Auroral Physics

David J. Knudsen • Joseph E. Borovsky •
Tomas Karlsson • Ryuho Kataoka • Noora Partamies
Editors

Auroral Physics

Previously published in *Space Science Reviews* in the Topical Collection "Auroral Physics"

Springer

Editors
David J. Knudsen
Dept. of Physics and Astronomy
University of Calgary
Calgary, Alberta, Canada

Joseph E. Borovsky
Space Science Institute
Boulder, CO, USA

Tomas Karlsson
Division of Space and Plasma Physics
KTH Royal Institute of Technology
Stockholm, Sweden

Ryuho Kataoka
National Institute of Polar Research
Tachikawa, Japan

Noora Partamies
The University Centre in Svalbard
Longyearbyen, Norway

ISSN 1385-7525 Space Sciences Series of ISSI
ISBN 978-94-024-2121-7

Cover Image: Artist's depiction of auroral forms covered in this volume, as described in the Preface. Coordinate circles are labelled with magnetic local time (hours) and magnetic latitude (degrees). Figure courtesy of D. Megan Gillies with contributions from Eric Donovan and David Knudsen.

Printed on acid-free paper

This Springer imprint is published by the registered company Springer Nature B.V.
The registered company address is: Van Godewijckstraat 30, 3311 GX Dordrecht, The Netherlands

Contents

Preface: Topical Collection on Auroral Physics

David J. Knudsen[1] · Joseph E. Borovsky[2] · Tomas Karlsson[3] · Ryuho Kataoka[4] · Noora Partamies[5]

Accepted: 10 January 2021 / Published online: 2 February 2021

1 Introduction

The term "aurora borealis", or "northern dawn" dates back centuries and refers to emissions of light from the otherwise-dark nighttime atmosphere, usually occurring in the polar regions except during highly disturbed periods. Aurora is distinct from airglow, which is a weak and relatively unstructured emission in the thermosphere caused by chemical reactions and ionization driven primarily by solar UV illumination during the day. In contrast, auroral emissions are the result of excitation of neutral atoms and molecules in the upper atmosphere by collisions with charged particles, typically which originate in the magnetosphere and precipitate along geomagnetic field lines with energies of hundreds of eV to tens of keV.

Popular descriptions of the aurora including in the media, dictionaries, encyclopedias and even textbooks often claim that the aurora is caused by "particles from the sun striking the upper atmosphere". It is well established in auroral science that such a description is not accurate except in very limited cases. While it is certainly true that particles from the sun – the solar wind – provide the energy that drives the aurora, the widely varying morphologies and behaviors of the aurora are the result of a complex chain of events that take place within Earth's magnetosphere or at its boundary, the magnetopause. The end result of this chain – excitation of neutrals by charged particles – is also well-established fact, as are

Auroral Physics
Edited by David Knudsen, Joe Borovsky, Tomas Karlsson, Ryuho Kataoka and Noora Partmies

✉ D.J. Knudsen
knudsen@ucalgary.ca

1 Department of Physics and Astronomy, University of Calgary, Calgary, Alberta, Canada

2 Center for Space Plasma Physics, Space Science Institute, Boulder, CO, USA

3 Space and Plasma Physics, KTH Royal Institute of Technology, 10405 Stockholm, Sweden

4 National Institute of Polar Research, 10-3 Midori-cho, Tachikawa, Tokyo 185-0031, Japan

5 Department of Arctic Geophysics, The University Centre in Svalbard, Longyearbyen, Norway

many other aspects described in this book volume. However, due to the vast region of the magnetosphere that is magnetically conjugate to the auroral ionosphere, and the difficulty in sampling it with a single or even multiple spacecraft, the nature of the magnetospheric generator responsible for driving individual auroral forms still remains one of the most elusive aspects of the aurora.

This book is designed to provide a comprehensive review of our current scientific understanding of the terrestrial aurora, both observational and theoretical, with an emphasis on developments since a previous collection devoted to the terrestrial aurora, "Auroral Plasma Physics" (*Space Science Reviews* volume *103*, 2002, by Paschmann, Haaland, and Treumann). That collection emphasizes introductory and background material which is not repeated in the current set.

Organization of this book is by type of aurora, summarized in Figure 1. State-of-the-art observations of each are summarized and compared to theories, while identifying open questions in obtaining a complete scientific understanding of the mechanisms responsible for their formation and behavior. As summarized below and documented in the individual reviews, while significant progress has been made over the past two decades, the remaining gaps in understanding are significant and in many cases fundamental.

(1) *Quiet discrete arcs* are featured in three separate reviews in this volume covering observations (Karlsson et al.), acceleration mechanisms (Lysak et al.), and generator theories (Borovsky et al.). These distinctive curtains are frequently visible in the evening and midnight sectors of the auroral zone (~60 – 70 degrees magnetic latitude), and are perhaps the most common and recognizable auroral form. Quiet arcs have characteristic time scales of several to tens of minutes and are associated with electric fields both perpendicular and parallel to the geomagnetic field, as well as field-aligned electric currents above the ionosphere and strong horizonal currents within it. At intermediate distances above the auroral ionosphere (up to ~2 earth radii), electrons are accelerated downward along the geomagnetic field to energies of a few to tens of keV, powered by quasi-static electric fields and currents imposed from still higher altitudes, by the so-called auroral generator. Presently, the magnetospheric auroral generator is the weakest link in terms of our understanding of this phenomenon, owing to the vast region of space magnetically conjugate to these arcs and the difficulty in sampling it with a single – or even multiple – spacecraft.

(2) *Small-scale dynamic aurora* (Kataoka et al. in this volume) is characteristic of auroral breakups, where energy stored in the stretched magnetotail is released episodically into the auroral zone, initially close to midnight but spreading rapidly poleward and to the west and east. Aurora associated with these transient events has characteristic spatial scales of the order of 0.1 – 10's of km, time scales of 0.1 – 10's of seconds, and includes rapidly evolving arcs and flickering aurora. Rapid developments in camera technology in the past two decades have brought vast improvements in our ability to resolve and study these structures and their causes. Electric and magnetic field fluctuations associated with these spatial and temporal scales are in the domain of inertial Alfvén waves (IAW), which can account for many properties of electron acceleration observed in situ. However, as with the case of quiet discrete arcs, the origin of the IAW and the basis of the distinctive morphologies they produce remain important open questions. In this chapter, a number of movie examples are included and are important to demonstrate the complexity and coherence of small-scale dynamic auroral features.

(3) *Diffuse and pulsating aurora* (Nishimura et al. in this volume) is generally less structured and dimmer than the discrete aurora but can be more widespread in local time, and in principle may deposit more energy into the neutral atmosphere overall. Unlike

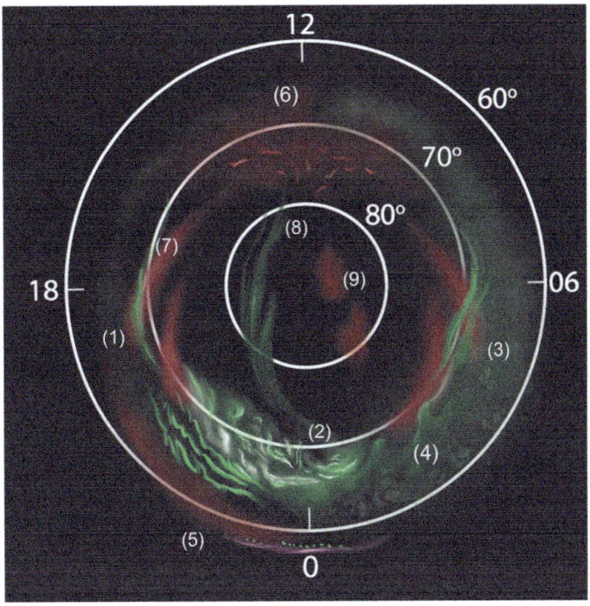

Fig. 1 A sketch of auroral forms reviewed in this collection, including (**1**) quiet discrete auroral arcs, (**2**) small-scale dynamic aurora, (**3**) diffuse and pulsating aurora, (**4**) mesoscale dynamical auroral forms, (**5**) proton aurora and sub-auroral forms, (**6**) dayside aurora, (**7**) field-line resonance arcs, (**8**) polar cap aurora, and (**9**) polar cap patches. Figure courtesy of D. Megan Gillies with contributions from Eric Donovan. After the original concept by Akasofu (*Space Science Reviews*, 19, 169, 1976)

their counterparts in the discrete aurora, electrons causing the diffuse aurora show no evidence of field-aligned acceleration, but rather are the result of pitch-angle scattering by plasma waves into the atmospheric loss cone by electron-cyclotron-resonance (ECH) or chorus whistler waves in the inner magnetosphere, on mostly dipolar field lines. This review documents significant advances in the past two decades including a direct connection of waves measured in situ with individual pulsating patches, providing, for the first time, definitive measurements of an auroral form as seen from the ground made simultaneously with measurements deep inside the magnetospheric source region. Still, many open issues remain, including the physical mechanisms that determine the spatial and temporal structure of pulsating auroras, the relative role of ECH and chorus waves, and the cause of small-scale sub-structure within the diffuse aurora.

(4) *Mesoscale dynamical auroral forms* (Forsyth et al. in this volume) have spatial scales of tens to hundreds of km, and include poleward boundary intensifications (PBI), streamers, beads, omega bands and torches, and giant undulations. These structures can be interpreted as visible maps of mesoscale and large-scale dynamics of the nightside magnetosphere. PBIs and streamers are associated with the poleward boundary of the auroral oval, and are interpreted in terms of reconnection in the distant magnetotail in the case of PBIs, and reconnected flux tubes and bursty bulk flows propagating earthward in the case of streamers. The connections between these phenomena have been strengthened considerably in the past twenty years due to an increase in the number of conjunctions between satellites deep in the magnetosphere, and closer to the ionosphere, along with ground-based and space-based imagers. Omega bands and giant undulations are structures that form in the equatorward part of the auroral zone, associated with the diffuse aurora. Kelvin-Helmholtz instability from shear flows in the inner magnetosphere likely plays a role, however many complexities are not yet explained, for example irregular evolution and the formation of poleward-pointing fingers (torches). The story is similar with beads, which are periodically-spaced structures that form along the thin "breakup arc" situated on the equatorward edge of the auroral zone immediately prior to substorm

onset. Beads are clearly a signature of a plasma instability in the magnetosphere associated with substorm onset, however there remains no scientific consensus on the specific mechanism or mechanisms responsible.

(5) *Proton aurora and subauroral forms* (Gallardo-Lacourt et al. in this volume) reviews morphologies just equatorward of the auroral zone that include detached arcs and spots (both ion and electron), stable auroral red (SAR) arcs, and the newly-identified phenomenon "STEVE". These phenomena map to and beyond the outer boundary of the plasmasphere, where dipolar field lines begin to transition to a more stretched configuration, and energetic ring current ions interact with low-energy plasma. These processes give rise to proton scattering and subsequent precipitation into the upper atmosphere, as well as heating of ionospheric electrons that is believed to the be direct cause of the sub-visual SAR arcs. A possible cause of STEVE's distinctive, mauve-coloured emission is the interaction between neutral particles and ions driven to hypersonic speeds by large convection electric fields originating in the inner magnetosphere. In this sense STEVE does not fit the usual definition of aurora in terms of charged particles precipitating from the magnetosphere, however it is often accompanied by a bright, spatially periodic feature in green-line emission known as the "picket fence", which may be associated with electron precipitation. Remaining open questions in this area include the specifics of proton scattering mechanisms, the source of electron heating in SAR arcs, and excitation and emission mechanisms within STEVE.

(6) *Dayside aurora* (Frey et al. in this volume) includes structures near and surrounding the dayside cusp region observed with space-borne auroral UV imagers capable of suppressing the strong background of solar illumination, or with ground-based cameras operated during the dark polar winter. Auroral phenomena in this region include high-latitude dayside auroral structures (HiLDAS), poleward-moving auroral forms (PMAFs), travelling convection vortices (TCVs), throat aurora, dayside diffuse aurora, and diffuse auroral spots. Many of these phenomena are associated with or driven directly by dayside magnetic reconnection and are used as a valuable diagnostic to complement in situ studies of the dayside magnetopause. As with other auroral types, significant progress in the past twenty years has been driven by advances in camera technology and availability of in situ observations. Remaining questions include the specific relation between dayside auroral forms and the morphology of the dayside X line (e.g. spatial extent, pulsed versus steady), and the role of the interplanetary magnetic field and other parameters.

(7) As reviewed by Rankin et al. in this volume, interhemispheric *field-line resonances* (FLRs) were proposed in the 1990's as a mechanism for the generation of discrete auroral arcs, however this explanation was not widely appreciated due to the relatively low energies (predominantly sub-keV) predicted by the theory of dispersive Alfvén waves, especially at such low frequencies, and because of the fact that only a minority of arcs were observed to oscillate at relevant frequencies. More recently, observations using a new generation of sensitive red-line (630 nm) cameras have been used to identify hundreds of examples of oscillating arcs. Many aspects of the overall morphology and behavior of FLR arcs can be explained by fluid simulations, however more sophisticated models of field-aligned electron acceleration are needed, and are being developed, in order to explain observed electron energies.

(8) *Polar cap aurora* (PCA) (Hosokawa et al. in this volume) tends to be sun-aligned on average, can drift in the dawn-dusk direction, and at times can extend across the entire extent of the polar cap to form the so-called theta aurora. PCA occurs almost exclusively during geomagnetically quiet periods when the interplanetary magnetic field is northward, and is generally interpreted as a tracer of large-scale magnetospheric topology. Recent scientific progress has been made possible due to rapid advances in both

ground and space-based camera technology, along with MHD-based modeling. Outstanding questions include the relation of PCA to open versus closed fields lines, magnetospheric source regions, generation mechanisms, cause of dawn-dusk motion, and the possible role of substorm triggering by polar cap arcs that encounter the poleward boundary of the nightside auroral zone.

(9) *Airglow Patches in the Polar Cap Region: A Review*, by Hosokawa et al. in this volume. Historically, the term "airglow patch" was based on the assumption of no particle precipitation, since polar cap patches, typically defined as two-fold increases in ionospheric plasma density, are formed by sunlit plasma on the dayside that convects antisunward into open field line regions, and in principle may not require an active source. However, recent findings based on conjunctions of the Swarm satellites with ground-based cameras have demonstrated that some polar cap patches are in fact associated with enhanced convection and field-aligned currents, and therefore presumably with electron precipitation. We note that the study of active driving of polar cap patches through precipitation is still in its early stages.

Fig. 2 Participants in the Auroral Physics Workshop in Bern, August 2018. Left to right: Rudolf von Steiger, Dan Whiter, David Knudsen, Jade Reidy, Laila Andersson, Shigeru Fujita, Toshi Nishimura, Marius Echim, Megan Gillies, Keisuke Hosokawa, Nithin Sivadas, Steve Milan, Noora Partamies, Ryuho Kataoka, Bob Strangeway, Joachim Birn, Joe Borovsky, Mike Henderson, Victor Sergeev, Colin Forsyth, Bob Lysak, Robert Rankin, Larry Kepko, Bea Gallardo-Lacourt, Marc Lessard, Harald Frey, Chris Chaston, Takeshi Sakanoi, Antonius Otto, Tomohiko Watanabe, Mizuki Fukizawa, Octav Marghitu, Tomas Karlsson, Jiashu Wu, Yuto Kato, Ying Zou, Kristina Lynch, Eric Grono, Anita Kullen, Yan Song. Photo credit: Saliba F. Saliba

Acknowledgements We are grateful to ISSI for facilitating this project, and for hosting the planning workshop held in Bern in August 2018. In particular we wish to express our sincere appreciation to Andrea Fischer, Jennifer Fankhauser, Alexandra Lehmann, Saliba F. Saliba, and Rudolf von Steiger.

Space Sci Rev (2020) 216:16
https://doi.org/10.1007/s11214-020-0641-7

Quiet, Discrete Auroral Arcs—Observations

T. Karlsson[1] · L. Andersson[2] · D.M. Gillies[3] · K. Lynch[4] · O. Marghitu[5] · N. Partamies[6,7] · N. Sivadas[8] · J. Wu[3]

Received: 13 June 2019 / Accepted: 28 January 2020 / Published online: 5 February 2020

Abstract Quiet, discrete auroral arcs are an important and fundamental consequence of solar wind-magnetosphere interaction. We summarize the current standing of observations of such auroral arcs. We review the basic characteristics of the arcs, including occurrence in time and space, lifetimes, width and length, as well as brightness, and the energy of the magnetospheric electrons responsible for the optical emission. We briefly discuss the connection between single and multiple discrete arcs. The acceleration of the magnetospheric electrons by high-altitude electric potential structure is reviewed, together with our current knowledge of these structures. Observations relating to the potential drop, altitude distribution and lifetimes are reviewed, as well as direct evidence for the parallel electric fields of the acceleration structures. The current closure in the ionosphere of the currents carried by the auroral electrons is discussed together with its impact on the ionosphere and thermosphere. The connection of auroral arcs to the magnetosphere and generator regions is briefly touched upon. Finally we discuss how to progress from the current observational status to further our understanding of auroral arcs.

Keywords Discrete auroral arcs · Auroral acceleration region · Ionospheric currents

Auroral Physics
Edited by David Knudsen, Joe Borovsky, Tomas Karlsson, Ryuho Kataoka and Noora Partmies

✉ T. Karlsson
tomas.karlsson@ee.kth.se

1 Space and Plasma Physics, KTH Royal Institute of Technology, 10405 Stockholm, Sweden

2 Laboratory for Atmospheric and Space Physics, University of Colorado Boulder, Boulder, CO, USA

3 Department of Physics and Astronomy, University of Calgary, Calgary, Alberta, Canada

4 Department of Physics and Astronomy, Dartmouth College, Hanover, NH, USA

5 Institute for Space Sciences, Bucharest, Romania

6 Department of Arctic Geophysics, The University Centre in Svalbard, Longyearbyen, Norway

7 Birkeland Centre for Space Science, Bergen, Norway

8 Boston University, Boston, MA, United States

1 Introduction

The bright, visible aurora (often >1 kR) is most frequently aligned with the local magnetic field, with sharp emission boundaries resulting in the notion of 'discrete' aurora (Davis 1978). This emission can be stable for tens of minutes or dance around the sky at high speed with different colors associated with it (Akasofu 1965; Hanna and Anger 1971). The more dynamic 'discrete' aurora is most of the time related to Alfvénic acceleration (Colpitts et al. 2013) while the more stable (minutes to hours) emissions are caused by electrons accelerated by a quasi-static electric potential structure. This stable emission is often also called the 'inverted-V' type of aurora due to the signature in the early satellite electron spectrograms. The focus of this chapter is the latest findings on the discrete, inverted-V, stable auroral arcs.

Optically, the diffuse and discrete aurora can be difficult to separate, especially since both can have sharp emission gradients. Depending on viewing angles, optical resolution, temporal cadence and observational wavelength, the cause of the precipitating particles can be identified. However, by not having all these observations, available studies in the literature might not have been able to separate the two and the reader should therefore be cautious of limitations of statistical studies from ground (note that recent advances in multispectral imaging are helping to mitigate these limitations.) Using satellite data, the processes can be most of the time be well separated but the horizontal extent and the dynamics can be lost. When interpreting single-satellite observations without imagery observation context, it is often assumed that the auroral arcs are in the east-west direction, while in reality the satellite might have crossed an auroral fold, spiral or other deviation from that simplified geometry. For this review we have decided to define a quiet discrete arc as an auroral phenomenon, which is associated with a upward field aligned current (FAC), emissions caused by accelerated precipitating particles (being more energetic than the thermal energy of the particles on the associated magnetic field line), and is temporarily stable (longer than the bounce time of an Alfvén wave on that magnetic field line, which is of the order of minutes, Vogt 2002).

The last decade has seen significant improvements in auroral imager performance and the number of all-sky stations. Using these improved observations, the knowledge of the characteristics of the quiet discrete arc has improved and is presented in Sect. 2. We now have access to high-resolution and often multiple satellite observations of the precipitation particles. This has led to improved understanding of the two dimensional aspects of the stable arc and its temporal evolution, which is presented in Sect. 3. With multiple point observation and using different instrumentation the latest understanding of the horizontal dynamics is presented in Sect. 4. Stable arcs can filament and evolve into multiple arcs. Recently significant progress has been made within this topic, which is discussed in Sect. 5. The aurora is not only a beautiful manifestation of the dynamics in the magnetosphere, but it also modifies the ionosphere. This effect is discussed in Sect. 6, while the driving magnetospheric sources are discussed in Sect. 7.

As demonstrated by this review, auroral physics is still a highly active research field and new types of satellite missions are needed, especially to increase our understanding of temporal and spatial variations. Only then can we fully understand the interaction between the ionosphere and magnetosphere. Section 8 highlights some of the measurements and questions that need to be addressed in the near future.

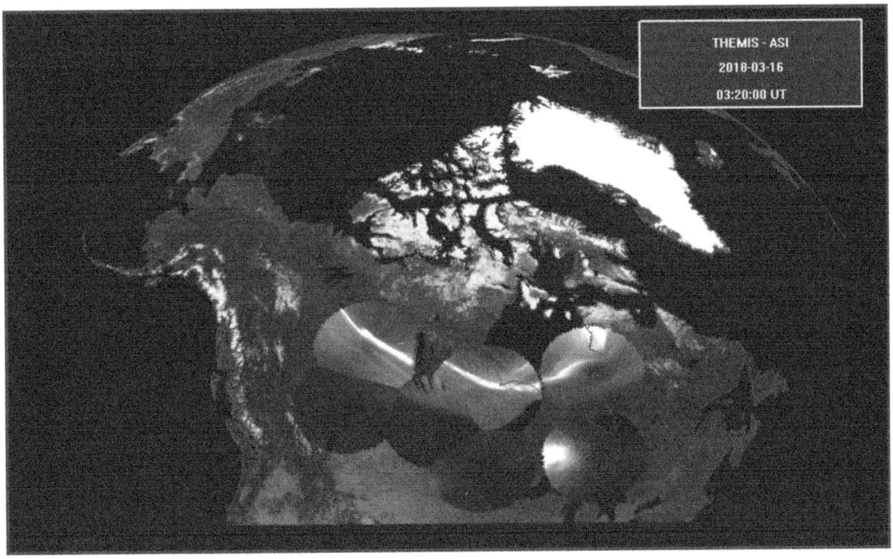

Fig. 1 An all-sky image mosaic of an example of typical quasi-stable arc captured by the panchromatic THEMIS cameras (Donovan et al. 2008) on 16 March, 2018, at 03:20 UT

2 Basic Characteristics

This section gives a comprehensive overview on recent findings framing the basic characteristics and appearance of quiet quasi-stable auroral arcs, such as where and when quiet arcs are likely to be found, their physical parameters (size, duration, brightness), and associated electron precipitation energies. Detailed understanding of these basic parameters is important as the knowledge of arc morphology places strong constraints on arc generation theories. Quiet arcs, as defined by Davis (1978), are highly magnetic east-west elongated discrete auroral forms with lengths of 100–1000 km. The arc luminosity experiences a sharp lower altitude boundary. In general, the emission within the arcs is limited to altitudes between 80 and 400 km, with a peak around 110 km (Davis 1978). Here we consider "quiet" to refer to arcs occurring under low, typically less than $\mathrm{Kp} = 4$, geomagnetic conditions. These can be growth phase arcs prior to the substorm onset (see Fig. 1) or high-latitude arcs, which map to the plasma sheet boundary layer.

2.1 Arc Occurrence

In order to accurately describe quiet auroral arcs it is helpful to first identify the large scale parameters such as when and where this type of aurora occurs.

Latitudinal Variation For the purposes of this review, the nightside auroral oval is defined as the region of closed geomagnetic field lines located equatorward of the "red-line shelf" and poleward of the optical "b2i" boundary. The "redline shelf" is used to identify the poleward boundary of soft-electron precipitation which generally marks the boundary of the polar cap (a region of open geomagnetic field lines, Blanchard et al. 1995, 1997; Wanliss et al. 2000; Johnsen and Lorentzen 2012). Alternately the optical "b2i" boundary is a visual manifestation of the equatorward boundary of the proton aurora corresponding to a transition from strong pitch angle scattering to bounce trapped particles (Jayachandran et al. 2002; Donovan et al. 2003). This is a reasonable definition in the context of this paper, since discrete auroral arcs are believed to be found exclusively poleward of the b2i boundary (Lyons et al. 1988; Newell et al. 1996). At present time, there is a lack of studies applying this definition to the dayside auroral oval. Thus, for the purpose of this review we refer to the nightside here. With these definitions in mind, quiet discrete arcs are generally found within the oval, between about 60–80 degrees geomagnetic latitude (MLat). Multiple studies have sought to capture the latitude range of quiet auroral arcs. Nevanlinna and Pulkkinen (2001) used an all-sky imager (ASI) network located in Finland to perform a survey of over 100,000 hours of all-sky data. The study extended to all forms of aurora with quiet auroral arcs as a subset of their data. They specifically looked at the percentage of auroral occurrence (the chance of seeing a particular type of aurora) and concluded that the occurrence of quiet arcs peaked near 70° MLat. The high latitude positions of the cameras used in this study meant that it was difficult to firmly establish upper and lower boundaries which potentially confine stable arcs. Gillies et al. (2014) used the THEMIS ASI network to study over 7500 quiet auroral arcs. Using ASIs spanning the auroral oval, they surveyed the latitude extent of single and multiple arc systems. They found that the 2687 observed single arcs occurred under lower geomagnetic activity levels and were confined to a narrower latitude band (67–75° MLat) as compared to the multiple arcs. The approximately 5400 images containing multiple arcs expanded across a broader range of latitudes of 62–75° MLat.

Pre-existing growth phase arcs (a subset of quiet arcs) were studied by Jiang et al. (2012) by comparing FAST spacecraft and ground-based optical observations. They determined that these arcs were located at, or very near, the boundary between the Region 1 and Region 2 field-aligned currents. In two events, Nishimura et al. (2012) found the upward FAC of preonset arcs located within the downward R2 FAC near the midnight. Using measurements from AMPERE, Motoba et al. (2015) further established that these growth phase arcs were observed to appear about 4.3 degrees equatorward of the R1/R2 boundary. This separation was found to decrease to about 1 degree as substorm breakup approached.

Diurnal Variation While quiet and growth phase arcs are confined to a relatively narrow latitude band, they can be found at nearly all magnetic local times (MLT). Growth phase arcs tend to have a strong preference for the midnight sector. Considering that this particular type of arc is a pre-substorm growth phase feature, it is unsurprising that they have a strong disposition to occurring at 20–03 MLT (Kozlovsky et al. 2001; Jiang et al. 2015a). Quiet arcs occur across all MLTs. Green emission dominated arc-like aurora has been observed frequently in the pre- and post-noon sectors on the dayside (Hu et al. 2009), while the noon region has found to contain fewest arcs (Qiu et al. 2016). The statistical study by Gillies et al. (2014) separated quiet arcs into single and multiple arc systems. They found that single (multiple) quiet arcs were seen between 14 and 8 MLT with a peak occurrence near (23–24 MLT) 22–23 MLT. This is in agreement with a study by Syrjäsuo and Donovan (2004) which used machine learning to identify and catalogue auroral structures. They found that the prevalence of arcs (of which quiet arcs was a subset) peaked near 20–21 MLT. This is also seen in a study by Partamies et al. (2008) which found a statistical preference for the inverted-V structures seen by FAST, confirmed by ground-based optical signatures, to occur around 21–23 MLT.

Alignment One of the unique features of the quiet auroral arcs is their alignment along lines of constant geomagnetic latitude. While arcs have frequently been observed to exhibit this behavior few studies have sought to quantify and further explore this property. Gillies

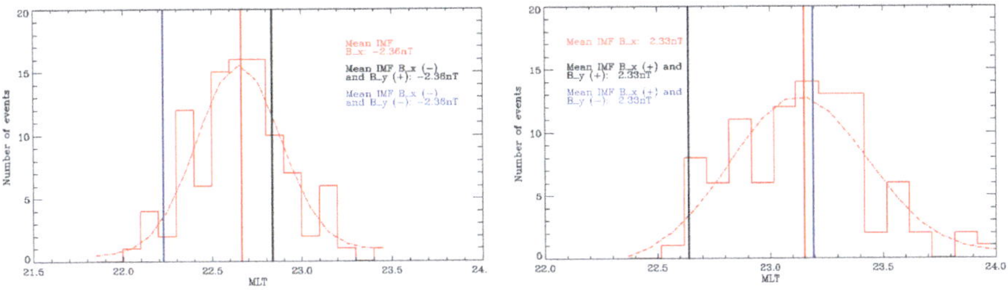

Fig. 2 Distributions of the single arc tilt reversal locations in MLT for subsets of about 100 events. IMF B_X negative (left) and positive (right) were investigated separately. The histograms give the observed reversal distribution to which the Gaussian (dashed red) was fitted to provide the MLT mean values of 22.6 and 23.1 (red vertical lines), respectively. Both datasets were further divided into cases of positive (black) and negative (blue) IMF B_Y. The same sign of B_X and B_Y moves the reversal duskward, while the opposite sign of B_X and B_Y twist the mean values dawnward. Figures from Gillies et al. (2014)

et al. (2014) performed one of the first large scale statistical studies quantifying the orientation of auroral arcs relative to the local magnetic east-west direction. Using five sites from the THEMIS ASI array, they identified over 7500 quiet arcs between the years 2007 and 2010 in images separated by at least 10 minutes. The sites chosen in this study provided a broad range of magnetic latitudes and longitudes designed to encompass a large portion of the auroral oval under different geomagnetic activity levels. One of the most notable results of this study was that stable auroral arcs are highly aligned with lines of constant geomagnetic latitude, indicating that their overall morphology is governed by the large-scale structure of the magnetosphere. In most cases the arcs were aligned (within $\pm 7.7^\circ$) with lines of constant geomagnetic latitude. Interestingly, there was a clear relationship between the direction and magnitude of the arc tilt and MLT seen in three of the five stations presented. The tilt reversal is the point where arc alignment, as viewed along the structure from west to east, changes from equatorial to poleward tilt, on average. The reversal point for three stations occurred near the magnetic midnight, with the reversal point shifting toward the morning sector at higher latitudes. Tilt reversal behaviour resembles that of Harang discontinuity, but the relationship between these two phenomena needs further investigations. The reversal point had a strong MLT dependence on the interplanetary magnetic field (IMF) orientation. The orientation of the B_x and B_y components shifted the arc reversal by as much as half an hour in most cases. For example, if both B_x and B_y were positive (black on the right panel of Fig. 2) or both were negative (blue on the left panel of Fig. 2), the shift would be towards dusk. A combination of positive and negative orientations resulted in a slight shift dawnward. This tilt effect has also been observed on dayside. The combination of positive B_x and negative B_y shifted the tilt reversal towards midday, whereas negative B_x together with positive B_y moved the reversal point dawnward (Qiu et al. 2016).

Seasonal and Solar Cycle Variation A frequently asked question is how quiet arcs vary by the solar cycle and the Earth's seasons. However, few studies have properly addressed this. One of the earliest studies was performed by Nevanlinna and Pulkkinen (2001). Under the hypothesis of a good statistical correlation between auroral activity and solar cycle, they used all-sky camera images in northern Finland to categorize auroral arcs (among other auroral forms). While the overall total number of auroral observations was only weakly tied to the solar cycle they did find the number of quiet arcs anti-correlated with the sunspot number. This indicated that while active auroral forms were less frequent during solar minima, quiet arcs were more common. Pulkkinen et al. (2011) recently furthered this idea by investigating the auroral electrojet evolution during the deep minimum at the end of the solar cycle 23 (2008–2009). One of the findings of this study was a confirmation of an increase in quiet arc occurrence during magnetically quiet periods. This is consistent with observations of downward accelerated electron beams (>5 mW/m^2), which occur twice as often during the solar minimum as compared to the solar maximum (Cattell et al. 2013). Partamies et al. (2014) used an automated auroral structure detection algorithm and performed a statistical study of the solar cycle dependence of arcs. Also in this study, quiet arcs were observed more frequently during the solar minimum years (specifically in 1996–1997 and in 2007).

From a ground-based perspective, seasonal studies are challenging as most auroral observations using ASIs require adequate dark conditions—a criterion fulfilled mostly in the winter season, particularly at high latitudes. The THEMIS array typically operates from late August to early April, similar to the high latitude stations of the Finnish all-sky imaging network. High-latitude stations, such as those on Svalbard, will observed the dark skies 24/7 from approximately November until February, but are still biased towards the winter season. Furthermore, dayside arc studies are very limited (Qiu et al. 2013; Hu et al. 2009). Thus, a vast majority of statistical studies are performed in the winter months, with September through March being preferred to avoid daylight complications. While not directly observing auroral arcs, there are a few studies of note which investigate the potential for auroral arc generation during winter and summer months. Newell et al. (1996a) used DMSP spacecraft to perform a survey of in-situ electron precipitation events and found that the occurrence rate of discrete aurora is lower during summer due to the increased probability of sunlit hours. Shortly after, Liou et al. (1997) analysed over 17,000 images acquired by the ultraviolet imager (UVI) on board the Polar satellite during the period of April–July 1996. Using these images, they calculated the N2 Lyman-Birge-Hopfield (LBH) auroral emission which can be used as a proxy for the total flux of precipitating electrons. Polar observations allowed monitoring of both hemispheres during the course of one orbit enabling the seasonal variation to be detected. This study concluded that discrete aurora was more common in the midnight sector in the dark hemisphere (winter) than in the sunlit hemisphere (summer). However, both of these studies included all discrete arcs, not just quiet arcs. One significant finding of this study was a link between seasons and geomagnetic activity levels. They attribute the gradual decrease in auroral emission on the nightside in the spring (April) to summer (July) months to this effect. While this study cannot distinguish between discreate and diffuse aurora, a later analysis of the same DMSP dataset (Newell et al. 2010) showed that the strongest monoenergetic precipitation comes to the pre-midnight sector, and that the winter time monoenergetic precipitation accounts for a higher energy flux compared to the summer scenario. More than the season, the precipitation energy flux was shown to be scaled by the solar wind. A further study of auroral electron precipitation was performed by Barth et al. (2004). They used the Student Nitric Oxide Explorer (SNOE) to determine the density of nitric oxide in the lower thermosphere—a marker of molecular nitrogen produced by auroral electron precipitation. An important result of their study was a clear seasonal dependence of precipitating auroral electron energy flux on the nightside at the energy of 4 keV marked by a minimum in electron flux (i.e. a reduction in NO production) as summer solstice was approached. They interpreted this to result from the high ionospheric conductivity which in the summer ionosphere suppresses the formation of auroral arcs. This seasonal dependence associated with auroral precipitation was investigated by Zheng et al. (2013) using a different approach: examining total hemispheric power (HP) calculated from satellite observations of the precipitating particle energy flux over the entire polar region. In this context, the precipitation related to quasi-static arcs is only part of the big picture. Nevertheless, they concluded that the winter hemisphere receives more auroral power than the summer hemisphere during low geomagnetic activity levels. They also reported that the hemispheric asymmetry has a weak positive association with the solar cycle.

The above studies on seasonal variation in auroral arcs relied solely upon space-based observations of UVI aurora (for which there is limited resolution for resolving small scale quiet arcs) or focused on ionospheric electron precipitation patterns. When looking at the seasonal dependence from a ground-based perspective, one must also consider the effect of weather patterns which can impede statistical studies. In 2018, an as yet unpublished statistical survey of cloudiness was performed by D.M. Gillies and D. Chaddock at the University of Calgary. Using three key sites in the THEMIS ASI network (Fort Simpson, Fort Smith and Gillam) the effect of the local weather on the observing capabilities was determined. Gillies and Chaddock sampled images every 10 minutes during the time period of January–December 2014. Each viewing hour (the time for which the camera was on and operating) was divided into segments of 10 minutes and each 10 minute 'bin' was assigned one of the following designations: cloudy, clear, or unclassifiable data. The 'clear' designation included events for which there were clear skies present—independent of the presence of aurora or light pollution effects from objects such as the moon. The cloudy classification was used when approximately 20% or more of the field-of-view of the camera contained clouds. Unclassifiable data contained images for which there was significant light pollution from external sources or weather conditions such as snow on the dome or rain precluded classification. Each site was classified independently to avoid bias in the results. In total, over 50,000 images were analysed. The results showed that over central Canada, the best viewing months from a cloud perspective, are November through April (Fig. 3). Unsurprisingly the winter months (November–February) contained the most images for which adverse weather conditions (such as snow obscuring the dome) precluded observations.

2.2 Arc Dimensions

Width One well-studied arc parameter is the latitudinal width/thickness: a feature easily identified in single ASI images. Knudsen et al. (2001) studied all-sky images of over 3000 auroral arcs found within 5 degrees of magnetic zenith. Limiting the study to a narrow band near the zenith is a necessary criterion in order to avoid elongating the width due to the altitude extent of the emission curtain. They concluded that a typical green emission arc has a width of 18 ± 9 km. This finding was further supported by a study of six evening sector arcs which had a similar median thickness of 18 km (Aikio et al. 2002), and by a set of about 400 thin pre-onset arcs at the equatorward boundary of the auroral oval (Nishimura et al. 2011). In contrast, a study of over 180 pre-onset arcs showed somewhat larger scales with widths of 20–50 km (Jiang et al. 2015a). Interestingly, as described by Aikio et al. (2002), the region of enhanced northward electric field on the equatorward side of the arc spanned 1–4 times the width of the arc, while a more recent study of midnight sector arcs showed arc widths (about 1.5 degrees in latitude) comparable to the size of the region with a strong electric field (50–150 km) (Aikio et al. 2018).

In their study, Knudsen et al. (2001) raised a question on whether the 1 km scale sizes do not exist or whether they had just not been measured. The optical resolution of the equipment used in the previous arc width studies had a minimum resolution of around 1 km. As a response, a set of 500 thin arcs with widths of 0.5–1.5 km were reported by Partamies et al. (2010) from an imager with a narrower field-of-view and higher spatial resolution (Fig. 4). They thus concluded that the spectrum of auroral arc widths is continuous.

From an in-situ observational point of view, inverted-V acceleration widths have been observed to be wider (>1 degree) than the ionospheric arc widths (<1 degree, Yago et al.

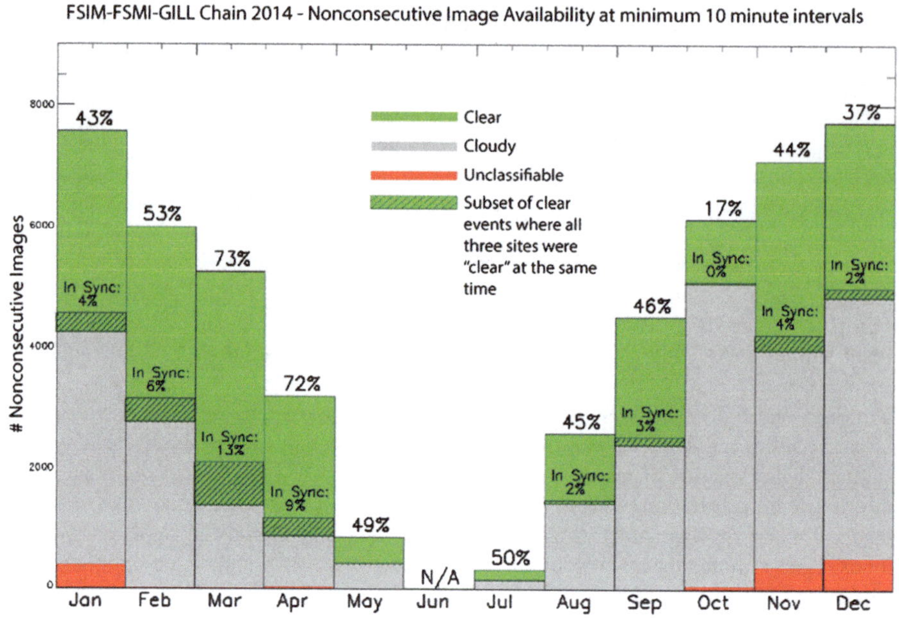

Fig. 3 Green indicates the time the camera was operating (as a percentage of the month), during which there were clear skies at an individual site to allow auroral observations. Grey represents cloudy conditions with zero chance of observing the aurora. The green hashed region demonstrates the percentage of time where the three adjacent sites (spanning approximately 3000 km) contain clear skies simultaneously, allowing large scale auroral observations

2005), suggesting that only higher energy electrons were able to produce observable emission. To further investigate this Partamies et al. (2008) conducted a statistical study of inverted-V structures as measured by the low-altitude orbiting FAST satellite. The typical acceleration potential was found 20–40 km wide, encompassing the ionospheric component of the arc. A more recent case study by Frey et al. (2010a) analysed an arc with an optical width of about 25 km and an inverted-V width of about 65 km. In agreement with the earlier results, they concluded that the optical emission with a limited width (25 km) was caused by the energy flux of the highest precipitation energies. The total energy flux input into the ionosphere, however, covered a larger region corresponding to the full width of the inverted-V precipitation.

Dayside arcs from the high-latitude Ny-Ålesund station were found to have comparable widths with the average of 18.5 km (Qiu et al. 2013), with the thinnest arcs closest to midday hours (at around 10 and 14 MLT). Post-midnight arcs, however, have been observed to have twice the scale (50–70 km) in association with low-energy precipitation of less than 200 eV (Aikio et al. 2004).

Length Less attention has been granted to the length of the auroral arcs, as that dimension is often limited by the camera field-of-view. The growth phase arcs have been concluded to extend over at least 0.75 hours in MLT (Lessard et al. 2007). Multiple studies have shown a soft upper limit of up to at least 2–5 hours in MLT (Donovan et al. 2008; Sergeev et al. 2012; Gillies et al. 2014; Nishimura et al. 2012).

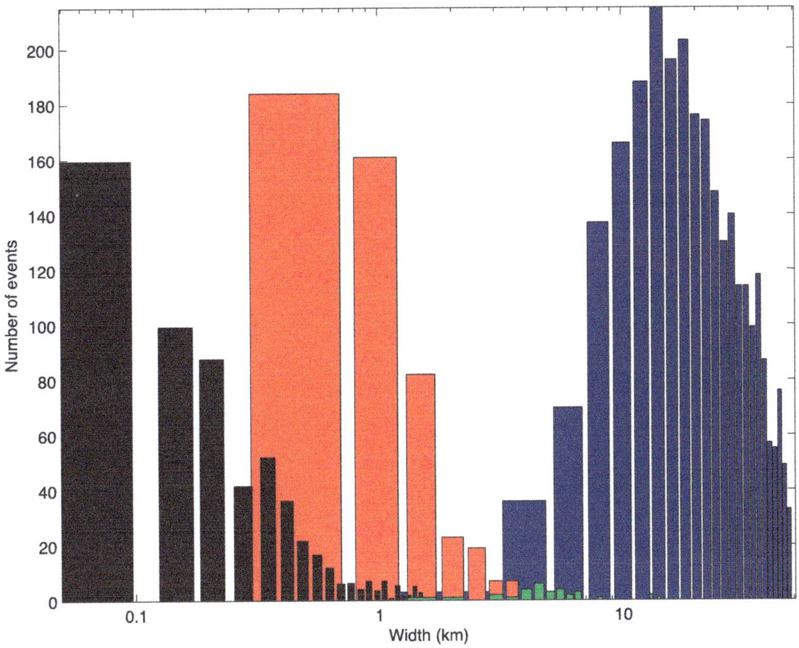

Fig. 4 A composite of observed arc widths based on ground-based studies. Auroral fine-scale observations by Maggs and Davis (1968) (black), meso-scale ASC observations by Knudsen et al. (2001) (blue) as well as widths captured by a 20-degree field-of-view by Partamies et al. (2010) (red). Some additional arc width observations are included from a short campaign through a 60-degree field-of-view (green). Figure from Partamies et al. (2010)

2.3 Brightness and Precipitation

The meso-scale (width of around 10–100 km, Partamies et al. 2010) arcs' peak intensities at the wavelength of 557.7 nm range from 0.2 to 200 kR with an average value of 15 kR (Knudsen et al. 2001). The peak brightness for an onset arc example by Frey et al. (2010a) at 427.8 nm was 3 kR. The brightness of the growth phase arcs has been observed to grow monotonically during the 4–5 minutes prior to the substorm onset (Lyons et al. 2002).

The precipitation energies of about 1 keV for high-latitude arcs (Safargaleev et al. 2003), 2–4 keV for inverted-V arcs (Partamies et al. 2008), 2.7 and 6 keV for quiet nightside arcs (Simon Wedlund et al. 2013), up to a few keV for pre-breakup arcs (Jiang et al. 2012) and several keV for pre-noon poleward moving arcs (Kozlovsky et al. 2003) have been reported. In the case of high-latitude arcs, the precipitation energy was seen to decrease with increasing distance to the auroral bulge (Safargaleev et al. 2003). These values are in a good agreement with the measured peak emission heights of 120–150 km (Sangalli et al. 2011; Qiu et al. 2013; Kozlovsky et al. 2003). The peak electron densities of evening sector arcs have been observed at the altitudes of 110–120 km, corresponding to a field aligned potential drop of about 1 kV (Aikio et al. 2002, 2018). In the events studied by Lessard et al. (2007), the inverted-V precipitation of the growth phase arc was gradually built up within the pre-existing diffuse precipitation (protons and electrons). In this case the peak energies stayed below 1 keV.

The typical electron energy fluxes of arcs were found to be 15–20 mW/m^2 (Partamies et al. 2004). Somewhat higher energies (12 keV) and energy fluxes (25 mW/m^2) were observed for an onset arc by Frey et al. (2010a), while lower energy fluxes (less than 10 mW/m^2) were modelled from intensities measured by a camera on board the Reimei spacecraft (Whiter et al. 2012). In contrast, tall rayed arcs, which are accelerated by a combination of inverted-V's and Alfvén waves, were reported to have an order of magnitude lower peak energy flux of 1–3 mW/m^2 in correlation with the optical intensity (Hallinan et al. 2001). Correspondingly, the lower boundaries of the tall rays were found at higher altitudes (130–170 km) in agreement with precipitation energies of 100–800 eV. Similar heights of about 130 km were also observed for late morning (10 MLT) discrete arcs, corresponding to a Total Electron Content (TEC) variation of 10 TEC units and the line-of-sight electron density of 1.25×10^6 cm^{-3} (Kintner et al. 2002). It is also interesting that while these examples of soft precipitation spectra provide lower energy fluxes, there are also regions of Alfvénically accelerated aurora (such as poleward boundary regions) where the soft precipitation delivers very large energy fluxes, as discussed in Forsyth et al. (this issue).

Equatorward of and often overlapping with a growth phase arc, a band of high-energy precipitation with energies of 10–300 keV appears some minutes before the breakup (Deehr and Lummerzheim 2001; Shiokawa et al. 2005; Lessard et al. 2007; Sergeev et al. 2012). This high-energy band of diffuse emission is related to the D region electron density enhancement and thus, cosmic radio noise absorption (CNA) (Jussila et al. 2004; McKay et al. 2018), and was shown to consist of mainly protons (Lessard et al. 2007) with an emission intensity of 150–1000 R at 486.1 nm (Deehr and Lummerzheim 2001). The diffuse band was also observed to move equatorward with the thinner growth phase arc with a roughly constant separation of some tens of km (Deehr and Lummerzheim 2001; McKay et al. 2018).

Recent sounding rocket studies (Grubbs et al. 2018b,a; Clayton et al. 2019b) have made conjunction observations between filtered ground-based imagery at several wavelengths, and nightside auroral sounding rocket in situ observations of electron precipitation. These conjunctions help to test and improve inversion calculations of precipitation based on multi-spectral imagery observations. The rapidly increasing capabilities of cameras makes this a valuable and growing field of study, as these inversions can provide time-dependent two-dimensional maps of electron precipitation (both energy flux and characteristic energy).

2.4 Stability

Quiet growth phase arcs are often seen as quasi-stable structures with lifetimes up to several tens of minutes (e.g. Kozlovsky et al. 2001; Shiokawa et al. 2005; Lessard et al. 2007; Nishimura et al. 2011; Jiang et al. 2015a). As an example, a quiet evening sector arc was observed for 6 minutes for triangulation of the peak emission height by Sangalli et al. (2011). They found an average peak emission height for green emission at 120–140 km with a variation *along* the arc structure of about 40 km.

Comparing arc lifetimes across the substorm phases, Partamies et al. (2015) found, as expected, that the longest durations take place during the growth phases. The lifetimes in this study accounted for periods when the arcs were continuously detectable with a maximum gap of 40 seconds (1-image gap in a 20-second cadence). This criterion is sensitive to intensity variations, such as fading, and resulted in a typical lifetime of only 1–2 minutes for any single arc, and with 90% of the arc events lasting for less than 10 minutes.

Similarly, a large range of lifetimes have been reported for the high-altitude acceleration potential structures. These are discussed in Sect. 3.4.

2.5 Arc Drift and Plasma Flows

In Sect. 4 below we will consider the controlling factors for horizontal structuring of discrete arcs. Here we touch briefly on the signatures of their motion, and of the motion of plasma near them. The equatorward drift speeds of the growth phase arcs have been measured on the order of the ionospheric convection speed, 70–170 m/s (Deehr and Lummerzheim 2001; Aikio et al. 2002; Kozlovsky et al. 2001; Jussila et al. 2004; McKay et al. 2018), but sometimes also significantly faster, e.g. 250–650 m/s (Kozlovsky et al. 2003, 2006; Safargaleev et al. 2003). No clear distinction is found between the arcs drifting along or faster than the ionospheric convection flow. Very fast plasma flow channels have been identified in the low conductance regions next to the arc emission (and high conductance). For instance, Aikio et al. (2018) reported eastward plasma flow speeds of 3.3 km/s on the poleward side of a midnight sector arc, corresponding to southward electric field. Recent observations by the Swarm mission have solidified these observations of fast flows adjacent to discrete arcs. The strong prevalent flows reported by Archer et al. (2017) are referred to as Birkeland current boundary flows, as they are seen to be largest (exceeding 1 km/s) at the boundary between upward and downward field aligned currents. The extensive Swarm observations allow the distinction between these auroral boundary flows, and other ionospheric features such as subauroral ion drifts (SAID) (Archer and Knudsen 2018).

The shear flow direction in the morning sector is westward on the equatorward side and eastward on the poleward side of the arc (Jiang et al. 2015a). In addition, the arcs have been observed to expand azimuthally at approximately the shear flow speeds of about 1–5 km/s (Kozlovsky et al. 2003; Safargaleev et al. 2003; Shiokawa et al. 2005; Yago et al. 2005). The shear flow over arc precipitation allows smaller-scale vortical structures to grow within the arcs. This may drive magnetic reconnection in the auroral acceleration region above the arc (Chaston 2015) leading to a breakup of an unstable arc at the substorm onset.

3 High-Altitude Observations, Along and Perpendicular to B

There is considerable direct observational evidence that quiet, meso-scale arcs are associated with precipitating, accelerated electrons, e.g. early observations of emission line intensity ratios, indicating electron acceleration up to several keV (e.g Mende and Eather 1976; Christensen et al. 1987). As described above combined observations of optical arcs and accelerated electrons is further evidence (McIlwain 1960; Stenbaek-Nielsen et al. 1998; Kozlovsky et al. 2003; Frey et al. 2010a; Jiang et al. 2012; Simon Wedlund et al. 2013). The REIMEI spacecraft observations, including an onboard auroral imager and electron precipitation sensors, provided a detailed view of these relationships, as, for instance, in Motoba and Hirahara (2016). The extensive THEMIS GBO array (Mende 2008) provides an enormous data archive of imagery allowing conjunction studies with many spacecraft. One nice example is found in Archer et al. (2017). Others combine REIMEI and Themis-GBO (Frey et al. 2010b), FAST and Themis-GBO (Haerendel and Frey 2014; Haerendel et al. 2012), and Swarm and Themis-GBO (Liu et al. 2018b; Wu et al. 2017; Gillies et al. 2015). Sounding rocket studies (Grubbs et al. 2018b,a; Clayton et al. 2019b) with dedicated groundbased imagery provide additional conjunction studies.

Early on it was suggested that the precipitating electrons described in Sect. 2 were accelerated by electric fields above the aurora (e.g. McIlwain 1960; Evans 1968, 1974; Carlqvist and Boström 1970; Hallinan and Davis 1970). Observations of large electric fields perpendicular to the geomagnetic field above the aurora were taken as confirmation of this idea,

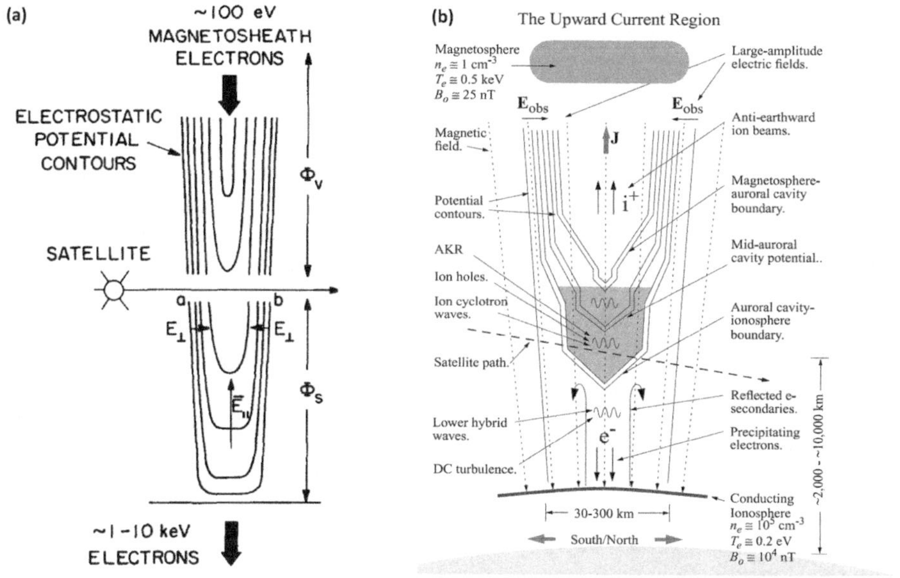

Fig. 5 Two interpretations of the U-shaped electric potentials above discrete auroral arcs. (**a**) Schematic adapted from Gurnett (1972) showing a potential structure with a relatively smoothly distributed $E_{\|}$. (**b**) Schematic from Ergun et al. (2004), where $E_{\|}$ to a large extent takes the form of localized double layers (see Sect. 3.4)

and were interpreted as a static, U-shaped electric potential structures, with an upward parallel electric field in the centre (Fig. 5a), based on measurements from the Injun-5 (Gurnett 1972) and S3-3 satellites (Mozer et al. 1977). These static electric potential structures are generally associated with the quiet arcs we discuss here (e.g. Marklund et al. 2011b; Li et al. 2013; Hull et al. 2016), and the central part with the upward-directed parallel electric field usually called the 'auroral acceleration region (AAR)'. More dynamic auroral forms can, on the other hand, be associated with time-varying electric fields associated with e.g. medium-frequency Alfvén waves. This type of acceleration is discussed in several companion papers in this issue. In this section we will discuss high-altitude observations associated with the static, U-shaped auroral acceleration potentials.

3.1 Overview

The basic configuration of the U-shaped electric potential structure, with converging perpendicular electric fields, has been confirmed by measurements from several satellites, such as DE-1 and 2 (Weimer et al. 1985), Viking (Block et al. 1987), FAST (McFadden et al. 1999), and Polar (Hull et al. 2003b). In particular the high-time resolution measurements from the FAST satellite have yielded detailed knowledge of the properties of the auroral acceleration region. An extensive review of results up until 2002 is given in the ISSI Auroral Plasma Physics book (Paschmann et al. 2003). We will mention some of them below, but will mostly focus on more recent results.

3.2 A Satellite Pass Above the Auroral Region

We will first briefly discuss a typical FAST pass above the auroral region, reproduced from Paschmann et al. (2003) and shown in Fig. 6. The current directions can be determined by

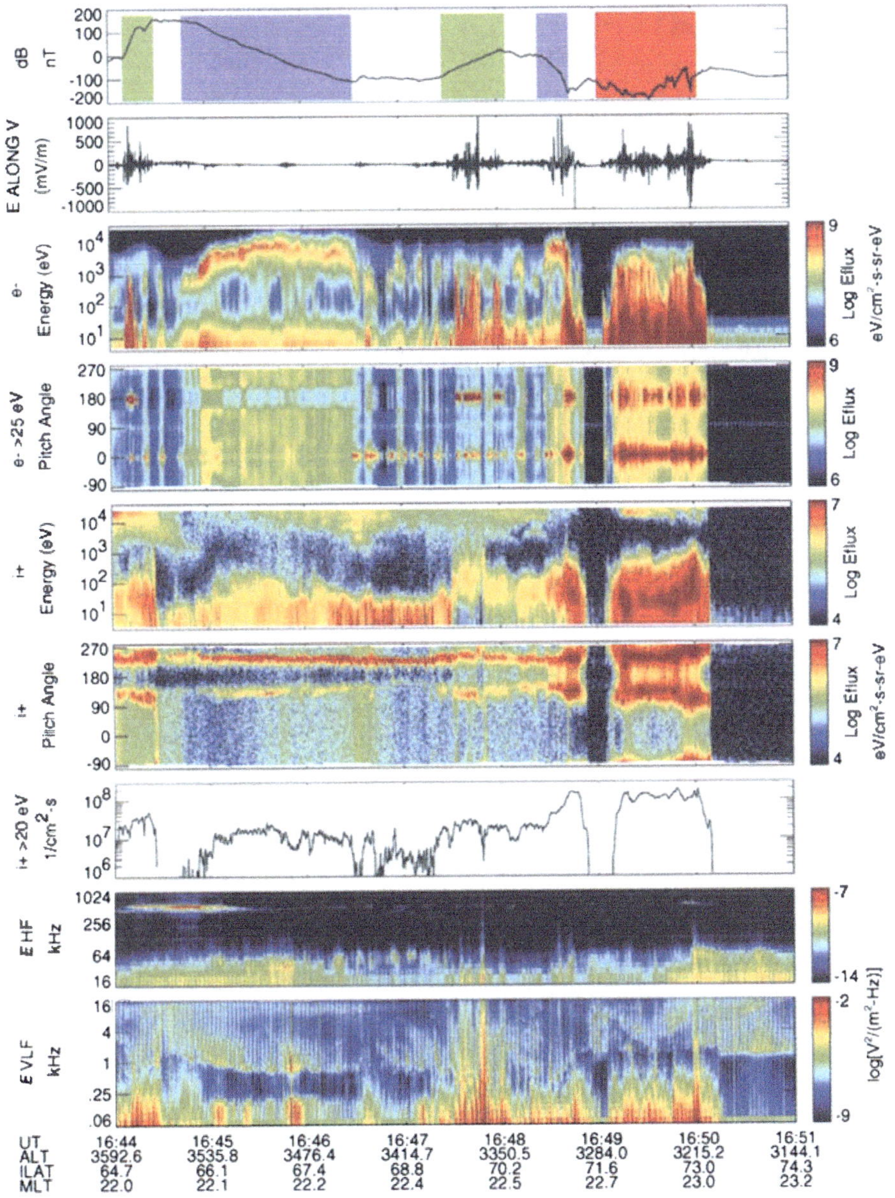

Fig. 6 A FAST satellite pass from the northern hemisphere auroral region. From top to bottom is shown east-west perturbation magnetic field *dB*, DC electric field, electron and ion spectrograms vs. energy and pitch angle, integrated upward ion flux, and electric field wave activity in two frequency intervals. Reproduced from Paschmann et al. (2003)

the gradient of *dB*, and are indicated in the figure: green for downward, purple for upward, and red for the changing current directions of the Alfvénic aurora, discussed by Forsyth et al. (this issue). The pass starts with a short traversal of a downward current region. Such regions will be discussed in Sect. 3.5. Between approximately 16:44:50 UT and 16:46:30

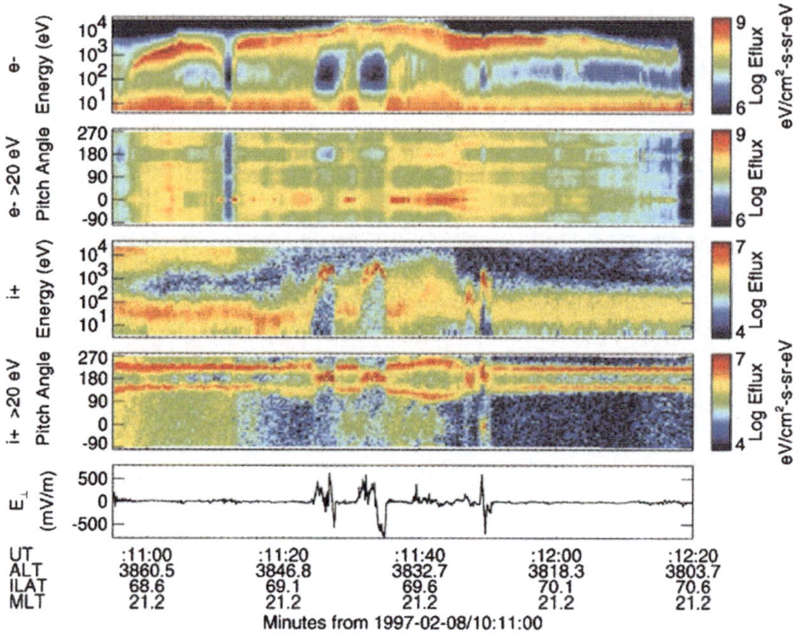

Fig. 7 A FAST pass across an upward current region. Adapted from (McFadden et al. 1999). From top to bottom is shown electron energy flux spectrograms versus energy and pitch angle, the same for ions, and the perpendicular (to the magnetic field) electric field

the spacecraft is inside a region of upward current. The electron measurements here show a typical signature of a large-scale inverted-V structure, being accelerated to maximum energies of close to 10 keV in the central current sheet. The lower energies of the electrons at the edges results in the typical inverted-V shape, and is consistent with electron acceleration by the U- (or 'V'-) shaped potentials shown in Fig. 5. This whole structure is also consistent with a large scale auroral arc, although the finer-scale structures in the electron inverted-V can correspond to individual smaller-scale arcs embedded inside the large arc (cf. Sect. 5). The high-energy electrons have a wide pitch-angle distribution, which reflects their plasma sheet origin. The low-energy population is a combination of ionospheric and atmospheric secondary electrons and photoelectrons emitted from the spacecraft. There is no sign of an upward accelerated ion beam in the upward current region, only ion conics created by transverse heating of ions and adiabatic transport along the field lines. This, together with the absence of large electric field signatures in association with the inverted-V boundaries, indicates that the spacecraft passes below the potential structure in this case. In contrast Fig. 7, from another FAST pass across an upward current region, shows clearly monoenergetically accelerated ion beams, also associated with very large electric fields. This is a consequence of the spacecraft being located above the lower border of the U-shaped potential structure for these times. Note that for these excursions into the auroral acceleration region, almost all of the low- and medium energy particles are absent. This is an example of the auroral density cavity, which will be discussed briefly below. These excursions into the AAR represent either small-scale spatial variations at its lower boundary ('auroral fingers', McFadden et al. 1999) or temporal variations of the position of the lower boundary.

3.3 Field-Aligned Current Sheets at High Altitude

There is long-standing proof that the auroral acceleration and precipitation takes place in regions of upward field-aligned current (Paschmann et al. 2003, and references therein). This is seen in terms of the large-scale Region-1 and -2 currents (Iijima and Potemra 1976), where the auroral oval emissions tend to take place in the Region-1 upward current in the premidnight region, and in the likewise upward-directed morningside Region-2 current (Paschmann et al. 2003). But above all, this is true on scales of individual auroral arcs and acceleration potential structures (where local small-scale upward currents can also be located within the general downward Region-1 and -2 current regions) (e.g. McFadden et al. 1999; Paschmann et al. 2003; Figueiredo et al. 2005). Johansson et al. (2007) studied the scales sizes of upward field-aligned currents and associated converging acceleration electric field structures, at an altitude of 4–7 R_E. They found scale sizes of 6.2 ± 2.6 km and 4.7 ± 0.45 km for the currents and electric fields, respectively, based on 85 Cluster observations of such structures.

The association between upward field-aligned current and upward-directed electric fields, accelerating electrons downward is consistent with the Knight relation (Knight 1973). This requires a parallel potential drop to allow the current carrying electrons overcome the mirror force and penetrate all the way to the ionosphere, where the currents can be closed by perpendicular ionospheric currents, maintaining current continuity. For a large range of field-aligned current densities this relation takes a linear form

$$j_{\parallel} = K U_{\parallel} \tag{1}$$

where $j_{\parallel}$ is the field-aligned current density, $U_{\parallel}$ is the total field-aligned potential drop, and K is the Knight or Lyons-Evans-Lundin constant (Knight 1973; Lundin and Sandahl 1978; Lyons et al. 1979). This is further discussed in the companion paper by Lysak et. al. (this issue).

The upward currents are often located close to and connecting to downward-directed field-aligned currents, constituting a balancing current, being connected via an ionospheric closure current, as described above (see also Sects. 4 and 3.5). Johansson et al. (2007) studied the scale sizes also of these currents at Cluster altitudes, and found similar mean and median scale sizes as for the upward currents. Notably this similarity of sizes is not seen in the lower ionosphere, where sounding rocket studies show downward current regions adjacent to arcs as being much narrower than their associated upward current neighbors (Lynch et al. 2015; Clayton et al. 2019b).

3.4 Structure of the Auroral Acceleration Potential

With single spacecraft measurements of the acceleration potential structures, it is difficult to disentangle temporal and spatial effects. With the Cluster spacecraft this is possible to some extent; a first example of this is the study by Marklund et al. (2011a), where the two-dimensional structure of the acceleration potential was revealed to consist of a two large U-potentials within a wider potential well, combined with an S-shaped potential (Fig. 8). One point made by the authors is that the scale sizes associated with the structure is very different when observed at different altitudes. Here we also introduce the S-shaped potential, which in contrast to the U-shaped one sustains a parallel electric field without uncoupling the perpendicular potential drop between the magnetosphere and the ionosphere (e.g. Mozer et al. 1980; Israelevich et al. 1988; Hwang et al. 2006b). The role of S-shaped potential structures in auroral acceleration is still unclear as their relative occurrence compared to U-shaped structures has not been established.

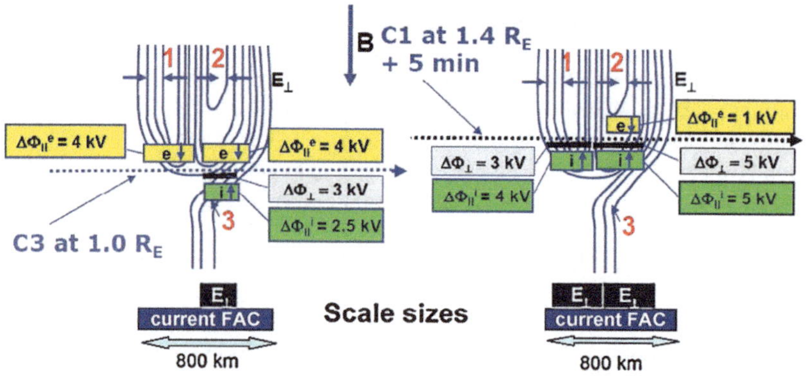

Fig. 8 Two-dimensional structure of the acceleration potential, based on measurements from Cluster 1 and 3 which passed through the potential structure with a time separation of five minutes, and a altitude separatin of 0.4 R_E. (Marklund et al. 2011a). Shown are equipotential structures (separation 1 kV), the widths of the down-going electron and up-going ion distributions (small yellow and green boxes), and estimates acceleration potentials (larger boxes). The potential estimates are based on upgoing ion energies (green), integrated perpendicular electric field (grey), and downgoing electrons (yellow). The width of the upward current region is indicated by the blue boxes, and the black lines and boxes indicate where the measured electric potential deviated appreciably from the ambient level

Many observations indicate that the potentials show considerable internal structure (e.g. Mozer and Kletzing 1998; McFadden et al. 1999; Hull et al. 2003b; Marklund et al. 2011a; Alm et al. 2015). Such internal structure may be due to spatial variations in the field-aligned current strength and/or magnetospheric source plasma, or temporal evolution of the plasma along the flux tube, as suggested by McFadden et al. (1999). This type of structure is potentially a strong observational test on theories of generation of auroral acceleration potentials, and should be studied in more detail.

Perpendicular Spatial Scales One expects the perpendicular (to the geomagnetic field) scale sizes of the acceleration potentials to match those of the shortest dimension (typically north-south) of the optical discrete arcs. We are aware of no direct comparison of a quiet arc optical observation and a high-altitude electric field measurement giving the acceleration region scale size, but Figueiredo et al. (2005) report on conjugacy between Cluster measurements of the acceleration region at high altitude and optical observations of a dynamic horn arc associated with an auroral substorm-related surge. Many observations exist, however, relating optical arc widths to the resulting accelerated precipitation of electrons from the closure of these acceleration structure (e.g. Hallinan and Stenbaek-Nielsen 2001; Grubbs et al. 2018b).

A comparison between arc sizes and potential structures can instead be based on statistical investigations of potential scale sizes, but there are very few such studies. Johansson et al. (2007) gives scale sizes of converging electric field structures of 1–10 km with a median of 5 km (mapped to ionospheric altitude), based on 85 measurements. Care is needed in comparing this result with optical measurements, since the authors define scale size as the half width of the electric field structure, and not of the potential. It is therefore likely that their result is more applicable to the boundary of the U potentials, and not the whole potential structure (cf. Fig. 5). Similar sizes were reported by Hull et al. (2003b), with the same reservations. More statistical studies, with a view to compare with optical and low-altitude electron measurements are desirable.

Altitude Distribution Another important observational constraint for acceleration mechanisms is the altitude distribution of the acceleration potential. Several studies have addressed this by various methods (Fig. 9). Statistical investigations of the change of the perpendicular electric field with altitude has been used by Lindqvist and Marklund (1990) and Weimer and Gurnett (1993). The former study showed that the acceleration region extended to an altitude of 1.7 R_E, but did not give a clear lower limit, due to the limitations of the spacecraft orbit. Using DE1 data Weimer and Gurnett (1993) found that most of the potential drop was found below 1 R_E, but that a minor part of it extended up to 3 R_E. A similar study based on statistics of ion beam energies measured by Polar placed most of the potential drop between 1 and 2 R_E (Mozer and Hull 2001). The latter result is consistent with a recent study by Sadeghi and Emami (2019), who report that 30% of the potential drop is located between 0.9 and 1.2 R_E. In contrast, based on Cluster measurements on ion and electron beam energies, Alm et al. (2015) reported that 50% of the potential drop is located above 2 R_E with 20% remaining at an altitude of 3.4 R_E. Above 4.6 R_E the remaining parallel potential quickly drops to zero, perhaps indicating the presence of a high-altitude double layer (cf. Sect. 3.4).

Statistical studies of course include a wide variation in plasma properties along the flux tubes depending on e.g. season and geomagnetic activity. E.g. in a study based on Akebono data, Morooka and Mukai (2003) show that the altitude extent of the acceleration potential shifts to higher altitudes during winter. But such studies do not say much about instantaneous distribution of the potential drops. Information on the instantaneous distribution of the parallel potential drop can be estimated by some different techniques. Marghitu et al. (2006) and Forsyth et al. (2012) use a technique to determine the top of the acceleration region based on the anisotropy of the electron distributions. For both case studies, the top altitude varied considerably during passes of wide auroral structures (0.6–1.5 R_E and 1.6–2.8 R_E, respectively), indicating either large temporal or internal spatial variations.

Simultaneous spacecraft measurements at different altitudes of a stable potential structure have been used in a few recent studies. Marklund et al. (2011a) used two Cluster spacecraft passing the same acceleration structure (using combined particle and electric field measurements to verify that the total potential drop remained constant), and showed that for one acceleration structure the whole acceleration potential of 4 kV was located between 1 and 1.4 R_E. Using the same method, Sadeghi et al. (2011) concluded that 18% of the total potential drop was located between 1.13 and 1.3 R_E.

In order to make progress in constraining the parallel potential distribution, we suggest that systematic studies combining the techniques of Marklund et al. (2011a) and Marghitu et al. (2006) be performed.

Within the auroral acceleration region the ionospheric plasma that originally could be present on that field-line is expelled, together with part of the magnetospheric plasma. This results in the low-density region of the 'auroral cavity'. It has been proposed that the auroral cavity is well collocated with the acceleration region (e.g. Ergun et al. 2002b), while recent results indicate that it can extend to altitudes above the top of the acceleration region (Alm et al. 2015).

Potential Drop Verification that the electric potential structures are responsible for auroral acceleration can be carried out by combined particle and electric field measurements. The total parallel potential drop within a U-shaped potential structure located below a spacecraft can be determined by integration of the perpendicular electric field measured from a point outside of the structure to the centre of the structure. This potential drop should be the same as the energy of the ion beam if the potential structure below the spacecraft is static on the time scale of the ion transit time. The total acceleration potential is then obtained by adding

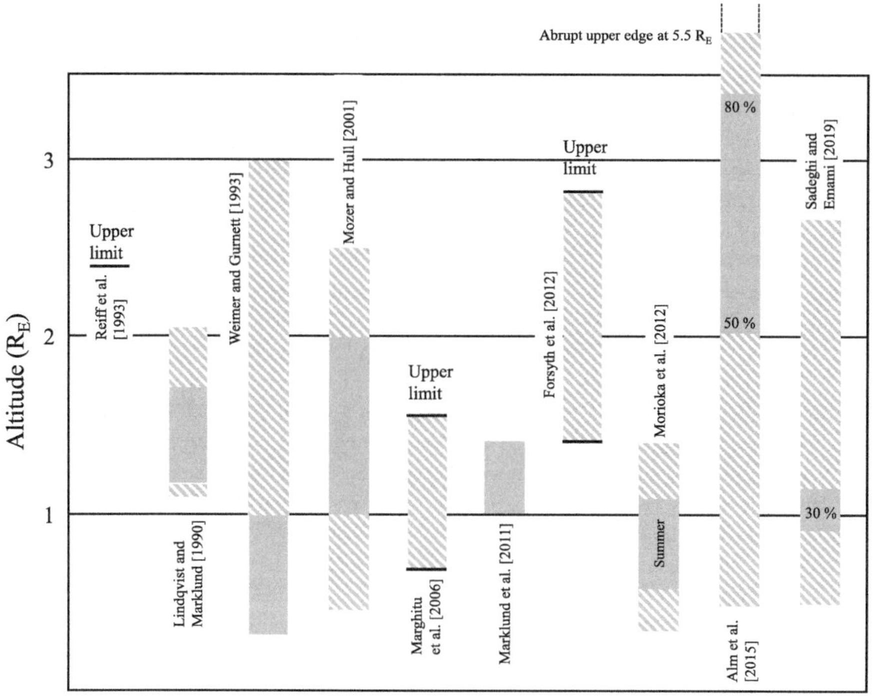

Fig. 9 Altitude distribution of the auroral acceleration region, reported by various authors. The hatched regions correspond to uncertain or varying results. Updated from Karlsson (2013)

the inferred acceleration potential of the electrons accelerated by the part of the potential structure extending above the spacecraft. Good agreement between ion beam energies and integrated electric potential drops was found using FAST measurements (McFadden et al. 1999), but there are few studies determining the total potential drop by this method. Case studies using Cluster passes through the acceleration region (additionally showing that the total potential drop stays constant on time scales similar to the satellite separation time) give expected potential drops of 4–7 kV (Marklund et al. 2011a), 3–3.3 kV (Sadeghi et al. 2011), 4.2–17.7 kV (Alm et al. 2013) (excluding a dayside event), and 2 kV (Forsyth et al. 2012). However, comprehensive statistical studies are missing.

Direct Measurements of $E_\parallel$ Direct observation of the parallel electric field $E_\parallel$ within the acceleration region by conventional double probe techniques is problematic due to several technical limitations, such as spacecraft shadowing of particle fluxes and asymmetric work functions of the probes (e.g. Karlsson 2012). However, when the parallel electric fields are large enough, detection may be possible. Mozer and Kletzing (1998) and Hull et al. (2003a) used the three-axis electric field measurements of the Polar satellite to demonstrate the likely existence of upward-directed parallel electric fields of a magnitude of 200–300 mV/m at an altitude of around 1 R_E. Ergun et al. (2002a) reported on a large number of FAST observations of upward-directed parallel electric fields in the form of strong, oblique double layers at the lower boundary of the auroral cavity. They concluded that these double layers contributed 10–50% of the total potential drop. There is also some evidence for the existence of double layers at somewhat higher altitudes, inside the auroral cavity (Ergun et al. 2004) (see

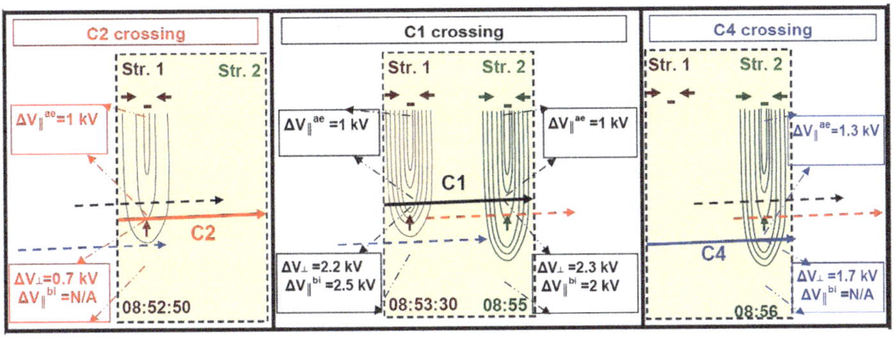

Fig. 10 Schematic of temporal evolution of two acceleration potentials based on measurements from Cluster 1, 2, and 4 (Sadeghi et al. 2011). The boxes below the spacecraft paths show the potentials below the spacecraft orbits, inferred from the electric potential drop and from the characteristic energy of the upgoing ions, while the boxes above the paths show the potential above the spacecraft, based on the energy of downgoing electrons. The crossing times of the structures are indicated at the bottom of the figure

also Fig. 5). For a more detailed discussion of double layers in the auroral region, see also Andersson and Ergun (2012).

Time Scales While a defining characteristic of the quiet arc is its relative temporal stability (cf. Sect. 2), the potential structures do change with time. The life time, and typical time scales for changes in the potential drop, size and location of the acceleration structures are important observational constraints to models of such structures. It is a striking fact that optical observations of quiet arcs show that their lifetimes can be tens of minutes (Knudsen et al. 2001), or even several hours (Galperin 2002). For acceleration structures, again, there exist no large statistical study, but a number of case studies based on multi-spacecraft observations exist.

Cluster multi-spacecraft observations show that acceleration structures can keep a clearly identifiable identity during up to 50 min (Figueiredo et al. 2005), and have very little change in total potential drop and width for around 5 minutes at an altitude of 1 R_E (Marklund et al. 2011a). Sadeghi et al. (2011) show, on the other hand, that an acceleration structure (structure 1 in Fig. 10) increased its potential drop significantly (from 1.7 kV to 3.5 kV) during 40 s. Interestingly, this increase took place exclusively below 1.3 R_E, while the potential drop stayed constant at 1 kV above this altitude. A similar result was reported by Forsyth et al. (2012) with an increase of the acceleration potential of 2 kV below an altitude of 4600 km during 150 s, while the potential above this altitude remained unchanged. While the field-aligned current density increased during this event, it remained constant in the Sadeghi et al. (2011) event, indicating changing plasma properties on the field line, as suggested by McFadden et al. (1999) in the context of auroral fingers.

While the study of temporal properties based on direct measurements of the electric potential structure is based on few cases, there is some supporting statistical studies of the resulting downward accelerated electrons. In an early small statistical study by Thieman and Hoffman (1985), the DE satellites were used to study the temporal stability of inverted-V electron precipitation. They concluded that well-identifiable inverted-V events could have life times of at least 18 minutes, and that the growth and decay of the events had similar time scales. A larger study by Boudouridis and Spence (2007), using two DMSP satellites showed that electron precipitation structures have a high correlation between the spacecraft up until around 2 min for scales sizes of 50–100 km, with lower correlation times for smaller scales.

More recently two studies have investigated the temporal stability of field-aligned currents, using the ST5 satellites. Gjerloev et al. (2011) reports that field-aligned currents with scale-sizes greater than 200 km (mapped to ionospheric altitudes) have life times of the order of a minute, consistent with the result of a smaller study by Le et al. (2009). The somewhat disparate results of these studies, as well as the unclear exact relation to discrete auroral arcs means that there is a need for more detailed statistical studies of temporal properties of structures clearly identified with discrete arcs.

When interpreting results on life times of aurora accelerations structures, as well as of optical signature of arcs, one must keep in mind that there is a certain uncertainty in its definition. A concept such as 'clearly identifiable identity' is somewhat vague and can be interpreted in different ways by different authors.

3.5 The Return Current Region

As mentioned in Sect. 3.3, the upward current region is often associated with a region of downward or 'return' current. Originally this region was thought to be a passive region, where the current was simply carried by thermal ionospheric electrons. The first indication that the situation was more complicated came from early observations of upward-directed strongly field-aligned fluxes of electrons accelerated up to energies of 500 eV (Klumpar and Heikkila 1982), as well as observations of extremely large, diverging electric fields at low altitudes (1400–1800 km) (Marklund et al. 1994; Karlsson and Marklund 1996).

Measurements from the FAST satellite greatly increased the observational knowledge of the return current region. Results from earlier than 2002 are covered in detail in the ISSI Auroral Plasma Physics book (Paschmann et al. 2003); we will here give a short recapitulation and update. Figure 6 shows a typical return current region from approximately 16:47:30 UT to 16:48 UT. The downward current (consistent with the gradient of the magnetic field) is carried by upgoing electrons with energies up to the order of 1 keV (Carlson et al. 1998b; Elphic et al. 1998). These electrons have a very low perpendicular temperature, which indicates their ionospheric origin (Carlson et al. 1998b). The integrated potential of the diverging electric fields collocated with the beams typically matches the beam maximum energy, which is strong evidence that the beam electrons have been accelerated by a quasi-static parallel electric field. However not all beams show a good correlation with the electric potential (Andersson and Ergun 2006), which may indicate acceleration by dynamic processes, e.g. Alfvén waves or evolving electric fields (Marklund et al. 2001; Andersson et al. 2008). Direct measurements of the parallel electric field have shown that it often takes the form of double layers with a parallel thickness of a few Debye lengths, moving upward along the field line approximately at the ion acoustic velocity (Andersson et al. 2002; Ergun et al. 2003).

The electron beam is strongly thermalized (Carlson et al. 1998b; Andersson and Ergun 2012), with a temperature comparable to the beam energy, which leads to a counter-streaming tail of the electron distribution. This thermalization is very likely connected to the presence of strong broadband electrostatic low frequency (BBELF) turbulence (e.g. Shen et al. 2018; Lynch et al. 2002), although the causal relation is not clear. The electron beams have at times been observed all the way up to an altitude of 3–4 R_E (Wright et al. 2008). At the low altitude end, sounding rocket observations in the F-region nightside often observe sharply localized return current signatures in magnetic field data, immediately adjacent to discrete arcs and tightly coupled to DC electric field structures (Lynch et al. 2015; Zettergren et al. 2014; Clayton et al. 2019b). As discussed in the next section, the horizontal current closure between adjacent upward and downward field aligned current structures demonstrates interesting ionospheric physics.

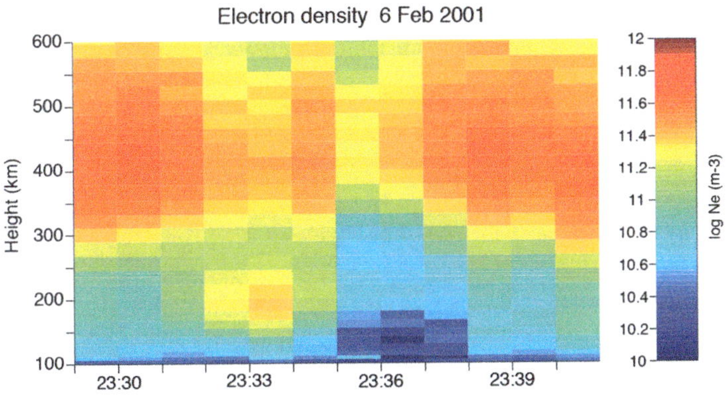

Fig. 11 EISCAT observations of ionospheric plasma densities, with a deep cavity (23:35 UT to 23:38 UT) in the downward current region. Adapted from Aikio et al. (2004)

The intense BBELF wave activity is believed to result in perpendicular heating of the ions in the return current region, which due to the conservation of the magnetic moment produces ion conics at higher altitudes (Carlson et al. 1998b; Lund et al. 1999; Lynch et al. 2002). Thus there is a considerable outflow of ions from the ionosphere in the return current region, something that considerably changes the plasma environment in the region.

Ions are also moving downwards in the form of accelerated beams (Andersson et al. 2002; Cattell et al. 2002; Hultqvist 2002), oppositely to the situation in the upward current region. However the combination of downward acceleration of ions and ion heating due the waves result in a 'pressure cooker' effect, and together with the requirement for the double layer to be in pressure balance, results in an anti-earthward motion of the double layer. The result is an efficient process to remove ions from the ionosphere (Hwang et al. 2008).

This asymmetry with the upward current region also extends to optical emissions; the return current region often is associated with the 'black aurora' (e.g. Trondsen and Cogger 1997; Peticolas et al. 2002), which are well-delimited regions of a lack of auroral emissions both of the diffuse and accelerated types. Both the ion beams and the lack of auroral emissions are consistent with a region of downward-directed $E_{\|}$, which can both accelerate ions downwards, and block access to the atmosphere and ionosphere for plasma sheet electrons (Archer et al. 2011). On the other hand, it has been suggested that rather than the electrons being blocked from access by en electric field, some magnetospheric mechanism may suppress the scattering of electrons into the loss cone in the first place (Peticolas et al. 2002; Blixt et al. 2005; Gustavsson et al. 2008; Obuchi et al. 2011; Sakanoi et al. 2012). This second type of black aurora probably takes place in the diffuse auroral region, and is not necessarily associated with a return current region. Similarly to the ordinary aurora, the black aurora exhibits not only elongated arcs, but also other types of geometrical shapes (Trondsen and Cogger 1997; Kimball and Hallinan 1998; Obuchi et al. 2011; Sakanoi et al. 2012), something that for the reurn current black aurora seems to also be reflected in variations in the shape of the electric field acceleration structures (Hwang et al. 2006a,b; Johansson et al. 2006).

An important consequence of the return current is the creation of deep density cavities in the E and F regions, created by a plasma outflow and increased recombination in the regions where the downward current closes to perpendicular currents (Doe et al. 1993, 1994; Aikio et al. 2004; Zettergren et al. 2014). Simulations by Zettergren and Semeter (2012), Karlsson et al. (2005) show that the depletion of E-region plasma occurs on a fast time-scale of 5–30 s depending on the current-system scale size. In the F-region, enhanced recombination from conversion of the plasma to molecular ions, a comparatively slower process (120–300 s), contributed to the depletion. The resulting low plasma densities and associated low conductivities may provide an important feedback to the magnetosphere (e.g. Karlsson et al. 2007). These structures are further discussed in the companion paper by Lysak et al.

4 Horizontal Dynamics, Ionospheric Current Continuity

A key element for auroral electrodynamics is the electric field, **E**, driven by magnetospheric convection and mapped to ionospheric altitudes along magnetic field lines. The ionospheric electric field differs from the magnetospheric electric field by the potential drop in the auroral acceleration region; it can also be a source/driver through local polarization charges and flows. Altogether, the electric field distribution and ionospheric convection are related to each other via the frozen-in condition, where $\mathbf{E} = -\mathbf{v} \times \mathbf{B}$ (down to the F-region), and thus flow fields around aurora, as read, e.g., by motion of visible structures and boundaries in TV images, can include information on the electric field structure. At the same time, the electric field and the ionospheric current, (which can be determined via Ohm's law, once the ionospheric conductance is known), are coupled through the continuity equation. The current system of the auroral arc, governed by ionospheric closure of the field-aligned current and Ohm's law, is summarized under Sect. 4.2.

4.1 Horizontal Dynamics

Studies of the auroral zone as seen by conjugate observations of SuperDarn and auroral imagery set the stage for a discussion of the horizontal dynamics of auroral activity. Zou et al. (2012, 2009a) and Bristow et al. (2003) show examples following the evolution of substorms and westward surges, showing the relationship between flow structures seen by the radars, and the motion of the arc systems as the oval evolves. Flows move around arcs and around the Harang discontinuity, and the oval itself moves equatorward and poleward, as the auroral system responds to the reconfiguration of current structures. The relationship between the flow structures and the visible signatures depends on the morphology: activity, local time, magnetic latitude, and sequence of events. More local studies using PFISR or EISCAT together with auroral imagery (Nishimura et al. 2014; Aikio et al. 2002; Semeter et al. 2010) explore the local signatures of flows near auroral arcs during the growth phase of substorms; the even more local studies of Archer et al. (2017) and Clayton et al. (2019b) show the finely detailed relationships between these features illustrated by conjunctions between in situ field observations and ground imagery. In all these cases, the underlying physics relating the auroral brightness—the precipitating energetic electrons—to the flows—the ionospheric DC electric field—is governed by the current continuity equation, as we discuss below.

Historically, explorations of the physics of these current continuity rules have focussed on idealized sheetlike arcs, with limited to no variation along the length of the arc. While sheetlike arcs—quiet discrete arcs, as the topic of this chapter—form a large portion of auroral signatures, their evolution and variability is of great interest; and, fully understanding the three dimensional structure of a simple sheetlike arc is critical for moving on to more complex situations. Three dimensional time dependent ionospheric models such as GEMINI (Zettergren and Semeter 2012; Zettergren and Snively 2015, 2019; Zettergren et al. 2015) and GITM (Ridley et al. 2006) allow us to explore these relationships in more complicated environments. Below we begin with the idealized arc and work through the implications of current continuity for increasingly complex morphologies. Even simple sheetlike arcs have interesting volumetric conductivity behaviour, from which much can be learned.

4.2 Ionospheric Current Continuity

The ionospheric closure of the field-aligned current can be addressed starting from the (quasi-)steady form of the current continuity equation, $\nabla \cdot \mathbf{J} = 0$, where time variation of the charge density is neglected. In a reference system where the field-aligned current, $j_\parallel$, is assumed perpendicular to ionosphere (roughly true in the auroral region), current continuity and Ohm's law can be worked out to

$$j_\parallel = \nabla \cdot \mathbf{J}_P + \nabla \cdot \mathbf{J}_H = \Sigma_P \nabla \cdot \mathbf{E} + \nabla \Sigma_P \cdot \mathbf{E} - (\nabla \Sigma_H \times \mathbf{E})_\parallel - \Sigma_H (\nabla \times \mathbf{E}))_\parallel \qquad (2)$$

where it is also assumed that magnetic field lines are equipotential and, consequently, the electric field, $\mathbf{E}$, is constant and maps through the F and E regions. Ionospheric Pedersen and Hall conductivities and currents can be integrated across the ionospheric height to the conductances Σ_P, Σ_H, and the currents $\mathbf{J}_P$, $\mathbf{J}_H$, respectively. The ionosphere is thus represented as a boundary layer and the divergence operators of the continuity equation operate in the 2-D plane of this layer. Nonetheless, the thick uniform arc model below takes into account also the thickness of the ionosphere; and, various altitudinally striated ionospheric models exist and are used for different studies (Lysak 1999; Ridley et al. 2006; Zettergren and Semeter 2012).

Even with an altitude-striated picture, the cumulative effect from height integrated conductances in the continuity equation still provides a valuable summary of the system. Equation (2) is written in the system of the neutral atmosphere, whose motion (the neutral wind) is often neglected, because it is rarely available, and the equivalent electric field is typically small compared to auroral electric fields. However, this is not always true, and a comprehensive examination of the neutral wind influence on auroral electrodynamics is yet to be performed. One nightside case study event of a discrete arc (from a sounding rocket event) explored, among other things, the effects of neutral winds using the GEMINI model (Lynch et al. 2015; Zettergren et al. 2014), showing the range of possible neutral wind effects on the terms of the continuity equation; for that event the effects could reach 10% depending on whether the neutral winds were sympathetic with the plasma motion.

Equation (2) is independent of the geometry of aurora and in general used by neglecting the inductive last term. While the last term is indeed negligible under (quasi-)steady conditions, it becomes important e.g. during substorms, when it accounts (Yoshikawa 2002b,a) for the inductive rise (and fall) of the divergence free part of the ionospheric current (not coupled to field-aligned current).

The geometry-independent form of current continuity and Ohm's law underlies a broad range of auroral studies, most of them based on ground observations (like magnetic field and optical data), addressing 2-D auroral features on meso- to large scale (see, e.g., the review by Vanhamäki and Amm 2011). More recently, 2-D investigations became possible also on small scale, by experiments like Auroral Structure and Kinetics (ASK, e.g., Lanchester and Gustavsson 2012). Next we explore more closely just the particular subset of elongated auroral forms, also known as auroral arcs, where the 2-D description can be further adapted to the (quasi)1-D arc geometry, with gradients (much) steeper across the arc than along the arc.

Equation (3) is adapted for this arc geometry, with ξ the direction across the arc (typically close to geomagnetic northward) and η the direction along the arc (typically close to geomagnetic eastward).

$$j_{\|} = \Sigma_P \frac{\partial \mathbf{E}_\xi}{\partial \xi} + \frac{\partial \Sigma_P}{\partial \xi}\mathbf{E}_\xi + \Sigma_P \frac{\partial \mathbf{E}_\eta}{\partial \eta} + \frac{\partial \Sigma_P}{\partial \eta}\mathbf{E}_\eta + \frac{\partial \Sigma_H}{\partial \eta}\mathbf{E}_\xi - \frac{\partial \Sigma_H}{\partial \xi}\mathbf{E}_\eta \tag{3}$$

The various terms on the right hand side are colored in blue (terms 1, 2), red (terms 3, 4, 5), or green (term 6), according to the related arc model, indicated with the same color at the top of Fig. 12. Thus, the blue terms are related to the thin uniform 1-D arc model, where the ionospheric thickness and gradients along the arc are neglected. The red terms are related to the thin non-uniform 2-D arc model, where ionospheric thickness is again neglected but gradients along the arc are allowed, even though the 1-D backbone is preserved. Finally, the green term is related to the Cowling effect (Cowling 1932; Chapman 1956), that can be understood by a thick uniform 2-D arc model, where the thickness of the ionosphere is no longer neglected, but gradients along the arc are ignored. Specific features of these three arc models are summarized below, following the review by Marghitu (2012).

In the case of the thin uniform 1-D arc, the field-aligned current is closed in the ionosphere by Pedersen current, driven by an electric field normal to the arc, while the Hall current electrojet (EJ) along the arc, driven by the same electric field, is divergence free. The 1-D symmetry is often extended to oval scales (e.g., Sugiura 1984), where gradients in the meridional direction dominate, under steady state, over gradients in the azimuthal direction.

Despite the seeming simplicity of the 1-D model, radar, rocket, and satellite observations exhibit a broad range of electric field profiles around the arc, addressed systematically by Marklund (1984) and shown to depend on the relative contributions to current closure by ionospheric polarization and field-aligned current. Subsequently, Chap. 6 of the ISSI book on Auroral Plasma Physics (Paschmann et al. 2003), included a follow-on examination of the electric field pattern around the arc, depending on the orientation of the background electric field with respect to the arc, under negligible (Fig. 6.2) and significant (Fig. 6.5) field-aligned current. In terms of (3), the field-aligned current above the arc (l.h.s. term) induces the enhanced conductance slab. This, in turn, is responsible for the current driven across the arc by the polarization electric field (first l.h.s term), generated by polarization charges accumulated at the arc edges. It also results in the current at these arc edges (second l.h.s. term), where the Pedersen conductance varies sharply. While these three terms balance each other under 1-D geometry, the field-aligned current on the l.h.s. is often small compared to the ionospheric current on the r.h.s.

Note the inherent limitation of this 'discrete' explanation of a continuous mechanism, explored further below in relation to Cowling channel, under the thick uniform 2-D arc model. Note as well that the relative weight of polarization and field-aligned current in providing ionospheric current continuity depends largely on the magnetospheric end of the auroral current circuit, whose structure is in general unknown, though the driving FAC can be measured. One cannot fully resolve the ionospheric end independent of the rest of the circuit, but a full picture of the ionospheric footpoint including its horizontally distributed current closure, as represented by (3), can be constraining. Recent advances in filtered groundbased imagery add data to this question, as the filtered imagery can provide 2D+time estimates of precipitating energy flux and characteristic energy, constraining possible magnetospheric sources (Grubbs et al. 2018; Clayton et al. 2019b). In situ observations of precipitation and fields have been used to infer magnetospheric drivers such as U-shaped potential structures (Hallinan et al. 2001).

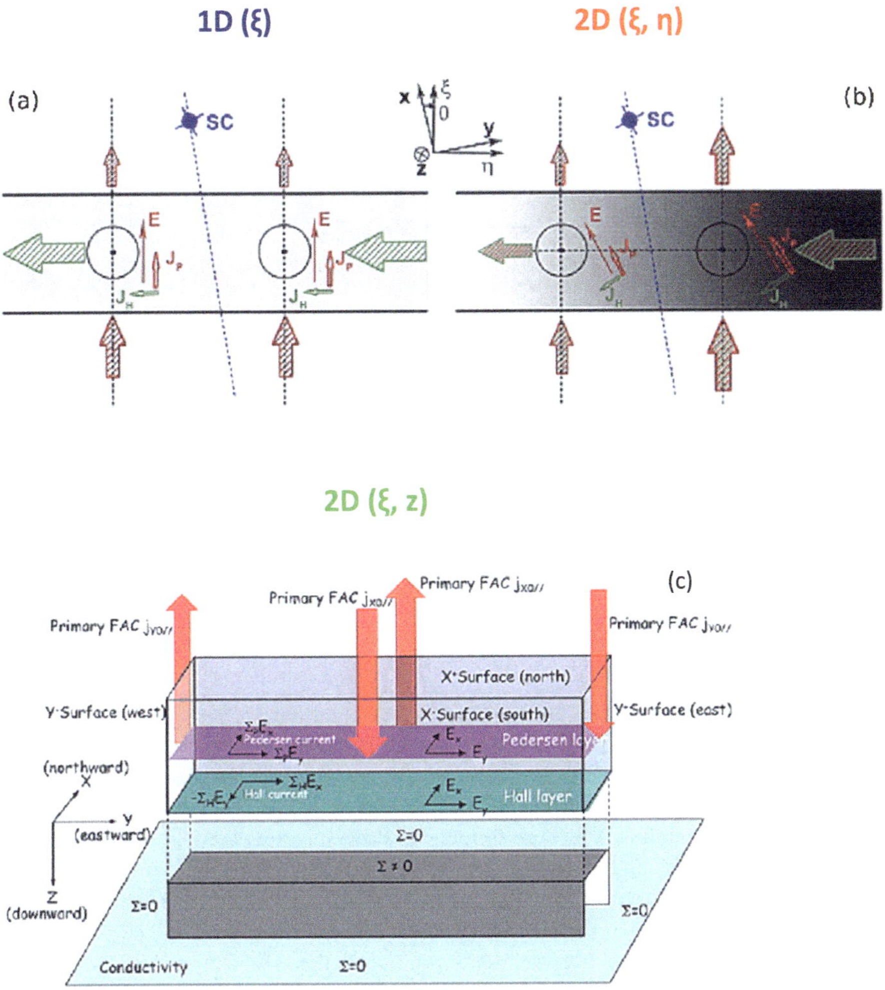

Fig. 12 (**a**) Thin uniform, (**b**) thin nonuniform, and (**c**) thick uniform (Cowling channel) arc models. Compound figure from Marghitu (2012), with panels (a) and (b) from Marghitu et al. (2011), and panel (c) from Fujii et al. (2011). In panels (**a**) and (**b**), the conductance, field-aligned current (FAC), ionospheric electric field, and ionospheric current are indicated by the gray shade, circles, solid arrows, and hatched arrows, respectively. Red and green arrows show the Pedersen and Hall components of the current. For simplicity, the conductance variation in normal direction is not shown (but is taken into account in the text). Note that in panels (**a**) and (**b**) the (ξ, η) frame is associated with the arc and (x, y) with the satellite overpass, while in panel (**c**), the (x, y) frame is associated with the arc. For consistency, x is replaced by ξ in the panel title

The use of a three-dimensional ionospheric model such as Gemini can help illustrate these relationships. An example is shown in Fig. 13, based on observations from PFISR and from filtered groundbased imagery of a quiet premidnight sheetlike arc over Poker Flat in winter 2017.

The case of thin non-uniform 2-D arc was substantiated by event studies of Marghitu et al. (2004, 2009, 2011), based on data observed by the FAST satellite (Carlson et al. 1998a). Data analysis by the newly introduced AuroraL Arc electroDYNamics (ALADYN) technique indicated that the electrojet divergence was significant at certain locations and,

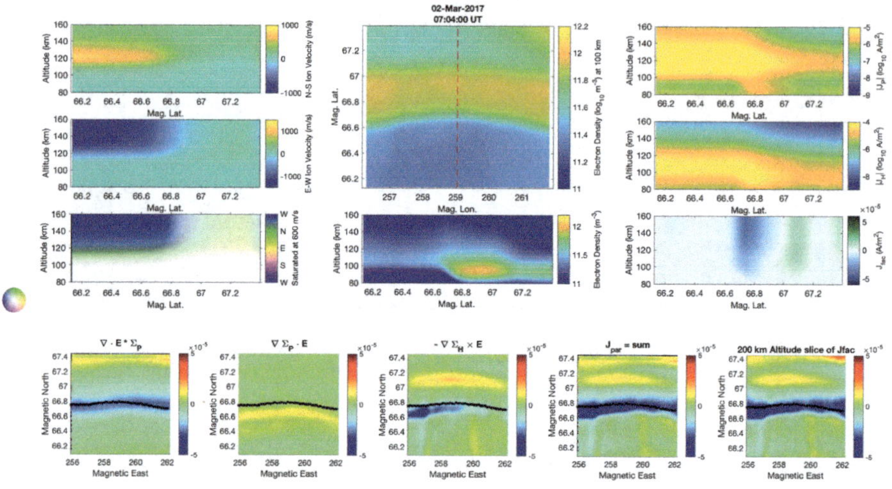

Fig. 13 Examples of GEMINI model output representing an event from the winter 2018 Poker Flat rocket campaign, as seen by PFISR and filtered groundbased imagery. The model is driven by precipitation maps created from the filtered imagery, and by two-dimensional electric field maps created by replicating the PFISR vvels (vector velocities) line of vectors along the extent of a sheetlike arc structure. The first set of panels show various cuts of flows, density, and current through the 3d modelled volume. The bottom set of panels show the pieces of (2) (Clayton et al. 2019a)

moreover, comparable to the FAC (field-aligned current) density, suggesting FAC closure also along the arc, by coupling to the electrojet. More recently, a statistical study by Jiang et al. (2015b), based on 181 growth phase arcs observed by FAST, provided further evidence for FAC–EJ coupling, by showing that the Pedersen current supposed to close the FAC across the arc is equal, on average, to just half of the field-aligned current (see Fig. 11 of this study). While the authors attributed the mismatch to underestimation of the ionospheric conductance, FAC–EJ coupling may play a role too.

Longitudinal gradients can become significant during the substorm growth phase, when the upcoming 1-D break-up arc 'prepares' to transform into 2-D aurora, and comparable to the meridional gradients at the substorm onset, when 1-D symmetry is lost. The mismatch found by Jiang et al. (2015b) can be regarded as a quantitative confirmation of this qualitative perspective, though a detailed examination of the data has not been performed so far.

The ALADYN technique can provide a local estimate of the electrojet divergence based on single-spacecraft data, but cannot indicate the respective weights of the longitudinal gradients in conductance and electric field (the red terms in (3)). In order to achieve such a goal, at least two satellites are needed, to sample the data at different locations along the auroral arc or oval, separated by a distance comparable to the electrojet length scale. Such a configuration is provided by the Swarm mission (Friis-Christensen et al. 2008), launched in November 2013. However, since precipitating particles are not directly measured, conductance proxies (e.g., like those of Robinson et al. (1987)) are more difficult to derive and the examination of the FAC closure in the longitudinal direction is yet to be done. Recent work however using imagery such as the THEMIS and REGO GBO databases (Mende 2008) can provide the necessary 2D+time context. An example case study is provided by the Isinglass sounding rocket study (Clayton et al. 2019b), where in situ flow data (along a rocket trajectory) is combined with ground imagery to replicate the trajectory at a variety of crossing points along the longitudinally-extended arc, in order to create a 2D flow field for driving the GEMINI model.

Until such a study is performed, a model based on a uniform FAC sheet and an electrojet divergence dominated by longitudinal gradients in conductance was suggested by Marghitu et al. (2009) (see in particular their Fig. 11), with the upward FAC above the arc fed by an electrojet of decreasing intensity—associated with decreasing conductance, related in turn to decreasing electron energy. The FAC can couple mainly to the Hall component of the electrojet for electron energies above some 5 keV, specific to active aurora, while coupling to both Pedersen and Hall component is possible at lower energies, as observed for quiet arcs. The model was further explored by Marghitu et al. (2011), who looked also at the direction across the arc, in the limit of low polarization. In this limit, supported for quiet aurora by the study of Amm et al. (2011), the results were found to be consistent with a divergence free Hall current, appropriate for quiet aurora. Moreover, (vanishingly) small tangential electric fields were found to characterize long arcs (as expected from the 1-D model), whereas substantial tangential electric fields, as observed, e.g., in the Harang region, were related to a higher probability of FAC coupling to Hall current, likely to happen during active times—in agreement with the dynamic character of the Harang region (Zou et al. 2009b). Many of these permutations can be explored with 3D ionospheric models such as GITM (Ridley et al. 2006) and GEMINI (Zettergren and Snively 2019). These models allow cuts along and across arc structures to illustrate how the different terms of the continuity equation interact, as illustrated in Fig. 13.

The thick uniform 2-D arc follows the Cowling channel model suggested by Amm et al. (2011) and Fujii et al. (2011), who addressed the competing contributions of polarization and FAC to ionospheric current continuity from a novel perspective. The textbook description of the Cowling channel emphasizes the multiplication effect on conductance and current along the channel, in a 1-D geometry: the primary electric field along the channel drives both a primary Pedersen current along the channel and a Hall current across the channel, which results in polarization charges at the channel boundaries (typically sharp gradients in conductance). The polarization electric field drives a secondary Pedersen current across the channel and a secondary Hall current along the channel, that adds to the primary Pedersen current and provides the multiplication effect, which scales with $(\Sigma_H/\Sigma_P)^2$. For a conductance ratio of around 3, which is large, but not unusual for active events, the current along the channel can be multiplied by one order of magnitude.

For simplicity, the textbook description concentrates on polarization and neglects the role of the FAC, that can discharge part of the polarization and reduce thus the multiplication effect. The result is a partial Cowling channel and relevant arc observations were discussed, e.g., in the review by Marklund (1984). More recently, Amm et al. (2011) formalized the efficiency of the Cowling effect, equivalent to the 'completeness' of the Cowling channel, in differential, geometry independent form, by a parameter C, with values between 0 and 1, indicating what fraction of the Hall current divergence is closed by the divergence of the polarization Pedersen current,

$$\nabla \cdot \mathbf{J}'_P = -C \nabla \cdot \mathbf{J}_H. \tag{4}$$

An efficiency of 1, equivalent to the complete Cowling channel, implies no FAC discharge of the polarization built up by the primary Hall current, while efficiency 0 is equivalent to no Cowling channel and full coupling of the Hall current For a 1D Cowling channel, the divergence of the Hall current reduces to the last (green) term in (3), while the divergence of the polarization Pedersen current is a fraction of the divergence of the total Pedersen current (the two blue terms), which includes also the primary Pedersen current, driven by the primary electric field, of magnetospheric origin. As noted already with respect to the 1D model, the weight of the polarization Pedersen current in the total Pedersen current, equivalent to knowing C, or the capability of the Hall current to build up polarization charges, (and not fully discharged via field-aligned currents), cannot be derived by just looking at the ionospheric end of the current circuit. Still, by analyzing electrojet data from ground magnetometers and radars, Amm et al. (2011) were also able to show that the probability of a high Cowling efficiency scales with geomagnetic activity.

In a companion paper, Fujii et al. (2011) advanced the Cowling channel model by giving up the approximation of a thin ionosphere and taking into account the different altitudinal profiles of the Pedersen and Hall conductivities (see Fig. 13 for an example of these altitudinal profiles.) Consequently, they suggested a model with Pedersen and Hall layers at different altitudes, as indicated in panel (c) of Fig. 12. The Hall current, at lower altitude, couples to FAC between the two layers, then part of it connects to the secondary Pedersen current, driven by the polarization electric field, while the other part escapes to magnetosphere. The Cowling channel is complete when the ionospheric loop of the Hall, FAC, and Pedersen current has no FAC leakage to the magnetosphere.

Fujii et al. (2011) gave up not only the thin ionosphere, but also the infinite longitudinal extension of the Cowling channel, assuming boundaries both across and along the channel, visible as well in panel (c) of Fig. 12. This has the important consequence that polarization charges can build up at the longitudinal ends of the channel, driving a polarization electric field also along the channel, though in other respects the channel remains uniform. Later on, Fujii et al. (2012) used the model to explain the equatorward motion of an afternoon arc, together with its current system, driven self-consistently by the polarization electric field along the Cowling channel associated with the arc. While the need for a height resolved ionosphere, in particular with respect to MIT coupling, has been identified for quite some time (e.g., Amm et al. 2008), studies like those mentioned above remain scarce, and likewise underlying in-situ observations. This may change in the future, e.g., if mission proposals like Daedalus (Sarris et al. 2019), recently selected by ESA for a detailed feasibility study, will materialize, and simultaneous ionospheric samplings at different altitudes will become available systematically.

An interesting problem raised by the Cowling channel model outlined above is related to the apparently increased likelihood that Cowling efficiency is high when geomagnetic activity is high, noticed by Amm et al. (2011). A (quasi) complete Cowling channel implies significant energy dissipation, via the secondary Pedersen current; but on the other hand the missing FAC connection to the magnetosphere raises the issue of missing energy input, typically delivered as Poynting flux, via the FAC system of the channel. This problem cannot be solved under steady state conditions and requires a time dependent framework, where energy delivery and energy dissipation are not necessarily simultaneous (and collocated). Such a formal framework was introduced by Yoshikawa et al. (2013) and checked on a numerical test case by Yoshikawa et al. (2013). A comprehensive review of recent Cowling channel developments, including those outlined above, is available in Yoshikawa and Fujii (2018).

To conclude this section and try to look a bit into the future, it seems clear that a complete understanding of the auroral arc current system requires a 3-D model, ideally time dependent, where length, width, and thickness are all taken into account. Recent work such as illustrated in Fig. 13 is helping to visualize this region. The 1-D backbone may still help to formalize such a model in a relatively simple framework, as yet not available, perhaps even tractable analytically. Future multi-point observations of the 3-D arc may help as well to unravel its structure. While radar, rocket, and single satellite observations were and still are the main tool for arc exploration, typically along a 1-D cross-section of the arc, this paradigm may change in the (near?) future, by easy access to space of CubeSat fleets—as planned, or already under development, by various research groups around the world. Upcoming ground facilities, like EISCAT 3D, may also shed new light on the complexity of aurora, including the 'simple' auroral arc.

5 Single and Multiple Arcs

Instead of a single narrow structure, auroral arcs often appear in multiple sets (Stenbaek-Nielsen et al. 1998), making up the bulk of discrete auroral observations (Gillies et al. 2014). A quasi-static multiple arc system in the nightside auroral oval consists of 2–5 discrete forms, aligned in the east-west direction and nearly parallel to each other. The latitudinal width of individual arcs is about 10–20 km (Sesiano and Cloutier 1976; Akasofu et al. 1966; Vondrak et al. 1983; Knudsen et al. 2001; Wu et al. 2017). A common separation distance between two adjacent arcs is between 30–50 km, and there are some arcs separated by more than 100 km. The orientation of arcs in multiple arc systems is similar to that seen for single arcs, including a small deviation from lines of constant geomagnetic latitude that reverse signs near midnight (Gillies et al. 2014). The multiple structures can occur during both quiet and active periods. However, they are more nearly parallel to each other and more easily identified during less active times, typically associated with slow equatorward motion of the aurora (Akasofu et al. 1966, 1967). Field-line resonance (FLR) arcs also display multiple structures, but with a temporally periodic morphology and poleward motion (Gillies et al. 2018).

The field-aligned current (FAC) system associated with multiple auroral arcs has attracted much attention in both satellite and rocket studies (e.g. Armstrong et al. 1975; Sesiano and Cloutier 1976; Kamide and Akasofu 1976; Aikio et al. 2002; Juusola et al. 2016a). Stenbaek-Nielsen et al. (1998) reported multiple arcs observed both by an aircraft-borne all-sky camera and from electron energy spectrometers on FAST. The arcs are found to be associated with upward field-aligned current, as described in Sect. 3.3, and are identified by multiple peaks in energy and flux spectra of down-going electrons (inverted-V events). However, the relation between arcs and current sheets is not one-to-one, since an arc indicates an enhancement in electron energy flux but not necessarily in upward field-aligned current. Wu et al. (2017) investigate the large-scale FACs observed by the Swarm satellites and show the 'unipolar' and 'multipolar' FACs associated with multiple arc systems. In 'unipolar' FAC events, each arc is an intensification within a broad, unipolar current sheet and downward return currents occur outside of this broad sheet (Fig. 14). In 'multipolar' FAC events, multiple arc systems represent a collection of multiple up and down current pairs. They also found 'unipolar' FAC pattern is more common for quasi-static multiple arcs in the pre-midnight, pre-substorm auroral zone.

Of smaller width than the above meso-scale arc set is the multiple arc system composed of elements each having a width of the order of several hundred meters (Maggs and Davis 1968; Davis 1978; Borovsky 1993; Trondsen and Cogger 1997). Such an array of arcs can appear in diffuse aurora (Stenbaek-Nielsen et al. 1998) or blend together to give the impression of one single arc when not viewed in the magnetic zenith. When the system is active,

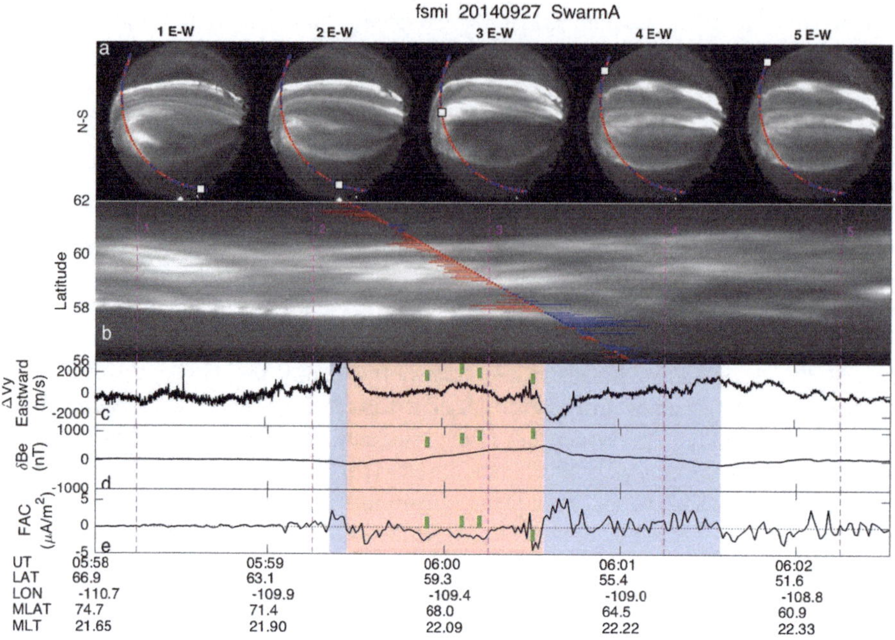

Fig. 14 Multiple arcs observed by the Fort Smith ASI and Swarm A with a "unipolar" FAC system. (**a**) The Swarm A trajectory and multiple arcs in the ASI images at 1-minute time intervals. Red and blue colors indicate upward and downward FACs. (**b**) Satellite-aligned keogram of Swarm A as a function of geomagnetic latitude and time. The values of upward (red) and downward (blue) FACs are displayed together with the flight path. (**c**) Horizontal cross-track (approximately eastward) ion velocity in m/s. (**d**) Eastward component of residual magnetic field. (**e**) Field-aligned current density $j_\parallel$ ($\mu A/m^2$) derived from the single satellite method. The red and blue areas in panels (**c**)–(**e**) are large-scale upward and downward FAC regions identified interactively by neglecting small-scale fluctuations. The positions of peak brightness of arcs are indicated by green lines (Wu et al. 2017)

motions along individual arcs frequently occur, and it is common to see opposite directions of motion along adjacent arcs (Hallinan and Davis 1970; Trondsen and Cogger 1997; Vogt et al. 1999b; Miles et al. 2018).

Multiple arcs are also a common feature in the dayside auroral oval, especially in the local time sector between 14 and 16 hours (Lundin and Evans 1985; Meng and Lundin 1986; Elphinstone et al. 1996; Sandholt et al. 2006). The precipitating electrons with energies of 0.3-2 keV in this local time sector usually appear as multiple narrow structures in latitude (Evans 1985). Moen et al. (1994) showed examples of postnoon discrete arcs with a thickness of 10-20 km and a lifetime of about 2 to 5 min. Kozlovsky et al. (2009) demonstrates these arcs were located in the region of the postnoon convection reversal, and the background upward FAC had a density of about 1 $\mu A/m^2$.

One interesting question about multiple arc systems is the source of multiplicity. Currently proposed theories include a structured magnetospheric generator (Akasofu and Chapman 1961; Coppi et al. 1966), instability in the ionosphere (Atkinson 1970), multiple reflection of Alfvén waves (e.g. Maltsev et al. 1977; Mallinckrodt and Carlson 1978; Haerendel 1994; Lysak 1990) and the internal structuring of the M-I coupling medium (stationary inertial Alfvén waves) (Knudsen 1996). The observations discussed in this section are meant to provide constraints for candidate theories.

6 Impact on Ionosphere and Thermosphere

Discrete auroral arcs enhance the E-region plasma density and increase ionospheric conductance within the arc by virtue of precipitating electrons accelerated within the auroral acceleration zone. Arcs that are not polarization arcs (Paschmann et al. 2003) (Chap. 6) are sustained by parallel currents that generate electric fields within the downward current region causing ionospheric heating via ion-neutral collisions. The heating may result in density cavities due to evacuation of current carriers and ion outflow. The enhanced electric fields and plasma density alter the E and F region neutral wind velocities through ion drag.

Discrete arcs contribute to about 16% of the total quiet time hemispherical precipitating energy flux (Newell et al. 2009). Electron density enhancements caused by precipitation within an arc can be greater than 2–3 times the background with typical densities ranging between 10^{11}–10^{12} m^{-3} peaking at about 105–120 km (Marklund et al. 1994; Kirkwood et al. 1988). Broadly speaking, the energy spectra associated with discrete arcs can be categorized into monoenergetic precipitation caused by electrostatic potential structures or broadband precipitation produced by Alfvénic acceleration. The peak energy of precipitating electrons within a discrete arc is around 1–10 keV, with electron energies increasing towards the center of the arc (Marklund et al. 1982). The precipitating energy flux is $\sim 1\ \mathrm{mW/m^2}$ and can reach up to $\sim 15\ \mathrm{mW/m^2}$ for Alfvénic acceleration (Colpitts et al. 2013). Precipitating electrons of energies up to 100 keV have also been observed on field lines conjugate with discrete auroral arcs (Swift and Gorney 1989). Higher mean electron energies can lead to a descent in the altitudes of maximum ionization and conductivity within the upward current region. An increase in the precipitating energy flux can lead to an increase in the maximum ionization and conductivity observed in this region. A statistical study by Deehr et al. (2005) has shown that the mean energy of precipitating electrons is higher in the day side of the dawn-dusk boundary, suggesting that a higher initial conductance (due to dayside photoionization) leads to greater acceleration. On the contrary, discrete arcs are frequently observed in the dusk to midnight sector. The reason may be that this region has low enough background ionospheric conductance and ion temperature for the ionospheric feedback instability (IFI) to develop and sustain them (Newell et al. 1996b, 2001).

Discrete arcs that are bounded by upward and downward field-aligned currents can be closed by a perpendicular Pedersen current in the horizontal plane. Observations from the CLUSTER satellites suggests the downward current region is dominated by current filaments of 10–20 km scale size in the ionosphere, with field-aligned current densities of a few $\mu\mathrm{A/m^2}$ consistent with other studies (Wright et al. 2008). As mentioned in Sect. 3.5, the downward current region develops a depletion region, since the Pedersen current carries ions away and electrons flow upwards along the field line. The decreased plasma density results in a lower conductance that forces an increased electric field to sustain the same Pedersen current. The upward current region lies within the arc, and is 1–4 times narrower than the downward current-region. (Note that for the black aurora, satellite measurements show narrow downward current regions (Elphic et al. 1998; Marklund et al. 2001).) Due to electron precipitation the region has higher conductance, with field-aligned current densities reaching tens of $\mu\mathrm{A/m^2}$ (Aikio et al. 2002). Smaller arc scale sizes may have higher field-aligned current densities (Juusola et al. 2016b; Stasiewicz and Potemra 1998).

Median Hall and Pedersen conductance ranges from 1–8 S and 1–6 S across MLT respectively, with minima located at dawn and maximum at $\sim$1–2 MLT (Aikio and Selkälä 2009). Within a stable discrete arc, the Hall and Pedersen conductance is higher than the background and can vary from $\sim$20–60 S and $\sim$8–29 S respectively (Kirkwood et al. 1988; Aikio et al. 2002). The magnitude of Hall and Pedersen currents carried within these arcs can reach up to ~1000 mA/m and ~500 mA/m respectively (Aikio et al. 2002). In the dayside, since the background conductance is already high due to photoionization, arc formation is less favored and the conductance within the arc is only weakly increased from 0.5–1 S to 2–3 S (Nilsson et al. 2005). Discrete aurora mapped to the plasma-sheet boundary layer have high conductivities (Kirkwood et al. 1988), and their energy spectra fit the kappa distribution observed within the high-latitude plasma sheet (Kaeppler et al. 2014). The E-region electric fields within the discrete arcs are close to zero, while much stronger ($\gtrsim$50 mV/m) adjacent to them (Kirkwood et al. 1988).

The electric field enhancement excites the two stream instability leading to meter-scale electron density irregularities that backscatter radio waves called a Radar aurora (Bahcivan et al. 2006). Through ion-neutral collisions the enhanced electric field leads to frictional Joule heating. Median Joule heating rates are low during magnetically quiet times: 0–2 mW/m^2 with minimum in the dayside and maximum near 2 MLT (Aikio and Selkälä 2009). However, energy dissipation by both kinetic energy of particles and Joule heating can reach ~100 mW/m^2 each (Marklund et al. 1982). The heating can cause the conversion of F region O^+ into NO^+ and result in faster recombination loss, since molecular ions have shorter chemical lifetimes. This results in additional loss of plasma density in the downward current region apart from upwelling or transport of current carriers out of the region (Zettergren and Semeter 2012). Since ion mobility is inversely proportional to collision frequency, and collision frequency increases with temperature, Joule heating causes the ion mobility to decrease. According to simulations by Streltsov (2008) this can suppress the IFI effectively. The result points to an additional reason for discrete arcs to be prevalent during night-time and the winter seasons: arcs attributed to IFI develop when ions in the ionosphere are cold.

Plasma flows of ~340 to 720 m/s directed eastward were observed for two auroral arc filaments corresponding to southward electric fields of 18 and 40 mV/m using optical observations of afterglow of O^+ emissions (Dahlgren et al. 2009). Such flows can affect thermospheric wind velocities through enhanced ion drag due to the increased plasma density and localized ionospheric electric field associated with the auroral arc. In the E-region, the background neutral wind can rotate from the background wind field to align itself parallel along the discrete arc and revert back to the background flow direction when the arc disappears. In the particular case observed by Kosch et al. (2010), the wind rotations occurred within 7–16 minutes. In the F-region the effect is much weaker, but Zou et al. (2018) observed that the winds still exhibit a temporal evolution similar to the plasma flows but with a delay of $\lesssim$20 minutes, which is much faster than what was previously thought (a few hours). The neutral wind and plasma flow velocities are likely balanced by inertia and ion-drag with F-region neutral wind gaining about 19% of the plasma flow velocity gain (~950 m/s) adjacent to the quasi-steady discrete arc studied in the above work. Arrays of Fabry-Perot interferometers can provide maps of neutral wind behavior for assimilation with radars and visible auroral imagery (Conde et al. 2001; Nicolls et al. 2012).

7 Generator and Magnetospheric Impact

The auroral arc (and aurora in general) is the low altitude effect of a complex magnetosphere/ionosphere/thermosphere coupling chain, sometimes called the auroral current circuit. The first suggestion on how this circuit looks came from Boström (1964), who identified two possible configurations, Type 1 and Type 2, presented in Fig. 15. Both Type 1 and Type 2 are discussed in more detail by Lysak et al. (this issue). Here we just note that, apart from the different topologies, in both cases electromagnetic energy is supplied by a

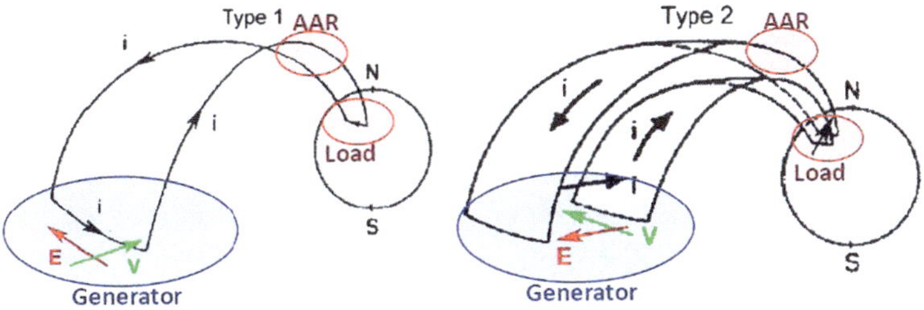

Fig. 15 The two basic configurations of the auroral current circuit suggested by Boström (1964): Type 1 (left) and Type 2 (right). Generator and load regions are indicated by blue and red shading, respectively. See text for more details. Adapted from Boström (1964)

magnetospheric generator region, where $\mathbf{E} \cdot \mathbf{i} < 0$, with **E** the electric field and **i** the current density. The mechanical energy of the plasma motion (**V** in Fig. 15) is converted into Poynting flux, transmitted along the auroral field lines to the ionospheric load, $\mathbf{E} \cdot \mathbf{i} > 0$, where it is converted back to mechanical energy by Joule heating.

After the original suggestion of Boström (1964), the understanding of the auroral current circuit was progressively consolidated and refined. A key milestone was the discovery and first systematic investigation of the AAR (indicated in Fig. 15), based on data from the S3-3 satellite (e.g. Mozer et al. 1980; Chiu et al. 1983). The acceleration of precipitating electrons and of outflowing ion beams (see more on ion beams below) occurs as well at the expense of the Poynting flux. Investigations based on Polar data indicated that the downward Poynting flux above the AAR is strong enough to feed auroral dissipation (e.g. Wygant et al. 2000), while statistical studies based on Polar (Keiling et al. 2002) and FAST (Chaston et al. 2002) data showed that energy flux is indeed dominated by Poynting flux above the AAR and by electron energy flux below the AAR (see the comparison by Chaston (2006), Fig. 16).

Among the three key regions of the auroral current circuit, the auroral ionosphere and the auroral acceleration region were explored far more intensively by satellite missions than the magnetospheric generator. Even though theoretical models and numerical simulations of auroral generator regions are well represented in the topical literature (see further details in Lysak et al., this issue) and Rostoker and Boström (1976), Lysak (1985), Vogt et al. (1999a), Birn and Hesse (2005), experimental evidence is scarce. Unlike the electrical circuits of daily life, where the various elements, including the generator, are localized (lumped), the key elements of the auroral current circuit are distributed over vast regions of space. While the power fed to the circuit is the integral over the generator volume and typically in the GW range, a satellite mission can only sample the local power density, $\mathbf{E} \cdot \mathbf{i}$, which can well be within the error margin. Moreover, one-spacecraft missions cannot infer the full current density vector, **i**. With the launch of the four satellite Cluster mission (Escoubet et al. 2001), it became possible both to fully derive **i** and to improve the accuracy of the electric field estimates, due to the redundancy of the available data, such that a systematic search for generator regions became possible. The first results of an investigation along this line were published in two companion papers by Marghitu et al. (2006) and Hamrin et al. (2006), where conjugacy with low altitude inverted-V events observed by FAST provided further evidence on the connection of the observed generator regions to the aurora (Fig. 17).

Later on, this investigation was extended to a comprehensive survey of energy conversion regions (ECRs) in the plasma sheet, including concentrated generator regions (CGRs, $\mathbf{E} \cdot \mathbf{i} < 0$) and concentrated load regions (CLRs, $\mathbf{E} \cdot \mathbf{i} > 0$), where the term 'concentrated' indicates a local power density above the error margin of $\sim 1\ \mathrm{pW/m^3}$. An outstanding result of

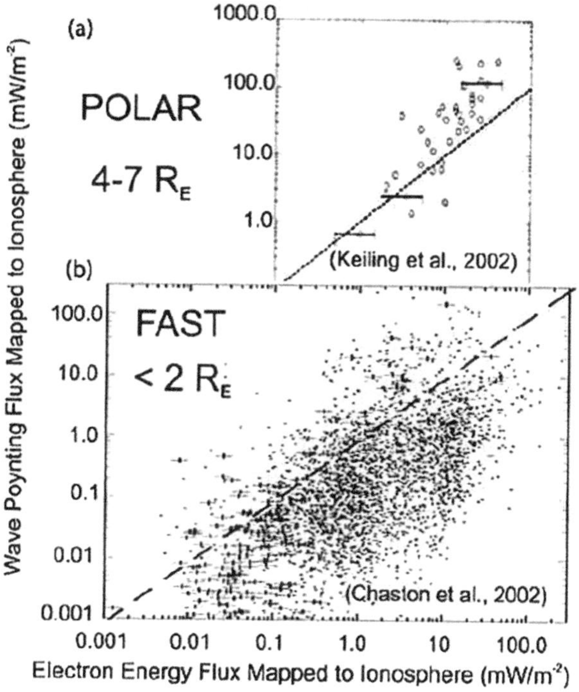

Fig. 16 Comparison of the Poynting flux and electron energy flux, as observed by the Polar satellite above the AAR (Keiling et al. 2002) and by the FAST satellite below the AAR (Chaston et al. 2002). After Chaston (2006)

this survey, summarized in the review by Hamrin et al. (2011), was that the observed ECRs were typically associated with bursty bulk flows (BBFs, Angelopoulos et al. 1992, 1994). Most of the ECRs were CLRs (as expected, considering the large scale load character of the plasma sheet), but the fraction of CGRs, of about one fifth, was significant too. In view of their statistical association with BBFs, it is likely that many of the observed CGRs actually fed their power to auroral streamers (not to auroral arcs). A closer examination of the energy conversion matter as related to the BBF structure and age was performed subsequently by Hamrin et al. (2013, 2014). Further details on BBFs and on their coupling to ionospheric streamers can be found in (Forsyth et al., this issue), as well as, e.g., in a recent review by Liu et al. (2018a).

The MI coupling chain described so far emphasizes the active role of the magnetosphere, that drives the auroral current circuit via generator regions, and suggests a rather passive role of the ionosphere, where energy is merely dissipated. However, energy dissipation generates in turn processes that feed back onto the magnetosphere and have an impact on the overall evolution of the MI system. An outstanding example of this feedback is the ion outflow, whose origin is in the ionospheric heating by Joule effect and electron precipitation. The gravitational binding of the heated ions is thus weakened and further heating processes, at higher altitudes, end up by ionospheric ion outflow injected into magnetosphere (e.g. Strangeway et al. 2005). The impact of the ionospheric ion outflow on the magnetosphere dynamics is diverse, ranging from the influence on the reconnection rate, through the speed of magnetospheric convection, to the rise and fall of the ring current (see e.g. Moore et al. 2014; Kronberg et al. 2014).

The ionospheric dissipation serves as well to relax the magnetic shear stress of the twisted flux tubes, again feeding back on the magnetospheric end, where the twisting proceeds. During active events the twisting prevails (note similarity with Boström Type 1), as reflected by

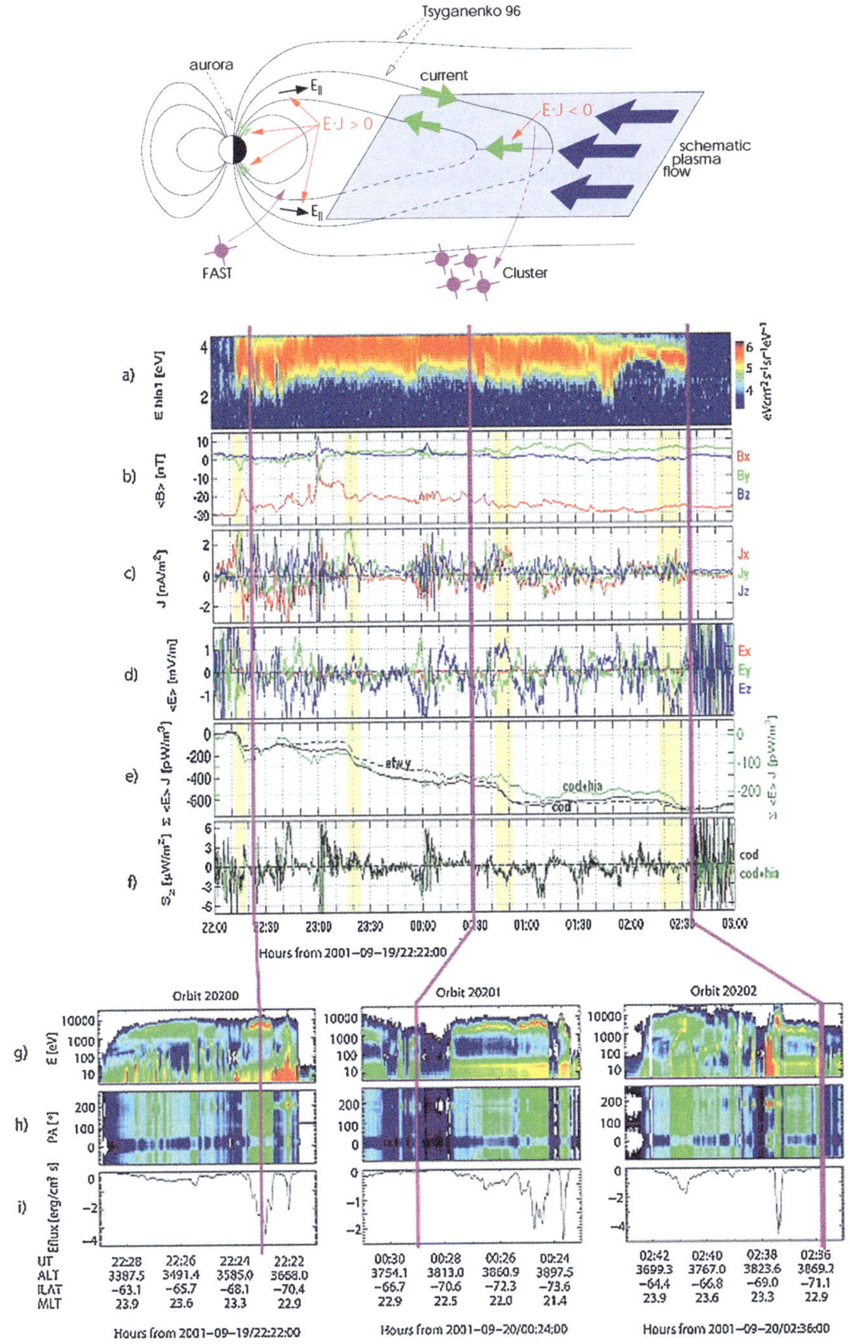

Fig. 17 Top: Schematic representation of the auroral current circuit, with the magnetospheric generator region sampled by the Cluster spacecraft and the low altitude end observed by the FAST satellite. From Hamrin et al. (2006). Bottom: Generator regions observed by Cluster in the magnetospheric plasma sheet (yellow stripes) conjugate (magenta lines) to inverted-V events observed by FAST at some 3000 km altitude above the auroral ionosphere. From the review by Hamrin et al. (2011), adapted after Marghitu et al. (2006) and Hamrin et al. (2006)

the large scale filamentary FAC of the substorm current wedge (McPherron et al. 1973; Keiling et al. 2009), or by meso-scale filamentary FACs associated with BBF-streamer current wedgelets (Birn et al., 2004; Liu et al., 2018). De-twisting dominates, e.g., during the (late) recovery stage of auroral substorms, when omega bands, threaded by meso-scale filamentary FACs, relax back to quiet arcs, threaded by (possibly undulated) sheet-like FACs (note, this time, similarity with Boström Type 2). A recent event study by Bunescu et al. (2017), based on Cluster observations conjugate to ground and FAST data, captured the progress of this process and suggested a relaxation time scale of about 10 min. The relevance of this time scale for the related magnetospheric processes and the specific auroral current circuit is yet to be explored.

The interhemispheric (a)symmetry of aurora provides another example where observations of low altitude features help understanding and, subsequently, diagnosing the coupled MI system, as well as the forcing exerted by the solar wind. Even though, at first sight, one may expect that auroras in the northern and southern hemispheres are merely symmetric images of each other, in reality this is not the case (e.g. Laundal and Østgaard 2009). As a matter of fact, as shown by both ground (e.g. Sato et al. 1998) and spacecraft (e.g. Østgaard et al. 2005) observations, asymmetry is rather the rule than the exception. As noted by Sato et al. (2015), perfect symmetry of conjugate auroras would require symmetric conditions both in the solar wind forcing and in the MI coupling chain. Considering the large number of potential asymmetries between the two hemispheres, many of which are essentially related to the asymmetric configuration of the magnetic field (Laundal et al. 2017), 'highly similar conjugate auroral events (...) are likely to be exceptionally rare in nature' (Sato et al. 2015). Specific drivers of interhemispheric asymmetries between conjugate auroras include the IMF components Bx and By, the difference in ionospheric conductance, and the difference in auroral acceleration, detailed, e.g., in the reviews by Østgaard et al. (2015) and Sato et al. (2015).

While the ionospheric footprint of the auroral region forms only one part of this complex magnetospheric/ionospheric/thermospheric coupling chain, its behaviour contains much information about the entire system. The visible signatures provide a movie of ongoing events; the electrodynamic signatures—**E** and **B**—connect the players, and the ionosphere and thermosphere, as revealed by radars, in situ observations, and neutral wind mappers, tell us the state of the footpoint. If we understand the physics of the auroral system, we can learn to read these signatures to constrain what can be happening in the far-dispersed magnetospheric regions.

8 Measurements Needed to Progress

Much of our understanding of auroral processes comes through a view of the auroral arc as an idealized sheetlike structure. This is an excellent startpoint that has provided a rich baseline for studying auroral plasma processes, particularly the details of both auroral acceleration processes, and of the effects of precipitation on the ionosphere and thermosphere. Many questions remain, however, as we move into system studies of the non-ideal aurora. Even sheetlike aurora have unaddressed physics in the longitudinal direction. And there is far more to the auroral system than sheetlike quiet arcs. We need to develop a firm understanding of the three-dimensional structure and time evolution of quiet arcs before we can tackle more complex morphologies such as westward surges and the interplay between discrete arcs and Alfvénic systems.

Moving beyond one-dimensional cuts of sheetlike arcs, what questions do we encounter? How does plasma flow into an arc region perpendicular to the magnetic field and perpendicular to the length of the arc? What magnetospheric generators create the need for auroral current systems? We can explore the ionospheric footprint of auroral current closure within a localized region, but examining the magnetospheric end will require development of closed theories of auroral physics. What allows an arc to remain relatively stable for many tens of minutes, and then abruptly change morphologies? What controls these shifts: magnetospheric drivers or ionospheric responses? What controls the plasma density profile both along and across the field line in the auroral zone? What are the rules of plasma transport at play here?

Various tools will help us move forward toward a more complete auroral theory. Modelling tools such as GITM and GEMINI allow us to decompose the three-dimensional volumetric region near an auroral arc into its various parameters and responses. These tools allow us to move beyond height-integrated idealized sheetlike systems into more complex morphologies. What happens to current closure as a sheetlike arc begins to bend? We can explore the responses we understand, in this complex environment, using volumetric models at a variety of resolutions. The new observations which are presently being planned also will move us in this direction. A variety of them involve the non-sheetlike system of the auroral zone, studied in a variety of multipoint and multidata fashions. The EISCAT3D will provide volumetric imaging (McCrea et al. 2015), and the continuation of the THEMIS GBO into multispectral information will be the means of extracting far more quantitative information from auroral imagery than simple whitelight images can provide. Other groundbased arrays and networks are designed to study not auroral physics at one point, but over a large field of view at high resolution. Finally, localized swarms of in situ high-resolution observations, following on and growing from the success of the recent ESA Swarm mission, and increasingly done in conjugacy with groundbased arrays, will fundamentally change our view from the idealized sheet to the real auroral arc.

Acknowledgements The authors are grateful for the kind hospitality of the International Space Science Institute (ISSI), Bern, Switzerland.

Figures 4, and 11, and part of Fig. 17 were reproduced under the Creative Commons Attribution 3.0 Unported License (https://creativecommons.org/licenses/by/3.0/).

References

A.T. Aikio, A. Selkälä, Statistical properties of Joule heating rate, electric field and conductances at high latitudes. Ann. Geophys. **27**(7), 2661–2673 (2009). https://doi.org/10.5194/angeo-27-2661-2009. https://www.ann-geophys.net/27/2661/2009/

A.T. Aikio, T. Lakkala, A. Kozlovsky, P.J.S. Williams, Electric fields and currents of stable drifting auroral arcs in the evening sector. J. Geophys. Res. Space Phys. **107**(A12), 3–1314 (2002). https://doi.org/10.1029/2001JA009172. https://agupubs.onlinelibrary.wiley.com/doi/abs/10.1029/2001JA009172

A.T. Aikio, K. Mursula, S. Buchert, F. Forme, O. Amm, G. Marklund, M. Dunlop, D. Fontaine, A. Vaivads, A. Fazakerley, Temporal evolution of two auroral arcs as measured by the cluster satellite and coordinated ground-based instruments. Ann. Geophys. **22**(12), 4089–4101 (2004). https://doi.org/10.5194/angeo-22-4089-2004. https://www.ann-geophys.net/22/4089/2004/

A. Aikio, K. Mursula, S. Buchert, F. Forme, O. Amm, G. Marklund, M. Dunlop, D. Fontaine, A. Vaivads, A. Fazakerley, Temporal evolution of two auroral arcs as measured by the Cluster satellite and coordinated ground-based instruments, in *Annales Geophysicae*, vol. 22 (2004), pp. 4089–4101

A.T. Aikio, H. Vanhamäki, A.B. Workayehu, I.I. Virtanen, K. Kauristie, L. Juusola, S. Buchert, D. Knudsen, Swarm satellite and EISCAT radar observations of a plasma flow channel in the auroral oval near magnetic midnight. J. Geophys. Res. Space Phys. **123**(6), 5140–5158 (2018). https://doi.org/10.1029/2018JA025409. https://agupubs.onlinelibrary.wiley.com/doi/abs/10.1029/2018JA025409

S.-I. Akasofu, Dynamic morphology of auroras. Space Sci. Rev. **4**(4), 498–540 (1965)

S.-I. Akasofu, S. Chapman, A neutral line discharge theory of the aurora polaris. Philos. Trans. R. Soc. Lond. Ser. A, Math. Phys. Sci. **253**(1031), 359–406 (1961). https://doi.org/10.1098/rsta.1961.0004

S.-I. Akasofu, D. Kimball, C.-I. Meng, Dynamics of the aurora–VII equatorward motions and the multiplicity of auroral arcs. J. Atmos. Terr. Phys. **28**, 627–635 (1966)

S.-I. Akasofu, S. Chapman, P. Kendall, *The Significance of the Multiple Structure of the Auroral Arc* (1967)

L. Alm, G.T. Marklund, T. Karlsson, A. Masson, Pseudo altitude: a new perspective on the auroral density cavity. J. Geophys. Res. Space Phys. **118**(7), 4341–4351 (2013)

L. Alm, B. Li, G. Marklund, T. Karlsson, Statistical altitude distribution of the auroral density cavity. J. Geophys. Res. Space Phys. **120**(2), 996–1006 (2015)

L. Alm, G.T. Marklund, T. Karlsson, Electron density and parallel electric field distribution of the auroral density cavity. J. Geophys. Res. Space Phys. **120**(11), 9428–9441 (2015). https://doi.org/10.1002/2015JA021593. https://agupubs.onlinelibrary.wiley.com/doi/abs/10.1002/2015JA021593

O. Amm, A. Aruliah, S.C. Buchert, R. Fujii, J. Gjerloev, A. Ieda, T. Matsuo, C. Stolle, H. Vanhamäki, A. Yoshikawa, Towards understanding the electrodynamics of the 3-dimensional high-latitude ionosphere: present and future, in *Annales Geophysicae: Atmospheres, Hydrospheres and Space Sciences*, vol. 26 (2008), p. 3913

O. Amm, R. Fujii, K. Kauristie, A. Aikio, A. Yoshikawa, A. Ieda, H. Vanhamäki, A statistical investigation of the Cowling channel efficiency in the auroral zone. J. Geophys. Res. Space Phys. **116**(A2), A02304 (2011)

L. Andersson, R. Ergun, Acceleration of antiearthward electron fluxes in the auroral region. J. Geophys. Res. Space Phys. **111**(A7), A07203 (2006)

L. Andersson, R. Ergun, The search for double layers in space plasmas, in *Auroral Phenomenology and Magnetospheric Processes: Earth and Other Planets*. Geophys. Monogr. Ser., vol. 197 (2012)

L. Andersson, R. Ergun, D. Newman, J. McFadden, C. Carlson, Y.-J. Su, Characteristics of parallel electric fields in the downward current region of the aurora. Phys. Plasmas **9**(8), 3600–3609 (2002)

L. Andersson, D. Newman, R. Ergun, M. Goldman, C. Carlson, J. McFadden, Influence of suprathermal background electrons on strong auroral double layers: observations. Phys. Plasmas **15**(7), 072901 (2008)

V. Angelopoulos, W. Baumjohann, C. Kennel, F. Coroniti, M. Kivelson, R. Pellat, R. Walker, H. Lühr, G. Paschmann, Bursty bulk flows in the inner central plasma sheet. J. Geophys. Res. Space Phys. **97**(A4), 4027–4039 (1992)

V. Angelopoulos, C. Kennel, F. Coroniti, R. Pellat, M. Kivelson, R. Walker, C. Russell, W. Baumjohann, W. Feldman, J. Gosling, Statistical characteristics of bursty bulk flow events. J. Geophys. Res. Space Phys. **99**(11), 257–280 (1994). Paper 94ja01263

W. Archer, D.J. Knudsen, Distinguishing subauroral ion drifts from Birkeland current boundary flows. J. Geophys. Res. Space Phys. **123**, 819–826 (2018). https://doi.org/10.1002/2017JA024577

J. Archer, H. Dahlgren, N. Ivchenko, B. Lanchester, G. Marklund, Dynamics and characteristics of black aurora as observed by high-resolution ground-based imagers and radar. Int. J. Remote Sens. **32**(11), 2973–2985 (2011)

W.E. Archer, D.J. Knudsen, J.K. Burchill, B. Jackel, E. Donovan, M. Connors, L. Juusola, Birkeland current boundary flows. J. Geophys. Res. Space Phys. **122**(4), 4617–4627 (2017)

J.C. Armstrong, S.-I. Akasofu, G. Rostoker, A comparison of satellite observations of Birkeland currents with ground observations of visible aurora and ionospheric currents. J. Geophys. Res. **80**(4), 575–586 (1975)

G. Atkinson, Auroral arcs: result of the interaction of a dynamic magnetosphere with the ionosphere. J. Geophys. Res. **75**(25), 4746–4755 (1970). https://doi.org/10.1029/JA075i025p04746. https://agupubs.onlinelibrary.wiley.com/doi/abs/10.1029/JA075i025p04746

H. Bahcivan, D.L. Hysell, D. Lummerzheim, M.F. Larsen, R.F. Pfaff, Observations of colocated optical and radar aurora. J. Geophys. Res. **111**(A12), 12308 (2006). https://doi.org/10.1029/2006JA011923. http://doi.wiley.com/10.1029/2006JA011923

C.A. Barth, D.N. Baker, S.M. Bailey, Seasonal variation of auroral electron precipitation. Geophys. Res. Lett. **31**(4), L04809 (2004). https://doi.org/10.1029/2003GL018892. https://agupubs.onlinelibrary.wiley.com/doi/abs/10.1029/2003GL018892

J. Birn, M. Hesse, Energy release and conversion by reconnection in the magnetotail, in *Annales Geophysicae*, vol. 23 (2005), pp. 3365–3373

G.T. Blanchard, L.R. Lyons, J.C. Samson, F.J. Rich, Locating the polar cap boundary from observations of 6300 Å auroral emission. J. Geophys. Res. Space Phys. **100**(A5), 7855–7862 (1995). https://doi.org/10.1029/94JA02631. https://agupubs.onlinelibrary.wiley.com/doi/abs/10.1029/94JA02631

G.T. Blanchard, L.R. Lyons, J.C. Samson, Accuracy of using 6300 Å auroral emission to identify the magnetic separatrix on the nightside of Earth. J. Geophys. Res. Space Phys. **102**(A5), 9697–9703 (1997). https://doi.org/10.1029/96JA04000. https://agupubs.onlinelibrary.wiley.com/doi/abs/10.1029/96JA04000

E. Blixt, M. Kosch, J. Semeter, Relative drift between black aurora and the ionospheric plasma, in *Annales Geophysicae*, vol. 23 (2005), pp. 1611–1621

L.P. Block, C.-G. Fälthammar, P.-A. Lindqvist, G. Marklund, F. Mozer, A. Pedersen, T. Potemra, L. Zanetti, Electric field measurements on viking: first results. Geophys. Res. Lett. **14**(4), 435–438 (1987)

J.E. Borovsky, Auroral arc thicknesses as predicted by various theories. J. Geophys. Res. Space Phys. **98**(A4), 6101–6138 (1993)

R. Boström, A model of the auroral electrojets. J. Geophys. Res. **69**(23), 4983–4999 (1964)

A. Boudouridis, H.E. Spence, Separation of spatial and temporal structure of auroral particle precipitation. J. Geophys. Res. Space Phys. **112**(A12), A12217 (2007)

W. Bristow, G. Sofko, H. Stenbaek-Nielsen, S. Wei, D. Lummerzheim, A. Otto, Detailed analysis of substorm observations using superDARN, UVI, ground-based magnetometers, and all-sky imagers. J. Geophys. Res. Space Phys. **108**(A3), 1124 (2003). https://doi.org/10.1029/2002JA009242

P. Carlqvist, R. Boström, Space-charge regions above the aurora. J. Geophys. Res. **75**(34), 7140–7146 (1970)

C. Carlson, R. Pfaff, J. Watzin, The FAST Auroral SnapshoT (FAST) mission. Geophys. Res. Lett. **25**(12), 2013–2016 (1998a)

C. Carlson, J. McFadden, R. Ergun, M. Temerin, W. Peria, F. Mozer, D. Klumpar, E. Shelley, W. Peterson, E. Moebius, et al., Fast observations in the downward auroral current region: energetic upgoing electron beams, parallel potential drops, and ion heating. Geophys. Res. Lett. **25**(12), 2017–2020 (1998b)

C. Cattell, L. Johnson, R. Bergmann, D. Klumpar, C. Carlson, J. McFadden, R. Strangeway, R. Ergun, K. Sigsbee, R. Pfaff, FAST observations of discrete electrostatic waves in association with down-going ion beams in the auroral zone. J. Geophys. Res. Space Phys. **107**(A9), 1238 (2002)

C. Cattell, J. Dombeck, L. Hanson, Solar cycle effects on parallel electric field acceleration of auroral electron beams. J. Geophys. Res. Space Phys. **118**(9), 5673–5680 (2013). https://doi.org/10.1002/jgra.50546. https://agupubs.onlinelibrary.wiley.com/doi/abs/10.1002/jgra.50546

S. Chapman, The electrical conductivity of the ionosphere: a review. Nuovo Cimento (1955–1965) **4**, 1385–1412 (1956)

C. Chaston, ULF waves and auroral electrons, in *Magnetospheric ULF Waves: Synthesis and New Directions*, vol. 169 (2006), pp. 239–257

C.C. Chaston, Magnetic reconnection in the auroral acceleration region. Geophys. Res. Lett. **42**(6), 1646–1653 (2015). https://doi.org/10.1002/2015GL063164. https://agupubs.onlinelibrary.wiley.com/doi/abs/10.1002/2015GL063164

C. Chaston, J. Bonnell, C. Carlson, M. Berthomier, L. Peticolas, I. Roth, J. McFadden, R. Ergun, R. Strangeway, Electron acceleration in the ionospheric Alfven resonator. J. Geophys. Res. Space Phys. **107**(A11), 41 (2002)

Y. Chiu, J. Cornwall, J. Fennell, D. Gorney, P. Mizera, Auroral plasmas in the evening sector: satellite observations and theoretical interpretations. Space Sci. Rev. **35**(3), 211–257 (1983)

A. Christensen, L. Lyons, J. Hecht, G. Sivjee, R. Meier, D. Strickland, Magnetic field-aligned electric field acceleration and the characteristics of the optical aurora. J. Geophys. Res. Space Phys. **92**(A6), 6163–6167 (1987)

R. Clayton, M. Burleigh, K. Lynch, M. Zettergren, M. Conde, T. Evans, G. Grubbs, D. Hampton, D. Hysell, M. Lessard, R. Michell, A. Reimer, T.M. Roberts, M. Samara, R. Varney, 3d modeling results of the GEMINI model using reconstructed 2d maps of auroral data (2019a), in preparation

R. Clayton, K. Lynch, M. Zettergren, M. Burleigh, M. Conde, G. Grubbs, D. Hampton, D. Hysell, M. Lessard, R. Michell, A. Reimer, T.M. Roberts, M. Samara, R. Varney, Two-dimensional maps of in situ ionospheric plasma flow data near auroral arcs using auroral imagery. J. Geophys. Res. Space Phys. **124**(4), 3036–3056 (2019b). https://doi.org/10.1029/2018JA026440. https://agupubs.onlinelibrary.wiley.com/doi/abs/10.1029/2018JA026440

C.A. Colpitts, S. Hakimi, C.A. Cattell, J. Dombeck, M. Maas, Simultaneous ground and satellite observations of discrete auroral arcs, substorm aurora, and Alfvénic aurora with FAST and THEMIS GBO. J. Geophys. Res. Space Phys. **118**(11), 6998–7010 (2013). https://doi.org/10.1002/2013JA018796. http://doi.wiley.com/10.1002/2013JA018796

M. Conde, J.D. Craven, T. Immel, E. Hoch, H. Stenbaek-Nielsen, T. Hallinan, R.W. Smith, J. Olson, W. Sun, L.A. Frank, J. Sigwarth, Assimilated observations of thermospheric winds, the aurora, and ionospheric currents over Alaska. J. Geophys. Res. Space Phys. **106**(A6), 10493–10508 (2001). https://doi.org/10.1029/2000JA000135. https://agupubs.onlinelibrary.wiley.com/doi/abs/10.1029/2000JA000135

B. Coppi, G. Laval, R. Pellat, Dynamics of the geomagnetic tail. Phys. Rev. Lett. **16**, 1207–1210 (1966). https://doi.org/10.1103/PhysRevLett.16.1207. https://link.aps.org/doi/10.1103/PhysRevLett.16.1207

T. Cowling, Magnetism, solar: the electrical conductivity of an ionised gas in the presence of a magnetic field. Mon. Not. R. Astron. Soc. **93**, 90 (1932)

H. Dahlgren, N. Ivchenko, B.S. Lanchester, M. Ashrafi, D. Whiter, G. Marklund, J. Sullivan, First direct optical observations of plasma flows using afterglow of O+ in discrete aurora. J. Atmos. Sol.-Terr. Phys. **71**(2), 228–238 (2009). https://doi.org/10.1016/J.JASTP.2008.11.015. https://www.sciencedirect.com/science/article/pii/S1364682608003751?via%3Dihub

T.N. Davis, Observed characteristics of auroral forms. Space Sci. Rev. **22**(1), 77–113 (1978). https://doi.org/10.1007/BF00215814.

C. Deehr, D. Lummerzheim, Ground-based optical observations of hydrogen emission in the auroral substorm. J. Geophys. Res. Space Phys. **106**(A1), 33–44 (2001). https://doi.org/10.1029/2000JA002010. https://agupubs.onlinelibrary.wiley.com/doi/abs/10.1029/2000JA002010

C.S. Deehr, M.H. Rees, A.E.H. Belon, G.J. Romick, D. Lummerzheim, Influence of the ionosphere on the altitude of discrete auroral arcs. Ann. Geophys. **23**(3), 759–766 (2005). https://doi.org/10.5194/angeo-23-759-2005. http://www.ann-geophys.net/23/759/2005/

R.A. Doe, M. Mendillo, J.F. Vickrey, L.J. Zanetti, R.W. Eastes, Observations of nightside auroral cavities. J. Geophys. Res. Space Phys. **98**(A1), 293–310 (1993)

R. Doe, M. Mendillo, J. Vickrey, J. Ruohoniemi, R. Greenwald, Coordinated convection measurements in the vicinity of auroral cavities. Radio Sci. **29**(1), 293–309 (1994)

E.F. Donovan, B.J. Jackel, I. Voronkov, T. Sotirelis, F. Creutzberg, N.A. Nicholson, Ground-based optical determination of the b2i boundary: a basis for an optical MT-index. J. Geophys. Res. Space Phys. **108**(A3), 1115 (2003). https://doi.org/10.1029/2001JA009198. https://agupubs.onlinelibrary.wiley.com/doi/abs/10.1029/2001JA009198

E. Donovan, W. Liu, J. Liang, E. Spanswick, I. Voronkov, M. Connors, M. Syrjäsuo, G. Baker, B. Jackel, T. Trondsen, M. Greffen, V. Angelopoulos, C.T. Russell, S.B. Mende, H.U. Frey, A. Keiling, C.W. Carlson, J.P. McFadden, K.-H. Glassmeier, U. Auster, K. Hayashi, K. Sakaguchi, K. Shiokawa, J.A. Wild, I.J. Rae, Simultaneous THEMIS in situ and auroral observations of a small substorm. Geophys. Res. Lett. **35**(17), L17S18 (2008). https://doi.org/10.1029/2008GL033794. https://agupubs.onlinelibrary.wiley.com/doi/abs/10.1029/2008GL033794

R. Elphic, J. Bonnell, R. Strangeway, L. Kepko, R. Ergun, J. McFadden, C. Carlson, W. Peria, C. Cattell, D. Klumpar, et al., The auroral current circuit and field-aligned currents observed by fast. Geophys. Res. Lett. **25**(12), 2033–2036 (1998)

R. Elphinstone, J. Murphree, L. Cogger, What is a global auroral substorm? Rev. Geophys. **34**(2), 169–232 (1996)

R. Ergun, L. Andersson, D. Main, Y.-J. Su, C. Carlson, J. McFadden, F. Mozer, Parallel electric fields in the upward current region of the aurora: indirect and direct observations. Phys. Plasmas **9**(9), 3685–3694 (2002a)

R. Ergun, L. Andersson, D. Main, Y.-J. Su, D. Newman, M. Goldman, C. Carlson, J. McFadden, F. Mozer, Parallel electric fields in the upward current region of the aurora: numerical solutions. Phys. Plasmas **9**(9), 3695–3704 (2002b)

R. Ergun, L. Andersson, C. Carlson, D. Newman, M. Goldman, Double layers in the downward current region of the aurora. Nonlinear Process. Geophys. **10**(1/2), 45–52 (2003)

R. Ergun, L. Andersson, D. Main, Y.-J. Su, D. Newman, M. Goldman, C. Carlson, A. Hull, J. McFadden, F. Mozer, Auroral particle acceleration by strong double layers: the upward current region. J. Geophys. Res. Space Phys. **109**(A12), A12220 (2004)

C. Escoubet, M. Fehringer, M. Goldstein, The cluster mission—introduction, in *Annales Geophysicae*, vol. 19, (Eur. Math. Soc., Lindau, 2001), pp. 1197–1200

D.S. Evans, The observations of a near monoenergetic flux of auroral electrons. J. Geophys. Res. **73**(7), 2315–2323 (1968)

D.S. Evans, Precipitating electron fluxes formed by a magnetic field aligned potential difference. J. Geophys. Res. **79**(19), 2853–2858 (1974)

D.S. Evans, The characteristics of a persistent auroral arc at high latitude in the 1400 MLT sector, in *The Polar Cusp*, ed. by J.A. Holtet, A. Egeland (Springer, Dordrecht, 1985), pp. 99–109. ISBN 978-94-009-5295-9

S. Figueiredo, G. Marklund, T. Karlsson, T. Johansson, Y. Ebihara, M. Ejiri, N. Ivchenko, P.-A. Lindqvist, H. Nilsson, A. Fazakerley, Temporal and spatial evolution of discrete auroral arcs as seen by cluster. Ann. Geophys. **23**(7), 2531–2557 (2005)

C. Forsyth, A. Fazakerley, A. Walsh, C.E. Watt, K. Garza, C. Owen, D. Constantinescu, I. Dandouras, K.-H. Fornaçon, E. Lucek, et al., Temporal evolution and electric potential structure of the auroral acceleration region from multispacecraft measurements. J. Geophys. Res. Space Phys. **117**(A12), A12203 (2012)

H.U. Frey, O. Amm, C.C. Chaston, S. Fu, G. Haerendel, L. Juusola, T. Karlsson, B. Lanchester, R. Nakamura, N. Østgaard, T. Sakanoi, E. Séran, D. Whiter, J. Weygand, K. Asamura, M. Hirahara, Small and meso-scale properties of a substorm onset auroral arc. J. Geophys. Res. Space Phys. **115**(A10), A10209 (2010a). https://doi.org/10.1029/2010JA015537. https://agupubs.onlinelibrary.wiley.com/doi/abs/10.1029/2010JA015537

H. Frey, O. Amm, C. Chaston, S. Fu, G. Haerendel, L. Juusola, T. Karlsson, B. Lanchester, R. Nakamura, N. Østgaard, et al., Small and meso-scale properties of a substorm onset auroral arc. J. Geophys. Res. Space Phys. **115**(A10), A10209 (2010b). https://doi.org/10.1029/2010JA015537

E. Friis-Christensen, H. Lühr, D. Knudsen, R. Haagmans, Swarm–an Earth observation mission investigating geospace. Adv. Space Res. **41**(1), 210–216 (2008)

R. Fujii, O. Amm, A. Yoshikawa, A. Ieda, H. Vanhamäki, Reformulation and energy flow of the Cowling channel. J. Geophys. Res. Space Phys. **116**(A2), A02305 (2011)

R. Fujii, O. Amm, H. Vanhamäki, A. Yoshikawa, A. Ieda, An application of the finite length Cowling channel model to auroral arcs with longitudinal variations. J. Geophys. Res. Space Phys. **117**(A11), A11217 (2012)

Y.I. Galperin, Multiple scales in auroral plasmas. J. Atmos. Sol.-Terr. Phys. **64**(2), 211–229 (2002)

D.M. Gillies, D.J. Knudsen, E.F. Donovan, E.L. Spanswick, C. Hansen, D. Keating, S. Erion, A survey of quiet auroral arc orientation and the effects of the interplanetary magnetic field. J. Geophys. Res. Space Phys. **119**(4), 2550–2562 (2014). https://doi.org/10.1002/2013JA019469. https://agupubs.onlinelibrary.wiley.com/doi/abs/10.1002/2013JA019469

D. Gillies, D. Knudsen, E. Spanswick, E. Donovan, J. Burchill, M. Patrick, Swarm observations of field-aligned currents associated with pulsating auroral patches. J. Geophys. Res. Space Phys. **120**(11), 9484–9499 (2015). https://doi.org/10.1002/2015JA021416

D.M. Gillies, D. Knudsen, R. Rankin, S. Milan, E. Donovan, A statistical survey of the 630.0 nm optical signature of periodic auroral arcs resulting from magnetospheric field line resonances. Geophys. Res. Lett. **45**, 4648–4655 (2018)

J. Gjerloev, S. Ohtani, T. Iijima, B. Anderson, J. Slavin, G. Le, Characteristics of the terrestrial field-aligned current system. Ann. Geophys. **29**(10) 1713–1729 (2011)

G. Grubbs II, R. Michell, M. Samara, D. Hampton, J.-M. Jahn, Predicting electron population characteristics in 2-d using multispectral ground-based imaging. Geophys. Res. Lett. **45**(1), 15–20 (2018). https://doi.org/10.1002/2017GL075873. https://agupubs.onlinelibrary.wiley.com/doi/abs/10.1002/2017GL075873

G. Grubbs, R. Michell, M. Samara, D. Hampton, J. Hecht, S. Solomon, J.-M. Jahn, A comparative study of spectral auroral intensity predictions from multiple electron transport models. J. Geophys. Res. Space Phys. **123**(1), 993–1005 (2018a)

G. Grubbs, R. Michell, M. Samara, D. Hampton, J.-M. Jahn, Predicting electron population characteristics in 2-d using multispectral ground-based imaging. Geophys. Res. Lett. **45**(1), 15–20 (2018b)

D.A. Gurnett, Electric Field and Plasma Observations in the Magnetosphere, Technical report, Iowa Univ. Iowa City Dept. Of Physics and Astronomy, 1972

B. Gustavsson, M.J. Kosch, A. Senior, A.J. Kavanagh, B. Brändström, E. Blixt, Combined EISCAT radar and optical multispectral and tomographic observations of black aurora. J. Geophys. Res. Space Phys. **113**(A6), A06308 (2008)

G. Haerendel, Acceleration from field-aligned potential drops. Int. Astron. Union Colloq. **142**, 765–774 (1994). https://doi.org/10.1017/S0252921100078076

G. Haerendel, H. Frey, Role and origin of the poleward Alfvénic arc. J. Geophys. Res. Space Phys. **119**(4), 2945–2962 (2014). https://doi.org/10.1002/2014JA019786

G. Haerendel, H.U. Frey, C.C. Chaston, O. Amm, L. Juusola, R. Nakamura, E. Seran, J. Weygand, Birth and life of auroral arcs embedded in the evening auroral oval convection: a critical comparison of observations with theory. J. Geophys. Res. Space Phys. **117**(A12), A12220 (2012). https://doi.org/10.1029/2012JA018128

T.J. Hallinan, T.N. Davis, Small-scale auroral arc distortions. Planet. Space Sci. **18**(12), 1735–1744 (1970)

T. Hallinan, H. Stenbaek-Nielsen, The connection between auroral acceleration and auroral morphology. Phys. Chem. Earth, Part C, Sol.-Terr. Planet. Sci. **26**(1–3), 169–177 (2001)

T.J. Hallinan, J. Kimball, H.C. Stenbaek-Nielsen, K. Lynch, R. Arnoldy, J. Bonnell, P. Kintner, Relation between optical emissions, particles, electric fields, and Alfvén waves in a multiple rayed arc. J. Geophys. Res. **106**, 15445 (2001)

T.J. Hallinan, J. Kimball, H.C. Stenbaek-Nielsen, K. Lynch, R. Arnoldy, J. Bonnell, P. Kintner, Relation between optical emissions, particles, electric fields, and Alfvén waves in a multiple rayed arc. J. Geophys. Res. Space Phys. **106**(A8), 15445–15454 (2001). https://doi.org/10.1029/2000JA000321. https://agupubs.onlinelibrary.wiley.com/doi/abs/10.1029/2000JA000321

M. Hamrin, O. Marghitu, K. Rönnmark, B. Klecker, M. André, S. Buchert, L. Kistler, J. McFadden, H. Reme, A. Vaivads, Observations of concentrated generator regions in the nightside magnetosphere by Cluster/FAST conjunctions, in *Annales Geophysicae*, vol. 24 (2006), pp. 637–649

M. Hamrin, O. Marghitu, P. Norqvist, S. Buchert, M. André, B. Klecker, L.M. Kistler, I. Dandouras, Energy conversion regions as observed by cluster in the plasma sheet. J. Geophys. Res. Space Phys. **116**(A1), A00K08 (2011)

M. Hamrin, P. Norqvist, T. Karlsson, H. Nilsson, H. Fu, S. Buchert, M. André, O. Marghitu, T. Pitkänen, B. Klecker, et al., The evolution of flux pileup regions in the plasma sheet: cluster observations. J. Geophys. Res. Space Phys. **118**(10), 6279–6290 (2013)

M. Hamrin, T. Pitkänen, P. Norqvist, T. Karlsson, H. Nilsson, M. André, S. Buchert, A. Vaivads, O. Marghitu, B. Klecker, et al., Evidence for the braking of flow bursts as they propagate toward the Earth. J. Geophys. Res. Space Phys. **119**(11), 9004–9018 (2014)

P. Hanna, C. Anger, Auroral colour variations. Planet. Space Sci. **19**(4), 399–411 (1971)

Z.-J. Hu, H. Yang, D. Huang, T. Araki, N. Sato, M. Taguchi, E. Seran, H. Hu, R. Liu, B. Zhang, D. Han, Z. Chen, Q. Zhang, J. Liang, S. Liu, Synoptic distribution of dayside aurora: multiple-wavelength all-sky observation at yellow river station in ny-Ålesund, svalbard. J. Atmos. Sol.-Terr. Phys. **71**(8), 794–804 (2009). https://doi.org/10.1016/j.jastp.2009.02.010. http://www.sciencedirect.com/science/article/pii/S1364682609000339

A. Hull, J. Bonnell, F. Mozer, J. Scudder, C. Chaston, Large parallel electric fields in the upward current region of the aurora: evidence for ambipolar effects. J. Geophys. Res. Space Phys. **108**(A6), 1265 (2003a)

A. Hull, J. Bonnell, F. Mozer, J. Scudder, A statistical study of large-amplitude parallel electric fields in the upward current region of the auroral acceleration region. J. Geophys. Res. Space Phys. **108**(A1), 1007 (2003b)

A. Hull, C. Chaston, H. Frey, M. Fillingim, M. Goldstein, J. Bonnell, F. Mozer, The "Alfvénic surge" at substorm onset/expansion and the formation of "Inverted Vs": Cluster and IMAGE observations. J. Geophys. Res. Space Phys. **121**(5), 3978–4004 (2016)

B. Hultqvist, Downward ion acceleration at auroral latitudes: cause of parallel electric field, in *Annales Geophysicae*, vol. 20 (2002), pp. 1117–1136

K.-J. Hwang, K. Lynch, C. Carlson, J. Bonnell, W. Peria, Fast auroral snapshot observations of perpendicular dc electric field structures in downward auroral current regions: morphology. J. Geophys. Res. Space Phys. **111**(A9), A09205 (2006a)

K.-J. Hwang, K. Lynch, C. Carlson, J. Bonnell, W. Peria, Fast auroral snapshot observations of perpendicular DC electric field structures in downward current regions: implications. J. Geophys. Res. Space Phys. **111**(A9), A09206 (2006b)

K.-J. Hwang, R. Ergun, L. Andersson, D. Newman, C. Carlson, Test particle simulations of the effect of moving DLs on ion outflow in the auroral downward-current region. J. Geophys. Res. Space Phys. **113**(A1), A01308 (2008)

T. Iijima, T.A. Potemra, The amplitude distribution of field-aligned currents at northern high latitudes observed by triad. J. Geophys. Res. **81**(13), 2165–2174 (1976)

P. Israelevich, I. Podgorny, A. Kuzmin, N. Nikolaeva, E. Dubinin, Convection and field-aligned currents, related to polar cap arcs, during strongly northward IMF (11 January 1983). Planet. Space Sci. **36**(12), 1317–1328 (1988)

P.T. Jayachandran, E.F. Donovan, J.W. MacDougall, D.R. Moorcroft, J.-P. St. Maurice, P. Prikryl, Superdarn e-region backscatter boundary in the dusk-midnight sector—tracer of equatorward boundary of the auroral oval. Ann. Geophys. **20**(12), 1899–1904 (2002). https://doi.org/10.5194/angeo-20-1899-2002. https://www.ann-geophys.net/20/1899/2002/

F. Jiang, R.J. Strangeway, M.G. Kivelson, J.M. Weygand, R.J. Walker, K.K. Khurana, Y. Nishimura, V. Angelopoulos, E. Donovan, In situ observations of the "preexisting auroral arc" by themis all sky imagers and the fast spacecraft. J. Geophys. Res. Space Phys. **117**(A5) (2012). https://doi.org/10.1029/2011JA017128. https://agupubs.onlinelibrary.wiley.com/doi/abs/10.1029/2011JA017128

F. Jiang, M.G. Kivelson, R.J. Strangeway, K.K. Khurana, R. Walker, Ionospheric flow shear associated with the preexisting auroral arc: a statistical study from the fast spacecraft data. J. Geophys. Res. Space Phys. **120**(6), 5194–5213 (2015a). https://doi.org/10.1002/2013JA019255. https://agupubs.onlinelibrary.wiley.com/doi/abs/10.1002/2013JA019255

F. Jiang, M.G. Kivelson, R.J. Strangeway, K.K. Khurana, R. Walker, Ionospheric flow shear associated with the preexisting auroral arc: a statistical study from the FAST spacecraft data. J. Geophys. Res. Space Phys. **120**(6), 5194–5213 (2015b)

T. Johansson, G. Marklund, T. Karlsson, S. Liléo, P.-A. Lindqvist, A. Marchaudon, H. Nilsson, A. Fazakerley, On the profile of intense high-altitude auroral electric fields at magnetospheric boundaries, in *Annales Geophysicae*, vol. 24 (2006), pp. 1713–1723

T. Johansson, G. Marklund, T. Karlsson, S. Liléo, P.-A. Lindqvist, H. Nilsson, S. Buchert, Scale sizes of intense auroral electric fields observed by Cluster, in *Annales Geophysicae*, vol. 25 (2007), pp. 2413–2425

M.G. Johnsen, D.A. Lorentzen, A statistical analysis of the optical dayside open/closed field line boundary. J. Geophys. Res. Space Phys. **117**(A2), A02218 (2012). https://doi.org/10.1029/2011JA016984. https://agupubs.onlinelibrary.wiley.com/doi/abs/10.1029/2011JA016984

J.R.T. Jussila, A.T. Aikio, S. Shalimov, S.R. Marple, Cosmic radio noise absorption events associated with equatorward drifting arcs during a substorm growth phase. Ann. Geophys. **22**(5), 1675–1686 (2004). https://doi.org/10.5194/angeo-22-1675-2004. https://www.ann-geophys.net/22/1675/2004/

L. Juusola, K. Kauristie, H. Vanhamäki, A. Aikio, M. Kamp, Comparison of auroral ionospheric and field-aligned currents derived from swarm and ground magnetic field measurements. J. Geophys. Res. Space Phys. **121**(9), 9256–9283 (2016a)

L. Juusola, W.E. Archer, K. Kauristie, J.K. Burchill, H. Vanhamäki, A.T. Aikio, Ionospheric conductances and currents of a morning sector auroral arc from Swarm-A electric and magnetic field measurements. Geophys. Res. Lett. **43**(22), 11519–11527 (2016b). https://doi.org/10.1002/2016GL070248. http://doi.wiley.com/10.1002/2016GL070248

S.R. Kaeppler, M.J. Nicolls, A. Strømme, C.A. Kletzing, S.R. Bounds, Observations in the E region ionosphere of kappa distribution functions associated with precipitating auroral electrons and discrete aurorae. J. Geophys. Res. Space Phys. **119**(12), 10164–10183 (2014). https://doi.org/10.1002/2014JA020356. http://doi.wiley.com/10.1002/2014JA020356

Y. Kamide, S.-I. Akasofu, The location of the field-aligned currents with respect to discrete auroral arcs. J. Geophys. Res. **81**(22), 3999–4003 (1976)

T. Karlsson, The acceleration region of stable auroral arcs, in *Chapman Conference on the Relationship Between Auroral Phenomenology and Magnetospheric Processes*, Feb. 27–Mar. 04, 2011, Fairbanks, AK (Am. Geophys. Union, Washington, 2012), pp. 227–239

T. Karlsson, *The Acceleration Region of Stable Auroral Arcs* (Am. Geophys. Union, Washington, 2013), pp. 227–240. https://doi.org/10.1029/2011GM001179. https://agupubs.onlinelibrary.wiley.com/doi/abs/10.1029/2011GM001179

T. Karlsson, G. Marklund, A statistical study of intense low-altitude electric fields observed by freja. Geophys. Res. Lett. **23**(9), 1005–1008 (1996)

T. Karlsson, G. Marklund, N. Brenning, I. Axnäs, On enhanced aurora and low-altitude parallel electric fields. Phys. Scr. **72**(5), 419 (2005)

T. Karlsson, N. Brenning, O. Marghitu, G. Marklund, S. Buchert, High-altitude signatures of ionospheric density depletions caused by field-aligned currents. arXiv preprint, arXiv:0704.1610 (2007)

A. Keiling, J. Wygant, C. Cattell, W. Peria, G. Parks, M. Temerin, F. Mozer, C. Russell, C. Kletzing, Correlation of Alfvén wave poynting flux in the plasma sheet at 4–7 RE with ionospheric electron energy flux. J. Geophys. Res. Space Phys. **107**(A7), 24 (2002)

A. Keiling, V. Angelopoulos, A. Runov, J. Weygand, S. Apatenkov, S. Mende, J. McFadden, D. Larson, O. Amm, K.-H. Glassmeier, et al., Substorm current wedge driven by plasma flow vortices: themis observations. J. Geophys. Res. Space Phys. **114**(A1), A00C22 (2009)

J. Kimball, T. Hallinan, Observations of black auroral patches and of their relationship to other types of aurora. J. Geophys. Res. Space Phys. **103**(A7), 14671–14682 (1998)

P.M. Kintner, H. Kil, C. Deehr, P. Schuck, Simultaneous total electron content and all-sky camera measurements of an auroral arc. J. Geophys. Res. Space Phys. **107**(A7), 13–1136 (2002). https://doi.org/10.1029/2001JA000110. https://agupubs.onlinelibrary.wiley.com/doi/abs/10.1029/2001JA000110

S. Kirkwood, H. Opgenoorth, J.S. Murphree, Ionospheric conductivities, electric fields and currents associated with auroral substorms measured by the EISCAT radar. Planet. Space Sci. **36**(12), 1359–1380 (1988). https://doi.org/10.1016/0032-0633(88)90005-0. https://www.sciencedirect.com/science/article/pii/0032063388900050?via%3Dihub

D. Klumpar, W. Heikkila, Electrons in the ionospheric source cone: evidence for runaway electrons as carriers of downward Birkeland currents. Geophys. Res. Lett. **9**(8), 873–876 (1982)

S. Knight, Parallel electric fields. Planet. Space Sci. **21**(5), 741–750 (1973)

D.J. Knudsen, Spatial modulation of electron energy and density by nonlinear stationary inertial Alfvén waves. J. Geophys. Res. Space Phys. **101**(A5), 10761–10772 (1996). https://doi.org/10.1029/96JA00429

D. Knudsen, E. Donovan, L. Cogger, B. Jackel, W. Shaw, Width and structure of mesoscale optical auroral arcs. Geophys. Res. Lett. **28**(4), 705–708 (2001)

M.J. Kosch, C. Anderson, R.A. Makarevich, B.A. Carter, R.A.D. Fiori, M. Conde, P.L. Dyson, T. Davies, First *E* region observations of mesoscale neutral wind interaction with auroral arcs. J. Geophys. Res. Space Phys. **115**(A2) A02303, (2010). https://doi.org/10.1029/2009JA014697. http://doi.wiley.com/10.1029/2009JA014697

A. Kozlovsky, T. Lakkala, J. Kangas, A. Aikio, Response of the quiet auroral arc motion to ionospheric convection variations. J. Geophys. Res. Space Phys. **106**(A10), 21463–21473 (2001). https://doi.org/10.1029/2001JA900043. https://agupubs.onlinelibrary.wiley.com/doi/abs/10.1029/2001JA900043

A.E. Kozlovsky, V.V. Safargaleev, J.R.T. Jussila, A.V. Koustov, Pre-noon high-latitude auroral arcs as a manifestation of the interchange instability. Ann. Geophys. **21**(12), 2303–2314 (2003). https://doi.org/10.5194/angeo-21-2303-2003. https://www.ann-geophys.net/21/2303/2003/

A.E. Kozlovsky, H. Nilsson, V.V. Safargaleev, Complex study of the auroral arc dynamics and ionospheric plasma convection in prenoon hours. Geomagn. Aeron. **46**(4), 473–484 (2006). https://doi.org/10.1134/S0016793206040098.

A. Kozlovsky, T. Turunen, S. Massetti, Field-aligned currents of postnoon auroral arcs. J. Geophys. Res. Space Phys. **114**(A3), A03301 (2009)

E.A. Kronberg, M. Ashour-Abdalla, I. Dandouras, D.C. Delcourt, E.E. Grigorenko, L.M. Kistler, I.V. Kuzichev, J. Liao, R. Maggiolo, H.V. Malova, et al., Circulation of heavy ions and their dynamical effects in the magnetosphere: recent observations and models. Space Sci. Rev. **184**(1-4), 173–235 (2014)

B. Lanchester, B. Gustavsson, Imaging of aurora to estimate the energy and flux of electron precipitation, in *Auroral Phenomenology and Magnetospheric Processes: Earth and Other Planets*, vol. 197 (2012), pp. 171–182

K. Laundal, N. Østgaard, Asymmetric auroral intensities in the Earth's northern and southern hemispheres. Nature **460**(7254), 491 (2009)

K.M. Laundal, I. Cnossen, S.E. Milan, S. Haaland, J. Coxon, N. Pedatella, M. Förster, J.P. Reistad, North–South asymmetries in Earth's magnetic field. Space Sci. Rev. **206**(1–4), 225–257 (2017)

G. Le, Y. Wang, J. Slavin, R. Strangeway, Space Technology 5 multipoint observations of temporal and spatial variability of field-aligned currents. J. Geophys. Res. Space Phys. **114**(A8), A08206 (2009)

M.R. Lessard, W. Lotko, J. LaBelle, W. Peria, C.W. Carlson, F. Creutzberg, D.D. Wallis, Ground and satellite observations of the evolution of growth phase auroral arcs. J. Geophys. Res. Space Phys. **112**(A9), A09304 (2007). https://doi.org/10.1029/2006JA011794. https://agupubs.onlinelibrary.wiley.com/doi/abs/10.1029/2006JA011794

B. Li, G. Marklund, T. Karlsson, S. Sadeghi, P.-A. Lindqvist, A. Vaivads, A. Fazakerley, Y. Zhang, E. Lucek, T. Sergienko, et al., Inverted-v and low-energy broadband electron acceleration features of multiple auroras within a large-scale surge. J. Geophys. Res. Space Phys. **118**(9), 5543–5552 (2013)

P.-A. Lindqvist, G.T. Marklund, A statistical study of high-altitude electric fields measured on the viking satellite. J. Geophys. Res. Space Phys. **95**(A5), 5867–5876 (1990). https://doi.org/10.1029/JA095iA05p05867

K. Liou, P.T. Newell, C.-I. Meng, M. Brittnacher, G. Parks, Synoptic auroral distribution: a survey using polar ultraviolet imagery. J. Geophys. Res. Space Phys. **102**(A12), 27197–27205 (1997). https://doi.org/10.1029/97JA02638. https://agupubs.onlinelibrary.wiley.com/doi/abs/10.1029/97JA02638

J. Liu, V. Angelopoulos, Z. Yao, X. Chu, X.-Z. Zhou, A. Runov, The current system of dipolarizing flux bundles and their role as wedgelets in the substorm current wedge, in *Electric Currents in Geospace and Beyond*, vol. 235, (Wiley Online Library, New York, 2018a), pp. 323–337

J. Liu, L. Lyons, W. Archer, B. Gallardo-Lacourt, Y. Nishimura, Y. Zou, C. Gabrielse, J. Weygand, Flow shears at the poleward boundary of omega bands observed during conjunctions of swarm and THEMIS ASI. Geophys. Res. Lett. **45**(3), 1218–1227 (2018b). https://doi.org/10.1002/2017GL076485

E. Lund, E. Möbius, R. Ergun, C. Carlson, Mass-dependent effects in ion conic production: the role of parallel electric fields. Geophys. Res. Lett. **26**(24), 3593–3596 (1999)

R. Lundin, D.S. Evans, Boundary layer plasmas as a source for high-latitude, early afternoon, auroral arcs. Planet. Space Sci. **33**(12), 1389–1406 (1985)

R. Lundin, I. Sandahl, Some characteristics of the parallel electric field acceleration of electrons over discrete auroral arcs as observed from two rocket flights, in *European Sounding Rocket, Balloon and Related Research, with Emphasis on Experiments at High Latitudes*, vol. 135 (1978)

K. Lynch, J. Bonnell, C. Carlson, W. Peria, Return current region aurora: $E_{\|}$, j_Z, particle energization, and broadband ELF wave activity. J. Geophys. Res. Space Phys. **107**(A7), 1115 (2002)

K. Lynch, D. Hampton, M. Zettergren, T. Bekkeng, M. Conde, P. Fernandes, P. Horak, M. Lessard, R. Miceli, R. Michell, et al., Mica sounding rocket observations of conductivity-gradient-generated auroral ionospheric responses: small-scale structure with large-scale drivers. J. Geophys. Res. Space Phys. **120**(11), 9661–9682 (2015)

L. Lyons, D. Evans, R. Lundin, An observed relationship between magnetic field-aligned electric fields and downward electron energy fluxes in the vicinity of auroral forms. J. Geophys. Res. **84**, 457 (1979)

L. Lyons, J. Fennell, A. Vampola, A general association between discrete auroras and ion precipitation from the tail. J. Geophys. Res. Space Phys. **93**(A11), 12932–12940 (1988)

L.R. Lyons, I.O. Voronkov, E.F. Donovan, E. Zesta, Relation of substorm breakup arc to other growth-phase auroral arcs. J. Geophys. Res. Space Phys. **107**(A11), 26–12610 (2002). https://doi.org/10.1029/2002JA009317. https://agupubs.onlinelibrary.wiley.com/doi/abs/10.1029/2002JA009317

R.L. Lysak, Auroral electrodynamics with current and voltage generators. J. Geophys. Res. Space Phys. **90**(A5), 4178–4190 (1985)
R.L. Lysak, Electrodynamic coupling of the magnetosphere and ionosphere. Space Sci. Rev. **52**(1), 33–87 (1990). https://doi.org/10.1007/BF00704239
R.L. Lysak, Propagation of Alfvén waves through the ionosphere: dependence on ionospheric parameters. J. Geophys. Res. **104**, 10017–10030 (1999). https://doi.org/10.1029/1999JA900024
J. Maggs, T. Davis, Measurements of the thicknesses of auroral structures. Planet. Space Sci. **16**(2), 205–209 (1968)
A.J. Mallinckrodt, C.W. Carlson, Relations between transverse electric fields and field-aligned currents. J. Geophys. Res. Space Phys. **83**(A4), 1426–1432 (1978). https://doi.org/10.1029/JA083iA04p01426. https://agupubs.onlinelibrary.wiley.com/doi/abs/10.1029/JA083iA04p01426
Y.P. Maltsev, W.B. Lyatsky, A.M. Lyatskaya, Currents over the auroral arc. Planet. Space Sci. **25**(1), 53–57 (1977). https://doi.org/10.1016/0032-0633(77)90117-9. http://www.sciencedirect.com/science/article/pii/0032063377901179
O. Marghitu, Auroral arc electrodynamics: review and outlook, in *Auroral Phenomenology and Magnetospheric Processes: Earth and Other Planets*, vol. 197 (2012), pp. 143–158
O. Marghitu, B. Klecker, G. Haerendel, J. McFadden, ALADYN: a method to investigate auroral arc electrodynamics from satellite data. J. Geophys. Res. Space Phys. **109**(A11), A11305 (2004)
O. Marghitu, M. Hamrin, B. Klecker, A. Vaivads, J. McFadden, S. Buchert, L.M. Kistler, I. Dandouras, M. André, H. Rème, Experimental investigation of auroral generator regions with conjugate Cluster and FAST data, in *Annales Geophysicae*, vol. 24 (2006), pp. 619–635
O. Marghitu, B. Klecker, J. McFadden, The anisotropy of precipitating auroral electrons: a fast case study. Adv. Space Res. **38**(8), 1694–1701 (2006)
O. Marghitu, T. Karlsson, B. Klecker, G. Haerendel, J. McFadden, Auroral arc and oval electrodynamics in the Harang region. J. Geophys. Res. Space Phys. **114**(A3), A03214 (2009)
O. Marghitu, C. Bunescu, T. Karlsson, B. Klecker, H.C. Stenbaek-Nielsen, On the divergence of the auroral electrojets. J. Geophys. Res. Space Phys. **116**(A1), A00K17 (2011)
G. Marklund, Auroral arc classification scheme based on the observed arc-associated electric field pattern. Planet. Space Sci. **32**(2), 193–211 (1984)
G. Marklund, I. Sandahl, H. Opgenoorth, A study of the dynamics of a discrete auroral arc. Planet. Space Sci. **30**(2), 179–197 (1982). https://doi.org/10.1016/0032-0633(82)90088-5. https://www.sciencedirect.com/science/article/pii/0032063382900885?via%3Dihub
G. Marklund, L. Blomberg, C.-G. Fälthammar, P.-A. Lindqvist, On intense diverging electric fields associated with black aurora. Geophys. Res. Lett. **21**(17), 1859–1862 (1994)
G.T. Marklund, N. Ivchenko, T. Karlsson, A. Fazakerley, M. Dunlop, P.-A. Lindqvist, S. Buchert, C. Owen, M. Taylor, A. Vaivalds, et al., Temporal evolution of the electric field accelerating electrons away from the auroral ionosphere. Nature **414**(6865), 724 (2001)
G.T. Marklund, S. Sadeghi, T. Karlsson, P.-A. Lindqvist, H. Nilsson, C. Forsyth, A. Fazakerley, E.A. Lucek, J. Pickett, Altitude distribution of the auroral acceleration potential determined from cluster satellite data at different heights. Phys. Rev. Lett. **106**(5), 055002 (2011a)
G.T. Marklund, S. Sadeghi, J.A. Cumnock, T. Karlsson, P.-A. Lindqvist, H. Nilsson, A. Masson, A. Fazakerley, E. Lucek, J. Pickett, et al., Evolution in space and time of the quasi-static acceleration potential of inverted-V aurora and its interaction with Alfvénic boundary processes. J. Geophys. Res. Space Phys. **116**(A1), A00K13 (2011b)
I. McCrea, A. Aikio, L. Alfonsi, E. Belova, S. Buchert, M. Clilverd, N. Engler, B. Gustavsson, C. Heinselman, J. Kero, et al., The science case for the EISCAT_3d radar. Prog. Earth Planet. Sci. **2**(1), 21 (2015)
J.P. McFadden, C.W. Carlson, R.E. Ergun, Microstructure of the auroral acceleration region as observed by fast. J. Geophys. Res. Space Phys. **104**(A7), 14453–14480 (1999). https://doi.org/10.1029/1998JA900167. https://agupubs.onlinelibrary.wiley.com/doi/abs/10.1029/1998JA900167
C.E. McIlwain, Direct measurement of particles producing visible auroras. J. Geophys. Res. **65**(9), 2727–2747 (1960)
D. McKay, N. Partamies, J. Vierinen, Pulsating aurora and cosmic noise absorption associated with growth-phase arcs. Ann. Geophys. **36**(1), 59–69 (2018). https://doi.org/10.5194/angeo-36-59-2018. https://www.ann-geophys.net/36/59/2018/
R.L. McPherron, C.T. Russell, M.P. Aubry, Satellite studies of magnetospheric substorms on August 15, 1968: 9. Phenomenological model for substorms. J. Geophys. Res. **78**(16), 3131–3149 (1973)
S. Mende, The THEMIS array of ground-based observatories for the study of auroral substorms. Space Sci. Rev. **141**, 357 (2008). https://doi.org/10.1007/s11214-008-9380-x
S. Mende, R. Eather, Monochromatic all-sky observations and auroral precipitation patterns. J. Geophys. Res. **81**(22), 3771–3780 (1976)

C.-I. Meng, R. Lundin, Auroral morphology of the midday oval. J. Geophys. Res. Space Phys. **91**(A2), 1572–1584 (1986)

D. Miles, I.R. Mann, I. Pakhotin, J.K. Burchill, A.D. Howarth, D.J. Knudsen, R.L. Lysak, D. Wallis, L. Cogger, A. Yau, Alfvénic dynamics and fine structuring of discrete auroral arcs: swarm and e-pop observations. Geophys. Res. Lett. **45**(2), 545–555 (2018)

J. Moen, P. Sandholt, M. Lockwood, A. Egeland, K. Fukui, Multiple, discrete arcs on sunward convecting field lines in the 14-15 MLT region. J. Geophys. Res. Space Phys. **99**(A4), 6113–6123 (1994)

T. Moore, M.-C. Fok, K. Garcia-Sage, The ionospheric outflow feedback loop. J. Atmos. Sol.-Terr. Phys. **115**, 59–66 (2014)

M. Morooka, T. Mukai, Density as a controlling factor for seasonal and altitudinal variations of the auroral particle acceleration region. J. Geophys. Res. Space Phys. **108**(A7), 1306 (2003)

T. Motoba, M. Hirahara, High-resolution auroral acceleration signatures within a highly dynamic onset arc. Geophys. Res. Lett. **43**(5), 1793–1801 (2016). https://doi.org/10.1002/2015GL067580. https://agupubs.onlinelibrary.wiley.com/doi/abs/10.1002/2015GL067580

T. Motoba, S. Ohtani, B.J. Anderson, H. Korth, D. Mitchell, L.J. Lanzerotti, K. Shiokawa, M. Connors, C.A. Kletzing, G.D. Reeves, On the formation and origin of substorm growth phase/onset auroral arcs inferred from conjugate space-ground observations. J. Geophys. Res. Space Phys. **120**(10), 8707–8722 (2015). https://doi.org/10.1002/2015JA021676. https://agupubs.onlinelibrary.wiley.com/doi/abs/10.1002/2015JA021676

F.S. Mozer, A. Hull, Origin and geometry of upward parallel electric fields in the auroral acceleration region. J. Geophys. Res. Space Phys. **106**(A4), 5763–5778 (2001). https://doi.org/10.1029/2000JA900117

F. Mozer, C. Kletzing, Direct observation of large, quasi-static, parallel electric fields in the auroral acceleration region. Geophys. Res. Lett. **25**(10), 1629–1632 (1998)

F.S. Mozer, C.W. Carlson, M. Hudson, R. Torbert, B. Parady, J. Yatteau, M.C. Kelley, Observations of paired electrostatic shocks in the polar magnetosphere. Phys. Rev. Lett. **38**(6), 292 (1977)

F.S. Mozer, C. Cattell, M. Hudson, R. Lysak, M. Temerin, R. Torbert, Satellite measurements and theories of low altitude auroral particle acceleration. Space Sci. Rev. **27**(2), 155–213 (1980)

H. Nevanlinna, T.I. Pulkkinen, Auroral observations in Finland: results from all-sky cameras, 1973–1997. J. Geophys. Res. Space Phys. **106**(A5), 8109–8118 (2001). https://doi.org/10.1029/1999JA000362. https://agupubs.onlinelibrary.wiley.com/doi/abs/10.1029/1999JA000362

P.T. Newell, Y.I. Feldstein, Y.I. Galperin, C.-I. Meng, Morphology of nightside precipitation. J. Geophys. Res. Space Phys. **101**(A5), 10737–10748 (1996)

P.T. Newell, K.M. Lyons, C.-I. Meng, A large survey of electron acceleration events. J. Geophys. Res. Space Phys. **101**(A2), 2599–2614 (1996a). https://doi.org/10.1029/95JA03147. https://agupubs.onlinelibrary.wiley.com/doi/abs/10.1029/95JA03147

P.T. Newell, C.-I. Meng, K.M. Lyons, Suppression of discrete aurorae by sunlight. Nature **381**(6585), 766–767 (1996b). https://doi.org/10.1038/381766a0. http://www.nature.com/articles/381766a0

P.T. Newell, R.A. Greenwald, J.M. Ruohoniemi, The role of the ionosphere in aurora and space weather. Rev. Geophys. **39**(2), 137–149 (2001)

P.T. Newell, T. Sotirelis, S. Wing, Diffuse, monoenergetic, and broadband aurora: the global precipitation budget. J. Geophys. Res. Space Phys. **114**(A9) A09207, (2009). https://doi.org/10.1029/2009JA014326. http://doi.wiley.com/10.1029/2009JA014326

P.T. Newell, T. Sotirelis, S. Wing, Seasonal variations in diffuse, monoenergetic, and broadband aurora. J. Geophys. Res. Space Phys. **115**(A3), A03216 (2010). https://doi.org/10.1029/2009JA014805. https://agupubs.onlinelibrary.wiley.com/doi/abs/10.1029/2009JA014805

M.J. Nicolls, S.L. Vadas, J.W. Meriwether, M.G. Conde, D. Hampton, The phases and amplitudes of gravity waves propagating and dissipating in the thermosphere: application to measurements over Alaska. J. Geophys. Res. Space Phys. **117**(A5), A03216 (2012). https://doi.org/10.1029/2012JA017542. https://agupubs.onlinelibrary.wiley.com/doi/abs/10.1029/2012JA017542

H. Nilsson, A. Kozlovsky, T. Sergienko, A. Kotikov, Radar observations in the vicinity of pre-noon auroral arcs. Ann. Geophys. **23**(5), 1785–1796 (2005). https://doi.org/10.5194/angeo-23-1785-2005. http://www.ann-geophys.net/23/1785/2005/

Y. Nishimura, L.R. Lyons, V. Angelopoulos, T. Kikuchi, S. Zou, S.B. Mende, Relations between multiple auroral streamers, pre-onset thin arc formation, and substorm auroral onset. J. Geophys. Res. Space Phys. **116**(A9), A09214 (2011). https://doi.org/10.1029/2011JA016768. https://agupubs.onlinelibrary.wiley.com/doi/abs/10.1029/2011JA016768

Y. Nishimura, L.R. Lyons, T. Kikuchi, V. Angelopoulos, E.F. Donovan, S.B. Mende, H. Lühr, Relation of substorm pre-onset arc to large-scale field-aligned current distribution. Geophys. Res. Lett. **39**(22), L22101 (2012). https://doi.org/10.1029/2012GL053761. https://agupubs.onlinelibrary.wiley.com/doi/abs/10.1029/2012GL053761

Y. Nishimura, L.R. Lyons, M.J. Nicolls, D.L. Hampton, R.G. Michell, M. Samara, W.A. Bristow, E.F. Donovan, E. Spanswick, V. Angelopoulos, S.B. Mende, Coordinated ionospheric observations indicating coupling between preonset flow bursts and waves that lead to substorm onset. J. Geophys. Res. Space Phys. **119**(5), 3333–3344 (2014). https://doi.org/10.1002/2014JA019773. https://agupubs.onlinelibrary.wiley.com/doi/abs/10.1002/2014JA019773

Y. Obuchi, T. Sakanoi, K. Asamura, A. Yamazaki, Y. Kasaba, M. Hirahara, Y. Ebihara, S. Okano, Fine-scale dynamics of black auroras obtained from simultaneous imaging and particle observations with the reimei satellite. J. Geophys. Res. Space Phys. **116**(A1), A00K07 (2011)

N. Østgaard, N. Tsyganenko, S. Mende, H. Frey, T. Immel, M. Fillingim, L. Frank, J. Sigwarth, Observations and model predictions of substorm auroral asymmetries in the conjugate hemispheres. Geophys. Res. Lett. **32**(5) (2005). https://doi.org/10.1029/2004GL022166

N. Østgaard, J.P. Reistad, P. Tenfjord, K.M. Laundal, K. Snekvik, S. Milan, S. Haaland, Mechanisms that produce auroral asymmetries in conjugate hemispheres, in *Auroral Dynamics and Space Weather*, vol. 215 (2015), p. 133

N. Partamies, P. Janhunen, K. Kauristie, S. Mäkinen, T. Sergienko, Testing an inversion method for estimating electron energy fluxes from all-sky camera images. Ann. Geophys. **22**, 1961–1971 (2004). https://doi.org/10.5194/angeo-22-1961-2004

N. Partamies, E. Donovan, D. Knudsen, Statistical study of inverted-V structures in FAST data. Ann. Geophys. **26**, 1439–1449 (2008). www.ann-geophys.net/26/1439/2008/

N. Partamies, M. Syrjäsuo, E. Donovan, M. Connors, D. Charrois, D. Knudsen, Z. Kryzanowsky, Observations of the auroral width spectrum at kilometre-scale size. Ann. Geophys. **28**, 711–718 (2010). www.ann-geophys.net/28/711/2010/

N. Partamies, D. Whiter, M. Syrjäsuo, K. Kauristie, Solar cycle and diurnal dependence of auroral structures. J. Geophys. Res. Space Phys. **119**(10), 8448–8461 (2014). https://doi.org/10.1002/2013JA019631. https://agupubs.onlinelibrary.wiley.com/doi/abs/10.1002/2013JA019631

N. Partamies, L. Juusola, D. Whiter, K. Kauristie, Substorm evolution of auroral structures. J. Geophys. Res. Space Phys. **120**(7), 5958–5972 (2015). https://doi.org/10.1002/2015JA021217. https://agupubs.onlinelibrary.wiley.com/doi/abs/10.1002/2015JA021217

G. Paschmann, S. Haaland, R. Treumann, *Auroral Plasma Physics* (2003)

L. Peticolas, T. Hallinan, H. Stenbaek-Nielsen, J. Bonnell, C. Carlson, A study of black aurora from aircraft-based optical observations and plasma measurements on fast. J. Geophys. Res. Space Phys. **107**(A8), 1217 (2002)

T.I. Pulkkinen, E.I. Tanskanen, A. Viljanen, N. Partamies, K. Kauristie, Auroral electrojets during deep solar minimum at the end of solar cycle 23. J. Geophys. Res. Space Phys. **116**(A4), A04207 (2011). https://doi.org/10.1029/2010JA016098. https://agupubs.onlinelibrary.wiley.com/doi/abs/10.1029/2010JA016098

Q. Qiu, H.-G. Yang, Q.-M. Lu, Q.-H. Zhang, D.-S. Han, Z.-J. Hu, Widths of dayside auroral arcs observed at the Chinese yellow river station. J. Atmos. Sol.-Terr. Phys. **102**, 222–227 (2013). https://doi.org/10.1016/j.jastp.2013.06.002. http://www.sciencedirect.com/science/article/pii/S1364682613001752

Q. Qiu, H.-G. Yang, Q.-M. Lu, Z.-J. Hu, D.-S. Han, Q. Wang, Orientation variation of dayside auroral arc alignments obtained from all-sky observation at yellow river station, svalbard. J. Atmos. Sol.-Terr. Phys. **142**, 20–24 (2016). https://doi.org/10.1016/j.jastp.2016.02.019. http://www.sciencedirect.com/science/article/pii/S1364682616300505

A.J. Ridley, Y. Deng, G. Tóth, The global ionosphere-thermosphere model. J. Atmos. Sol.-Terr. Phys. **68**(8), 839–864 (2006). https://doi.org/10.1016/j.jastp.2006.01.008. http://www.sciencedirect.com/science/article/pii/S1364682606000071

R. Robinson, R. Vondrak, K. Miller, T. Dabbs, D. Hardy, On calculating ionospheric conductances from the flux and energy of precipitating electrons. J. Geophys. Res. Space Phys. **92**(A3), 2565–2569 (1987)

G. Rostoker, R. Boström, A mechanism for driving the Gross Birkeland current configuration in the auroral oval. J. Geophys. Res. **81**(1), 235–244 (1976)

S. Sadeghi, M. Emami, Large-scale altitude distribution profile of auroral parallel electric potentials: a statistical analysis of cluster data. Adv. Space Res. **64**(2), 378–384 (2019)

S. Sadeghi, G.T. Marklund, T. Karlsson, P.-A. Lindqvist, H. Nilsson, O. Marghitu, A. Fazakerley, E.A. Lucek, Spatiotemporal features of the auroral acceleration region as observed by cluster. J. Geophys. Res. Space Phys. **116**(A1) (2011). https://doi.org/10.1029/2011JA016505

V.V. Safargaleev, A.E. Kozlovsky, S.V. Osipenko, V.R. Tagirov, Azimuthal expansion of high-latitude auroral arcs. Ann. Geophys. **21**(8), 1793–1805 (2003). https://doi.org/10.5194/angeo-21-1793-2003. https://www.ann-geophys.net/21/1793/2003/

T. Sakanoi, Y. Obuchi, Y. Ebihara, Y. Miyoshi, K. Asamura, A. Yamazaki, Y. Kasaba, M. Hirahara, T. Nishiyama, S. Okano, *Fine-Scale Characteristics of Black Aurora and Its Generation Process*. Geophysical Monograph Series, vol. 197 (Am. Geophys. Union, Washington, 2012)

P.E. Sandholt, H.C. Carlson, A. Egeland, *Dayside and Polar Cap Aurora*, vol. 270 (Springer, Berlin, 2006)

L. Sangalli, B. Gustavsson, N. Partamies, K. Kauristie, Estimating the peak auroral emission altitude from all-sky images. Opt. Pura Apl. **44**, 593–598 (2011)

T.E. Sarris, E. R. Talaat, M. Palmroth, I. Dandouras, E. Armandillo, G. Kervalishvili, S. Buchert, D. Malaspina, A. Jaynes, N. Paschalidis, J. Sample, J. Halekas, S. Tourgaidis, V. Lappas, M. Clilverd, Q. Wu, I. Sandberg, A. Aikio, P. Pirnaris, Daedalus: a low-flying spacecraft for the exploration of the lower thermosphere—Ionosphere. Geoscientific Instrumentation, Methods and Data Systems Discussions, 1–59 (2019). https://doi.org/10.5194/gi-2019-3

N. Sato, T. Nagaoka, K. Hashimoto, T. Saemundsson, Conjugacy of isolated auroral arcs and nonconjugate auroral breakups. J. Geophys. Res. Space Phys. **103**(A6), 11641–11652 (1998)

N. Sato, A. Kadokura, T. Motoba, K. Hosokawa, G. Björnsson, T. Saemundsson, Interhemispheric symmetries and asymmetries of aurora from ground-based conjugate observations, in *Auroral Dynamics and Space Weather*, vol. 215 (2015), p. 145

J. Semeter, T. Butler, M. Zettergren, C. Heinselman, M. Nicolls, Composite imaging of auroral forms and convective flows during a substorm cycle. J. Geophys. Res. Space Phys. **115**(A8) (2010). https://doi.org/10.1029/2009JA014931

V. Sergeev, Y. Nishimura, M. Kubyshkina, V. Angelopoulos, R. Nakamura, H. Singer, Magnetospheric location of the equatorward prebreakup arc. J. Geophys. Res. Space Phys. **117**(A1), A01212 (2012). https://doi.org/10.1029/2011JA017154. https://agupubs.onlinelibrary.wiley.com/doi/abs/10.1029/2011JA017154

J. Sesiano, P. Cloutier, Measurements of field-aligned currents in a multiple auroral arc system. J. Geophys. Res. **81**(1), 116–122 (1976)

Y. Shen, D.J. Knudsen, J.K. Burchill, A.D. Howarth, A.W. Yau, D.M. Miles, H.G. James, G.W. Perry, L. Cogger, Low-altitude ion heating, downflowing ions, and BBELF waves in the return current region. J. Geophys. Res. Space Phys. **123**(4), 3087–3110 (2018). https://doi.org/10.1002/2017JA024955

K. Shiokawa, K. Yago, K. Yumoto, D.G. Baishev, S.I. Solovyev, F.J. Rich, S.B. Mende, Ground and satellite observations of substorm onset arcs. J. Geophys. Res. Space Phys. **110**(A12), A12225 (2005). https://doi.org/10.1029/2005JA011281. https://agupubs.onlinelibrary.wiley.com/doi/abs/10.1029/2005JA011281

C. Simon Wedlund, H. Lamy, B. Gustavsson, T. Sergienko, U. Brändström, Estimating energy spectra of electron precipitation above auroral arcs from ground-based observations with radar and optics. J. Geophys. Res. Space Phys. **118**(6), 3672–3691 (2013). https://doi.org/10.1002/jgra.50347. https://agupubs.onlinelibrary.wiley.com/doi/abs/10.1002/jgra.50347

K. Stasiewicz, T. Potemra, Multiscale current structures observed by Freja. J. Geophys. Res. Space Phys. **103**(A3), 4315–4325 (1998). https://doi.org/10.1029/97JA02396. http://doi.wiley.com/10.1029/97JA02396

H.C. Stenbaek-Nielsen, T.J. Hallinan, D.L. Osborne, J. Kimball, C. Chaston, J. McFadden, G. Delory, M. Temerin, C.W. Carlson, Aircraft observations conjugate to fast: auroral are thicknesses. Geophys. Res. Lett. **25**(12), 2073–2076 (1998). https://doi.org/10.1029/98GL01058. https://agupubs.onlinelibrary.wiley.com/doi/abs/10.1029/98GL01058

R. Strangeway, R. Ergun, Y.-J. Su, C. Carlson, R. Elphic, Factors controlling ionospheric outflows as observed at intermediate altitudes. J. Geophys. Res. Space Phys. **110**(A3), A03221 (2005)

A.V. Streltsov, Effects of ionospheric heating on feedback-unstable electromagnetic waves. J. Geophys. Res. Space Phys. **113**(A9) A09211, (2008). https://doi.org/10.1029/2008JA013199. http://doi.wiley.com/10.1029/2008JA013199

M. Sugiura, A fundamental magnetosphere-ionosphere coupling mode involving field-aligned currents as deduced from DE-2 observations. Geophys. Res. Lett. **11**(9), 877–880 (1984)

D.W. Swift, D.J. Gorney, Production of very energetic electrons in discrete aurora. J. Geophys. Res. **94**(A3), 2696 (1989). https://doi.org/10.1029/JA094iA03p02696. http://doi.wiley.com/10.1029/JA094iA03p02696

M.T. Syrjäsuo, E.F. Donovan, Diurnal auroral occurrence statistics obtained via machine vision. Ann. Geophys. **22**(4), 1103–1113 (2004). https://doi.org/10.5194/angeo-22-1103-2004. https://www.ann-geophys.net/22/1103/2004/

J.R. Thieman, R.A. Hoffman, Determination of inverted-v stability from dynamics explorer satellite data. J. Geophys. Res. Space Phys. **90**(A4), 3511–3516 (1985)

T. Trondsen, L. Cogger, High-resolution television observations of black aurora. J. Geophys. Res. Space Phys. **102**(A1), 363–378 (1997)

H. Vanhamäki, O. Amm, Analysis of ionospheric electrodynamic parameters on mesoscales-a review of selected techniques using data from ground-based observation networks and satellites, in *Annales Geophysicae*, vol. 29 (2011), p. 467. Copernicus GmbH

J. Vogt, Alfvén wave coupling in the auroral current circuit. Surv. Geophys. **23**(4), 335–377 (2002)

R. Vondrak, S. Harris, S. Mende, Ground-based observations of subauroral energetic-electron arcs. Geophys. Res. Lett. **10**(7), 557–560 (1983)

J. Vogt, G. Haerendel, K. Glassmeier, A model for the reflection of Alfvén waves at the source region of the Birkeland current system: the tau generator. J. Geophys. Res. Space Phys. **104**(A1), 269–278 (1999a)

J. Vogt, H. Frey, G. Haerendel, H. Höfner, J. Semeter, Shear velocity profiles associated with auroral curls. J. Geophys. Res. Space Phys. **104**(A8), 17277–17288 (1999b)

J.A. Wanliss, J.C. Samson, E. Friedrich, On the use of photometer data to map dynamics of the magnetotail current sheet during substorm growth phase. J. Geophys. Res. Space Phys. **105**(A12), 27673–27684 (2000). https://doi.org/10.1029/2000JA000178. https://agupubs.onlinelibrary.wiley.com/doi/abs/10.1029/2000JA000178

D.R. Weimer, D.A. Gurnett, Large-amplitude auroral electric fields measured with DE 1. J. Geophys. Res. Space Phys. **98**(A8), 13557–13564 (1993). https://doi.org/10.1029/93JA00793

D.R. Weimer, C. Goertz, D. Gurnett, N. Maynard, J. Burch, Auroral zone electric fields from DE 1 and 2 at magnetic conjunctions. J. Geophys. Res. Space Phys. **90**(A8), 7479–7494 (1985)

D.K. Whiter, B.S. Lanchester, T. Sakanoi, K. Asamura, Estimating high-energy electron fluxes by intercalibrating reimei optical and particle measurements using an ionospheric model. J. Atmos. Sol.-Terr. Phys. **89**, 8–17 (2012). https://doi.org/10.1016/j.jastp.2012.06.014. http://www.sciencedirect.com/science/article/pii/S136468261200168X

A.N. Wright, C.J. Owen, C.C. Chaston, M.W. Dunlop, Downward current electron beam observed by cluster and fast. J. Geophys. Res. Space Phys. **113**(A6), A06202 (2008)

J. Wu, D. Knudsen, D. Gillies, E. Donovan, J. Burchill, Swarm observation of field-aligned currents associated with multiple auroral arc systems. J. Geophys. Res. Space Phys. **122**(10), 10145–10156 (2017)

J. Wygant, A. Keiling, C. Cattell, I.M. Johnson, R. Lysak, M. Temerin, F. Mozer, C. Kletzing, J. Scudder, W. Peterson, et al., Polar spacecraft based comparisons of intense electric fields and poynting flux near and within the plasma sheet-tail lobe boundary to uvi images: an energy source for the aurora. J. Geophys. Res. Space Phys. **105**(A8), 18675–18692 (2000)

K. Yago, K. Shiokawa, K. Hayashi, K. Yumoto, Auroral particles associated with a substorm brightening arc. Geophys. Res. Lett. **32**(6), L06104 (2005). https://doi.org/10.1029/2004GL021894. https://agupubs.onlinelibrary.wiley.com/doi/abs/10.1029/2004GL021894

A. Yoshikawa, Excitation of a Hall-current generator by field-aligned current closure, via an ionospheric, divergent Hall-current, during the transient phase of magnetosphere–ionosphere coupling. J. Geophys. Res. Space Phys. **107**(A12), 1445 (2002a). https://doi.org/10.1029/2001JA009170

A. Yoshikawa, How does the ionospheric rotational Hall current absorb the increasing energy from the field-aligned current system? Geophys. Res. Lett. **29**(7), 37 (2002b)

A. Yoshikawa, R. Fujii, Earth's ionosphere: theory and phenomenology of Cowling channels, in *Electric Currents in Geospace and Beyond*, vol. 235 (Wiley Online Library, New York, 2018), pp. 427–443

A. Yoshikawa, O. Amm, H. Vanhamäki, R. Fujii, Illustration of Cowling channel coupling to the shear Alfven wave. J. Geophys. Res. Space Phys. **118**(10), 6405–6415 (2013)

A. Yoshikawa, O. Amm, H. Vanhamäki, A. Nakamizo, R. Fujii, Theory of Cowling channel formation by reflection of shear Alfven waves from the auroral ionosphere. J. Geophys. Res. Space Phys. **118**(10), 6416–6425 (2013)

M. Zettergren, J. Semeter, Ionospheric plasma transport and loss in auroral downward current regions. J. Geophys. Res. Space Phys. **117**(A6) A06306, (2012). https://doi.org/10.1029/2012JA017637. http://doi.wiley.com/10.1029/2012JA017637

M. Zettergren, J. Semeter, Ionospheric plasma transport and loss in auroral downward current regions. J. Geophys. Res. Space Phys. **117**(A16), 6306 (2012). https://doi.org/10.1029/2012JA017637

M. Zettergren, J. Semeter, Ionospheric plasma transport and loss in auroral downward current regions. J. Geophys. Res. Space Phys. **117**(A6), A06306 (2012)

M.D. Zettergren, J.B. Snively, Ionospheric response to infrasonic-acoustic waves generated by natural hazard events. J. Geophys. Res. Space Phys. **120**(9), 8002–8024 (2015). https://doi.org/10.1002/2015JA021116. https://agupubs.onlinelibrary.wiley.com/doi/abs/10.1002/2015JA021116

M.D. Zettergren, J.B. Snively, Latitude and longitude dependence of ionospheric tec and magnetic perturbations from infrasonic-acoustic waves generated by strong seismic events. Geophys. Res. Lett. **46**(3), 1132–1140 (2019). https://doi.org/10.1029/2018GL081569. https://agupubs.onlinelibrary.wiley.com/doi/abs/10.1029/2018GL081569

M. Zettergren, K. Lynch, D. Hampton, M. Nicolls, B. Wright, M. Conde, J. Moen, M. Lessard, R. Miceli, S. Powell, Auroral ionospheric f region density cavity formation and evolution: MICA campaign results. J. Geophys. Res. Space Phys. **119**(4), 3162–3178 (2014)

M.D. Zettergren, J.L. Semeter, H. Dahlgren, Dynamics of density cavities generated by frictional heating: formation, distortion, and instability. Geophys. Res. Lett. **42**(23), 10120–10125 (2015). https://doi.org/10.1002/2015GL066806. https://agupubs.onlinelibrary.wiley.com/doi/abs/10.1002/2015GL066806

L. Zheng, S. Fu, Q. Zong, G. Parks, C. Wang, X. Chen, Solar cycle dependence of the seasonal variation of auroral hemispheric power. Chin. Sci. Bull. **58**(4), 525–530 (2013). https://doi.org/10.1007/s11434-012-5378-6

S. Zou, L.R. Lyons, M.J. Nicolls, C.J. Heinselman, S.B. Mende, Nightside ionospheric electrodynamics associated with substorms: PFISR and THEMIS ASI observations. J. Geophys. Res. Space Phys. **114**(A12), A12301 (2009a). https://doi.org/10.1029/2009JA014259. https://agupubs.onlinelibrary.wiley.com/doi/abs/10.1029/2009JA014259

S. Zou, L. Lyons, C.-P. Wang, A. Boudouridis, J. Ruohoniemi, P. Anderson, P. Dyson, J. Devlin, On the coupling between the Harang reversal evolution and substorm dynamics: a synthesis of SuperDARN, DMSP, and IMAGE observations. J. Geophys. Res. Space Phys. **114**(A1), A01205 (2009b)

S. Zou, L.R. Lyons, Y. Nishimura, *Mutual Evolution of Aurora and Ionospheric Electrodynamic Features Near the Harang Reversal During Substorms*. Geophysical Monograph Series, vol. 197 (Am. Geophys. Union, Washington, 2012). https://doi.org/10.1029/2011GM001163

Y. Zou, Y. Nishimura, L. Lyons, M. Conde, R. Varney, V. Angelopoulos, S. Mende, Mesoscale *F* region neutral winds associated with quasi-steady and transient nightside auroral forms. J. Geophys. Res. Space Phys. **123**(9), 7968–7984 (2018). https://doi.org/10.1029/2018JA025457. http://doi.wiley.com/10.1029/2018JA025457

Space Sci Rev (2020) 216:92
https://doi.org/10.1007/s11214-020-00715-5

Quiet, Discrete Auroral Arcs: Acceleration Mechanisms

R. Lysak[1] · **M. Echim**[2,3] · **T. Karlsson**[4] · **O. Marghitu**[3] ·
R. Rankin[5] · **Y. Song**[1] · **T.-H. Watanabe**[6]

Received: 11 June 2019 / Accepted: 27 June 2020 / Published online: 9 July 2020

Abstract The theory of the acceleration of auroral particles is reviewed, focusing on developments in the last 15 years. We discuss elementary plasma physics processes leading to acceleration of electrons to energies compatible with emission observed for quiet, discrete auroral arcs, defined as arcs that have time scales of minutes or more and spatial scales ranging from less than 1 km to tens of kilometers. For context, earlier observations are first described briefly. The theoretical fundamentals of auroral particle acceleration are based on the kinetic theory of plasmas, in particular the development of parallel electric fields. These parallel electric fields can either be distributed along the magnetic field lines, often associated with the mirror geometry of the geomagnetic field, or concentrated into narrow regions of charge separation known as double layers. Observations have indicated that the acceleration process depends on whether the field-aligned currents are directed away from the Earth, toward the Earth, or in mixed regions of currents often associated with the propagation of Alfvén waves. Recent observations from the NASA Fast Auroral SnapshoT (FAST) satellite, the ESA satellite constellation Cluster, and the Japanese Reimei satellite have provided new insights into the auroral acceleration process and have led to further refinements to the theory of auroral particle acceleration.

Keywords Aurora · Particle acceleration · Magnetosphere · Ionosphere

Auroral Physics
Edited by David Knudsen, Joe Borovsky, Tomas Karlsson, Ryuho Kataoka and Noora Partmies

✉ R. Lysak
lysak001@umn.edu

1 School of Physics and Astronomy, University of Minnesota, Minneapolis, MN 55455, USA

2 Institut Royale d'Aeronomie Spatiale de Belgique, Bruxelles, Belgium

3 Institute for Space Sciences, Bucharest, Romania

4 School of Electrical Engineering and Computer Science, KTH Royal Institute of Technology, Stockholm, Sweden

5 Department of Physics, The University of Alberta, Edmonton, Alberta, Canada T6G 2G7

6 Department of Physics, Nagoya University, Nagoya, Japan

1 Historical Introduction

Observations of the discrete aurora have been made since ancient times, leading to many fanciful descriptions speculating about their source. Much of this folklore has been summarized in the book of Eather (1980). Scientific study of the phenomenon began during the Renaissance, with Galileo being credited with introducing the term *aurora borealis*, or "northern dawn." Gauss noted that strong aurorae were associated with deflections of the compass needle, tying the aurora to magnetic phenomena. With the coming of the 20th century, Kristian Birkeland studied the aurora extensively from northern Scandinavia, making "terrella" experiments to show that magnetic fields could funnel charged particles into the polar regions (e.g., Jago 2001). Birkeland postulated that the aurora was associated with a system of field-aligned currents, which now bears his name.

The International Geophysical Year (1957-1958) led to more detailed measurements tying the auroral emissions to the acceleration of charged particles. Observations of auroral X-rays from sounding rockets (Van Allen 1957) and balloons (Winckler et al. 1958) indicated that the aurora was associated with high energy particles, with Winckler et al. (1958) noting that the results seemed consistent with an acceleration mechanism close to Earth. A direct measurement of the auroral particles from a sounding rocket (McIlwain 1960), who noted that the aurora was associated with "nearly monoenergetic electrons" with an energy of 6 keV, concluding that "the presence of the monoenergetic electrons strongly suggests an electrostatic acceleration mechanism." On the theoretical side, Alfvén (1958) and Alfvén and Fälthammar (1963) proposed that electric fields parallel to the magnetic field could exist in a magnetic mirror geometry when electrons and protons had different pitch angle distributions. Alfvén (1958) proposed the term "double layer" for the localized parallel electric field associated with constrictions in a laboratory plasma column, and suggested that such structures could form at the interface between hot and cold plasmas, such as the plasma sheet and the ionosphere (see review by Fälthammar 1979).

A further period of progress in the understanding of auroral acceleration came with the advent of polar-orbiting satellites. Particle observations from the low-altitude Injun-5 satellite showed that the "monoenergetic" peak in the electron distribution rose in energy to a peak and then fell, showing a structure that in energy-time spectrograms looked like an "inverted-V" (Frank and Ackerson 1971; Gurnett and Frank 1973). This led Gurnett (1972) to postulate U-shaped equipotential structures to explain this acceleration mode. This type of structure was confirmed by the S3-3 satellite, which had a higher orbit with an apogee of 8000 km (Mozer et al. 1977, 1980). Particle measurements from S3-3 were found to be consistent with quasi-static acceleration (Mizera and Fennell 1977; Croley et al. 1978). The electric field experiment on S3-3 measured large (>100 mV/m) perpendicular electric fields above the aurora. The electrostatic nature of these structures was verified by Temerin et al. (1981) who noted that for higher altitude cases, the potential found by integrating the perpendicular electric field observed by the satellite was equal to the energy of the upgoing ion beam at the center of the structure.

Theoretical models of the acceleration process also were developed in the 1970's. Knight (1973; see also Lemaire and Scherer 1973; Fridman and Lemaire 1980; Lyons 1980) considered the motion of electrons in magnetic mirror geometry in the presence of a parallel potential drop and found that for typical auroral energies (1-10 keV), there was a linear relationship between the current density and the total potential drop, which can be written as $j_{\|} = K\Phi$. Considering this relationship and assuming that the field-aligned currents were closed by Pedersen currents in the ionosphere, gave a characteristic scale length of $\sqrt{\Sigma_P/K}$. This scale length is about 100 km for typical parameters (e.g., $\Sigma_P = 10$ mho and

$K = 10^{-9}$ mho/m^2). This is at the upper end of the scale sizes observed in auroral structures (Karlsson et al. 2020). The Knight relation was supported by the sounding rocket observations of Lyons et al. (1979). This type of model was further developed by Chiu and Schulz (1978) and Chiu and Cornwall (1980) who made self-consistent models satisfying Poisson's equation. Models based on the concept of anomalous resistivity due to wave-particle interactions were considered (e.g., Papadopoulos 1977), although definitive evidence of this effect remains elusive. The theory of double layers in terms of BGK modes was developed by Block (1972) and Knorr and Goertz (1974), and double layers in which the electric field was oblique to the magnetic field were considered by Swift (1975, 1979), who called them electrostatic shocks. The distinction between double layers and electrostatic shocks was debated, with Goertz (1979) concluding that double layers were stationary with respect to the plasma while electrostatic shocks moved; however, this distinction has largely faded from the literature and such structures are now mostly called double layers.

All of these models were electrostatic in nature and did not address the formation and evolution of the parallel electric fields. The enhancement of the field-aligned current required for the development of parallel electric fields is largely due to the electromagnetic nature of wave energy propagation along the field lines, generally carried by Alfvén waves. Mallinckrodt and Carlson (1978) considered an Alfvén wave generated by a source moving with respect to the plasma and showed that the reflection of this wave from the ionosphere produced paired electric field structures like those observed by S3-3 without invoking a parallel electric field. Hasegawa (1976) considered mode conversion at the plasma sheet boundary layer from a surface Alfvén wave to a so-called kinetic Alfvén wave by including electron pressure and finite ion gyroradius effects, the former of which leads to parallel electric fields. Goertz and Boswell (1979) noted that the effect of electron inertia was more important than electron pressure in cold plasmas like the topside ionosphere. This effect would later be invoked to explain the broadband electron distributions observed by Freja (Louarn et al. 1994) and FAST (Chaston et al. 2002). Furthermore, Lysak and Dum (1983) showed that in the presence of a localized anomalous resistivity, a propagating Alfvén wave could evolve into a quasi-static potential structure.

The effects of kinetic Alfvén waves propagating obliquely to the magnetic field was considered by Haerendel (1983), who developed a "fracture" model that included the proper motion of the arc with respect to the plasma, or equivalently the motion of the plasma across the arc. In his model, a parallel electric field developed due to an anomalous resistivity region that diffused across the plasma. A steady-state oblique arc model was further developed by Seyler (1988, 1990), who considered the nonlinear evolution of the flux tube in which electron inertia gave rise to parallel electric fields. He invoked a tearing instability operating on the perturbed magnetic field to structure auroral currents. A similar steady-state convecting arc model was developed by Knudsen (1996), who showed that structures on the order of tens of electron inertial lengths could be self-consistently established. However, this model did not include the effects of ionospheric coupling that should play an important role in the evolution of auroral arcs.

Greater understanding of the nature of the auroral acceleration region has been provided by a number of satellites, including Dynamics Explorer (Burch et al. 1981), Viking (Hultqvist 1987), Freja (Lundin et al. 1994), Akebono (Tsuruda and Oya 1993) and FAST (Carlson et al. 1998a, 1998b). This work is described in more detail later in this manuscript. This review focuses on the acceleration mechanisms linked to quiet, discrete auroral arcs. This class of auroral forms is widely encountered and can be considered as a "prototype", fundamental paradigm for all other types of aurora. In coherence with Borovsky et al.

(2020), we define a quiet arc as the one "not associated with dynamic events such as substorm breakup, and which has lifetimes ranging from several minutes to hours." The remainder of this article will detail the advances in our understanding of the auroral acceleration mechanism since the publication of the previous book (Paschmann et al. 2003) as well as some key aspects not fully discussed in the previous review. As in the previous article (Karlsson et al. 2020), we will consider discrete arcs as being associated with the precipitation of accelerated particles that are stable on time scales of minutes or longer.

2 Theoretical Framework for Quasi-Static Acceleration Processes

One of the most challenging issues in theory of auroral acceleration is how to sustain the parallel electric field with spatiotemporal scales relevant to auroral phenomena in collisionless plasma (e.g., Song and Lysak 2006, 2015). The parallel electrostatic field is shielded by thermal electrons in spatial scales longer than the Debye length λ_D, while the Debye shielding is established in a time scale of electron plasma oscillation. Therefore, electrostatic phenomena involving charge separation such as double layers are characterized by short spatial and temporal scales. On the longer spatiotemporal scales, the electrostatic potential energy leaking out of the shielding is smaller than the electron thermal energy (i.e., $|e\varphi/k_B T_e| < 1$), where the quasi-neutrality (or, simply, the charge neutrality) condition of $\sum_s q_s n_s = 0$ is well satisfied. Here, q_s and n_s mean the charge and the number density of particle species of s. Note that quasi-neutrality does not imply that the parallel electric fields are strictly zero. In the presence of current flows, a weak parallel electric field can exist over larger scales, as is discussed in Sect. 3.

In the kinetic Vlasov approach one adopts a statistical description of plasma dynamics and considers the microscopic structure of auroral (magnetospheric and ionospheric) plasmas at the level of velocity distribution functions (VDFs) of electrons and ions, f_e and f_i. In the general case the VDFs are functions of time and of the six components of the phase space (three position and three velocity coordinates, respectively). The Vlasov equation describes the spatio-temporal evolution of f_e and f_i and is derived from the general Boltzmann equation by neglecting the close range (binary) interactions between particles and by considering electromagnetic forces.

$$\frac{\partial f_\alpha}{\partial t} + \mathbf{v} \cdot \frac{\partial f_\alpha}{\partial \mathbf{r}} + q_\alpha \left(\mathbf{E} + \mathbf{v} \times \mathbf{B}\right) \cdot \frac{\partial f_\alpha}{\partial \mathbf{v}} = 0 \tag{1}$$

where f_s is the VDF of species s, $\mathbf{E}$ and $\mathbf{B}$ are the electric and magnetic field. The Vlasov equation is "coupled" to the full Maxwell set of equations describing the electromagnetic field. The coupling is mutual in the sense that the solutions of Maxwell's equations are considered in the force term of the Vlasov equation while the charge and current densities of each species, the "sources" of the Maxwell set, are evaluated from the moments of the VDF:

$$Q_\alpha^{(ijk)} = \iiint v_x^i v_y^j v_z^k f_\alpha \left(v_x, v_y, v_z\right) dv_x dv_y dv_z \tag{2}$$

where $q_\alpha Q_\alpha^{(000)}$ gives the charge density of species a, $q_\alpha Q_\alpha^{(100)}$ gives the J_x component of the current density, etc. The above expression for the moments is given in a canonical representation as an integral in the velocity space; note, however, that a representation in the space of constants of motion (e.g. total energy, canonical momenta) is fully equivalent (Lemaire and Scherer 1970; see also Khazanov et al. 1998). The kinetic Vlasov treatment of the particle dynamics in mirroring magnetic fields, as is typical for the auroral regions, allows for a self-consistent estimation of the relationship between the moments of the VDF and the electric potential. An example is the well-known current-voltage relationship provided by various models for upward and downward field-aligned currents. Non-thermal effects, like the role of kappa VDFs can also be quantified. Note also that the Vlasov equation gives a valid description of plasma processes with characteristic time and space scales shorter than the mean free time and the mean free path. A linearization of the Vlasov equation allows for study of various class of waves and their respective dispersion relation in magnetoactive plasmas (as is the case for the auroral regions) with or without thermal effects.

While the Vlasov approach gives the most complete and self-consistent description of the plasma, application of this approach is generally restricted to time-independent situations (e.g., Echim et al. 2007). This is due to the fact that Vlasov models should resolve the Debye length and plasma frequency time scales, which are 74 m and 0.1 ms, respectively, for a plasma with a temperature of 100 eV and a density of 1 cm^{-3}. Global scale hybrid-Vlasov models are now being developed (e.g., Palmroth et al. 2015) which treat the ions by a Vlasov model while the electrons are a neutralizing fluid. However, these results are still preliminary. The difficulty of these models to model the global dynamics has led many researchers to adopt fluid models of the magnetosphere.

The magnetohydrodynamic (MHD) equations are the most popular set of equations for describing low frequency and macro-scale plasma phenomena where the charge neutrality is assumed. Since MHD is based on a single fluid approximation, it does not directly describe the auroral electron acceleration; however, it is an important theoretical framework to describe Alfvén waves in the magnetosphere. The Ohm's law in the MHD equation tells us that the parallel electric field vanishes in the ideal MHD limit with no resistivity (that is, $\boldsymbol{E} = -\boldsymbol{V} \times \boldsymbol{B}$ where $\boldsymbol{E}$, $\boldsymbol{V}$, and $\boldsymbol{B}$ denote the electric field, flow velocity and the magnetic field, respectively). The Ohm's law derived from the equation of motion for electrons is generalized by taking into account higher order terms neglected in the ideal MHD, such as the electron inertia, the electron pressure gradient, and the Hall terms as well as a finite ion gyroradius effect. By means of the generalized Ohm's law, one can discuss generation of the inductive parallel electric field associated with the kinetic and/or inertial Alfven waves.

Further reduction of the MHD equation is also useful in application to the auroral region with the strong background magnetic field B_0. The reduced MHD equations (e.g., Hazeltine and Meiss 1992) are often used to decouple the shear Alfven waves from the fast and slow compressional modes, where only the parallel component of the vorticity and the vector potential (flux function) are retained to describe the shear Alfvén dynamics. The derivation is done by assuming the flute ordering of

$$\frac{L_\perp}{L_\parallel} \sim \frac{B_1}{B_0} \sim \frac{V}{V_A} \sim O(\varepsilon) \tag{3}$$

where $L_\perp$ and $L_\parallel$ are the perpendicular and parallel scale lengths of fluctuations, B_1 is the perturbed magnetic field, V means the flow velocity, V_A denotes the Alfvén speed, and ϵ is a smallness parameter. The reduced MHD equations are used to describe MHD instabilities (such as the kink, ballooning, and tearing instabilities) as well as the Alfvénic coupling of the magnetosphere and the ionosphere (e.g., Lysak 1991; Watanabe 2010). The parallel component of the generalized Ohm's law can also be incorporated into the reduced MHD equations.

In order to investigate interactions of electromagnetic fields and electron motions, one may return to the two-fluid equations before taking summation over particle species in derivation of the MHD equation where assumption of the charge neutrality is still useful to avoid the high-frequency plasma oscillation. In the two-fluid model, the displacement current term ($\epsilon_0 \partial \boldsymbol{E}/\partial t$) is often retained, and plays a role in a low density region where V_A/c is non-negligible (e.g., Lysak 1993; Streltsov and Lotko 2003). This condition is often met in the auroral density cavity at altitudes of 1 R_E or less where the magnetic field is strong and the density is low. The displacement current term related to the parallel electric field is often neglected; however, it is critical for the development of the charge separation necessary to form structures like double layers (Song and Lysak 2001a, 2001b, 2006).

Here, we must recall that it may not be straightforward to derive two-fluid equations from the set of Vlasov-Poisson (or Vlasov-Maxwell) equations by taking fluid moments of the kinetic equation unless suitable closure relations are given. Hammett and Perkins (1990) developed a kinetic fluid closure model (called the Hammett-Perkins or the Landau closure) which mimics the phase relation between the parallel heat flux and the temperature perturbation given by the linearized Vlasov-Poisson system so that the Landau damping is effectively included.

Any fluid equations, however, cannot describe the particle acceleration process self-consistently with formation of the parallel electric field, while a test particle calculation in the electromagnetic fields given by the fluid models may be fruitful to estimate the amount of possible acceleration (e.g., Chaston et al. 2002; Lysak and Song 2005). This is why a kinetic description is essential to the auroral electron acceleration. The gyrokinetic equations (Rutherford and Frieman 1968; Antonsen and Lane 1980; Frieman and Chen 1982; a recent review for the modern gyrokinetic equations is found in Brizard and Hahm 2007) applicable to the low frequency phenomena have been derived in studies of drift wave instabilities and turbulence in magnetically confined plasma under the ordering of

$$\frac{\rho}{L} \sim \frac{L_\perp}{L_\parallel} \sim \frac{\delta f}{F_0} \sim \frac{e\Phi}{k_B T} \sim \frac{B_1}{B_0} \sim O(\varepsilon) \tag{4}$$

where L, ρ, δf, F_0, ϕ, and T denote the scale length of the system, thermal gyroradius, perturbed and equilibrium distribution functions, electrostatic potential, and temperature, respectively. The perturbed distribution function is averaged over the gyromotion, and is defined as a function on the five-dimensional phase space, such as $\delta f(\boldsymbol{X}, v_\parallel, \mu)$, where $\boldsymbol{X}$, $v_\parallel$, and μ represent the gyrocenter position, the parallel velocity component, and the magnetic moment, respectively. The gyrokinetic equations describe time-evolution of δf along trajectories of charged particles with $\boldsymbol{E} \times \boldsymbol{B}$, curvature, and ∇B drift motions and the mirror force. Perturbations of the electromagnetic fields are given by solving the Poisson equation and the Ampere's law with effects of finite gyroradius and polarization. It is noteworthy that the gyrokinetic ordering given above has a similarity to that used in derivation of the reduced MHD equation. It means that the gyrokinetic equations are a kinetic extension of the reduced MHD equations relevant to low frequency phenomena. Actually, the gyrokinetic theory has been applied to auroral electron acceleration by the kinetic Alfvén waves amplified through the feedback instability in the magnetosphere-ionosphere coupling (Watanabe 2014). In a limit of small $\rho/L_\perp \ll 1$ (that is, a limit of long perpendicular wavelength), one finds that the drift kinetic equations are derived from the gyrokinetic equations. The former is often applied to study of electron dynamics in low frequency waves such as the inertial (or dispersive) Alfvén waves (Watt et al. 2004; Watt and Rankin 2009).

Finally, it is remarked that velocity-space integrals of the gyrokinetic equations lead to the gyrofluid equations for each particle species. Relevant closure relations are required to truncate the moment hierarchy, not only for the parallel heat flux but also for

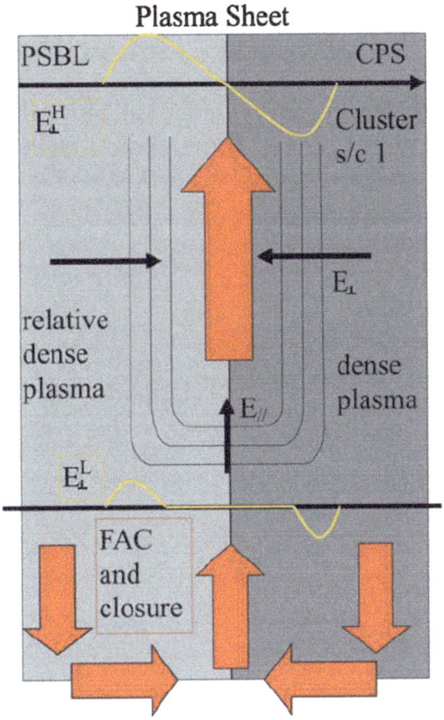

Fig. 1 A diagram of the auroral current circuit sustained by a magnetospheric generator at the interface between the plasma sheet and the plasma sheet boundary layer. The diagram also illustrates the current closure in the upper ionospheric layers (adapted from Marklund et al., 2007)

the magnetic drift term and the finite gyroradius correction (Dorland and Hammett 1993; Beer and Hammett 1996). Application of the gyrofluid equations for the auroral particle acceleration is, however, still immature so far.

3 Quasi-Static Acceleration in the Upward Current Region

A quiet discrete arc is connected by magnetic field lines to a high-altitude generator that supplies an electromotive force in the electric circuit feeding the arc. Such an equivalent circuit is a simplified view of the magnetosphere-ionosphere coupling in auroral regions. Nevertheless, it allows for a description of the main features relevant for a quiet, discrete arc: (i) a magnetospheric generator, (ii) an upward current branch, (iii) an ionospheric load corresponding to the arc itself, (iv) a downward current region to close the circuit (see Fig. 1). In this section we discuss the acceleration of particles in the upward current leg of the auroral circuit. Magnetospheric electrons moving downward along auroral magnetic field lines are accelerated and precipitate into the outer ionospheric layers where they dissipate the energy (mainly through collisions with the ambient neutral gas) and trigger optical auroral emissions. As mentioned in the historical introduction of this chapter, the observed kinetic properties of the precipitating electrons can give a hint on the acceleration mechanism. Indeed, it is believed that inverted-V spectra result from acceleration by a quasi-static parallel electric field (see details on the experimental evidence in the last section of this chapter).

The existence of an electric field component parallel to the magnetic field in a collisionless magnetized plasma is intriguing at a first sight since the electrons and ions have large mobility (virtually infinite) along magnetic field lines, thus is tempting to think that any electric charge imbalance would be immediately neutralized. Nevertheless, a kinetic treatment of particle dynamics in the Vlasov approach reveals the internal plasma mechanisms leading to the formation of such electric field structures and, for some simplified cases, provides a quantitative estimation of the parallel electric field. Let us follow a recent review of auroral acceleration (Birn et al. 2012) and discuss the main mechanisms for quasi-static parallel electric fields in the region of auroral upward field-aligned currents (see also Boström 2003, 2004).

3.1 Acceleration by Distributed, Charge Neutral, Large Scale Parallel Electric Fields

A distributed large scale parallel electric field is set-up in an auroral magnetic flux tube as a result of the effect of the mirroring magnetic field on the dynamics of electrons and ions moving upward/downward along the magnetic field lines (Boström 2003). As mentioned in the introductory historical notes, Alfvén and Fälthammar (1963) were the first to estimate the parallel component of the electric field due to electrons and ions moving in such a mirroring geometry. They found an analytic expression for $E_\parallel$ that depends on (i) the pitch angle anisotropy of ions and electrons and (ii) the rate of variation of the magnetic field intensity with the altitude. Nevertheless the validity of the expression postulated by Alfvén and Fälthammar (1963) is limited by the assumption that magnetospheric electrons and ions are monoenergetic (i.e. their velocity distribution function is a delta function). The Vlasov equation allows for an estimation of the parallel component of the electric field. In the case of a one-dimensional mirroring magnetic field configuration a general expression of $E_\parallel$ is given by (Persson 1966):

$$E_\parallel = \frac{m_\alpha}{e}\left[\frac{v_\perp}{2B}\frac{dB}{ds}\left(v_\perp \frac{\partial f_\alpha}{\partial v_\parallel} - v_\parallel \frac{\partial f_\alpha}{\partial v_\perp}\right) - v_\parallel \frac{\partial f_\alpha}{\partial s}\right]\left(\frac{\partial f_\alpha}{\partial v_\parallel}\right)^{-1} \tag{5}$$

where s is the coordinate in the direction parallel to the magnetic field. Obviously one needs to obtain the same value of $E_\parallel$ from the Vlasov treatment of each species α, electrons and ions. An additional constraint is quasi-neutrality, i.e. charge balance between electrons and ions.

A general solution for the parallel component of the electric field is not easy to find analytically. In fact, as noted by Persson (1966), if f_α does not depend explicitly on s but implicitly through the electric potential, $\Phi(s)$, then equation (5) is a tautology. General solutions of (5) for two or three dimensional configurations are not known. Under additional assumptions, such as the monotonicity of Φ as a function of s, one can attempt solutions for the one-dimensional case of a mirroring field. However, the physics of auroral acceleration is often discussed in terms of a "current-voltage relationship (CVR)", linking the field-aligned current density, $J_\parallel$, with the potential drop, $\Delta\Phi$, at altitude s. Various expressions of the CVR have been proposed for different auroral configurations, at microscopic and macroscopic level (for a review see Boström 2003; Pierrard et al. 2007).

As noted by Boström (2003) the effect of orbital dynamics of charged particles in the auroral flux tubes is similar in many ways with the classical plasma physics problem of estimating the flux of electrons between an emitter and a collector in electric discharges (Mott-Smith and Langmuir 1926). Large scale distributed parallel electric fields are formed due to the mutual interaction of ions and electrons moving upward/downward along the auroral flux tube from the magnetospheric (generator) and ionospheric (load) ends. Their orbital motion in the mirroring field leads to corresponding altitude-dependent variations of the phase space density. Various regions of the phase space are populated by particles with different origins and properties, magnetospheric and/or ionospheric. The fundamental dynamics of such particles and the effects on their velocity distribution function and its moments (e.g. $J_{\parallel}$) is similar in many ways, including the mathematical apparatus, with the dynamics of particles as treated in the exospheric models of polar and solar wind (Lemaire and Scherer 1970, 1971).

The exospheric approach was applied to plasma sheet particle precipitation by Lemaire and Scherer (1973) and Knight (1973) who provided the first expressions for an auroral current-voltage relationship. In this approach the key-issue is the phase space accessibility. Indeed, depending on their kinetic properties at the origin (magnetospheric or ionospheric end of the auroral flux tube) and the local value of the electric and magnetic field, auroral particles can populate only restricted sub-regions of the phase space (see also Whipple 1977; Miller and Khazanov 1993; Liemohn and Khazanov 1998; Khazanov et al. 1998). The moments of the VDF are then obtained from piece-wise integration of the VDF over these subregions of the phase space. This approach for solving the Vlasov equation can be effective for a quasi-static quasi-neutral description of auroral particle dynamics and fields. However, it should be noted that these models generally assume a quasi-static potential structure and then determine the distributions that can support it and do not address the development of such a potential structure (e.g., Song and Lysak 2001a, 2001b, 2006).

3.2 Acceleration by Confined, Space-Charge Driven Parallel Electric Fields: Double Layers

Acceleration of downward going auroral electrons can be achieved by small scale electric potential structures that, although globally neutral, have an internal structure organized as two layers of net charges of opposite sign – *a double layer* (*DL*). The existence of double layers is an inherent feature of plasmas, as shown by numerous laboratory experiments (see, e.g., Hershkowitz 1985; Raadu and Rasmussen 1988; Singh et al. 2011). Advocated by Alfven (1986) to have a crucial role in astrophysical particle acceleration, double layers were confirmed as efficient quasi-static auroral accelerations by experimental evidence in spite some theoretical debate (Bryant et al. 1992; Borovsky 1992; see also Eriksson and Boström 1993). Indeed, DLs were detected indirectly for hundreds of FAST orbits (see, e.g., Ergun et al. 2002). The difference with respect to the auroral large scale distributed parallel E-field discussed above is that the double layer introduces a sharp jump of the electric potential (and other plasma parameters) and the parallel electric field is confined within spatial scales of the order of the Debye length.

The origin of these sharp potential structures and fine scale parallel electric fields is at the contact between plasmas with different macroscopic temperatures, ionospheric and magnetospheric, similar to thermoelectric fields in thermodynamics. This similarity was first discussed and described quantitatively in an auroral context by Hultqvist (1971). Experimental evidence from Dynamic Explorer and FAST spacecraft indicates the formation of low-density cavity in the upward field-aligned current region (Persoon et al. 1988; McFadden et al. 1999). FAST data also shows that the low altitude boundary of the auroral cavity, at the contact with the higher density ionospheric plasma, is the site of a strong double layer and a corresponding parallel electric field accelerating downward going electrons (see, e.g., McFadden et al. 1999; Ergun et al. 2002). As argued by McFadden et al. (1999), the formation of the DL may be a consequence of the requirement to maintain quasineutrality on a global scale in the cavity itself that is devoid of cold ionospheric plasma; thus a sharp DL prevents the more energetic secondary ionospheric electrons from entering the cavity.

On the theoretical side, quasi-static double layer solutions are obtained from solving the Vlasov equation together with the Poisson equation:

$$\varepsilon_0 \nabla^2 \Phi = -\left[q_i Q_i^{(000)} + q_e Q_e^{(000)}\right] \tag{6}$$

where the densities are determined as moments of the VDF, as defined by equation (2). The problem of finding stable DL solutions in one spatial dimension is solved by integrating the Vlasov-Poisson system of equations (e.g., Schamel and Bujabarua 1983; Robertson 2014). The solution is described in terms of $V(\Phi)$, the so-called Sagdeev potential, which is equal to within an additive constant to the negative of the total particle pressure (Raadu and Rasmussen 1988):

$$-\frac{1}{2}\varepsilon_0 E^2 + \sum_{i,e} \int_{-\infty}^{\infty} m_{i,e} v^2 f_{i,e}(x, v)\, dv = -\frac{1}{2}\varepsilon_0 E^2 + V(\Phi) = -\Pi \tag{7}$$

where Π is a constant. Two existence conditions must be met by a stable DL: (i) the total particle pressure must be in balance across the DL (*the Langmuir condition*) and (ii) *the generalized Bohm criterion* stating that, assuming the parallel electric field is zero at the edges of the DL, the net charge on the low and high potential sides are negative and positive, respectively. The latter criterion leads to a condition in terms of velocity distribution functions in close relationship to the dispersion relation for electrostatic waves in a uniform plasma (Raadu and Rasmussen 1988). However, testing of the DL existence criteria with auroral experimental data is difficult and practically never fully achieved.

The conditions for existence of DL solutions in the time independent form are well understood, at least for simplified geometries; however, their stability and therefore the importance for quasi-static acceleration of auroral electrons in the upward field-aligned current region remain questionable. Nevertheless, observations from FAST (Ergun et al. 2002) show evidence that the auroral DL are stable over time intervals of the order of tens of seconds to minutes. Numerical simulations show examples of DL solutions that are stable over time scales compatible with observations of quiet auroral arcs. Vlasov simulations in one and two dimensions (Main et al. 2006) initialized with conditions similar to observations of DL by FAST, show that the DL is stable over time intervals longer than 30000 electron plasma periods, $(\omega_{pe})^{-1}$, which is tens of seconds assuming typical auroral parameters. These simulations show formation of ion phase space holes at the upper boundary of the cavity but their interaction with the DL does not destabilize its configuration although relatively diminishes its strength. The simulation run is however rather short (of the order of second) compared to real life observations (tens of seconds to minutes). Particle-in-Cell simulations (Main et al. 2010) do confirm the stability of the DL during the entire short time span of the run. As conditions for DL stability the PIC simulations find that the "density of the cold ionospheric electrons to be greater that 30% and the temperature to be less than 100 eV" (Main et al. 2010).

Vlasov simulations provide additional insight compared to PIC. It is shown that the electron acceleration is practically achieved in the double layer itself, when a total potential drop, $\Delta\Phi$, is assumed (Gunell et al. 2013). Nevertheless, a third of the final acceleration energy is gained above the DL; also, the DL altitude decreases with increasing $\Delta\Phi$. Vlasov simulations also confirm that the structure of the accelerating potential is monotonic with the altitude when $\Delta\Phi$ is stable in time but can show local minima/maxima when $\Delta\Phi$ is time dependent (Gunell et al. 2015).

However, the question of how double layers are formed and maintained remains open. The charge separation necessary for a double layer occurs due to the inability of the auroral electrons to carry the current required by the magnetic shear. It has been commonly believed that the non-ideal terms in the parallel components of the generalized Ohm's law, electron inertia, anomalous resistivity, or the electron pressure gradient, inhibit the electron motion and can give rise to parallel electric fields, causing auroral particle acceleration. However it has been pointed out (Song and Lysak 2000, 2001a, 2001b, 2006) that the three non-ideal terms do not describe the generation of parallel electric fields, but simply describe a force balance between the electric force and other forces. An analysis of the full set of Maxwell's equations and Newton's laws indicates that the parallel displacement current, which is often neglected in descriptions of low-frequency phenomena in dense plasmas, is crucial in the generation of parallel electric fields when the plasma density is low (Song and Lysak 2006). As the full Ampere's Law states, when the magnetic shear cannot be balanced by charged particle flow, the displacement current term will lead to an enhancement of the parallel electric field, forming what has been termed an Alfvénic double layer (Song and Lysak 2015).

The generation of a sustained parallel electric field, which requires a sustained energy supply, is associated with the temporal changes and spatial gradients of magnetic and velocity shears. It has been proposed (Song and Lysak 2001a, 2001b) that the nonlinear interaction between incident and reflected Alfven wave packets in the auroral acceleration regions can create local double layers or charge holes. The final state of the interaction of these wave packets can be obtained by considering the approximate conservation of angular momentum and magnetic helicity carried by the two discrete wave packets before and after the collision (Song and Lysak 1994).

It is well known that the reflection coefficient for Alfvén waves interacting with the ionosphere has been given by (Mallinckrodt and Carlson 1978):

$$R = \frac{E_{ref}}{E_{inc}} = \frac{\Sigma_A - \Sigma_P}{\Sigma_A + \Sigma_P} \tag{8}$$

The Pedersen conductance is generally greater than 1 mho while the Alfvén conductance, defined by $\Sigma_A = 1/\mu_0 V_A$, is typically 0.1 mho or less (e.g., Lysak 1990). Therefore, in the usual situation, the reflected wave has a perpendicular electric field in the opposite direction as the incident electric field. In this situation, the incident and reflected wave packets will have the same direction of field-aligned current, and the enhanced magnetic shear in this interaction will lead to an enhancement in the displacement current and the generation of a double layer (Song and Lysak 2001a, 2006). On the other hand, Alfvén waves can also be reflected from strong gradients in the Alfvén speed such as those that exist above the ionosphere. Since the effective conductivity decreases when the Alfvén speed is high, i.e., when the plasma density is low, reflection from the low-density plasma in the acceleration region will produce a reflection with the reflected electric field in the same direction as in the incident electric field. Such reflections will be conducive to the generation of localized electric fields observed as charge holes. Thus, the nonlinear interaction between the incident and reflected Alfvén wave packets in the auroral particle acceleration region can either convert the kinetic energy into electromagnetic energy to generate and support a parallel potential drop, or convert the magnetic energy into kinetic energy to generate a charge condensation (or charge hole), depending on the reflection coefficients. Since upgoing and downgoing shear Alfvén wave packets reflect multiple times from the ionosphere and from Alfvén speed gradients, multiple localized potential drops or charge holes will be generated.

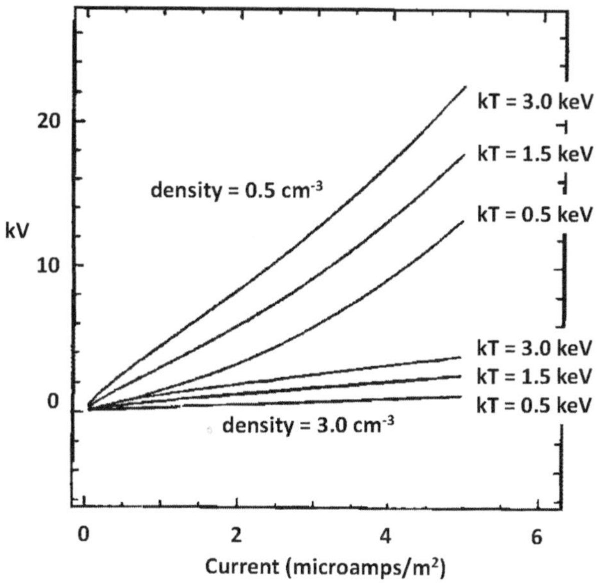

Fig. 2 Current-voltage relation in the downward current region for varying magnetospheric electron density and temperatures for the quasi-neutrality model used by Temerin and Carlson (1998)

4 The Downward Current Region

There are three main observations in the downward, or 'return' current region that need to be addressed by models: acceleration of upward-moving electrons and downward ions, ion conics, and the formation of low-altitude ionospheric densities cavities.

4.1 Particle Acceleration, Parallel Electric Field, and Ion Conics

The existence of both upward accelerated electrons and downward accelerated ions in the return current region indicates acceleration by a downward-directed parallel electric field (Hultqvist 2002), and theories for particle acceleration have concentrated on modelling such a field. One of the first tries (based on earlier work by Stern (1981), and Chiu et al. (1981)) to model the potential drop in the return current region was made by Temerin and Carlson (1998). They invoked quasi-neutrality along a magnetic field line populated by four plasma species; cold ionospheric electrons and ions, and hot magnetospheric electrons and ions. By prescribing an ion density altitude distribution and a field-aligned current, the quasi-neutrality requirement forces an electric field to accelerate the electrons upward in the region above the ionosphere, where the ion density decreases to low values. This results in a current-voltage relation of the return current region, which depends on the density and temperature of the magnetospheric electrons, similar to the Knight relation of the upward current region (see Fig. 2). However, the potential drop in the return current region is typically substantially smaller than the corresponding value in the upward current region for the same value of current density.

Similar models have been developed by Cran-McGreehin and Wright (2005a, 2005b), and Vedin and Rönnmark (2005). Vedin and Rönnmark extended the model to high altitudes and showed that above the maximum of B/n_I, where n_I is the ionospheric electron number density, the electric field changes sign, and points downward. Cran-McGreehin and Wright also attach importance to the B/n peak and argue that the potential drop for a given field-aligned current density is determined by the detailed properties of the ion density profile in the immediate vicinity of that point. It has long been recognized (e.g., Lysak and Hudson 1979) that the peak in the B/n ratio determines the location at which the drift velocity of electrons relative to ions is strongest for a steady-state field-aligned current.

None of the above models takes into account the modification of the ionospheric plasma resulting from the parallel electric field, which is necessary for a self-consistent model of the downward current region. Observations show that the return current region is associated with intense broadband electrostatic low frequency (BBELF) turbulence, heated ions (also in the form of upflowing conics), and a strongly thermalized upward electron beam. These modified plasma populations will affect the solutions to the quasi-neutrality requirement. The exact causal relation between these elements of the downward current region physics are not exactly known and have been modelled in various different ways.

Jasperse (1998), Jasperse and Grossbard (2000), and Jasperse et al. (2010), in a kinetic model (again based on quasi-neutrality) prescribe an ion heating rate due to an ion cyclotron resonance. They assume that BBELF waves exist on the downward current flux tube and are excited by the accelerated electron beam or the field-aligned current without specifying the exact mechanism. The solutions obtained show (in addition to a downward directed electric field) ion distributions identifiable as ion conics. These conics are consistent with the idea of the 'pressure cooker' mechanism suggested by Gorney et al. (1985), where the ions are contained at the low-altitude side of the parallel potential drop and are continuously heated until their perpendicular energy is high enough that the mirror force will overcome the potential drop. In this way the ions will escape the ionosphere and under further adiabatic motion will decrease their pitch angle, and form ion beam-like outflows, in spite of the downward-directed electric field.

In the above models, the resulting electric field is smoothly distributed in altitude. Observations indicate that the electric field at times takes the form of double layers, moving upward along the field line at the ion acoustic speed (Andersson et al. 2002). This has been modelled by several authors (Newman et al. 2001, 2008; Liu and Liao 2011; Singh 2003; Singh et al. 2009). Newman et al. (2001, 2008) have used a 1-D Vlasov code to study current driven double layers. A snapshot of the resulting electron phase space is shown in Fig. 3 (which also includes a qualitative comparison to observational data). It can also be seen that the resulting accelerated electron beam is unstable and generates strong electrostatic turbulence, which indicates that the thermalization of the electron beam is a candidate for generating at least a subset of the BBELF waves. Non-linear interaction of the electrostatic turbulence and the ambient plasma can also produce electron phase-space holes, consistent with observations (Goldman et al. 1999, 2007; Ergun et al. 2003; Singh et al. 2009). The idea that the BBELF is located at the high-altitude side of the double layers has also lead to suggestions of an alternative to the pressure-cooker model for the ion heating, involving a self-consistent feedback between the double layer and the heated ion population (Hwang et al. 2008, 2009).

So far, modelling of the downward current region has been focused on the 1-D properties, and the altitude distribution of the parallel electric field. One notable exception is a two-fluid MHD model (Streltsov and Marklund 2006), describing the temporal evolution of the downward current region, where a modification of the ionospheric plasma results in creation of perpendicular, diverging electric fields of the type observed. Similar simulations also show that the ionospheric feedback results in structuring of the downward current region in the perpendicular direction into small-scale current elements or electromagnetic structures (Streltsov and Karlsson 2008; Streltsov and Lotko 2003, 2008). Streltsov (2018) has also addressed the self-consistent formation of narrow, intense regions of downward current, as a consequence of the ionospheric feedback instability (e.g. Atkinson 1970; Sato 1978; Lysak and Song 2002).

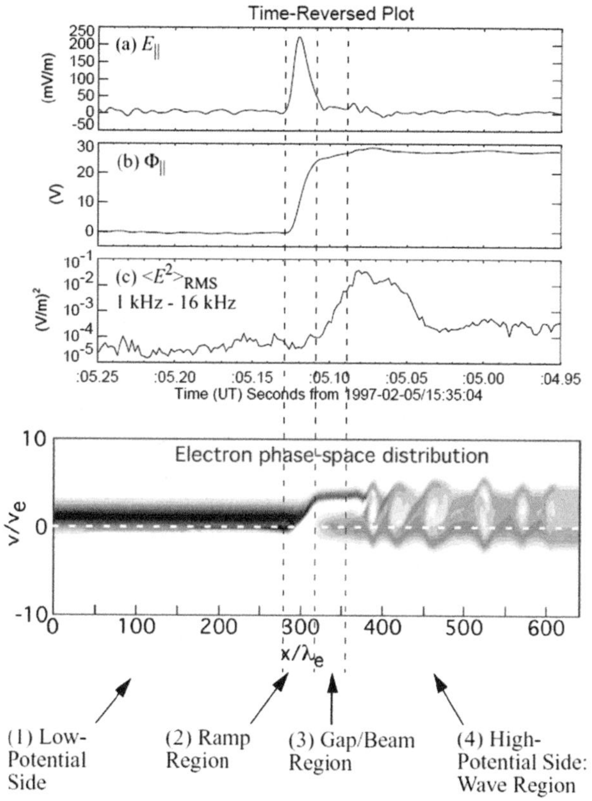

Fig. 3 Comparison between observation of a double layer in the downward current region by the FAST spacecraft, and a numerical simulation by Newman et al. (2001). (From Ergun et al. 2003)

The ionospheric modification resulting from the downward currents is further described in the next section.

4.2 Ionospheric Density Cavities

The strong downward current densities in the return current region will result in the creation of a drastic reduction of plasma density when the downward current closes to perpendicular currents in the ionosphere, resulting in a density cavity. This has been modelled in several studies, and it has been shown that the reason for the cavity formation is three-fold.

The first mechanism is an outflow of plasma, resulting from the fact that the field-aligned currents are mainly carried by electrons, due to their high mobility, while the perpendicular Pedersen currents closing the upward and downward currents are mainly carried by ionospheric ions (Doe et al. 1995; Blixt and Brekke 1996). This leads to a net outflow of current carriers in regions where the downward current connects to the perpendicular currents, which can evacuate the main part of the E region plasma in tens of seconds (Doe et al. 1995; Karlsson and Marklund 1998; Karlsson et al. 2005), see Fig. 4.

For the F region, current closure is less efficient in evacuating the ionospheric plasma, and increased recombination due to ohmic heating of the plasma becomes important (Zettergren et al. 2010; Zettergren and Semeter 2012). This process includes temperature-dependent chemical conversion of ionospheric atomic ions to molecular ions, with subsequent fast recombination. At the top-side ionosphere also increased vertical transport becomes important. This transport should perhaps also be taken into account in calculations of the parallel

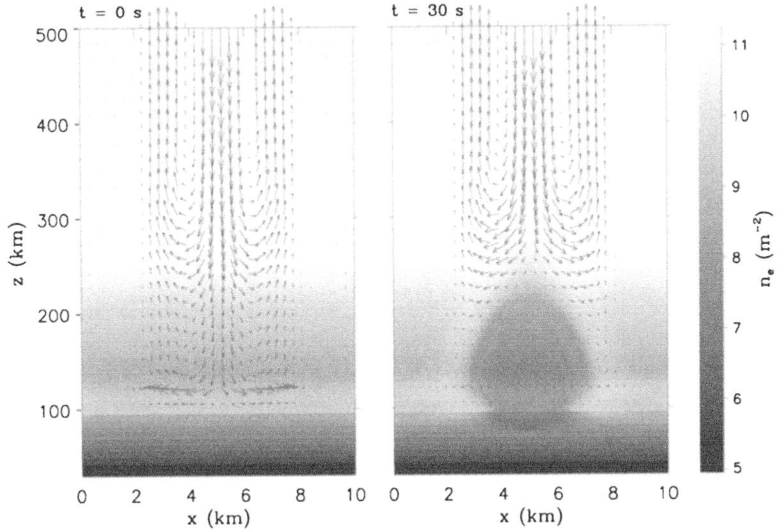

Fig. 4 Ionospheric E-region cavity formation in the downward current region due to an outflow of plasma. (Karlsson et al. 2005)

electric field at higher altitudes. Recent modelling also shows that the resulting cavities can be severely deformed, due to plasma drift and instabilities triggered by the large gradients produced by the cavity formation (Zettergren et al. 2015), or by re-distribution of the currents in the ionosphere (Karlsson and Marklund 1998), or their self-consistent magnetospheric response (Cran-McGreehin et al. 2007).

An important consequence of the cavity formation is the resulting low ionospheric conductivity in that region (Karlsson et al. 2007), which will feed back to the interaction with the magnetosphere. This was mentioned at the end of the previous subsection, and has been further studied by e.g. Russell et al. (2013, 2015). Another suggested consequence is that the large, ohmic parallel electric field formed within the E region cavity can create a suprathermal electron population, which may produce optical emissions similar to those in subauroral red arcs and in sprites (Karlsson et al. 2005). This is a possible mechanism for the phenomenon of enhanced auroral emissions (e.g. Hallinan et al. 1985).

5 Signatures of Alfvénic Acceleration

It was first suggested by Hasegawa (1976) that small perpendicular wavelength Alfvén waves carry a frequency-dependent parallel electric field capable of accelerating electrons to energies sufficient to produce aurora. A fraction of electrons accelerated along the geomagnetic field by these waves will enter the loss cone at a range of altitudes depending on their initial energy and pitch angle, while electrons that cannot overcome the mirror force at low altitudes will form conics as they reverse their direction and move upward along field lines (André and Eliasson 1992). Over the past few decades, numerous observations from rockets (e.g., Arnoldy et al. 1985; McFadden et al. 1986; Lynch et al. 1994, 1999; Semeter et al. 2001) and scientific satellites (e.g., Clemmons et al. 1994; Knudsen et al. 1998; Chaston et al. 2002, 2003a, 2003b) have confirmed this "Alfvénic acceleration" process as viable for producing aurora by elucidating the complex interplay between parallel electric fields, electron conics, and structuring of discrete arcs. While this Alfvénic acceleration process will likely play a role in more dynamic auroral structures (e.g., see the companion article by Kataoka et al. 2020), it can also be associated with field line resonances that have periods of minutes (e.g., Tikhonchuk and Rankin 2000, 2002), and so the resulting auroral structures will appear to be quasi-static. Gillies et al. (2018) have made observations of recurring auroral arcs with periods consistent with field line resonances using a red line (630 nm) imager, which has a response that is dominated by energies < 1 keV; however, they did not measure the electron spectrum directly. As noted above, Alfvén waves also play a major role in the development of quasi-static parallel electric fields, as was suggested in the early work of Lysak and Dum (1983) and in more detail in the work of Vedin and Rönnmark (2006). If the Alfvénic structure is standing in the flow (e.g., Knudsen 1996), the resulting aurora will also appear quasi-static. Such a situation may occur at the plasma sheet boundary layer, where the Alfvénic acceleration may be associated with mode conversion at the Alfvén speed gradients in the boundary layer (e.g., Lysak and Song 2011). Here, we review a subset of observations and theory that summarize the current state of knowledge on the relationship between small scale Alfvén waves and the aurora.

5.1 Two-Fluid Theory

Parallel electric fields are supported by small perpendicular scale Alfvén waves propagating in the inertial or kinetic regime, which depends on the background plasma conditions in which waves propagate. At altitudes of a few Earth radii above the auroral zone, electron pressure gradients support parallel electric fields, with the relevant perpendicular scale being the ion acoustic gyro-radius, $\rho_s = c_s/\Omega_i$, where the ion sound speed and the ion cyclotron frequency are expressed as $c_s = \sqrt{T_e/m_i}$ and $\Omega_i = eB_0/m_i$, respectively, and T_e is the electron temperature in eV. Finite ion Larmor radius effects set a limiting perpendicular scale for waves when $\lambda_e < \rho_s < \rho_i$, where $\rho_i = v_{Ti}/\Omega_i$ is the ion gyro-radius, $v_{Ti} = \sqrt{T_i/m_i}$ is the ion thermal speed, $\lambda_e = c/\omega_{pe}$ is the electron skin depth, and $\omega_{pe} = \sqrt{ne^2/m_e\varepsilon_0}$ is the electron plasma frequency. At scales approaching ρ_i, wave dispersion is altered, but there is no associated contribution to the parallel electric field. When the condition $\lambda_e > \max(\rho_s, \rho_i)$ is satisfied, which is generally true in the auroral acceleration region, the dispersion relation and expression for the parallel electric field are different from the kinetic regime. Within the two-fluid formalism of dispersive Alfvén waves, the parallel electric field, $E_\parallel$, in the inertial regime is given by (e.g., Lysak and Song 2005):

$$\frac{E_\parallel}{E_\perp} = -\frac{k_\parallel k_\perp \lambda_e^2}{1 + k_\perp^2 \lambda_e^2} \tag{9}$$

while in the kinetic regime it is given by,

$$\frac{E_\parallel}{E_\perp} = k_\parallel k_\perp \rho_s^2 \tag{10}$$

Both expressions for the parallel electric field assume an infinite homogeneous plasma with the second equation, which is valid for $\lambda_e < \rho_i < \rho_s$, commonly referred to as the kinetic regime. In the above expressions, $E_\perp$ denotes the perpendicular electric field, and $(k_\parallel, k_\perp)$ represent parallel and perpendicular wavenumbers, respectively. It is clear that for dispersive Alfvén waves, $E_\parallel$ in the inertial and kinetic regimes has opposite sign and that in warm plasma the strength of the field increases in proportion to v_{Te}^2/v_A^2, which can easily reach values on the order of ten in the equatorial plasma sheet. The dependence of the strength of the parallel electric field on increasing plasma temperature with altitude suggests that electron acceleration by dispersive Alfvén waves may be efficient at high altitudes as well as in the "auroral acceleration region." The assessment of this possibility requires a kinetic treatment in which self-consistent Landau damping in hot plasma sheet plasma is accounted for, as is discussed in the next section.

Although two-fluid theory excludes Landau damping effects, it has nevertheless aided the interpretation of distribution functions modified by dispersive Alfvén waves. In particular, because direct measurements of $E_{||}$ are not generally available, observations of wave perpendicular fields are used along with the two-fluid dispersion relation to provide evidence of electron acceleration by $E_{||}$ (Wygant et al. 2002). Test particle simulations and linear wave two-fluid theory successfully also reproduces observations of electron energy-time dispersion expected of small scale Alfvén waves (Thompson and Lysak 1996; Chaston et al. 2002). Nonlinear theory of two-fluid dispersive Alfvén waves has been developed to explain observations attributed to colliding wave packets (Song and Lysak 2001a, 2001b, 2006), ponderomotive effects (Sydorenko et al. 2008), and the formation of kinetic Alfvén wave solitons (e.g., Wu et al. 1995).

In homogeneous plasma, the inertial and kinetic regimes of Alfvén wave propagation do not overlap. Some studies have combined the two-fluid inertial and kinetic regimes by merely adding expressions for the parallel electric field for the inertial and kinetic regimes. The motivation for this is that Alfvén waves propagating along geomagnetic field lines naturally transition from the kinetic regime at high altitude to the inertial regime at low altitude (e.g., Lysak and Carlson 1981). However, there is no formal justification for combining the two limits of the dispersion relation, which is in any case easily avoided by solving the full kinetic dispersion relation (Lysak and Lotko 1996). On geomagnetic field lines with changing density, temperature, and magnetic field strength the kinetic dispersion relation can be used to describe the characteristics of propagating Alfvén waves provided the parallel wavelength is smaller than the characteristic length scale of variation of field line plasma parameters.

Two-fluid theory reveals that at a certain altitude the parallel electric field of a propagating or standing two-fluid dispersive Alfvén wave changes sign at a location where counteracting effects of inertial and kinetic wave dispersion are in balance. The transition from one regime to the other has important implications for standing Alfvén waves, which can become trapped on geomagnetic field lines into a narrow range of L-shell (Streltsov and Lotko 1995) with little or no losses, except at the ionospheres. In a driven system under these circumstances, phase mixing on auroral field lines will lead to a spatial structure consistent with a discrete arc, i.e., a field-aligned current system confined to a very narrow range of latitude, and with a correspondingly strong parallel electric field (Lysak and Song 2008, 2011). For dispersive standing Alfvén waves, this focusing effect leads naturally to a concentration of wave energy and parallel electric fields on geomagnetic field lines where this minimum perpendicular wavelength occurs.

5.2 Kinetic Dispersion and Simulations

Theoretical and numerical studies on field-aligned electron acceleration processes by the Alfvén waves have largely been advanced in the last two decades. The kinetic dispersion relation incorporating ρ_s and λ_e effects can be derived following (Nakamura 2000) and is written for a Lorentzian distribution function,

$$\frac{E_{\|}}{E_{\perp}} = \frac{v_{T\kappa}^2}{V_A^2} k_{\|} k_{\perp} \lambda_e^2 \frac{1}{2\left[1 - 1/2\kappa + \zeta Z_\kappa(\zeta)\right]} \tag{11}$$

where $v_{T\kappa} = [2\kappa - 3]^{1/2} v_{Th}$ is an effective thermal speed of electrons, κ is the so-called "Kappa" parameter, $\zeta = \omega / k_{\|} v_{T\kappa}$, and $Z_\kappa(\zeta)$ is the modified dispersion function introduced by Summers and Thorne (1992). For large values of κ the expression for $E_{||}$ obtained from the kinetic dispersion relation has cold and warm plasma limits corresponding to the inertial and kinetic regimes described earlier. However, in addition to these effects, Landau damping is included in the kinetic description associated with the imaginary part of the dispersion function. There have been several studies to assess the consequences of the kinetic regime of electron acceleration including the Landau damping (Damiano et al. 2007; Seyler and Liu 2007; Swift 2007; Watt and Rankin 2009; Lysak and Song 2011). Interactions of kinetic Alfvén waves and field-aligned electron motions have been investigated by means of the one-dimensional drift kinetic model (see, for example, Watt et al. 2004; Watt and Rankin 2009) or by use of the particle-in-cell (Clark and Seyler 1999; Seyler and Liu 2007) or a hybrid-MHD simulations (Damiano et al. 2003; Damiano and Wright 2005).

The effect of electron trapping in kinetic Alfvén waves has also been considered (Watt and Rankin 2009, 2010, 2012; Damiano et al. 2015, 2016). This trapping mechanism is much more efficient than the electrostatic mechanism considered by Damiano et al. (2016) and is supported by Vlasov-kinetic simulations (Watt and Rankin 2009). Watt and Rankin (2010) have shown that this mechanism can accelerate electrons to the 1-10 keV range with sufficient energy flux to power discrete auroral emissions. Their simulations considered waves that had a characteristic period of 2 seconds and a wavelength mapped to the ionosphere of 4 km. Damiano et al. (2015) found that ion temperature effects reduce parallel currents and electron energization all along geomagnetic field lines as expected for a kinetic Alfvén wave. In a subsequent study by the same authors (Damiano et al. 2016) it was found that as the ion temperature increases the ability of kinetic Alfvén waves to carry and energize trapped electrons is reduced by more significant wave energy dispersion perpendicular to the ambient magnetic field. However, the trapping mechanism evaluated by these authors differs from the mechanism considered by Artemyev et al. (2015), which can be realized only in a system with magnetic field inhomogeneity. In this case, an effective potential well into which electrons become trapped is generated by the wave parallel electric field and mirror force.

The local dispersion relation used in all of these kinetic theories is strictly valid in infinite homogeneous plasma. When long-period field line resonances (FLRs) reach latitude widths at which dispersive effects are important, the electron transit time along geomagnetic field lines has to be considered. The latter can be much shorter than a typical FLR wave period, in which case Ohm's law breaks down. A non-local theory of the interaction of auroral electrons that accounts for electron bounce motion has been developed by Rankin et al. (1999) and Tikhonchuk and Rankin (2000) and exhibits enhanced parallel electric fields in FLRs. More detail on these types of interaction in FLRs and the limitations of two-fluid and local kinetic theories are given in the companion paper by Rankin et al. (2020). A final aspect of Alfvénic acceleration of auroral electrons is formulated in terms of the feedback magnetosphere-ionosphere (M-I) coupling, which has long been considered to be a mechanism for quiet auroral arcs (Atkinson 1970; Sato 1978). These studies consider the ionization produced by particle precipitation, which can cause conductivity gradients that can give rise to kinetic Alfvén waves when the phase relations between up and down going waves is favorable (Miura and Sato 1980; Lysak 1991). Such feedback can occur due to reflections in the conjugate ionosphere, in which case the resulting structures are slowly evolving on time scales of minutes, or by reflections in the ionospheric Alfvén resonator (e.g., Lysak and Song 2002; Streltsov and Lotko 2008). A feedback model where the kinetic Alfvén waves are described by means of the gyrokinetic equations, are amplified through the M-I interactions, and accelerate electrons self-consistently based on reflections from the conjugate ionosphere has been presented (Watanabe 2014). On the other hand, Sydorenko and Rankin (2017) have questioned the existence of this instability; however, this conclusion has been challenged by Streltsov and Mishin (2018). Thus, this question remains open.

6 Observations of Auroral Acceleration and of the Auroral Acceleration Region

Following the suggestion by Alfvén (1958), that (quasi)static parallel electric fields / double layers could develop on auroral field lines and decouple the ionospheric from the magnetospheric plasma motion, a broad range of signatures was observed, supporting Alfvén's early intuition. As described earlier in this chapter, rocket experiments in the 1960's showed monoenergetic peaks in the electron data, indicative for acceleration by a quasi-static potential drop along the magnetic field line, while later on, the inverted-V signature was discovered in electron spectrograms from the Injun-5 satellite. Gurnett (1972a, 1972b) found that inverted-V signatures were associated with electric field reversals and interpreted this observation in terms of a 'U' (or 'V') shaped of the electron acceleration region (Fig. 5, top left). A similar conclusion was reached independently by Carlqvist and Boström (1970), based on high speed (well above typical ionospheric speeds) counter-streaming flows at the arc boundaries, observed by a low luminosity TV camera.

The discovery "almost by chance" (according to Fälthammar 1983), of the auroral acceleration region (AAR) by the S3-3 satellite, confirmed the suggestion of Gurnett (1972a, 1972b), and was based on systematic observations of specific features such as electrostatic shocks (Mozer et al. 1977), ion beams (Shelley et al. 1976; Ghielmetti et al. 1978) and ion conics (Sharp et al. 1977; Gorney et al. 1981). Signatures of the ion and electron phase space consistent with quasi-static acceleration by parallel electric fields (Mizera and Fennell 1977), as well as double layers and solitary waves (Temerin et al. 1982), were also observed by S3-3. Comprehensive reviews of the S3-3 discoveries and investigations were provided e.g. by Mozer et al. (1980) and Chiu et al. (1983). Figure 5 shows an updated perspective of the AAR as derived from S3-3 observations (Mizera et al. 1982).

After S3-3, several spacecraft missions on polar or near polar orbits explored the AAR with progressively better resolution and accuracy, among them Dynamics Explorer (DE), Viking, Akebono, Freja, Interball-Aurora, and Polar. The DE mission, consisting of two satellites at high and low altitude (DE-1, apogee at $\sim$23000 km, and DE-2, apogee at $\sim$1000 km), was designed to provide repeated conjunctions between the two spacecraft, which enabled Reiff et al. (1988) to perform consistent comparisons of the field-aligned potential drop inferred by DE-1 (electron and ion data) and DE-2 (electron data).

A comprehensive investigation of the AAR was made possible by the Fast Auroral SnapshoT (FAST) mission (1996–2009, apogee at 4100 km, see Carlson et al. 1998a), whose orbit and high resolution payload turned out to be optimal to explore both quasi-static and transient features of the auroral acceleration. By showing systematically that the integral of the electric field along the satellite path is essentially equal to the average energy of the beam ions (McFadden et al. 1998), FAST provided strong evidence for the electrostatic character of the auroral acceleration in the upward current region. A similar relationship was found in the downward current region, this time between the integral and of the electric field and the average energy of the beam electrons (Carlson et al. 1998b), providing systematic support

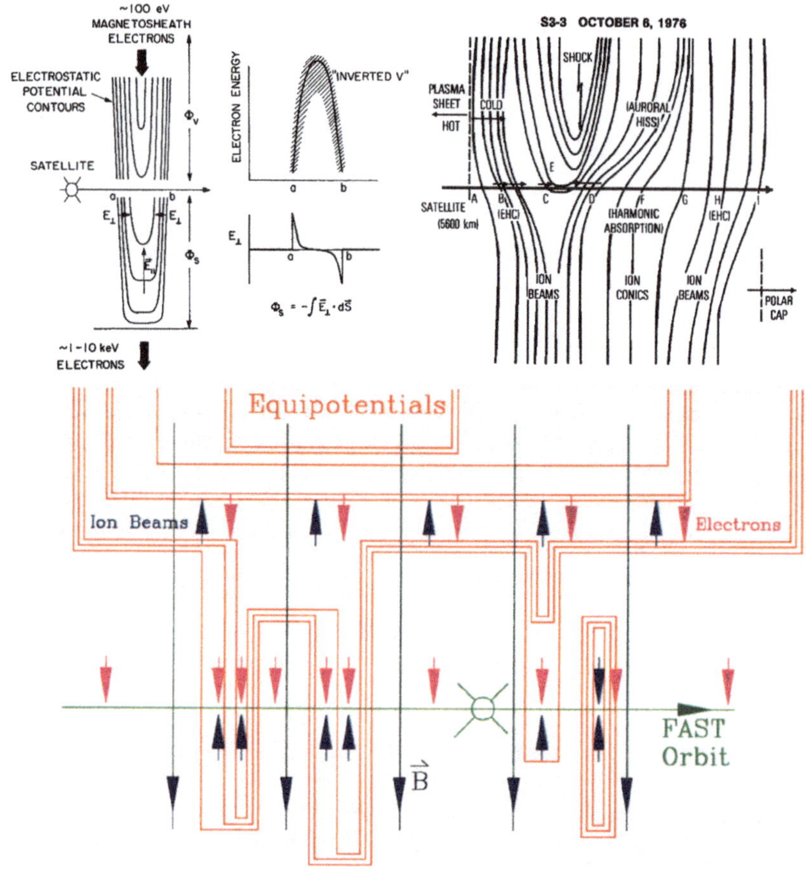

Fig. 5 The configuration of the auroral acceleration region as inferred from Injun-5 (Gurnett 1972b, top left), S3-3 (Mizera et al. 1982, top right), and FAST (McFadden et al. 1999, bottom) data

for quasi-static acceleration over 'black aurora', first observed in the Freja data by Marklund et al. (1994).

A summary of AAR FAST findings was provided by McFadden et al. (1999). Figure 5 (bottom) shows a sketch of the FAST perspective on the AAR structure in the upward current region. Subsequently, a broad range of FAST observations was used to illustrate various AAR features in the ISSI monograph on Auroral Plasma Physics (Eds. Paschmann et al. 2003), in particular in Chap. 4.

With less payload but a high resolution narrow angle camera, actively pointed to the ionospheric footprint of the magnetic field line, the Reimei satellite (launched in 2005), in an almost circular orbit at $\sim$650 km (see Saito et al. 2005), was able to complement FAST data and provide insight into the relationship between quasi-static and Alfvénic acceleration, e.g. in the case of an onset arc (Frey et al. 2010). Reimei data were particularly useful to advance the understanding of highly dynamic aurora, whose small scale structuring was found to be driven by Alfvénic turbulence cascade (Chaston et al. 2011).

Until the launch of the four-spacecraft Cluster mission in July-August 2000, the AAR was explored mainly by single spacecraft missions, with the notable exception of DE, providing conjugacy along the magnetic field between the low and high altitude satellite, as well as of Polar and Reimei, equipped with cameras able to take auroral images conjugate to the in-situ observations. However, until Cluster, no mission was able to separate unequivocally temporal from spatial variations and, consequently, to investigate the AAR evolution in time. One early example for this capability of Cluster was provided by an event study of the divergent electric field evolution in time (Marklund et al. 2001a, 2001b). Later on, when the gradual change in orbit due to moon influence resulted in frequent Cluster crossings of the AAR topside, several studies explored both the altitudinal distribution of the Cluster potential and its development in time (e.g. Marklund et al. 2001a, 2001b, 2012; Sadeghi et al. 2011). More details on Cluster observations are available in the previous article in this collection (Karlsson et al. 2020), as well as, e.g., in the AAR review by Karlsson (2012).

As mentioned earlier, an important feature of the field-aligned potential drop is the so-called current-voltage relationship (Knight 1973), based on energy conservation and on the adiabatic particle motion in a converging magnetic field. Irrespective of the altitudinal distribution of the electric potential (subject to conditions of accessibility, see, e.g., Boström 2003, 2004), the current density can be shown to vary linearly with the field-aligned potential drop, for a broad range of potential drops typical of the AAR. This relationship, which is particularly important for including the AAR in M–I coupling models, was first verified by rocket data (Lundin and Sandahl 1978; Lyons et al. 1979) and later on by satellite data, e.g. Weimer et al. (1985, 1987) and Lu et al. (1991) based on DE data, Sakanoi et al. (1995) based on Akebono data, Frey et al. (1998) and Olsson et al. (1998) based on Freja data, and Elphic et al. (1998) based on FAST data.

Finally, the altitudinal distribution of the potential drop and the altitudes of the lower and upper AAR boundaries were investigated by various statistical and event-oriented approaches. An overview of statistical results was provided by Karlsson (2012), completed, in the meanwhile, by a Cluster based investigation (Alm et al. 2015a). Among the event oriented techniques, Marghitu et al. (2006) and Forsyth et al. (2012) used the anisotropy of the observed electron distributions to infer the altitude of the upper AAR boundary, while more recently, Hatch et al. (2019) estimated the same altitude by fitting moment-voltage relationships parameters (including magnetic field ratio) over a range of different voltages. Detailed features of the electron distribution function can be used as well, e.g. to find properties of concentrated electric fields / double layers located above the satellite (Vedin et al. 2007). Naturally, multi-point observations provide better constraints to the altitudinal distribution of the electric potential, as shown by Cluster events (e.g. Sadeghi et al. 2011; Alm et al. 2015b).

7 Summary and Conclusions

While a great deal of progress has been made on the theory of quiet, discrete arcs, a number of important questions remain. One issue involves the predicted scale sizes of arcs. As noted in the previous chapter (Karlsson et al. 2020) observed scale sizes appear to have a rather continuous spectrum from less than 100 m up to about 100 km (Maggs and Davis 1968; Knudsen et al. 2001; Partamies et al. 2010). The larger scale size is consistent with estimates based on the Knight relation and current closure through the ionosphere (Lyons 1980), while smaller scale sizes may be associated with the electron inertial length in the acceleration region (e.g., Goertz and Boswell 1979), typically a few kilometers. Ionospheric feedback tends to produce very small scale structures, which are likely limited by the effect of collisional resistivity in the ionosphere (Forget et al. 1991; Lessard and Knudsen 2001), which can give scales of less than a kilometer (Lysak and Song 2002). It may be the case that these smaller-scale structures are mostly associated with dynamic arcs rather than the quiet arcs that are the main focus here. However, the rather continuous spectrum in scale sizes raises the question of whether the underlying dynamics is much different for quiet and more dynamic arcs.

Predicted time scales for auroral arcs also constitute a broad spectrum. Of course, the steady-state models described in Sect. 2 above assume that the system is not evolving, and so can shed little information about time scales. As noted above, double layer simulations have produced stable structures over the length of the runs, with periods of up to tens of seconds. It is not clear how long a double layer could ultimately exist if the simulations were carried out to larger times. Multi-spacecraft investigations of how a quasi-static acceleration region develops, similar to observations from the Cluster mission (Marklund et al. 2001a, 2001b, 2012), are also needed in the future to establish this process. It should also be noted that the time scale of quiet discrete arcs may also be determined by the generation mechanism. Potential generator mechanisms are discussed in the companion paper by Borovsky et al. (2020).

One clear distinction in time scales can be seen in the distinction between monoenergetic and broadband electron acceleration. Broadband acceleration would be favored if the transit time of electrons through the acceleration region, time scale of a few seconds for an acceleration region the order of 1 R_E in length, is longer than the time scale of the evolution. Thus, Alfvénic acceleration in the ionospheric Alfvén resonator would produce a broadband spectrum of electron energies, while a similar acceleration process in a field line resonance that has a period of a few minutes would give a more monoenergetic spectrum that would be stable over this time scale.

A final question is how the energy of the accelerated electrons depends on the acceleration mechanism. Observations generally show that monoenergetic electrons generally have energies above 1 keV, while broadband electrons are often of lower energies. However, test particle models and observations of auroral acceleration (e.g., Chaston et al. 2002, 2003a, 2003b) show that inertial Alfvén waves can produce broadband acceleration up to 10 keV, although typical energies are 1 keV or less. Thus, while quiet arcs are generally associated with monoenergetic electrons, it is possible that in some cases they may be produced by a broadband electron distribution.

Further observations, combined with relevant theory, could shed new light on these structures. For example, simultaneous observations at different altitudes along the field line could address the question of whether the parallel electric fields are distributed over large distances, as was discussed above in Sect. 3A, or if they are concentrated in double layers as in Sect. 3B. Further studies relating observed auroral arcs with electron precipitation, like those from the Reimei satellite (Chaston et al. 2011; Fukuda et al. 2014) would help establish what types of particle acceleration are associated with different auroral emissions. In addition, studies such as those from the Swarm and ePOP missions (e.g., Miles et al. 2018; Pakhotin et al. 2018) that combine observations with model results can address the structure of electric and magnetic fields in the auroral region.

Therefore, despite the fact that the basic ingredients of auroral acceleration processes are reasonably well understood, many questions as to how these ingredients work together remain to be answered. Hopefully, future missions, likely with multiple point measurements, will be planned to investigate the remaining issues in auroral acceleration processes.

Acknowledgements The authors wish to thank the International Space Science Institute (ISSI), Bern, for organization of the reviews and for support of the team meeting. RLL and YS have been supported by NSF grant AGS 1558134. MME acknowledges support from the Romanian Ministry of Research (PCCDI Grant VESS), the Romanian Space Agency (STAR project 182-OANA), the Belgian Solar Terrestrial Center of Excellence (STCE), the BRAIN-BE project MOMA BR/175/A2/MOMA. OM acknowledges support by SIFACIT contract 4000118383/16/I–EF with ESA and STAR EXPRESS contract 119/2017 with Romanian Space Agency. THW is supported by JSPS KAKENHI Grant Number JP16H04086 and JP17H01177.

Publisher's Note Springer Nature remains neutral with regard to jurisdictional claims in published maps and institutional affiliations.

References

H. Alfvén, On the theory of magnetic storms and aurorae. Tellus **10**, 104 (1958)

H. Alfven (1986). Double layers and circuits in astrophysics (TRITA-EPP–86-04). Sweden

H. Alfvén, C.-G. Fälthammar, *Cosmic Electrodynamics* (Clarendon Press, Oxford, 1963)

L. Alm, B. Li, G.T. Marklund, T. Karlsson, Statistical altitude distribution of the auroral density cavity. J. Geophys. Res. Space Phys. **120**, 996–1006 (2015a). https://doi.org/10.1002/2014JA020691

L. Alm, G.T. Marklund, T. Karlsson, Electron density and parallel electric field distribution of the auroral density cavity. J. Geophys. Res. Space Phys. **120**, 9428–9441 (2015b). https://doi.org/10.1002/2015JA021593

L. Andersson, R.E. Ergun, D.L. Newman, J.P. McFadden, C.W. Carlson, Y.J. Su, Characteristics of parallel electric fields in the downward current region of the aurora. Phys. Plasmas **9**(8), 3600–3609 (2002)

M. André, L. Eliasson, Electron acceleration by low frequency electric field fluctuations: electron conics. Geophys. Res. Lett. **19**, 1073 (1992)

T.M. Antonsen Jr., B. Lane, Kinetic equations for low frequency instabilities in inhomogeneous plasma. Phys. Fluids **23**, 1205–1214 (1980)

R.L. Arnoldy, T.E. Moore, L.J. Cahill, Low-altitude field-aligned electrons. J. Geophys. Res. **90**, 8445 (1985)

A.V. Artemyev, R. Rankin, M. Blanco, Electron trapping and acceleration by kinetic Alfvén waves in the inner magnetosphere. J. Geophys. Res. Space Phys. **120**, 10,305–10,316 (2015). https://doi.org/10.1002/2015JA021781

G. Atkinson, Auroral arcs: result of the interaction of a dynamic magnetosphere with the ionosphere. J. Geophys. Res. **75**(25), 4746–4755 (1970)

M.A. Beer, G.W. Hammett, Toroidal gyrofluid equations for simulations of tokamak turbulence. Phys. Plasmas **3**, 4046–4064 (1996)

J. Birn, A.V. Artemyev, D.N. Baker, M. Echim, M. Hoshino, L.M. Zelenyi, Particle acceleration in the magnetotail and aurora. Space Sci. Rev. **173**, 49 (2012)

E.M. Blixt, A. Brekke, A model of currents and electric fields in a discrete auroral arc. Geophys. Res. Lett. **23**(18), 2553–2556 (1996)

L.P. Block, Potential double layers in the ionosphere. Cosm. Electrodyn. **3**, 349 (1972)

J.E. Borovsky, Double layers do accelerate particles in the auroral zone. Phys. Rev. Lett. **69**, 1054 (1992)

J.E. Borovsky, J. Birn, M.M. Echim, S. Fujita, R.L. Lysak, D.J. Knudsen, O. Marghitu, A. Otto, T.-H. Watanabe, T. Tanaka, Quiescent discrete auroral arcs: a review of magnetospheric generator mechanisms. Space Sci. Rev. **216**, 1 (2020). https://doi.org/10.1007/s11214-019-0619-5

R. Boström, Kinetic and space charge control of current flow and voltage drops along magnetic flux tubes: kinetic effects. J. Geophys. Res. **108**(A4), 8004 (2003). https://doi.org/10.1029/2002JA009295

R. Boström, Kinetic and space charge control of current flow and voltage drops along magnetic flux tubes: 2. Space charge effects. J. Geophys. Res. **109**(A18), A01208 (2004). https://doi.org/10.1029/2003JA010078

A.J. Brizard, T.S. Hahm, Foundations of nonlinear gyrokinetic theory. Rev. Mod. Phys. **79**, 421–468 (2007)

D.A. Bryant, R. Bingham, U. de Angelis, Double layers are not particle accelerators. Phys. Rev. Lett. **68**, 37 (1992)

J.L. Burch, J.D. Winningham, V.A. Blevins, N. Eaker, W.C. Gibson, R.A. Hoffman, High altitude plasma instrument for Dynamics Explorer-A. Space Sci. Instrum. **5**, 455 (1981)

P. Carlqvist, R. Boström, Space-charge regions above the aurora. J. Geophys. Res. **75**(34), 7140–7146 (1970)

C.W. Carlson, R.P. Pfaff, J.G. Watzin, The Fast Auroral SnapshoT (FAST) mission. Geophys. Res. Lett. **25**, 2013 (1998a)

C. Carlson et al., FAST observations in the downward auroral current region: energetic upgoing electron beams, parallel potential drop, and ion heating. Geophys. Res. Lett. **25**, 2017–2021 (1998b)

C.C. Chaston, J.W. Bonnell, L.M. Peticolas, C.W. Carlson, J.P. McFadden, R.E. Ergun, Driven Alfven waves and electron acceleration: a FAST case study. Geophys. Res. Lett. **29**(11), 1535 (2002). https://doi.org/10.1029/2001GL013842

C.C. Chaston, J.W. Bonnell, C.W. Carlson, J.P. McFadden, R.E. Ergun, R.J. Strangeway, Properties of small-scale Alfvén waves and accelerated electrons from FAST. J. Geophys. Res. **108**, 8003 (2003a). https://doi.org/10.1029/2002JA009420

C.C. Chaston, J.W. Bonnell, C.W. Carlson, J.P. McFadden, R.J. Strangeway, R.E. Ergun, Kinetic effects in the acceleration of auroral electrons in small scale Alfven waves: a FAST case study. Geophys. Res. Lett. **30**(6), 1289 (2003b). https://doi.org/10.1029/2002GL015777

C.C. Chaston, K. Seki, T. Sakanoi, K. Asamura, M. Hirahara, C.W. Carlson, Cross–scale coupling in the auroral acceleration region. Geophys. Res. Lett. **38**, L20101 (2011). https://doi.org/10.1029/2011GL049185

Y.T. Chiu, J.M. Cornwall, Electrostatic model of a quiet auroral arc. J. Geophys. Res. **85**, 543 (1980)

Y.T. Chiu, M. Schulz, Self-consistent particle and parallel electrostatic field distributions in the magnetospheric-ionospheric auroral region. J. Geophys. Res. **83**, 629 (1978)

Y.T. Chiu, A.L. Newman, J.M. Cornwall, On the structure and mapping of auroral electrostatic potentials. J. Geophys. Res. **86**, 10,029 (1981)

Y. Chiu, J. Cornwall, J. Fennell, D. Gorney, P. Mizera, Auroral plasma in the evening sector: satellite observations and theoretical interpretations. Space Sci. Rev. **35**, 211–257 (1983)

A.E. Clark, C.E. Seyler, Electron beam formation by small-scale oblique inertial Alfvén waves. J. Geophys. Res. **104**, 17 233–17 249 (1999)

J.H. Clemmons, M.H. Boehm, G.E. Paschmann, G. Haerendel, Signatures of energy-time dispersed electron fluxes observed by Freja. Geophys. Res. Lett. **21**, 1899 (1994)

A.P. Cran-McGreehin, A.N. Wright, Electron acceleration in downward auroral field-aligned currents. J. Geophys. Res. Space Phys. **110**, A10S15 (2005a)

A.P. Cran-McGreehin, A.N. Wright, Current-voltage relationship in downward field-aligned current region. J. Geophys. Res. Space Phys. **110**, A10S10 (2005b)

A.P. Cran-McGreehin, A.N. Wright, A.W. Hood, Ionospheric depletion in auroral downward currents. J. Geophys. Res. Space Phys. **112**, A10309 (2007)

D.R. Croley Jr., P.F. Mizera, J.F. Fennell, Signature of a parallel electric field in ion and electron distributions in velocity space. J. Geophys. Res. **83**, 2701 (1978)

P.A. Damiano, A.N. Wright, Two-dimensional hybrid MHD-kinetic electron simulations of an Alfvén wave pulse. J. Geophys. Res. **110**, A01201 (2005). https://doi.org/10.1029/2004JA010603

P.A. Damiano, R.D. Sydora, J.C. Samson, Hybrid magnetohydrodynamic-kinetic model of standing shear Alfvén waves. J. Plasma Phys. **69**, 277–304 (2003)

P.A. Damiano, A.N. Wright, R.D. Sydora, J.C. Samson, Energy dissipation via electron energization in standing shear Alfvén waves. Phys. Plasmas **14**, 062904 (2007)

P.A. Damiano, J.R. Johnson, C.C. Chaston, Ion temperature effects on magnetotail Alfvén wave propagation and electron energization. J. Geophys. Res. Space Phys. **120**, 5623–5632 (2015). https://doi.org/10.1002/2015JA021074

P.A. Damiano, J.R. Johnson, C.C. Chaston, Ion gyroradius effects on particle trapping in kinetic Alfvén waves along auroral field lines. J. Geophys. Res. Space Phys. **121**, 10,831–10,844 (2016). https://doi.org/10.1002/2016JA022566

R.A. Doe, J.F. Vickrey, M. Mendillo, Electrodynamic model for the formation of auroral ionospheric cavities. J. Geophys. Res. Space Phys. **100**(A6), 9683–9696 (1995)

W. Dorland, G.W. Hammett, Gyrofluid turbulence models with kinetic effects. Phys. Fluids B **5**, 812–835 (1993)

R.H. Eather, *Majestic Lights: The Aurora in Science, History, and the Arts* (American Geophysical Union, Washington, 1980)

M.M. Echim, M. Roth, J. DeKeyser, Sheared magnetospheric plasma flows and discrete auroral arcs: a quasi-static coupling model. Ann. Geophys. **25**, 317 (2007)

R.C. Elphic et al., The auroral current circuit and field-aligned currents observed by FAST. Geophys. Res. Lett. **25**, 2033–2036 (1998)

R.E. Ergun, L. Andersson, D.S. Main, Y.J. Su, C.W. Carlson, J.P. McFadden, F.S. Mozer, Parallel electric fields in the upward current region of the aurora: indirect and direct observations. Phys. Plasmas **9**(9), 3685–3694 (2002)

R.E. Ergun, L. Andersson, C.W. Carlson, D.L. Newman, M.V. Goldman, Double layers in the downward current region of the aurora. Nonlinear Process. Geophys. **10**, 45–52 (2003). https://doi.org/10.5194/npg-10-45-2003

A.I. Eriksson, R. Boström, Are weak double layers important for auroral particle acceleration? in *Auroral Plasma Dynamics*, ed. by R.L. Lysak. Geophysical Monograph, vol. 80 (American Geophysical Union, Washington, 1993), p. 105

C.-G. Fälthammar, Non-resistive potential drops in cosmical plasmas, in *Particle Acceleration in Astrophysics*, ed. by J. Arons, C. Max, C. McKee (American Institute of Physics, New York, 1979), p. 27

C.-G. Fälthammar, Magnetic-field-aligned electric fields. ESA J. **7**, 385–404 (1983)

B. Forget, J.-C. Cerisier, A. Berthelier, J.-J. Berthelier, Ionospheric closure of small-scale Birkeland currents. J. Geophys. Res. **96**, 1843 (1991)

C. Forsyth et al., Temporal evolution and electric potential structure of the auroral acceleration region from multispacecraft measurements. J. Geophys. Res. **117**, A12203 (2012). https://doi.org/10.1029/2012JA017655

L.A. Frank, K.L. Ackerson, Observations of charged particle precipitation into the auroral zone. J. Geophys. Res. **76**, 3612 (1971)

H.U. Frey et al., Freja and ground-based analysis of inverted-V events. J. Geophys. Res. **103**, 4303–4314 (1998)

H.U. Frey et al., Small and meso-scale properties of a substorm onset auroral arc. J. Geophys. Res. **115**, A10209 (2010). https://doi.org/10.1029/2010JA015537

M. Fridman, J. Lemaire, Relationship between auroral electron fluxes and field aligned electric potential differences. J. Geophys. Res. **85**, 664 (1980)

E.A. Frieman, L. Chen, Nonlinear gyrokinetic equations for low-frequency electromagnetic waves in general plasma equilibria. Phys. Fluids **25**, 502–508 (1982)

Y. Fukuda, M. Hirahara, K. Asamura, T. Sakanoi, Y. Miyoshi, T. Takada, A. Yamazaki, K. Seki, Y. Ebihara, Electron properties in inverted-V structures and their vicinities based on Reimei observations. J. Geophys. Res. Space Phys. **119**, 3650–3663 (2014). https://doi.org/10.1002/2013JA018938

A. Ghielmetti, R. Johnson, R. Sharp, E. Shelley, The latitudinal, diurnal, and altitudinal distributions of upward flowing energetic ions of ionospheric origin. Geophys. Res. Lett. **5**, 59–62 (1978)

D.M. Gillies, D. Knudsen, R. Rankin, S. Milan, E. Donovan, A statistical survey of the 630.0-nm optical signature of periodic auroral arcs resulting from magnetospheric field line resonances. Geophys. Res. Lett. **45**, 4648–4655 (2018). https://doi.org/10.1029/2018GL077491

C.K. Goertz, Double layers and electrostatic shocks in space. Rev. Geophys. Space Phys. **17**, 418 (1979)

C.K. Goertz, R.W. Boswell, Magnetosphere-ionosphere coupling. J. Geophys. Res. **84**, 7239 (1979)

M.V. Goldman, M.M. Oppenheim, D.L. Newman, Theory of localized bipolar wave-structures and nonthermal particle distributions in the auroral ionosphere. Nonlinear Process. Geophys. **6**(3/4), 221–228 (1999)

M.V. Goldman, D.L. Newman, A. Mangeney, Theory of weak bipolar fields and electron holes with applications to space plasmas. Phys. Rev. Lett. **99**(14), 145002 (2007)

D.J. Gorney, A. Clarke, D. Croley, J. Fennell, J. Luhmann, P. Mizera, The distribution of ion beams and conics below 8000 km. J. Geophys. Res. **86**, 83 (1981)

D.J. Gorney, Y.T. Chiu, D.R. Croley, Trapping of ion conics by downward parallel electric fields. J. Geophys. Res. Space Phys. **90**(A5), 4205–4210 (1985)

H. Gunell, J. De Keyser, E. Gamby, I. Mann, Vlasov simulations of parallel potential drops. Ann. Geophys. **31**(7), 1227–1240 (2013). https://doi.org/10.5194/angeo-31-1227-2013. 2013

H. Gunell, L. Andersson, J. De Keyser, I. Mann, Vlasov simulations of trapping and loss of auroral electrons. Ann. Geophys. **33**, 279–293 (2015). https://doi.org/10.5194/angeo-33-279-2015

D.A. Gurnett, INJUN-5 observations of magnetospheric electric fields and plasma convection, in *Earth's Magnetospheric Processes*, ed. by B.M. McCormac (1972a), p. 253

D.A. Gurnett, Electric field and plasma observations in the magnetosphere, in *Critical Problems of Magnetospheric Physics*, ed. by E.R. Dyer (1972b), pp. 123–138

D.A. Gurnett, L.A. Frank, Observed relationships between electric fields and auroral particle precipitation. J. Geophys. Res. **78**, 145 (1973)

G. Haerendel, An Alfvén wave model of auroral arcs, in *High-Latitude Space Plasma Physics*, ed. by B. Hultqvist, T. Hagfors (Plenum, New York, 1983), p. 515

T.J. Hallinan, H.C. Stenbaek-Nielsen, C.S. Deehr, Enhanced aurora. J. Geophys. Res. Space Phys. **90**(A9), 8461–8475 (1985)

G.W. Hammett, F.W. Perkins, Fluid moment models for Landau damping with application to the ion-temperature-gradient instability. Phys. Rev. Lett. **64**, 3019–3022 (1990)

A. Hasegawa, Particle acceleration by MHD surface wave and formation of aurora. J. Geophys. Res. **81**, 5083 (1976)

S.M. Hatch, J. LaBelle, C.C. Chaston, Inferring source properties of monoenergetic electron precipitation from kappa and Maxwellian moment-voltage relationships. J. Geophys. Res. Space Phys. **124**, 1548–1567 (2019). https://doi.org/10.1029/2018JA026158

R.D. Hazeltine, J.D. Meiss, *Plasma Confinement* (Addison-Wesley, Redwood City, 1992)

N. Hershkowitz, Review of recent laboratory double layer experiments. Space Sci. Rev. **41**, 351–391 (1985). https://doi.org/10.1007/BF00190655

B. Hultqvist, On the production of a magnetic-field-aligned electric field by the interaction between the hot magnetospheric plasma and the cold ionosphere. Planet. Space Sci. **19**, 749 (1971)

B. Hultqvist, The Viking project. Geophys. Res. Lett. **14**, 379 (1987)

B. Hultqvist, Downward ion acceleration at auroral latitudes: cause of parallel electric field. Ann. Geophys. **20**(8), 1117–1136 (2002)

K.J. Hwang, R.E. Ergun, L. Andersson, D.L. Newman, C.W. Carlson, Test particle simulations of the effect of moving DLs on ion outflow in the auroral downward-current region. J. Geophys. Res. Space Phys. **113**(A1) (2008). https://doi.org/10.1029/2007JA012640

K.J. Hwang, R.E. Ergun, D.L. Newman, J.B. Tao, L. Andersson, Self-consistent evolution of auroral downward-current region ion outflow and moving double layer. Geophys. Res. Lett. **36**(21) (2009). https://doi.org/10.1029/2009GL040585

L. Jago, *The Northern Lights* (Vintage Books, New York, 2001)

J.R. Jasperse, Ion heating, electron acceleration, and the self-consistent parallel E-field in downward auroral current regions. Geophys. Res. Lett. **25**(18), 3485–3488 (1998)

J.R. Jasperse, N.J. Grossbard, The Alfven-Falthammar formula for the parallel E-field and its analogue in downward auroral-current regions. IEEE Trans. Plasma Sci. **28**(6), 1874–1886 (2000)

J.R. Jasperse, B. Basu, E.J. Lund, N. Grossbard, The self-consistent parallel electric field due to electrostatic ion-cyclotron turbulence in downward auroral-current regions of the Earth's magnetosphere. IV. Phys. Plasmas **17**(6), 062904 (2010)

T. Karlsson, The acceleration region of stable auroral arcs, in *Auroral Phenomenology and Magnetospheric Processes: Earth and Other Planets*, ed. by A. Keiling, E. Donovan, F. Bagenal, T. Karlsson. AGU Monograph, vol. 197 (American Geophysical Union, Washington, 2012)

T. Karlsson, G. Marklund, Simulations of effects of small-scale auroral current closure in the return current region. Phys. Space Plasmas **15**, 401 (1998)

T. Karlsson, G. Marklund, N. Brenning, I. Axnäs, On enhanced aurora and low-altitude parallel electric fields. Phys. Scr. **72**(5), 419 (2005)

T. Karlsson, N. Brenning, O. Marghitu, G. Marklund, S. Buchert, High-altitude signatures of ionospheric density depletions caused by field-aligned currents (2007). ArXiv preprint. arXiv:0704.1610

T. Karlsson, L. Andersson, D.M. Gilies, K. Lynch, O. Marghitu, N. Paramies, N. Sivadas, J. Wu, Quiet, discrete aurora arcs: observations. Space Sci. Rev. **216**, 16 (2020). https://doi.org/10.1007/s11214-020-0641-7

R. Kataoka, C. Chaston, D. Knudsen, K.A. Lynch, R. Lysak, Y. Song, K. Murase, T. Sakanoi, J. Semeter, T.-H. Watanabe, D. Whiter, Small-scale dynamic aurora. Space Sci. Rev. (2020), submitted

G.V. Khazanov, M.W. Liemohn, E.N. Krivorutsky, T.E. Moore, Generalized kinetic description of a plasma in an arbitrary field-aligned potential energy structure. J. Geophys. Res. **103**, 6871–6890 (1998)

S. Knight, Parallel electric fields. Planet. Space Sci. **21**, 741 (1973)

G. Knorr, C.K. Goertz, Existence and stability of strong potential double layers. Astrophys. Space Sci. **31**, 209 (1974)

D.J. Knudsen, Spatial modulation of electron energy and density by nonlinear stationary inertial Alfvén waves. J. Geophys. Res. **101**, 10,761 (1996)

D.J. Knudsen, J.H. Clemmons, J.-E. Wahlund, Correlation between core ion energization, suprathermal electron beams, and broadband ELF plasma waves. J. Geophys. Res. **103**, 4171 (1998)

D.J. Knudsen, E.F. Donovan, L.L. Cogger, B. Jackel, W.D. Shaw, Width and structure of mesoscale optical auroral arcs. Geophys. Res. Lett. **28**, 705 (2001)

J. Lemaire, M. Scherer, Model of the polar ion-exosphere. Planet. Space Sci. **18**, 103–120 (1970)

J. Lemaire, M. Scherer, Kinetic models of the solar wind. J. Geophys. Res. **76**, 7479 (1971)

J. Lemaire, M. Scherer, Plasma sheet particle precipitation: a kinetic model. Planet. Space Sci. **21**, 281–289 (1973)

M.R. Lessard, D.J. Knudsen, Ionospheric reflection of small-scale Alfvén waves. Geophys. Res. Lett. **28**, 3573 (2001)

M.W. Liemohn, G.V. Khazanov, Collisionless plasma modeling in an arbitrary potential energy distribution. Phys. Plasmas **5**, 580–589 (1998)

S. Liu, J. Liao, Numerical analysis of double layers in the downward current region of the aurora. J. Plasma Phys. **77**(3), 345–356 (2011)

P. Louarn, J.-E. Wahlund, T. Chust, H. deFeraudy, A. Roux, B. Holback, P.O. Dovner, A.I. Eriksson, G. Holmgren, Observation of kinetic Alfvén waves by the Freja spacecraft. Geophys. Res. Lett. **21**, 1847 (1994)

G. Lu et al., On the auroral current-voltage relationship. J. Geophys. Res. **96**, 3523–3531 (1991)

R. Lundin, I. Sandahl, Some characteristics of the parallel electric field acceleration of electrons over discrete auroral arcs as observed from two rocket flights, in *European Sounding-Rocket, Balloon and Related Research, with Emphasis on Experiments at High Latitudes*, vol. SP–135 (1978), pp. 125–136. ESA

R. Lundin, G. Haerendel, S. Grahn, The Freja project. Geophys. Res. Lett. **21**, 1823 (1994)

K.A. Lynch, R.L. Arnoldy, P.M. Kintner, J.L. Vago, Electron distribution function behavior during localized transverse ion acceleration events in the topside auroral zone. J. Geophys. Res. **99**, 2227 (1994)

K.A. Lynch, D. Pietrowski, R.B. Torbert, N. Ivchenko, G. Marklund, F. Primdahl, Multiple-point electron measurements in a nightside auroral arc: Auroral Turbulence II particle observations. Geophys. Res. Lett. **26**, 3361 (1999)

L.R. Lyons, Generation of large-scale regions of auroral currents, electric potentials, and precipitation by the divergence of the convection electric field. J. Geophys. Res. **85**, 17 (1980)

L. Lyons, D. Evans, R. Lundin, An observed relation between magnetic field aligned electric fields and downward energy fluxes in the vicinity of auroral forms. J. Geophys. Res. **84**, 457–461 (1979)

R.L. Lysak, Electrodynamic coupling of the magnetosphere and ionosphere. Space Sci. Rev. **52**, 33 (1990)

R.L. Lysak, Feedback instability of the ionospheric resonant cavity. J. Geophys. Res. **96**, 1553–1568 (1991)

R.L. Lysak, Generalized model of the ionospheric Alfven resonator, in *Auroral Plasma Dynamics*, ed. by R.L. Lysak. AGU Monograph, vol. 80 (1993), p. 121

R.L. Lysak, C.W. Carlson, Effect of microscopic turbulence on magnetosphere-ionosphere coupling. Geophys. Res. Lett. **8**, 269 (1981)

R.L. Lysak, C.T. Dum, Dynamics of magnetosphere-ionosphere coupling including turbulent transport. J. Geophys. Res. **88**, 365 (1983)

R.L. Lysak, M.K. Hudson, Coherent anomalous resistivity in the region of electrostatic shocks. Geophys. Res. Lett. **6**, 661 (1979)

R.L. Lysak, W. Lotko, On the kinetic dispersion relation for shear Alfvén waves. J. Geophys. Res. **101**, 5085 (1996)

R.L. Lysak, Y. Song, Energetics of the ionospheric feedback interaction. J. Geophys. Res. Space Phys. **107**(A8), SIA-6 (2002)

R.L. Lysak, Y. Song, Non-local interactions between electrons and Alfvén waves on auroral field lines. J. Geophys. Res. **110**, A10S06 (2005). https://doi.org/10.1029/2004JA010803

R.L. Lysak, Y. Song, Propagation of kinetic Alfvén waves in the ionospheric Alfvén resonator in the presence of density cavities. Geophys. Res. Lett. **35**, L20101 (2008). https://doi.org/10.1029/2008GL035728

R.L. Lysak, Y. Song, Development of parallel electric fields at the plasma sheet boundary layer. J. Geophys. Res. **116**, A00K14 (2011). https://doi.org/10.1029/2010JA016424

J.E. Maggs, T.N. Davis, Measurements of the thickness of auroral structures. Planet. Space Sci. **16**, 205 (1968)

D.S. Main, D.L. Newman, R.E. Ergun, Double layers and ion phase-space holes in the auroral upward-current region. Phys. Rev. Lett. **97**, 185001 (2006)

D.S. Main, D.L. Newman, R.E. Ergun, Conditions for establishing quasistable double layers in the Earth's auroral upward current region. Phys. Plasmas **17**, 122901 (2010)

A.J. Mallinckrodt, C.W. Carlson, Relations between transverse electric fields and field-aligned currents. J. Geophys. Res. **83**, 1426 (1978)

O. Marghitu, B. Klecker, J.P. McFadden, The anisotropy of precipitating auroral electrons: a FAST case study. Adv. Space Res. **38**, 1694–1701 (2006). https://doi.org/10.1016/j.asr.2006.03.028

G. Marklund, L. Blomberg, C.-G. Fälthammar, P.-A. Lindqvist, On intense diverging electric fields associated with black aurora. Geophys. Res. Lett. **21**, 1859–1862 (1994)

G.T. Marklund, N. Ivchenko, T. Karlsson et al., Temporal evolution of the electric field accelerating electrons away from the auroral ionosphere. Nature **414**, 724–727 (2001a). https://doi.org/10.1038/414724a

G.T. Marklund, S. Sadeghi, J.A. Cumnock, T. Karlsson, P.-A. Lindqvist, H. Nilsson, A. Masson, A. Fakerley, E. Lucek, J. Pickett, Y. Zhang, Evolution in space and time of the quasi-static acceleration potential of inverted-V aurora and its interaction with Alfvénic boundary processes. J. Geophys. Res. **116**, A00K13 (2001b). https://doi.org/10.1029/2011JA016237

G. Marklund, T. Johansson, S. Lileo, T. Karlsson, Cluster observations of an auroral potential and associated field-aligned current reconfiguration during thinning of the plasma sheet boundary layer. J. Geophys. Res. **112**, A01208 (2007). https://doi.org/10.1029/2006JA011804

G.T. Marklund et al., Cluster multipoint study of the acceleration potential pattern and electrodynamics of an auroral surge and its associated horn arc. J. Geophys. Res. **117**, A10223 (2012). https://doi.org/10.1029/2012JA018046

J.P. McFadden, C.W. Carlson, M.H. Boehm, Field-aligned electron precipitation at the edge of an arc. J. Geophys. Res. **91**, 1723 (1986)

J.P. McFadden et al., Spatial structure and gradients of ion beams observed by FAST. Geophys. Res. Lett. **25**, 2021–2024 (1998)

J.P. McFadden, C.W. Carlson, R.E. Ergun, Microstructure of the auroral acceleration region as observed by FAST. J. Geophys. Res. **104**, 14453–14480 (1999)

C. McIlwain, Direct measurement of the particles producing visible aurora. J. Geophys. Res. **65**, 2727 (1960)

D.M. Miles, I.R. Mann, I.P. Pakhotin, J.K. Burchill, A.D. Howarth, D.J. Knudsen, R.L. Lysak, D.D. Wallis, L.L. Cogger, A.W. Yau, Alfvénic dynamics and fine structuring of discrete auroral arcs: Swarm and e-POP observations. Geophys. Res. Lett. **45**, 545–555 (2018). https://doi.org/10.1002/2017GL076051

R.H. Miller, G.V. Khazanov, Self-consistent electrostatic potential due to trapped plasma in the magnetosphere. Geophys. Res. Lett. **20**, 1331 (1993)

A. Miura, T. Sato, Numerical simulation of the global formation of auroral arcs. J. Geophys. Res. **85**, 73 (1980)

P. Mizera, J. Fennell, Signatures of electric fields from high and low altitude particles distributions. Geophys. Res. Lett. **4**, 311–314 (1977)

P.F. Mizera, D.J. Gorney, J.F. Fennell, Experimental verification of an S-shaped potential structure. J. Geophys. Res. **87**, 1535–1539 (1982)

H.M. Mott-Smith, I. Langmuir, The theory of collectors in gaseous discharges. Phys. Rev. **28**, 727–763 (1926). https://doi.org/10.1103/PhysRev.28.727

F.S. Mozer, C.W. Carlson, M.K. Hudson, R.B. Torbert, B. Parady, J. Yatteau, M.C. Kelley, Observations of paired electrostatic shocks in the polar magnetosphere. Phys. Rev. Lett. **38**, 292 (1977)

F.S. Mozer, C.A. Cattell, M.K. Hudson, R.L. Lysak, M. Temerin, R.B. Torbert, Satellite measurements and theories of auroral particle acceleration. Space Sci. Rev. **27**, 155 (1980)

T.K. Nakamura, Parallel electric field of a mirror kinetic Alfvén wave. J. Geophys. Res. **105**, 10,729 (2000)

D.L. Newman, M.V. Goldman, R.E. Ergun, A. Mangeney, Formation of double layers and electron holes in a current-driven space plasma. Phys. Rev. Lett. **87**(25), 255001 (2001)

D.L. Newman, L. Andersson, M.V. Goldman, R.E. Ergun, N. Sen, Influence of suprathermal background electrons on strong auroral double layers: laminar and turbulent regimes. Phys. Plasmas **15**(7), 072903 (2008)

A. Olsson et al., Freja studies of the current-voltage relation in substorm-related events. J. Geophys. Res. **103**, 4285–4301 (1998)

I.P. Pakhotin, I.R. Mann, R.L. Lysak, D.J. Knudsen, J.W. Gjerloev, I.J. Rae, C. Forsyth, K.R. Murphy, D.M. Miles, L.G. Ozeke, G. Balasis, Diagnosing the role of Alfvén waves in magnetosphere-ionosphere coupling: Swarm observations of large amplitude nonstationary magnetic perturbations during an interval of northward IMF. J. Geophys. Res., Space Phys. **123** (2018). https://doi.org/10.1002/2017JA024713

M. Palmroth et al., ULF foreshock under radial IMF: THEMIS observations and global kinetic simulation Vlasiator results compared. J. Geophys. Res. Space Phys. **120**, 8782–8798 (2015). https://doi.org/10.1002/2015JA021526

K. Papadopoulos, A review of anomalous resistivity for the ionosphere. Rev. Geophys. Space Phys. **15**, 113 (1977)

N. Partamies, M. Syrjäsuo, E. Donovan, M. Connors, D. Charrois, D. Knudsen, Z. Kryzanowsky, Observations of the auroral width spectrum at kilometre-scale size. Ann. Geophys. **28**, 711 (2010)

G. Paschmann, S. Haaland, R. Treumann (eds.), *Auroral Plasma Physics* (Kluwer Academic Publishers, Dordrecht, 2003)

A.M. Persoon, D.A. Gurnett, W.K. Peterson, J.H. Waite, J.L. Burch, J.L. Green, Electron density depletions in the nightside auroral zone. J. Geophys. Res. **93**, 1871 (1988)

H. Persson, Electric field parallel to the magnetic field in a low-density plasma. Phys. Fluids **9**, 1090–1098 (1966). https://doi.org/10.1063/1.1761807

V. Pierrard, G.V. Khazanov, J.F. Lemaire, Current voltage relationship. J. Atmos. Sol.-Terr. Phys. **69**, 2048–2057 (2007). https://doi.org/10.1016/j.jastp.2007.08.005

M.A. Raadu, J.J. Rasmussen, Dynamical aspects of electrostatic double layers. Astrophys. Space Sci. **144**, 43–71 (1988)

R. Rankin, J.C. Samson, V.T. Tikhonchuk, Parallel electric fields in dispersive shear Alfvén waves in the dipolar magnetosphere. Geophys. Res. Lett. **26**, 3601 (1999)

R. Rankin, D. Gillies, A.W. Degeling, On the relationship between shear Alfven waves, auroral electron acceleration, and field line resonances. Space Sci. Rev. (2020), submitted

P. Reiff, H. Collin, J. Craven, J. Burch, J. Winningham, E. Shelley, L. Frank, M. Friedman, Determination of auroral electrostatic potentials using high- and low-altitude particle distributions. J. Geophys. Res. **93**, 7441–7465 (1988)

S. Robertson, Kinetic model for an auroral double layer that spans many gravitational scale heights. Phys. Plasmas **21**, 122901 (2014). https://doi.org/10.1063/1.4903337

A.J. Russell, A.N. Wright, A.V. Streltsov, Production of small-scale Alfvén waves by ionospheric depletion, nonlinear magnetosphere-ionosphere coupling and phase mixing. J. Geophys. Res. Space Phys. **118**(4), 1450–1460 (2013)

A.J. Russell, T. Karlsson, A.N. Wright, Magnetospheric signatures of ionospheric density cavities observed by Cluster. J. Geophys. Res. Space Phys. **120**(3), 1876–1887 (2015)

P.H. Rutherford, E.A. Frieman, Drift instabilities in general magnetic field configurations. Phys. Fluids **11**, 569–585 (1968)

S. Sadeghi, G.T. Marklund, T. Karlsson, P.-A. Lindqvist, H. Nilsson, O. Marghitu, A. Fazakerley, E.A. Lucek, Spatiotemporal features of the auroral acceleration region as observed by Cluster. J. Geophys. Res. **116**, A00K19 (2011). https://doi.org/10.1029/2011JA016505

H. Saito et al., An overview and initial in-orbit status of "INDEX" satellite. IEIC Tech. Rep. **105**, 29–34 (2005)

T. Sakanoi et al., Relationship between field-aligned currents and inverted-V parallel potential drops observed at midaltitudes. J. Geophys. Res. **100**, 19343–19360 (1995)

T. Sato, A theory of quite auroral arcs. J. Geophys. Res. **83**, 1042–1048 (1978)

H. Schamel, S. Bujabarua, Analytical double layers. Phys. Fluids **26**, 190 (1983). https://doi.org/10.1063/1.864006

J. Semeter, J. Vogt, G. Haerendel, K. Lynch, R. Arnoldy, Persistent quasiperiodic precipitation of suprathermal ambient electrons in decaying auroral arcs. J. Geophys. Res. **106**(A7), 12,863–12,874 (2001)

C.E. Seyler, Nonlinear 3-d evolution of bounded kinetic Alfvén waves due to shear flow and collisionless tearing instability. Geophys. Res. Lett. **15**, 756 (1988)

C.E. Seyler, A mathematical model of the structure and evolution of small-scale discrete auroral arcs. J. Geophys. Res. **95**, 17,199 (1990)

C.E. Seyler, K. Liu, Particle energization by oblique inertial Alfvén waves in the auroral region. J. Geophys. Res. **112**, A09302 (2007). https://doi.org/10.1029/2007JA012412

R. Sharp, R. Johnson, E. Shelley, Observation of an ionospheric acceleration mechanism producing energetic (keV) ions primarily normal to the geomagnetic field direction. J. Geophys. Res. **82**, 3324–3328 (1977)

E. Shelley, R. Sharp, R. Johnson, Satellite observations of an ionospheric acceleration mechanism. Geophys. Res. Lett. **3**, 654–656 (1976)

N. Singh, Dynamically evolving double layers and density depletions. J. Geophys. Res. Space Phys. **108**(A8), 1322 (2003)

N. Singh, K. Arcot, B.E. Wells, Parallel electric fields in mixing hot and cold plasmas in the auroral downward current region: Double layers and ambipolar fields. J. Geophys. Res. Space Phys. **114**(A3) (2009). https://doi.org/10.1029/2008JA013591

N. Singh, S. Araveti, E.B. Wells, Mesoscale PIC simulation of double layers and electron holes affecting parallel and transverse accelerations of electrons and ions. J. Geophys. Res. **116**, A00K09 (2011). https://doi.org/10.1029/2010JA016323

Y. Song, R.L. Lysak, Alfvénon, driven reconnection and the direct generation of field-aligned current. Geophys. Res. Lett. **21**, 1755 (1994)

Y. Song, R.L. Lysak, Paradigm transition in cosmic plasma physics, magnetic reconnection and the generation of field-aligned current, in *Magnetospheric Current Systems*, ed. by S.-I. Ohtani et al. AGU Monograph, vol. 118 (American Geophysical Union, Washington, 2000), p. 11

Y. Song, R.L. Lysak, The physics in the auroral dynamo regions and auroral particle acceleration. Phys. Chem. Earth **26**, 33 (2001a)

Y. Song, R.L. Lysak, Towards a new paradigm: from a quasi-steady description to a dynamical description of the magnetosphere. Space Sci. Rev. **95**, 273 (2001b)

Y. Song, R.L. Lysak, The displacement current and the generation of parallel electric fields. Phys. Rev. Lett. **96**, 145002 (2006)

Y. Song, R.L. Lysak, Dynamical generation of quasi-stationary alfvenic double layers and charge holes and unified theory of quasi-static and alfvenic auroral arc formation. Abstract SM23B-2555, in *Fall Meeting*, San Francisco, Dec. 14-18 (2015)

D.P. Stern, One-dimensional models of quasi-neutral parallel electric fields. J. Geophys. Res. Space Phys. **86**(A7), 5839–5860 (1981)

A.V. Streltsov, On the Asymmetry Between Upward and Downward Field-Aligned Currents Interacting with the Ionosphere. J. Geophys. Res. Space Phys. **123** (2018). https://doi.org/10.1029/2018JA025826

A.V. Streltsov, T. Karlsson, Small-scale, localized electromagnetic waves observed by Cluster: result of magnetosphere-ionosphere interactions. Geophys. Res. Lett. **35**, L22107 (2008)

A. Streltsov, W. Lotko, Dispersive field line resonances on auroral field lines. J. Geophys. Res. **100**, 19,457 (1995)

A.V. Streltsov, W. Lotko, Small-scale electric fields in downward auroral current channels. J. Geophys. Res. Space Phys. **108**(A7), 1289 (2003)

A.V. Streltsov, W. Lotko, Coupling between density structures, electromagnetic waves and ionospheric feedback in the auroral zone. J. Geophys. Res. Space Phys. **113**(A5) (2008). https://doi.org/10.1029/2007JA012594

A.V. Streltsov, G.T. Marklund, Divergent electric fields in downward current channels. J. Geophys. Res. Space Phys. **111**(A7) (2006). https://doi.org/10.1029/2005JA011196

A.V. Streltsov, E.V. Mishin, On the existence of ionospheric feedback instability in the Earth's magnetosphere-ionosphere system. J. Geophys. Res. Space Phys. **123**, 8951–8957 (2018). https://doi.org/10.1029/2018JA025942

D. Summers, R.M. Thorne, A new tool for analyzing microinstabilities in space plasmas modeled by a generalized Lorentzian (kappa) distribution. J. Geophys. Res. **97**, 16,827 (1992)

D.W. Swift, On the formation of auroral arcs and the acceleration of auroral electrons. J. Geophys. Res. **80**, 2096 (1975)

D.W. Swift, An equipotential model for auroral arcs: the theory of two-dimensional laminar electrostatic shocks. J. Geophys. Res. **84**, 6427 (1979)

D.W. Swift, Simulation of auroral electron acceleration by inertial Alfven waves. J. Geophys. Res. **112**, A12207 (2007). https://doi.org/10.1029/2007JA012423

D. Sydorenko, R. Rankin, The stabilizing effect of collision-induced velocity shear on the ionospheric feedback instability in Earth's magnetosphere. Geophys. Res. Lett. **44**, 6534–6542 (2017). https://doi.org/10.1002/2017GL073415

D. Sydorenko, R. Rankin, K. Kabin, Nonlinear effects in the ionospheric Alfvén resonator. J. Geophys. Res. **113**, A10206 (2008). https://doi.org/10.1029/2008JA013579

M. Temerin, C.W. Carlson, Current-voltage relationship in the downward auroral current region. Geophys. Res. Lett. **25**(13), 2365–2368 (1998)

M. Temerin, C. Cattell, R. Lysak, M. Hudson, R.B. Torbert, F.S. Mozer, R.D. Sharp, P.M. Kintner, The small scale structure of electrostatic shocks. J. Geophys. Res. **86**, 11,278 (1981)

M. Temerin, K. Cerny, W. Lotko, F. Mozer, Observations of double layers and solitary waves in the auroral plasma. Phys. Rev. Lett. **48**, 1175–1179 (1982)

B.J. Thompson, R.L. Lysak, Electron acceleration by inertial Alfvén waves. J. Geophys. Res. **101**, 5359 (1996)

V.T. Tikhonchuk, R. Rankin, Electron kinetic effects in standing shear Alfvén waves in the dipolar magnetosphere. Phys. Plasmas **7**, 2630 (2000)

V.T. Tikhonchuk, R. Rankin, Parallel potential driven by a kinetic Alfvén wave on geomagnetic field lines. J. Geophys. Res. **107**(A7), 1104 (2002). https://doi.org/10.1029/2001JA000231

K. Tsuruda, H. Oya, Introduction to the EXOS-D (Akebono) project. Geophys. Res. Lett. **18**, 293 (1993)

J.A. Van Allen, Proc. Natl. Acad. Sci. **43**, 57 (1957)

J. Vedin, K. Rönnmark, Electrostatic potentials in the downward auroral current region. J. Geopyhs. Res. **110** (2005). https://doi.org/10.1029/2005JA011083

J. Vedin, K. Rönnmark, Particle-fluid simulation of the auroral current circuit. J. Geophys. Res. **111**, A12201 (2006). https://doi.org/10.1029/2006JA011826

J. Vedin, K. Rönnmark, C. Bunescu, O. Marghitu, Estimating properties of concentrated parallel electric fields from electron velocity distributions. Geophys. Res. Lett. **34**, L16107 (2007). https://doi.org/10.1029/2007GL030162

T.-H. Watanabe, Feedback instability in the magnetosphere-ionosphere coupling system: revisited. Phys. Plasmas **17**, 022904 (2010)

T.-H. Watanabe, A unified model of auroral arc growth and electron acceleration in the magnetosphere-ionosphere coupling. Geophys. Res. Lett. **41** (2014). https://doi.org/10.1002/2014GL061166

C.E.J. Watt, R. Rankin, Electron trapping in shear Alfvén waves that power the aurora. Phys. Rev. Lett. **102**, 045002 (2009)

C.E.J. Watt, R. Rankin, Do magnetospheric shear Alfvén waves generate sufficient electron energy flux to power the aurora? J. Geophys. Res. **115**, A07224 (2010). https://doi.org/10.1029/2009JA015185

C.E.J. Watt, R. Rankin, Alfvén wave acceleration of auroral electrons in warm magnetospheric plasma, in *Auroral Phenomenology and Magnetospheric Processes: Earth and Other Planets*, ed. by A. Keiling, E. Donovan, F. Bagenal, T. Karlsson. AGU Monograph, vol. 197 (American Geophysical Union, Washington, 2012), p. 251

C.E.J. Watt, R. Rankin, R. Marchand, Kinetic simulations of electron response to shear Alfvén waves in magnetospheric plasmas. Phys. Plasmas **11**, 1277–1284 (2004)

D.R. Weimer et al., Auroral zone electric field from DE 1 and 2 at magnetic conjunctions. J. Geophys. Res. **90**, 7479–7494 (1985)

D.R. Weimer et al., The current-voltage relationship in auroral current sheets. J. Geophys. Res. **92**, 187–194 (1987)

E.C. Whipple, The signature of parallel electric fields in a collisionless plasma. J. Geophys. Res. **82**, 1525 (1977)

J.R. Winckler, L. Peterson, R. Arnoldy, R. Hoffman, X-rays from visible aurorae at Minneapolis. Phys. Rev. **110**, 1221 (1958)

D.-J. Wu, D.Y. Sang, C.-G. Fälthammar, An analytical solution of finite-amplitude solitary kinetic Alfvén waves. Phys. Plasmas **2**, 4476 (1995)

J.R. Wygant, A. Keiling, C.A. Cattell, R.L. Lysak, M. Temerin, F.S. Mozer, C.A. Kletzing, J.D. Scudder, A.V. Streltsov, W. Lotko, C.T. Russell, Evidence for kinetic Alfvén waves and parallel electron energization at 4–6 R_E altitudes in the plasma sheet boundary layer. J. Geophys. Res. **107**(A8), 1201 (2002). https://doi.org/10.1029/2001JA900113

M. Zettergren, J. Semeter, Ionospheric plasma transport and loss in auroral downward current regions. J. Geophys. Res. Space Phys. **117**, A06306 (2012)

M. Zettergren, J. Semeter, B. Burnett, W. Oliver, C. Heinselman, P.L. Blelly, M. Diaz, Dynamic variability in F-region ionospheric composition at auroral arc boundaries. Ann. Geophys. **28**(2) (2010). https://doi.org/10.5194/angeo-28-651-2010

M.D. Zettergren, J.L. Semeter, H. Dahlgren, Dynamics of density cavities generated by frictional heating: formation, distortion, and instability. Geophys. Res. Lett. **42**(23), 10–120 (2015)

Space Sci Rev (2020) 216:1
https://doi.org/10.1007/s11214-019-0619-5

Quiescent Discrete Auroral Arcs: A Review of Magnetospheric Generator Mechanisms

Joseph E. Borovsky[1] · Joachim Birn[1] · Marius M. Echim[2,3] · Shigeru Fujita[4,5] · Robert L. Lysak[6] · David J. Knudsen[7] · Octav Marghitu[3] · Antonius Otto[8] · Tomo-Hiko Watanabe[9] · Takashi Tanaka[10]

Received: 10 June 2019 / Accepted: 12 November 2019 / Published online: 16 December 2019

Abstract One of the longstanding questions of space science is: How does the Earth's magnetosphere generate auroral arcs? A related question is: What form of energy is extracted from the magnetosphere to drive auroral arcs? Not knowing the answers to these questions hinders our ability to determine the impact of auroral arcs on the magnetospheric system. Magnetospheric mechanisms for driving quiescent auroral arcs are reviewed. Two types of quiescent arcs are (1) low-latitude non-Alfvénic (growth-phase) arcs magnetically connecting to the electron plasma sheet and (2) high-latitude arcs magnetically connecting near the plasma-sheet boundary layer. The reviews of the magnetospheric generator mechanisms are separated for the two types of quiescent arcs. The driving of auroral-arc currents in large-scale computer simulations is examined. Predicted observables in the magnetosphere and in the ionosphere are compiled for the various generator mechanisms.

Keywords Aurora · Magnetosphere · Plasma physics · Discrete arcs

Auroral Physics
Edited by David Knudsen, Joe Borovsky, Tomas Karlsson, Ryuho Kataoka and Noora Partmies

✉ J.E. Borovsky
jborovsky@spacescience.org

1 Center for Space Plasma Physics, Space Science Institute, Boulder, CO, USA

2 Institut Royale d'Aeronomie Spatiale de Belgique, Bruxelles, Belgium

3 Institute of Space Science, Bucharest, Romania

4 Meteorological College, Kashiwa, Japan

5 National Institute of Polar Research, Tachikawa, Japan

6 School of Physics and Astronomy, University of Minnesota, Minneapolis, MN, USA

7 Department of Physics and Astronomy, University of Calgary, Calgary, Alberta, Canada

8 Geophysical Institute, University of Alaska, Fairbanks, AK, USA

9 Department of Physics, Nagoya University, Nagoya, Japan

10 International Center for Space Weather Science and Education, Kyushu University, Fukuoka, Japan

1 Introduction

The mechanisms that generate auroral arcs are not fully understood; neither is the form of energy conversion that occurs when the arc is powered by the magnetosphere. In various models the energy powering the auroral arc comes from magnetic-field energy, ion pressure, electron pressure, flow kinetic energy, solar-wind-driven waves, etc. For magnetospheric physics the generation of auroral arcs by the magnetosphere has been an outstanding question for decades, as testified in the reviews of auroral-arc generator mechanisms (e.g. Swift 1978; Atkinson 1978; Borovsky 1993; De Keyser and Echim 2010; Haerendel 2011). (For a review of pre-space-age theories of the cause of aurora, see Eather 1980.)

An auroral arc is a geomagnetic east-west aligned curtain of optical emission in the upper atmosphere produced by the impacts of magnetospheric electrons that were accelerated downward along the Earth's magnetic field. The emission is associated with an east-west aligned sheet of upward magnetic-field-aligned electrical current that connects into the magnetosphere, with the current carried largely by the downward-accelerated electrons.

The purpose of this review is to discuss contemporary ideas about the magnetospheric generators of quiescent auroral arcs: what mechanisms provide the power for the arcs, what mechanisms divert the current from the magnetosphere, what mechanisms produce the perpendicular electric fields and related E-cross-B flows. By "quiescent" we mean arcs not associated with dynamic events such as substorm breakup, and which have lifetimes ranging from several minutes to hours. These quiescent arcs are typically associated with electron energies of keV to tens of keV. Because they may be driven by different magnetospheric processes, quiescent arcs are split into two types in this review: (1) high-latitude quiescent arcs magnetically connected to the plasma sheet boundary layer (PSBL) and associated with Alfvén waves in the magnetosphere and/or with magnetospheric interfaces (e.g. Burke et al. 1994; Keiling et al. 2006) and (2) low-latitude quiescent arcs magnetically connected into the electron plasma sheet and associated with mild geomagnetic activity prior to substorm onsets ("growth-phase arcs") or without substorms occurring (Feldstein and Galperin 1985; Galperin and Feldstein 1996; Motoba et al. 2015). Low-latitude field-line-resonance (FLR) arcs (e.g. Samson et al. 1996; Gillies et al. 2017), which also are associated with Alfvén waves, are not discussed in this review.

Auroral arcs do not necessarily have a one-to-one relation to upward current sheets; in some cases single or multiple arcs are current-sheet intensifications within broader regions of upward field-aligned current (Wu et al. 2017). Growth-phase quiescent arcs have optical widths of 0.2–30 km (Kim and Volkman 1963; Knudsen et al. 2001; Partamies et al. 2010), often appear in multiple sets (Akasofu 1976; Wu et al. 2017), and can appear simultaneously in conjugate hemispheres (Sato et al. 1998; Motoba et al. 2012) but favor the winter hemisphere (Newell et al. 1996a). Auroral arcs are associated with geomagnetic activity: the stronger the activity (and the strong the solar-wind driving), the more intense the auroral-arc activity (Sandford 1968; Newell et al. 2009). Auroral arcs are associated with perpendicular electric fields (Marklund 1984) and with vorticity (flow shear) (Kelley and Carlson 1977). During very quiet times, discrete arcs can be absent, though it is believed that some level of auroral emission is essentially always present somewhere. The properties of quiescent discrete auroral arcs are reviewed in Karlsson et al. (2019).

From a systems science point of view, auroral arcs can be considered to be an "emergent phenomenon" in the magnetosphere-ionosphere system (Borovsky and Valdivia 2018). As such, understanding how an arc is produced might require a global (e.g. whole-magnetotail) discussion. Because of their energy extraction, the emergence of arcs has an impact on the system behavior.

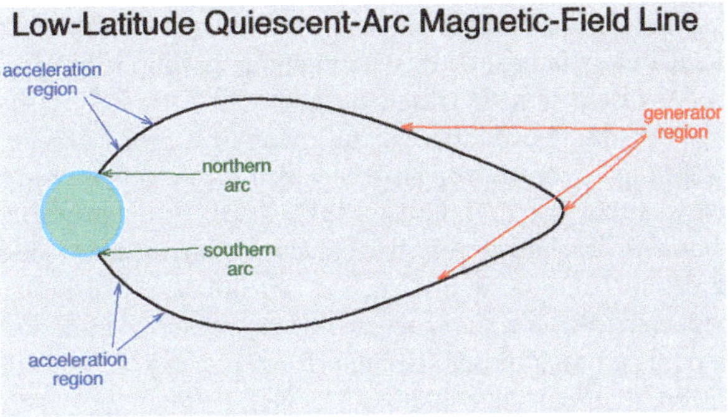

Fig. 1 A sketch of the magnetic-field line connecting a quiescent arc in the atmosphere to the plasma sheet in the nightside magnetosphere

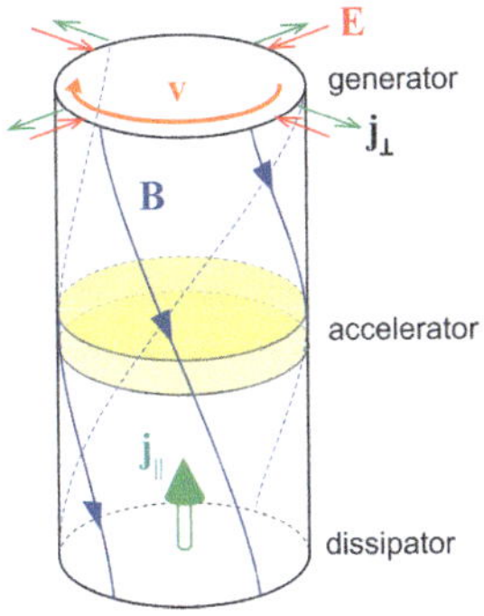

Fig. 2 A simplified sketch of the field, currents, and flows of an auroral arc (in an idealized cylindrical geometry for illustrative purposes) showing the generator region, an accelerator region, and a dissipator region. Magnetic fields are in blue, currents are in green, and electric fields are in red. (Cf. Fig. 3.7.c of Paschmann et al. 2002)

The arc is the site of intense energy extraction from the magnetosphere, with the magnetospheric energy powering the electron acceleration and the flow of current in the dissipative ionosphere. An auroral arc can be envisioned (see Fig. 1) as being comprised of three key regions that are magnetically connected: (1) a generator region in the magnetosphere that supplies power and current to the arc, (2) an accelerator region above the ionosphere where magnetospheric electrons are accelerated downward toward the Earth, and (3) an ionospheric dissipation region that contains the atmospheric-emission curtain.

Figure 2 illustrates these regions in the simplest, cylindrically symmetric, form for the fields and currents that are typically associated with an arc. (1) The generator region in the magnetosphere consists of a shear flow or twisting motion associated with a converging electric field. During a buildup phase this causes a magnetic shear or twist of a magnetic flux tube, corresponding to an increase of upward field-aligned current. For long lasting arcs this would lead into a steady state. By current continuity, the field-aligned current must become converted into (or from) perpendicular current $j_\perp$ in the magnetosphere, with $j_\perp$ being opposite to the electric field, thus constituting the generator ($\underline{J} \bullet \underline{E} < 0$). (This perpendicular current that feeds the field-aligned current is associated with a field-aligned gradient of the field-aligned vorticity Song and Lysak 2006.) The perpendicular electric field together with the perturbed azimuthal magnetic-field component also provide the downward Poynting flux that powers the arc. (2) The perpendicular electric field is known to be stronger at high altitudes than near the ionosphere (Weimer et al. 1985); in a quasi-static state this implies an upward electric field parallel to $\underline{B}$ in the acceleration region, which causes the electron acceleration. In the acceleration region this also represents a load ($\underline{J} \bullet \underline{E} > 0$), which converts part of the power. The parallel electric field may also be visualized as a partial field line slippage, which reduces the twisting motion. (3) In the ionosphere, the field-aligned current is again closed through perpendicular current, which together with the remaining perpendicular electric field (and corresponding motion) represents a load ($\underline{J} \bullet \underline{E} > 0$), which now causes dissipation via Ohmic heating.

This simple cylindrical picture, however, needs to be modified in light of several facts. One is the geometry of an arc, which is strongly elongated in the east-west direction in the ionosphere (cf. Hallinan 1981). In the magnetosphere, this may be associated with a geometry in which the electric field, associated with the shear motion, may be aligned in one direction, but the current diversion is in the transverse direction. And the vectors need not be collocated. Also in the magnetotail, the perturbed currents that connect to the field-aligned currents are superposed onto unperturbed currents, particularly the cross tail current, which distorts the generator and load picture. Furthermore, the current diversion and generator regions, which appear closely connected in the simple picture of Fig. 2, may be separated substantially, as found in MHD simulations to be discussed in Sect. 5.

The simple picture of Fig. 2 also leaves out the ultimate energy source in the magnetosphere. While the twisting or shear motion appears to be essential this does not mean that the energy is supplied by the kinetic motional energy. Most detailed energy release and conversion investigations in magnetotail models and observations have found a strong dominance of enthalpy (that is, essentially thermal energy) flux (Birn and Hesse 2005; Aunai et al. 2011; Eastwood et al. 2013; Tyler et al. 2016). But how and where this is converted to the Poynting flux that eventually powers the arcs is still uncertain. The picture illustrated in Fig. 2 applies not only to ion or plasma flows but also, on smaller scales potentially relevant for arcs, to electron flows.

Determining the generator mechanism of low-latitude quiescent auroral arcs (and the impact that those arcs have on the magnetosphere) is hindered by uncertainties as to where the visible arcs map to in the nightside magnetosphere. One school has low-latitude arcs magnetically connected into the plasma sheet in the outer dipolar portion of the nightside magnetosphere (e.g. McIlwain 1975; Meng et al. 1979; Kremser et al. 1988; Elphinstone et al. 1991; Pulkkinen et al. 1991; Mauk and Meng 1991; Lu et al. 2000; Yang et al. 2013; Motoba et al. 2015), a second school has those arcs magnetically connected to the plasma sheet in the stretched magnetotail (e.g. Yahnin et al. 1997, 1999; Birn et al. 2004a,b, 2012; Sergeev et al. 2012; Hsieh and Otto 2014), and a third school has the low-latitude arcs magnetically connected into the mid magnetotail but with the arc current generated at the plasma-sheet/lobe boundary (e.g. Tanaka 1995, 2015; Tanaka et al. 2017; Ebihara and Tanaka 2017). In mapping quiescent auroral arcs into the nightside magnetosphere, it is unlikely that they magnetically connect to the equatorial region of the magnetotail beyond about 15 R_E: even under quiet geomagnetic conditions those regions of the plasma sheet are characterized by turbulent flow (Borovsky et al. 1997; Voros et al. 2007; Stepanova et al. 2011; El-Alaoui et al. 2012) and irregular magnetic field (Borovsky and Funsten 2003; Voros et al. 2004). An example is given in Fig. 3 where the flow and field measured at $X = -20\ R_E$ are shown for an interval of normal geomagnetic activity. It is difficult to see how a laminar, large-scale quiescent arc could be driven from such an irregular dynamic region.

This review is organized as follows. Section 2 reviews generator mechanisms for high-latitude quiescent arcs associated with the plasma sheet boundary layer: Sect. 2.1 discusses arcs driven by Alfvén waves and Sect. 2.2 discusses arcs driven by plasma interfaces. Section 3 reviews generator mechanisms for low-latitude quiescent arcs: Sect. 3.1 reviews the driving of arcs by pressure gradients in the plasma sheet, Sect. 3.2 reviews an arc model based on plasma pressure and magnetic stress in the global magnetospheric convection, Sect. 3.3 reviews arcs that are driven by thin current sheets in the Earth's plasma sheet, Sect. 3.4 reviews the stationary-Alfvén-wave picture of arcs driven by magnetic energy, and Sect. 3.5 reviews ionospheric-feedback models for auroral arcs. Section 4 discusses the

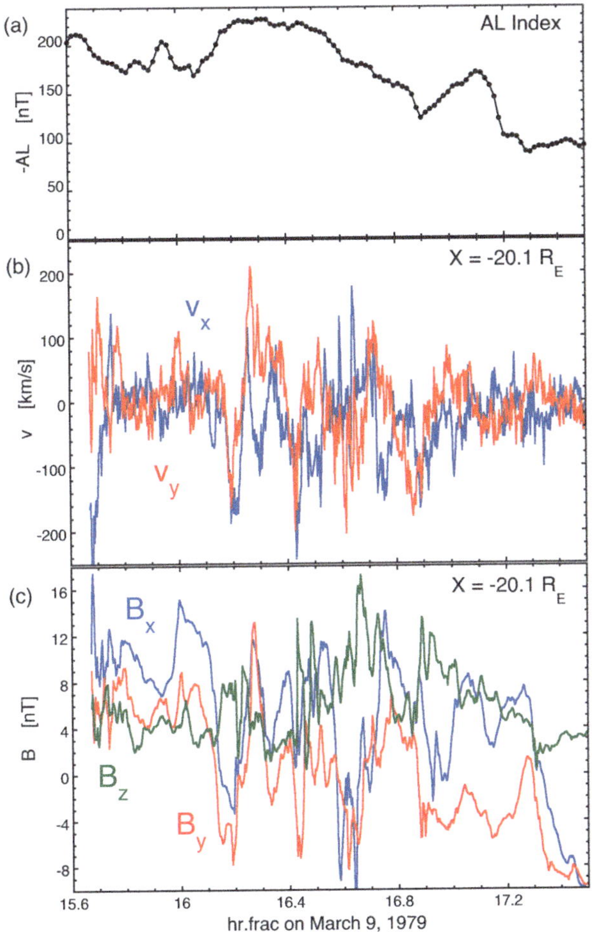

Fig. 3 ISEE-2 measurements in the plasma sheet 20 R_E downtail. Average value of -AL for years 1966–2013 is 128 nT

driving of arcs located near the Harang discontinuity. Section 5 looks at what large-scale numerical simulations say about the driving of quiescent auroral arcs. Section 6 summarizes the review and provides an overview of unknown issues associated with the generation of quiescent auroral arcs. A number of arc-generator models from the literature that are in disuse are briefly discussed in the Appendix.

2 High-Latitude Quiescent Arcs: The PSBL and LLBL

High-latitude quiescent auroral arcs have been associated with the polar-cap boundary and plasma sheet boundary layer (PSBL) (e.g. Burke et al. 1994; Onsager and Mukai 1996). On the duskside, these arcs might be consistent with magnetic connection into the low-latitude boundary layer (e.g. Echim et al. 2008). High-latitude quiescent arcs tend to involve the precipitation of lower-energy electrons than do low-latitude arcs (Boyd et al. 1971; Safargaleev et al. 2003; Karlsson et al. 2019)

In equilibrium models of the magnetotail (in GSE coordinates), the Y-direction magnetic-field flaring of the plasma sheet is larger than the Y-flaring of the lobe magnetic field (cf. Fig. 4 of Birn 1989 or Fig. 1 of Birn and Hesse 1991): this means that there is a magnetic shear between the lobe and the plasma sheet, and a Region-I sense field-aligned current associated with that magnetic shear. At $X = -100\ R_E$, the plasma sheet flares about $5\ R_E$ further in the Y direction than the lobe does (cf. Figs. 2–5 of Birn and Hesse 1991). This means that the magnetic shear angle is about 5/100 radians, or about 2.9° at this downtail distance. If the magnetic-field strength is about 20 nT, then the change in B_y from the lobe to the plasma sheet is about 1.0 nT, which corresponds to a field-aligned current per unit length $J_\| = 8 \times 10^{-4}$ A/m in the shear, where the length is in the Y-direction. At $X = -100\ R_E$, the premidnight shear layer is about $20\ R_E$ in extent in the Y-direction (cf. Figs. 2–5 of Birn and Hesse 1991), giving a total field-aligned current in the PSBL shear layer of 1×10^5 A going into one hemisphere. For a 15-km-wide high-latitude arc at the PSBL with an energy flux of 20 mW/m^2 carried by 3 keV electrons (numbers from Vaivads et al. 2003), the current of the arc is 1×10^5 A if the arc is 1000-km long in the ionosphere. In a sense, the high-latitude quiescent arc is part of the equilibrium configuration of the magnetotail.

The Earth's low-latitude boundary layer (LLBL) also magnetically connects to the high-latitude portion of the auroral oval near the polar-cap boundary, although the connection may be predominantly to the dayside ionosphere (e.g. Elphinstone et al. 1991). The velocity shear between the tailward-moving flow of the LLBL relative to the more-stationary plasma sheet has a vorticity $\underline{\omega} = \nabla \times \underline{v}$ in the equatorial plane that has $\underline{\omega} \bullet \underline{B} > 0$ on the dusk side of the Earth, which corresponds to a negative space charge (cf. Eq. (11) of Borovsky and Birn 2014) and converging perpendicular electric fields as in auroral arcs. And on the dusk side the twist of the magnetic field produced by the tailward flow of the LLBL relative to the plasma sheet corresponds to a field-aligned current that is out of the ionosphere: a Region-I sense current as is the current associated with the PSBL. This mapping is pertinent to the model discussed in Sect. 2.2.

2.1 Alfvén-Wave-Driven Arcs

Alfvén-wave Earthward Poynting flux is seen for these high-latitude arcs (e.g. Wygant et al. 2000; Chaston et al. 2003; Keiling et al. 2006). Time-varying Alfvén waves with small perpendicular scale sizes (a few km or less in the auroral zone) are able to accelerate electrons to energies as high as few keV (Goertz and Boswell 1979; Lysak and Song 2011) and in transient bursts (~ 1 s in duration) as high as 10 keV (Watt and Rankin 2010). While these spatial and temporal scales are at or below the low end of those relevant to quiescent arcs, Alfvén waves are involved in the transient stage leading to quiescent arcs, and may be important in maintaining "red-line" arcs produced by lower-energy electrons (Gillies et al. 2017; Liang et al. 2019). In a generator-accelerator-ionosphere picture of the arc (cf. Fig. 4), the Alfvén waves play the role of the accelerator: to determine the magnetospheric cause (generator) of the arc, one must determine what drives the Alfvén waves near the PSBL and where the energy for the waves comes from.

Zelenyi et al. (2004) and Grigorenko et al. (2007) have argued that the Alfvén waves (which have linear polarizations with the wave's velocity and magnetic-field vectors in the Y direction in the magnetotail) are driven by firehose instabilities powered by Earthward-traveling beams of energetic ions in the PSBL, known as velocity dispersed ion beams (e.g. Takahashi and Hones 1988; Zelenyi et al. 1990). Since the gradients in the beam intensity are stronger in the Z direction than the Y direction it was argued (Zelenyi et al. 2004; Grigorenko et al. 2007) that the firehose would produce the observed Y-direction polarization of the Alfvén waves. In this picture, the source of the ion beams is the source (generator) of the arc. If the Earthward ion beams are produced by Earthward plasma flow from magnetotail

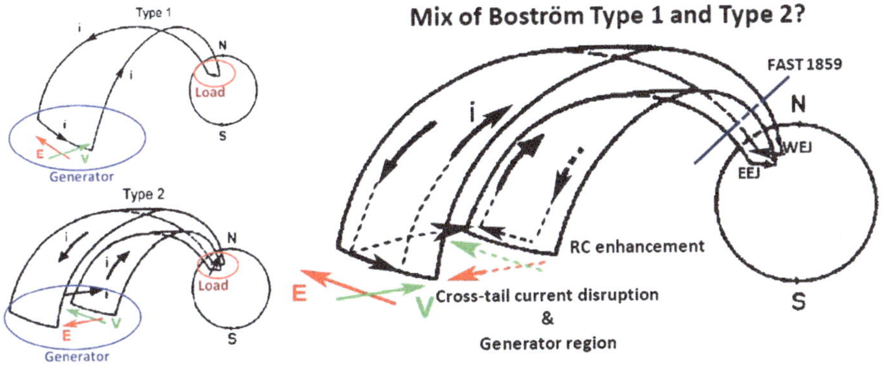

Fig. 4 Left: Type 1 and Type 2 auroral current circuit. Adapted after Boström (1964). Right: Mix of Type 1 and Type 2 (see text). Figure based on Marghitu et al. (2009) and on subsequent FAST and Swarm work on longitudinal gradients in the auroral region, as yet unpublished

reconnection process (e.g. Onsager et al. 1991; Birn et al. 2015), then the energy source of the high-latitude Alfvénic arc is magnetic-field annihilation associated with the distant reconnection site. If the Earthward ion beams are produced by Speiser-orbiting ions escaping from the cross-tail current sheet (e.g. Lyons and Speiser 1982; Bosqued et al. 1993) then the beams have obtained their energization in a cross-tail electric field, which is consistent with a coherent large-scale convection in the magnetotail. In the turbulent regions of the magnetotail (Borovsky et al. 1997; Voros et al. 2004; Weygand et al. 2005) beyond $\sim 15\ R_E$ where Earthward transport is dominated by bursty bulk flows (Angelopoulos et al. 1993, 1994; Cao et al. 2013) and where the electric field is spatially and temporally chaotic (Cattell and Mozer 1982; Pedersen et al. 1985; Borovsky et al. 1998a), the concept of a global cross-tail electric field may not be valid.

It has been argued (Wright and Allan 2008) that if there were a source of fast magnetosonic-mode waves in the plasma sheet (such as from a substorm or the movement of a reconnection plasmoid), the fast mode waves could mode convert into shear Alfvén waves in the plasma sheet boundary layer; similarly it has been argued (Dmitrienko 2011) that fast magnetosonic waves entering the magnetotail from the solar wind could mode convert into shear Alfvén waves in the plasma sheet boundary layer. To make a quiescent high-latitude arc during quiet geomagnetic times, the magnetosonic-wave source in the solar wind is more promising than is the (impulsive) source in the plasma sheet. This is similar to the idea that solar-wind-driven Kelvin-Helmholtz waves on the magnetotail magnetopause could mode convert into shear (and kinetic) Alfvén waves in the plasma sheet boundary layer (Smith et al. 1986; Harrold et al. 1990; Hanson and Harrold 1994) (see also Ruderman and Wright 1998). In these pictures, the power source for the arc is the solar-wind driving wave energy into the magnetotail.

Another potential source of shear Alfvén waves in the PSBL is associated with distant reconnection and the flaring of the magnetotail. As noted above, in the equilibrium configuration of the magnetotail there is a change in flaring angle between the open field lines of the lobes and the closed field lines of the plasma sheet, with a Region-I sense field-aligned current at the PSBL. As a northern-lobe magnetic-field line reconnects at the distant X line, it convects downward in Z to join the plasma sheet. The Z velocity of the field line and its plasma is approximately $0.1\ v_A$ for collisionless reconnection (Birn et al. 2001), where v_A is the local Alfvén speed at the reconnection site. Owing to the change in flaring angle between magnetic-field lines in the lobe and magnetic-field lines in the plasma sheet, the field line and its plasma must also flow in the Y-direction outward from local midnight after it crosses the separatrix. With the plasma carried on the field line, this Y flow will have some inertia. If the inertia carries the field line past the equilibrium flaring angle of the plasma sheet, there may be a $v_y - B_y$ spatio-temporal oscillation set up in the Z flow entering the PSBL from the lobe (i.e. in the PSBL). This oscillation will take the form of a shear Alfvén wave, with the linear polarization as observed for Alfvén waves in the PSBL (Wygant et al. 2000). Whether or not there is an Alfvénic oscillation will depend on how rapidly the field line makes the transition in flaring angle.

Although not for a quiescent arc, Lysak et al. (2009) have pointed out that the Earthward plasma injection from a near-Earth neutral line at substorm onset compresses plasma and launches fast-mode waves that can propagate from the equatorial region of the plasma sheet to become mode converted into Alfvén waves at the PSBL to produce poleward boundary intensifications.

Electron acceleration by dispersive Alfvén waves is limited to velocities of less than 2 v_A (Kletzing 1994), corresponding to energies of less than ~ 2 keV, and resulting in relatively dim emissions dominated by the red line. Furthermore, Alfvénic acceleration is inherently transient, which makes it seemingly incompatible with quiescent arcs, although the long lifetime of red-line emissions will tend to smooth out dynamical variations. Therefore, Alfvén waves may explain electron acceleration within low-energy arcs such as high-latitude (PSBL) arcs. They also clearly play a role in field-line resonance arcs, which occur most frequently at dawn and at dusk (Gillies et al. 2018), in contrast to the brighter and more energetic growth-phase arcs that are most common near midnight.

2.2 Model of Magnetospheric Interface Driven Quiescent Arcs: PSBL and LLBL

Motivated by the works of Lyons et al. (1979), Lyons (1980, 1981) and Lyons and Evans (1984), Roth et al. (1993) suggested that magnetospheric plasma interfaces provide the electromotive force needed to sustain the magnetic-field-aligned potential difference that accelerates auroral electrons. One key feature is the polarization electric field across the plasma interface sustained by pressure gradients or shears of bulk velocity. Roth et al. (1993) portray this magnetospheric DC generator with an analogy to the "contact potential" difference produced at the interface between two metallic conductors at different temperatures" and show examples of kinetic Vlasov solutions computed for interfaces formed in the Earth's plasma sheet. The ultimate energy of the coupled auroral arc is provided by the polarization electric field across the generator. This polarization electric field is stronger when the temperature and/or density gradient, and/or the velocity shear across the plasma interface is stronger.

Roth et al. (1996), Echim et al. (2007, 2008), De Keyser and Echim (2010, 2013), De Keyser et al. (2010) further developed these ideas and evaluated qualitatively and quantitatively the auroral effects due to the magnetic coupling between the strong electric fields formed at magnetospheric interfaces and the auroral ionosphere. It is argued that plasmas with different macroscopic properties (e.g. density, temperature, bulk velocity) are in contact in various regions of the magnetosphere (e.g. at the inner edge of the Low Latitude Boundary Layer (LLBL) or the Plasma Sheet Boundary Layer (PSBL)) and lead to the formation of transition layers/interfaces with scale lengths of the order of the electron and/or proton gyroradius or larger. Extensive earlier work focused on the driving of arcs by the velocity shear between the plasma sheet and the tailward-flowing LLBL (e.g. Sonnerup 1980; Lundin and Evans 1985; Lotko et al. 1987; Wei et al. 1996).

The interaction between particles with different macroscopic properties from the two sides of the magnetospheric interface leads to electric polarization and formation of a normal electric field. A Vlasov kinetic treatment provides the profile of variation across the interface of the plasma moments (e.g., current and charge density, temperature, bulk velocity, etc.) and of the electromagnetic field. The interface is connected to the auroral ionosphere by magnetic-field lines such that magnetospheric/ionospheric particles can move downwards/upwards in the magnetic flux tube connecting the DC generator to the auroral ionosphere. The field-aligned dynamics of these particles is affected by the magnetic mirroring and electrostatic forces. Particles from the magnetospheric interface are (locally) scattered into the loss cone by wave particle interactions sustained by inherent magnetospheric turbulence and/or instabilities due to the strong gradients across the interface. The model assumes that these magnetospheric processes are sufficiently efficient to replenish permanently the loss cone. The loss cone particles will then contribute to a field-aligned current density. The mirror force impedes the downward motion of magnetospheric electrons leading to a field-aligned variation of their current density and consequently to a $\Delta\Phi$, the field-aligned potential difference between the magnetosphere and ionosphere. The latter however will accelerate electrostatically the downgoing electrons. The mutual effect of the mirror force and electrostatic acceleration determines the energy flux of the precipitating electrons. If the electric potential varies monotonically with the field-aligned distance analytical expressions can be found for the field-aligned current density of precipitating electrons (Knight 1973; Lemaire and Scherer 1973) and for the flux of precipitating energy (Lundin and Sandahl 1978) as a function of $\Delta\Phi$.

When mapped to ionospheric altitudes these expressions give the magnetospheric inflow of particles and energy. The model of Echim et al. (2007, 2008) considers also the contributions to the net field-aligned current density due to upgoing electrons and ions and downgoing ions. At the topside ionosphere the upward field-aligned current must be compensated by (perpendicular) ionospheric currents such that the current continuity $\nabla \bullet \underline{j} = 0$ is satisfied. The precipitating flux of electrons, the field-aligned current density, and the magnetospheric electrostatic potential (Φ_m) mapped to the ionospheric altitude have a spatial variation imprinted by the profile of plasma parameters and fields in the DC generator in the magnetosphere. The current continuity is the mathematical kernel that allows the computation of Φ_i, the ionospheric electrostatic potential as a function of the coordinate normal to the discrete auroral arc. It is then straightforward to compute $\Delta\Phi = \Phi_i - \Phi_m$ and all the auroral arc parameters that depend on it, like the field-aligned current density, the influx of precipitating energy, and the Pedersen conductance (see Fig. 5). The multiplicity of arcs can occur when the velocity shears and/or kinetic pressure gradients across the interface are so strong that an equilibrium configuration cannot form such that the interface splits in several smaller scale interfaces—the roots of smaller scale auroral arcs (De Keyser and Echim 2013).

These models are quasi-static and relevant for stable, discrete auroral arcs. They provide the main properties of the discrete auroral arc (like thickness, field-aligned current density, flux of precipitating energy, Pedersen conductance) as a function of the DC generator strength. This quasi-static description of the arc allows for parametric studies and leads to estimation of the effects of various types of drivers on the properties of the auroral arc. For instance it was shown that increased levels of shears at the inner edge of the LLBL tend to produce brighter arcs while increased kinetic pressure gradients lead to narrower auroral structures (Echim et al. 2008). The LLBL and the PSBL are sites where cold-hot interfaces do form. Note, however, that other types of interfaces (e.g. hot-hot) can occur in various magnetospheric regions, as for instances in the plasma sheet (De Keyser and Echim 2013).

The plasma-interface generator model and its coupling with the auroral ionosphere was validated with satellite observations. Conjugate observations of the magnetospheric generator and of the auroral arc by Cluster and DMSP spacecraft (Vaivads et al. 2003) was a

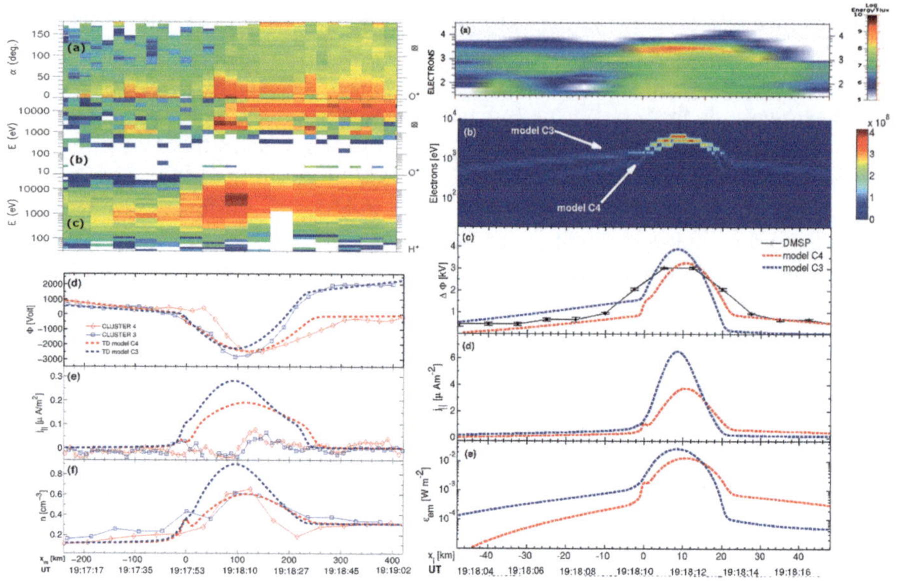

Fig. 5 The left panel shows Cluster data and model results for a magnetospheric interface generator located at the lobe-PSBL interface; (**a**) pitch angle distribution of O^+ ions, from Cluster 4 CIS (Rème et al. 2001); (**b**) O^+ and (**c**) H^+ energy spectrum from Cluster 4; (**d**) electrostatic potential from Cluster data (Vaivads et al. 2003) and model, (**e**) field-aligned current density, positive for upward current, (**f**) number density. The right panel shows model results and DMSP data for the arc connected to the generator described by the left panel; (**a**) energy of precipitating electrons from DMSP-F14; (**b**) model energy spectra obtained for Maxwellian electrons injected at 4.7 R_E; (**c**) the field-aligned potential drop, $\Delta\Phi$, derived from DMSP F14 electron spectra (black symbols) and from current continuity (dashed lines); (**d**) the field-aligned current density; (**e**) the flux of precipitating energy. Figures after Echim et al. (2009)

fortunate occasion when virtually all the model observables mentioned in Table 1 could be compared with experimental data. Echim et al. (2009) showed that their model estimates well the plasma properties of the generator (the polarization electric field normal to the interface and its V-shaped profile and spatial scale, the gradients of density, temperature and the shear of the bulk velocity), as well as the properties of the coupled auroral (the thickness, the accelerating potential, the flux of precipitating energy). A comparison between the model and the optical (by TIMES spacecraft) and in-situ (by CLUSTER) observations of a polar cap arc shows that the model estimates well the thickness of the polar cap arc, the accelerating potential and the optical emission rates (Maggiolo et al. 2012). A statistical analysis of the Region 1 field-aligned currents from DMSP data (Wing et al. 2011) confirms the trends observed from the parametric study of plasma interface generator model (Echim et al. 2008): in the afternoon sector, (12–16 MLT), an increasing density gradient across the generator interface results in an increase of the field aligned current density, a decrease of the maximum field-aligned potential drop, and has little effect on the maximum of the precipitating energy flux.

3 Low-Latitude (Growth-Phase) Quiescent Arcs

In this section various ideas about the magnetospheric mechanisms that drive quiescent low-latitude arcs (a.k.a. growth-phase arcs) are reviewed. Predictions about magnetospheric observables for the various mechanisms are collected into Table 1.

Table 1 Predicted observables for the various non-Alfvénic models for quiescent auroral-arc generation: the final column denotes where the observable should be seen, in the magnetosphere (M) or ionosphere (I)

Magnetospheric-Interface-Driven Arcs	
• Polarization electric field normal to the interface	M
• V-shaped electric potential profile across the interface	M
• Gradients of density and/or temperature and/or shears of bulk velocity	M
• Total pressure equilibrium across the interface	M
• Interface scales with the Larmor scale of dominant species	M
• Arcs's field-aligned current density and the accelerating potential increase with increasing velocity shear and density gradient in the generator	M + I
• Precipitating energy flux increases with the magnetospheric velocity shear, decreases with the generator temperature gradient, shows no dependence on the density gradient	M + I
• Arc thickness related to the Larmor radius of the species dominant in the generator and on the altitude of the generator; decreases with the density gradient in the generator	M + I
Pressure-Driven Arcs	
• $(\nabla P_i + \nabla P_e) \times \nabla B$ is southward in the equatorial region	M
• $(\nabla P_i + \nabla P_e) \times \nabla B$ must be the dominant term in the current-diversion equation	M
• Arc thickness related to radial gradient scale	M + I
• $\lvert B \rvert$ change across arc (from $\nabla_r P$ pressure equilibrium or from $\nabla_r B$ in $\nabla_a P \times \nabla_r B$)	M
Embedded Arcs	
• Pressure "ledge" profile in "radial" direction	M
• $\nabla_r P \times \nabla_a B$ southward at equator	M
• Arc thickness related to pressure-ledge thickness	M + I
• Azimuthal westward flow inside arc or tailward of arc	M + I
• Normal flow through arc	M + I
• Alfvén bounce dynamics; decay times	I
Stationary Inertial Alfvén Arcs	
• Spatial oscillations of n, $j_\parallel$, electron energy, $E_\perp$	I
• Normal flow through arc	M + I
• Reduced azimuthal magnetic twist in flow exiting arc	M + I
• Growth times	I
Thin Cross-Tail-Current-Sheet Arcs	
• Arc magnetically connects to Earthward edge of cross-tail current sheet	M
• $\underline{v}_i \neq \underline{v}_e$ in the cross-tail current sheet	M
• There is a reduction of B_{mag} in the current sheet	M
• On the field lines connecting to the cross-tail sheet there is a duskward ($j_\perp$ or $E_\perp$)	M
• Across the field lines connecting to the sheet, B_{mag} is greater on the high-latitude side than the low-latitude side (dB_{mag}/dz)	M
Feedback-Instability Arcs	
• Normal flow through arc	M + I
• Spatial oscillations of $j_\parallel$ and ionospheric conductivity in accordance with models	M + I
• Growth time and temporal oscillation period compared with models	I
All Models	
• vorticity $\underline{\omega} \bullet \underline{B} > 0$	M + I
• magnetic shear from $j_\parallel$	M + I
• upflowing ions (not Alfvénic arcs)	M

3.1 Pressure-Driven Arcs

It is well known that the magnetosphere can drive field-aligned currents into the ionosphere via particle pressure gradients that are perpendicular to the magnetospheric magnetic field: this process is described by the "Vasyliunas formula" (e.g. Eq. (3.21) of Schindler and Birn 1978, Eq. (5) of Sato and Iijima 1979, Eq. (15) of Hasegawa and Sato 1979, and Eq. (12) of Strangeway 2012)

$$\begin{aligned}(1/L_{\parallel})(j_{\parallel}/B) &= (2/B^3)\big([\nabla P_i + \nabla P_e] \times \nabla B\big)_{\parallel} + (2\rho/B^3)\big((d\underline{v}/dt) \times \nabla B\big)_{\parallel} \\ &\quad - (1/B^2)\big((d\underline{v}/dt) \times \nabla\rho\big)_{\parallel} + (\rho/B^3)\underline{\omega} \bullet d\underline{B}/dt \\ &\quad + (\omega_{\parallel}/eB^3)(\nabla k_B T \times \nabla\rho)_{\parallel} + (\rho/B^2)d\omega_{\parallel}/dt, \qquad (1)\end{aligned}$$

where P_i and P_e are the ion and electron pressure, B is the magnetic-field strength, ρ is the plasma mass density, and $\underline{\omega} = \nabla \times \underline{v}$ is the vorticity. On the left hand side $L_{\parallel}$ is the length along a magnetospheric field line away from the equator that the generator mechanism acts. The left-hand side of expression (1) describes the strength of field-aligned current and the six terms on the right-hand side are driver (current-diversion) terms in the magnetosphere for that field-aligned current. The first term on the right-hand side represents pressure-gradient driving of parallel currents: for pressure driven arcs the first term on the right-hand side must dominate over all other terms on the right hand side, yielding $\nabla_{\parallel}(j_{\parallel}/B) = (2/B^3)(\nabla_{\perp}P \times \nabla_{\perp}B)$ (Grad 1964; Vasyliunas 1970). The other terms on the right-hand side of expression (1) represent flow braking, vorticity and magnetic-flux change, baroclinic flow ($\nabla T \times \nabla\rho$), and advected vorticity. In dynamical MHD simulations of field-aligned currents driven in the magnetotail, Birn et al. (1999, 2011) find that the $\nabla_{\perp}P \times \nabla_{\perp}B$ term tends to be dominant. For driving auroral arcs (with upward current coming out of the ionosphere), the vector $\nabla_{\perp}P \times \nabla_{\perp}B$ evaluated in the equatorial plane of the magnetosphere should be pointing southward. The driving of auroral arcs by pressure gradients in the Earth's plasma sheet has been suggested several times, focusing on various pressure configurations that may form in the nightside magnetosphere. In the plasma sheet typically $T_i > T_e$ and the ion pressure dominates over the electron pressure, so the focus has usually been on the ion-pressure configuration.

Three prominent applications of plasma-sheet pressure configurations driving arcs are by Stasiewicz (1985), Galperin et al. (1992), and Coroniti and Pritchett (2014). Stasiewicz (1985) hypothesized that premidnight quiescent arcs could be driven by elongated streams of lower-pressure plasma advecting from the magnetotail sunward into the duskside dipolar magnetosphere, with the auroral arc magnetically mapping to the duskward edge of the low-pressure stream where the pressure gradient $\nabla_{\perp}P$ points in the duskward direction. (Stasiewicz 1985 also discussed the possibility of electron pressure gradients caused by spatial variations of the electron temperature.) Galperin et al. (1992) (see also Galperin and Bosqued 1999) looked at an Earthward-pointing pressure gradient $\nabla_{\perp}P$ at the Earthward edge of a strong cross-tail current at a location where $|B_z|$ has a localized minimum in the magnetotail. A feedback in the cross-tail current sheet is described wherein ion stochasticity leads to a strengthened current which leads to a reduced $|B_z|$ and more stochasticity. Examining the Tsyganenko T87 magnetic-field model (Tsyganenko 1987), they argued that there is a slight misalignment of the directions of $\nabla_{\perp}P$ and $\nabla_{\perp}B$ in this $|B_z|$ minimum that make $\nabla_{\perp}P \times \nabla_{\perp}B$ nonzero. Coroniti and Pritchett (2014) analyzed a configuration wherein there is an Earthward-directed radial gradient $\nabla_r P$ duskward of midnight where the magnetic-field strength B decreases going away from midnight. They argued that if this maximum $\nabla_r P$ occurs at a location where $\nabla_y P = 0$ then field-aligned current will be maximal.

Tverskoi (cf. Antonova et al. 1998; Stepanova et al. 2003) suggested a mechanism whereby the hot-plasma pressure profile in the plasma sheet can become stratified to form multiple quiescent auroral arcs. In upward-current regions, they identified a plasma pressure feedback with the electrostatic potential (Tverskoi 1983) wherein stratified perturbation potentials in the magnetosphere grow and trap ions to produce pressure stratification; the result is a stratification of the magnetospheric plasma, the plasma pressure, field-aligned current, and magnetospheric convection. Two assumptions for this stratification instability to work are (1) that magnetospheric potential perturbations are linearly proportional to magnetospheric pressure perturbations and (2) that the magnetospheric spatial scales of the stratification perturbations are on the order of the magnetospheric ion gyroradius. The stratified plasma pressure in the magnetosphere drives stratified upward field-aligned current via the $\nabla_\perp P \times \nabla_\perp B$ current-diversion mechanism.

For pressure-driven currents, the width of the arc in the atmosphere is governed by the magnetic mapping of the scalesizes of the pressure gradient and/or magnetic-field-strength gradient from the magnetosphere to the atmosphere.

It should be noted that there are electron pressure gradients in the nightside magnetosphere that can be very sharp. The inner edge of the electron plasma sheet is often very sharp in the radial direction (Gussenhoven et al. 1981; Borovsky et al. 1998b) with the gradient scale of the hot-electron density being comparable to an ion gyroradius or an ion inertial length. The location of the electron inner edge depends on the level of magnetospheric convection (Gussenhoven et al. 1983; Elphic et al. 1999), with the edge moving deeper into the nightside dipole as global magnetospheric convection becomes stronger, e.g. during the growth phase of a substorm; as the inner edge moves deeper into the dipole its magnetic mapping to the ionosphere moves equatorward. Owing to temporal changes in the level of magnetospheric convection, multiple edges can form.

3.2 Model of Embedded Arcs

Haerendel (2007, 2008, 2009, 2010, 2011, 2012a, 2012b) has put forth an extensive model of auroral arcs "embedded" in the convection of the plasma sheet. In the model Earthward flow from the magnetotail sets up a local pressure maximum in the near-Earth magnetotail; in particular a local maximum in PV is set up, where P is the plasma particle pressure and V is the flux-tube volume. The maximum is in both the radial and azimuthal directions, with the azimuthal maximum near local midnight. To drive strong arcs, the plasma beta at the pressure maximum must be somewhat above unity. As plasma and plasma pressure build up on the nightside, the azimuthal pressure gradient pushes the plasma azimuthally around the dipolar obstacle. The magnetospheric plasma is magnetically connected to the ionosphere where ion-neutral collisions oppose the azimuthal flow in the ionosphere, leading to a twist (magnetic stress) of the plasma-sheet magnetic-field lines as the magnetospheric azimuthal flow gets ahead of the ionospheric azimuthal flow. This twist represents a magnetic stress.

At a location where the radial pressure gradient is toward the Earth (tailward of the pressure maximum) it is conjectured that a sharp "ledge" forms in the broad radial pressure gradient, the ledge being a steepened pressure gradient. This is sketched in Fig. 6. The auroral arc magnetically maps to this pressure ledge, with the width of the arc set by the width of the ledge. (Although structuring of the optical signature of the arc to smaller scales caused by Alfvén wave transits and reflections cannot be ruled out (e.g. Lysak 1985; Haerendel 1994).) Because of the sharp pressure gradients at the ledge, an intense upward field-aligned current is driven from the magnetosphere into the arc.

Because of the intense field-aligned current, a field-aligned voltage forms above the ionosphere on the arc field lines: associated with the field-aligned potential drop, the field lines

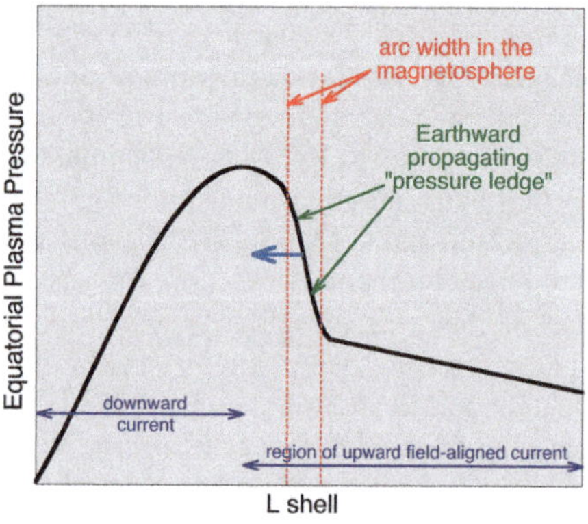

Fig. 6 A sketch of the "pressure ledge" in the plasma sheet in the embedded-arc model of quiescent auroral arcs

are no longer equipotential and there is a partial disconnection of the strong perpendicular electric field in the magnetosphere from a weaker perpendicular electric field in the ionosphere. This allows the magnetosphere's field lines to partially slip relative to the ionospheric flow. This azimuthal slippage reduces the magnetic stress on the magnetospheric field lines in the azimuthal flow in the arc. The azimuthal magnetospheric flow must do work against the opposition of the ionospheric with its ion-neutral collisions and the azimuthal magnetospheric flow must do work to power electron acceleration through the field-aligned voltage drop: the energy for this work comes from both (a) a reduction of the magnetic stress in the magnetosphere and (b) a reduction of the particle pressure in the magnetosphere (reduction of internal energy). (Earlier papers (Haerendel 2007, 2008, 2009) explicitly discussed the release of internal energy (nK_BT thermal energy) from the magnetosphere in the arc, the so-called "auroral pressure valve" (Haerendel 2000), whereas later papers (Haerendel 2010, 2011, 2012a, 2012b) discussed only the magnetic stress release supplying power to the arc.)

In the model it is hypothesized that after the local energy in the arc (and pressure ledge) is depleted, the arc will advect Earthward into the pressure gradient and utilize new magnetic-stress energy and thermal energy to keep the arc powered, tapping stress and pressure as it radially propagates through the magnetospheric plasma. The Earthward migration of the pressure ledge in the magnetosphere corresponds to an equatorward drift of the arc in the atmosphere.

The relief of magnetic stress by the decoupling of the magnetic-field connection to the ionosphere yields an azimuthal acceleration of magnetospheric plasma away from midnight at the leading edge (Earthward edge) of the arc. It is proposed that this azimuthal flow overshoots its equilibrium and begins to flow azimuthally back toward midnight, forming a return flow on the trailing edge of the arc. These forward and backward flows along the arc are consistent with $\underline{E} \times \underline{B}$ drift in a converging perpendicular electric field, i.e. a U-shaped potential around the magnetospheric portion of the arc. In the evening sector, with a vorticity $\underline{\omega} = \nabla \times \underline{v}$ that is northward in the equatorial plane, $\underline{\omega} \bullet \underline{B} > 0$ and hence the arc's vorticity layer in the magnetosphere contains a negative space charge (cf. Eq. (11) of Borovsky and Birn 2014 where the $j \bullet \underline{v}$ term inside the arc is smaller than the $\underline{\omega} \bullet \underline{B}$ term by a factor of $2B/\Delta B$ where B is the equatorial magnetic-field strength and ΔB is the diamagnetic change in the magnetic-field strength across the ledge pressure gradient at the equator).

It is stated (Haerendel 2012a) that the ultimate source of energy for the arc is the magnetospheric pressure gradient in the azimuthal direction.

Fig. 7 Schematic of a U-shaped potential associated with the auroral acceleration region. Magnetic field lines are shown in blue, potential contours in red and electric field vectors in green. Modified after Fig. 1 of Birn et al. (2012)

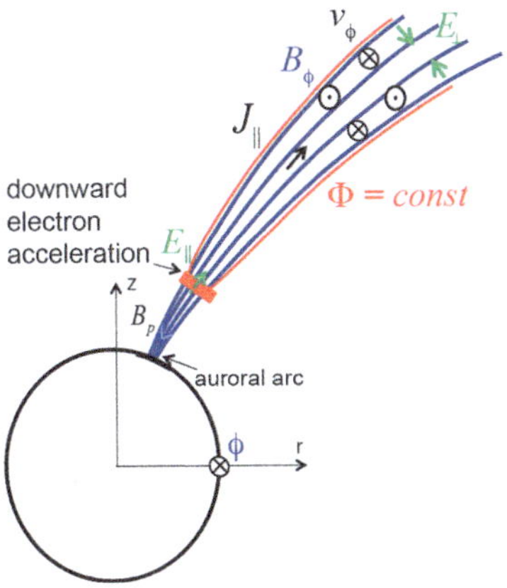

3.3 Arcs Associated with Cross-Tail Current Sheets

The parallel electric field, which is the essence of electron acceleration above the atmosphere in auroral arcs, is part of a U- or S-shaped potential connected with perpendicular electric fields, as illustrated in Fig. 7. While the parallel electric field is confined to the acceleration region the perpendicular electric field maps out into the magnetosphere, although it is not fully established how far. As illustrated in Fig. 7, the converging perpendicular electric field is associated with an $\underline{E} \times \underline{B}$ drift, which would cause a shear in the magnetic field with a B_ϕ component appropriate for the upward field-aligned current in auroral arcs. Thus, the main features connecting the auroral acceleration region with the source region in the magnetosphere are the upward field-aligned current and the perpendicular electric field.

The typical scale size of quiet arcs of ~ 10 km in the ionosphere becomes close to an ion inertia length when mapped into the magnetotail. At this scale, ions are no longer frozen into the magnetic-field lines whereas the electrons are. Thus, the perpendicular electric field is associated with electron $\underline{E} \times \underline{B}$ drift, assuming that the picture is quasi-steady and the electric field maps out into the magnetosphere as a potential field. Since the ions do not (fully) participate in $\underline{E} \times \underline{B}$ drifting, the electron $\underline{E} \times \underline{B}$ drift corresponds to a Hall current, confined to a thin sheet or sheets. This has led several authors (Paschmann et al. 2002; Schindler and Birn 2002; Birn et al. 2004a, 2012; Coroniti and Pritchett 2014) to suggest a connection between auroral arcs and thin current sheets in the magnetotail. As outlined in Sect. 7.7.5 of Paschmann et al. (2002), there are several potential thin current sheet regions in the magnetotail that have locations suitable for an association with auroral arcs: the inner edge of the electron plasma sheet, the thin growth phase current sheet, the near-Earth reconnection site, the plasma sheet boundary layer, and bursty bulk flows. The association of the latter with field-aligned currents and perpendicular electric fields will be discussed in Sect. 5. Here we are particularly interested in the possible source regions of quiet arcs: the inner edge of the electron plasma sheet (cf. Sect. 3.1) and the thin growth phase current sheet.

Extreme current-sheet thinning in the inner magnetotail prior to substorm onset has been documented by observations by many authors (e.g., McPherron et al. 1987; Sergeev et al. 1993; Pulkkinen et al. 1994; Sanny et al. 1994). The main features that support theoretically the association of the thin growth-phase current sheet with the quiet arc are derived from self-consistent models or particle simulations of thin current sheets at and below the ion-inertial scales (e.g., Pritchett and Coroniti 1994; Hesse et al. 1996, 1998): (a) the fact that the current in such current sheets is carried by electron $\underline{E} \times \underline{B}$ flow, and (b) that these sheets are

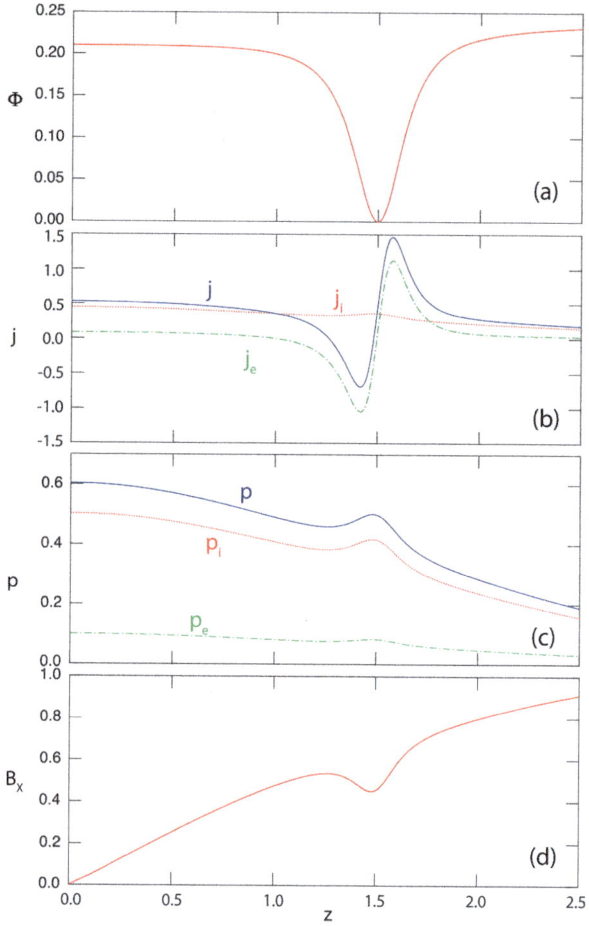

Fig. 8 Vlasov model of a thin double current sheet, suitable for the generation of converging perpendicular electric fields: (**a**) electric potential, (**b**) current density contributions, (**c**) pressure contributions, and (**d**) magnetic field component B_x as functions of z. Modified after Fig. 10 of Birn et al. (2004a)

associated with field-aligned potentials and perpendicular electric fields that could provide the connection to the auroral acceleration region. On the basis of 2-D Vlasov equilibrium theory (Schindler and Birn 2002), Birn et al. (2004a) constructed models of thin current sheets with potentials suitable for the connection to a quasi-static arc. Figure 8 shows an example with a potential that corresponds to the converging perpendicular electric field of a U-shaped auroral potential. In reproducing this figure, we noticed that the labels of p_i and p_e were misplaced in Fig. 10 of Birn et al. (2004a). It is actually common in these thin-current-sheet models that the ion pressure dominates despite the fact that the current is carried primarily by the electrons (in a frame in which the electric field vanishes outside the current sheet).

Since these models are typically two-dimensional, they do not include the field-aligned current that is a necessary ingredient for auroral arcs. However, qualitative arguments based on embedding the 2-D models in a 3-D environment indicate that the azimuthal electron flow may well be associated with field-aligned currents, as illustrated in Fig. 9. In fact, a similar field bending is part of the kinetic structure of the thin current sheet at a near-tail reconnection site and has been confirmed by the quadrupolar B_y field associated with Hall currents in particle simulations (e.g., Sonnerup 1979; Terasawa 1983; Pritchett 2001).

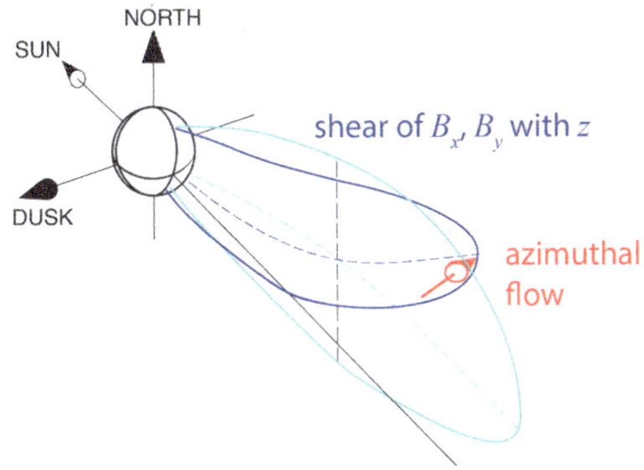

Fig. 9 Schematic of an azimuthal flow causing field-line bending and magnetic shear. Modified from part of Fig. 3.4 of Paschmann et al. (2002). The schematic was originally used to illustrate the effects of plasma flow but can also be applied to electron flow if the ions are decoupled

An auroral arc and precipitation model closely related to properties of the plasma sheet and current sheet thinning is based on so-called isotropy boundaries (Sergeev et al. 1983, 2012). The basic concept for such boundaries is the isotropy or lack of isotropy of ions and electrons observed by low orbiting spacecraft in the auroral region. The fact that the loss cone is filled is taken as an indication for effective scattering of particles into the loss cone and has been associated with chaotic particle motion in the plasma sheet (Buechner and Zelenyi 1986, 1989; Sergeev et al. 1990). (Note that in the magnetotail plasma sheet there is an absence of waves that would pitch angle scatter ions (e.g. Usanova et al. 2012) and, unless reconnection is ongoing, an absence of waves that would pitch-angle scatter electrons Zhang et al. 1999.) The relevant parameter for chaotic motion is the ratio $\kappa^2 = R_c/r_g$, where R_c is the radius of curvature of the magnetic field in the center of the tail current sheet and r_g is the (90°-pitch-angle) gyroradius in the center field. When this ratio κ^2 is sufficiently small particle motion is strongly nonadiabatic and scattering can easily fill a loss cone caused by auroral precipitation. It is obvious that more energetic particles with larger gyroradii are more easily nonadiabatic, and ions are less adiabatic than electrons. The dependence on the radius of curvature implies that nonadiabaticity increases for thinner current sheets.

The radius-of-curvature scattering mechanism has been employed to categorize types of precipitation based on low orbiting spacecraft (e.g., Newell et al. 1996b; Wing et al. 2005), to identify ground based precipitation boundaries (Donovan et al. 2003), and to improve mapping between the plasma sheet and the ionosphere (Shevchenko et al. 2010; Kubyshkina et al. 2011). Using ground, low altitude, and plasma sheet THEMIS observations, Sergeev et al. (2012) used an adaptive magnetic-field model to identify the magnetospheric location of the pre-breakup (growth-phase) arc. Figure 10 presents the comparison of optical and low orbiting energetic particle data for parallel and perpendicular energy fluxes, illustrating the isotropy boundaries and the good agreement between the isotropy boundary of the energetic electrons with the optical arc location. Comparison with the adaptive magnetic-field model implies an arc location at the most-earthward edge of a thin cross-tail current sheet.

In separate computational studies Yang et al. (2013) and Hsieh and Otto (2014) examined the formation of a very thin cross-tail current sheet based on adiabatic convection from the midnight tail region to the dayside and its mapping into the auroral ionosphere. In both studies, the imposed convection generated a strong magnetic-flux depletion in the midnight sector causing an equatorward expansion of all auroral signatures/proxies. Figure 11 (top) from Hsieh and Otto (2014) presents the field-aligned current pattern mapped into the ionosphere at the start of the simulation, corresponding to the currents from the Tsyganenko model used to initialize the simulation. The bottom left of Fig. 11 shows the field-aligned currents 45 minutes into the simulation with the maximum current density almost an order

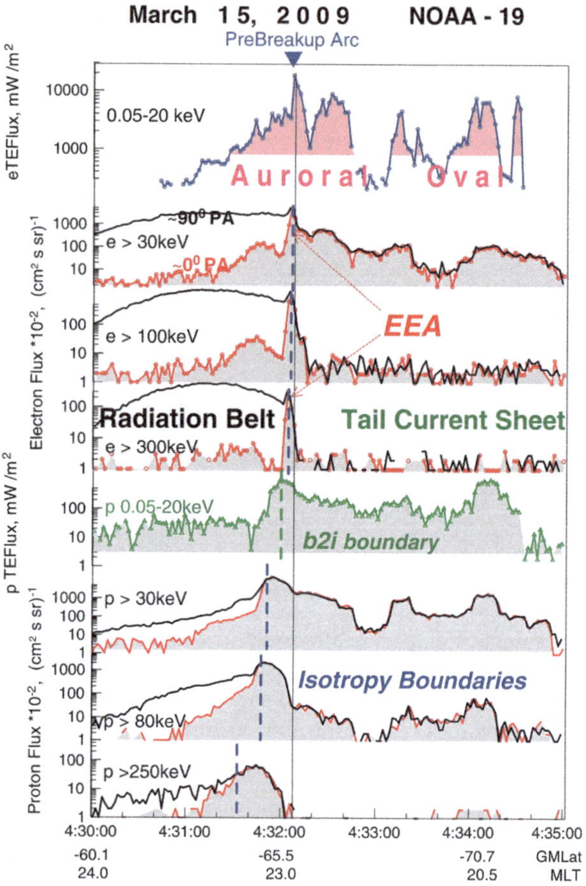

Fig. 10 Comparison of auroral energy flux and parallel (red) and perpendicular (black) particle flux for different electron and ion energy ranges from Sergeev et al. (2012)

of magnitude higher, much narrower, and by about 2 degrees further equatorward. Using a value of $\kappa = 8$ the yellow and black solid lines show the mapping of the 100-keV electron and 30-keV ion isotropy boundaries with the electron isotropy boundary at the equatorward edge of the thin region 1 field-aligned current. The bottom right of Fig. 11 shows the equatorial magnetic-field strength mapped as a reference to the two isotropy boundaries. While Yang et al. (2013) used the modified Rice Convection Model, Hsieh and Otto (2014) used a mesoscale MHD simulation. However, results are quantitatively very similar regarding field-aligned current evolution, motion, and isotropy boundary as a proxy for precipitation energy. Note, that the mechanism for the field aligned current evolution is the same as illustrated in Fig. 9.

3.4 Stationary Alfvén-Wave Arc Picture

Auroral arcs are a signature of magnetosphere-ionosphere coupling via fields and particles. For quasi-static arcs, the relevant timescales are from minutes to tens of minutes. This is in the domain of Alfvén waves, which are defined as having a characteristic frequency ω much less than the relevant ion cyclotron frequencies. In principle, ω can extend to zero, meaning quasi-static fields can be treated as Alfvén waves after all transients have died out. However the question then arises as to why a wave treatment is necessary. The answer is

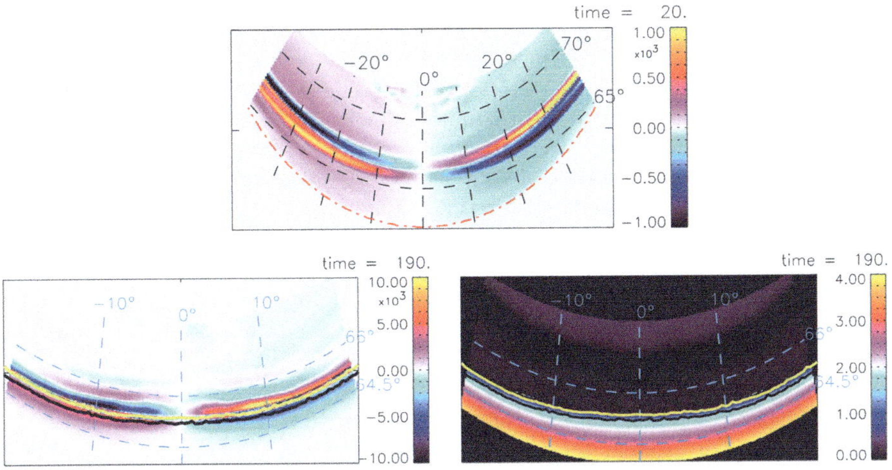

Fig. 11 Mapping of field-aligned currents into the ionosphere at the beginning of the simulation (top) and after 45 minutes (bottom left). Yellow and black traces show the isotropy boundaries for 100 keV electron s and 30 keV ions. The lower left plot represents the equatorial tail magnetic field magnitude mapped into the ionosphere (1 unit corresponding to 20 nT). From Hsieh and Otto (2014)

Fig. 12 In a non-drifting reference frame (e.g. channels of enhanced conductivity in the ionosphere) the current sheet coupling the magnetosphere and ionosphere can be susceptible to spatial structuring of electron density (plotted here) and field-aligned energy powered by advection of magnetic energy. After Knudsen (1996)

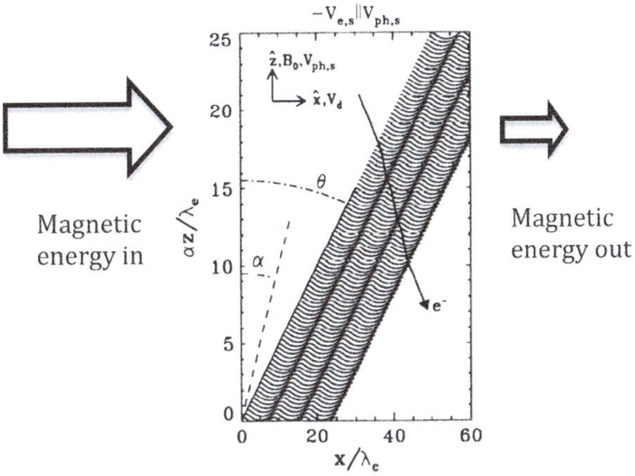

that various wave-like behaviors arise in quasi-static situations where sources of perturbed fields or currents move relative to the background plasma (Jupiter's moon Io is an example), or equivalently when plasma convects past stationary sources.

The term "stationary Alfvén wave" was used by Maltsev et al. (1977) and Mallinckrodt and Carlson (1978), who considered plasma convection across a localized ionospheric conductivity enhancement and showed that the resulting field-aligned currents propagate upward as a localized shear Alfvén wave with a wavefront that is inclined with respect to the magnetic field by the small Mach angle $\alpha = \text{Arctan}(v_{\text{drift}}/v_{\text{A}})$. A stationary Alfvén wave is depicted in Fig. 12. (Such structures can also be described as "Alfvén wings" (e.g. Drell et al. 1965).) This wave can reflect in the conjugate ionosphere, however the net effect of convection and the relatively slow Alfvén wave speed (relative to c) is to displace the return wave in latitude relative to its source (Gurnett and Goertz 1981). Multiple bounces can occur, and are put forward by Maltsev et al. (1977) and Mallinckrodt and Carlson (1978) as a possible explanation of multiple auroral arcs. Once a quasi-static steady state is reached, the Alfvénic character of the structures is maintained by polarization currents resulting from plasma convection across stationary field structures, which are then closed via field-aligned currents.

These original stationary-Alfvén-wave models do not provide a self-consistent mechanism for the creation of parallel electric fields and particle acceleration. Furthermore, auroral arcs and multiple arc systems are observed to form on timescales of seconds, much faster than the inter-hemispheric Alfvén-wave bounce period of many minutes. Lysak (1991) explored wave reflections within the topside ionospheric Alfvén resonator, which has a fundamental resonant frequency of the order of 1 Hz, as a means of reducing the growth time and creating small-scale structure through multiple interfering reflections. Lysak and Song (2002) argued that the energy for such structures derives from a reduction of Joule heating associated with the large-scale convection electric field; in the case of quiet-time background convection this is insufficient to power auroral arcs.

Knudsen (1996) developed a model of stationary inertial Alfvén (SIA) waves based on a two-fluid (ion and electron) model that included several new effects including (a) finite electron inertia allowing finite $E_{\|}$, (b) a source of $E_{\|}$ of the order of $E_c \delta B/B_0$ arising through the interaction between the perturbed magnetic field δB and the convection electric field E_c (Seyler 1990), and (c) self-consistent inclusion of advection and associated nonlinearities. Solutions were then sought in the form of stationary wavefronts resulting (after all transients have died away) from propagation along B at speed of the order of (but not necessarily equal to) the Alfvén speed, with a superimposed uniform background convection electric field E_y which is normal to all density and field variations. In other words, the model allows for two-dimensional arc-like structures having a uniform background convection flow in the cross-arc direction (x) and no variation in y. Finnegan et al. (2008) added the effects of finite electron temperature and collisions.

An important additional element of the Knudsen (1996) model is an imposed, large-scale background field-aligned currents sheet, the source of which is unspecified. Within such a sheet, the model admits solutions of in the form of nonlinear spatially-periodic oscillations in density, field-aligned electron energy, field-aligned currents, and electric field (perpendicular and parallel to B). The allowed wavelengths of these structures across B are several to tens of electron inertial lengths λ_e, more than an order of magnitude of larger than the $\lambda_\perp$ values associated with electron acceleration by time-varying inertial Alfvén waves. For example, taking nominal auroral acceleration region densities in the range 10–1000 cm^{-3} at an altitude of 4,000 km gives λ_e mapped to the ionosphere of 0.8–8 km, leading to SIA wavelengths across B of several to tens of km. Unlike the case of time-varying inertial Alfvén waves, electron acceleration by SIA waves is not limited by the Alfvén speed and electrons can reach many times V_A, corresponding to energies much greater than 1 keV. However, the model of Knudsen (1996) cannot produce such energies within the observed field-aligned extent of the auroral acceleration region ($\sim$ 10,000 km). Furthermore, that model does not account for several potentially important effects such as an ionospheric boundary and variations in the background plasma (e.g. density and magnetic field). For this reason, the model does not make predictions that are specific to the terrestrial M-I coupling region, for example the altitude of the electron acceleration region. Addressing these issues will require a proper numerical simulation as opposed to the analytical models used by Knudsen (1996) and Finnegan et al. (2008).

SIA waves can be thought of as the electromagnetic analog of perturbations in a fast-flowing river. They are not standing waves in the sense that there is no sinusoidal variation in time. Rather, they are stationary spatial structures resulting from the flowing fluid. The assumed background field-aligned currents, in addition to being necessary to give rise to the spatially periodic solutions described above, also provides a source of energy, namely advection of magnetic energy, that is comparable to observed values (Knudsen et al. 2011). The concept of magnetic energy as a source of energy for auroral arcs was proposed by Haerendel et al. (1993) for the case of time-varying (shrinking) current sheets.

3.5 Feedback-Instability-Driven Arcs

The feedback instability in magnetosphere-ionosphere (M-I) coupling was proposed as a possible formation mechanism of quiet auroral arcs (Atkinson 1970; Sato and Holzer 1973; Sato 1978): the feedback is between field-aligned currents and the conductivity variations produced by those currents. The simplest M-I coupling model for the feedback instability is described by the magnetohydrodynamic equations for the magnetosphere and the height-integrated two-fluid equations for the ionosphere, where the spatiotemporal scales are assumed to be much longer than the ion gyroradius and gyroperiod. The ionospheric equations consist of the continuity of particles and currents, along with charge neutrality. In the steady state with constant ionospheric density, the latter is simply given by the divergence free condition of the electric current, suggesting a current circuit in the M-I coupling.

To explain the physical mechanism of the feedback instability, let us suppose first that a large-scale $\underline{E} \times \underline{B}$ convection flow is set up in the equilibrium state of the magnetosphere, and the perpendicular electric field $\underline{E}$ is mapped onto the ionosphere where Pedersen and Hall currents are driven by the electric field. The Joule dissipation in the ionosphere is balanced with the input Poynting flux from the magnetosphere. If a density perturbation arises in the ionosphere, it causes a polarization electric field $\underline{E}$ as well as a perturbation of the ionospheric and the field-aligned current $j_{\parallel}$. Continuity of $\underline{E}$ and $j_{\parallel}$ between the ionosphere and the magnetosphere demands propagation of electromagnetic perturbations in the magnetosphere, that is, shear Alfvén waves.

Consider the same process occurring in both hemispheres, or suppose a symmetry or anti-symmetry condition at the magnetic equator. This scenario is supported by the conjugacy of auroral appearance on closed field lines. Then, there exist upward and downward propagating waves carrying $j_{\parallel}$. If the upward and downward components of $j_{\parallel}$, respectively, coincide with the ionospheric density increase and depletion, the perturbation may be enhanced. This is because the electron precipitation increases the plasma density (or conductivity) in the ionosphere, and vice-versa. In contrast, if the electron precipitation carrying the upward $j_{\parallel}$ is in phase with the density depletion, the perturbation may be damped. This situation is possible only when one takes into account the time-evolution of the ionospheric density. Then, the perturbations of density, $\underline{E}$, and $j_{\parallel}$ satisfying the unstable phase relation can grow exponentially in time in the linear regime, leading to the feedback instability in the M-I coupling.

The free energy for the feedback instability comes from reduction of the ionospheric Joule heating (Lysak and Song 2002). As discussed above, the background convection electric field leads to steady Joule heating in the equilibrium state, if the ionospheric density is constant in time. When the ionospheric density perturbation arises, a local reduction of the Joule heating releases energy to the Alfvén waves reflected at the ionosphere, which can be characterized as an over-reflection of the Alfvén waves. The same conclusion can be derived from the energy balance equation for the reduced MHD and the ionospheric equations, where the input Poynting flux provided by the equilibrium-scale electric field and the field-aligned current (inducing the magnetic field) turns out to be the energy source.

The feedback instability has been shown to be excited under realistic parameters when the convection electric field exceeds a critical value related to the stabilization effects of recombination loss or diffusion in the ionosphere. The linear theory predicts the most unstable wave length of fluctuations once the ionospheric and magnetospheric parameters are given. This provides an important implication to the theory of auroral arcs. The feedback instability potentially explains typical spatiotemporal scales of auroral arcs, which are selected by the M-I coupling system itself (rather than external conditions) if the equilibrium state with characteristic parameters are given.

The most unstable fluctuation propagates in the direction of the ionospheric current with a phase speed characterized by the Pedersen and the Hall mobility (Watanabe 2010), while the real frequency is roughly scaled by that of the field line resonance. If the Alfvén velocity is non-uniform, and has a sharp gradient along the field line, a partial reflection of the Alfvén waves enables stronger feedback M-I coupling with higher frequencies and growth rate than those of the entire field line resonance (Lysak 1991). The feedback instability is further enhanced when one introduces the ionization effect of neutrals due to precipitating electrons (Sato 1978).

Morphological features of the feedback instability are described below. Local density enhancements and depletions are generated by the feedback instability in the ionosphere. A pair of upward and downward field aligned current, that is, a local current circuit, is also spontaneously formed, where the upward current $j_{\parallel}$ carried by precipitating electrons generates auroral arcs with density increase. Simultaneously, the electric field perturbation drives the local $\underline{E} \times \underline{B}$ flow with a perpendicular velocity shear not only in the ionosphere but also in the magnetosphere. The sheared flow adjacent to auroral structures may induce a secondary (nonlinear) instability if the primary feedback instability can grow to a large amplitude. Recent nonlinear simulation of the M-I feedback coupling reveals the secondary excitation of the Kelvin-Helmholtz type instability, which leads to roll-up of auroral vortex structures (Watanabe 2010; Watanabe et al. 2016). While direct observational evidence for the feedback instability is rare, recent work analyzing sounding rocket data has shown characteristic consistent with the feedback instability (Cohen et al. 2013).

The above discussions are limited to the simplest model of M-I coupling with uniform magnetic field and the reduced MHD model for the magnetosphere. A variety of extensions of the feedback instability has also been discussed in the literatures, such as the global distribution of arc structures (Miura and Sato 1980; Watanabe and Sato 1988; Watanabe et al. 1993, 1994), introduction of the two-fluid effects (Streltsov and Lotko 2003, 2004, 2008), application to the dipole geometry (Lu et al. 2007, 2008; Hiraki and Watanabe 2011, 2012), fine structure formation (Jia and Streltsov 2014; Watanabe et al. 2016), and gyrokinetic model for the magnetosphere (Watanabe 2014). Sydorenko and Rankin (2017) have argued that height variation of the ion-neutral collision frequency in the ionosphere inhibits the feedback instability; the generality of this conclusion is disputed by Watanabe and Maeyama (2018) and Streltsov and Mishin (2018). Overall, further elaboration is necessary for a complete understanding of ionospheric feedback in the M-I coupled system, including parallel electric field generation and auroral particle acceleration, coupling with pressure perturbations, nonlinear dynamics and turbulence transition, and a realistic three-dimensional geometry that can resolve auroral fine structures.

4 Considerations for Arc and Oval Current System Near the Harang Region

As detailed in Sect. 1.4 of Paschmann et al. (2002), observations near the low-altitude end of the auroral-current circuit suggest that one-dimensional planar symmetry, which is often assumed for quiet auroral arcs, may not hold at specific locations and times, notably in the

Harang region during the substorm growth phase. Likewise, the auroral oval may not be azimuthally symmetric. Here we briefly address the effects of the modified arc and oval symmetry on the auroral current circuit and its magnetospheric generator.

Two basic configurations of the auroral current circuit were proposed by Boström (1964), remarkably enough, at a time when the existence of field-aligned currents had not been confirmed yet (cf. Fig. 4).

Boström's "Type 1" auroral current circuit relies on a pair of filamentary field-aligned currents that are connected to each other by a Cowling channel in the ionosphere driven by a primary electric field in the approximately east-west direction. The Type-1 auroral circuit is powered by a magnetospheric generator dominated by an electric field and current in the azimuthal direction. The ionospheric electrojet includes a Pedersen component that is associated with the primary east-west electric field and a Hall component that is associated with a secondary north-south polarization electric field. Subsequently, the Type-1 circuit was found to be realized on a large scale by the substorm current wedge (McPherron et al. 1973; Birn and Hesse 1991; Scholer and Otto 1991; Keiling et al. 2009) and on smaller scales by the current "wedgelets" (Rostoker 1998) related to high speed plasma flows in the magnetotail (Sergeev et al. 1996, 2000; Birn et al. 2004b; Liu et al. 2013).

Boström's "Type 2" auroral current circuit relies on a pair of sheet field-aligned currents connected to each other by a north-south Pedersen current. The Type 2 circuit is powered by a magnetospheric generator dominated by an electric field and current in the radial direction. The east-west Hall electrojet driven by the primary north-south ionospheric electric field is divergence free. Type-2 circuits are realized by the auroral arc current system (e.g. Elphic et al. 1998) and on auroral-oval spatial scales by the Region 1/Region 2 currents (Iijima and Potemra 1978). As such, Type 2 geometries underlie the one-dimensional aurora-arc and oval models mentioned above.

Since aurora can be quite complex, in particular during substorms, Type-1 and Type-2 configurations can co-exist, possibly embedded in each other. An example is auroral arcs embedded in the westward travelling surge, related to the upward field-aligned current filament of the substorm current wedge (e.g. Marklund et al. 2012). With due care to real-world complexity, the Type-2 configuration can be regarded as typical for the current system of the quiet arc and oval, while the Type-1 configuration appears to describe the auroral current circuit under more dynamic conditions.

The events investigated by Marghitu et al. (2004, 2009, 2011) at the low-altitude end of the auroral current circuit show a different mix of Type-1 and Type-2 circuits, in some sense a hybrid between the two with field-aligned current sheet geometry specific to Type 2 and field-aligned current-electrojet coupling specific to Type 1. Statistical evidence gathered by Jiang et al. (2015) appears to confirm that field-aligned current-electrojet coupling may indeed make a significant contribution to the ionospheric field-aligned current closure of the growth phase auroral arc prior to break up at substorm onset. Other studies (e.g. Zou et al. 2009) indicate that such arcs are often located in the Harang region.

The effect of the topology of the ionospheric field-aligned-current closure currents on the magnetospheric generator end of the auroral current circuit was addressed by Marghitu et al. (2009). In the Harang region poleward of the convection reversal boundary the coupling of the upward field-aligned current to the westward electrojet suggests that there is azimuthal field-aligned current closure in the magnetosphere with a current opposite to the cross-tail current and contributing to the current disruption at the substorm onset. The magnetospheric current of the westward-electrojet loop is also opposite to the dawn-to-dusk magnetospheric electric field, denoting a magnetospheric generator that drives the ionospheric dissipation in the upward current region (Fig. 4). Depending on how much of the field-aligned current closure is achieved by the Pedersen and Hall component of the ionospheric current, energy dissipation can be dominated by Joule heating when Pedersen closure prevails or by accelerated particle heating when Hall closure prevails.

Likewise, equatorward of the convection reversal boundary the coupling of the downward field-aligned current to the eastward electrojet indicates that the magnetospheric segment of this current loop has the same sense as the ring current and may contribute to the evening-sector asymmetry of the ring current as inferred from ground magnetic data (e.g. Newell and Gjerloev 2012) and energetic-neutral-atom imaging (e.g. Brandt et al. 2018). Future work is needed to investigate the relationship between the time variation of the ring current asymmetry and the substorm cycle.

5 What Global and Magnetotail Computer Simulations Say About Driving Auroral Arcs

At first, the topic of this section appears to be an oxymoron. In Sect. 3.3 we have argued that the arc size investigated in this review paper involves scales short enough for a breakdown of the frozen-in condition when mapped to the presumed magnetotail source region, implying a separation of ion and electron motion. However, the two important elements in arc generation are the establishment of field-aligned currents that connect the source region with the arc dissipation region and the generation of perpendicular electric fields that allow the closure through the parallel potentials essential for the auroral acceleration region. These elements are essential ingredients not only on electron scales but also on larger ion or fluid scales treated in MHD simulations, such that conclusions from the MHD simulations may also apply to shorter scale mechanisms. Arc models of the stationary Alfvén waves and the feedback instability are the examples that relate the MHD dynamics to auroral fine structuring.

In this section we address mechanisms of field-aligned current generation on the basis of local magnetotail simulations, which are confined to the magnetotail source region and involve a detailed study of source mechanisms, and through global simulations, which also include a self-consistent coupling with the inner magnetosphere and ionosphere.

5.1 Local Magnetotail Simulations

The details described in this section are largely based on a simulation (Birn et al. 2011) that models the onset and dynamic evolution of magnetic reconnection in the magnetotail together with the generation of flow bursts and dipolarization fronts, limited to a region tailward of about $x = -7.5\ R_E$. This and similar simulations have been used to investigate field-aligned current generation applicable to the substorm current wedge (SCW) as a whole (Birn and Hesse 1991; Scholer and Otto 1991; Birn et al. 2004b; Keiling et al. 2009), as well as to individual flow bursts, which may contribute to the SCW as "wedgelets" (Liu et al. 2013; Birn and Hesse 2013, 2014).

The main mechanism is illustrated in Fig. 13, modified after Birn et al. (2004b). The vorticity associated with a flow from the magnetotail, its braking and diversion, causes buildup of magnetic shear or twist, that is, field-aligned current. Figure 13 illustrates only the twist on the western or duskward edge, which causes the buildup of outward current. A similar but opposite twist is generated at the downward edge, causing earthward field-aligned current, thus completing the standard current system of the substorm current wedge. However, from

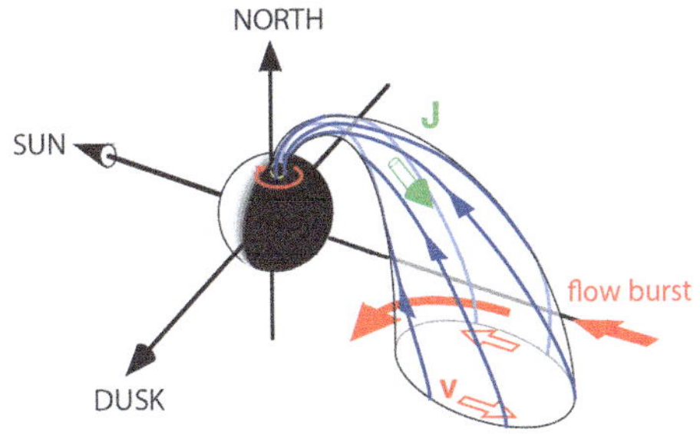

Fig. 13 Mechanism of field-aligned current generation or intensification by flow vorticity. Modified after Birn et al. (2004b)

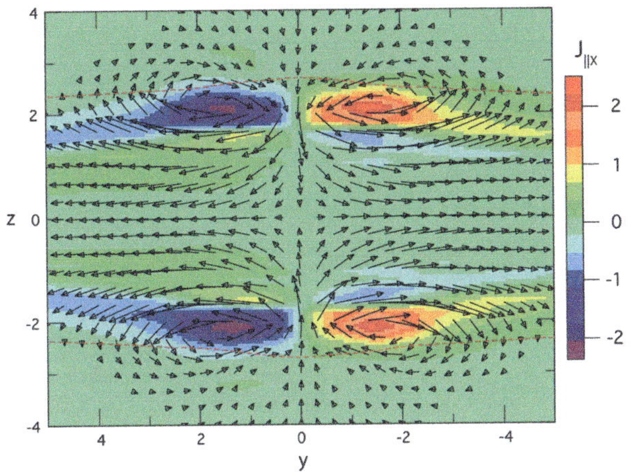

Fig. 14 Field-aligned currents and $\underline{E} \times \underline{B}$ flow vectors near the inner boundary of an MHD simulation of tail reconnection, dipolarization, and flow bursts (Birn et al. 2011). Color indicates the x component of $\underline{J}_{\|}$ with red indicating earthward currents and blue tailward currents

an auroral-arc point of view the outward current, corresponding to precipitating electrons, is of primary interest.

It should be noted that this fluid picture can also be applied on smaller scales to the electron fluid, when ions are no longer frozen in the fluid but electrons are. In that case, that difference between electron and ion flow represents a Hall current, such that Hall current layers may be considered as possible source regions for field-aligned currents on sub-ion scales (e.g., Birn et al. 2012; Coroniti and Pritchett 2014) as discussed in Sect. 3.1.

The buildup of field-aligned currents by flow shear includes self-consistently another crucial element for auroral arcs. The flow, which on small scales is carried only by electrons, is associated with a converging perpendicular electric field, which is necessary for the closure through parallel potentials. (Because there is a nonzero vorticity in the perpendicular-to-$\underline{B}$ flow $\nabla_\perp \times \underline{v}_\perp \neq 0$ with $\underline{v}_\perp = -\underline{E}_\perp \times \underline{B}/B^2$, vector identities yield $(\nabla_\perp \bullet \underline{E}_\perp)\underline{B}/B^2 \neq 0$.) This flow and perpendicular electric field exist not only near the equatorial plane, where the driver might be, but extends to higher latitudes. This is illustrated by Fig. 14, which shows the field-aligned current near the inner Earthward boundary of the magnetotail simulation of Birn et al. (2011) together with $\underline{E} \times \underline{B}$ velocity vectors in the y, z plane.

So far, we have discussed the flow pattern as a mechanism of build-up of field-aligned currents. In principle, if ionospheric dissipation and magnetospheric relaxation were neglected, the field-aligned currents would keep flowing without further need of a flow driver. In that case, it might not be possible to infer the field-aligned currents from a flow pattern and it is common to deduce them from force balance and current continuity via the Vasyli-

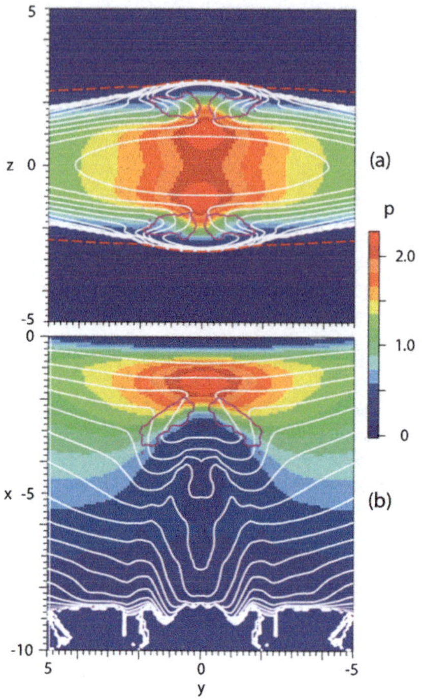

Fig. 15 (**a**) Plasma pressure (color) and contours of constant flux tube volume V (white), defined by (2), at $x = -1$, (**b**) same at $z = 0$, evaluated at $t = 130$. Purple contours indicate enhanced field-aligned currents evaluated from Vasyliunas' equation (2). The red dashed line (top) represents the open/closed boundary. After Birn and Hesse (2013)

unas formula (cf. expression (1)), here generalized to include the inertia term,

$$j_{\parallel} = -(\underline{\boldsymbol{B}}/B) \bullet (\nabla P + \rho dv/dt) \times \nabla V, \tag{2}$$

where V is the volume of a magnetic flux tube of unit flux, defined by

$$V = \int B^{-1} ds, \tag{3}$$

with s being the distance along the flux-tube axis. We note that expression (2) does not describe pressure gradients or inertia as a "driver" in a causal sense but merely an association between the different terms. An evaluation of Eq. (2) on the basis of the Birn et al. (2011) simulation showed that the dominant "driver" of the field-aligned currents is the pressure gradient term, even under dynamic conditions. This is demonstrated by Fig. 15, taken from Birn and Hesse (2013).

5.2 Global MHD Simulations

The connection between magnetotail dynamics and auroral features has been investigated in a number of global MHD simulations, modeling the interaction of the magnetosphere with the driving solar wind as well as the dissipation and current closure in the ionosphere. In the global simulations nightside field-aligned currents resembling auroral currents are identified, the currents are traced and their origins established, and their roles in the global magnetosphere are investigated. Of course global MHD simulations lack the kinetic physics and field-aligned potential drops to fully simulate auroral arcs.

Ge et al. (2011) simulated an event observed by THEMIS and were able to reproduce several of the observed features, such as auroral breakup, poleward expansion and the westward traveling surge. The breakup in particular was found to be related to a flow burst in the tail, producing strong flow shear at the edges of the flow channel, as illustrated in Fig. 13. Raeder et al. (2012) demonstrated a connection between ballooning instability in the tail and auroral beads, which are frequently observed in preexisting arcs prior to substorm onset (e.g., Liang et al. 2008; Henderson 2009). With global simulations Wiltberger et al. (2015) investigated flow bursts from reconnection in a global MHD simulation, confirming the details discussed in Sect. 5.1: an enhancement in B_Z and a decrease in density preceding a peak in the flow velocity, and a reduction of field line entropy. They also showed that the fast flow perpendicular to the magnetic field at the equator was turned into field-aligned flow off the equator.

The magnetotail connection with aurora was investigated in a series of global simulation studies of substorms done by Tanaka's group (Tanaka et al. 2010, 2017; Tanaka 2015; Ebihara and Tanaka 2015a,b, 2016), who did not aim at modeling particular substorms but rather investigated the physical mechanisms governing the interaction of magnetospheric and ionospheric processes. The initial state of the magnetosphere-ionosphere system was set up for stationary convection under northward IMF (Tanaka 2007). This was followed by a southward turning of the imposed IMF. Upward field-aligned currents, used as proxy of electron precipitation, were interpreted to be related to the possible quiet arc formation in the substorm growth phase, the onset arc brightening, and into the development and evolution of the westward traveling surge, observed, e.g., by Kadokura et al. (2002).

Ebihara and Tanaka (2016) identified field-aligned current systems, which they related to N-S arcs and the quiet preonset arc. The quiescent-arc intensification in the substorm growth phase was found to be related to an off-equatorial concentration of a high-pressure region under enhanced convection, occurring when the IMF in the global simulation was changed from weakly northward to strongly southward. This simulation of Tanaka et al. (2017) showed that the current attributed to quiet growth-phase arcs was found not to come from the equatorial region of the plasma sheet, but rather from the flow shear at the plasma sheet-lobe boundary, which is a part of the magnetospheric convection in the growth phase (cf. Sect. 5.1). The growth-phase arc was magnetically connected to the equatorial mid-tail plasma sheet; however, current lines do not follow magnetic-field lines and the current of the arc mapped to the outer edge of the plasma sheet. This current system, belonging to the Region 1 system, is shown in Fig. 16. An ultimate generator location for the current was attributed to the mantle region (Tanaka 1995) and it was noted that the current did not pass through the equatorial region of the plasma sheet.

In contrast to the setup of the quiet arc system, Ebihara and Tanaka (2015a) concluded that the initial brightening of the arc at substorm onset was generated by a near-Earth dynamo. Similar to the picture drawn from the local simulations, this dynamo was found to be related to high-speed earthward flow in the plasma sheet, resulting from the release of the magnetic tension due to reconnection at a near-Earth neutral line (Tanaka et al. 2017). The dynamo was driven by the squeezing of plasma along the field lines by the field-aligned Earthward flow driven by a field-aligned pressure gradient from the high-pressure plasma in the magnetotail plasma sheet in front of the Earthward flow burst in the equatorial plane. The upward field-aligned current attributed to the substorm-onset brightening in the simulation was located equatorward of the quiet auroral arc at a location consistent with observations by Akasofu (1964).

Similar pressure enhancement in the Earthward portion of a collapsing magnetic flux tube was also found in magnetotail simulations (Birn et al. 2004b). The pressure enhancement is associated with a local depression of the magnetic-field strength, corresponding to

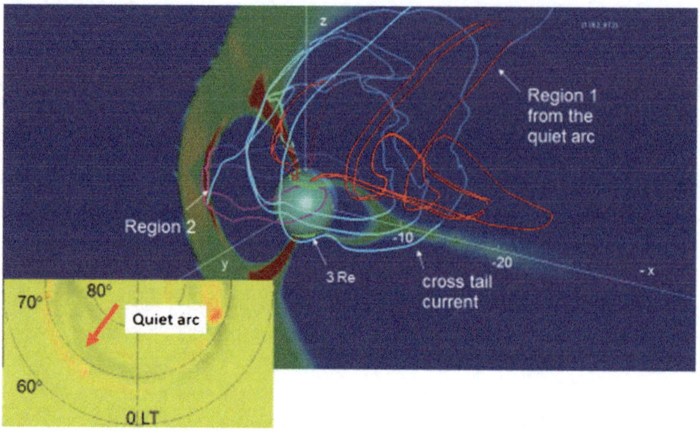

Fig. 16 The current system of the quiet arcs (red). The color shading stands for thermal pressure in the noon-midnight meridian. The left-bottom insert shows ionospheric distribution of the field-aligned current in the growth phase

a current perturbation (counterclockwise in the northern hemisphere; Birn et al. 2004a; Ebihara and Tanaka 2015a) that reverses the cross tail current below the field depression but enhances it above. The downward current below together with the duskward electric field in the collapsing flux tube constitute the dynamo interpreted by Ebihara and Tanaka (2015a) and Tanaka et al. (2017) as the source of the arc brightening.

Birn et al. (2004a) also noted that the perturbed current system is equivalent to the Hall currents associated with the perpendicular electric fields above a V-shaped auroral potential, if driven by electron flows at scales below the characteristic ion scales, as discussed in Sect. 3.1. However, it has not been shown yet that these current perturbations can be setup also at sub ion scales.

The Tanaka et al. (2017) simulation reproduced the two-step onset (initial brightening and the westward traveling surge) observed in a substorm by Kadokura et al. (2002). However, the MHD simulation cannot reproduce the actual enhancement of the auroral luminosity nor the electron precipitation, and because the mesh size in the ionosphere was as large as 100 km, the thin arc structure of the quiet arc could not be resolved. However, the upward field-aligned current, used as proxy of electron precipitation and auroral arc brightening, may be a seed of the thin auroral form generated by the feedback instability (Watanabe 2014).

It is important to note that the investigation of dynamo action, emphasized by Tanaka, is different from, but not necessarily contrary to, the consideration of flow shear and current diversion stressed particularly in Sect. 5.1. It concerns the conversion of energy (primarily from thermal transport) into magnetic energy and Poynting flux, which is the ultimate energy driver of auroral arcs. This energy conversion typically takes place closer to Earth and off the equatorial plane (e.g. Hamrin et al. 2009, 2011; Marghitu et al. 2010) in contrast to the flow shear that might be driven at or near the equatorial plane.

6 Summary and Outstanding Issues

The physical processes that act in the magnetosphere to drive quiescent auroral arcs are not known; several candidate mechanisms that have been put forth in the literature were reviewed in Sect. 2 for high-latitude quiescent arcs and in Sect. 3 for low-latitude quiescent arcs. Predictions for the various driver mechanisms are collected in Table 1. Strengths, weaknesses, and accuracies of the various models can be judged via future evaluations of the predictions in Table 1. Presently, the form of the energy that is extracted from the magnetosphere to power the arcs is not known, and so the impact of auroral arcs on the dynamics of the magnetospheric system is not known.

Much of the uncertainty about the magnetospheric mechanisms that drive low-latitude quiescent arcs comes from an uncertainty about the magnetic connection of visible atmospheric arcs out into the nightside magnetosphere: there are on one hand arc observations and low-altitude measurements of auroral arcs and on the other hand data sets of equatorial magnetospheric measurements, but unambiguously connecting them together has not been possible. This will remain true in the near future making evaluations of Table 1 uncertain.

Spacecraft mission designs have been suggested to connect equatorial magnetospheric measurements to visible auroral forms using energetic electron beams fired from magnetospheric spacecraft into the atmospheric loss cone (Borovsky et al. 1998c; Borovsky 2002; NASA 2003; National Research Council 2012; Delzanno et al. 2016; Borovsky and Delzanno 2019), but such missions have yet to fly.

In advancing the theoretical understanding of auroral-arc driving by the magnetosphere, large-scale kinetic plasma simulations are needed. Specifically, kinetic simulations that encompass the entire auroral-arc region from the equatorial magnetosphere to the resistive ionosphere, that included nonzero-beta plasma dynamics, Alfvén-wave transmission, charge-separation electric fields, collisionless Hall currents, ionospheric Pedersen and Hall currents, ionospheric dissipation, and ionospheric ion and electron outflows would be helpful tools.

Another potential tool for discerning the driver mechanisms of arcs is machine learning and data analytics (Bobra and Ilonidis 2016; Camporeale et al. 2018; McGranaghan et al. 2018), statistically connecting up auroral observations and magnetospheric spacecraft measurements.

Acknowledgements The authors wish to thank Johan De Keyser, Mike Henderson, Michael Hesse, Joseph Lemaire, Bill Lotko, Romain Maggiolo, Noora Partamies, Michael Roth, and Michelle Thomsen for helpful conversations. JEB was supported by the NASA Heliophysics LWS TRT program via grants NNX16AB75G and NNX14AN90G, by the NSF GEM Program via award AGS-1502947, by the NASA Heliophysics Guest Investigator Program via grants NNX17AB71G, by the NSF SHINE program via award AGS-1723416. JB acknowledges support by NASA grants 80NSSC18K0834 and 80NSSC18K1452 and NSF grant 1602655. MME acknowledges support from the Romanian Ministry of Research (PCCDI Grant VESS), the Romanian Space Agency (STAR project 182-OANA) and the Belgian Solar Terrestrial Center of Excellence (STCE). SF is supported by JSPS KAKENHI Grant Number JP17K05671. DJK is supported by the National Sciences and Engineering Council of Canada. RLL is supported by NSF grant AGS-1558134. OM acknowledges support by SIFACIT contract 4000118383/16/I–EF with ESA and STAR EXPRESS contract 119/2017 with Romanian Space Agency. THW is supported by JSPS KAKENHI Grant Number JP16H04086 and JP17H01177. The authors also wish to thank the International Space Science Institute ISSI-Bern for organization of this review and for financial support of the team meeting.

Publisher's Note Springer Nature remains neutral with regard to jurisdictional claims in published maps and institutional affiliations.

Appendix: Some Theories of Auroral Arc Generation that Are in Disuse

Some auroral-arc generator models that were reviewed previously (Borovsky 1993) are no longer actively discussed in the literature. These theories are briefly outlined below.

Arcs Generated at Reconnection X-Lines Atkinson (1992) argued that auroral arcs magnetically connect to reconnection X-lines where they are driven by charge-separation electric fields produced by the differences in electron and ion inertia in the plasma flow through the reconnection diffusion region (i.e. collisionless Hall effects). He further argued (Atkinson et al. 1989) that multiple auroral arcs map to multiple reconnection X-lines in the magnetotail (see also Safargaleev et al. 1997).

Arcs Generated on a Resonant Absorption Layer The idea was discussed (Hasegawa 1976; Goertz 1984) wherein kinetic Alfvén waves could be driven by mode conversion on a resonant layer in the magnetotail and that the kinetic Alfvén waves could accelerate electrons into the atmosphere to produce an auroral arc. The energy source most discussed was MHD surface waves (e.g. Kelvin-Helmholtz waves) on the magnetopause driven by the solar wind, with the evanescent surface waves mode converting on density boundaries in the plasma sheet. One drawback with this idea is that the magnetospheric surface waves are traveling antisunward and they will mode convert to produce kinetic Alfvén waves that are propagating antisunward, opposite to the Alfvén waves observed to accelerate electrons sunward (Earthward) to produce aurora. Chen and Kivelson (1991) evaluated the amplitudes of ULF MHD waves in the lobe (between the magnetopause surface wave and the plasma sheet) and found the power transport to be weak (see also the comments of Keiling (2009)). Note that ideas about mode conversion from solar-wind-driven waves to Alfvén waves that drive the aurora are active (cf. Sect. 2.1), with the solar-wind-driven waves being magnetosonic waves instead of Kelvin-Helmholtz waves.

Arcs Generated by Earthward Ion or Plasma Streams Driving Electrostatic Shocks Ideas were presented (e.g. Kan 1975; Kan and Akasofu 1976) wherein observed Earthward streams of ions (or plasma flows) in the high-latitude portions of the plasma sheet could drive electrostatic shocks (e.g. Swift 1976) near the Earth that stopped the Earthward ion flow and accelerated electrons Earthward to produce auroral arcs. In this picture the power for the auroral arc came from the ram kinetic energy of the Earthward moving ions. The conversion of the ion flow energy into electron beam energy by an electrostatic potential structure has not recently been discussed in the literature, although the conversion of ion-beam energy into Alfvén waves via firehose-type instabilities has been discussed (cf. Sect. 2.1).

Arcs Generated by Earthward Ion Streams Driving Lower-Hybrid Waves Lower-hybrid waves can transfer energy between ions and electrons. A model for high-latitude auroral arcs was suggested in which Earthward streams of ions in the plasma sheet boundary layer could drive a turbulence of lower-hybrid waves, which in turn stochastically accelerated electrons toward the atmosphere to produce auroral arcs (Bingham et al. 1984, 1988; Roy and Lakhina 1985; Bryant 1990; Bryant et al. 1991). Analogous to the prior model, this idea has not been discussed recently in the literature.

Plasma-Sheet Flow Turbulence Inverse Cascading to Form an Arc For MHD turbulence restricted to two dimensions there was a plasma-physics research focus on an inverse cascade of turbulent energy to large spatial scales (Fyfe et al. 1977; Pouquet 1978). In magnetospheric physics several calculations and simulations focused on the idea of two-dimensional MHD or electrostatic $\underline{E} \times \underline{B}$ turbulence in the plasma sheet undergoing an inverse energy cascade to form a coherent structure, mapping to an east-west-aligned auroral arc (Swift 1977, 1979, 1981; Lotko and Schultz 1988; Song and Lysak 1988).

References

S.-I. Akasofu, The development of the auroral substorm. Planet. Space Sci. **12**, 273 (1964)
S.-I. Akasofu, Recent progress in studies of DMSP auroral photographs. Space Sci. Rev. **19**, 169 (1976)
V. Angelopoulos, C.F. Kennell, F.V. Coroniti, R. Pellat, H.E. Spence, M.G. Kivelson, R.J. Walker, W. Baumjohann, W.C. Feldman, J.T. Gosling, C.T. Russell, Characteristics of ion flow in the quiet state of the inner plasma sheet. Geophys. Res. Lett. **20**, 1711 (1993)
V. Angelopoulos, C.F. Kennell, F.V. Coroniti, R. Pellat, M.G. Kivelson, R.J. Walker, C.T. Russell, W. Baumjohann, W.C. Feldman, J.T. Gosling, Statistical characteristics of bursty bulk flow events. J. Geophys. Res. **99**, 21257 (1994)
E.E. Antonova, M.V. Stepanova, M.V. Teltzov, B.A. Tverskoy, Multiple inverted-V structures and hot plasma pressure gradient mechanism of plasma stratification. J. Geophys. Res. **103**, 9317 (1998)
G. Atkinson, Auroral arcs: result of the interaction of a dynamic magnetosphere with the ionosphere. J. Geophys. Res. **75**, 4746 (1970)
G. Atkinson, Review of auroral currents and auroral arcs. J. Geomagn. Geoelectr. **30**, 435 (1978)
G. Atkinson, Mechanism by which merging at X lines causes discrete auroral arcs. J. Geophys. Res. **97**, 1337 (1992)
G. Atkinson, F. Creutzberg, R.L. Gattinger, J.S. Murphree, Interpretation of complicated discrete arc structure and behavior in terms of multiple X lines. J. Geophys. Res. **94**, 5292 (1989)
N.G. Aunai, G. Belmont, R. Smets, Energy budgets in collisionless magnetic reconnection: ion heating and bulk acceleration. Phys. Plasmas **18**, 122901 (2011)
R. Bingham, D.A. Bryant, D.S. Hall, A wave model for the aurora. Geophys. Res. Lett. **11**, 327 (1984)
R. Bingham, D.A. Bryant, D.S. Hall, Auroral electron acceleration by lower-hybrid waves. Ann. Geophys. **6**, 159 (1988)
J. Birn, Three-dimensional equilibria for the extended magnetotail and the generation of field-aligned current sheets. J. Geophys. Res. **94**, 252 (1989)
J. Birn, M. Hesse, The substorm current wedge and field-aligned currents in MHD simulations of magnetotail reconnection. J. Geophys. Res. **96**, 1611 (1991)
J. Birn, M. Hesse, Energy release and conversion by reconnection in the magnetotail. Ann. Geophys. **23**, 3365 (2005)
J. Birn, M. Hesse, The substorm current wedge in MHD simulations. J. Geophys. Res. **118**, 3364 (2013)
J. Birn, M. Hesse, The substorm current wedge: further insights from MHD simulations. J. Geophys. Res. Space Phys. **119**, 3503 (2014)
J. Birn, M. Hesse, G. Haerendel, W. Baumjohann, K. Shiokawa, Flow braking and the substorm current wedge. J. Geophys. Res. **104**, 19895 (1999)
J. Birn, J.F. Drake, M.A. Shay, B.N. Rogers, R.E. Denton, M. Hesse, M. Kuznetsova, Z.W. Ma, A. Bhattacharjee, A. Otto, P.L. Pritchett, Geospace Environment Modeling (GEM) magnetic reconnection challenge. J. Geophys. Res. **106**, 3715 (2001)
J. Birn, K. Schindler, M. Hesse, Thin electron current sheets and their relation to auroral potentials. J. Geophys. Res. **109**, A02217 (2004a)
J. Birn, J. Raeder, Y.L. Wang, R.A. Wolf, M. Hesse, On the propagation of bubbles in the geomagnetic tail. Ann. Geophys. **22**, 1773 (2004b)
J. Birn, R. Nakamura, E. Panov, M. Hesse, Bursty bulk flows and dipolarization in MHD simulations of magnetotail reconnection. J. Geophys. Res. **116**, A01210 (2011)
J. Birn, K. Schindler, M. Hesse, Magnetotail aurora connection: the role of thin current sheets. Geophys. Monogr. Ser. **197**, 337 (2012)
J. Birn, M. Hesse, A. Runov, X.-Z. Zhou, Ion beams in the plasma sheet boundary layer. J. Geophys. Res. **120**, 7522 (2015)
M.G. Bobra, S. Ilonidis, Predicting coronal mass ejections using machine learning methods. Astrophys. J. **821**, 127 (2016)
J.E. Borovsky, Auroral arc thicknesses as predicted by various theories. J. Geophys. Res. **98**, 6101 (1993)
J.E. Borovsky, The Magnetosphere-Ionosphere Observatory (MIO). Los Alamos National Laboratory (2002). https://www.lanl.gov/csse/MIOwriteup.pdf
J.E. Borovsky, J. Birn, The solar-wind electric field does not control the dayside reconnection rate. J. Geophys. Res. **119**, 751 (2014)
J.E. Borovsky, J.L. Delzanno, Space active experiments: the future. Front. Astron. Space Sci. **6**, 31 (2019)
J.E. Borovsky, H.O. Funsten, MHD turbulence in the Earth's plasma sheet: dynamics, dissipation, and driving. J. Geophys. Res. **108**, 1284 (2003)
J.E. Borovsky, J.A. Valdivia, The Earth's magnetosphere: a systems science overview and assessment. Surv. Geophys. **39**, 817 (2018)

J.E. Borovsky, R.C. Elphic, H.O. Funsten, M.F. Thomsen, The Earth's plasma sheet as a laboratory for turbulence in high-beta MHD. J. Plasma Phys. **57**, 1 (1997)
J.E. Borovsky, M.F. Thomsen, R.C. Elphic, The driving of the plasma sheet by the solar wind. J. Geophys. Res. **103**, 17617 (1998a)
J.E. Borovsky, M.F. Thomsen, D.J. McComas, T.E. Cayton, D.J. Knipp, Magnetospheric dynamics and mass flow during the November-1993 storm. J. Geophys. Res. **103**, 26373 (1998b)
J.E. Borovsky, R.A. Greenwald, T.J. Hallinan, J.L. Horwitz, M.C. Kelley, D.M. Klumpar, R.L. Lysak, B.H. Mauk, T.M. Moore, G.D. Reeves, H.J. Singer, M.F. Thomsen, The magnetosphere-ionosphere facility: a satellite cluster in geosynchronous orbit connected to ground-based observatories. Eos **79**(45), F744 (1998c)
J.M. Bosqued, M. Ashour-Abdalla, M. El Alaoui, V. Peroomian, L.M. Zelenyi, C.P. Escoubet, Dispersed ion structures at the poleward edge of the auroral oval: low-altitude observations and numerical modeling. J. Geophys. Res. **98**, 19181 (1993)
R. Boström, A model of the auroral electrojets. J. Geophys. Res. **69**, 4983 (1964)
J.S. Boyd, A.E. Belon, G.J. Romick, Latitude and time variations in precipitated electron energy inferred from measurements of auroral heights. J. Geophys. Res. **76**, 7694 (1971)
P.C. Brandt, S.Y. Hsieh, R. DeMajistre, D.G. Mitchell, ENA imaging of planetary ring currents. Geophys. Monogr. Ser. **235**, 95 (2018)
D.A. Bryant, Auroral electron acceleration. Phys. Scr. T **30**, 215 (1990)
D.A. Bryant, A.C. Cook, Z.-S. Wang, U. de Angelis, C.H. Perry, Turbulent acceleration of auroral electrons. J. Geophys. Res. **96**, 13829 (1991)
J. Buechner, L.M. Zeleny, Deterministic chaos in the dynamics of charged particles near a magnetic field reversal. Phys. Lett. A **118**, 395 (1986)
J. Buechner, L.M. Zelenyi, Regular and chaotic charged particle motion in magnetotaillike field reversals. I—Basic theory of trapped motion. J. Geophys. Res. **94**, 11,821 (1989)
W.J. Burke, J.S. Machuzak, N.C. Maynard, E.M. Basinski, G.M. Erickson, R.A. Hoffman, J.A. Slavin, W.B. Hanson, Auroral ionospheric signatures of the plasma sheet boundary layer in the evening sector. J. Geophys. Res. **99**, 2489 (1994)
E. Camporeale, S. Wing, J. Johnson, *Machine Learning Techniques for Space Weather* (Elsevier, Amsterdam, 2018)
J. Cao, Y. Ma, G. Parks, H. Reme, I. Dandouras, T. Zhang, Kinetic analysis of the energy transport of busty bulk flows in the plasma sheet. J. Geophys. Res. **118**, 313 (2013)
C.A. Cattell, F.S. Mozer, Electric fields measured by ISEE-1 within and near the neural sheet during quiet and active times. Geophys. Res. Lett. **9**, 1041 (1982)
C.C. Chaston, L.M. Peticolas, J.W. Bonnell, C.W. Carlson, R.E. Ergun, J.P. McFadden, R.J. Strangeway, Width and brightness of auroral arcs driven by inertial Alfvén waves. J. Geophys. Res. **198**, 1091 (2003)
S.-H. Chen, M.G. Kivelson, On ultralow frequency waves in the lobes of the Earth's magnetotail. J. Geophys. Res. **96**, 15711 (1991)
I.J. Cohen et al., Auroral Current and Electrodynamics Structure (ACES) observations of ionospheric feedback in the Alfvén resonator and model responses. J. Geophys. Res. Space Phys. **118**, 3288 (2013)
F.V. Coroniti, P.L. Pritchett, The quiet evening auroral arc and the structure of the growth phase near-Earth plasma sheet. J. Geophys. Res. **119**, 1827 (2014)
National Research Council, Magnetosphere-to-ionosphere field-line tracing technology, in *Solar and Space Physics: A Science for a Technological Society* (National Academies Press, Washington, D.C., 2012), p. 333
J. De Keyser, M. Echim, Auroral and sub-auroral phenomena: an electrostatic picture. Ann. Geophys. **28**, 633 (2010)
J. De Keyser, M. Echim, Electric potential differences across auroral generator interfaces. Ann. Geophys. **31**, 251 (2013)
J. De Keyser, R. Maggiolo, M. Echim, Monopolar and bipolar auroral electric fields and their effects. Ann. Geophys. **28**, 2027 (2010)
G.L. Delzanno, J.E. Borovsky, M.F. Thomsen, B.E. Gilchrist, E. Sanchez, Can an electron gun solve the outstanding problem of magnetosphere-ionosphere connectivity? J. Geophys. Res. **121**, 6769 (2016)
I.S. Dmitrienko, Formation of accelerated ion flows in Alfvén disturbances of the magnetotail. Geomagn. Aeron. **51**, 1160 (2011)
E.F. Donovan, B.J. Jackel, I. Voronkov, T. Sotirelis, F. Creutzberg, N.A. Nicholson, Ground-based optical determination of the b2i boundary: a basis for an optical MT-index. J. Geophys. Res. **108**, 1115 (2003)
S.D. Drell, H.M. Foley, M.A. Ruderman, Drag and propulsion of large satellites in the ionosphere: an Alfvén propulsion engine in space. Phys. Rev. Lett. **14**, 171 (1965)
J.P. Eastwood, T.D. Phan, J.F. Drake, M.A. Shay, A.L. Borg, B. Lavraud, M.G.G.T. Taylor, Energy partition in magnetic reconnection in Earth's magnetotail. Phys. Rev. Lett. **110**, 225001 (2013)

R.H. Eather, *Majestic Lights: The Aurora in Science, History, and the Arts* (American Geophysical Union Press, Washington, 1980)
Y. Ebihara, T. Tanaka, Substorm simulation: insight into the mechanisms of initial brightening. J. Geophys. Res. Space Phys. **120**, 7270 (2015a)
Y. Ebihara, T. Tanaka, Substorm simulation: formation of westward traveling surge. J. Geophys. Res. Space Phys. **120**, 10466 (2015b)
Y. Ebihara, T. Tanaka, Substorm simulation: quiet and N-S arcs preceding auroral breakup. J. Geophys. Res. Space Phys. **121**, 1201 (2016)
Y. Ebihara, T. Tanaka, Energy flow exciting field-aligned current at substorm expansion onset. J. Geophys. Res. **122**, 12288 (2017)
M.M. Echim, M. Roth, J. de Keyser, Sheared magnetospheric plasma flows and discrete auroral arcs: a quasi-static coupling model. Ann. Geophys. **25**, 317 (2007)
M.M. Echim, M. Roth, J. de Keyser, Ionospheric feedback effects on the quasi-stationary coupling between LLBL and postnoon/evening discrete auroral arcs. Ann. Geophys. **26**, 913 (2008)
M.M. Echim, R. Maggiolo, M. Roth, J. De Keyser, A magnetospheric generator driving ion and electron acceleration and electric currents in a discrete auroral arc observed by Cluster and DMSP. Geophys. Res. Lett. **36**, L12111 (2009)
M. El-Alaoui, R.L. Richard, M. Ashour-Abdalla, R.J. Walker, M.L. Goldstein, Turbulence in a global magnetohydrodynamic simulation of the Earth's magnetosphere during northward and southward interplanetary magnetic field. Nonlinear Process. Geophys. **19**, 165 (2012)
R.C. Elphic et al., The auroral current circuit and field-aligned currents observed by FAST. Geophys. Res. Lett. **25**, 2033 (1998)
R.C. Elphic, M.F. Thomsen, J.E. Borovsky, D.J. McComas, Inner edge of the electron plasma sheet: empirical models of boundary location. J. Geophys. Res. **104**, 22679 (1999)
R.D. Elphinstone, D. Hearn, J.S. Murphree, L.L. Cogger, Mapping using the Tsyganenko Long magnetospheric model and its relationship to Viking auroral images. J. Geophys. Res. **96**, 1467 (1991)
Y.I. Feldstein, Y.I. Galperin, The auroral luminosity structure in the high-latitude upper atmosphere: its dynamics and relationship to the large-scale structure of the Earth's magnetosphere. Rev. Geophys. **23**, 217 (1985)
S.M. Finnegan, M.E. Koepke, D.J. Knudsen, The dispersive Alfvén wave in the time-stationary limit with a focus on collisional and warm-plasma effects. Phys. Plasmas **15**, 052108 (2008)
D. Fyfe, D. Montgomery, G. Joyce, Dissipative, forced turbulence in two-dimensional magnetohydrodynamics. J. Plasma Phys. **17**, 369 (1977)
Y.I. Galperin, J.M. Bosqued, Stationary magnetospheric convection on November 24, 1981. 1. A case study of "pressure gradient/minimum-B" auroral arc generation. Ann. Geophys. **17**, 358 (1999)
Y.I. Galperin, Y.I. Feldstein, Mapping the precipitation regions to the plasma sheet. J. Geomagn. Geoelectr. **48**, 857 (1996)
Y.I. Galperin, A.V. Volosevich, L.M. Zelenyi, Pressure gradient structures in the tail neutral sheet as "roots of the arcs" with some effects of stochasticity. Geophys. Res. Lett. **19**, 2163 (1992)
Y.S. Ge, J. Raeder, V. Angelopoulos, M.L. Gilson, A. Runov, Interaction of dipolarization fronts within multiple bursty bulk flows in global MHD simulations of a substorm on 27 February 2009. J. Geophys. Res. **116**, A00123 (2011)
D.M. Gillies, D. Knudsen, E. Donovan, B. Jackel, R. Gillies, E. Spanswick, Identifying the 630 nm auroral arc emission height: a comparison of the triangulation, FAC profile, and electron density methods. J. Geophys. Res. **122**, 8181 (2017)
D.M. Gillies, D. Knudsen, R. Rankin, S. Milan, E. Donovan, A statistical survey of the 630.0-nm optical signature of periodic auroral arcs resulting from magnetospheric field line resonances. Geophys. Res. Lett. **45**, 4648 (2018)
C.K. Goertz, Alfvén waves on auroral field lines. Planet. Space Sci. **32**, 1387 (1984)
C.K. Goertz, R.W. Boswell, Magnetosphere-ionosphere coupling. J. Geophys. Res. **84**, 7239 (1979)
H. Grad, Some new variational properties of hydromagnetic equilibria. Phys. Fluids **7**, 1283 (1964)
E.E. Grigorenko, J.-A. Sauvaud, L.M. Zelenyi, Spatial-temporal characteristics of ion beamlets in the plasma sheet boundary layer of magnetotail. J. Geophys. Res. **112**, A05218 (2007)
D.A. Gurnett, C.K. Goertz, Multiple Alfvén wave reflections excited by Io: origin of the Jovain decametric arcs. J. Geophys. Res. **86**, 717 (1981)
M.S. Gussenhoven, D.A. Hardy, W.J. Burke, DMSP/F2 electron observations of equatorward auroral boundaries an their relationship to magnetospheric electric fields. J. Geophys. Res. **86**, 768 (1981)
M.S. Gussenhoven, D.A. Hardy, N. Heinemann, Systematics of the equatorward diffuse auroral boundary. J. Geophys. Res. **88**, 5692 (1983)
G. Haerendel, Acceleration from field-aligned potential drops. Astrophys. J. Suppl. Ser. **90**, 765 (1994)

G. Haerendel, Outstanding issues in understanding the dynamics of the inner plasma sheet and ring current during storms and substorms. Adv. Space Res. **25**, 2379 (2000)

G. Haerendel, Auroral arcs as sites of magnetic stress release. J. Geophys. Res. **112**, A09214 (2007)

G. Haerendel, Auroral arcs as current transformers. J. Geophys. Res. **113**, A07205 (2008)

G. Haerendel, Poleward arcs of the auroral oval during substorms and the inner edge of the plasma sheet. J. Geophys. Res. **114**, A06214 (2009)

G. Haerendel, Equatorward moving arcs and substorm onset. J. Geophys. Res. **115**, A07212 (2010)

G. Haerendel, Six auroral generators: a review. J. Geophys. Res. **116**, A00K05 (2011)

G. Haerendel, A tool for characterizing and evaluating Type II auroral arcs. J. Geophys. Res. **117**, A06214 (2012a)

G. Haerendel, Auroral generators: a survey. Geophys. Monogr. Ser. **197**, 347 (2012b)

G. Haerendel, S. Buchert, C. La Hoz, B. Raaf, E. Rieger, On the proper motion of auroral arcs. J. Geophys. Res. **98**, 6087 (1993)

T.J. Hallinan, *The Distribution of Vorticity in Auroral Arcs*. Geophys. Res. Monogr., vol. 25, (1981), p. 42

M. Hamrin, P. Norqvist, O. Marghitu, S. Buchert, B. Klecker, L.M. Kistler, I. Dandouras, Occurrence and location of concentrated load and generator regions observed by Cluster in the plasma sheet. Ann. Geophys. **27**, 4131 (2009)

M. Hamrin, O. Marghitu, P. Norqvist, S. Buchert, M. Andre, B. Klecker, L.M. Kistler, I. Danouras, Energy conversion regions as observed by Cluster in the plasma sheet. J. Geophys. Res. **116**, A00K8 (2011)

P.J. Hanson, B.G. Harrold, Parallel inhomogeneity and the Alfvén resonance 1. Open field lines. J. Geophys. Res. **99**, 2429 (1994)

B.G. Harrold, C.K. Goertz, R.A. Smith, P.J. Hansen, Resonant Alfvén wave heating of the plasma sheet boundary layer. J. Geophys. Res. **95**, 15039 (1990)

A. Hasegawa, Particle acceleration by MHD surface wave and formation of aurora. J. Geophys. Res. **81**, 5083 (1976)

A. Hasegawa, T. Sato, Generation of field aligned current during substorm, in *Dynamics of the Magnetosphere*, ed. by S.-I. Akasofu (D. Reidel Publishing, Dordrecht, 1979), p. 529

M. Henderson, Observational evidence for an inside-out substorm onset scenario. Ann. Geophys. **27**, 2120 (2009)

M. Hesse, D. Winske, M. Kuznetsova, J. Birn, K. Schindler, Hybrid modeling of the formation of thin current sheets in magnetotail configurations. J. Geomagn. Geoelectr. **48**, 749 (1996)

M. Hesse, D. Winske, J. Birn, On the ion-scale structure of thin current sheets in the magnetotail. Phys. Scr. T **74**, 63 (1998)

Y. Hiraki, T. Watanabe, Feedback instability analysis for dipole configuration with ionospheric and magnetospheric cavities. J. Geophys. Res. **116**, A11220 (2011)

Y. Hiraki, T.-H. Watanabe, Hybrid Alfvén resonant mode generation in the magnetosphere-ionosphere coupling system. Phys. Plasmas **19**, 102904 (2012)

M.-S. Hsieh, A. Otto, The influence of magnetic flux depletion on the magnetotail and auroral morphology during the substorm growth phase. J. Geophys. Res. **119**, 3430 (2014)

T. Iijima, T.A. Potemra, Large-scale characteristics of field-aligned currents associated with substorms. J. Geophys. Res. **83**, 599 (1978)

N. Jia, A.V. Streltsov, Ionospheric feedback instability and active discrete auroral forms. J. Geophys. Res. **119**, 2243 (2014)

F. Jiang, M.G. Kivelson, R.J. Strangeway, K.K. Khurana, R. Walker, Ionospheric flow shear associated with the preexisting auroral arc: a statistical study from the FAST spacecraft data. J. Geophys. Res. Space Phys. **120**, 5194 (2015)

A. Kadokura, A.-S. Yukimatu, M. Ejiri, T. Oguti, M. Pinnock, M.R. Hairston, Detailed analysis of a substorm event on 6 and 7 June 1989, 1, growth phase evolution of nightside auroral activities and ionospheric convection toward expansion phase onset. J. Geophys. Res. **107**, 1479 (2002)

J. Kan, Energization of auroral electrons by electrostatic shock-waves. J. Geophys. Res. **80**, 2089 (1975)

J.R. Kan, S.-I. Akasofu, Energy source and mechanisms for accelerating the electrons and driving the field-aligned currents for the discrete auroral arc. J. Geophys. Res. **81**, 5123 (1976)

T. Karlsson, L. Andersson, M. Gillies, K. Lynch, O. Marghitu, N. Partamies, N. Sivadas, J. Wu, Quiet, discrete arcs—observations. Space Sci. Rev. (2019, submitted)

A. Keiling, Alfvén waves and their roles in the dynamics of the Earth's magnetotail: a review. Space Sci. Rev. **142**, 73 (2009)

A. Keiling, G.K. Parks, H. Reme, I. Dandouras, M. Wilber, L. Distler, C. Owen, A.N. Fazakerley, E. Lucek, M. Maksimovic, N. Cornilleau-Wehrlin, Energy-dispersed ions in the plasma sheet boundary layer and associated phenomena: ion heating, electron acceleration, Alfvén waves, broadband waves, perpendicular electric field spikes, and auroral emissions. Ann. Geophys. **24**, 2685 (2006)

A. Keiling et al., Substorm current wedge driven by plasma flow vortices: THEMIS observations. J. Geophys. Res. **114**, A00C22 (2009)
M.C. Kelley, C.W. Carlson, Observations of intense velocity shear and associated electrostatic waves near an auroral arc. J. Geophys. Res. **82**, 2343 (1977)
J.S. Kim, R.A. Volkman, Thickness of zenithal auroral arc over Fort Churchill, Canada. J. Geophys. Res. **68**, 3187 (1963)
C.A. Kletzing, Electron acceleration by kinetic Alfvén waves. J. Geophys. Res. **99**, 11095 (1994)
L. Knight, Parallel electric fields. Planet. Space Sci. **21**, 741 (1973)
D.J. Knudsen, Spatial modulation of electron energy and density by nonlinear stationary inertial Alfvén waves. J. Geophys. Res. **101**, 10761 (1996)
D.J. Knudsen, E.F. Donovan, L.L. Cogger, B. Jackel, W.D. Shaw, Width and structure of mesoscale optical auroral arcs. Geophys. Res. Lett. **28**, 705 (2001)
D.J. Knudsen, J.K. Burchill, E.F. Donovan, V.M. Uritsky, Advection of magnetic energy as a source of power for auroral arcs. Geophys. Res. Lett. **38**, L24103 (2011)
G. Kremser, A. Korth, S.L. Ullaland, S. Perraut, A. Roux, A. Pedersen, R. Schmidt, P. Tanskanen, Field-aligned beams of energetic electrons (16 keV $< E <$ 80 keV) observed at geosynchronous orbit at substorm onset. J. Geophys. Res. **93**, 14453 (1988)
M. Kubyshkina, V. Sergeev, N. Tsyganenko, V. Angelopoulos, A. Runov, E. Donovan, H. Singer, U. Auster, W. Baumjohann, Time-dependent magnetospheric configuration and breakup mapping during a substorm. J. Geophys. Res. **116**, A00I27 (2011)
J. Lemaire, M. Scherer, Plasma sheet particle precipitation: a kinetic model. Planet. Space Sci. **21**, 281 (1973)
J. Liang, E.F. Donovan, W.W. Liu, B. Jackel, M. Syrjäsuo, S.B. Mende, H.U. Frey, V. Angelopoulos, M. Connors, Intensification of preexisting auroral arc at substorm expansion phase onset: wave-like disruption during the first tens of seconds. Geophys. Res. Lett. **35**, L17S19 (2008)
J. Liang, Y. Shen, D. Knudsen, E. Spanswick, J. Burchill, E. Donovan, e-POP and red line optical observations of Alfvénic aurora. J. Geophys. Res. **124**, 4672 (2019)
J. Liu, V. Angelopoulos, A. Runov, X.-Z. Zhou, On the current sheets surrounding dipolarizing flux bundles in the magnetotail: the case for wedgelets. J. Geophys. Res. Space Phys. **118**, 2000 (2013)
W. Lotko, C.G. Schultz, Internal shear layers in auroral dynamics. Geophys. Monogr. Ser. **44**, 121 (1988)
W. Lotko, B.U.O. Sonnerup, R.L. Lysak, Nonsteady boundary layer flow including ionospheric drag and parallel electric fields. J. Geophys. Res. **92**, 8635 (1987)
G. Lu, M. Brittnacher, G. Parks, D. Lummerzheim, On the magnetospheric source regions of substorm-related field-aligned currents and auroral precipitation. J. Geophys. Res. **105**, 18483 (2000)
J.Y. Lu, R. Rankin, R. Marchand, I.J. Rae, W. Wang, S.C. Solomon, J. Lei, Electrodynamics of magnetosphere-ionosphere coupling and feedback on magnetospheric field line resonances. J. Geophys. Res. **112**, A10219 (2007)
J.Y. Lu, W. Wang, R. Rankin, R. Marchand, J. Lei, S.C. Solomon, I.J. Rae, J.-S. Wang, G.-M. Le, Electromagnetic waves generated by ionospheric feedback instability. J. Geophys. Res. **113**, A05206 (2008)
R. Lundin, D.S. Evans, Boundary layer plasmas as a source for high-latitude, early afternoon, auroral arcs. Planet. Space Sci. **32**, 1389 (1985)
R. Lundin, I. Sandahl, Some characteristics of the parallel electric field acceleration of electrons over discrete auroral arcs as observed from two rocket flights. Tech. Rep. SP-135, European Space Agency (1978)
L.R. Lyons, Generation of large-scale regions of auroral currents, electric potentials and precipitation by the divergence of the convection electric field. J. Geophys. Res. **85**, 17 (1980)
L.R. Lyons, Discrete aurora as the direct result of an inferred high-altitude generating potential distribution. J. Geophys. Res. **86**, 1 (1981)
L.R. Lyons, D.S. Evans, An association between discrete aurora and energetic particle boundaries. J. Geophys. Res. **89**, 2395 (1984)
L.R. Lyons, T.W. Speiser, Evidence for current sheet acceleration in the geomagnetic tail. J. Geophys. Res. **87**, 2276 (1982)
L.R. Lyons, D.S. Evans, R. Lundin, An observed relation between magnetic field aligned electric fields and downward electron energy fluxes in the vicinity of auroral forms. J. Geophys. Res. **84**, 457 (1979)
R.L. Lysak, Auroral electrodynamics with current and Voltage generators. J. Geophys. Res. **90**, 4178 (1985)
R.L. Lysak, Feedback instability of the ionospheric resonant cavity. J. Geophys. Res. **96**, 1553–1568 (1991)
R.L. Lysak, Y. Song, Energetics of the ionospheric feedback interaction. J. Geophys. Res. **107**, 1160 (2002)
R.L. Lysak, Y. Song, Development of parallel electric fields at the plasma sheet boundary layer. J. Geophys. Res. **116**, A00K14 (2011)
R.L. Lysak, Y. Song, T.W. Jones, Propagation of Alfvén waves in the magnetotail during substorms. Ann. Geophys. **27**, 2237 (2009)

R. Maggiolo, M. Echim, C.S. Wedlund, Y. Zhang, D. Fontaine, G. Lointier, J.-G. Trotignon, Polar cap arcs from the magnetosphere to the ionosphere: kinetic modelling and observations by Cluster and TIMED. Ann. Geophys. **30**, 283 (2012)

A.J. Mallinckrodt, C.W. Carlson, Relations between transverse electric fields and field-aligned currents. J. Geophys. Res. **83**, 1426 (1978)

Y.P. Maltsev, W.B. Lyatsky, A.M. Lyatskaya, Currents over the Auroral Arc. Planet. Space Sci. **25**, 53 (1977)

O. Marghitu, B. Klecker, G. Haerendel, J. McFadden, ALADYN: a method to investigate auroral arc electrodynamics from satellite data. J. Geophys. Res. **109**(A11), 305 (2004)

O. Marghitu, T. Karlsson, B. Klecker, G. Haerendel, J. McFadden, Auroral arc and oval electrodynamics in the Harang region. J. Geophys. Res. **114**, A03214 (2009)

O. Marghitu, M. Hamrin, B. Klecker, K. Ronnmak, S. Buchert, L.M. Kistler, M. Andre, H. Reme, Cluster observations of energy conversion regions in the plasma sheet, in *The Cluster Active Archive*, ed. by H. Laakso et al. (Springer, Berlin, 2010)

O. Marghitu, C. Bunescu, T. Karlsson, B. Klecker, On the divergence of the auroral electrojets. J. Geophys. Res. **116**, A00K17 (2011)

G. Marklund, Auoral arc classification scheme based on the observed arc-associated electric field pattern. Planet. Space Sci. **32**, 193 (1984)

G.T. Marklund et al., Cluster multipoint study of the acceleration potential pattern and electrodynamics of an auroral surge and its associated horn arc. J. Geophys. Res. **117**, A10223 (2012)

B.H. Mauk, C.-I. Meng, The aurora and middle magnetospheric processes, in *Auroral Physics*, ed. by C.-I. Meng, M.J. Rycroft, L.A. Frank (Cambridge Press, Cambridge, 1991), p. 223

R.M. McGranaghan, A.J. Manucci, B. Wilson, C.A. Mattmann, R. Chadwick, New capabilities for prediction of high-latitude ionospheric scintillation: a novel approach with machine learning. Space Weather **16**, 1817 (2018)

C.E. McIlwain, Auroral electron beams near the magnetic equator, in *Physics of the Hot Plasma in the Magnetosphere*, ed. by B. Hultqvist, L. Stenflo (Plenum, New York, 1975), p. 91

R.L. McPherron, C.T. Russell, M.P. Aubry, Satellite studies of magnetospheric substorms on August 15, 1968, 9, phenomenological model for substorms. J. Geophys. Res. **78**, 3131 (1973)

R.L. McPherron, A. Nishida, C.T. Russell, Is near-Earth current sheet thinning the cause of substorm onset? in *Quantitative Modeling of Magnetosphere-Ionosphere Coupling Processes*, ed. by Y. Kamide, R.A. Wolf (Kyoto Sangyo University, Kyoto, 1987), p. 252

C.-I. Meng, B. Mauk, C.E. McIlwain, Electron precipitation of evening diffuse aurora and its conjugate electron fluxes near the magnetospheric equator. J. Geophys. Res. **84**, 2545 (1979)

A. Miura, T. Sato, Numerical simulation of global formation of auroral arcs. J. Geophys. Res. **85**, 73 (1980)

T. Motoba, K. Hosokawa, A. Kadodura, N. Sato, Magnetic conjugacy of northern and southern auroral beads. Geophys. Res. Lett. **39**, L08101 (2012)

T. Motoba, S. Ohtani, B.J. Anderson, H. Korth, D. Mitchell, L.J. Lanzerotti, K. Shiokawa, M. Connors, C.A. Kletzing, G.D. Reeves, On the formation and origin of substorm growth phase/onset auroral arcs inferred from conjugate space-ground observations. J. Geophys. Res. **120**, 8707 (2015)

NASA, Sun-Earth Connection Roadmap 2003–2028, p. 92, http://www.dept.aoe.vt.edu/~cdhall/courses/aoe4065/NASADesignSPs/SEC_2003_roadmap_full.pdf (2003)

P.T. Newell, J.W. Gjerloev, SuperMAG-based partial ring current indices. J. Geophys. Res. **117**, A05215 (2012)

P.T. Newell, C.-I. Meng, K.M. Lyons, Suppression of discrete aurorae by sunlight. Nature **381**, 766 (1996a)

P.T. Newell, Y.I. Feldstein, Y.I. Galperin, C.-I. Meng, Morphology of nightside precipitation. J. Geophys. Res. **101**, 10,737 (1996b)

P.T. Newell, T. Sotirelis, S. Wing, Diffuse, monoenergetic, and broadband aurora: the global precipitation budget. J. Geophys. Res. **114**, A09207 (2009)

T.G. Onsager, T. Mukai, The structure of the plasma sheet and its boundary layers. J. Geomagn. Geoelectr. **48**, 687 (1996)

T.G. Onsager, M.F. Thomsen, R.C. Elphic, J.T. Gosling, Model of electron and ion distributions in the plasma sheet boundary layer. J. Geophys. Res. **96**, 20999 (1991)

N. Partamies, M. Syrjasuo, E. Donovan, M. Conners, D. Charrois, D. Knudsen, Z. Kryzanowsky, Observations of the auroral width spectrum at kilometer-scale size. Ann. Geophys. **29**, 711 (2010)

G. Paschmann, S. Haaland, R. Treumann, *Auroral Plasma Physics*. Space Sci. Rev., vol. 103 (2002)

A. Pedersen, C.A. Cattell, C.-G. Falthammar, K. Knott, P.-A. Lindqvist, R.H. Manka, F.S. Mozer, Electric fields I the plasma sheet and plasma sheet boundary layer. J. Geophys. Res. **90**, 1231 (1985)

A. Pouquet, On two-dimensional magnetohydrodynamic turbulence. J. Fluid Mech. **88**, 1 (1978)

P.L. Pritchett, Geospace Environment Modeling magnetic reconnection challenge: simulations with a full particle electromagnetic code. J. Geophys. Res. **106**, 3783 (2001)

P.L. Pritchett, F.V. Coroniti, Convection and the formation of thin current sheets in the near-Earth plasma sheet. Geophys. Res. Lett. **21**, 1587 (1994)
T.I. Pulkkinen, H.E.J. Koskinen, R.J. Pellinen, Mapping of auroral arcs during substorm growth phase. J. Geophys. Res. **96**, 21087 (1991)
T. Pulkkinen, D.N. Baker, D.G. Mitchell, R.L. McPherron, C.Y. Huang, L.A. Frank, Thin current sheets in the magnetotail during substorms: CDAW6 revisited. J. Geophys. Res. **99**, 5793 (1994)
J. Raeder, P. Zhu, Y. Ge, G. Siscoe, Auroral signatures of ballooning mode near substorm onset: open geospace general circulation model simulations. Geophys. Monogr. Ser. **197**, 389 (2012)
H. Rème, C. Aoustin, J.M. Bosqued, I. Dandouras, B. Lavraud, J.A. Sauvaud et al., First multispacecraft ion measurements in and near the Earth's magnetosphere with the identical Cluster Ion Spectrometry (CIS) experiment. Ann. Geophys. **19**, 1303 (2001)
G. Rostoker, On the place of the pseudo-breakup in a magnetospheric substorm. Geophys. Res. Lett. **25**, 217 (1998)
M. Roth, D.S. Evans, J. Lemaire, Theoretical structure of a magnetospheric plasma boundary: application to the formation of discrete auroral arcs. J. Geophys. Res. **98**, 11411 (1993)
M. Roth, J. De Keyser, M. Kuznetsova, Vlasov theory of the equilibrium structure of tangential discontinuities in space plasmas. Space Sci. Rev. **76**, 251 (1996)
M. Roy, G.S. Lakhina, Lower hybrid wave model for aurora. Astrophys. Space Sci. **117**, 111 (1985)
M.S. Ruderman, A.N. Wright, Excitation of resonant Alfvén waves in the magnetosphere by negative energy surface waves on the magnetopause. J. Geophys. Res. **103**, 26573 (1998)
V. Safargaleev, W. Lyatsky, V. Tagirov, Luminosity variations in several parallel auroral arcs before auroral breakup. Ann. Geophys. **15**, 959 (1997)
V.V. Safargaleev, A.E. Kozlovsky, S.V. Osipenko, V.R. Tagirov, Azimuthal expansion of high-latitude auroral arcs. Ann. Geophys. **21**, 1793 (2003)
J.C. Samson, L.L. Gogger, Q. Pao, Observations of field line resonances, auroral arcs, and auroral vortex structures. J. Geophys. Res. **101**, 17373 (1996)
P.B. Sandford, Variations of auroral emissions with time, magnetic activity and the solar cycle. J. Atmos. Terr. Phys. **30**, 1921 (1968)
J. Sanny, R.L. McPherron, C.T. Russell, D.N. Baker, T.I. Pulkkinen, A. Nishida, Growth-phase thinning of the near-Earth current sheet during the CDAW-6 substorm. J. Geophys. Res. **99**, 5805 (1994)
T. Sato, A theory of quiet auroral arcs. J. Geophys. Res. **83**, 1042 (1978)
T. Sato, T.E. Holzer, Quiet auroral arcs and electrodynamic coupling between the ionosphere and the magnetosphere, 1. J. Geophys. Res. **78**, 7314 (1973)
T. Sato, T. Iijima, Primary sources of large-scale Birkeland currents. Space Sci. Rev. **24**, 347 (1979)
N. Sato, T. Nagaoka, K. Hashimoto, T. Saemundsson, Conjugacy of isolated auroral arcs and nonconjugate auroral breakups. J. Geophys. Res. **103**, 11641 (1998)
K. Schindler, J. Birn, Magnetospheric physics. Phys. Rep. **47**, 109 (1978)
K. Schindler, J. Birn, Models of two-dimensional embedded thin current sheets from Vlasov theory. J. Geophys. Res. **107**, SMP20 (2002)
M. Scholer, A. Otto, Magnetotail reconnection: current diversion and field-aligned currents. Geophys. Res. Lett. **18**, 7331 (1991)
V.A. Sergeev, E.M. Sazhina, N.A. Tsyganenko, J.A. Lundblad, F. Soraas, Pitch-angle scattering of energetic protons in the magnetotail current sheet as the dominant source of their isotropic precipitation into the nightside ionosphere. Planet. Space Sci. **31**, 1147 (1983)
V.A. Sergeev, P. Tanskanen, K. Mursula, A. Korth, R.C. Elphic, Current sheet thickness in the near-Earth plasma sheet during substorm growth phase. J. Geophys. Res. **95**, 3819 (1990)
V.A. Sergeev, D.G. Mitchell, C.T. Russell, D.J. Williams, Structure of the tail plasma/current sheet at 11 R_E and its changes in the course of a substorm. J. Geophys. Res. **98**, 17345 (1993)
V.A. Sergeev, R.J. Pellinen, T.I. Pulkkinen, Steady magnetospheric convection: a review of recent results. Space Sci. Rev. **75**, 551 (1996)
V.A. Sergeev, J.-A. Sauvaud, D. Popescu, R.A. Kovrazhkin, K. Liou, P.T. Newell, M. Brittnacher, G. Parks, R. Nakarnura, T. Mukai, G.D. Reeves, Multiple-spacecraft observation of a narrow transient plasma jet in the Earth's plasma sheet. Geophys. Res. Lett. **27**, 851 (2000)
V. Sergeev, Y. Nishimura, M. Kubyshkina, V. Angelopoulos, R. Nakamura, H. Singer, Magnetospheric location of the equatorial prebreakup arc. J. Geophys. Res. **117**, A01212 (2012)
C.E. Seyler, A mathematical model of the structure and evolution of small-scale discrete auroral arcs. J. Geophys. Res. **95**, 17199 (1990)
I.G. Shevchenko, V. Sergeev, M. Kubyshkina, V. Angelopoulos, K.H. Glassmeier, H.J. Singer, Estimation of magnetosphere-ionosphere mapping accuracy using isotropy boundary and THEMIS observations. J. Geophys. Res. **115**, A11206 (2010)

R.A. Smith, C.K. Goertz, W. Grossmann, Thermal catastrophe in the plasma sheet boundary layer. Geophys. Res. Lett. **13**, 1380 (1986)

Y. Song, R.L. Lysak, Turbulent generation of auroral currents and fields—a spectral simulation of two-dimensional MHD turbulence. Geophys. Monogr. Ser. **44**, 197 (1988)

Y. Song, R.L. Lysak, Displacement current and the generation of parallel electric fields. Phys. Rev. Lett. **96**, 145002 (2006)

B.U.O. Sonnerup, Magnetic field reconnection, in *Solar System Plasma Physics*, vol. III, ed. by L.T. Lanzerotti, C.F. Kennel, E.N. Parker (North-Holland, New York, 1979), p. 45

B.U.O. Sonnerup, Theory of the low-latitude boundary layer. J. Geophys. Res. **85**, 2017 (1980)

K. Stasiewicz, Generation of magnetic-field aligned currents, parallel electric field, and inverted-V structures by plasma pressure inhomogeneities in the magnetosphere. Planet. Space Sci. **9**, 1037 (1985)

M.V. Stepanova, E.E. Antonova, G. Stanev, N. Bankov, N.V. Isaev, Study of stratification of magnetospheric convection using Intercosmos-Bulgaria-1300 electric field observations. Adv. Space Res. **31**, 1419 (2003)

M. Stepanova, V. Pinto, J.A. Valdivia, E.E. Antonova, Spatial distribution of the eddy diffusion coefficients in the plasma sheet during quiet time and substorms from THEMIS satellite data. J. Geophys. Res. **116**, A00I24 (2011)

R.J. Strangeway, The relationship between magnetospheric processes and auroral field-aligned current morphology. Geophys. Monogr. Ser. **197**, 355 (2012)

A.V. Streltsov, W. Lotko, Small-scale electric fields in the downward auroral current channels. J. Geophys. Res. **108**, 1289 (2003)

A.V. Streltsov, W. Lotko, Multiscale electrodynamics of the ionosphere-magnetosphere system. J. Geophys. Res. **109**, A09214 (2004)

A.V. Streltsov, W. Lotko, Coupling between density structures, electromagnetic waves and ionospheric feedback in the auroral zone. J. Geophys. Res. **113**, A05212 (2008)

A.V. Streltsov, E.V. Mishin, On the existence of ionospheric feedback instability in the Earth's magnetosphere-ionosphere system. J. Geophys. Res. **123**, 8951 (2018)

D.W. Swift, An equipotential model for auroral arcs 2. Numerical solutions. J. Geophys. Res. **81**, 3935 (1976)

D.W. Swift, Turbulent generation of electrostatic fields in the magnetosphere. J. Geophys. Res. **82**, 5143 (1977)

D.W. Swift, Mechanisms for the discrete aurora—a review. Space Sci. Rev. **22**, 35 (1978)

D.W. Swift, On the structure of auroral arcs: the results of numerical simulations. J. Geophys. Res. **84**, 469 (1979)

D.W. Swift, Numerical simulations of the generation of electrostatic turbulence in the magnetotail. J. Geophys. Res. **86**, 2273 (1981)

D. Sydorenko, R. Rankin, The stabilizing effect of collision-induced velocity shear on the ionospheric feedback instability in Earth's magnetosphere. Geophys. Res. Lett. **44**, 6534 (2017)

K. Takahashi, E.W. Hones, ISEE 1 and 2 observations of ion distributions at the plasma sheet-tail lobe boundary. J. Geophys. Res. **93**, 8558 (1988)

T. Tanaka, Generation mechanisms for magnetosphere—ionosphere current systems deduced from a three-dimensional MHD simulation of the solar wind-magnetosphere-ionosphere coupling processes. J. Geophys. Res. **100**, 12057 (1995)

T. Tanaka, Magnetosphere-ionosphere convection as a compound system. Space Sci. Rev. **133**, 1 (2007)

T. Tanaka, Substorm auroral dynamics reproduced by advanced global magnetosphere-ionosphere (M-I) coupling simulation. Geophys. Monogr. Ser. **215**, 177 (2015)

T. Tanaka, A. Nakamizo, A. Yoshikawa, S. Fujita, H. Shinagawa, H. Shimazu, T. Kikuchi, K.K. Hashimoto, Substorm convection and current system deduced from the global simulation. J. Geophys. Res. **115**, A05220 (2010)

T. Tanaka, Y. Ebihara, M. Watanabe, M. Den, S. Fujita, T. Kikuchi, K.K. Hashimoto, R. Kataoka, Global simulation study for the time sequence of events leading to the substorm onset. J. Geophys. Res. Space Phys. **122**, 6210 (2017)

T. Terasawa, Hall current effect on tearing mode instability. Geophys. Res. Lett. **10**, 475 (1983)

N.A. Tsyganenko, Global quantitative models of the geomagnetic field in the cislunar magnetosphere for different disturbance levels. Planet. Space Sci. **35**, 1347 (1987)

B.A. Tverskoi, Magnetosphere-ionosphere interaction and polar auroras. Sov. Phys. Usp. **26**, 383 (1983)

E. Tyler, C. Cattell, S. Thaller, J. Wygant, C. Gurgiolo, M. Goldstein, C. Mouikis, Partitioning of integrated energy fluxes in four tail reconnection events observed by Cluster. J. Geophys. Res. Space Phys. **121**, 11798 (2016)

M.E. Usanova, I.R. Mann, J. Bortnik, L. Shao, V. Angelopoulos, THEMIS observations of electromagnetic ion cyclotron wave occurrence: dependence on AE, SYMH, and solar wind dynamic pressure. J. Geophys. Res. **117**, A10218 (2012)

A. Vaivads, M. Andre, S. Buchert, A.I. Eriksson, A. Olsson, J.-E. Wahlund, P. Janhunen, G. Marklund, L.M. Kistler, C. Mouikis, D. Winningham, A. Fazakerley, P. Newell, What high altitude observations tell us about the auroral acceleration: a Cluster/DMSP conjunction. Geophys. Res. Lett. **30**, 1106 (2003)

V.M. Vasyliunas, Mathematical models of magnetospheric convection and its coupling to the ionosphere, in *Particles and Fields in the Magnetosphere*, ed. by B.M. McCormac (D. Reidel, Dordrecht, 1970), p. 60

Z. Voros, W. Baumjohann, R. Nakamura, M. Volwerk, A. Runov, T.L. Zhang, H.U. Eichelberger, R. Treumann, E. Georgescu, A. Balogh, B. Klecker, H. Reme, Magnetic turbulence in the plasma sheet. J. Geophys. Res. **109**, A11215 (2004)

Z. Voros, W. Baumjohann, R. Nakamura, A. Runov, M. Volwek, T. Takada, E.A. Lucek, H. Reme, Spatial structure of plasma flow associated turbulence in the Earth's plasma sheet. Ann. Geophys. **23**, 13 (2007)

T.-H. Watanabe, Feedback instability in the magnetosphere-ionosphere coupling system: revisited. Phys. Plasmas **17**, 022904 (2010)

T.-H. Watanabe, A unified model of auroral arc growth and electron acceleration in the magnetosphere-ionosphere coupling. Geophys. Res. Lett. **41**, 6071 (2014)

T.-H. Watanabe, S. Maeyama, Unstable eigenmodes of the feedback instability with collision-induced velocity shear. Geophys. Res. Lett. **45**, 10043 (2018)

K. Watanabe, T. Sato, Geophys. Res. Lett. **15**, 717 (1988)

T. Watanabe, H. Oya, K. Watanabe, T. Sato, Comprehensive simulation study on local and global development of auroral arcs and field-aligned potentials. J. Geophys. Res. **98**, 21391 (1993)

T. Watanabe, H. Oya, K. Watanabe, T. Sato, Correction. J. Geophys. Res. **99**, 6151 (1994)

T.-H. Watanabe et al., Generation of auroral turbulence through the magnetosphere-ionosphere coupling. New J. Phys. **18**, 125010 (2016)

C.E.J. Watt, R. Rankin, Do magnetospheric shear Alfvén waves generate sufficient electron energy flux to power the aurora? J. Geophys. Res. **115**, A07224 (2010)

C.Q. Wei, B.U.O. Sonnerup, W. Lotko, Model of the low-latitude boundary layer with finite field-aligned potential drops and nonconstant mapping factors. J. Geophys. Res. **101**, 21463 (1996)

D.R. Weimer, C.K. Goertz, D.A. Gurnett, N.C. Maynard, J.L. Burch, Auroral zone electric fields from DE 1 and 2 at magnetic conjunctions. J. Geophys. Res. **90**, 7479 (1985)

J.M. Weygand, M.G. Kivelson, K.K. Khurana, H.K. Schwartzl, S.M. Thompson, R.L. McPherron, A. Balogh, L.M. Kistler, M.L. Goldstein, J. Borovsky, D.A. Roberts, Plasma sheet turbulence observed by Cluster II. J. Geophys. Res. **110**, A01205 (2005)

M. Wiltberger, V. Merkin, J.G. Lyon, S. Ohtani, High-resolution global magnetohydrodynamic simulation of bursty bulk flows. J. Geophys. Res. Space Phys. **120**, 4555 (2015)

S. Wing, J.R. Johnson, P.T. Newell, C.-I. Meng, Dawn-dusk asymmetries, ion spectra, and sources in the northward interplanetary magnetic field plasma sheet. J. Geophys. Res. **110**, A08205 (2005)

S. Wing, S. Ohtani, J.R. Johnson, M. Echim, P.T. Newell, T. Higuchi, G. Ueno, G.R. Wilson, Solar wind driving of dayside field-aligned currents. J. Geophys. Res. **116**, A08208 (2011)

A.N. Wright, W. Allan, Simulations of Alfvén waves in the geomagnetic tail and their auroral signatures. J. Geophys. Res. **113**, A02206 (2008)

J. Wu, D.J. Knudsen, D.M. Gilles, E.F. Donovan, J.K. Burchill, Swarm observation of field-aligned currents associated with multiple auroral arc systems. J. Geophys. Res. **122**, 10145 (2017)

J.R. Wygant, A. Keiling, C.A. Cattell, M. Johnson, R.L. Lysak, M. Temerin, F.S. Mozer, C.A. Kletzing, J.D. Scudder, W. Petersen, C.T. Russell, G. Parks, M. Brittnacher, G. Germany, J. Spann, Polar spacecraft based comparisons of intense electric fields and Poynting flux near and within the plasma sheet-tail lobe boundary to UVI images: an energy source for the aurora. J. Geophys. Res. **105**, 18675 (2000)

A.G. Yahnin, V.A. Sergeev, B.B. Gvozdevsky, S. Vennerstrom, Magnetospheric source region of discrete auroras inferred from their relationship with isotropy boundaries of energetic particles. Ann. Geophys. **15**, 943 (1997)

A.G. Yahnin, V.A. Sergeev, B.B. Gvozdevsky, S. Vennerstrom, Reply. Ann. Geophys. **17**, 42 (1999)

J. Yang, R.A. Wolf, F.R. Toffoletto, S. Sazykin, RCE-E simulation of substorm growth phase are associate with large-scale adiabatic convection. Geophys. Res. Lett. **40**, 6017 (2013)

L.M. Zelenyi, R.A. Kovrazkhin, J.M. Bosqued, Velocity-dispersed ion beams in the nightside auroral zone: AUREOL 3 observations. J. Geophys. Res. **95**, 12119 (1990)

L.M. Zelenyi, E.E. Grigorenko, A.O. Fedorov, Spatial-temporal ion structures in the Earth's magnetotail: beamlets as a result of nonadiabatic impulse acceleration of the plasma. JETP Lett. **80**, 663 (2004)

Y. Zhang, H. Matsumoto, H. Kojima, Whistler mode waves in the magnetotail. J. Geophys. Res. **104**, 28633 (1999)

S. Zou, L. Lyons, C.-P. Wang, A. Boudouridis, J. Ruohoniemi, P. Anderson, P. Dyson, J. Devlin, On the coupling between the Harang reversal evolution and substorm dynamics: a synthesis of SuperDARN, DMSP, and IMAGE observations. J. Geophys. Res. **114**, A01205 (2009)

Space Sci Rev (2021) 217:17
https://doi.org/10.1007/s11214-021-00796-w

Small-Scale Dynamic Aurora

Ryuho Kataoka[1,2] · Christopher C. Chaston[3] · David Knudsen[4] · Kristina A. Lynch[5] · Robert L. Lysak[6] · Yan Song[6] · Robert Rankin[7] · Kiyoka Murase[2] · Takeshi Sakanoi[8] · Joshua Semeter[9] · Tomo-Hiko Watanabe[10] · Daniel Whiter[7]

Received: 11 June 2020 / Accepted: 13 January 2021 / Published online: 1 February 2021

Abstract Small-scale dynamic auroras have spatial scales of a few km or less, and temporal scales of a few seconds or less, which visualize the complex interplay among charged particles, Alfvén waves, and plasma instabilities working in the magnetosphere-ionosphere coupled regions. We summarize the observed properties of flickering auroras, vortex motions, and filamentary structures. We also summarize the development of fundamental theories,

Auroral Physics
Edited by David Knudsen, Joe Borovsky, Tomas Karlsson, Ryuho Kataoka and Noora Partmies

Supplementary Information The online version contains supplementary material available at https://doi.org/10.1007/s11214-021-00796-w.

✉ R. Kataoka
kataoka.ryuho@nipr.ac.jp

C.C. Chaston
ccc@ssl.berkeley.edu

D. Knudsen
knudsen@ucalgary.ca

K.A. Lynch
kristina.a.lynch@dartmouth.edu

R.L. Lysak
lysak001@umn.edu

Y. Song
yan@umn.edu

R. Rankin
rrankin@ualberta.ca

K. Murase
murase.kiyoka@nipr.ac.jp

T. Sakanoi
tsakanoi@pparc.geophys.tohoku.ac.jp

J. Semeter
jls@bu.edu

such as dispersive Alfvén waves (DAWs), plasma instabilities in the auroral acceleration region, ionospheric feedback instabilities (IFI), and the ionospheric Alfvén resonator (IAR).

Keywords Magnetosphere-ionosphere coupled region · Auroral phenomena · Auroral breakup · Flickering aurora · Dispersive Alfven waves · Ionospheric feedback instabilities · Ionospheric Alfven resonator

1 Introduction

For the purposes of this review, we define small-scale, dynamic aurora to include optical and plasma structure having characteristic spatial scales of a few km or less, and time scales of a few seconds or less. On these scales, spatial and temporal effects are often tightly intertwined. The justification of these choices is as follows.

Until the 1960's, scientific study of auroral morphology depended on eyewitness accounts and drawings, and later, film-based cameras having limited resolution in space and time, as exposure times of many tens of seconds were required due to the relatively low light levels of most auroral phenomena (Stormer 1955). Despite these limitations, many key types of auroral phenomena were categorized and given names such as luminous bands, curtains, omega bands, pillars (Vallance-Jones 1974). One of the most common and recognizable forms apparent in conventional film-based photographs of the period is now known as a stable auroral arc, which often dominates the evening-time and pre-midnight auroral sky, is highly elongated in the east-west direction, and has characteristic north-south widths of the order of 10's of km (Kim and Volkman 1963). Using a sounding rocket, McIlwain (1960) established that such arcs are due to magnetic field-aligned electron beams having energies of several keV. Later these were termed "inverted-V" (also "discrete") arcs due to their signatures in energy-time spectrograms recorded by polar-orbiting satellites (Frank and Ackerson 1971). Early satellite observations based on spinning photometers on ISIS-II also established a broad region of unstructured emission (at least down to the resolution of those

T.-H. Watanabe
watanabe.tomohiko@nagoya-u.jp

D. Whiter
d.whiter@soton.ac.uk

[1] National Institute of Polar Research, 10-3 Midori-cho, Tachikawa, Tokyo 185-0031, Japan

[2] SOKENDAI, 10-3 Midori-cho, Tachikawa, Tokyo 185-0031, Japan

[3] Space Sciences Laboratory, University of California, Berkeley, CA 94720, USA

[4] Dept of Physics and Astronomy, University of Calgary, Calgary, AB T2N 1N4 Canada

[5] Dept of Physics and Astronomy, Dartmouth College, Hanover, NH 03755 USA

[6] School of Physics and Astronomy, University of Minnesota, Minneapolis, MN, USA

[7] Physics & Astronomy, University of Southampton, SO17 1BJ, Southampton, UK

[8] Planetary Plasma and Atmospheric Research Center, Aramaki-aza-Aoba 6-3, Aoba, Sendai, Miyagi 980-8578, Japan

[9] Department of Electrical and Computer Engineering and Center for Space Physics, Boston University, Boston, MA, USA

[10] Department of Physics, Nagoya University, Nagoya 464-8602, Japan

composite images) at the equatorward edge of the auroral zone, termed the "diffuse" aurora (Lui and Anger 1973).

Through naked-eye observations, it was known in those early days that the aurora contains features which are more highly structured in space (scales of ~1 km and smaller) and time (several seconds or less) than could be recorded with conventional film cameras, for example flickering aurora, which oscillates at frequencies of the order of 10 Hz (Beach et al. 1968). In a landmark paper, Maggs and Davis (1968) used an image-intensified TV camera with a telescopic lens to record fine structure within the aurora having characteristic widths as small as 70 m. Borovsky and Suszcynsky (1993) confirmed the existence of these small-scale features, and Borovsky (1993) pointed out that, at the time, no auroral theories could account for them. A reanalysis of the Maggs and Davis (1968) observations by Stenbaek-Nielsen et al. (1999) concluded that those particular fine structures were in fact not associated with discrete arcs, but instead were embedded within the diffuse aurora. More recent observations (Dahlgren et al. 2012) have since shown that even discrete arcs may contain fine structure down to scales of 0.09 km.

The renewed interest in small-scale auroral structures following Borovsky and Suszcynsky (1993) led some to view ~100 m as the "true" scale of the aurora, and that mesoscale arcs were in fact assemblages of such small structure. However, Knudsen et al. (2001) demonstrated that in the case of mesoscale arcs observed by more modern, CCD-based all-sky cameras, north-south structure of stable arcs is dominated by Gaussian-like profiles with a characteristic widths of 10-30 km, and with little sub-structure down to the camera resolution of 1.7 km. In other words, small-scale (km or sub-km scale) aurora is distinct from quiet, mesoscale phenomena such as discrete arcs.

As will be described below, rapid advancements in camera technology in the past two to three decades have led to extensive observations of small-scale and dynamic auroral features, and have established that small-scale aurora is also a common feature of auroral breakup following substorm onset.

The spatiotemporal properties of small-scale auroras were reviewed in Sandahl et al. (2008) and Sandahl et al. (2011). After the successful ground-based observations utilizing EMCCD cameras (e.g., Semeter et al. 2008) and sCMOS cameras (e.g., Dahlgren et al. 2013), new types of small-scale auroras have also been found by these fine resolution observations. The ASK project was the most successful in revealing the energy flux associated with fine structures using three identical EMCCD cameras with multiple wavelengths. For example, Dahlgren et al. (2008) reported that "filaments" or "curls" were the energetic part of surrounding auroras. Dahlgren et al. (2011) reported that thin arcs are related to increase in number flux without an increase in energy.

A breakthrough was achieved by the Reimei satellite observations. The Reimei satellite has capabilities of simultaneous auroral imaging and auroral electron measurement with a spatial resolution of ~1 x 1 km and a time resolution of 120 ms (Asamura et al. 2003; Sakanoi et al. 2003), and revealed the precise relationship between auroral arcs and precipitating electrons (Asamura et al. 2009; Frey et al. 2010, Chaston et al. 2010, 2011, Fukuda et al. 2014, Motoba and Hirahara 2016). Figure 1 shows an event of inverted-V electrons and auroral arcs observed by Reimei. In this case, Reimei observed auroral band structure with a latitudinal width of about 100 km. Note that there are multiple small-scale auroral arcs showing sheared motions (Fig. 1, see also Movie 1) in the poleward half of the auroral band, and rather uniform emission in the equatorward part, which is consistent with the previous finding that there is often a band-like emission of several tens of km in the background when a small-scale auroral arc appears (Haerendel 1999). In fact, electron data show the inverted-V electron precipitation with widths of ~100 km corresponding to the auroral band structure. In the poleward half of the inverted-V structure where the small-scale

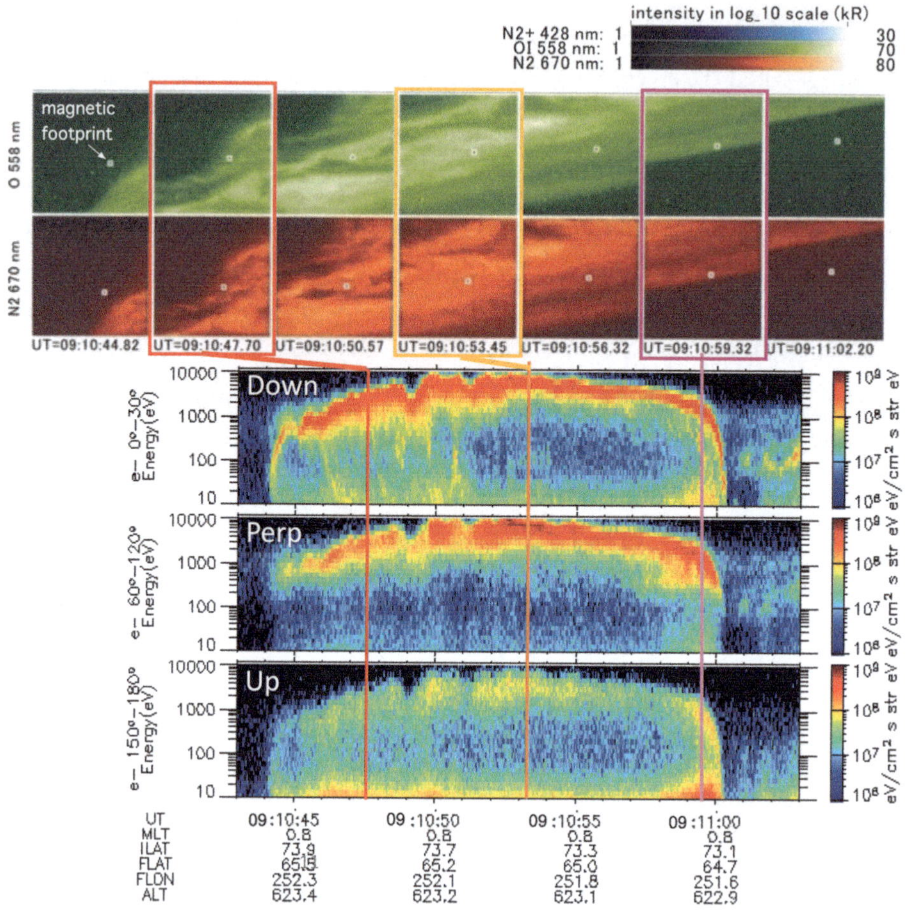

Fig. 1 Inverted-V electrons and auroral arcs observed by the Reimei satellite on 2005-12-26, moving across the auroral band structure with a latitudinal width of about 100 km. Multiple small-scale auroral arcs showed sheared motions as shown in Movie 1

auroral arcs appeared, there are small-scale fluctuations of the peak energies of electrons in the energy range of several keV, with time-dispersed low-energy electrons precipitating simultaneously, implying the existence of Alfvén wave acceleration (Whiter et al. 2012). It was concluded from this event that the small-scale auroral arcs are not generated by a single acceleration mechanism but produced by the fluctuations of peak energy of inverted-V type accelerated electrons as caused by the interaction of Alfvén waves.

As another example of the important role of Alfvén waves in the auroral acceleration region, Asamura et al. (2009) suggested that Alfvén waves are generated by the shear instability (Wu and Seyler 2003) using the Reimei data over auroral arcs with fast shear flows (Movie 2). In this case, Alfvén waves are a secondary process, generated by the shear flow in the magnetosphere-ionosphere coupled region.

Many researchers have considered that inertial Alfvén waves play an important role in auroral acceleration in the several thousands of km altitudes where a low-β condition is satisfied. When the wavelength of obliquely propagating inertial Alfvén waves $\lambda_\perp$ becomes comparable to the electron inertial length λ_e, a field-aligned acceleration electric field is formed (e.g., Stasiewicz et al. 2000a,b). Chaston et al. (2003b) estimated that the auroral width generated from inertial Alfvén waves can be 0.5-1.0 km. However, within this range, the acceleration energy is about 0.1-0.5 keV, which is an order of magnitude smaller than auroral precipitating electron energy that produces typical auroral arcs. Chen et al. (2005) demonstrated that Alfvén waves can produce modulation of inverted-V accelerated electrons in the keV energy range. Their previous studies suggested that the small-scale auroral structures are produced by coupling of multiple acceleration mechanisms including inertial Alfvén waves rather than a single acceleration mechanism. The vast majority of our discussion below will therefore be related to the important role of Alfvén waves in causing small-scale dynamic auroras.

In a recent study using suprathermal (100-350 eV) electron measurements on ePOP in addition to field measurements from Swarm, Wu et al. (2020b) carried out a statistical study of periodic electron bursts which recurred several times at periods consistent with the eigenfrequency of the ionospheric Alfvén resonator (IAR). These events are similar to those reported by Chaston et al. (2002a, 2002b) using the FAST satellite in the nightside auroral region, and Tanaka et al. (2005) using a sounding rocket in the dayside cusp, but extend the number of events to more than 50, and show that they are distributed over a wide range of magnetic local times (from pre-noon to post-midnight) and geomagnetic latitudes (64-78 degrees). In at least one case, the electron bursts observed by ePOP were associated with auroral rays, a connection reported previously by Lynch et al. (2012) based on sounding rocket observations, and indicating a possible coupling between rays and the IAR. Wu et al. (2020b) were also able to reproduce detailed properties of electron energy and pitch-angle dispersion observed within individual bursts by tracing test particles within a downward-propagating Alfvén wave.

Other sounding rocket observations have also contributed to the understanding of the IAR and ionospheric feedback. The Auroral Current and Electrodynamics Structure (ACES) mission observed small-scale electric and magnetic perturbations together with measurements of precipitating electrons to investigate the properties of the IAR (Cohen et al. 2013). They found results consistent with the ionospheric feedback model of Streltsov and Lotko (2008), including the structuring of the auroral currents and formation of a low density cavity. Studies of the IAR and feedback were also a focus of the Magnetosphere-Ionosphere Coupling in the Alfvén Resonator (MICA) sounding rocket program (Lynch et al. 2015). They compared their results with a simple electrostatic model based on both the in situ field measurements, and on the PFISR radar observations of the ionosphere (Zettergren et al. 2014), and showed that the model could not reproduce the small-scale structures, which they then interpreted as possibly being caused by feedback interactions in the IAR.

In Sect. 2, we review specific phenomena representing the small-scale dynamic aurora, such as vortices, filaments, and packets. We can classify the majority of the phenomena as a variety of localized Alfvénic interactions on multiple scales and morphologies. However, it is an open question as to how we can model the Alfvénic electron acceleration associated with the rapid formation and deformation of double layers, plasma instability, and turbulent cascading. In Sect. 3 we discuss recent developments in the essential theories, such as dispersive Alfvén waves, feedback instability, and the IAR. In Sect. 3.4, concluding remarks of this review article are summarized.

2 Specific Phenomena

2.1 Flickering

Bright auroras sometimes show localized flickering intensity modulation (Movies 3 and 4). The modulation frequency is typically 3-15 Hz and the typical size of the bright patch is 1-12 km across magnetic field lines at an emission altitude of 100 km (Kunitake and Oguchi 1984; Sakanoi and Fukunishi 2004; Michell et al. 2012). The modulation frequency is consistent with the O^+ ion-cyclotron frequency in the auroral acceleration region (Temerin et al. 1986; Lund et al. 1995) and the spatial structure is consistent with an interference pattern associated with dispersive Alfvén waves (Sakanoi et al. 2005; Gustavsson et al. 2008; Kataoka et al. 2011b; Yaegashi et al. 2011). Whiter et al. (2008) showed that flickering intensity increases when background non-flickering aurora becomes brighter.

Ground-based high-speed observations provide an important clue for understanding the possible generation mechanism. Whiter et al. (2010) identified "chirps" (variation in frequency over 1-2 s) in multi-spectral optical observations of flickering aurora, and found that the electron precipitation energy has an inverse correlation with the flickering frequency during these chirps, consistent with electron acceleration by Landau resonance with dispersive Alfvén waves. The leading edge of moving flickering patches was shown by Kataoka et al. (2011b) to be more energetic than the trailing edge, which could be a result of dispersive Alfvén waves losing energy by Landau damping. Recent advances in ground-based imaging techniques such as EMCCD cameras have allowed us to find zenith-view flickering auroras with frequencies above the oxygen ion-cyclotron frequency of auroral acceleration regions (Yaegashi et al. 2011). Using an sCMOS camera Fukuda et al. (2017) found 60-80 Hz flickering aurora, which can be modulated by proton-band electromagnetic ion cyclotron (EMIC) waves, coincident with 10 Hz flickering and therefore indicating the presence of multi-ion EMIC waves. The fastest-varying patch was smaller than the 10 Hz patch. The fastest variation of side-view flickering aurora was 180 Hz as observed with a photometer (McHarg et al. 1998).

Related phenomena include dispersive electrons called field-aligned bursts (FABs) as observed by sounding rockets above the ionosphere (Andersson et al. 2002), and EMIC and broad-band ELF (BBELF) waves as observed by satellites in the magnetosphere-ionosphere coupled regions (Erlandson et al. 1994, McFadden et al. 1987; Lund et al. 1998). The flickering auroras therefore manifest complex interactions among electrons and ions, and dispersive Alfvén waves nearby auroral acceleration regions.

2.2 Vortices

Km-scale vortex-like structures of the aurora are known as "curls" (Hallinan 1976; Trondsen and Cogger 1998; Vogt et al. 1999) (Fig. 2, see also Movies 5 and 6). More recently, smaller-scale structure was found along curls with inverse rotation and named "ruffs" (Dahlgren et al. 2010) (Fig. 2, see also Movie 7). Ivchenko et al. (2005) found that "curls" were caused by precipitation of energetic electrons without a low-energy population, while both high and low energy precipitation were present in the "rays." Another well-known small-scale vortex-like structure is "folds", which have typical scale of $\sim$10 km (Fig. 2, see also Movie 8). Recently, the formation and inverse cascade of folds were found just before auroral breakup (Kataoka et al. 2011a), which is similar to so-called auroral "beads" appearing at substorm onset (Motoba et al. 2012). Wavy structures which look different from typical folds were

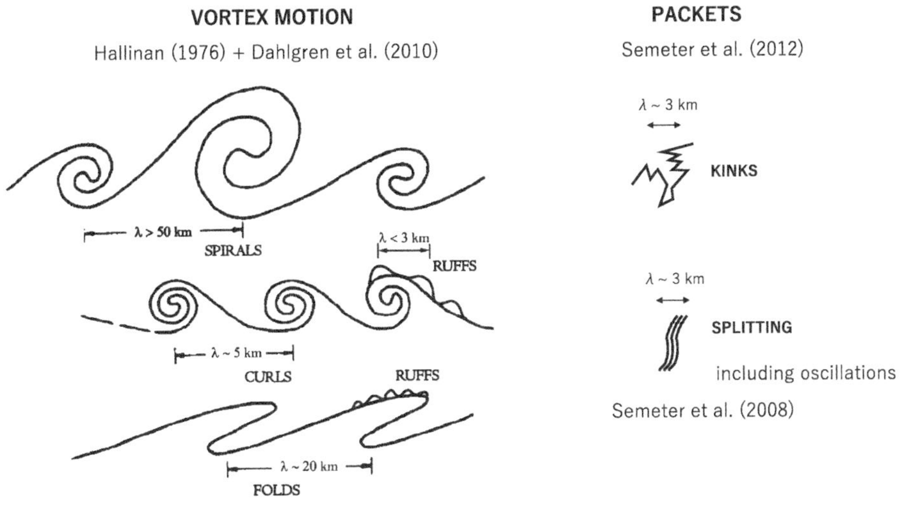

Fig. 2 Cartoon of the magnetic zenith-view of small scale auroral structures. (After Hallinan 1976; Dahlgren et al. 2010). The packet structures are compared against vortex motion (Semeter et al. 2008; Semeter 2012)

also reported at the poleward boundary of multiple arcs, along with eddies and flickering (Kataoka et al. 2015).

These small-scale vortex structures in the magnetic zenith view have long been considered as the cause of "rays" appearing in side-viewed auroral curtains. However, a different type of side-view appearance of "rays" was also found at poleward boundary intensification (PBI) by Lynch et al. (2012), where sounding rocket observations showed electrons of ionospheric origin accelerated by Alfvén waves. Sample images of these PBI rays are shown in Movie 9.

2.3 Filaments

Progressive advances in imaging technologies applied to auroral research have led to fundamental advances in our understanding of the connections between auroral morphology and the associated energy source. The goal of understanding "filamentary structure" in the aurora has arisen as a distinct quest in auroral physics, based partly on this technological progression. The lower minimum spatial scale perpendicular to the ambient magnetic field appears to be 70$\sim$90 m (Maggs and Davis 1968; Borovsky 1995; Dahlgren et al. 2012). This limit corresponds to the electron inertial length near the density minimum in the near-Earth magnetosphere, suggesting that inertial Alfvén wave dispersion plays an important role in producing these structures (e.g., Stasiewicz et al. 2000a, 2000b).

Bulk properties of the incident particle flux can be deduced from the auroral spectrum in the magnetic field-aligned direction, and recent observational programs have sought to bring such a multi-spectral imaging capability to auroral research (Dahlgren et al. 2008; Ivchenko et al. 2005; Grubbs et al. 2018). One notable result has been the discovery of spectral signatures consistent with higher energy (>8 keV) mono-energetic particle fluxes within filamentary features observed in the magnetic zenith (Dahlgren et al. 2012). This population was subsequently found to be co-mingled on the same flux tube with a higher energy $\sim$30 keV population (Dahlgren et al. 2015). This result is consistent with findings of Arnoldy et al. (1999) who suggested that broad-band field-aligned bursts may be correlated

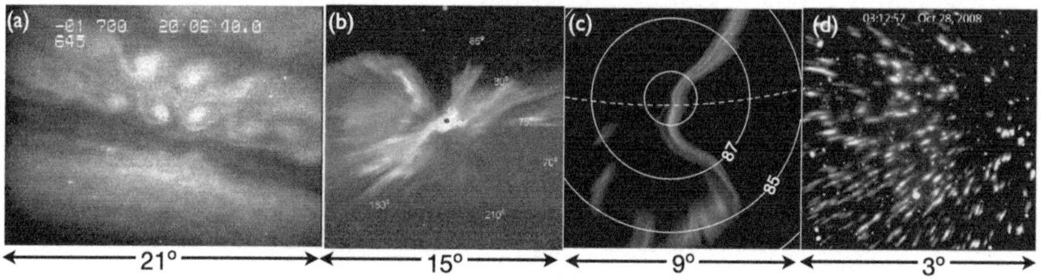

Fig. 3 Ground-observations of fine scale auroral phenomena with fields-of-view as shown: **a**) Auroral curls (Vogt et al. 1999), **b**) "Flaming" auroral filaments (Dahlgren et al. 2013), **c**) Auroral "packets" (Semeter et al. 2008), **d**) Decameter filaments produced by HAARP (Kendall et al. 2010)

with modulation of the entire inverted-V potential drop. Such co-mingling has also been recently observed for filament structure embedded within the diffuse aurora (Sivadas et al. 2019), and may also explain observation of flickering observed in sub-kilometer features (Whiter et al. 2008). These results do not call into question the importance of Alfvén waves in modulating the aurora, but they suggest an interplay among energization mechanisms powering the aurora that may not be fully understood.

Ionospheric heating experiments have been shown to create filamentary aurora-like emissions with spatiotemporal characteristics similar to natural auroras (Kendall et al. 2010) (Fig. 3, see also Movie 10). Furthermore, for heater facilities co-located near an ISR facility, these features are found to be accompanied by enhanced ion-acoustic backscatter, similar to NEIALs (Kosch et al. 2007). These results of heater experiments such as HAARP and EISCAT HF facility have contributed substantially to our understanding of small-scale M-I coupling (Streltsov et al. 2018) and have guided simulation studies (Akbari et al. 2016).

2.4 Packets

An auroral breakup is synonymous with the impulsive release of magnetic stress in the magnetosphere-ionosphere system. The resultant energy influx leads to creation and modulation of field and particle fluxes embodying a rich variety of spatial and temporal scales (e.g., Kataoka, 2011). Although auroral displays are highly variable during these events, coherent patterns can nonetheless be identified in observations of sufficient resolution. One intriguing and repeatable phenomenon is the bifurcation of narrow arcs into systems of parallel forms (Trondsen et al. 1997; Semeter and Blixt 2006; Semeter et al. 2008). This dynamic is often superposed on a moving reference frame, giving the appearance of a translating wave packet (Fig. 3, see attached Movie 11). Semeter et al. (2008) have termed such motion "dispersive" (as opposed to "fluid"), in that there is a distinctly different phase velocity (motion of individual arc elements) and group-velocity (motion of the packet) in the observer frame of reference. The packet formation is qualitatively consistent with the auroral projection of an Alfvén resonant cone (Singh 1999).

Such patterns in auroral breakup are not confined to arc-like features. Dahlgren et al. (2013) observed periodic cylindrical filaments racing upward along the magnetic field. Termed "flaming aurora," a careful analysis of these signatures using a high frame-rate sCMOS sensor revealed consistency with dispersive field-aligned bursts (FABs) commonly observed by particle detectors in regions of Alfvénic turbulence (Lynch et al. 1994). The appearance of these Alfvénic relationships between these features and the particle source populations remains poorly understood. Borovsky (1995) once concluded that "*Existing auroral-arc theories do not explain the finescale structures because none of these theories will allow a structure as narrow as the observed finescale structures*". Indeed, the thesis put forth by Borovsky (1995) appears to still hold.

2.5 Effects on the Ionospheric State

The fields and precipitation patterns associated with small-scale auroral phenomena produce small-scale irregularities in the ionosphere, affecting radar backscatter, trans-ionospheric radio propagation, and the global-scale M-I energy budget. Electron distribution functions associated with filamentary aurora are spread broadly in energy, extending from several keV's down to thermal levels, as discussed in Sect. 3. The low-energy portion can interact resonantly with Langmuir and ion-acoustic modes in the ionosphere (Akbari et al. 2012). Such interactions are detected by Incoherent Scatter Radar (ISR) as "Naturally Enhanced Ion Acoustic Line" (NEIAL) (Grydeland et al. 2003). Coordinated observations with optical sensors have been used to correlate NEIAL features with specific types of auroral structure (Blixt et al. 2005; Michell and Samara 2010; Akbari et al. 2012). The region of destabilization presents a source of anomalous resistivity, which has been implicated in regulating macro-scale M-I processes such as ion outflow (Forme and Fontaine 1999).

Irregularities induced by small-scale auroral processes also affect ground-space communication links and Global Navigation Satellite Systems (GNSS). Recent work has found GPS signal disruptions, in the form of scintillation and complete loss-of-lock, to be clustered at the trailing edge of an auroral surge, suggesting a structuring mechanism involving the interplay between precipitation-induced gradients, and field-induced drift (i.e., gradient drift instability) (Semeter et al. 2017). Such technological effects of fine-scale auroral dynamics may have important implications for space weather forecasting in newly opened arctic shipping routes.

3 Theories and Discussions

3.1 Dispersive Alfvén Waves and Field-Aligned Electron Acceleration

While specific auroral forms are distinguished by their morphology and evolution, in-situ measurements of the electron acceleration processes are distinguished by the form of the electron spectrum (Paschmann et al. 2003). Figure 4 presents measurements returned from the Reimei spacecraft showing electron spectra (a) and conjunctive imagery (b and c). On the left the electron spectrum is broad and characteristic of what is often termed the 'Alfvénic aurora' while on the right there is a distinct peak at $\sim$10 keV representing what is termed an 'inverted-V' (Lyons 1981). The latter is often identified as the 'quasi-static' aurora because the electron spectrum is largely invariant over many Alfvén bounce times between the acceleration region and ionosphere. While the physics of Alfvén waves is central to the dynamics and structure of both the auroral forms shown in Figs. 2b and 2c in Chaston et al. (2010), in the case of 'Alfvénic aurora' dispersive Alfvén waves are responsible for the electron acceleration itself. In fact, the correlation of broad-spectrum electromagnetic field fluctuations in the ELF range with bursts of broad spectrum field-aligned as shown in Fig. 4a electrons is an established feature of the in-situ observations above the auroral oval (Andersson et al. 2002; Chaston et al. 2003a; Lynch et al. 1996; Knudsen et al. 1998). Measurements show that the relative variation of the electric and magnetic spectra of these fluctuations with spacecraft frame frequency are in general consistent with the properties of a broad k-spectrum of dispersive Alfvén waves (Wahlund et al. 1998, 2003; Stasiewicz

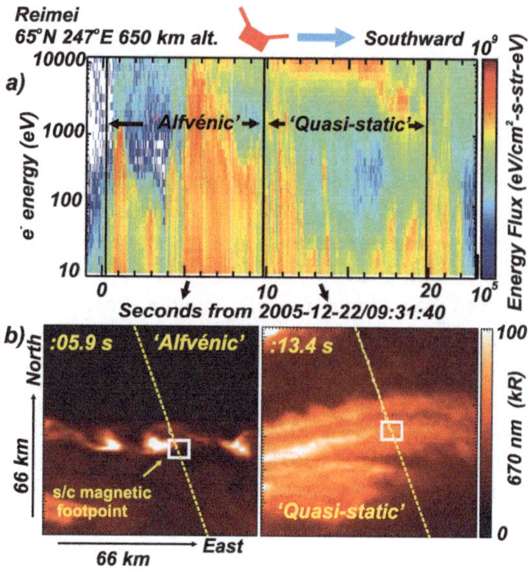

Fig. 4 Auroral electron and magnetically conjugate camera observations of rapidly evolving auroral forms returned from the Remei spacecraft. (**a**) Electron energy spectrogram revealing the electron energy fluxes responsible for auroral emission at 670 nm shown for (**b**) Alfvénic and (**c**) Quasi-static aurora (After Chaston et al. 2010)

et al. 2000a 2000b; Chaston et al. 1999, 2008; Hull et al. 2010, 2016). Dispersive Alfvén waves have wavelengths transverse to the background magnetic field of the order of characteristic plasma scales including the electron inertial length ($\lambda_e = c/\omega_{pe}$), ion gyro-radius ($\rho_i = (T_i/m_i)^{1/2}/\Omega_i$) and ion acoustic gyro-radius ($\rho_s = (T_e/m_i)^{1/2}/\Omega_i$). Here ω_{pe} is the electron plasma frequency, Ω_i is the ion cyclotron frequency, T_i is the ion temperature, T_e is the electron temperature, and m_i the ion mass. Unlike MHD Alfvén waves the propagation of these modes through the plasma requires the existence of a significant wave electric field component parallel to the background magnetic field ($E_\parallel$) (Lysak and Lotko 1996). It is this property that is responsible for the field-aligned electron distributions commonly observed within these waves. Detailed reviews of the properties of these waves and the acceleration processes that operate within them can be found in Stasiewicz et al. (2000b), Chaston (2006), Birn et al. (2012) and Mottez (2014).

Field-aligned electron bursts observed within dispersive Alfvén waves above the auroral oval are typically impulsive, broad in energy ('broadband' or supra-thermal bursts), or time dispersed features on sub-second timescales extending from at most a few keV down to the eV range (Andersson et al. 2002). These features can occur in isolation or in combination with classical inverted-V type electron spectra (e.g. Kletzing and Hu 2001; Chaston et al. 2002a; Asamura et al. 2009). Efforts to explain the characteristics of these features have in general relied on a linear fluid description of the perpendicular field and plasma dynamics coupled to the electron momentum equation (Thompson and Lysak 1996) or Vlasov equation (Watt and Rankin 2012) along the geomagnetic field. The relationship between the parallel electric field and the plasma can be most easily understood by considering the force balance described by the electron momentum equation where,

$$E_\parallel = \mu_o \lambda_e^2 \left(\frac{\partial J_\parallel}{\partial t} + v_\perp \cdot \nabla_\perp J_\parallel + J_\parallel \nabla_\parallel \frac{J_\parallel}{q_e n_e}\right) - \frac{\nabla_\parallel P_{e\parallel}}{q_e n_e} - \frac{(P_{e\parallel} - P_{e\perp}) \nabla_\parallel B_o / B_o}{q_e n_e}. \tag{1}$$

Here $J_\parallel = n_e q_e v_\parallel$ is the field-aligned current, P_e is the electron pressure, $v_\perp$ and $v_\parallel$ are velocities perpendicular and parallel to the geomagnetic field, q_e is the electron charge, n_e is the electron density and B_o is the geomagnetic field strength. For cold electrons and small-scale waves at low altitudes ($\lesssim 2$ R_E above the surface) where $\lambda_e > \rho_s$ the first term on the RHS of Eq. (1) dominates and the waves are generally termed inertial Alfvén waves (Goertz and Boswell 1979; Lysak and Carlson 1981; Stasiewicz et al. 2000a,b). At higher altitudes $\rho_s > \lambda_e$ the second term becomes more important and the waves are known as kinetic Alfvén waves (Hasegawa 1976; Lysak and Lotko 1996). The third term describes the parallel electric field balancing the mirror force acting on primarily hot plasma sheet current carrying electrons. For the zero-frequency case the integrated contribution of this term along the geomagnetic field corresponds to the well-known Knight relation (Knight 1973). For time varying fields this contribution has been evaluated analytically using Vlasov or gyrokinetic approaches (Rankin et al. 1999a,b; Nakamura 2000; Lysak and Song 2003; Watanabe 2014) and via numerical simulation (Watt 2010; Damiano and Johnson 2013). For the highly field-aligned distributions characteristic of electron bursts in dispersive Alfvén waves this contribution is likely not the dominant driver of field aligned electron bursts. Consequently, efforts to model the electron acceleration mechanism leading to these bursts have largely focused on the inertial and electron pressure gradient contributions (Chaston et al. 2003a, 2003b, 2003c). Conjunctive observations from spacecraft separated along the geomagnetic field and via imaging suggest that these effects drive field-aligned electron acceleration over many Earth radii along auroral field-lines (Wygant et al. 2000, 2002; Keiling et al. 2002; Chaston et al. 2005; Dombeck et al. 2005; Damiano et al. 2018)

Test particle simulations have been widely performed to understand the generation of field-aligned electron bursts in model wavefields propagating along realistic altitude dependent phase speed profiles. These have shown how the macro-scale features of the geomagnetic field and plasma distribution together with the micro-physics of the wave-particle interaction conspire to define the observed form and evolution of the electron distribution in velocity space (Thompson and Lysak 1996; Chaston et al. 2000, 2002a, 2002b; Kletzing and Hu 2001; Andersson et al. 2002; Su et al. 2004; Chen et al. 2005). How this works can be qualitatively understood from a consideration of the field-aligned potential (ϕ) in the wavefield, the electron velocity ($v_\parallel$) and the parallel wave phase speed ($\frac{\omega}{k_\parallel}$) expressed as,

$$\frac{\omega}{k_\parallel} - \sqrt{\frac{2q_e\phi}{m_e}} < v_\parallel < \frac{\omega}{k_\parallel}. \tag{2}$$

Here ω is the wave frequency, $k_\parallel$ the wavenumber along B_0 and m_e is the electron mass. Any electron with $v_\parallel$ in this range can be picked up by the wave and carried along with speed of the order of $\frac{\omega}{k_\parallel}$. The fact that the geomagnetic field varies as the inverse cube of the radial distance from the Earth, while the mass density of the plasma is relatively constant above a few thousand kilometers altitude, means that an Alfvén wave propagating from a magnetospheric source will speed-up as it approaches Earth. This provides an accelerator for electrons which become resonant with the wave according to Eq. (2) at high altitudes where the Alfvén speed is less than the electron thermal speed and the resonant location in phase space occurs where significant phase space densities are found. Consequently, large fractions of the source electron distribution may become resonant with the wave. These resonant electrons are then driven Earthward with progressively increasing speed as the phase speed of the wave increases (Chaston 2006; Watt and Rankin 2009). The convergence of the geomagnetic field over this path focusses both the wave energy density and electron flux to account for appreciable energy deposition (Keiling et al. 2002, 2003, 2019; Chaston et al. 2003a, 2007a,b; Hatch et al. 2016, 2017, 2018; Dombeck et al. 2018)

The broad flat-top like distributions that this interaction generates are a consequence of the differences in energization due to stochastic interaction with different wave phases, trapping (Clark and Seyler 1999; Su et al. 2004; Watt and Rankin 2009; Damiano et al. 2015), and variations in energy gain depending on the source altitude of the accelerated an electron (Chaston et al. 2000, 2002a; Andersson et al. 2002). The special case of time dependent energy dispersed bursts observed at low altitudes as part of this process occurs specifically through the action of large amplitude impulsive coherent field variations. The source altitude that the dispersion infers corresponds to the upper edge of the topside ionosphere where the acceleration process terminates due to increasing electron densities (Kletzing and Hu 2001; Chaston et al. 2002b).

While test particle simulations provide basic insight into the manner in which the acceleration occurs, the fact that the field-aligned potential in the wavefield is often comparable to the temperature of the supporting plasma means that a self-consistent non-linear approach is required. In a series of papers Watt et al. (2004, 2005, 2006, 2009), Watt and Rankin (2010, 2012) using a Vlasov model have shown how the bulk of the electron population remains non-resonant with the wave and carries the field-aligned wave current while a fraction are resonantly accelerated to energies of the order of twice the Alfvén speed. It has been suggested that these two processes naturally provide a dispersive burst terminated by a non-dispersive 'broadband' or supra-thermal burst as the wave passes over the spacecraft. Features having this form are commonly observed (Andersson et al. 2002). Hybrid fluid-kinetic approaches have also been implemented (Hui and Seyler 1992; Clark and Seyler 1999; Swift 2007a,b; Damiano et al. 2015, 2016). The results from Damiano et al. have for example explored the dependency of the electron energization on the relative ion to electron temperature ratio and finite ion gyro-radius effects along the high-altitude portion of an auroral field-line. This work along with that by Swift (2007a) has revealed the importance of wave dispersion across the geomagnetic field as a constraint on the energization process. This dispersion is largest in the presence of hot ions where finite gyro-radii effects spread wave energy across the geomagnetic field and can significantly reduce the effectiveness of the acceleration process. This suggests that the Alfvén wave acceleration process will be most effective on auroral field-lines connected to the cold dense plasma sheet and is consistent with the observation of the most energetic field-aligned electron bursts occur preferentially at high latitudes.

These simulation results provide basic physical understanding of why field-aligned electron distributions are observed in association with dispersive Alfvén waves. However further efforts are required to more credibly connect these processes to visible aurora and to incorporate the action of small-scale kinetic processes observed within the Alfvén wavefield. Imaging from the Reimei spacecraft show that the form of aurora driven by dispersive Alfvén waves are not consistent with the simplified 1-D or even 2-D model geometries assumed. Filaments and vortices are characteristic of Alfvénic aurora (Chaston et al. 2010). These features require a 3-D model and represent the action of non-linear instabilities in the perpendicular plane (Seyler 1990). While the parallel dynamics described above still apply it is likely that the mechanisms through which the acceleration is energetically supported is not just through field aligned Poynting flux at a particular wave scale as modelled. Seyler and Liu (2007) instead have performed 3-D particle in cell simulations that capture this morphology while also examining small scale kinetic interactions associated with wave breaking. This process is found to drive field-aligned electron acceleration while simultaneously driving ion acoustic waves. The later may be related to observations of large amplitude impulsive parallel electric fields in dispersive Alfvén waves (Stasiewicz et al. 1997; Chust et al. 1998; Chaston et al. 1999, 2007a,b; Ergun et al. 2005) suggestive of the formation of double layers (Lysak and Hudson 1987; Silberstein and Otani 1994) or more generally Debye scale regions of charge separation within larger scale Alfvén wave fields. Genot et al. (2004) have explored the coupling of the Alfvén wave to small scale kinetics to demonstrate how the focusing of wave energy into density cavities leads to the formation of phase space holes. These features have been observationally linked to the generation of supra-thermal electron bursts (Chaston et al. 2006). These PIC simulations that allow an examination of the detailed kinetics in the wave are necessarily performed on small scales so how they couple with, and perhaps modify, the larger scale parallel acceleration dynamics along the geomagnetic field remains undetermined.

3.2 Dynamics of Small Scale Structured Aurora

Dispersive Alfvén waves and the electron acceleration discussed in the previous section propose a possible source of electron precipitation in regard to small scale structures of auroras. In addition to the particle source, one demands a physics mechanism elucidating structure formation and dynamics of small-scale auroras. The dynamics and structuring of small-scale features in auroral forms is in broad terms determined by the ionospheric response to current closure and ionization, phase mixing and non-linear evolution of Alfvén waves, and instabilities acting on auroral current sheets and potential structures through the acceleration region. For the general description of those processes that lead to curls, folds and filamentation, the strength of the geomagnetic field (B_0) along an auroral fluxtube conveniently allows the relevant physics to be formulated in terms of the dynamics perpendicular and parallel to B_0 (Seyler 1988; Song and Lysak 2006). The former can be approximated by the single fluid (ion) momentum equation describing force balance in the perpendicular plane and the latter by the parallel electron momentum equation (Eq. (1)) describing the self-consistent distribution of parallel electric fields along auroral flux tubes that drive field-aligned electron acceleration to form aurorae. The occurrence of double layers along auroral field-lines (Ergun et al. 2004) tells us that in principle the parallel dynamics should be characterized using a fully kinetic model, however achieving this in 3-D while faithfully incorporating the macroscale variation in plasma parameters extending several Earth radii from the auroral ionosphere remains computationally challenging. Solutions to these equations are constrained by boundary conditions prescribed by current closure through the ionosphere (Seyler 1990; Watanabe 2010), and driven at the high-altitude end by applied current and voltage profiles representing magnetospheric drivers often referred to as 'generators' (Lysak 1990). The momentum equation describing the perpendicular dynamics (e.g. Biskamp 2003; Eq. 2.7) can be expressed as

$$\rho\left(\frac{\partial v_\perp}{\partial t}+v_\perp\cdot\nabla_\perp v_\perp\right)\approx J\times B-\nabla_\perp P\approx J_\parallel\times b_\perp+\frac{B_o}{\mu_o}\frac{\partial b_\perp}{\partial z},\tag{3}$$

where ρ is the mass density, J is the total current, $b_\perp$ is the perturbed magnetic field,and P is the plasma pressure. The very low values for plasma β through the auroral acceleration region allow the contribution from $\nabla_\perp P$ to be ignored. Advection of structure with $v_\perp$ through the acceleration region (Seyler 1990) suggest that the dynamics described by this equation are projected onto upper atmosphere by precipitating electrons (Hallinan 1981). Consequently, the observed motion and structuring of auroral forms qualitatively reflect that of auroral acceleration structures. Rotational vorticity in auroral luminosity therefore represents that of $v_\perp$ through the acceleration region requiring that the convective non-linearity ($\rho v_\perp\cdot\nabla_\perp v_\perp$) on the LHS side on Eq. (3) be non-zero. On the other hand, the concentration of $J_\parallel$ into filaments suggests that the magnetic non-linearity ($J_\parallel\times b_\perp$) on the RHS of

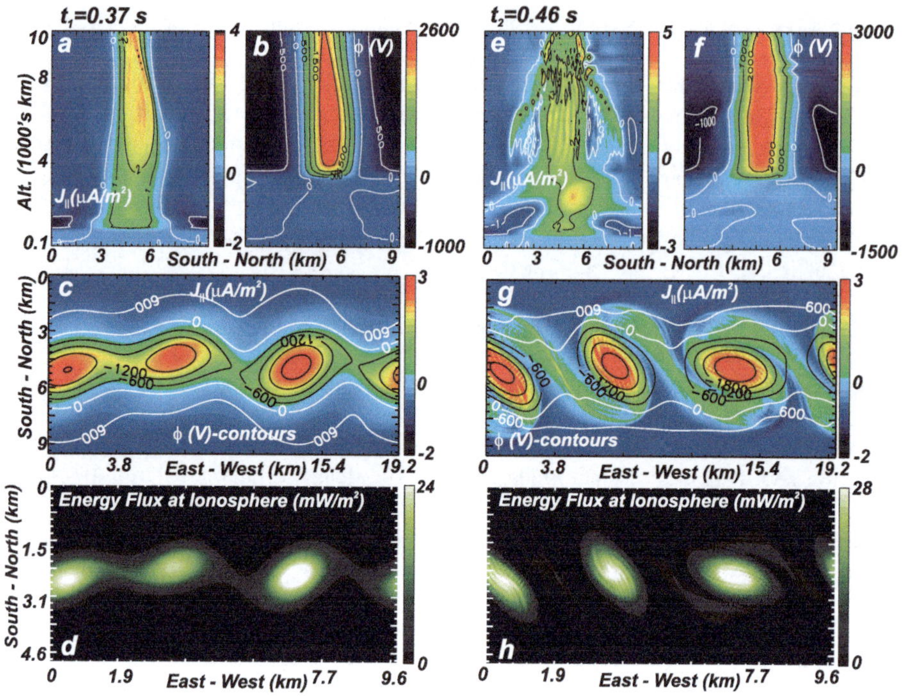

Fig. 5 Snapshots in the evolution of an auroral current sheet for plasma parameters that provide M_sub_A > 1 and auroral 'curls' (After Chaston and Seki 2010). **a**) and **e**) show vertical structure, **c**) and **d**) show horizontal slices of the current and potential through the acceleration region while **g**) and **h**) show energy flux at the ionosphere. Spatial scales vary with altitude

Eq. (3) is significant. For an 'Alfvénic' arc with $\omega \sim k_\parallel v_A$, where v_A is the Alfvén speed, and $v_\perp \sim b_\perp v_A / B_o$, then Ampere's law provides the ratio of the non-linear terms to the linear terms in Eq. (3) as,

$$\sim k_\perp b_\perp / k_\parallel B_0. \tag{4}$$

This relationship shows that as an auroral arc becomes narrower ($\sim k_\perp b_\perp$ increases) so too does the importance of the non-linear force terms. If ω is defined by an Alfvén transit time to the ionosphere from the acceleration region at $\sim$1 Earth radii, then using models for the Alfvén speed profiles above the aurorae (e.g. Stasiewicz et al. 2000a,b) the dynamics of kilometer scale arcs with $J_\parallel \gtrsim 1\ \mu\text{A/m}^2$ will be dominated by the non-linearities leading to instabilities (Seyler 1988). This recognition explains why small scale auroral features are usually observed to be distorted from laminar form.

A consideration of the relative magnitudes of the convective and magnetic non-linearities in Eq. (3) provides further insight into what drives the distortion of auroral forms in the auroral acceleration region. The ratio of these terms reduces to the Alfvén Mach number $M_A = (\Delta v / \Delta v_{A\perp})$ where $\Delta v_{A\perp} = \Delta b_\perp / \sqrt{\mu_o \rho}$, and Δv and $\Delta b_\perp$ are respectively the change in velocity and magnetic field across the arc (Chaston and Seki 2010). Values greater than one may give rise to instabilities driven by flow shear (e.g. Kelvin Helmholtz; Fig. 5) thought to lead to the features described above as 'folds' and 'curls' (Hallinan and Davis 1970) while values less than one lead to instabilities driven by magnetic shear (e.g. tearing, Seyler 1990) identified in vortices observed from the Reimei spacecraft (Chaston 2015a,b;

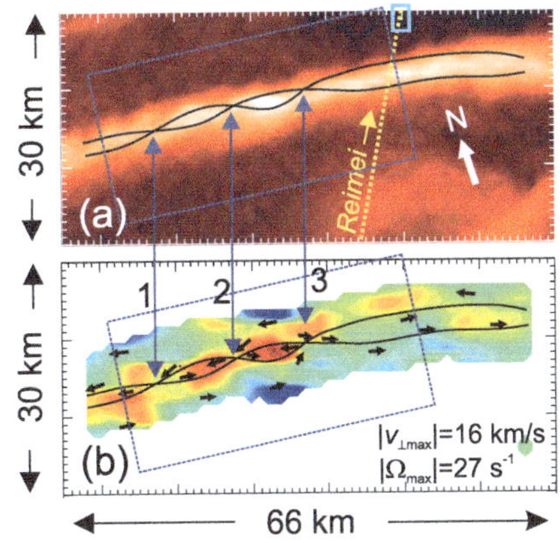

Fig. 6 Observations of likely tearing instability in a discrete feature from the Reimei spacecraft (After Chaston 2015a,b). **a**) shows intensity at 670 nm while **b**) shows the corresponding optical vorticity with arrows indicating the direction of optical flow

Fig. 6). For Alfvénic arcs on kilometer scales electron inertial effects in the acceleration region provide $M_A > 1$ favoring flow shear. This local picture is however complicated by reflections from the conducting ionosphere, and/or density gradients at the base of the acceleration region that alter the relationships between the field and flows along the flux tube (Lysak and Dum 1983; Seyler 1988). Alternatively, for 'quasi-static' arcs, as described by the current voltage relation $J_{\|} = -K\phi$, we find $M_A = 1/\mu_o \Delta x^2 v_A K$ where Δx is the arc width and K is the field-line conductivity defined for example by Knights relation (Knight 1973). In this model the cross-over between magnetic-shear or flow-shear dominance for kilometer scale features and acceleration region values for v_A occurs at $K \sim 10^{-9}$ (ohm m^2)$^{-1}$. It is interesting to note that $K \sim 10^{-9}$ (ohm m^2)$^{-1}$ is a typical value derived from in-situ observations above discrete aurora (Lyons 1981; Sakanoi et al. 1995; Elphic et al. 1998). This suggests that both flow shear and magnetic shear are intrinsic to the observed evolution of discrete auroral forms. Moreover, the dependence on K is indicative of the importance of the electron kinetics in defining the evolution of auroral arcs and from a macroscopic perspective represents of the importance of the properties of the electron source region that in turn define K. Qualitatively, hot tenuous magnetospheric source regions will tend to drive the evolution through flow shears while cooler more dense source region plasmas will provide evolutionary sequences that are more likely controlled by magnetic shear.

A detailed exposition of various instabilities active along an auroral flux-tube that can provide small scale structuring of auroral forms is provided in Seyler and Wu (2001) and Wu and Seyler (2003). Other relevant works not already cited include Wagner et al. (1983); Lotko et al. (1987); Chmyrev et al. (1992), Otto and Birk (1993), Rankin et al. (1993), Lysak and Song (1996), Shukla and Stenflo (1999), Peñano and Ganguli (2000) and Chaston et al. (2011) among others. How the processes described in these works, which operate largely through the acceleration region connected to a passive ionosphere, couple to an active and self-consistently evolving ionosphere is required for understanding the evolution of auroral displays on time-scales longer than an Alfvén bounce time. In the following section the representative case of coupling ionospheric feedback to the Kelvin-Helmholtz instability which is active through the acceleration region is described.

The feedback instability in the magnetosphere-ionosphere (M-I) coupling was first proposed as a mechanism of self-excitation of quiet auroral arcs as briefly summarized in the review of discrete arcs (Borovsky et al. 2020), and has been developed with extensions introducing a variety of physics models.

In the feedback M-I coupling model, the shear Alfvén waves propagating along auroral field lines carry the field aligned current $\tilde{j}_{\|}$ and the plasma motion $\tilde{v}$ given by the $E \times B$

drift, where tilde ($\tilde{}$) denotes a perturbed quantity. Continuity of $\tilde{j}_\parallel$ and $\tilde{E}$ between the magnetosphere and the ionosphere constructs the M-I coupling, and induces auroral structuring with the ionospheric density change $\tilde{n}$. In the original feedback instability theory, the typical time scale (τ) of auroras is characterized by periods of low-order Alfvén harmonics of field line resonance (say, order of a minute), and the horizontal scale at the ionospheric altitude is given by $V_d\tau$, where V_d means the Pedersen or Hall drift speed in the auroral region. If $V_d \sim 100$ m/sec and $\tau \sim 100$ sec, one finds a simple estimate of the perpendicular (horizontal) wave length of fluctuations $\lambda \sim 10$ km which is comparable to latitudinal scales of auroral arcs. The feedback instability spontaneously generates auroral structures of $\tilde{j}_\parallel$, $\tilde{B}$, $\tilde{E}$, $\tilde{v}$, and $\tilde{n}$ in the form of a two-dimensional (obliquely propagating) plane wave, where $\tilde{B}$ and $\tilde{E}$ mean the perturbed electromagnetic fields with the same space-time scales as $\tilde{j}_\parallel$, $\tilde{v}$, and $\tilde{n}$. In the following, we discuss possible mechanisms or extensions of the feedback instability, leading to smaller-scale structures of auroras.

Now, let us consider extensions of the above model in a framework of the linear instability. In case with inhomogeneity of the Alfvén speed along the auroral field line, such as the IAR discussed below, a shorter time scale for the feedback instability is brought by the shear (possibly dispersive) Alfvén waves with higher frequencies (of the order of $\tau \sim 1$ sec; see the following subsection). It leads to excitation of auroral structures with smaller horizontal spatial scales of $\lambda \sim 100$ m, if we take into account the feedback resulting from M-I coupling with the IAR (Lysak 1991). It is remarked that the horizontal scale of auroras considered here stems from the field-aligned inhomogeneity of the Alfvén speed, and not from the perpendicular one resulting from phase mixing and wave propagation.

Extensions of the MHD equations with effects of the finite gyroradius and the electron inertia are essential to incorporate the parallel electric field $\tilde{E}_\parallel$ of the dispersive Alfvén waves and particle acceleration, and simultaneously introduce additional spatial scales such as the ion acoustic gyroradius or the electron skin depth. The small scales related to kinetic dynamics of magnetospheric plasma often lead to stabilization of short wavelength components and a cut off wavenumber for the feedback instability as found in the M-I coupling analysis based on the gyrokinetic theory (Watanabe 2014).

Introduction of pressure perturbations and field line curvature provides an additional instability in the M-I coupling. The ballooning instability in the MHD or kinetic regimes has widely been investigated as a possible mechanism for auroral beading (or wavy) structures. Since the cut-off wavenumber of the ballooning instability is given by extended MHD effects such as the finite ion gyroradius, small-scale structures can be generated even in the linear instability. An integrated modelling of the feedback and ballooning instabilities is recently developed by means of the reduced MHD framework of the three-dimensional M-I coupling (Watanabe 2020).

Nonlinear interactions of shear (or dispersive) Alfvén waves bring other possible mechanisms to create auroral fine structures through the feedback M-I coupling. Let us recall the feedback instability generates $\tilde{v}$, that is, the $E \times B$ flow driven by the perturbed electric field and the static magnetic field. Since Alfvén waves are accompanied by $\tilde{v}$, a horizontal sheared flow with a spatial scale of λ is induced in the ionosphere. As the feedback instability grows, the flow speed as well as the velocity shear is amplified. If the flow shear exceeds a critical value, the Kelvin-Helmholtz (K-H) type instability is secondarily is switched on as shown in Fig. 7, if the nonlinear wave interaction is properly included in the M-I coupling (Watanabe 2010 and Watanabe et al. 2016). This is because the K-H instability is driven by the Reynolds stress coming from the advection term $\tilde{v} \cdot \nabla \tilde{v}$ in the equation of motion of a fluid. Here, the K-H instability secondary induced by the feedback instability leads to deformation of the primary arc structure, and generates vortex structures in the M-I coupling

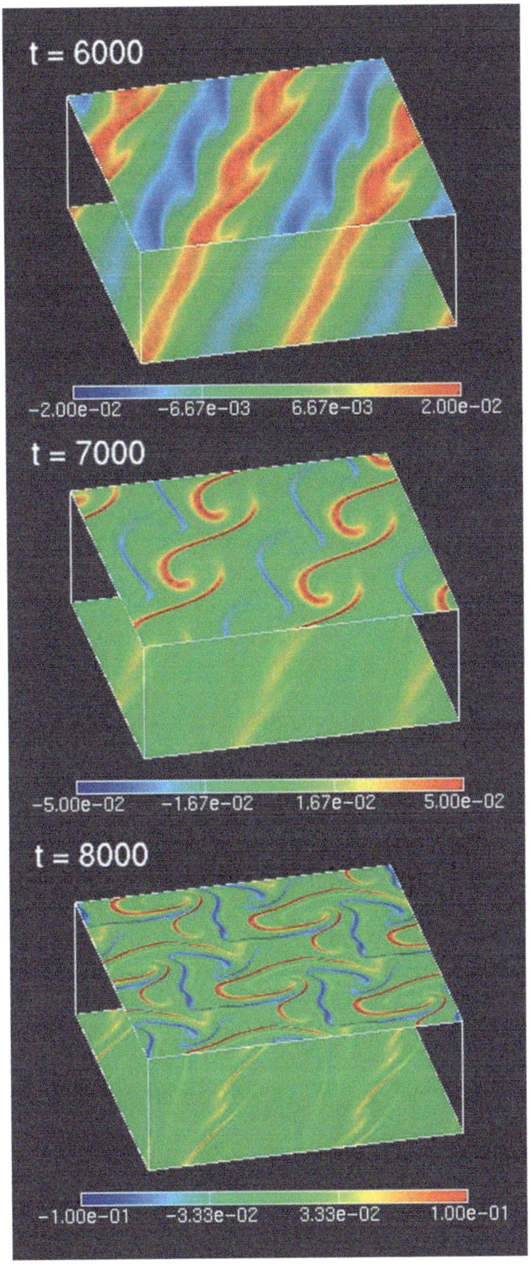

Fig. 7 Color contours of the vorticity distributions on the ionosphere (lower plane) and the magnetic equator (upper plane) at three different time steps of the nonlinear simulation, where the vertical scale is shortened just for clarity of the plots. (After Watanabe 2010)

system, of which pattern has a similarity to folds observed in auroral breakup arc (Kataoka et al. 2011a,b). It is noteworthy to mention that the primary arc structure causing the K-H instability is also a result of spontaneous growth of the feedback instability, and is not *a priori* given by the initial set up of the system. Direct numerical simulation using the three-dimensional nonlinear M-I coupling model clearly captures the secondary K-H instability growth and transition to turbulence (Watanabe et al. 2016).

If the nonlinear interaction of the Alfvén waves propagating in the upward and downward directions is strong enough, the feedback instability develops into a fully turbulent state via the K-H instability growth. Interactions of counter propagating shear (or kinetic) Alfvén waves have widely been investigated in studies of the MHD turbulence (Goldreich and Sridhar 1995). In the quasi two-dimensional MHD turbulence with a strong external magnetic field, fluctuations with higher perpendicular wavenumbers are produced through interactions of the counter propagating shear Alfvén waves, leading to the energy cascade and the power spectrum of $k_\perp^{-5/3}$, where $k_\perp$ denotes the perpendicular wavenumber. This is the case that we consider for the feedback M-I coupling with nonlinearity. Indeed, a continuous power spectrum of the Alfvénic turbulence is found in the nonlinear regime of the feedback instability (Watanabe et al. 2016). It is qualitatively consistent with FAST observations of the power law spectrum of dispersive Alfvén waves (Chaston et al. 2008). More detailed theoretical studies and quantitative comparisons with observations are desired to determine what type of auroral structures are caused through turbulent Alfvénic wave interactions of the magnetosphere and the ionosphere.

Much of the previous analysis of the feedback instability has assumed a slab ionosphere described by height-integrated Pedersen and Hall conductances. However, the ionosphere has a finite thickness and an inhomogeneous profile of the conductivities and the Alfvén speed. An analysis of the feedback instability in a model including a full altitude profile of the Pedersen conductivity was conducted by Sydorenko and Rankin (2017). They noted that the inhomogeneity in the ion-neutral collision frequency produced a large shear in the Pedersen drift, and considered that this shear tends to smear out the density perturbations produced by the feedback interaction, preventing this interaction from going unstable. Later, Watanabe and Maeyama (2018) performed an eigenmode analysis of the instability in a model with an inhomogeneous ionosphere and found that unstable eigenmodes exists for cases with long parallel wavelengths in field line resonances, where the effect of inhomogeneity in the ionosphere is much less significant than in the IAR. Further work will be necessary to clarify under which conditions, if any, the feedback instability exists.

However, the feedback interaction may be effective at generating small-scale currents even in the absence of an instability. For example, Russell et al. (2013) performed numerical experiments in which the magnetosphere was assumed to be uniform, so that the reflection of waves from the ionospheric density gradient or the conjugate ionosphere, which is required for the feedback instability, was not present. Despite this, they found that in the downward current region, the upward motion of the electrons led to a depletion of the ionospheric density. This created conditions favorable to the small-scale structuring of the field-aligned currents. Sounding rocket results from Lynch et al. (2015) have shown that conductivity enhancements caused by the precipitation of electrons can also lead to structuring of the field-aligned currents. In addition, they point to the role of the neutral wind dynamo as an important effect in the structuring of currents. These results make it clear that the further investigations of the magnetosphere-ionosphere interaction are required to understand the structuring of auroral currents.

3.3 Ionospheric Alfvén Resonator

In the ionosphere, a resonant cavity known as the Ionospheric Alfvén Resonator (IAR) can form when the plasma density drops more steeply with altitude than the decrease in strength of the dipole magnetic field. The IAR has an Alfvén speed profile that increases to a peak value in the range 4000-8000 km before dropping more slowly at higher altitudes. It was first identified by Polyakov and Rapoport (1981) during ionospheric heating experiments.

Subsequently, Trakhtengertz and Feldstein (1984, 1991), and Lysak (1991, 1993), discussed the potential importance of the IAR in the context of auroral region physics. They showed that the sharp gradient in the Alfvén speed in the IAR results in partial reflection of Alfvén waves, trapping a fraction of their energy between the bottom of the ionosphere and the Alfvén speed peak. As will be shown below, the IAR is associated with electron acceleration on spatial and temporal scales characteristic of small-scale, dynamic aurora. For typical ionospheric parameters, the fundamental IAR resonance frequency is in the 0.1-1.0 Hz range.

A comprehensive theory of the IAR was derived by Lysak and Yoshikawa (2006), who suggested the trapped resonant wave modes could perhaps explain observations of broadband precipitating electrons with energies ranging from 100 eV to over one keV. The particle distributions of these precipitating electrons were identified as superthermal electron bursts (e.g., Johnstone and Winningham 1982; McFadden et al. 1986, 1987); however, definitive identification of the association between these types of bursts and Alfvén waves in the IAR came from NASA FAST satellite observations (e.g., Chaston et al. 2000, 2002a,b, 2003a,b). In a series of publications, Chaston demonstrated that the Alfvénic aurora is consistent with the particle distributions measured by FAST. The evidence lies in the fact that a broadband wave distribution means that electrons see a fluctuating parallel electric field during their passage through the acceleration region. An electron of energy 100 eV travels at roughly 1 $R_E\ s^{-1}$, and thus for an acceleration length of around 1 R_E, fluctuations on time scales of seconds are required to produce the observed particle distributions. Test particle modeling established this process is possible (Thompson and Lysak 1996), while later more detailed test particle simulations compare well with FAST data (Chaston et al. 2002a,b), confirming the Alfvénic nature of the acceleration process observed by FAST.

The focus here is on developments since the previous ISSI book on the aurora (Paschmann et al. 2003). Models of the IAR developed before 2003 generally considered the linearized MHD equations, with the addition of two-fluid effects of finite electron inertia and electron thermal pressure. Later, these two-dimensional models were extended to include the nonlinear Lorentz force due to field-aligned currents on auroral flux tubes (Sydorenko et al. 2008; Streltsov and Lotko 2008). These authors considered a geometry in which the z-direction is along geomagnetic field lines, the y-direction is azimuthal, and the x-direction completes the right-hand system. Since Alfvén waves have an electric field in the x-direction, and a magnetic perturbation in the y-direction, this leads to a nonlinear ponderomotive force, $j_x B_y$ directed upward along the geomagnetic field. Sydorenko et al. (2008) and Streltsov and Lotko (2008) described how this force accelerates ions away from the acceleration region to form a density cavity that enhances the ability of the plasma to support a parallel electric field. The generation of parallel electrostatic fields together with low density cavities is a positive feedback process in auroral acceleration regions, since the increased parallel electric field also contributes to the acceleration of ions and electrons out of the acceleration region, deepening the density cavity (Song and Lysak 2006). This process is a result of Alfvénic interactions in the solar wind-magnetosphere-ionosphere coupling system.

Phase mixing at the boundaries of these density cavities leads to a decrease in the perpendicular wavelength, further enhancing the parallel electric field, as described by Lysak and Song (2008), who prescribed a density cavity rather than having it form self-consistently. The structuring of waves perpendicular to the geomagnetic field was confirmed using observations from the FAST satellite (Chaston et al. 2006), which showed that density depletion was associated with ion outflow produced by Alfvén waves. Similar physics operates in field line resonances, where low frequency standing waves produce density cavities, perpendicular structuring and enhancement of parallel electric fields, and ion outflow (Rankin et al. 1999a,b; Lu et al. 2003a,b).

While early models of the IAR assumed that the field lines were vertical, measurements at low latitude (Simões et al. 2012) from the C/NOFS satellite have also shown IAR signatures. Lysak et al. (2013) developed a model in non-orthogonal dipole coordinates to investigate the development of Alfvén waves in the IAR at lower latitudes. This work has shown the existence of the IAR when magnetic dip angle is less than 90 degrees, although not to the equatorial latitudes described by Simões et al. (2012).

The modeling work of Lysak et al. (2013) also confirmed earlier suggestions (e.g., Greifinger and Greifinger 1968; Fraser 1975a,b; Fujita and Tamao 1988; Knudsen et al. 1992; Neudegg et al. 1995) that in the presence of Hall conductivity, shear Alfvén waves can mode convert to the fast wave. While the shear Alfvén mode propagates along the field line, the fast mode can propagate across magnetic field lines, and so the sharp increase in the Alfvén speed leads to an ionospheric waveguide. Thus, the incident shear Alfvén wave can produce ground signatures 100's of kilometers from the source. The process was modeled by Woodroffe and Lysak (2012) who showed that the polarization pattern from such a localized source is consistent with the observations (Fraser 1975b).

A great deal of work on the IAR in the last 15 years has focused on verifying the expected signatures of this structure with satellites and sounding rockets as well as ground observations. Much of this work has considered the ratio of the perpendicular electric field to the orthogonal component of the magnetic perturbation and their phase difference. For a static sheet-like current system, field-aligned currents are closed by the presence of Pedersen conductance, for which this ratio is

$$\delta E_x/\delta B_y = 1/\mu_0\Sigma_{\mathrm{P}}. \tag{5}$$

This expression uses the same coordinates defined earlier. On the other hand, for an inertial Alfvén wave, this ratio is

$$\delta E_x/\delta B_y = V_A[(1+k_\perp{}^2\lambda_e{}^2)(1+k_\perp{}^2\rho_i{}^2))^{1/2}. \tag{6}$$

Here, we have included the finite ion gyroradius term but have neglected the electron kinetic term (often written in terms of the ion acoustic gyroradius) since it is not usually important at low altitudes in the IAR. For brevity, we will refer to this ratio as the "E/B ratio" in the discussion that follows. For typical parameters, the ratio indicated by Eq. (6) is usually much less than that given by Eq. (5). For both of these cases, the phase shift between δE_x and δB_y is 0 or 180 degrees, depending on the direction of the geomagnetic field. On the other hand, the IAR is characterized by the interference of up and down going waves and so the phase shift for a strong gradient in Alfvén speed oscillates around ± 90 degrees, switching over when the mode structure of the IAR has an approximate node in either the electric or magnetic field at the point of observation (e.g., Lysak 1991). In addition, the E/B ratio becomes a function of frequency, becoming quite large, for example, when the standing wave pattern has a node in the magnetic field at the point of observation. In general, for realistic (Chapman-layer-like) ionospheric density profiles the magnitude of the reflection coefficient for Alfvén waves at the lower boundary of the IAR is less than unity, and often much less, leading to $E-\delta B$ phase shifts much smaller than ± 90 degrees, as shown by Knudsen et al. (1992), who verified these predictions with sounding rocket observations. Furthermore, the ionospheric reflection coefficient varies strongly with frequency in the same range as the characteristic frequencies of the IAR (of the order of 1 Hz). This is due to the decrease of the effective conductivity due to the collisional skin depth effect, which becomes smaller than the ionospheric thickness at these frequencies (Lysak et al. 2013), as well as coupling to the fast mode.

Using Akebono data, Hirano et al. (2005) investigated the properties of waves in the IAR at three different altitudes, roughly 2000 km, 4000 km and 7500 km. They showed that at the lowest altitude range the E/B ratio is lower, with E/B approaching $1/\mu_0\Sigma_P$, while at higher altitudes they were more consistent with E/B $= V_A$. In addition, they showed that the phase relationship between E and B was consistent with the interference of waves in the IAR. One interesting observation is that the ratio at 2000 km altitude approached the inertial Alfvén wave case discussed above for perpendicular wavelengths shorter than 1 km. This may be because these short wavelength waves can be absorbed in the ionosphere (e.g., Forget et al. 1991; Lessard and Knudsen 2001).

Similar observations have been carried out more recently using the enhanced Polar Outflow Probe (ePOP) and the Swarm constellation of satellites (Miles et al. 2018). They observed that the E/B ratio approached the ionospheric values for lower frequency, consistent with the interpretation that the ionospheric E/B ratio should hold within a fraction of one wavelength of the ionosphere, which would be a larger distance for low frequency waves. Pakhotin et al. (2018, 2020) used the multi-point measurements from Swarm to verify the time-dependent nature of the wave structures, and also considered the field ratio and phase difference as a function of frequency to indicate correspondence with the standing wave model. Moreover, they pointed out that the low-frequency E/B ratio could be used as an independent measure of the Pedersen conductance in the ionosphere.

Wu et al. (2020a) carried out a statistical study of regions of intense E and δB field fluctuations in the 0-8 Hz range, selecting for enhanced correlations between the two fields as an indication of Alfvénic activity. Those authors searched for a consistent pattern between Alfvénic fluctuations, field-aligned currents, and auroral arcs observed with white-light all-sky cameras, and found no one-to-one relation between them. Rather, regions of enhanced Alfvén waves occurred within both upward and downward field-aligned current regions, and at their boundaries, and at varying distances away from bright auroral arcs. One conclusion that can be drawn from that study is that the ultimate sources of Alfvénic (and IAR) energy remain to be identified.

This combination of new model developments and more detailed observations have firmly established that the ionospheric Alfvén resonator is an important structure for the development of auroral field-aligned currents and the resulting particle acceleration. The IAR is particularly important in the evolution of small-scale structures when coupling with the ionospheric feedback interactions discussed in the previous section of this publication.

3.4 Conclusions

In recent decades, our understanding of small-scale dynamic auroras has been rapidly advanced via both improved observations and theories. Advancements in camera technology have been parallelled by developments of in situ plasma measurements, which demonstrate magnetic field-aligned electron beams that appear bursty in the spacecraft frame during active periods such as auroral breakup. However, such measurements are often challenging to interpret given the space-time ambiguity. A suborbital sounding rocket crosses a 1-km-wide structure in 1 s, which is also the characteristic lifetime of such a structure as observed by ground-based cameras. Thus it is difficult to simultaneously resolve event lifetimes and spatial scales, both of which are needed to constrain auroral theories. Camera imaging can be used to resolve this ambiguity when available; another approach is the use of multiple spacecraft. More complete specification of the dynamic structure of small-scale aurora in space and time remains an important research goal. New development will also be valuable in carefully resolving the vertical luminosity distribution of small-scale dynamic auroras (Ivchenko et al. 2005; Dahlgren et al. 2013; Tuttle et al. 2014; Kataoka et al. 2016), which is complementary information to that of the multi-spectral observations (Dahlgren et al. 2008; Kataoka et al. 2011a,b).

Future observations of the various types of dynamic auroral forms should be put into context if the community is to understand what causes them and what their impact on the system might be. Future reports of the occurrence of these forms should state, as appropriate, (1) the geomagnetic activity level, (2) the phase of substorms, (3) the local time, (4) the relative latitude in the auroral zone (e.g. high-latitude or low-latitude), and (5) what other auroral forms are present and where is the dynamic form located relative to those other aurora.

On the theory side, Alfvén waves play an essential role in the formation of small-scale dynamic auroras. Several different cascading modes of Alfvén waves can form a variety of fine structures. For example, K-H and tearing instabilities of an arc lead to "folds" and "curls" (Hallinan and Davis 1970; Wagner et al. 1983; Wu and Seyler 2003; Chaston and Seki 2010), although reconnection in the M-I region also contributes to a particular type of vortical motions of auroras (Chaston 2015b). DAWs play the essential role for causing "packets" (Semeter et al. 2008) and "rays" (Lynch et al. 2012). The interference of resonant EMIC waves cause "flickering" (Temerin et al. 1986; Sakanoi et al. 2005). Consistent theories have been therefore proposed for small-scale dynamic auroras. However, predictive simulations have not been tested yet. In other words, it is still an open question how we can predict the occurrence of small-scale dynamic auroras such as curls, folds, flickering, and packets.

In future, 3-D kinetic simulations that extend from the equator to the ionosphere should be made to understand the driving of the Alfvénic perturbations and their dynamics in the presence of self-consistently formed parallel electric fields, based on the future investigations of the Alfvénic energy source.

What is the essential role of the filamentation of auroras for global energetics? Cross-scale coupling has been poorly investigated by observations. Continuous international effort is therefore needed for systematic monitoring observations of both global and local auroras at the same time, which would be necessary to build a firm empirical relationship among global-scale, mesoscale, and the small-scale dynamic auroras as reviewed in this article. Systematic surveys have also been conducted to identify the large-scale context of the small-scale structures, such as large-scale upward current or downward current (Liang et al. 2019; Wu et al. 2020a, 2020b). Next generation high-resolution global imaging of aurora from spacecraft would be a dream, but for the time being, rapidly increasing resolution of ground-based all-sky cameras, the networked distributions of imagery and small spacecraft, and timely data sharing combined with realistic simulations will renovate this aspect. Citizen scientists can play a key role in providing high quality opportunistic observations, as they have in the study of STEVE (MacDonald et al. 2018) and its mysterious filamentary structures (Semeter et al. 2020). Quantitative measurement of the filamentary structures using ground-based facilities combined with cutting-edge cameras would be essential to accumulate the quantitative knowledge of small-scale dynamic auroras (Lanchester et al. 1994, 1997, 1998, 2001), which has been taken over to the ongoing EISCAT 3D project.

Acknowledgements The work of RK was supported by JSPS Kakenhi (25302006, 15H02628, 16H06286). C.C. Chaston was supported by ISSI and through NASA grant NNX17AI55G. T.-H. Watanabe was supported by ISSI and by JSPS KAKENHI Grant Number JP16H04086 and JP17H01177. D.K. Whiter was supported by ISSI and NERC (UK) through grant NE/S015167/1. Movies 4, 6 and 7 are from the ASK instrument which has been funded by PPARC, STFC (PP/C502614/1) and NERC (NE/H024433/1, NE/N004051/1) of the United Kingdom and Vetenskapsrådet of Sweden. The authors wish to thank the International Space Science Institute ISSI-Bern for organization of this review and for financial support of the team meeting.

References

H. Akbari, J.L. Semeter, H. Dahlgren, M. Diaz, M. Zettergren, A. Stromme, M.J. Nicolls, C. Heinselman, Anomalous ISR echoes preceding auroral breakup: evidence for strong Langmuir turbulence. Geophys. Res. Lett. **39**, L03102 (2012). https://doi.org/10.1029/2011GL050288

H. Akbari, P. Guio, M.A. Hirsch, J.L. Semeter, Zakharov simulations of beam-induced turbulence in the auroral ionosphere. J. Geophys. Res. Space Phys. **121**(5), 4811–4825 (2016). https://doi.org/10.1002/2016JA022605

L. Andersson, N. Ivchenko, J. Clemmons, A.A. Namgaladze, B. Gustavsson, J.-E. Wahlund, L. Eliasson, R.Y. Yurik, Electron signatures and Alfvén waves. J. Geophys. Res. **107**(A9), 1244 (2002). https://doi.org/10.1029/2001JA900096

R.L. Arnoldy, K.A. Lynch, J.B. Austin, P.M. Kintner, Energy and pitch angle-dispersed auroral electrons suggesting a time-variable, inverted-V potential structure. J. Geophys. Res. **104**(A10), 22613–22621 (1999). https://doi.org/10.1029/1999JA900219

K. Asamura, D. Tsujita, H. Tanaka, Y. Saito, T. Mukai, M. Hirahara, Auroral particle instrument onboard the index satellite. Adv. Space Res. **32**(3), 375–378 (2003). https://doi.org/10.1016/S0273-1177(03)90275-4

K. Asamura, C.C. Chaston, Y. Itoh, M. Fujimoto, T. Sakanoi, Y. Ebihara, A. Yamazaki, M. Hirahara, K. Seki, Y. Kasaba, M. Okada, Sheared flows and small-scale Alfvén waves generation in the auroral acceleration region. Geophys. Res. Lett. **36**, L05105 (2009). https://doi.org/10.1029/2008GL036803

G. Atkinson, Auroral arcs: result of the interaction of a dynamic magnetosphere with the ionosphere. J. Geophys. Res. **75**(25), 4746–4755 (1970). https://doi.org/10.1029/JA075i025p04746

R. Beach, G.R. Cresswell, T.N. Davis, T.J. Hallinan, L.R. Sweet, Flickering, a 10 cps fluctuation within bright auroras. Planet. Space Sci. **16**, 1525 (1968)

A.V.A. Birn, A.V. Artemyev, D.N. Baker, M. Echim, M. Hoshino, L.M. Zelenyi, Particle acceleration in the magnetotail and aurora. Space Sci. Rev. **173**, 49–102 (2012). https://doi.org/10.1007/s11214-012-9874-4

D. Biskamp, *Magnetohydrodynamic Turbulence* (Cambridge University Press, Cambridge, 2003). ISBN 0521810116

E.M. Blixt, T. Grydeland, N. Ivchenko, T. Hagfors, C. La Hoz, B.S. Lanchester, U.P. Lvhaug, T.S. Trondsen, Dynamic rayed aurora and enhanced ion-acoustic radar echoes. Ann. Geophys. **23**, 3–11 (2005)

J.E. Borovsky, Auroral arc thicknesses as predicted by various theories. J. Geophys. Res. Space Phys. **98**(A4), 6101–6138 (1993)

J.E. Borovsky, Fine-scale structures in auroral arcs: an unexplained phenomenon, in *Space Plasmas: Coupling Between Small and Medium Scale Processes*, ed. by M. Ashour-Abdalla, T. Chang, P. Dusenbery. Geophys. Monogr. Ser., vol. 86 (Am. Geophys. Union, Washington, 1995), pp. 255–267

J.E. Borovsky, D.M. Suszcynsky, Optical measurements of the fine structure of auroral arcs, in *Auroral Plasma Dynamics*, ed. by R.L. Lysak (1993). https://doi.org/10.1029/GM080p0025

J.E. Borovsky, J. Birn, M.M. Echim et al., Quiescent discrete Auroral arcs: a review of magnetospheric generator mechanisms. Space Sci. Rev. **216**, 1 (2020). https://doi.org/10.1007/s11214-019-0619-5

C.C. Chaston, ULF waves and auroral electrons, in *ULF Waves*. AGU Geophysical Monograph Series, vol. 169 (2006), p. 239

C.C. Chaston, Magnetic reconnection in the auroral acceleration region. Geophys. Res. Lett. **42**, 1646–1653 (2015a). https://doi.org/10.1002/2015GL063164

C.C. Chaston, Inverted-v auroral arcs and Alfvén waves, in *Auroral Dynamics and Space Weather*, ed. by e.Y. Zhang, L.J. Paxton (Wiley, Hoboken, 2015b). https://doi.org/10.1002/9781118978719.ch3

C.C. Chaston, K. Seki, Small-scale auroral current sheet structuring. J. Geophys. Res. **115**, A11221 (2010). https://doi.org/10.1029/2010JA015536

C.C. Chaston, C.W. Carlson, W.J. Peria, R.E.E.J.P. McFadden, FAST observations of inertial Alfvén waves in the dayside aurora. Geophys. Res. Lett. **26**(6), 647–650 (1999)

C.C. Chaston, C.W. Carlson, R.E. Ergun, J.P. McFadden, Alfvén waves, density cavities and electron acceleration observed from the FAST spacecraft. Phys. Scr. T **84**, 64 (2000)

C.C. Chaston, J.W. Bonnell, L.M. Peticolas, C.W. Carlson, R.E. Ergun, J.P. McFadden, Driven Alfvén waves and electron acceleration: a FAST case study. Geophys. Res. Lett. **29**, 30 (2002a). https://doi.org/10.1029/2001GL013842

C.C. Chaston, J.W. Bonnell, C.W. Carlson, M. Berthomier, L.M. Peticolas, I. Roth, J.P. McFadden, R.J. Strangeway, Electron acceleration in the ionospheric Alfvén resonator. J. Geophys. Res. **107**, 1413 (2002b)

C.C. Chaston, C.W. Carlson, R.E. Ergun, J.P. McFadden, R.J. Strangeway, Properties of small-scale Alfvén waves and accelerated electrons from FAST. J. Geophys. Res. **108**, 8003 (2003a)

C.C. Chaston, L.M. Peticolas, J.W. Bonnell, C.W. Carlson, R.E. Ergun, J.P. McFadden, Width and brightness of auroral arcs driven by inertial Alfvén waves. J. Geophys. Res. **108**, 1091 (2003b)

C.C. Chaston, J. Bonnell, C.W. Carlson, J.P. McFadden, R.E. Ergun, R.J. Strangeway, Kinetic effects in the acceleration of electrons by small scale Alfvén waves: a FAST case study. Geophys. Res. Lett. **30**, 1289 (2003c)

C.C. Chaston, L.M. Peticolas, C.W. Carlson, J.P. McFadden, M. Wilber, G.K. Parks, R.E. Ergun, R.J. Strangeway, M. Andre, Y. Khoyaintsev, M. Goldstein, M. Acuna, E.J. Lund, H. Reme, I. Dandouras, A.N. Fazakerely, A. Balogh, Energy deposition by Alfvén waves into the dayside auroral oval: cluster and FAST observations. J. Geophys. Res. **110**, A02211 (2005)

C.C. Chaston, V. Genot, J.W. Bonnell, C.W. Carlson, J.P. McFadden, R.E. Ergun, R.J. Strangeway, E.J. Lund, K.J. Hwang, Ionospheric erosion by Alfvén waves. J. Geophys. Res. **111**, A03206 (2006). https://doi.org/10.1029/2005JA011367

C.C. Chaston, C.W. Carlson, J.P. McFadden, R.E. Ergun, R.J. Strangeway, How important are dispersive Alfvén waves for auroral particle acceleration? Geophys. Res. Lett. **34**, L07101 (2007a). https://doi.org/10.1029/2006GL029144

C.C. Chaston, A.J. Hull, J.W. Bonnell, C.W. Carlson, R.E. Ergun, R.J. Strangeway, J.P. McFadden, Large parallel electric fields, currents and density cavities in dispersive Alfvén waves above the aurora. J. Geophys. Res. **112**, A05215 (2007b). https://doi.org/10.1029/2006JA012007

C.C. Chaston, C. Salem, J.W. Bonnell, C.W. Carlson, R.E. Ergun, R.J. Strangeway, J.P. McFadden, The turbulent Alfvénic aurora. Phys. Rev. Lett. **100**, 175003 (2008)

C.C. Chaston, K. Seki, T. Sakanoi, K. Asamura, M. Hirahara, Motion of aurorae. Geophys. Res. Lett. **37**, L08104 (2010). https://doi.org/10.1029/2009GL042117

C.C. Chaston, K. Seki, T. Sakanoi, K. Asamura, M. Hirahara, C.W. Carlson, Cross-scale coupling in the auroral acceleration region. Geophys. Res. Lett. **38**, L20101 (2011). https://doi.org/10.1029/2011GL049185

L.-J. Chen, C.A. Kletzing, S. Hu, S.R. Bounds, Auroral electron dispersion below inverted-V energies: resonant deceleration and acceleration by Alfvén waves. J. Geophys. Res. **110**, A10S13 (2005). https://doi.org/10.1029/2005JA011168

V.M. Chmyrev, V.A. Marchenko, O.A. Pokhotelov, P.K. Shukla, L. Stenflo, A.V. Streltsov, The development of discrete active auroral forms. IEEE Trans. Plasma Sci. **20**(6), 764 (1992). https://doi.org/10.1109/27.199525

T. Chust, P. Louarn, M. Volwerk, H. Feraudy, A. Roux, J.-E. Wahlund, B. Holback, Electric fields with a large parallel component observed by the Freja spacecraft: artifacts or real signals? J. Geophys. Res. **103**(A1), 215–224 (1998). https://doi.org/10.1029/97JA02587

A.E. Clark, C.E. Seyler, Electron beam formation by small-scale oblique inertial Alfvén waves. J. Geophys. Res. **104**(A8), 17233–17249 (1999). https://doi.org/10.1029/1999JA900212

I.J. Cohen et al., Auroral current and electrodynamics structure (ACES) observations of ionospheric feedback in the Alfvén resonator and model responses. J. Geophys. Res. Space Phys. **118**, 3288–3296 (2013). https://doi.org/10.1002/jgra.50348

H. Dahlgren, N. Ivchenko, J. Sullivan, B.S. Lanchester, G. Marklund, D. Whiter, Morphology and dynamics of aurora at fine scale: first results from the ASK instrument. Ann. Geophys. **26**, 1041–1048 (2008). https://doi.org/10.5194/angeo-26-1041-2008

H. Dahlgren, A. Aikio, K. Kaila, N. Ivchenko, B.S. Lanchester, D.K. Whiter, G.T. Marklund, Simultaneous observations of small multi-scale structures in an auroral arc. J. Atmos. Sol.-Terr. Phys. **72**, 633 (2010). https://doi.org/10.1016/j.jastp.2010.01.014
H. Dahlgren, B. Gustavsson, B.S. Lanchester, N. Ivchenko, U. Brändström, D.K. Whiter, T. Sergienko, I. Sandahl, G. Marklund, Energy and flux variations across thin auroral arcs. Ann. Geophys. **29**, 1699–1712 (2011). https://doi.org/10.5194/angeo-29-1699-2011
H. Dahlgren, N. Ivchenko, B.S. Lanchester, Monoenergetic high-energy electron precipitation in thin auroral filaments. Geophys. Res. Lett. **39**, L20101 (2012). https://doi.org/10.1029/2012GL053466
H. Dahlgren, J. Semeter, R. Marshall, M. Zettergren, The optical manifestation of field-aligned bursts in the aurora. J. Geophys. Res. **118**, 4572–4582 (2013). https://doi.org/10.1002/jgra.50415
H. Dahlgren, B.S. Lanchester, N. Ivchenko, Coexisting structures from high- and low-energy precipitation in fine-scale aurora. Geophys. Res. Lett. **42**, 1290–1296 (2015). https://doi.org/10.1002/2015GL063173
P.A. Damiano, J.R. Johnson, Mirror force induced wave dispersion in Alfvén waves. Phys. Plasmas **20**, 062901 (2013). https://doi.org/10.1063/1.4810788
P.A. Damiano, J.R. Johnson, C.C. Chaston, Ion temperature effects on magnetotail Alfvén wave propagation and electron energization. J. Geophys. Res. Space Phys. **120**, 5623–5632 (2015). https://doi.org/10.1002/2015JA021074
P.A. Damiano, J.R. Johnson, C.C. Chaston, Ion gyroradius effects on particle trapping in kinetic Alfvén waves along auroral field lines. J. Geophys. Res. Space Phys. **121**, 10,831–10,844 (2016). https://doi.org/10.1002/2016JA022566
P.A. Damiano, C.C. Chaston, A.J. Hull, J.R. Johnson, Electron distributions in kinetic scale field line resonances: a comparison of simulations and observations. Geophys. Res. Lett. **45**, 5826–5835 (2018). https://doi.org/10.1029/2018GL077748
J. Dombeck, C. Cattell, J.R. Wygant, A. Keiling, J. Scudder, Alfvén waves and Poynting flux observed simultaneously by polar and FAST in the plasma sheet boundary layer. J. Geophys. Res. **110**, A12S90 (2005). https://doi.org/10.1029/2005JA011269
J. Dombeck, C. Cattell, N. Prasad, E. Meeker, E. Hanson, J. McFadden, Identification of auroral electron precipitation mechanism combinations and their relationships to net downgoing energy and number flux. J. Geophys. Res. Space Phys. **123**, 10,064–10,089 (2018). https://doi.org/10.1029/2018JA025749
R.C. Elphic, J.W. Bonnell, R.J.S.L. Kepko, R.E.E.J.P. McFadden, C.W. Carlson, W. Peria, C.A. Cattell, D. Klumpar, E. Shelley, W. Peterson, E. Moebius, L. Kistler, R. Pfaff, The auroral current circuit and field-aligned currents observed by FAST. Geophys. Res. Lett. **25**, 2033 (1998). https://doi.org/10.1029/98GL01158
R.E. Ergun, L. Andersson, D. Main, Y.-J. Su, D.L. Newman, M.V. Goldman, C.W. Carlson, A.J. Hull, J.P. McFadden, F.S. Mozer, Auroral particle acceleration by strong double layers: the upward current region. J. Geophys. Res. **109**, A12220 (2004). https://doi.org/10.1029/2004JA010545
R.E. Ergun, L. Andersson, Y.-J. Su, D.L. Newman, M.V. Goldman, W. Lotko, C.C. Chaston, C.W. Carlson, Localized parallel electric fields associated with inertial Alfvén waves. Phys. Plasmas **12**, 072901 (2005)
R.E. Erlandson, L.J. Zanetti, M.H. Acuña, A.I. Eriksson, L. Eliasson, M.H. Boehm, L.G. Blomberg, Freja observations of electromagnetic ion cyclotron ELF waves and transverse oxygen ion acceleration on auroral field lines. Geophys. Res. Lett. **21**, 1855 (1994)
B. Forget, J.-C. Cerisier, A. Berthelier, J.-J. Berthelier, Ionospheric closure of small-scale Birkeland currents. J. Geophys. Res. **96**, 1843 (1991)
F.R.E. Forme, D. Fontaine, Enhanced ion acoustic fluctuations and ion outflows. Ann. Geophys. **17**, 182–189 (1999). https://doi.org/10.1007/s00585-999-0182-6
L.A. Frank, K.L. Ackerson, Observations of charged particle precipitation into the auroral zone. J. Geophys. Res. **76**(16), 3612–3643 (1971)
B.J. Fraser, Ionospheric duct propagation and Pc1 pulsation source. J. Geophys. Res. **80**, 2790 (1975a)
B.J. Fraser, Polarization of Pc1 pulsations and high and middle latitudes. J. Geophys. Res. **80**, 279 (1975b)
H.U. Frey et al., Small and meso-scale properties of a substorm onset auroral arc. J. Geophys. Res. **115**, A10209 (2010). https://doi.org/10.1029/2010JA015537
S. Fujita, T. Tamao, Duct propagation of hydromagnetic waves in the upper ionosphere, 1, electromagnetic field disturbances in high latitudes associated with localized incidence of a shear Alfvén wave. J. Geophys. Res. **93**(14), 665 (1988)
Y. Fukuda, M. Hirahara, K. Asamura, T. Sakanoi, Y. Miyoshi, T. Takada, A. Yamazaki, K. Seki, Y. Ebihara, Electron properties in inverted-V structures and their vicinities based on Reimei observations. J. Geophys. Res. Space Phys. **119**, 3650–3663 (2014). https://doi.org/10.1002/2013JA018938
Y. Fukuda, R. Kataoka, H.A. Uchida, Y. Miyoshi, D. Hampton, K. Shiokawa, Y. Ebihara, D. Whiter, N. Iwagami, K. Seki, First evidence of patchy flickering aurora modulated by multi-ion electromagnetic ion cyclotron waves. Geophys. Res. Lett. **44**, 3963–3970 (2017). https://doi.org/10.1002/2017GL072956

V. Genot, P. Louarn, F. Mottez, Alfvén wave interaction with inhomogeneous plasmas: acceleration and energy cascade toward smaller scales. Ann. Geophys. **22**, 931 (2004)

C.K. Goertz, R.W. Boswell, Magnetosphere-ionosphere coupling. J. Geophys. Res. **84**(A12), 7239–7246 (1979). https://doi.org/10.1029/JA084iA12p07239

P. Goldreich, S. Sridhar, Toward a theory of interstellar turbulence II. Strong Alfvénic turbulence. Astrophys. J. **438**, 763–775 (1995)

C. Greifinger, P. Greifinger, Theory of hydromagnetic propagation in the ionospheric waveguide. J. Geophys. Res. **73**, 7473 (1968)

G. Grubbs II., R. Michell, M. Samara, D. Hampton, J.-M. Jahn, Predicting electron population characteristics in 2-D using multispectral ground-based imaging. Geophys. Res. Lett. **44**, 15–20 (2018). https://doi.org/10.1002/2017GL075873

T. Grydeland, C. La Hoz, T. Hagfors, E.M. Blixt, S. Saito, A. Stromme, A. Brekke, Interferometric observations of filamentary structures associated with plasma instability in the auroral ionosphere. Geophys. Res. Lett. **30**, 1338 (2003). https://doi.org/10.1029/2002GL016362

B. Gustavsson, J. Lunde, E.M. Blixit, Optical observations of flickering aurora and its spatiotemporal characteristics. J. Geophys. Res. **113**, A12317 (2008). https://doi.org/10.1029/2008JA013515

G. Haerendel, Origin and dynamics of thin auroral arcs. Adv. Space Res. **23**(10), 1637–1645 (1999). https://doi.org/10.1016/S0273-1177(99)00374-9

T.J. Hallinan, Auroral spirals 2. Theory J. Geophys. Res. **81**(22), 3959–3965 (1976)

T.J. Hallinan, The distribution of vorticity in auroral arcs, in *Physics of Auroral Arc Formation*, ed. by S.-I. Akasofu, J.R. Kan. Geophys. Monogr. Ser., vol. 25 (Am. Geophys. Union, Washington, 1981). https://doi.org/10.1029/GM025p0042. 42 pp

T.J. Hallinan, T.N. Davis, Small scale auroral arc distortions. Planet. Space Sci. **18**, 1735–1736 (1970). https://doi.org/10.1016/0032-0633(70)90007-3

A. Hasegawa, Particle acceleration by MHD surface wave and formation of aurora. J. Geophys. Res. **81**, 5083 (1976)

S.M. Hatch, C.C. Chaston, J. LaBelle, Alfvén wave-driven ionospheric mass outflow and electron precipitation during storms. J. Geophys. Res. Space Phys. **121**, 7828–7846 (2016). https://doi.org/10.1002/2016JA022805

S.M. Hatch, J. LaBelle, W. Lotko, C.C. Chaston, B. Zhang, IMF control of Alfvénic energy transport and deposition at high latitudes. J. Geophys. Res. Space Phys. **122**, 12,189–12,211 (2017). https://doi.org/10.1002/2017JA024175

S.M. Hatch, J. LaBelle, C.C. Chaston, Storm phase-partitioned rates and budgets of global Alfvénic energy deposition, electron precipitation, and ion outflow. J. Atmos. Sol.-Terr. Phys. **167**, 1–12 (2018). https://doi.org/10.1016/j.jastp.2017.08.009

Y. Hirano, H. Fukunishi, R. Kataoka, T. Hasunuma, T. Nagatsuma, W. Miyake, A. Matsuoka, Evidence for the resonator of inertial Alfvén waves in the cusp topside ionosphere. J. Geophys. Res. **110**, A07218 (2005). https://doi.org/10.1029/2003JA010329

C.-H. Hui, C.E. Seyler, Electron acceleration by Alfvén waves in the magnetosphere. J. Geophys. Res. **97**(A4), 3953–3963 (1992). https://doi.org/10.1029/91JA03101

A.J. Hull, M. Wilber, C.C. Chaston, J.W. Bonnell, J.P. McFadden, F.S. Mozer, M. Fillingim, M.L. Goldstein, Time development of field-aligned currents, potential drops, and plasma associated with an auroral poleward boundary intensification. J. Geophys. Res. **115**, A06211 (2010). https://doi.org/10.1029/2009JA014651

A.J. Hull, C.C. Chaston, H.U. Frey, M.O. Fillingim, M.L. Goldstein, J.W. Bonnell, F.S. Mozer, The "Alfvénic surge" at substorm onset/expansion and the formation of "inverted vs": cluster and IMAGE observations. J. Geophys. Res. Space Phys. **121**, 3978–4004 (2016). https://doi.org/10.1002/2015JA022000

N. Ivchenko, E.M. Blixt, B.S. Lanchester, Multispectral observations of auroral rays and curls. Geophys. Res. Lett. **32**, L18106 (2005). https://doi.org/10.1029/2005GL022650

A.D. Johnstone, J.D. Winningham, Satellite observations of suprathermal electron bursts. J. Geophys. Res. **87**, 2321 (1982)

R. Kataoka, Y. Miyoshi, T. Sakanoi, A. Yaegashi, K. Shiokawa, Y. Ebihara, Turbulent microstructures and formation of folds in auroral breakup arc. J. Geophys. Res. **116**, A00K02 (2011a). https://doi.org/10.1029/2010JA016334

R. Kataoka, Y. Miyoshi, T. Sakanoi, A. Yaegashi, Y. Ebihara, K. Shiokawa, Ground-based multispectral high-speed imaging of flickering aurora. Geophys. Res. Lett. **38**, L14106 (2011b). https://doi.org/10.1029/2011GL048317

R. Kataoka, Y. Fukuda, Y. Miyoshi, H. Miyahara, S. Itoya, Y. Ebihara, D. Hampton, H. Dahlgren, D. Whiter, N. Ivchenko, Compound auroral micromorphology: ground-based high-speed imaging. Earth Planets Space **67**, 23 (2015). https://doi.org/10.1186/s40623-015-0190-6

R. Kataoka, Y. Fukuda, H.A. Uchida, H. Yamada, Y. Miyoshi, Y. Ebihara, H. Dahlgren, D. Hampton, High-speed stereoscopy of aurora. Ann. Geophys. **34**, 41–44 (2016). https://doi.org/10.5194/angeo-34-41-2016

A. Keiling, J.R. Wygant, C. Cattell, W. Peria, G. Parks, M. Temerin, F.S. Mozer, C.T. Russell, C.A. Kletzing, Correlation of Alfvén wave Poynting flux in the plasma sheet at 4–7 Re with ionospheric electron energy flux. J. Geophys. Res. **107**(A7), 1132 (2002). https://doi.org/10.1029/2001JA900140

A. Keiling, J.R. Wygant, C.A. Cattell, F.S. Mozer, C.T. Russell, The global morphology of wave Poynting flux: Powering the aurora. Science **299**, 383–386 (2003)

A. Keiling, S. Thaller, J. Wygant, J. Dombeck, Assessing the global Alfven wave power flow into and out of the auroral acceleration region during geomagnetic storms. Science Advances **5**(6) (2019) https://doi.org/10.1126/sciadv.aav8411

E. Kendall, R. Marshall, R.T. Parris, A. Bhatt, A. Coster, T. Pedersen, P. Bernhardt, C. Selcher, Decameter structure in heater-induced airglow at the high frequency active auroral research program facility. J. Geophys. Res. **115**, A08306 (2010). https://doi.org/10.1029/2009JA015043

J.S. Kim, R.A. Volkman, Thickness of zenithal auroral arc over Fort Churchill, Canada. J. Geophys. Res. **68**(10), 3187–3190 (1963)

C.A. Kletzing, S. Hu, Alfvén wave generated electron time dispersion J. Geophys. Res. Lett. **28**(4), 693–696 (2001). https://doi.org/10.1029/2000GL012179

S. Knight, Parallel electric fields. Planet. Space Sci. **21**, 741–750 (1973). https://doi.org/10.1016/0032-0633(73)90093-7

D.J. Knudsen, M.C. Kelley, J.F. Vickrey, Alfvén waves in the auroral ionosphere: a numerical model compared with measurements. J. Geophys. Res. **97**(A1), 77–90 (1992). https://doi.org/10.1029/91JA02300

D.J. Knudsen, J.H. Clemmons, J.E. Wahlund, Correlation between core ion energization, suprathermal electron bursts, and broadband ELF plasma waves. J. Geophys. Res. **103**(A3), 4171–4186 (1998)

D.J. Knudsen, E.F. Donovan, L.L. Cogger, B. Jackel, W.D. Shaw, Width and structure of mesoscale optical auroral arcs. Geophys. Res. Lett. **28**, 705–708 (2001)

M.J. Kosch, T. Pedersen, M.T. Rietveld, B. Gustavsson, S.M. Grach, T. Hagfors, Artificial optical emissions in the high-latitude thermosphere induced by powerful radio waves: an observational review. Adv. Space Res. **40**, 365–376 (2007). https://doi.org/10.1016/j.asr.2007.02.061. ISSN 0273-1177

M. Kunitake, T. Oguchi, Spatial-temporal characteristics of flickering spots in flickering auroras. J. Geomagn. Geoelectr. **36**, 121–138 (1984)

B.S. Lanchester, J.R. Palmer, M.H. Rees, D. Lummerzheim, K. Kaila, T. Turunen, Energy flux and characteristic energy of an elemental auroral structure. Geophys. Res. Lett. **21**, 2789–2792 (1994)

B.S. Lanchester, M.H. Rees, D. Lummerzheim, A. Otto, H.U. Frey, K.U. Kaila, Large fluxes of auroral electrons in filaments of 100 m width. Geophys. Res. Lett. **102**, 9741–9748 (1997)

B.S. Lanchester, M.H. Rees, K.J.F. Sedgemore et al., Ionospheric response to variable electric fields in small-scale auroral structures. Ann. Geophys. **16**, 1343–1354 (1998). https://doi.org/10.1007/s00585-998-1343-8

B.S. Lanchester, M.H. Rees, D. Lummerzheim, A. Otto, K.J.F. Sedgemore-Schulthess, H. Zhu, I.W. McCrea, Ohmic heating as evidence for strong field-aligned currents in filamentary aurora. J. Geophys. Res. **106**(A2), 1785? 1794 (2001). https://doi.org/10.1029/1999JA000292

M.R. Lessard, D.J. Knudsen, Ionospheric reflection of small-scale Alfvén waves. Geophys. Res. Lett. **28**, 3573 (2001)

J. Liang, Y. Shen, D. Knudsen, E. Spanswick, J. Burchill, E. Donovan, e-POP and red line optical observations of Alfvénic auroras. J. Geophys. Res. Space Phys. **124**, 4672–4696 (2019). https://doi.org/10.1029/2019JA026679

W. Lotko, B.U. Sonnerup, R.L. Lysak, Nonsteady boundary layer flow including ionospheric drag and parallel electric fields. J. Geophys. Res. **92**(A8), 8635–8648 (1987). https://doi.org/10.1029/JA092iA08p08635

J.Y. Lu, R. Rankin, R. Marchand, V.T. Tikhonchuk, Nonlinear acceleration of dispersive effects in field line resonances. Geophys. Res. Lett. **30**(10), 1540 (2003a). https://doi.org/10.1029/2003GL016929

J.Y. Lu, R. Rankin, R. Marchand, V.T. Tikhonchuk, J. Wanliss, Finite element modeling of nonlinear dispersive field line resonances: trapped shear Alfvén waves inside field-aligned density structures. J. Geophys. Res. Space Phys. **108**(A11), 0148 (2003b). https://doi.org/10.1029/2003JA010035

A.T. Lui, C.D. Anger, A uniform belt of diffuse auroral emission seen by the ISIS-2 scanning photometer. Planet. Space Sci. **21**(5), 799–809 (1973)

E.J. Lund, J. LaBelle, R.B. Torbert, K. Liou, W. Peria, C.A. Kletzing, M.C. Kelley, S.D. Baker, F. Primdahl, H.C. Stenbaek-Nielsen, A. Ranta, G. Haerendel, H.U. Frey, Observation of electromagnetic oxygen cyclotron waves in a flickering aurora. Geophys. Res. Lett. **22**(18), 2465–2468 (1995). https://doi.org/10.1029/95GL02409

E.J. Lund, E. Möbius, L. Tang, L.M. Kistler, M.A. Popecki, D.M. Klumpar, W.K. Peterson, E.G. Shelley, B. Klecker, D. hovestadt, M. Temerin, R.E. Ergun, J.P. McFadden, C.W. Carlson, F.S. Mozer, R.C.

Elphic, R.J. Strangeway, C.A. Cattell, R.F. Pfaff, FAST observations of preferentially acceleration He^+ in association with auroral electromagnetic ion cyclotron waves. Geophys. Res. Lett. **25**, 2049 (1998)
K.A. Lynch, R.L. Arnoldy, P.M. Kintner, J.L. Vago, Electron distribution function behavior during localized transverse ion acceleration events in the topside auroral zone. J. Geophys. Res. **99**(A2), 2227–2241 (1994). https://doi.org/10.1029/93JA02959
K.A. Lynch, R.L. Arnoldy, P.M. Kintner, J. Bonnell, The AMICIST auroral sounding rocket: a comparison of transverse ion acceleration mechanisms. Geophys. Res. Lett. **23**, 3293 (1996)
K.A. Lynch, D. Hampton, M. Mella, B. Zhang, H. Dahlgren, M. Disbrow, P.M. Kintner, M. Lessard, E. Lundberg, H.C. Stenbaek-Nielsen, Structure and dynamics of the nightside poleward boundary: sounding rocket and ground-based observations of auroral electron precipitation in a rayed curtain. J. Geophys. Res. **117**, A11202 (2012). https://doi.org/10.1029/2012JA017691
K.A. Lynch et al., MICA sounding rocket observations of conductivity-gradient-generated auroral ionospheric responses: small-scale structure with large-scale drivers. J. Geophys. Res. Space Phys. **120**, 9661–9682 (2015). https://doi.org/10.1002/2014JA020860
L.R. Lyons, The field aligned current versus electric potential relation and auroral electrodynamics, in *Physics of Auroral Arc Formation*, ed. by S.-I. Akasofu, J.R. Kan. Geophys. Monogr. Ser., vol. 25 (Am. Geophys. Union Washington, 1981). 252 pp
R.L. Lysak, Electrodynamic coupling of the magnetosphere and the ionosphere. Space Sci. Rev. **52**, 33 (1990). https://doi.org/10.1007/BF00704239
R.L. Lysak, Feedback instability of the ionospheric resonant cavity. J. Geophys. Res. **96**, 1553 (1991). https://doi.org/10.1029/90JA02154
R.L. Lysak, Generalized model of the ionospheric Alfvén resonator, in *Auroral Plasma Dynamics*, ed. by R.L. Lysak. AGU Monograph, vol. 80 (1993), p. 121
R.L. Lysak, C.W. Carlson, The effect of microscopic turbulence on magnetosphere-ionosphere coupling. Geophys. Res. Lett. **8**(3), 269–272 (1981). https://doi.org/10.1029/GL008i003p00269
R.L. Lysak, C.T. Dum, Dynamics of magnetosphere ionosphere coupling including turbulent transport. J. Geophys. Res. **88**, 365–380 (1983). https://doi.org/10.1029/JA088iA01p00365
R.L. Lysak, M. Hudson, Effect of double layers on magnetosphere ionosphere coupling. Laser Part. Beams **5**, 351 (1987)
R.L. Lysak, W. Lotko, On the kinetic dispersion relation for shear Alfvén waves. J. Geophys. Res. **101**, 5085 (1996)
R.L. Lysak, Y. Song, Coupling of Kelvin-Helmholtz and current sheet instabilities to the ionosphere: a dynamic theory of auroral spirals. J. Geophys. Res. **101**, 15,411–15,422 (1996). https://doi.org/10.1029/96JA00521
R.L. Lysak, Y. Song, Nonlocal kinetic theory of Alfvén waves on dipolar field lines. J. Geophys. Res. **108**(A8), 1327 (2003). https://doi.org/10.1029/2003JA009859
R.L. Lysak, Y. Song, Propagation of kinetic Alfvén waves in the ionospheric Alfvén resonator in the presence of density cavities. Geophys. Res. Lett. **35**, L20101 (2008). https://doi.org/10.1029/2008GL035728
R.L. Lysak, A. Yoshikawa, Resonant cavities and waveguides in the ionosphere and atmosphere, in *Magentospheric ULF Waves: Synthesis and New Directions*, ed. by K. Takahashi, P.J. Chi, R.E. Denton, R.L. Lysak. AGU Geophysical Monograph, vol. 169 (Am. Geophys. Union, Washington, 2006), p. 289
R.L. Lysak, C.L. Waters, M.D. Sciffer, Modeling of the ionospheric Alfvén resonator in dipolar geometry. J. Geophys. Res. Space Phys. **118**, 1514–1528 (2013). https://doi.org/10.1002/jgra.50090
E.A. MacDonald, E. Donovan, Y. Nishimura, N.A. Case, D.M. Gillies, B. Gallardo-Lacourt et al., New science in plain sight: citizen scientists lead to the discovery of optical structure in the upper atmosphere. Sci. Adv. **4**(3), eaaq0030 (2018). https://doi.org/10.1126/sciadv.aaq0030. https://advances.sciencemag.org/content/4/3/eaaq0030
J.E. Maggs, T.N. Davis, Measurements of the thicknesses of auroral structures. Planet. Space Sci. **16**(2), 205–209 (1968)
J.P. McFadden, C.W. Carlson, M.H. Boehm, Field-aligned electron precipitation at the edge of an arc. J. Geophys. Res. **91**, 1723 (1986)
J.P. McFadden, C.W. Carlson, M.H. Boehm, T.J. Hallinan, Field-aligned electron flux oscillations that produce flickering aurora. J. Geophys. Res. **92**, 11,133 (1987)
M.G. McHarg, D.L. Hampton, H.C. Stenbaek-Nielsen, Fast photometry of flickering in discrete auroral arcs. Geophys. Res. Lett. **25**, 2637–2640 (1998). https://doi.org/10.1029/98GL01972
C.E. McIlwain, Direct measurement of particles producing visible auroras. J. Geophys. Res. **65**(9), 2727–2747 (1960)
R.G. Michell, M. Samara, High-resolution observations of naturally enhanced ion acoustic lines and accompanying auroral fine structures. J. Geophys. Res. **115**, A03310 (2010). https://doi.org/10.1029/2009JA014661

R.G. Michell, M.G. McHarg, M. Samara, D.L. Hampton, Spectral analysis of flickering aurora. J. Geophys. Res. **117**, A03321 (2012). https://doi.org/10.1029/2011JA016703
D.M. Miles, I.R. Mann, I.P. Pakhotin, J.K. Burchill, A.D. Howarth, D.J. Knudsen, R.L. Lysak, D.D. Wallis, L.L. Cogger, A.W. Yau, Alfvénic dynamics and fine structuring of discrete auroral arcs: Swarm and e-POP observations. Geophys. Res. Lett. **45**, 545–555 (2018). https://doi.org/10.1002/2017GL076051
T. Motoba, M. Hirahara, High-resolution auroral acceleration signatures within a highly dynamic onset arc. Geophys. Res. Lett. **43**, 1793? 1801 (2016). https://doi.org/10.1002/2015GL067580
T. Motoba, K. Hosokawa, A. Kadokura, N. Sato, Magnetic conjugacy of northern and southern auroral beads. Geophys. Res. Lett. **39**, L08108 (2012). https://doi.org/10.1029/2012GL051599
F. Mottez, Alfvén wave acceleration of particles in the aurora. Plasma Phys. Control. Fusion **57**(1), 014011 (2014)
T.K. Nakamura, Parallel electric field of a mirror kinetic Alfvén wave. J. Geophys. Res. **105**(A5), 10729–10737 (2000). https://doi.org/10.1029/1999JA900494
D.A. Neudegg, B.J. Fraser, F.W. Menk, H.J. Hansen, G.B. Burns, R.J. Morris, M.J. Underwood, Source and velocities of Pc1-2 ULF waves at high latitudes. Geophys. Res. Lett. **22**, 2965 (1995)
A. Otto, G.T. Birk, Formation of thin auroral arcs by current striation. Geophys. Res. Lett. **20**, 2833–2836 (1993). https://doi.org/10.1029/93GL02492
I.P. Pakhotin, I.R. Mann, R.L. Lysak, D.J. Knudsen, J.W. Gjerloev, I.J. Rae, C. Forsyth, K.R. Murphy, D.M. Miles, L.G. Ozeke, G. Balasis, Diagnosing the role of Alfvén waves in magnetosphere-ionosphere coupling: swarm observations of large amplitude nonstationary magnetic perturbations during an interval of northward IMF. J. Geophys. Res. Space Phys. **123**, 326–340 (2018). https://doi.org/10.1002/2017JA024713
I.P. Pakhotin, I.R. Mann, D.J. Knudsen, R.L. Lysak, J.K. Burchill, Diagnosing the role of Alfvén waves in global field-aligned current system dynamics during southward IMF: swarm observations. J. Geophys. Res. Space Phys. **125**, e2019JA027277 (2020). https://doi.org/10.1029/2019JA027277
G. Paschmann, S. Haaland, R. Treumann (eds.), *Auroral Plasma Physics* (Kluwer Academic, Dordrecht, 2003)
J.R. Peñano, G. Ganguli, Generation of ELF electromagnetic waves in the ionosphere by localized transverse DC electric fields: subcycloton regime. J. Geophys. Res. **105**, 7441–7457 (2000). https://doi.org/10.1029/1999JA000303
S.V. Polyakov, V.O. Rapoport, Ionospheric Alfvén resonator. Geomagn. Aeron. **21**, 816 (1981)
R. Rankin, B.G. Harrold, J.C. Samson, P. Frycz, The nonlinear evolution of field line resonances in the Earth's magnetosphere. J. Geophys. Res. **98**(A4), 5839–5853 (1993). https://doi.org/10.1029/92JA01606
R. Rankin, J.C. Samson, V.T. Tikhonchuk, Parallel electric fields in dispersive shear Alfvén waves in the dipolar magnetosphere. Geophys. Res. Lett. (1999a). https://doi.org/10.1029/1999GL010715
R. Rankin, J.C. Samson, V.T. Tikhonchuk, I. Voronkov, Auroral density fluctuations on dispersive field line resonances. J. Geophys. Res. **104**, 4399 (1999b)
A.J.B. Russell, A.N. Wright, A.V. Streltsov, Production of small-scale Alfvén waves by ionospheric depletion, nonlinear magnetosphere-ionosphere coupling and phase mixing. J. Geophys. Res. Space Phys. **118**, 1450–1460 (2013). https://doi.org/10.1002/jgra.50168
K. Sakanoi, H. Fukunishi, Temporal and spatial structures of flickering aurora derived from high-speed imaging photometer observations at Syowa Station in the Antarctic. J. Geophys. Res. **109**, A01221 (2004). https://doi.org/10.1029/2003JA010081
T. Sakanoi, H. Fukunishi, T. Mukai, Relationship between field-aligned currents and inverted-V parallel potential drops observed at midaltitudes. J. Geophys. Res. **100**(A10), 19343–19360 (1995). https://doi.org/10.1029/95JA01285
T. Sakanoi, S. Okano, Y. Obuchi, T. Kobayashi, M. Ejiri, K. Asamura, M. Hirahara, Development of the multi-spectral auroral camera onboard the INDEX satellite. Adv. Space Res. **32**(3), 379–384 (2003). https://doi.org/10.1016/S0273-1177(03)00273-4
K. Sakanoi, H. Fukunishi, Y. Kasahara, A possible generation mechanism of temporal and spatial structures of flickering aurora. J. Geophys. Res. **110**, A03206 (2005). https://doi.org/10.1029/2004JA010549
I. Sandahl, T. Sergienko, U. Brandstrom, Fine structure of optical aurora. J. Atmos. Sol.-Terr. Phys. **70**, 2275–2292 (2008). https://doi.org/10.1016/j.jastp.2008.08.016
I. Sandahl, U. Brändström, T. Sergienko, Fine structure of aurora. Int. J. Remote Sens. **32**(11), 2947–2972 (2011). https://doi.org/10.1080/01431161.2010.541507
J. Semeter, Coherence in auroral fine structure, in *Auroral Phenomenology and Magnetospheric Processes: Earth and Other Planets*. Geophysical Monograph Series, vol. 197 (2012)
J. Semeter, E.M. Blixt, Evidence for Alfvén wave dispersion identified in high-resolution auroral imagery. Geophys. Res. Lett. **33**, L13106 (2006). https://doi.org/10.1029/2006GL026274

J. Semeter, M. Zettergren, M. Diaz, S. Mende, Wave dispersion and the discrete aurora: new constraints derived from high-speed imagery. J. Geophys. Res. **113**, A12208 (2008). https://doi.org/10.1029/2008JA013122

J. Semeter, S. Mrak, M. Hirsch, J. Swoboda, H. Akbari, G. Starr, V. Pankratius, GPS signal corruption by the discrete Aurora: Precise measurements from the Mahali experiment. Geophys. Res. Lett. **44**, 9539–9546 (2017). https://doi.org/10.1002/2017GL073570

J. Semeter, E. MacDonald, M. Hunnekuhl, M. Hirsch, N. Zeller, The mysterious green streaks below steve. Earth Space Sci. Open Archive **21** (2020). https://doi.org/10.1002/essoar.10502878.1

C.E. Seyler Jr., Nonlinear 3-D evolution of bounded kinetic Alfvén waves due to shear flow and collisionless tearing instability. Geophys. Res. Lett. **15**, 756–759 (1988). https://doi.org/10.1029/GL015i008p00756

C.E. Seyler, A mathematical model of the structure and evolution of small-scale discrete auroral arcs. J. Geophys. Res. **95**(A10), 17199–17215 (1990). https://doi.org/10.1029/JA095iA10p17199

C.E. Seyler, K. Liu, Particle energization by oblique inertial Alfvén waves in the auroral region. J. Geophys. Res. **112**, A09302 (2007). https://doi.org/10.1029/2007JA012412

C.E. Seyler, K. Wu, Instability at the electron inertial scale. J. Geophys. Res. **106**, 21,623–21,644 (2001). https://doi.org/10.1029/2000JA000410

P.K. Shukla, L. Stenflo, Nonlinear phenomena involving dispersive Alfvén waves, in *Nonlinear MHD Waves and Turbulence*, ed. by T. Passot, P.-L. Sulem (Springer, Berlin, 1999). https://doi.org/10.1007/3-540-47038-7_1. 1 pp.

M. Silberstein, N.F. Otani, Computer simulation of Alfvén waves and double layers along auroral magnetic field lines. J. Geophys. Res. **99**(A4), 6351–6365 (1994). https://doi.org/10.1029/93JA02963

F. Simões et al., Detection of ionospheric Alfvén resonator signatures in the equatorial ionosphere. J. Geophys. Res. **117**, A11305 (2012). https://doi.org/10.1029/2012JA017709

N. Singh, Field patterns of Alfvén wave resonance cones. J. Geophys. Res. **104**(A4), 6999–7009 (1999)

N. Sivadas, J. Semeter, Y.T. Nishimura, S. Mrak, Optical signatures of the outer radiation belt boundary. Geophys. Res. Lett. **46**, 8588–8596 (2019). https://doi.org/10.1029/2019GL083908

Y. Song, R.L. Lysak, The displacement current and the generation of parallel electric fields. Phys. Rev. Lett. **96**, 145002 (2006)

K. Stasiewicz, G. Gustafsson, G. Marklund, P.-A. Lindqvist, J. Clemmons, L. Zanetti, Cavity resonators and Alfvén resonance cones observed on Freja. J. Geophys. Res. **102**(A2), 2565–2575 (1997). https://doi.org/10.1029/96JA03462

K. Stasiewicz, Y. Khotyaintsev, M. Berthomier, J-E. Wahlund, Identification of widespread turbulence of dispersive Alfvén waves. Geophys. Res. Lett. (2000a). https://doi.org/10.1029/1999GL010696

K. Stasiewicz, P.M. Bellan, C.C. Chaston, C. Kletzing, R. Lysak, J. Maggs, O. Pokhotelov, C. Seyler, P. Shukla, L. Stenflo, A. Streltsov, J-E. Wahlund, Small scale Alfvénic structure in the aurora. Space Sci. Rev. **92**, 423 (2000b)

H.C. Stenbaek-Nielsen, T.J. Hallinan, L. Peticolas, Why do auroras look the way they do? Eos Trans. AGU **80**(17), 193–199 (1999). https://doi.org/10.1029/99EO00138

C. Stormer, *The Polar Aurora* (Clarendon, Oxford, 1955)

A.V. Streltsov, W. Lotko, Coupling between density structures, electromagnetic waves and ionospheric feedback in the auroral zone. J. Geophys. Res. **113**, A05212 (2008). https://doi.org/10.1029/2007JA012594

A.V. Streltsov, J. Berthelier, A.A. Chernyshov et al., Past, present and future of active radio frequency experiments in space. Space Sci. Rev. **214**, 118 (2018). https://doi.org/10.1007/s11214-018-0549-7

Y.-J. Su, S.T. Jones, R.E. Ergun, S.E. Parker, Modeling of field-aligned electron bursts by dispersive Alfvén waves in the dayside auroral region. J. Geophys. Res. **109**, A11201 (2004). https://doi.org/10.1029/2003JA010344

D.W. Swift, A two-dimensional particle code simulation of inertial Alfvén waves and auroral electron acceleration. J. Geophys. Res. **112**, A01204 (2007a). https://doi.org/10.1029/2006JA011998

D.W. Swift, Simulation of auroral electron acceleration by inertial Alfvén waves. J. Geophys. Res. **112**, A12207 (2007b). https://doi.org/10.1029/2007JA012423

D. Sydorenko, R. Rankin, The stabilizing effect of collision-induced velocity shear on the ionospheric feedback instability in Earth's magnetosphere. Geophys. Res. Lett. **44**, 6534–6542 (2017). https://doi.org/10.1002/2017GL073415

D. Sydorenko, R. Rankin, K. Kabin, Nonlinear effects in the ionospheric Alfvén resonator. J. Geophys. Res. **113**, A10206 (2008). https://doi.org/10.1029/2008JA013579

H. Tanaka, Y. Saito, K. Asamura, S. Ishii, T. Mukai, High time resolution measurement of multiple electron precipitations with energy-time dispersion in high-latitude part of the cusp region. J. Geophys. Res. **110**, A07204 (2005). https://doi.org/10.1029/2004JA010664

M. Temerin, J. McFadden, M. Boehm, C.W. Carlson, W. Lotko, Production of flickering aurora and field-aligned electron flux by electromagnetic ion cyclotron waves. J. Geophys. Res. **91**(A5), 5769–5792 (1986). https://doi.org/10.1029/JA091iA05p05769

B.J. Thompson, R.L. Lysak, Electron acceleration by inertial Alfvén waves. J. Geophys. Res. **101**, 5359 (1996)
V.Yu. Trakhtengertz, A.Ya. Feldstein, Quiet auroral arcs: ionospheric effect of magnetospheric convection stratification. Planet. Space Sci. **32**, 127 (1984)
V.Yu. Trakhtengertz, A.Ya. Feldstein, Turbulent Alfvén boundary layer in the polar ionosphere, I, excitation conditions and energetics. J. Geophys. Res. **96**, 19,363 (1991)
T.S. Trondsen, L.L. Cogger, A survey of small-scale spatially periodic distortions of auroral forms. J. Geophys. Res. **103**(A5), 9405–9415 (1998)
T.S. Trondsen, L.L. Cogger, J.C. Samson, Asymmetric multiple auroral arcs and inertial Alfvén waves. Geophys. Res. Lett. **24**(22), 2945–2948 (1997). https://doi.org/10.1029/97GL52855
S. Tuttle, B. Gustavsson, B. Lanchester, Temporal and spatial evolution of auroral electron energy spectra in a region surrounding the magnetic zenith. J. Geophys. Res. Space Phys. **119**, 2318–2327 (2014). https://doi.org/10.1002/2013JA019627
A. Vallance-Jones, Aurora, in *Geophysics and Astrophysics Monograph*, vol. 9, ed. by D. McCormac (Reidel, Dordrecht, 1974)
J. Vogt, H.U. Frey, G. Haerendel, H. Hofner, J.L. Semeter, Shear velocity profiles associated with auroral curls. J. Geophys. Res. **104**(A8), 17277–17288 (1999). https://doi.org/10.1029/1999JA900148
J.S. Wagner, R.D. Sydora, T. Tajima, T. Hallinan, L.C. Lee, S.-I. Akasofu, Small-scale auroral arc deformations. J. Geophys. Res. **88**(A10), 8013–8019 (1983). https://doi.org/10.1029/JA088iA10p08013
J.-E. Wahlund et al., Broadband ELF plasma emission during auroral energization: 1. Slow ion acoustic waves. J. Geophys. Res. **103**(A3), 4343–4375 (1998). https://doi.org/10.1029/97JA02008
J.-E. Wahlund et al., Observations of auroral broadband emissions by CLUSTER. Geophys. Res. Lett. **30**(11), 1563 (2003). https://doi.org/10.1029/2002GL016335.
T.-H. Watanabe, Feedback instability in the magnetosphere-ionosphere coupling system, revisited. Phys. Plasmas **17**, 022904 (2010). https://doi.org/10.1063/1.3304237
T.-H. Watanabe, A unified model of auroral arc growth and electron acceleration in the magnetosphere-ionosphere coupling. Geophys. Res. Lett. **41**, 6071–6077 (2014). https://doi.org/10.1002/2014GL061166
T.-H. Watanabe, Feedback and ballooning instabilities in the magnetosphere-ionosphere coupling. Geophys. Res. Lett. **47**, e2020GL088233 (2020). https://doi.org/10.1029/2020GL088233
T.-H. Watanabe, S. Maeyama, Unstable eigenmodes of the feedback instability with collision-induced velocity shear. Geophys. Res. Lett. **45**, 10,043–10,049 (2018). https://doi.org/10.1029/2018GL079715
T.-H. Watanabe, H. Kurata, S. Maeyama, Generation of auroral turbulence through the magnetosphere-ionosphere coupling. New J. Phys. **18**, 125010 (2016)
C.E.J. Watt, R. Rankin, Electron trapping in shear Alfvén waves that power the aurora. Phys. Rev. Lett. **102**(4), 045002 (2009). https://doi.org/10.1103/PhysRevLett.102.045002
C.E.J. Watt, R. Rankin, Do magnetospheric shear Alfven waves generate sufficient electron energy flux to power the aurora? J. Geophys. Res. **115**, A07224 (2010). https://doi.org/10.1029/2009JA015185
C.E.J. Watt, R. Rankin, Alfvén wave acceleration of auroral electrons in warm magnetospheric plasma, in *Auroral Phenomenology and Magnetospheric Processes: Earth and Other Planets*. Geophys. Monogr. Ser., vol. 197 (Am. Geophys. Union, Washington, 2012), pp. 251–260
C.E.J. Watt, R. Rankin, R. Marchand, Kinetic simulations of electron response to shear Alfvén waves in magnetospheric plasmas. Phys. Plasmas **11**, 1277 (2004). https://doi.org/10.1063/1.1647140
C.E.J. Watt, R. Rankin, I.J. Rae, D.M. Wright, Self-consistent electron acceleration due to inertial Alfvén wave pulses. J. Geophys. Res. **110**, A10S07 (2005). https://doi.org/10.1029/2004JA010877
C.E.J. Watt, R. Rankin, I.J. Rae, D.M. Wright, Inertial Alfvén waves and acceleration of electrons in nonuniform magnetic fields. Geophys. Res. Lett. **33**, L02106 (2006). https://doi.org/10.1029/2005GL024779
D.K. Whiter, B.S. Lanchester, B. Gustavsson, N. Ivchenko, J.M. Sullivan, H. Dahlgren, Small-scale structures in flickering aurora. Geophys. Res. Lett. **35**, L23103 (2008). https://doi.org/10.1029/2008GL036134
D.K. Whiter, B.S. Lanchester, B. Gustavsson, N. Ivchenko, H. Dahlgren, Using multispectral optical observations to identify the acceleration mechanism responsible for flickering aurora. J. Geophys. Res. **115**, A12315 (2010). https://doi.org/10.1029/2010JA015805
D.K. Whiter, B.S. Lanchester, T. Sakanoi, K. Asamura, Estimating high-energy electron fluxes by intercalibrating Reimei optical and particle measurements using an ionospheric model. J. Atmos. Sol.-Terr. Phys. **89**, 8–17 (2012). https://doi.org/10.1016/j.jastp.2012.06.014
J.R. Woodroffe, R.L. Lysak, Ultra-low frequency wave coupling in the ionospheric Alfvén resonator: characteristics and implications for the interpretation of ground magnetic fields. J. Geophys. Res. **117**, A03223 (2012). https://doi.org/10.1029/2011JA017057
K. Wu, C.E. Seyler, Instability of inertial Alfvén waves in transverse sheared flow. J. Geophys. Res. **108**(A6), 1236 (2003). https://doi.org/10.1029/2002JA009631

J. Wu, D.J. Knudsen, D.M. Gillies, J.K. Burchill, Swarm survey of Alfvénic fluctuations and their relation to nightside field-aligned current and auroral arc systems. J. Geophys. Res. Space Phys. **125**, e2019JA027220 (2020a). https://doi.org/10.1029/2019JA027220

J. Wu, D.J. Knudsen, Y. Shen, D.M. Gillies, e-POP observations of suprathermal electron bursts in the ionospheric Alfvénic resonator. J. Geophys. Res. (2020b, in revision)

J.R. Wygant et al., Polar spacecraft based comparisons of intense electric fields and Poynting flux near and within the plasma sheet-tail lobe boundary to UVI images: an energy source for the aurora. J. Geophys. Res. **105**(A8), 18675–18692 (2000). https://doi.org/10.1029/1999JA900500

J.R. Wygant et al., Evidence for kinetic Alfvén waves and parallel electron energization at 4–6 Re altitudes in the plasma sheet boundary layer. J. Geophys. Res. **107**(A8), 1201 (2002). https://doi.org/10.1029/2001JA900113

A. Yaegashi, T. Sakanoi, R. Kataoka, K. Asamura, Y. Miyoshi, M. Sato, S. Okano, Spatial-temporal characteristics of flickering aurora as seen by high-speed EMCCD imaging observations. J. Geophys. Res. **116**, A00K04 (2011). https://doi.org/10.1029/2010JA016333

M. Zettergren, K. Lynch, D. Hampton, M. Nicolls, B. Wright, M. Conde, J. Moen, M. Lessard, R. Miceli, S. Powell, Auroral ionospheric F region density cavity formation and evolution: MICA campaign results. J. Geophys. Res. Space Phys. **119**, 3162–3178 (2014). https://doi.org/10.1002/2013JA019583

Space Sci Rev (2020) 216:4
https://doi.org/10.1007/s11214-019-0629-3

Diffuse and Pulsating Aurora

Yukitoshi Nishimura[1] · Marc R. Lessard[2] · Yuto Katoh[3] · Yoshizumi Miyoshi[4] · Eric Grono[5] · Noora Partamies[6,7] · Nithin Sivadas[1] · Keisuke Hosokawa[8] · Mizuki Fukizawa[3] · Marilia Samara[9] · Robert G. Michell[10,9] · Ryuho Kataoka[11,12] · Takeshi Sakanoi[3] · Daniel K. Whiter[13] · Shin-ichiro Oyama[4,14,15] · Yasunobu Ogawa[11,12] · Satoshi Kurita[4]

Received: 16 June 2019 / Accepted: 14 December 2019 / Published online: 8 January 2020

Abstract This chapter reviews fundamental properties and recent advances of diffuse and pulsating aurora. Diffuse and pulsating aurora often occurs on closed field lines and involves energetic electron precipitation by wave-particle interaction. After summarizing the definition, large-scale morphology, types of pulsation, and driving processes, we review observation techniques, occurrence, duration, altitude, evolution, small-scale structures, fast modulation, relation to high-energy precipitation, the role of ECH waves, reflected and sec-

Auroral Physics
Edited by David Knudsen, Joe Borovsky, Tomas Karlsson, Ryuho Kataoka and Noora Partmies

Electronic supplementary material The online version of this article (https://doi.org/10.1007/s11214-019-0629-3) contains supplementary material, which is available to authorized users.

✉ Y. Nishimura
toshi16@bu.edu

M.R. Lessard
marc.lessard@unh.edu

Y. Katoh
yuto@stpp.gp.tohoku.ac.jp

Y. Miyoshi
miyoshi@isee.nagoya-u.ac.jp

E. Grono
emgrono@ucalgary.ca

N. Partamies
noora.partamies@unis.no

N. Sivadas
nithin@bu.edu

K. Hosokawa
keisuke.hosokawa@uec.ac.jp

M. Fukizawa
fukizawa.m@pparc.gp.tohoku.ac.jp

M. Samara
marilia.samara@nasa.gov

ondary electrons, ionosphere dynamics, and simulation of wave-particle interaction. Finally we discuss open questions of diffuse and pulsating aurora.

Keywords Pulsating aurora · Diffuse aurora · Wave-particle interaction · Whistler-mode chorus · Energetic electron precipitation

1 Introduction

This chapter reviews diffuse aurora, with a particular emphasis on recent advances in pulsating aurora in the past 20 years. Section 1.1 describes definition and basic properties. Section 1.2 reviews a historical context of pulsating aurora. Section 2 describes recent observations of pulsating aurora over the past 20 years, including measurement techniques (Sect. 2.1) and observational findings (Sect. 2.2). Section 3 states recent advances in simulation. Section 4 briefly summarizes those findings and discusses open questions.

1.1 Definition and Basic Properties

Diffuse aurora is caused by precipitation from the plasma sheet without additional acceleration in the low-altitude magnetosphere. Unlike discrete or Alfvenic aurora, diffuse aurora is a region of glow without large-scale structures (Lui et al. 1973) and does not form rays or show shear or rotational motion (Yamamoto 1988). Pulsating aurora is a type of diffuse aurora that appears as irregular patches with quasi-periodic on-off switching of its intensity.

R.G. Michell
robert.g.michell@nasa.gov

R. Kataoka
kataoka.ryuho@nipr.ac.jp

T. Sakanoi
tsakanoi@pparc.gp.tohoku.ac.jp

D.K. Whiter
d.whiter@soton.ac.uk

S.-i. Oyama
soyama@isee.nagoya-u.ac.jp

Y. Ogawa
yogawa@nipr.ac.jp

S. Kurita
kurita@isee.nagoya-u.ac.jp

[1] Department of Electrical and Computer Engineering and Center for Space Physics, Boston University, Boston, MA, USA

[2] Space Science Center, University of New Hampshire, Durham, NH USA

[3] Department of Geophysics, Graduate School of Science, Tohoku University, Sendai, Japan

[4] Institute for Space Earth Environmental Research, Nagoya University, Nagoya, Japan

[5] Department of Physics and Astronomy, University of Calgary, Calgary, Alberta, Canada

[6] Department of Arctic Geophysics, The University Centre in Svalbard, Longyearbyen, Norway

[7] Birkeland Centre for Space Science, Bergen, Norway

The horizontal size and periodicity are typically ∼10–200 km of ∼2–20 s (Royrvik and Davis 1977; Yamamoto 1988). Auroral intensity modulation of a similar size and period can also occur over discrete aurora but such modulation is not included in this definition.

Pulsating aurora is dominated by the green color at 557.7 nm wavelength. The intensity is somewhat dimmer than typical discrete aurora but is visible in human eyes as blinking patches over the sky (some hundreds of R up to tens of kR at 557.7 nm; a few hundred R to ∼10 kR at 427.8 nm) (McEwen et al. 1981). The intensity modulation relative to background ranges over ∼50–100%.

The emission and ionization occurs in the lower E-region ionosphere at ∼90–107 km (Stenbaek-Nielsen and Hallinan 1979; Kataoka et al. 2016; Jones et al. 2009). The vertical extent of pulsating aurora is a few to a few tens of km. The altitude is nearly constant during individual pulses but gradually increases with MLT (Stenbaek-Nielsen and Hallinan 1979).

Early research associated with pulsating aurora is described in detail by Lessard (2013) and goes as far back as Størmer (1955) and references therein. Although a limited number of publications on pulsating aurora appeared in the literature during these times, interest in the topic increased, somewhat, during the International Geophysical Year (1957/1958) and then significantly with a joint Canadian/Japanese Pulsating Aurora Campaign in Saskatchewan in 1980 (McEwan and Duncan 1981). Throughout the period between that campaign and the last decade, research on pulsating aurora continued at a steady, though perhaps reduced, rate. However, community interest in the topic has grown recently and has resulted in significant strides in our understanding of the process.

Large-Scale Morphology A connection between pulsating aurora and magnetospheric substorms was described long ago, by Akasofu (1968). During a substorm expansion phase, bright and dynamic arcs race poleward. Those arcs are typically driven by quasi-static or Alfvenic acceleration of electrons. Following this activity, aurora is characterized by the presence of dim, diffuse auroral patches. After a period of perhaps 10–20 min, these patches can appear to pulsate and continue to pulsate for several 10 s of minutes or more. Figure 1 shows a typical pulsating aurora event during a substorm. Pulsating aurora starts ∼10–30 min after substorm poleward expansion and lasts throughout the rest of the substorm. The overall activity typically lasts for ∼1.5 hour, while events can last for many hours for long-lasting enhanced geomagnetic activity (Jones et al. 2011, 2013). Pulsating aurora occupies the equatorward portion of the auroral oval (also illustrated in Fig. 2), and the occurrence region expands with geomagnetic activity. For weak geomagnetic activity ($Kp \lesssim 2$), pulsating aurora is localized to the midnight-dawn sector at ∼65–70 deg latitude. Pulsating aurora extends to the dayside for moderate activity and to lower latitudes ($Kp \sim 3$), and can be seen at all local times at more active conditions (Oguti et al. 1981a; Royrvik and Davis 1977).

8 Department of Communication Engineering and Informatics, University of Electro-Communications, Tokyo, Japan

9 NASA Goddard Space Flight Center, Greenbelt, MD, USA

10 Department of Astronomy, University of Maryland, College Park, MD, USA

11 National Institute of Polar Research (NIPR), Tachikawa, Japan

12 SOKENDAI, Tachikawa, Japan

13 Department of Physics and Astronomy, University of Southampton, Southampton, UK

14 National Institute of Polar Research, Tokyo, Japan

15 Space Physics and Astronomy Research Unit, University of Oulu, Oulu, Finland

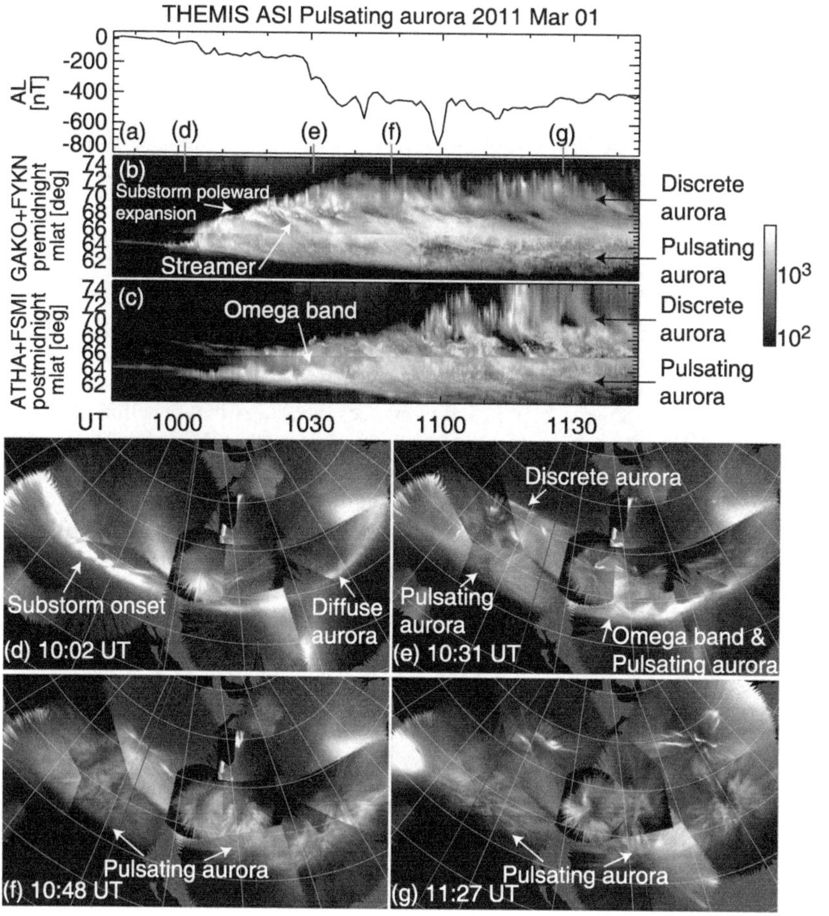

Fig. 1 Pulsating aurora distribution and evolution during the 10–11 UT, 1 March 2011 substorm, detected by the THEMIS all-sky imagers

Fig. 2 Schematic illustration of pulsating auroral distribution in substorm aurora

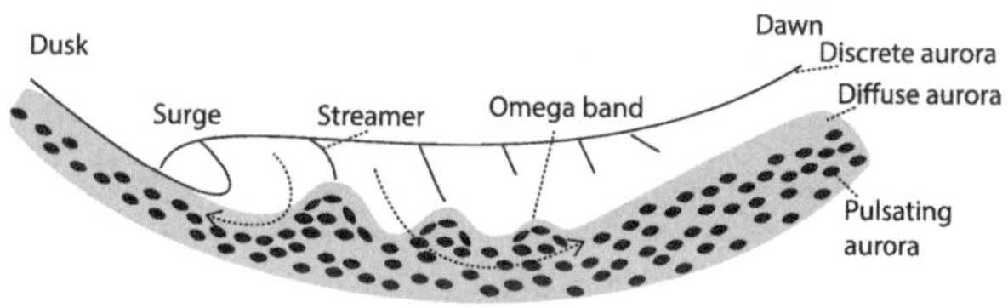

The poleward portion of the auroral oval is dominated by discrete aurora (Figs. 1–2). Transient auroral brightenings along the poleward boundary of the substorm surge (poleward boundary intensifications or PBIs) occur repetitively and propagate equatorward (streamers). Once the streamers reach the equatorward portion of the auroral oval, pulsating aurora intensifies and its occurrence region extends poleward. At post-midnight, this interaction can be seen as poleward-stretching torch-like structures or Omega bands containing pulsating patches that emerge at the poleward boundary of the diffuse auroral region (Oguti et al. 1981b; Sato et al. 2015), often followed by streamers (Henderson 2012).

The whole pattern of pulsating aurora at post-midnight drifts eastward with a speed of a few hundred meters/sec up to a few km/s and extends toward the dayside. The drift is dawnward after midnight and duskward before midnight (Nakamura and Oguti 1987). The patch drift speed has been reported to follow the $E \times B$ drift speed (Yang et al. 2015). The westward drift before midnight suggests that pulsating auroral structures are determined by distribution of cold plasma in the magnetosphere rather than by eastward magnetic drift of energetic electrons. However, some types of pulsating aurora, such as the streaming and propagating types (see below), propagate much faster than a typical convection speed. This suggests that convection of cold plasma is not the only process that controls pulsating aurora dynamics. Some patches indeed do not follow the $E \times B$ drift speed (Grono et al. 2017; Grono and Donovan 2018; Humberset et al. 2018).

Types of Pulsation Each patch generally shows a series of pulsations that typically lasts for several minutes, and the pulse amplitude and shape stay constant for the duration (Yamamoto 1988; Humberset et al. 2018). Pulsation of each patch is incoherent with neighboring patches; the pulsation period and phase of a patch are independent of those at other patches (Scourfield et al. 1972). Pulsating aurora can also co-exist with non-pulsating diffuse auroral patches (Shiokawa et al. 2014; Grono and Donovan 2018).

Pulsating aurora has various types of the spatial structure of intensity modulation. Commonly known types are (1) pure pulsation: pulsation at a fixed size and shape, (2) expansion: pulsation expanding from a core and receding back, (3) streaming: a patch moving away from its original position usually along diffuse auroral boundaries including Omega band (at tens of km/s), (4) propagating: recurrent waves synchronous over hundreds of km in size sweeping north–south (at ∼10 km/s), (5) flaming (field-aligned motion), (6) flashing (∼0.2–0.5 s rapid pulsation, including 3-Hz modulation), and (7) flooding (east-west elongated patches equatorward of streaming patches, moving in east-west) (Oguti et al. 1981a; Yamamoto and Oguti 1982). The propagating pulsation tends to occur at dawnside and is correlated with geomagnetic pulsation (Oguti et al. 1981b). Figure 3 show selected snapshots of Movies 1–6, which show high-speed imaging of pure/expanding pulsation, streaming pulsation, and non-pulsating substructures. See Sect. 2.2.3 for details.

The on-off temporal sequence of pulsations can be classified into three types: (1) Positive pulsation has "on" time longer than "off" time; (2) negative pulsation has "on" time shorter than "off" time; (3) irregular burstlike pulsation and complex mixtures of (1) and (2) (Yamamoto 1988). The "on" time has two primary populations (∼0.2–0.5 s and 2–6 s). The pulsation period has a wide range as mentioned above, but the pulsation "on" time is fairly constant across different types of patches, while the pulsation "off" time is variable (between nearly zero and longer than 1 min) (Yamamoto 1988). Pure, streaming and propagating pulsation has ∼2–6 s of "on" time. A more rapid pulsation (∼0.2–0.5 s) is found for flash and 3-Hz modulation. Those rapid pulsations do not necessarily occur coherently over a patch but occur at substructures of patches (Nishiyama et al. 2016). The ∼3 Hz modulation appears in more than 50% of all pulsating auroras in the midnight and morning sectors, and the amplitude of modulation using panchromatic imaging ranges up to 20% (Royrvik and Davis 1977).

Dayside Diffuse and Pulsating Aurora Most pulsating auroral studies concern the nightside, and much of this chapter covers nightside pulsating aurora. But here we review dayside diffuse and pulsating aurora, which is also associated with substorms. Typically, these observations have been at high latitudes (i.e., near cusp latitudes), as opposed to the traditional occurrences, which tend to be near the equatorward edge of the auroral zone.

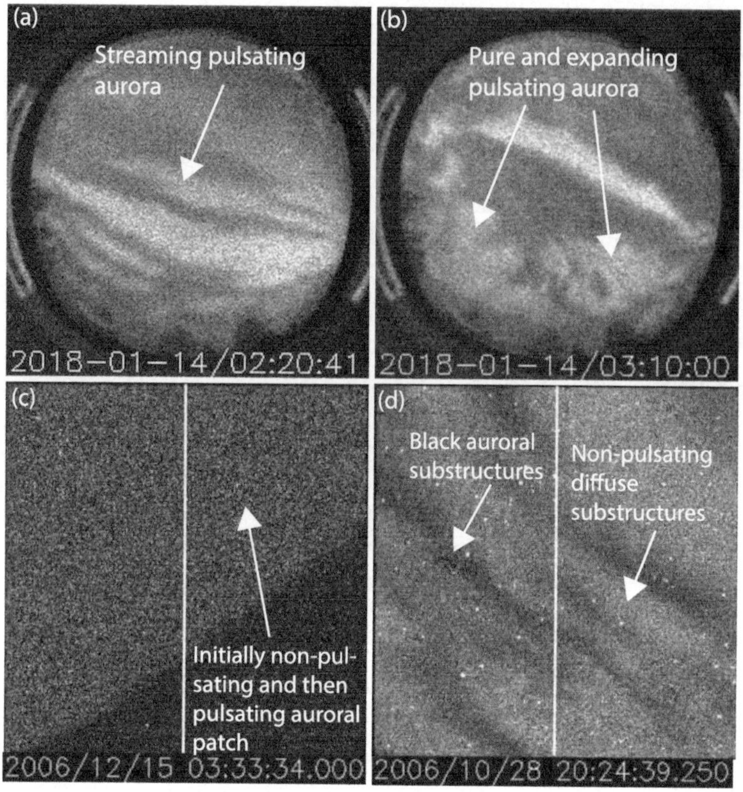

Fig. 3 Selected snapshots of diffuse and pulsating aurora. (**a**) Streaming pulsation (Movies 1 and 3), (**b**) pure and expanding pulsation (Movies 2 and 4), (**c**) pure pulsation and its moving boundary (Movie 5), and (**d**) non-pulsating substructures within pulsating aurora (Movie 6)

There have been, however, at least a couple of notable exceptions. First, Royrvik and Davis (1977) noted that "all-sky camera data from Byrd Station (near the equatorward boundary of the auroral oval) demonstrate that the pulsating aurora can extend eastward from the darkside auroral oval around the noon meridian or even beyond". Note that, because of the relative orientations of the magnetic and geographic poles in the southern hemisphere, stations in the auroral zone remain in darkness 24 hours per day.

In addition, Berkey (1978) presented photometric and riometer observations obtained just after twilight (1300–1600 MLT) at College, Alaska, and showed the occurrence of pulsating aurora in the afternoon sector. The events were detected by simultaneously observing the sky with a 428 nm photometer and a riometer. They compared 428 nm emissions to sunlight (actually, to twilight). As the sun set, a pulsating aurora signature was observed corresponding to absorptions that were present before the optical signatures could be observed, due to sunlit conditions. With this arrangement, only a half dozen events were observed during the winter of 1967–1968. Still, the results suggest that the pulsating aurora generation mechanism can operate on nearly global scales.

At high latitudes (on the dayside), pulsating aurora was first reported by Brekke and Pettersen (1971), using data from Spitzbergen. In that report, they described this aurora as seldom being more than 5% of the total 427.8 nm emission and having pulsating periods

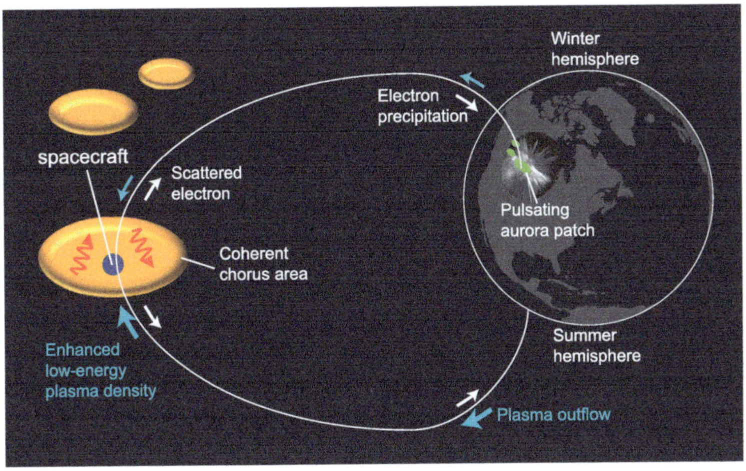

Fig. 4 Schematic illustration of chorus and pulsating aurora connection (Nishimura et al. 2015)

from 25–40 s. Occurrence rates varied with magnetic local time, with a maximum being slightly before magnetic noon.

Craven and Burns (1990) examined data from Davis, Antarctica (ILAT of −74.5 degrees), using optical emissions at 557.7 nm. At this wavelength, the authors typically found an enhancement of 250 R (or less) on a 1 kR background and also noted that pulsating periods tended to be 20–30 s, quite similar to Brekke and Pettersen (1971). Occurrence rates peaked near 14.5% with a peak in occurrences near 1330–1400 magnetic local time (i.e., postnoon).

Wu and Rosenberg (1992) addressed the apparent discrepancy reported by these authors, regarding whether the peak in occurrence comes before or after magnetic noon. These authors examined 427.8 nm emissions from South Pole station and found a single-peaked distribution at magnetic noon during quiet times and a double-peaked during moderate activity at 1000–1030 MLT and 1400–1600 MLT. They note a number of variables that could account for ow their results differ from that of Craven and Burns (1990) and Wu and Rosenberg (1992), emphasizing differences in MLT of the observing times at these stations.

1.2 Driving Processes

The mechanism that drives pulsating aurora had long been thought to be associated with VLF waves near the equator that scattered energetic electrons in that region (Tsuruda et al. 1981). Rocket observations indeed found that pulsating aurora is primarily driven by modulating electron precipitation of energies above a few keV (Johnstone 1978; McEwan and Duncan 1981). As described in Sect. 2 and in recent review papers by Li et al. (2013) and Ni et al. (2016), quasi-periodic precipitation into the upper atmosphere is caused by a group of chorus waves in the equatorial magnetosphere that turn on and off quasi-periodically in $\sim$10 s periodicity and scatter plasma sheet electrons into the loss cone. Thorne et al. (2010) was able to show that, specifically, lower-band chorus is most effective for electron precipitation at higher energies ($>$7 keV) and that upper-band chorus is more important for energies below $\sim$3 keV. Figure 4 illustrates this process.

Confirmation of this process was provided by Nishimura et al. (2010), who showed evidence of lower-band chorus waves, measured by THEMIS spacecraft, in conjunction with ground-based optical observations of pulsating aurora. Correlation between chorus and pulsating aurora last for hours and are highest among other types of waves (Nishimura et al. 2018), giving a firm evidence that the chorus-pulsating aurora connection is not spurious. The chorus-pulsating aurora connection allows to test accuracy of magnetic field models and evolution of magnetic field geometry (Nishimura et al. 2010, 2018, Kawamura et al. 2019).

Evidence of equatorial electron scattering above pulsating aurora was presented by Jaynes et al. (2013), who showed data from the Magnetospheric Electron Detector (MAGED) on the GOES 13 satellite while it was located in a position where its magnetic footprint mapped to the ground in a region with pulsating aurora. These data included a direct correlation between diffuse luminosity fluctuation periods and particle pulsation periods over an energy range of 30 to 50 keV and, to a lesser degree, 50–100 keV.

Finally, Kasahara et al. (2018) showed data from the Japanese ARASE satellite as it passed over The Pas (in Manitoba, Canada) where pulsating aurora was observed, with chorus waves being highly correlated with energetic electrons spanning from 10 keV to 30 keV. Individual elements of chorus waves would create more rapid modulation of precipitating electrons and have been shown to explain the 3-Hz modulation (Ozaki et al. 2015). Such a comprehensive measurement is both unique and important, as it provides direct confirmation of the theory.

The cause of the chorus modulation is, of course, of interest and a number of authors have determined that the periodicity of pulsating aurora has a latitudinal dependence. Such a situation might arise, for example, if chorus waves modulation is driven by field-line resonances (Thomas and Stenbaek-Nielsen 1981; Duncan et al. 1981; Creutzberg et al. 1981). Much more recently, Spanswick et al. (2005) showed that pulsations in the cosmic noise absorption (indicative of energetic electron precipitation) correlate with field-line resonances, and Jaynes et al. (2015) showed in-situ observations of structured chorus waves in conjunction with an apparent field-line resonance, along with ground-based observations of pulsating aurora, although the modulation of chorus waves occurs at half the period (twice the frequency) of the ULF waves. The resonance in this case is a poloidal mode, which would include compressional wave power that might drive the oscillations.

A similar result was obtained by Motoba et al. (2017), but with regard to dayside aurora. These authors correlated all-sky imager observations of pulsating aurora with compressional Pc3 waves, which they suggest are driven by upstream waves.

As mentioned above, patch motion can be characterized by the $E \times B$ drift, and magnetic drifts of energetic electrons cannot explain its motion. These properties suggest that patch motion reflects structure of cold plasma in the magnetosphere rather than drifts of energetic electrons. While it is still unknown how such cold plasma regions form and regulate auroral pulsation, satellite observations have shown modulating low-energy plasma fluxes that suggest to originate as plasma outflows from the ionosphere and to modulate wave growth rates (Nishimura et al. 2015; Liang et al. 2015). Substantial ionization and heating in the ionosphere are associated with pulsating aurora (Hosokawa and Ogawa 2015; Liang et al. 2018), and such processes may be the source of plasma outflow. Waves going to the southern (northern) hemisphere interact with electrons moving to the northern (southern) hemisphere. Waves do not necessarily propagate toward both hemispheres at the same time, and this can explain incoherence of pulsations in the two hemispheres, although overall morphology of pulsating aurora has conjugacy (Fujii et al. 1987; Watanabe et al. 2007).

Spectroscopy of pulsating aurora also supports that the typical energy of pulsating aurora is above a few keV (McEwen et al. 1981). Pulsating aurora is dominated by emission at the 557.7 nm wavelength, and the intensity ratio between 557.7 and 427.8 is about 4–5. Emission at 630.0 nm is generally weak and unchanged during pulsations. However, some events are associated with discernible 630.0 nm emission coincident with the pulsations, although the level of intensity modulation is much smaller than at the other wavelengths. In such cases, pulsating aurora involves a small amount of < keV electrons (Eather 1969; Liang et al. 2016).

2 Recent Advances

2.1 Observation Technique

Diffuse aurora appears quite faint in ground-based all-sky images except during westward travelling surges (Lui et al. 1977). Pulsating aurora occurs within regions of diffuse aurora and can be measured by the same instruments but with a high cadence to capture high frequency pulsations. Historically, these forms of aurora have been measured using direct visual observations and photographic images (Lessard 2013 and references therein), all-sky imagers, television and image intensifier camera systems, meridian scanning photometers, and spectrometers. Apart from the optical signatures, measurements have been made of alternate signatures such as ionospheric electron density using radars (Kirkwood et al. 1988), cosmic noise absorption with riometers (Arnoldy et al. 1982; Berkey et al. 1980), ionospheric currents with magnetometers (Arnoldy et al. 1982; Oguti and Hayashi 1984; Michell and Samara 2015), electron and ion precipitation using particle detectors or electrostatic analyzers onboard sounding rockets (Bryant et al. 1969) and spacecraft (Evans et al. 1987), and X-ray bremsstrahlung emissions using X-ray detectors on air balloons (Sørensen et al. 1973).

Technological advances in the past couple of decades have enhanced the capabilities of the above measurement techniques. All-sky imager networks that had used photographic films, are now entirely digital cameras with solid-state optical detectors. Scientific grade CCDs, that can measure both weak and intense emissions simultaneously, are used in camera networks such as MIRACLE in Scandinavia (Sangalli et al. 2011), and in THEMIS Ground-based Observatory (GBO) in North America (Mende et al. 2008). Electron multiplication CCD (EMCCD) detectors are being used in more recent experiments, which are highly sensitive and have better frame rates. Most cameras use narrow-band filters that allow quantitative measurement of intensities from particular auroral emissions. Auroral Structure and Kinetics (ASK) is a multi-spectral imager consisting of three EMCCD cameras with identical, narrow (3° or 6°) fields-of-view centered on magnetic zenith (Dahlgren et al. 2008). Each ASK camera is fitted with a different narrow passband interference filter, in a combination designed to image the energy and flux of the electron precipitation in fine-scale auroral features at up to 32 frames per second (Lanchester et al. 2009; Dahlgren et al. 2016). The Multispectral Observatory Of Sensitive EMCCDs (MOOSE) suite of imagers was deployed in Poker Flat, AK as a pathfinder for future larger arrays for exploring the electrodynamics of the aurora from ground multi-spectral imaging. It consists of five Andor Ixon DU-888 EMCCD imagers that have a 1024×1024 pixel chip each, with internal binning and sub-framing capabilities that allow trade-offs between temporal and spatial resolution (details in Michell et al. 2014). These imagers are well suited to high spatial and temporal resolution auroral morphology studies (such as, Michell et al. 2012). However, the driving force behind this observatory was the simultaneous use of these imagers with filters of different wavelengths for a multi-spectral view of the aurora in real time. To that end, each imager has telecentric optics to allow the use of narrowband pass filters for two dimensional auroral photometry studies which are actively used to infer the average energy and total energy flux of the precipitating electron distribution, such as Grubbs et al. (2018).

Modern auroral TV-cameras use image-intensified charge coupled device (ICCD) detectors, and are capable of obtaining 25–30 frames per second, and are better at measuring faint aurora than their older counterparts. Scientific complementary metal-oxide semiconductor (sCMOS) cameras, that allow higher frame rates with similar sensitivities as CCDs, have been used to observe the fine-scale morphology of aurora. An NIPR-CMOS system, installed in February 2014 at Poker Flat Research Range, is capable of measuring auroral emissions at a high resolution of 2048 × 2048 pixels, at 100 frames per second, within a field of view corresponding to 25.6 × 25.6 km at 100 km altitude (Kataoka et al. 2015).

Photometers measure intensity of light, through a tube consisting of a lens, filter and a detector. Meridian scanning photometers, with filters, scan the magnetic meridian and produce elevation profiles of auroral emissions at different wavelengths typically: O^+ 557.7 nm and 630.0 nm, N_2^+ 427.8 nm, H_β 486.1 nm (Deehr and Lummerzheim 2001; Johnston 1989; Unick et al. 2017). Spectrographs, measure the spectrum of the aurora with light collected through a narrow slit, and focused through a grating onto a detector. An imaging spectrograph, such as the CEDAR Optical Tomographic Imaging Facility (COTIF), measures all elevations simultaneously, with a time resolution identical for all emission wavelengths dictated by the instrument sensitivity (Semeter et al. 1999).

One area of substantial interest has been to study the altitude profiles of auroral emissions. These can be estimated using a few different techniques: (1) multiple cameras viewing the same volume of sky from different directions (Whiter et al. 2013) in combination with tomography-like methods (Enell et al. 2012; Kataoka et al. 2016), (2) spectral measurements of the same region of the sky (Semeter et al. 1999), (3) space-based visible and UV tomography (Comberiate et al. 2007; Solomon et al. 1984). The Auroral Large Imaging System in Scandinavia is specifically built for the first technique and is the first system in the world that can produce three-dimensional luminosity distribution of aurora (Brändström 2003; Tanaka et al. 2011). ALIS is capable of measuring scale-sizes down to 100 m at a 5 sec temporal resolution, and its imagers are equipped with filter wheels that study visual and sub-visual aurora. The system, which can be remotely-controlled via internet can be made to point in any direction.

Identifying diffuse and pulsating aurora automatically is challenging, and several automatic auroral image classification techniques have arisen over the last two decades using techniques from computer vision, pattern recognition, and machine vision. Over the last few years, advancements in machine learning and access to predefined algorithms have enabled the application of deep neural networks on large auroral-image datasets. Using this technique, Clausen and Nickisch (2018) were able to classify images into no aurora, cloudy, moon, arc, diffuse and discrete with about 82% accuracy. Tracking the convection of auroral forms has also been a topic of increasing interest in the past decade. Grono et al. (2017) used a technique based on wavelet filtering to track the velocity of pulsating aurora patches.

The ionospheric effects of pulsating and diffuse aurora have been studied using a few instrument types. Fabry-Perot interferometers (FPI) are widely used to estimate ionospheric wind velocities and temperatures by measuring the atomic line profiles of low-light emissions. These instruments have been employed to study the effect of pulsating aurora on the variation of wind velocities (Oyama et al. 2010, 2016). A novel experiment was carried out using an eight-microphone array of ground-based infrasonic receivers at Fairbanks, Alaska, to correlate atmospheric acoustic pulsations with pulsating aurora (Wilson et al. 2005).

Another extensively used tool to study the signature of aurora in the ionosphere are the Incoherent Scatter Radars (ISRs). ISRs can estimate a number of different quantities varying in altitude by using model estimates of the neutral atmospheric composition and ionospheric chemistry. The EISCAT radar system in Northern Europe and the Advanced Modular ISR (AMISR) systems in North America are the two main ISR systems currently used to study the ionosphere in the auroral regions. The following quantities are derived from the power spectra of the backscattered radar waves: the altitude profiles of electron density, electron and ion temperature, and line of sight velocity (Robinson 2004). The AMISR systems have the ability to probe multiple points in the ionosphere simultaneously using the electronically steerable phased-array system (Nicolls and Heinselman 2007). As a result, with appropriate assumptions they can estimate the ion velocities in the F-region, and neutral-wind velocities in the E-region. Using electron density profiles from ISRs it is possible to estimate Hall and Pedersen conductivities in the ionosphere (Hosokawa et al. 2010; Kirkwood et al. 1988), energy spectra of precipitating electrons (Fang et al. 2010; Semeter and Kamalabadi 2005; Sivadas et al. 2017), ionospheric electron heating (Liang et al. 2018), and Hall and Pedersen currents. Atmospheric radars such as PANSY in Antarctica (Sato et al. 2014) can be used to identify deep mesospheric ionization during diffuse or pulsating aurora (Kataoka et al. 2019). SuperDARN, a global network of high-frequency radars in the Arctic and Antarctic, uses coherent scatter signals to study ionospheric plasma convection. Recently, it was shown that attenuation in the background atmospheric radio noise during solar proton events can be used to estimate ionospheric absorption similar to riometers (Bland et al. 2018), and the same technique can be used to detect energetic precipitation within diffuse or pulsating aurora (Bland et al. 2019). Riometers are passive receivers operating in the range of 30–40 MHz, and are used to measure the extent to which comic radio noise is absorbed by the ionosphere. The absorption is primarily a result of the D-region ionosphere, ionized by high energy precipitation, UV and X-rays. In the last two decades, imaging riometers have been built that can be used to image the spatial structure of energetic precipitation (Honary et al. 2011; Kosch et al. 2001; McKay et al. 2018).

Besides ground-based instruments, space- and air-borne experiments have also contributed substantially to our understanding of the diffuse and pulsating aurora. Of particular importance are observations of particle energy spectra and waves by THEMIS and GOES-13 in the magnetosphere, and optical images by the low-earth orbiting satellite Reimei and e-POP (Lui et al. 2015). Electric field and electron temperature measurements from Swarm (Knudsen et al. 2017), measurements of Chorus, Hiss, and EMIC waves in the magnetosphere using spacecraft such as CRRES, THEMIS and RBSP (Nishi et al. 2018), particle measurements by low-earth orbiting satellites such as DMSP, FAST, NOAA, and ARASE (Kasahara et al. 2018), X-ray images from the PIXIE on-board the low-earth orbiting POLAR satellite, ultra-violet images from WIC on-board IMAGE (Mende et al. 2002) and GUVI on-board TIMED (Zhang et al. 2005), are among the other datasets available to study the source and signature of pulsating and diffuse aurora. Though rare, early rocket-borne experiments recorded energy flux variations of precipitating electrons that correlate with optical observations of pulsating aurora (McEwen et al. 1981).

2.2 Observations

2.2.1 Pulsating Auroral Properties

Large-Scale Occurrence and Duration Jones et al. (2011) reported that the most probable duration of pulsating aurora events is between 90 and 120 min yet events persisting for several or more hours are common. Within their set of 74 events spread across 119 days of good optical data from Gillam THEMIS all-sky imager, the likelihood of observing pulsating aurora at pre and post-midnight is 14% and 54%, respectively, indicating that this is predominantly a morning-sector phenomenon. Jones et al. (2013) highlighted a particularly long-lived event which covered more than 10 hours magnetic local time and endured for at least 15 hours until the cameras shut down at sunrise. Cameras are often unable to observe the end of morning-sector pulsating aurora events, making it impossible to accurately determine their duration or occurrence (See Sect. 1.1 for dayside measurements). Using a set of 398 events identified in a decade of MIRACLE ASC data, Partamies et al. (2017a) found the median lifetime of pulsating aurora to be 1.4 hours, in agreement with Jones et al. (2011). They highlighted the conservative nature of these estimates using all-sky images from Svalbard which captured pulsating aurora continuing beyond what could be observed by the MIRACLE ASCs in Lapland. In fact, the SuperDARN detection method by Bland et al. (2019) reported 1.16 hours longer median duration (2.83 h) as compared to optical observations (1.67 h). As the radar detection is independent of optical observing conditions, a seasonal occurrence variation could be revealed showing the highest occurrence rates during the winter months (up to ~50% at 4–5 MLT).

Altitude of Patches The altitude of aurora is primarily inferred through either stereoscopic imaging or the measurement of electron density profiles. Auroral stereoscopy is the triangulation of structures simultaneously observed from multiple locations. Electron density profiles can be used to approximate auroral altitude since peak emission and electron density are closely related. Apart from these, other methods have been used to less accurately establish the range of altitudes that pulsating aurora can originate from (Liang et al. 2016; Grandin et al. 2017).

Kataoka et al. (2013) demonstrated stereoscopic auroral imaging utilizing ground-based digital single-lens reflex (DSLR) cameras, which are cheaper and more portable than scientific-grade CCDs. Applied to two case studies, a discrete ray structure and pulsating aurora, they found broad altitude distributions which peaked above and below 100 km altitude, respectively. Kataoka et al. (2016) used this same technique with sCMOS cameras to perform the first-ever stereoscopic auroral imaging at 100 frames per second, determining the emission altitude of a pulsating patch to be 85–95 km and that it could vary by 10 km over 5 s. The overlapping fields-of-view of the MIRACLE all-sky imager network allowed Partamies et al. (2017a) to stereoscopically observe a decrease in peak emission height by approximately 8 km during the onset of a pulsating aurora event, lowering the hardest 10% of precipitating electrons to approximately 90 km. This was further found to apply primarily to the events occurring towards the end of substorm recovery phases when the patch sizes were systematically decreasing (Partamies et al. 2019).

While studying thermospheric wind variations within pulsating patches using a Fabry-Perot interferometer, Oyama et al. (2016) observed electron density peak heights in simultaneous EISCAT observations between 93 and 100 km with a maximum at 96 km. Using high-frequency EISCAT data Hosokawa and Ogawa (2015) separately measured the altitude profile of pulsating aurora during its bright and dim states and identified a 10 km difference that is lost when the measurement is more commonly performed with lower-resolution radar data.

Grandin et al. (2017) used the Sodankylä Ion-Neutral Chemistry model to estimate the amplitude of cosmic noise absorption (CNA) modulation during pulsating aurora, finding that uniform modulation of the ionization profile between 80 and 150 km matched their CNA observations. Close correlation was found between the temporal evolution of CNA and optical emission intensity. More recently, a good correlation between optical emission intensity and CNA was observed during amorphous pulsating aurora (Grono and Donovan 2018) but not during patchy pulsating aurora (Yang et al. 2019). Limited to a single event with an overpassing satellite their results suggest that amorphous pulsating aurora is associated with higher energy precipitation (peaking at ~20 keV) than patchy pulsating aurora (peaking at ~9 keV).

Liang et al. (2016) examined pulsating patches imaged by a 630 nm redline REGO all-sky imager and found that when compared to "white light" THEMIS all-sky imager data mapped to 110 km altitude, the redline pulsations overlapped best at an average altitude of 156 km. This is noticeably lower than steady-state redline auroras, which are conventionally mapped to 230 km (Jackel et al. 2003).

Patch Evolution Diffuse auroras also involve fine-structures at a few to a few tens of km (Sergienko et al. 2008; Shiokawa et al. 2009). These fine-scale structures were suggested to form along the western edges of eastward drifting pulsating auroral patches when the convection drift slows down, and to grow from Rayleigh-Taylor instability at the boundary of a drifting patch. Fukuda et al. (2016) reported rapid, often repetitive expansions around the edges of patches related to their intensity modulation. The expansion speed was typically 10s km/s, comparable to Alfvén speeds, indicating that this phenomenon could arise from the slow and fast-mode Alfvén waves. Samara and Michell (2010) observed that pulsating patch intensity is not symmetric in time, but rather increases more slowly than it decreases. Similarly asymmetric intensity rise and fall times were measured by Dahlgren et al. (2017), who reported an average of 1.6 and 1.4 s, respectively. Based on analysis of optical and radar data of individual patches, precipitation energy and energy flux were found to be 60% lower during off times mainly due to the reduction of high-energy particle flux. Despite this, the energy of precipitating particles remained greater than a few keV throughout off times. The presence of high energy precipitation during both on and off times led the authors to suggest that the same precipitation mechanism is in effect during both time periods. No high-frequency modulation was detected in the two studied events. The speed of patch evolution can be a useful criterion for differentiating pulsating aurora.

Pulsating Aurora Relation to Substorms Pulsating aurora has been recognized as a characteristic of post-onset aurora. While approximately 60% of pulsating aurora observations have been collected during substorm recovery phases (Partamies et al. 2017a), they are not uncommon in expansion phases. Pulsating auroral patches observed during substorm expansion phases have a tendency to grow in size, while the more common patches seen during the substorm recovery phases mainly decrease in size (Partamies et al. 2019). Pulsating auroral patches have recently been found within the growth phase auroral arc of a substorm (McKay et al. 2018).

It was noted by Weygand et al. (2015) and Partamies et al. (2017b) that every auroral omega band is accompanied by simultaneous, or nearly simultaneous, pulsating aurora either directly inside the auroral omega structure or further equatorward within the same band of diffuse aurora. This is not true of the converse, since the existence of pulsating aurora does not require particularly high geomagnetic activity (Partamies et al. 2017a) while the occurrence of omega bands does imply above average substorm activity (Partamies et al. 2017b). In particular, torch-like omega bands have been observed to consist of pulsating aurora (Sato et al. 2015). Their passage is typically followed by appearance of pulsating aurora, but can sometimes be preceded by pulsating aurora as well (Sato et al. 2017). The relation between omega bands and pulsating aurora was found to hold for eastward-expanding auroral surge (EEAS) events, which also take place shortly after substorm onsets (Tanaka et al. 2015).

Other Patch Characteristics The thickness of pulsating auroral patches can be estimated from electron density altitude profiles which reveal where particle precipitation has ionized neutral atmospheric particles. Utilizing electron density profiles measured by PFISR, Jones et al. (2009) observed a pulsating patch approximately 15–25 km thick. This result was corroborated by Hosokawa and Ogawa (2015) who observed similar thickness in EISCAT electron density profiles of four pulsating auroral patches.

The lifetimes of pulsating auroral patches are most commonly measured on an individual basis, and efforts that could allow their determination *en masse* have been few. While implementing an automatic algorithm to track pulsating patches, Grono et al. (2017) found that the number of images unique patches were identified in could function as a proxy for patch lifetime. It was observed that the lifetime of the shortest-lived structures was less than 1 min, while the longest-lasting could endure for 10s of minutes.

Humberset et al. (2018) quantitatively and qualitatively assessed the evolution of four patches over a portion of their lifetimes, measuring the stability of their shape and coherence of their pulsations. They found that 85–100% of the patch shape persist for 4.5–8.5 min. In addition, detailed characteristics of these patches were provided, including their emissions and apparent size which ranged approximately from 2.7–3.8 $\mathrm{kR\,km^{-2}}$ and 1000–4800 $\mathrm{km^2}$, respectively. Similar patch sizes were later reported by Partamies et al. (2019) and Ozaki et al. (2019).

It was noted by Grono et al. (2017) that while pulsating patches have been seen to follow ionospheric convection (Yang et al. 2015, 2017), all structures that could be defined as pulsating patches do not move in the same manner. Grono and Donovan (2018) proposed three categories of pulsating patches based on the stability of their shape and the spatial extent of their pulsations. Amorphous Pulsating Aurora (APA) describes rapidly evolving pulsating features that are generally difficult or impossible to track between successive images captured at a 3 second cadence. Patchy Pulsating Aurora (PPA) and Patchy Aurora (PA) are long-lived structures that can persist for 10s of minutes, but PA does not pulsate. These two types of aurora are clearly related, and appear to differ only by the existence of a modulating mechanism in their source region. While the apparent motion of APA appears unrelated to convection, Grono and Donovan (2018) revisited the events of Yang et al. (2015) and Grono et al. (2017) and found that they reported PPA and PA patches moving with convection.

Grono and Donovan (2019) found that pulsating aurora does not occur poleward of the proton aurora. Amorphous pulsating aurora appears within and equatorward of the proton aurora, while patchy and patchy pulsating auroras predominantly form equatorward of the optical b2i boundary (Donovan et al. 2003). The optical b2i is a variant of the b2i boundary (Newell et al. 1996), defined by the rapid decrease in downward proton fluxes. This boundary is the ionospheric marker of the isotropy boundary (e.g., Sergeev et al. 1983), indicating the transition between stably trapped protons and those scattered by magnetic field curvature. That patchy and patchy pulsating auroras primarily form equatorward of this point suggests that mechanism and conditions responsible for their structuring occurs in the region of mostly dipolar magnetic field.

Yang et al. (2019) reported evidence that APA was associated with higher energy electron precipitation than PPA. They investigated cosmic noise absorption (CNA) associated with APA and PPA, finding that CNA is systematically higher during APA than PPA and better correlated with its brightness. Furthermore, they found one conjunction between the FAST satellite and both APA and PPA. The APA feature had higher energy precipitation than the PPA feature.

Pulsating aurora is not generally coherent between the two hemispheres (Watanabe et al. 2007). Although pulsating aurora has the tendency to occur simultaneously on both hemispheres (Partamies et al. 2017a), the shape of the auroral form is generally different between the two hemispheres, and there is little or no interhemispheric correlation in the intensity variations. This property suggests that waves do not occur simultaneously in the northern and southern hemispheres but have different frequencies and sizes so that precipitating particles are modulated differently in different hemispheres.

2.2.2 *Morphology of Non-Pulsating Diffuse Aurora, Small-Scale Structures*

Fine small-scale structures are sometimes seen within non-pulsating diffuse aurora both in the dayside and nightside, while viewed through sensitive detectors. They are often termed structured diffuse aurora (SDA). In the dayside, they are most often observed at magnetic local noon. Two types of SDA have been identified to coexist within this region: stripy or patchy forms that drift towards higher latitudes and forms adjacent to discrete aurora that moves slowly along with it. Han et al. (2017) found that the latter has an increased He^{2+} concentration, which find their way into the magnetosphere from magneto-sheath, suggesting that the magneto-sheath particles may play an important role in producing the dayside diffuse aurora. The stripy and irregular forms of dayside aurora seem to be produced by electrons of energy ~1–10 keV (Han et al. 2015). Simultaneous measurements from an all-sky imager and THEMIS spacecraft by Nishimura et al. (2013) suggests that the SDA patterns reflect the spatial distribution of whistler-mode waves and ambient plasma density near the magnetic equatorial plane.

One of the earliest reports on the statistics of narrow discrete arcs by Maggs and Davis (1968), may have included a large fraction of nightside SDAs according to Stenbaek-Nielsen et al. (1999). Nightside SDA are regular, parallel auroral stripes brighter than the background luminosity observed mostly during the recovery phase of substorms (Sergienko et al. 2008). These aurorae have been shown to be associated with electrons ≥8 keV scattered into the loss-cone from trapped magnetospheric particles (Samara et al. 2010). A statistical analysis of about 500 diffuse stripe-like structures recorded during 5 events have found the scale sizes of these stripes to be on average about 13–14 km, and with a separation of about 5–23 km between each other (Axelsson et al. 2012). This separation maps to about 75–345 km in the equatorial plane of the magnetosphere. Using ionospheric equivalent current estimates from ground magnetometers, Axelsson et al. (2012) concluded that the stripes move equatorward at a speed close to zero in the plasma convection frame (~100–200 m/s). As in the case of the dayside aurora, the fine structures are understood to be a visual manifestation of cold plasma density or magnetic field structures in the equatorial magnetosphere (Axelsson et al. 2012; Ebihara et al. 2010). A further study on nightside SDAs by Sivadas et al. (2019) showed the diffuse structures within the energetic electron arc (EEA, Sergeev et al. 2012) equatorward of the growth phase arc. This precipitation was concluded to mark the outer radiation belt boundary of the 30–300 keV electrons and, despite the high energies, contribute to ~46% of the total 427.8 nm emission.

Apart from the SDA mentioned above, Dahlgren et al. (2012) reported an observation of an unusually narrow filament of ~70 m width, forming an irregular ring of ~3.8 km scale size within the diffuse aurora. The authors argue that the source mechanism of this filament is unlikely to be quasi-static potential structures, Alfvén waves, or pitch-angle scattering in the magnetosphere as the primary electron spectra is sharply mono-energetic at ~8 keV with a moderately high energy flux of ~230 mW/m^2, and confined to extremely narrow widths. Peticolas et al. (2002) analyzed six different nights of conjugate optical and satellite data that showed black aurora within uniform diffuse aurora. The black aurora corresponded to electron energy flux dropouts in FAST satellite measurements, with surrounding energy flux of electrons >2 keV in the diffuse aurora. The authors argue that pitch angle diffusion by chorus waves is suppressed in the localized magnetospheric regions leading to the observed black auroral features in the ionosphere.

More recently reported fast auroral transients (McKay et al. 2019) resemble non-pulsating auroral patches, which propagate azimuthally within the CNA arc on the equatorward side of the growth phase arc. The estimated propagation speed of the transients was ~10 km/s without a preference between eastward and westward motion. Similarly to the SDAs observed by Sivadas et al. (2019), the auroral transients map to the equatorial plane distance of about 9–12 Re.

2.2.3 Substructure and Fast Modulation

Two main frequency bands have been reported within pulsating aurora during one persistent, several hour event (Samara and Michell 2010). Lower frequency pulsations, within the range of 50–500 mHz, were present in large patches (10–50 km) seen by an all-sky imager, while fine-structures observed by a narrow field-of-view imager pulsated at higher frequencies, between 0.5–15 Hz. It was further suggested that pulsation frequency anti-correlates with the size of pulsating structures and correlates with the emission intensity of the pulsating structures. Thus, larger and dimmer auroral structures feature lower pulsation frequencies.

Much faster modulations with frequencies up to 54 Hz were documented by Kataoka et al. (2012) for one case using a state-of-the-art high-speed imager with a frame rate of 500 frames per second. These are the fastest variations observed to date, and they were seen at the edge of larger pulsating patches. In agreement with Samara and Michell (2010), the pulsation frequency appears to be related to structure size such that the smaller auroral features undergo faster pulsations. Ozaki et al. (2018) reported Particles and Waves in the INner magnetosphere using Ground-based network observations (PWING, Shiokawa et al. 2017) 100 Hz all-sky imager observations of a patch pulsating which demonstrated rapid evolution on the order of 10s of ms. This behaviour was attributed to successive chorus elements, or possibly to the chorus subpacket structures. It was further demonstrated that the spatiotemporal evolution of a single auroral patch experienced one-to-one correspondence with the individual chorus elements on the timescales of a few hundreds of milliseconds (Ozaki et al. 2019). This gives a more thorough insight into the processes controlling the patch sizes, but requires a theoretical foundation to understand the nature of their structuring.

Three independent and adjacent patches with periods between 4–7 s were observed by Nishiyama et al. (2012) with a high-speed camera (100 frames per second). One of the patches experienced high frequency modulation of about 1.5 Hz in a small region near the centre of the patch. Further, the small size of this patch supports the observation of Samara and Michell (2010) that smaller structures undergo higher frequency fluctuations.

Based on a statistical study of 53 events, Nishiyama et al. (2014) determined that quasi-periodic intensity modulation frequencies most commonly range between 1.5–3.3 Hz. Moreover, modulation frequency was found to correlate with the emission intensity (correlation coefficient 0.58), being particularly strong during the morning MLT hours (coefficient 0.67). However, the frequency of pulsating aurora on–off transitions shows no significant correlation with brightness. The high-frequency intensity modulations exist superimposed on the quasi-periodic on–off transitions whose frequencies are well below 1 Hz (2–16 s).

Nishiyama et al. (2016) used a principal component analysis to identify independent sub-structures within a pulsating auroral patch for the first time. They found that the most rapid fluctuations are localised on sub-structures smaller than 20 km in diameter which oscillate back and forth relative to the larger patch drifting along $E \times B$. The authors proposed that the oscillation related to the propagation of chorus wave packets. In agreement with Samara and Michell (2010), these substructures underwent higher frequency modulation than was exhibited in the larger patches.

Humberset et al. (2016) identified and tracked six well-defined patches and concluded that they did not pulsate with a regular period. The typical on-time for patches was found to vary between 3–5 s while the typical off-time was shorter, having a median of 0.6 s. Thus, they suggested that the term fluctuating be used instead of pulsating to describe these auroras.

Fritz et al. (2015) carried out a study of black aurora embedded within pulsating patches. They found that black aurora forms an apparent firm boundary between the auroral forms, much like the optical emissions as described by Humberset et al. (2018). This paper also shows evidence of black auroral curls, suggesting the presence of counterstreaming shear flows, perhaps in conjunction with field-aligned currents.

Six movies are available as electronic supplements as examples of fast variation of diffuse and pulsating aurora. Figure 3 shows representative snapshots and highlights the regions of interest. Movies 1 and 2 are obtained by the PWING EMCCD camera at Kevo, Finland at 100 frame/s on 14 January, 2018 at 02:20 and 03:10 UT. Movie 1 shows streaming pulsating aurora, where a series of auroral patches emerge, move rapidly away from original locations, and disappear. Movie 2 shows pure and expanding pulsating aurora, where each patch illuminates simultaneously over the patch or starting from its center. Motion of the pure/expanding pulsating aurora is much slower than for streaming pulsation, and each patch can be traced over several pulsations. When the images are down-sampled to 3-s (Movies 3 and 4), Movie 3 does not show the coherent motion seen in Movie 1, while Movie 4 still allows to trace each pulsating auroral patch over a few minutes. APA pulsations could arise from time-aliasing of rapidly-propagating pulsating aurora, and PPA pulsations are slowly drifting patches such as pure and expanding pulsations.

Similar but higher spatial resolution observations have been made with the ASK instrument introduced in Sect. 2.1, two samples of which are shown as Movies 5 and 6. The movie shows the rapid dynamics of a pulsating aurora patch seen above Tromsø, Norway, on 15 December 2006. The bottom panel shows a sequence of images in N_2 1PG (673.0 nm, a prompt emission from high energy precipitation). The top panel shows the absolute intensity along the cut highlighted by a white line on the bottom panel. In Movie 5, a sharp boundary defining the leading edge of the patch moves at an approximately constant velocity. Once this boundary has passed the field of view, the patch exhibits pure pulsation with "on" and "off" periods of highly varying lengths, and it is also clear that the patch "fills up" and empties from a specific direction rather than turning on and off uniformly. During the longer off periods the intensity drops to the noise floor. Movie 6 shows pulsating aurora with non-pulsating substructures. During the "on" period, the patch intensity is uniform, while bands of non-pulsating diffuse and black substructures appear during the "off" period. The substructures slowly drifted and changed shapes.

2.2.4 High-Energy Precipitation

Typical characteristic energy of the pulsating aurora electrons is in the range from a few keV to $\sim$100 keV (e.g., Bryant et al. 1975, Sandahl et al. 1981; Miyoshi et al. 2010). Since the energy is higher than discrete and diffuse aurora, the emission altitudes of the pulsating aurora is lower than other auroras (e.g., Brown et al. 1976, Partamies et al. 2017a).

If we consider that the characteristic energy of the pulsating aurora electrons is high, their cyclotron resonance with whistler mode chorus waves, lower-band chorus (LBC) waves especially, is a plausible candidate to cause the pitch angle scattering, while Electron Cyclotron Harmonic (ECH) waves are less likely to cause such high energy electron precipitations (e.g., Thorne et al. 2010). The relationship between LBC and the pulsating aurora has been confirmed by various authors (Nishimura et al. 2010; Jaynes et al. 2013; Miyoshi et al. 2015b). Kasahara et al. (2018) showed that the flux inside the loss cone increases simultaneously with the chorus waves by the Arase observations, and these electrons inside the loss cone correspond to the pulsating aurora on the ground. The characteristic energy of precipitating electrons depends on the magnetic local time. Hosokawa and Ogawa (2015) found that the ionization altitude, which indicates the precipitating energies of the pulsating aurora electrons, decreases in the morning side from the EISCAT observations. Oyama et al. (2017) indicated similar results.

Recently, it was found that sub-relativistic electrons precipitate into the middle atmosphere simultaneously during the pulsating aurora (Miyoshi et al. 2015a). Miyoshi et al. (2015a) proposed a model, in which wide energy electron simultaneously precipitate into the atmosphere during the pulsating aurora, if the chorus waves propagate to higher latitudes (Horne and Thorne 2003). The Miyoshi theory showed that chorus waves can cause tens keV electron scattering near the magnetic equator, which contributes to the optical emissions of the pulsating aurora. And then, several chorus waves can propagate to the higher-latitude along the field line as observed by Cluster (Santolik et al. 2014), which causes the scattering of sub-relativistic and relativistic electrons due to changes of the resonance condition. The Miyoshi theory suggested that the scattered sub-relativistic/relativistic electrons are observed as relativistic electron microbursts simultaneously with pulsating aurora.

This finding is important to consider the effects on the middle and upper atmosphere, because a few hundred keV electrons can ionize in the mesosphere and cause enhancements of NO_x and HO_x and resultant decrease of O_3 (Turunen et al. 2016).

Considering the results expected from the Miyoshi theory, it has been postulated that the microbursts of relativistic electrons occur in association with the pulsating aurora (Miyoshi et al. 2010; Saito et al. 2012). Quasi-3 Hz modulations embedded in the main modulation of the pulsating aurora and these internal modulations are caused by each rising tone elements of chorus waves (Miyoshi et al. 2015b). The time scale of the internal modulations of the pulsating aurora is similar to or a little bit longer than the microbursts of the relativistic electrons, suggesting a close relationship between the internal modulations of the pulsating aurora and the microbursts. Recent low-altitude observations such as SAMPEX and Firebird II suggested scale size of tens km, which may be the same scale size of the pulsating patch or smaller (Crew et al. 2016). It is interesting to investigate the relationship between the pulsating aurora and the microbursts of relativistic electrons, which contribute to more quantitative understanding of the pulsating aurora.

2.2.5 Role of ECH Waves

In the inner magnetosphere ($L \gtrsim 8$), diffuse aurora including pulsating aurora is mainly caused by lower-band chorus waves as demonstrated in Sect. 1, 2.2.3, 2.2.4. On the other hand, upper-band chorus and ECH waves can also, theoretically, scatter electrons through cyclotron resonance (e.g., Lyons 1974; Ni et al. 2008) and can account for lower energy portion of precipitating particles. Miyoshi et al. (2015b) showed that a close set of upper-band chorus waves causes the stable precipitations at $\sim$1 keV, while lower-band chorus bursts cause the main modulation of more energetic electrons, and a train of rising tone elements embedded in the lower-band chorus bursts drives the internal modulations. Notwithstanding the long radiative timescale of 630.0-nm emissions, pulsating auroral emissions at wavelength of 630.0 nm have been reported (Eather 1969; Liang et al. 2016). Their generation mechanism cannot be explained by gyroresonance with lower-band chorus waves because the gyroresonance energy of lower-band chorus waves is too high to cause 630.0 nm emissions in the ionospheric F region. Liang et al. (2016) suggested that upper-band chorus and ECH waves may scatter the pitch angle of low energy electrons and cause faint 630.0 nm pulsating aurora in addition to the 557.7 nm pulsating aurora. Fukizawa et al. (2018) performed a detailed event study using coordinated Arase satellite and ground-based optical observations and showed that ECH wave intensities had a positive correlation with pulsating auroral intensities whose period was a few dozen seconds, and that lower-band chorus wave intensities had correlation with pulsating auroral intensities whose period was a few seconds. It was suggested that lower-band chorus waves cause short-period pulsating aurora at low altitude, while ECH waves cause long-period pulsating aurora at high altitude.

In the outer magnetosphere ($L \gtrsim 8$), the occurrence rate of moderately intense chorus emissions ($\geq$10 pT) drops significantly, while moderately strong ECH emissions ($\sim$0.1–1.0 mV/m) have been reported to extend up to $\sim$12 R_E (e.g., Roeder and Koons 1989; Ni et al. 2011, 2017). These observations suggest that ECH waves may play a leading role in driving diffuse aurora at $L \gtrsim 8$. Liang et al. (2011) showed that intensities of ECH waves had an unambiguous positive correlation with those of diffuse auroral precipitation using the simultaneous in-situ wave and particle observations by THEMIS satellites and ground-based NORSTAR optical auroral observations at $L = 11.5$ during either an absence of chorus emissions or very weak chorus emissions. The scattering rate of central plasma sheet electrons by ECH waves, the subsequent ionospheric precipitation flux, and resulting auroral brightness were quantified by Ni et al. (2012), who concluded they were consistent with observations. ECH waves can play an important role in the outer magnetosphere to cause diffuse aurora.

2.2.6 *Reflected and Secondary Electrons*

Advances in imaging of aurora have opened up the exploration of auroral signatures to faster timescales, smaller spatial scales and, overall, scales that require wider dynamic range to enable more detailed morphological resolution. It can be said that a significant part of new insights come from such a change coupled with our ability to handle much larger volumes of data, a necessity when recording many hours a night of auroral activity. The recently observed optical signatures of inter-hemispheric electron reflections within pulsating aurora (Samara et al. 2017) are in that category. These had been inferred to be observable from prior theory and modeling work but even initial verification from optical signatures had been elusive partly because of the observational conditions needed.

Pulsating aurora is often observed against a non-pulsating background created by soft electrons (Smith et al. 1980; McEwen et al. 1981; Sandahl 1985). This background has been identified by Evans et al. (1987), who suggest that these electrons originate from the opposite hemisphere in conjunction with simultaneously occurring pulsations there. They determine that this population arises from secondary electrons and backscattered electrons produced by the primary high energy pulsating electrons. e-POP satellite observations over pulsating aurora have also detected upgoing low-energy electrons, which are considered to be backscattered secondary electrons (Knudsen et al. 2015).

Samara et al. (2017) highlighted a case where the temporal and intensity variations of pulsating auroral are in good agreement with the predictions of the SuperThermal Electron Transport (STET) code. A pulsating aurora event was observed at Poker Flat, AK with the MOOSE imager suite in multiple fields of view but it was the very narrow field of view imager (4 degree field of view) that was operated at 56 frames per second (16 ms exposure time) with no filter that enabled the observations of the secondary electron peaks. This type of analysis is only possible with a narrow field of view and high frame rates. In the optical signatures Samara et al. (2017) identified four primary peaks and assumed that they are caused by a pitch-angle scattering mechanism in the equatorial magnetosphere, which would send electrons into both loss-cones (northern and southern hemispheres) at the same time (Nishimura et al. 2010). This would produce primary peaks in both hemispheres at the same time. As part of this process a portion of these primary electrons get scattered back up and head toward the opposite hemisphere producing the secondary peaks observed. These faint secondary pulses were visible in the data because of the high frame rate, the high signal-to-noise ratio—maximized by the white-light imaging—and the morphology of the pulsations. Specifically, the temporal characteristics of the auroral event must be such that the "pulsation-on" time is short—less than the bounce time of a few seconds—and the spacing between pulses must be larger than at least several bounce periods. The beginning of the event analyzed in Samara et al. (2017) met both of those criteria with "pulsation-on" times of $\sim$1 s and times between the main pulses of 6–9 s. The spacing of the peaks (1.5 to 2 s) corresponds to electron energies of 6–10 keV, for a field line length of 14.26 Re, which is consistent with many observations of diffuse and more specifically pulsating auroral electron precipitation.

Modeling work using STET, such as Khazanov et al. (2017), has shown that the reflected primary and secondary electrons can lead to significant changes in the precipitating electron distributions in regions of diffuse aurorae. Due to the highly time-varying nature of these processes and the short bounce periods of 1–5 s, it has proven difficult to quantify any evidence of these reflected electrons in optical auroral observations. The Samara et al. (2017) work advanced our understanding of electron precipitation in two ways. First by offering a peak into a scale we do not often see and second by verifying predictions from consistent theoretical work. These results are consistent with the predictions of the STET model, both in terms of the production of a reflected electron population and the relative intensities of the reflected electron distributions. They are also consistent with early in-situ measurements and theoretical work suggesting that $\sim$10–20% of the precipitated electron fluxes get reflected back to the conjugate hemisphere (O'Brien 1964; Rees 1968).

There is a distinct variation in the reported measured electron energies associated with pulsating aurora (from a few keV to as high as 140 keV) suggesting both different generation mechanisms but also different resulting ionospheric effects. In all previous studies the secondary and backscattered electron populations that exist within the diffuse aurora have not been considered. As such the optical signatures of reflected electrons have broad implications because they highlight that the formation of the auroral electron distributions within regions of diffuse and pulsating aurora contain contributions from reflected primary and secondary electrons. These electrons, which can be the result of multiple reflections, can ultimately lead to larger fluxes than expected when considering only the primary injection of magnetospheric electrons. While they have largely been missing from the current theoretical studies of particle precipitation the inclusion of a reflected electron population which can have immediate implications for studies of magnetosphere-ionosphere coupling.

2.2.7 *Ionosphere Dynamics*

The effect of pulsating aurora on the ionospheric electrodynamics (current system) has been studied extensively by using ground-based radars and satellites in the last decade. Hosokawa et al. (2008) introduced an interval of pulsating aurora during which the ionospheric electric field, as detected by a SuperDARN radar in Iceland, oscillates in close correlation with the optical pulsation. They interpreted this oscillation as a result of quasi-periodic appearance of polarization electric field generated by a modulation of electron density (i.e., conductance) due to precipitation of pulsating aurora electrons (this idea is based on an earlier study by Oguti and Hayashi 1984)

Later, Hosokawa et al. (2010) demonstrated, by using the EISCAT radar in Tromsø, that the electron density and Hall conductance actually changes in harmony with the optical pulsation, which implies that pulsating aurora has a potential to modulate the ionospheric current system in the quasi-periodic manner. They also directly measured similar pulsations in the northward electric field by using one of the remote site radars in Kiruna, which should be a result of quasi-periodic appearance of electric field within the patches of pulsating aurora. However, to explain the origin of electric field oscillation by the current hypothesis in a quantitative manner, we need to consider an escape of charges accumulated at the boundary of pulsating aurora patches. Based on the calculation of Hosokawa et al. (2010), about 70% of charges should escape from the edges of pulsating aurora patches. Such a significant escape of charges could be associated with field-aligned current (FAC) flowing in/out along the boundary of pulsating aurora patches

An earlier study by Fujii et al. (1985) detected such small-scale structures of FAC in the vicinity of pulsating aurora patches by using a magnetometer onboard the MAGSAT satellite. They indicated that FACs are actually flowing near the edges of the patches. More recently, Gillies et al. (2015) have introduced similar observations of FACs in the vicinity of pulsating aurora patches by using Swarm satellite. They identified strong downward currents just poleward and equatorward of the pulsating patches. Weaker upward currents are observed throughout the interior of the patches, leading the authors to suggest that currents might close via horizontal currents in the ionosphere. Such a FAC near the boundary is basically consistent with the model of Oguti and Hayashi (1984). The above-mentioned set of studies indicates that electric field can be modified within the patches of pulsating aurora by the local ionospheric electrodynamics. Such a change in the ionospheric electric field may modify the large-scale magnetospheric electric field only within the source region of pulsating aurora patches. This process could characterize the shape of pulsating aurora patches by determining the spatial structure of cold plasma density in the magnetosphere. Not only that, the modification of electric field by the ionospheric feedback process may introduce a difference between the apparent patch motion in the ionosphere and the motion of cold source plasma in the magnetosphere, especially when the ambient large-scale electric field is relatively small (as in the case in Humberset et al. 2018).

The distinct difference in altitude of diffuse and pulsating aurora is important because it is indicative of the presence of two separate populations of precipitating electrons, with the higher-energy population being the pulsating one. In order to understand the relationship between diffuse and pulsating aurora, Brown et al. (1976) triangulated ground-based optical observations of diffuse and pulsating aurora from a 10 km baseline. This configuration provides, essentially, a measurement of the lower border of these emissions. They report that the diffuse aurora can have lower borders from 118 km to 375 km (based on 10 measurements). The median altitude of these events was near 150 km, which imply precipitation energies of up to $\sim$1 keV. The pulsating aurora (19 measurements), however, tended to have lower borders from 82 km to 105 km, with a median altitude of 92 km, implying 10 keV to 60 keV electron precipitation. Their Fig. 2 shows a schematic drawing illustrating the altitudinal relationship of pulsating aurora and the higher-altitude diffuse aurora in which it is embedded.

Partamies et al. (2017a) also triangulated ground-based optical measurements, an effort that included 400 intervals of pulsating aurora. They find that the peak in auroral emission height decreases by ∼8 km at the start of a pulsating aurora event, suggesting a corresponding increase in precipitating electron energy. While this result might appear to indicate a gradual transition from lower energies (with the diffuse aurora) to the higher energies of pulsating aurora, this is not the case. In fact, a distinct effort to exclude other types of aurora from this study, events were selected so that the start was when the pulsating aurora became the dominant auroral feature.

Using data from the Reimei satellite, Miyoshi et al. (2015b) showed optical observations of pulsating aurora that are well-correlated with electron measurements. They conclude that the electrons that coincide with the variations in brightness are those with energies above 3 keV (but up to the instrument limit of 12 keV). Lower energy electrons are also present, with energies of ∼1 keV but having stable signatures (i.e., producing diffuse aurora). They note a gap between these two populations, which they attribute to lower- and upper-band chorus waves. They also note that quasi-3 Hz modulation of only the higher energy portion of the electron precipitation and deem these to be driven by processes internal to the scattering mechanism.

Sandahl et al. (1980) provided a somewhat similar result. Using data from a rocket launched from the Esrange Space Center into pulsating aurora, with instrumentation to measure electron energies from near 0 to 200 keV, they found fluctuations in precipitating electron fluxes at energies above 5 keV, with a peak in energy flux at about 20 keV. Electron fluxes decrease significantly near 40 keV, but they show that weak pulsations were observed even up to 140 keV. As Miyoshi et al. (2015b) would later show, these authors also reported a ∼3 Hz modulation, but only at higher energies, above 25 keV in this case. Non-pulsating electrons are also measured below 5 keV, down to a few hundred eV, but with no gap, as described by Miyoshi et al. (2015b).

2.2.8 *Ion Outflow*

As explained by Strangeway et al. (2005), ion outflow (O^+ ions that have acquired gravitational escape speed) is a two-step process. At lower altitudes, ion upflow can occur via one of two very different mechanisms (Wahlund et al. 1992). In Type 1 ion upflow, O^+ ions are heated near 110–120 km altitude by Joule heating, a process that can lift this population up to perhaps several hundred kilometers. In Type 2 ion upflow, precipitating soft electrons collide with and heat ambient ionospheric electrons which then expand upwards adiabatically, thereby creating an ambipolar electric field that lifts the O^+ ions to altitudes of several hundred kilometers. In either case, once the O^+ ions are at these higher altitudes, they may be further energized by ambient waves that can accelerate them to escape velocity (then called ion outflow). If this secondary process does not develop, the ions will drift back downward to the ionosphere.

This process is most prevalent in the cusp region, where intense Joule heating and Alfvènic aurora (i.e., soft precipitation) are common. One would not expect to find ion upwelling in pulsating aurora because of the apparent lack of such processes in these cases, although the recent results described in the previous section may indicate otherwise.

Liang et al. (2015) sought to determine what processes define the shape and size of patches and highlighted a number of theories suggesting that flux tubes above the patches must contain enhanced cold plasma that determine the structure. These authors further suggested that ion outflow could provide the requisite cold plasma, though no observations of upflow were presented in that paper. Evidence of PFISR observations of ion upflow (not outflow) over pulsating aurora are presented by Liang et al. (2018), which they attribute to the deposition of heat flux from higher altitudes that lead to the formation of an ambipolar field and subsequent ion upflow.

Kenward et al. (2019) also used PFISR observations, in conjunction with all-sky imagers filtered at 557.7 and 630.0 nm and were able to show that the soft precipitation directly heated the electron population in the F-region. This heated population led to the formation of an observed ambipolar field, its resulting parallel electric field and the associated upflow. Still, as these authors point out, the upflow speeds are well below escape speeds (near 450 km altitude, where the data are acquired) and the question of whether a secondary process exists at higher altitudes that would further energize these ions is an open one.

2.2.9 Coupling to the Thermosphere

In this section, coupling of pulsating aurora to the thermosphere is addressed via a discussion of four disparate papers. Wilson et al. (2005) used an infrasonic array of 8 microphones to record acoustic pulses driven by pulsating aurora. They conclude that the acoustic pulses result from ionospheric heating due to the precipitating electron population, which then couples to the thermosphere to launch downward pulses. They show that their signatures originate from sources within 35 km of the zenith above the array (assuming a source height of 110 km).

Oyama et al. (2010) used data from a Fabry-Perot Interferometer (FPI) and an all-sky imager (both at a wavelength of 557.7 nm), as well as the EISCAT UHF radar, to study thermospheric winds in pulsating aurora. They found that in pulsating aurora, when convection electric fields were smaller than 15 mV/m, wind fluctuations were then isolated at the edges and/or in the darker areas of auroral patches. The largest vertical amplitudes they observed were up to 20 m/s, suggesting that the wind fluctuations were localized to the patches.

In terms of atmospheric effects of pulsating aurora, Takahashi et al. (2017) followed up on an earlier report by Tsuda et al. (2013) that described a decrease in mesospheric sodium in conjunction with pulsating aurora. Takahashi et al. (2017) examined data from a single event over Tromsø, Norway, observed using an all-sky imager and EISCAT Tromsø radar, with the radar showing enhanced ionization below 100 km. Simultaneous observations by the sodium lidar at Tromsø showed sodium depletions at those altitudes of 5–8%, lasting 8 min. Based on model results, the sequence they describe begins with pulsating auroral electron ionization producing N_2^+ and O_2^+, followed quickly by the transition of N_2^+ to NO^+. The environment of NO^+ and O_2^+ ultimately leads to the depletion of sodium via charge transfer reactions with these ions.

An entirely different atmospheric impact is described by Turunen et al. (2016), who sought to quantify the role of pulsating auroral electrons on ozone depletion, noting that the energetic precipitation produces odd nitrogen and odd hydrogen. Through catalytic reaction chains, these populations drive the ozone depletion either at the high altitudes where the initial ionization occurs, or at lower altitudes following transport of these populations. In their paper, they consider a single event that included >200 keV electron precipitation that led to model results showing depletion of odd oxygen by several tens of a percent.

3 Numerical Simulation

3.1 Wave-Particle Interaction, Chorus/ECH Generation, Scattering

Wave-particle interactions cause the pitch angle scattering of energetic particles in the magnetosphere, resulting in diffuse/pulsating aurora. Among a variety of wave modes in the magnetospheric plasma, whistler-mode waves play dominant roles in the pitch angle scattering of energetic electrons. Whistler-mode chorus emissions are typical example of whistler-mode waves in the magnetosphere. Chorus emissions are coherent electromagnetic waves observed mainly in the dawn side of the magnetosphere, characterized by their spectral features showing rising/falling tones and by the periodic generation of wave elements in the time scale closely related to pulsating aurora. Understanding of the chorus generation mechanism is essential for the thorough understanding of the mechanism controlling the periodicity of pulsating aurora.

The theory of the whistler-mode chorus generation has been substantially advanced since 2000's, based on the development of the nonlinear wave growth theory by Omura et al. (2008, 2009) supported by results of numerical experiments (Katoh and Omura 2007; Hikishima et al. 2009). By an electron hybrid code simulation in a spatially one-dimensional simulation system along a field line of the dipole field, Katoh and Omura (2007) reproduced that coherent wave elements with rising tones are emerged from a band of whistler-mode waves around the magnetic equator. Hikishima et al. (2009) also reproduced the whistler-mode chorus generation in the equatorial region by a plasma full particle code. Omura et al. (2008, 2009) proposed the nonlinear wave growth theory for the chorus generation from the magnetic equator.

The theory and simulation suggest that the generation process of chorus can be categorized into (1) linear phase and (2) non-linear phase. Under the presence of energetic electrons having a temperature anisotropy of the velocity distribution function, a band of whistler-mode waves is generated around the magnetic equator. The time scale of the wave growth can be estimated by the growth rate of waves based on the linear theory (e.g., Kennel and Petschek 1966; Summers et al. 2013). As the wave amplitude grows, a coherent whistler-mode wave appears from a band of waves, and then nonlinear effects become significant on the motion of resonant electrons. Thereby the wave generation process undergoes the non-linear phase.

Omura et al. (2008) suggested that an electromagnetic electron hole will be formed in the velocity phase space through nonlinear wave-particle interactions in the equatorial region of the magnetosphere. They also suggested that the most of resonant electrons will be untrapped from a coherent whistler-mode wave and that nonlinear trajectories of resonant untrapped electrons result in the formation of nonlinear resonant currents. Based on the nonlinear wave growth theory, the threshold amplitude required for the chorus generation has been proposed (Omura et al. 2009). A series of electron hybrid code simulations have been performed for different number density of energetic electrons (Katoh and Omura 2011), for different spatial gradient of the background magnetic field (Katoh and Omura 2013), and for different temperature anisotropy of energetic electrons at the magnetic equator (Katoh et al. 2018). Results of the simulations revealed that distinct chorus with rising tones appeared when the wave amplitude of a band of whistler-mode waves exceeds the theoretically estimated threshold amplitude. Simulation results in Katoh and Omura (2011) also revealed that the frequency sweep rate becomes high for large wave amplitude of chorus elements, as predicted by Omura et al. (2009). The proposed nonlinear wave growth theory well explains not only simulation results but also satellite observation. Kurita et al. (2012) analyzed THEMIS data of chorus elements without a gap at half the gyrofrequency. They showed that frequency profiles of the wave amplitudes of the chorus elements are well explained by the optimum wave amplitude proposed by Omura and Nunn (2011). Hikishima et al. (2010) revealed by a plasma full particle code simulation that microburst precipitation of 10–100 keV electrons accompanies the generation of discrete chorus emissions.

There are many reports from the low-altitude satellites and sounding rocket experiments that the precipitating electrons have clear energy dispersions, which have been used for the time-of-flight analysis to estimate the modulation region of the pulsating aurora. Miyoshi et al. (2010) proposed a model for the modulations of the pulsating aurora electrons, including the effect of the propagation of the chorus waves along the field line. They derived the theoretical curve of energy dispersion for precipitating electrons. The theoretical energy dispersion is fitted with the observed energy dispersion to estimate parameters of chorus waves. Later, Saito et al. (2012) extended this theoretical energy dispersion including relativistic effects.

More recently, Miyoshi et al. (2015b) conducted the GEMSIS-RBW test-particle simulations (Saito et al. 2012) to calculate cyclotron resonance between upper- and lower-band chorus waves and electrons bounced along the field line. In general, lower-band chorus bursts every a few seconds include a few rising tone elements, while continuous upper-band chorus include continuous elements. There is a half-gyro frequency gap between lower-band and upper-band chorus waves. As shown in Fig. 5, burst precipitations more than a few keV are observed with energy dispersion. Besides the burst precipitations, stable precipitations are found around 1 keV. From investigations of the simulation, the burst precipitations are caused by the lower-band chorus waves, while the stable precipitations are driven by the upper-band chorus waves. There is a precipitation gap between the burst precipitations and the stable precipitations, which corresponds to the resonance energy of the half-gyro frequency gap. Characteristics of energy spectrum of precipitating electrons are well consistent with the energy spectrum measured by the low-altitude satellite (Miyoshi et al. 2010, 2015b, Nishiyama et al. 2011a, 2011b). The study indicates that the origin of the internal modulations of the pulsating aurora is the train of the rising tone elements, while the main modulations is the lower-band chorus bursts. These relationships are confirmed by satellite-ground based conjugate observations; main modulations and chorus bursts: Nishimura et al. (2010), Kasahara et al. (2018), internal modulations and rising tone elements: Ozaki et al. (2018). Also, the background emissions of the pulsating aurora (Evans et al. 1987) is caused by the upper-band chorus waves.

It is interesting to note the possibility of tens keV and sub-relativistic/relativistic electron precipitations associated with the pulsating aurora. Miyoshi et al. (2010, 2015a) proposed a model that chorus waves propagating along the field line cause wide-energy electron precipitations from a few keV to multi-hundred keV and MeV electrons if chorus waves propagate to the higher latitudes, and they suggested that precipitations of multi-hundred keV and MeV electron simultaneously occur during the pulsating aurora. The theoretical energy dispersion curve derived from Saito et al. (2012) that includes precipitations from a few keV to more than MeV can be used to compare with the future observations on precipitating electrons for wide energy electrons associated with the pulsating aurora.

As we have mentioned in Sect. 2, ECH waves are another candidate to cause the pitch angle scattering to cause diffuse aurora (Ni et al. 2012; Zhang et al. 2015) and pulsating aurora (Liang et al. 2010). Ni et al. (2012) estimated scattering rate of electrons by ECH waves and claimed that ECH waves can play an important role to cause the diffuse aurora in higher L-shell region. Several studies estimated linear growth rate of ECH waves (Ashour-Abdalla and Kennel 1978; Zhou et al. 2017; Liu et al. 2018), and Zhou et al. (2017) and Liu et al. (2018) concluded that the loss-cone distribution should be a free energy source to drive ECH waves. Detail distribution function data have been observed by Van Allen Probes (e.g., Zhou et al. 2017) and Arase (Kazama et al. 2018) in the inner magnetosphere, comparisons between in-situ particle/wave observations and ground-based optical measurements are useful to identity generation of ECH waves and scattering of electrons.

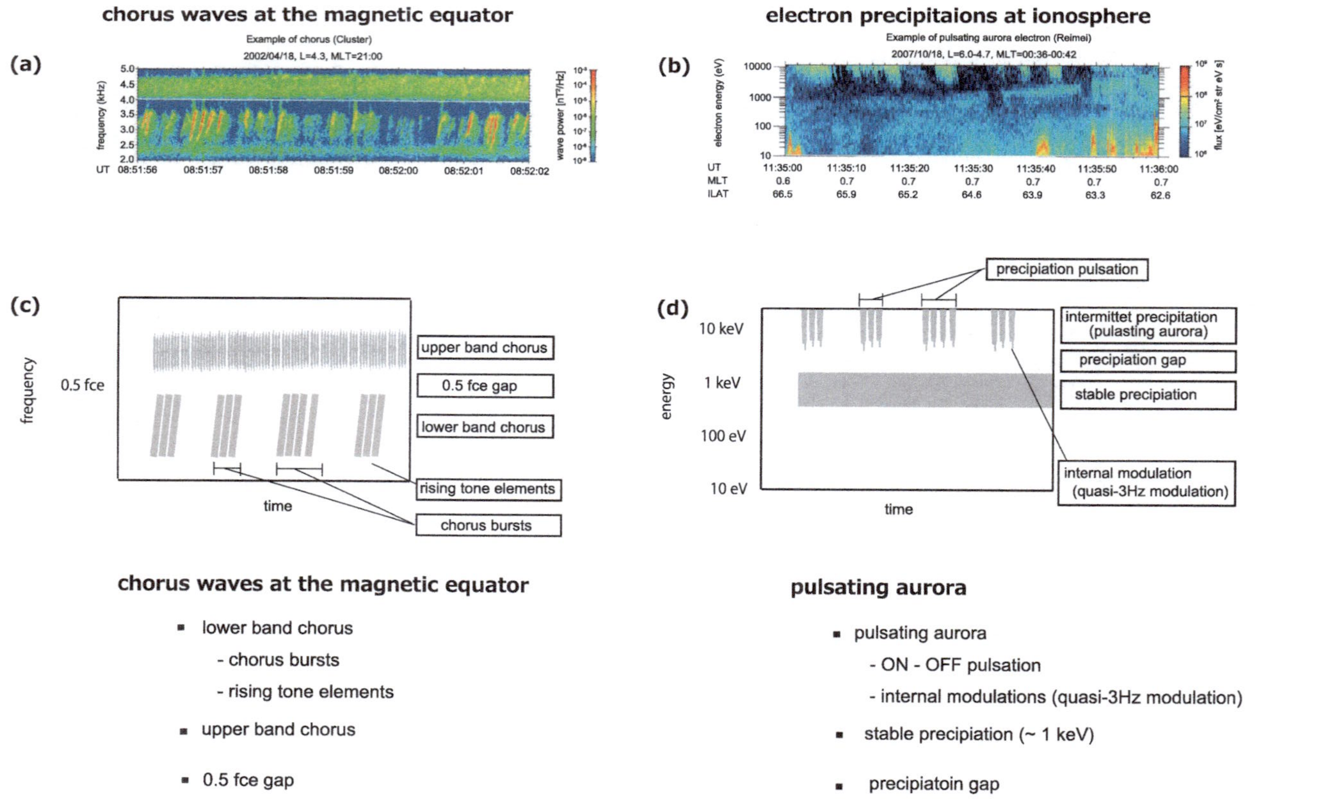

Fig. 5 (**a**) Example of chorus frequency spectrum observed by the Cluster satellite. (**b**) Example of pulsating aurora energy spectrum observed by the Reimei satellite. (**c** and **d**) Schematic diagram to indicate the correspondences between the frequency spectrum of chorus waves and energy spectrum of precipitating electrons for the pulsating aurora, together with typical frequency-time spectrogram of chorus waves and energy-time diagram of precipitating electrons for the pulsating aurora (Miyoshi et al. 2015b)

3.2 Spatial Extent, Modulation

The size of pulsating aurora patches has been explained by the spatial scale of wave-particle interactions occurring in the equatorial region of the magnetosphere. Demekhov and Trakhtengerts (1994) proposed a flow cyclotron maser model as a mechanism controlling both temporal and spatial scale of pulsating aurora. Assuming a plasma duct formed along a field line as a region where efficient wave-particle interaction occurs, they proposed that the pulsation 'on' time of pulsating patches corresponds to the duration of interactions in the duct and that the pulsating 'off' time is the time interval restoring free energy inside the duct through the transportation of energetic electrons. The modulation of auroral brightness inside the patch has been explained by the repetition period of individual chorus elements. The size of the duct has been related to the spatial structure of ambient plasma. Katoh (2014) performed a simulation study of the propagation of whistler-mode waves in the meridional plane of the inner magnetosphere. While the simulation results showed that the duct is formed by either enhancement or depletion of plasma density, the propagation property of waves inside the duct becomes different depending on the plasma density structure. The propagation property should affect the spatial modulation of auroral emissions inside pulsating patches.

3.3 Modeling Parameters, Consistency with Observation

Recent development of computational resources enables us to perform numerical experiments with realistic initial conditions as we observe in the magnetosphere. A self-consistent simulation of the whistler-mode chorus generation was performed in Katoh and Omura (2016) with the real spatial scale of the inner magnetosphere. They showed that the spectral properties of reproduced chorus elements are consistent with those observed by the Cluster spacecraft (Santolik et al. 2003). The consistency between the simulation and satellite observation has also been performed in Demekhov et al. (2016).

4 Summary and Open Questions

This chapter reviewed fundamental properties of diffuse and pulsating aurora. Section 1 summarized the definition, the large-scale morphology, types of pulsation, and driving process of diffuse/pulsating aurora. Section 2 reviewed recent advances. Section 2.1 covered observation techniques, Sect. 2.2 presented recent findings on occurrence, duration, altitude, evolution, small-scale structures, fast modulation, relation to high-energy precipitation, the role of ECH waves, reflected and secondary electrons, and ionosphere dynamics. Section 3 reviewed recent advances in simulation of wave-particle interaction.

Despite that major advances have been made in recent years, the community still faces outstanding questions on fundamental properties of diffuse and pulsating aurora as listed below:

- Origin of chorus modulation. Although the auroral pulsation has been found to originate in chorus modulation, it is still an open question what controls chorus modulation. Density modulation and non-linear processes have been suggested but no definitive conclusion has been made. Since whistler-mode waves in the plasma sheet do not always exhibit rising tones but also show falling tones or unstructured emission. Those different types of wave generation processes need to be investigated for understanding pulsating auroral modulation.

- Pulsating auroral patch size. The size of pulsating auroral patches reflect the size of coherent chorus size. This is also likely true of substructures of pulsating chorus and elements of chorus waves. However, it is unclear what determines the size of coherent chorus waves. It is much larger than electron gyroradius or chorus perpendicular wavelength, and thus collective processes beyond the electron scale are expected to play a role.
- Origin of fine-scale diffuse aurora including substructures. Diffuse aurora is not a structureless feature but involves fine-scale substructures. One fundamentally important issue has to do with the persistence of the shapes of pulsating patches that sustain over several minutes, even in cases where the patches appear to have disappeared but then reappeared. Although ideas such as pressure-driven processes have been proposed, it is not understood how such structures form in the magnetosphere and what determines the size and modulation. Since it is difficult to have satellite conjunctions with fine-scale aurora, knowledge of such structures has been severely limited. The implicit question is whether the precipitation source (i.e., near the equatorial region) or the ionosphere itself is responsible for the shape of patches. In the latter case where, for example, ionospheric conductivity plays a role, the question would then become how this might contribute to a feedback process.
- A related question is what determines the propagation speed of patches. Again, the answer may indicate that the ionosphere is intricately linked to the patches (i.e., if the patches are tied to convection), or may suggest that patches are simply signatures of source region dynamics.
- Role of ECH waves and upper-band chorus. While lower-band chorus has been identified as the dominant source of waves for pitch angle scattering, ECH waves and upper-band chorus have also been associated with pulsating aurora. A thorough study is desired to find the probability and conditions that those wave modes become important. Also the role of parallel potential drop and secondary electrons should also be investigated.

Acknowledgements The work of Y.N. was supported by NASA grant NNX17AL22G and 80NSSC18K0657, NSF grants PLR-1341359, AGS-1737823, and AFOSR FA9559-16-1-0364. The work by NP was supported by the Research Council of Norway under CoE contract 223252. Y.K., Y.M., and K.H. are supported by JSPS Kakenhi (15H05747). Y.K, and Y.M are supported by JSPS Kakenhi (15H05815). Y.M is supported by JSPS Kakenhi (16H06286). Y.K is supported by JSPS Kakenhi (18H03727). Y.K., Y.M., and K.H. are supported by the International Space Science Institutes Beijing (ISSI-BJ) International Team program. The work of NS was supported by NASA Earth & Space Science Fellowship Program grant 80NSSC17K0433. The THEMIS mission and all-sky imagers are supported by NASA contract NAS5-02099, CSA contract 9F007-046101 and NSF grant AGS-1004736. ML was supported by NASA grant 80NSSC18K0950. DKW was supported by the Natural Environment Research Council (NERC) of the UK under grant NE/N004051/1. The PWING imager was installed and operated by the pulsating aurora project (http://www.psa-research.org) and PWING project (http://www.isee.nagoya-u.ac.jp/dimr/PWING/en/).

References

S.I. Akasofu, *Polar and Magnetospheric Substorms* (Reidel, Dordrecht, 1968), p. 223

R.L. Arnoldy, K. Dragoon, L.J. Cahill, S.B. Mende, T.J. Rosenberg, Detailed correlations of magnetic field and riometer observations at $L = 4.2$ with pulsating aurora. J. Geophys. Res. **87**(A12), 10449 (1982). https://doi.org/10.1029/JA087iA12p10449

M. Ashour-Abdalla, C. Kennel, Nonconvective and convective electron cyclotron harmonic instabilities. J. Geophys. Res. **83**, 1531–1543 (1978). https://doi.org/10.1029/JA083iA04p01531

K. Axelsson, T. Sergienko, H. Nilsson, U. Brändstrm, Y. Ebihara, K. Asamura, M. Hirahara, Spatial characteristics of wave-like structures in diffuse aurora obtained using optical observations. Ann. Geophys. **30**(12), 1693–1701 (2012). https://doi.org/10.5194/angeo-30-1693-2012

F.T. Berkey, Observations of pulsating aurora in the day sector auroral zone. Planet. Space Sci. **26**, 635–650 (1978)

F.T. Berkey, C.D. Anger, S.-I. Akasofu, E.P. Rieger, The signature of large-scale auroral structure in radio wave absorption. J. Geophys. Res. **85**(A2), 593 (1980). https://doi.org/10.1029/JA085iA02p00593

E.C. Bland, E. Heino, M.J. Kosch, N. Partamies, SuperDARN radar-derived HF radio attenuation during the September 2017 solar proton events. Space Weather **16**(10), 1455–1469 (2018). https://doi.org/10.1029/2018SW001916

E.C. Bland, N. Partamies, E. Heino, A.S. Yukimatu, H. Miyaoka, Energetic electron precipitation occurrence rates determined using the Syowa East SuperDARN radar. J. Geophys. Res. Space Phys. **124**, 6253–6265 (2019). https://doi.org/10.1029/2018JA026437

U. Brändström, the auroral large imaging system-design, operation and scientific results. IRF Scientific Report 279, Kiruna (2003)

A. Brekke, H. Pettersen, Some observations of pulsating aurora at Spitzbergen. Planet. Space Sci. **19**, 536540 (1971)

N.B. Brown, T.N. Davis, T.J. Hallinan, H.C. Stenbaek-Nielsen, Altitude of pulsating aurora determined by a new instrumental technique. Geophys. Res. Lett. **3**, 403–404 (1976). https://doi.org/10.1029/GL003i007p00403

D. Bryant, G. Courtier, A. Johnstone, Modulation of auroral electrons at large distances from the Earth. J. Atmos. Terr. Phys. **31**(4), 579–592 (1969). https://doi.org/10.1016/0021-9169(69)90010-5

D. Bryant, M.J. Smith, G.M. Courtier, Distant modulation of electron intensity during the expansion phase of an auroral substorm. Planet. Space Sci. **23**, 867 (1975). https://doi.org/10.1016/0032-0633(75)90022-7

L.B.N. Clausen, H. Nickisch, Automatic classification of auroral images from the Oslo Auroral THEMIS (OATH) data set using machine learning. J. Geophys. Res. Space Phys. **123**(7), 5640–5647 (2018). https://doi.org/10.1029/2018JA025274

J.M. Comberiate, F. Kamalabadi, L.J. Paxton, A tomographic model for ionospheric imaging with the Global Ultraviolet Imager. Radio Sci. **42**(2), RS2011 (2007). https://doi.org/10.1029/2005RS003348

M. Craven, G.B. Burns, High latitude pulsating aurora. Geophys. Res. Lett. **17**(9), 1251–1254 (1990)

F. Creutzberg, R. Gattinger, F. Harris, A. Vallance Jones, Pulsating auroras in relation to proton and electron auroras. Can. J. Phys. **59**, 1124–1130 (1981)

A.B. Crew, H.E. Spence, J.B. Blake, D.M. Lumpar, B.A. Larsen, T.P. O'Brien, S. Driscoll, M. Handley, J. Legere, S. Longworth, K. Mashburn, E. Mosleh, N. Ryhajlo, S. Smith, L. Springer, M. Widholm, First multipoint in situ observations of electron microbursts: initial results from the NSF FIREBIRD II mission. J. Geophys. Res. **121**, 5272–5283 (2016). https://doi.org/10.1002/2016JA022485

H. Dahlgren, N. Ivchenko, J. Sullivan, B.S. Lanchester, G. Marklund, D. Whiter, Morphology and dynamics of aurora at fine scale: first results from the ASK instrument. Ann. Geophys. **26**, 1041–1048 (2008). https://doi.org/10.5194/angeo-26-1041-2008

H. Dahlgren, N. Ivchenko, B.S. Lanchester, Monoenergetic high-energy electron precipitation in thin auroral filaments. Geophys. Res. Lett. **39**(20), L20101 (2012). https://doi.org/10.1029/2012GL053466

H. Dahlgren, B.S. Lanchester, N. Ivchenko, D.K. Whiter, Electrodynamics and energy characteristics of aurora at high resolution by optical methods. J. Geophys. Res. **121**(6), 5966–5974 (2016). https://doi.org/10.1002/2016JA022446

H. Dahlgren, B.S. Lanchester, N. Ivchenko, D.K. Whiter, Variations in energy, flux, and brightness of pulsating aurora measured at high time resolution. Ann. Geophys. **35**(3), 493–503 (2017). https://doi.org/10.5194/angeo-35-493-2017. https://www.ann-geophys.net/35/493/2017/

C. Deehr, D. Lummerzheim, Ground-based optical observations of hydrogen emission in the auroral substorm. J. Geophys. Res. Space Phys. **106**(A1), 33–44 (2001). https://doi.org/10.1029/2000JA002010

A.G. Demekhov, V.Y. Trakhtengerts, A mechanism of formation of pulsating aurorae. J. Geophys. Res. **99**(A4), 5831–5841 (1994). https://doi.org/10.1029/93JA01804

A.G. Demekhov, U. Taubenschuss, O. Santolik, Simulation of VLF chorus emissions in the magnetosphere and comparison with THEMIS spacecraft data. J. Geophys. Res. Space Phys. **122**, 166–184 (2016). https://doi.org/10.1002/2016JA023057

E.F. Donovan, B.J. Jackel, I. Voronkov, T. Sotirelis, F. Creutzberg, N.A. Nicholson, Ground-based optical determination of the b2i boundary: a basis for an optical MT-index. J. Geophys. Res. Space Phys. **108**(A3), 1115 (2003). https://doi.org/10.1029/2001JA009198

E. Donovan, S. Mende, B. Jackel, H. Frey, M. Syrjasuo, I. Voronkov, T. Trondsen, L. Peticolas, V. Angelopoulos, S. Harris, M. Green, M. Connors, The THEMIS all-sky imaging array—system design and initial results from the prototype imager. J. Atmos. Sol.-Terr. Phys. **68**, 1472–1487 (2006). https://doi.org/10.1016/j.jastp.2005.03.027

C.N. Duncan, F. Creutzberg, R.L. Gattinger, F.R. Harris, A. Vallance Jones, Latitudinal and temporal characteristics of pulsating auroras. Can. J. Phys. **59**, 1063–1069 (1981)

R.H. Eather, Short-period auroral pulsations in λ6300 OI. J. Geophys. Res. **74**(21), 4998–5004 (1969). https://doi.org/10.1029/JA074i021p04998

Y. Ebihara, T. Sakanoi, K. Asamura, M. Hirahara, M.F. Thomsen, Reimei observation of highly structured auroras caused by nonaccelerated electrons. J. Geophys. Res. Space Phys. **115**(A8), n/a–n/a (2010). https://doi.org/10.1029/2009JA015009

C.-F. Enell, B. Gustavsson, B.U.E. Brändström, T.I. Sergienko, P.T. Verronen, P. Rydesäter, I. Sandahl, Tomography-like retrieval of auroral volume emission ratios for the 31 January 2008 Hotel Payload 2 event. Geosci. Instrum. Method. Data Syst. **2**(1), 1–21 (2012). https://doi.org/10.5194/gid-2-1-2012

D.S. Evans, G.T. Davidson, H.D. Voss, W.L. Imhof, J. Mobilia, Y.T. Chiu, Interpretation of electron spectra in morningside pulsating aurorae. J. Geophys. Res. **92**(A11), 12295 (1987). https://doi.org/10.1029/JA092iA11p12295

X. Fang, C.E. Randall, D. Lummerzheim, W. Wang, G. Lu, S.C. Solomon, R.A. Frahm, Parameterization of monoenergetic electron impact ionization. Geophys. Res. Lett. **37**(22), n/a–n/a (2010). https://doi.org/10.1029/2010GL045406

B.A. Fritz, M.L. Lessard, M.J. Blandin, P.A. Fernandes, Structure of black aurora associated with pulsating aurora. J. Geophys. Res. Space Phys. **120**, 10,096–10,106 (2015). https://doi.org/10.1002/2015JA021397

R. Fujii, T. Oguti, T. Yamamoto, Relationships between pulsating auroras and field-aligned electric currents. Mem. Natl. Inst. Polar Res. **36**, 95–103 (1985)

R. Fujii, N. Sato, T. Ono, H. Fukunishi, T. Hirasawa, S. Kokubun, T. Araki, T. Saemundsson, Conjugacies of pulsating auroras by all-sky TV observations. Geophys. Res. Lett. **14**, 115–118 (1987)

M. Fukizawa, T. Sakanoi, Y. Miyoshi, K. Hosokawa, K. Shiokawa, Y. Katoh et al., Electrostatic electron cyclotron harmonic waves as a candidate to cause pulsating auroras. Geophys. Res. Lett. **45**, 12,661–12,668 (2018). https://doi.org/10.1029/2018GL080145

Y. Fukuda, R. Kataoka, Y. Miyoshi, Y. Katoh, T. Nishiyama, K. Shiokawa, Y. Ebihara, D. Hampton, N. Iwagami, Quasi-periodic rapid motion of pulsating auroras. Polar Sci. **10**, 183–191 (2016). https://doi.org/10.1016/j.polar.2016.03.005

D.M. Gillies, D. Knudsen, E. Spanswick, E. Donovan, J. Burchill, M. Patrick, Swarm observations of field-aligned currents associated with pulsating auroral patches. J. Geophys. Res. Space Phys. **120**, 9484–9499 (2015). https://doi.org/10.1002/2015JA021416

M. Grandin, A. Kero, N. Partamies, D. McKay, D. Whiter, A. Kozlovsky, Y. Miyoshi, Observation of pulsating aurora signatures in cosmic noise absorption data. Geophys. Res. Lett. **44**, 5292–5300 (2017). https://doi.org/10.1002/2017GL073901

E. Grono, E. Donovan, Differentiating diffuse auroras based on phenomenology. Ann. Geophys. **36**(3), 891–898 (2018). https://doi.org/10.5194/angeo-36-891-2018

E. Grono, E. Donovan, Constraining the source regions of pulsating auroras. Geophys. Res. Lett. **46**, 10267–10273 (2019). https://doi.org/10.1029/2019GL084611

E. Grono, E. Donovan, K.R. Murphy, Tracking patchy pulsating aurora through all-sky images. Ann. Geophys. **35**(4), 777–784 (2017). https://doi.org/10.5194/angeo-35-777-2017

G. Grubbs, R. Michell, M. Samara, D. Hampton, J.-M. Jahn, Predicting electron population characteristics in 2-D using multispectral ground-based imaging. Geophys. Res. Lett. **45**, 15–20 (2018). https://doi.org/10.1002/2017GL075873

D.-S. Han, X.-C. Chen, J.-J. Liu, Q. Qiu, K. Keika, Z.-J. Hu et al., An extensive survey of dayside diffuse aurora based on optical observations at Yellow River Station. J. Geophys. Res. Space Phys. **120**(9), 7447–7465 (2015). https://doi.org/10.1002/2015JA021699

D.-S. Han, J.-X. Li, Y. Nishimura, L.R. Lyons, J. Bortnik, M. Zhou et al., Coordinated observations of two types of diffuse auroras near magnetic local noon by Magnetospheric Multiscale mission and ground all-sky camera. Geophys. Res. Lett. **44**(16), 8130–8139 (2017). https://doi.org/10.1002/2017GL074447

M.G. Henderson, Auroral substorms, poleward boundary activations, auroral streamers, omega bands, and onset precursor activity, in *Auroral Phenomenology and Magnetospheric Processes: Earth and Other Planets*. Geophysical Monograph Series, vol. 197 (Am. Geophys. Union, Washington, 2012). https://doi.org/10.1029/2011GM001165

M. Hikishima, S. Yagitani, Y. Omura, I. Nagano, Full particle simulation of whistler-mode rising chorus emissions in the magnetosphere. J. Geophys. Res. **114**, A01203 (2009). https://doi.org/10.1029/2008JA013625

M. Hikishima, Y. Omura, D. Summers, Microburst precipitation of energetic electrons associated with chorus wave generation. Geophys. Res. Lett. **37**, L07103 (2010). https://doi.org/10.1029/2010GL042678

F. Honary, S.R. Marple, K. Barratt, P. Chapman, M. Grill, E. Nielsen, Invited article: Digital beam-forming imaging riometer systems. Rev. Sci. Instrum. **82**(3), 031301 (2011). https://doi.org/10.1063/1.3567309

R.B. Horne, R.M. Thorne, Relativistic electron acceleration and precipiation during resonant interactions with whistler-mode chorus. Geophys. Res. Lett. **30**, 1527 (2003). https://doi.org/10.1029/2003GL016973

K. Hosokawa, Y. Ogawa, Ionospheric variation during pulsating aurora. J. Geophys. Res. Space Phys. **120**, 5943–5957 (2015). https://doi.org/10.1002/2015JA021401

K. Hosokawa, A. Kadokura, N. Sato, S.E. Milan, M. Lester, G. Bjornsson, T. Saemundsson, Electric field modulation behind pulsating aurora. J. Geophys. Res. **113**, A11322 (2008). https://doi.org/10.1029/2008JA013601

K. Hosokawa, Y. Ogawa, A. Kadokura, H. Miyaoka, N. Sato, Modulation of ionospheric conductance and electric field associated with pulsating aurora. J. Geophys. Res. Space Phys. **115**(A3), n/a–n/a (2010). https://doi.org/10.1029/2009JA014683

B.K. Humberset, J.W. Gjerloev, M. Samara, R.G. Michell, I.R. Mann, Temporal characteristics and energy deposition of pulsating auroral patches. J. Geophys. Res. Space Phys. **121**, 7087–7107 (2016). https://doi.org/10.1002/2016JA022921

B.K. Humberset, J.W. Gjerloev, I.R. Mann, R.G. Michell, M. Samara, On the persistent shape and coherence of pulsating auroral patches. J. Geophys. Res. Space Phys. **123**, 4272–4289 (2018). https://doi.org/10.1029/2017JA024405

B.J. Jackel, F. Creutzberg, E.F. Donovan, L.L. Cogger, Triangulation of auroral red-line emission heights, in *Proceedings of the 28th Annual European Meeting on Atmospheric Studies by Optical Methods*, vol. 92, ed. by K.U. Kaila, J.R.T. Jussila, H. Holma (Sodankyla Geophysical Observatory Publications, Oulu, 2003), pp. 97–100

A.N. Jaynes, M.R. Lessard, J.V. Rodriguez, E. Donovan, T.M. Loto'aniu, K. Rychert, Pulsating aurora electron flux modulations in the equatorial magnetosphere. J. Geophys. Res. Space Phys. **118**, 884–894 (2013). https://doi.org/10.1002/jgra.50434

A.N. Jaynes et al., Correlated Pc4–5 ULF waves, whistler-mode chorus and pulsating aurora observed by the Van Allen Probes and ground-based systems. J. Geophys. Res. Space Phys. **120**, 8749–8761 (2015). https://doi.org/10.1002/2015JA021380

S.F. Johnston, Autonomous meridian scanning photometer for auroral observations. Opt. Eng. **28**(1), 280120 (1989). https://doi.org/10.1117/12.7976895

A.D. Johnstone, Pulsating aurora. Nature **274**, 119–126 (1978)

S.L. Jones, M.R. Lessard, P.A. Fernandes, D. Lummerzheim, J.L. Semeter, C.J. Heinselman, K.A. Lynch, G. Michell, P.M. Kintner, H.C. Stenbaek-Nielsen, K. Asamura, PFISR and ROPA observations of pulsating aurora. J. Atmos. Sol.-Terr. Phys. **71**, 708–716 (2009). https://doi.org/10.1016/j.jastp.2008.10.004

S.L. Jones, M.R. Lessard, K. Rychert, E. Spanswick, E. Donovan, Large-scale aspects and temporal evolution of pulsating aurora. J. Geophys. Res. Space Phys. **116**, A03214 (2011). https://doi.org/10.1029/2010ja015840

S.L. Jones, M.R. Lessard, K. Rychert, E. Spanswick, E. Donovan, A.N. Jaynes, Persistent, widespread pulsating aurora: a case study. J. Geophys. Res. Space Phys. **118**, 2998–3006 (2013). https://doi.org/10.1002/jgra.50301

S. Kasahara, Y. Miyoshi, S. Yokota, T. Mitani, Y. Kasahara, S. Matsuda et al., Pulsating aurora from electron scattering by chorus waves. Nature **554**(7692), 337–340 (2018). https://doi.org/10.1038/nature25505

R. Kataoka, Y. Miyoshi, D. Hampton, T. Ishii, H. Kozako, Pulsating aurora beyond the ultra-low-frequency range. J. Geophys. Res. Space Phys. **117**, A8 (2012). https://doi.org/10.1029/2012JA017987.

R. Kataoka, Y. Miyoshi, K. Shigematsu, D. Hampton, Y. Mori, T. Kubo, A. Yamashita, M. Tanaka, T. Takahei, T. Nakai, H. Miyahara, K. Shiokawa, Stereoscopic determination of all-sky altitude map of aurora using two ground-based Nikon DSLR cameras. Ann. Geophys. **31**(9), 1543–1548 (2013). https://doi.org/10.5194/angeo-31-1543-2013. https://www.ann-geophys.net/31/1543/2013/

R. Kataoka, Y. Fukuda, Y. Miyoshi, H. Miyahara, S. Itoya, Y. Ebihara, D. Hampton, H. Dahlgren, D. Whiter, N. Ivchenko, Compound auroral micromorphology: ground-based high-speed imaging. Earth Planets Space **67**(1), 23 (2015). https://doi.org/10.1186/s40623-015-0190-6

R. Kataoka, Y. Fukuda, H.A. Uchida, H. Yamada, Y. Miyoshi, Y. Ebihara, D. Hampton, High-speed stereoscopy of aurora. Ann. Geophys. **34**(1), 41–44 (2016). https://doi.org/10.5194/angeo-34-41-2016

R. Kataoka, T. Nishiyama, Y. Tanaka, A. Kadokura, H.A. Uchida, Y. Ebihara, M.K. Ejiri, Y. Tomikawa, M. Tsutsumi, K. Sato, Y. Miyoshi, K. Shiokawa, S. Kurita, Y. Kasahara, M. Ozaki, K. Hosokawa, S. Matsuda, I. Shinohara, T. Takashima, T. Sato, T. Mitani, T. Hori, N. Higashio, Transient ionization of the mesosphere during auroral breakup: Arase satellite and ground-based conjugate observations at Syowa Station. Earth Planets Space **7**, 19 (2019). https://doi.org/10.1186/s40623-019-0989-7

Y. Katoh, A simulation study of the propagation of whistler-mode chorus in the Earth's inner magnetosphere. Earth Planets Space **66**, 6 (2014). https://doi.org/10.1186/1880-5981-66-6

Y. Katoh, Y. Omura, Computer simulation of chorus wave generation in the Earth's inner magnetosphere. Geophys. Res. Lett. **34**, L03102 (2007). https://doi.org/10.1029/2006GL028594

Y. Katoh, Y. Omura, Amplitude dependence of frequency sweep rates of whistler mode chorus emissions. J. Geophys. Res. **116**, A07201 (2011). https://doi.org/10.1029/2011JA016496

Y. Katoh, Y. Omura, Effect of the background magnetic field inhomogeneity on generation processes of whistler-mode chorus and hiss-like broadband emissions. J. Geophys. Res. Space Phys. **118**, 4189–4198 (2013). https://doi.org/10.1002/jgra.50395

Y. Katoh, Y. Omura, Electron hybrid code simulation of whistler-mode chorus generation with real parameters in the Earth's inner magnetosphere. Earth Planets Space **68**, 192 (2016). https://doi.org/10.1186/s40623-016-0568-0

Y. Katoh, Y. Omura, Y. Miyake, H. Usui, H. Nakashima, Dependence of generation of whistler-mode chorus emissions on the temperature anisotropy and density of energetic electrons in the Earth's inner magnetosphere. J. Geophys. Res. Space Phys. **123**, 1165–1177 (2018). https://doi.org/10.1002/2017JA024801

S. Kawamura, K. Hosokawa, S. Kurita, S. Oyama, Y. Miyoshi, Y. Kasahara et al., Tracking the region of high correlation between pulsating aurora and chorus: simultaneous observations with Arase satellite and ground-based all-sky imager in Russia. J. Geophys. Res. Space Phys. **124**, 2769–2778 (2019). https://doi.org/10.1029/2019JA026496

Y. Kazama, H. Kojima, Y. Miyoshi, Y. Kasahara, H. Usui, B.-J. Wang et al., Density depletions associated with enhancements of electron cyclotron harmonic emissions: an ERG observation. Geophys. Res. Lett. **45**, 10,075–10,083 (2018). https://doi.org/10.1029/2018GL080117

C.F. Kennel, H.E. Petschek, Limits on stably trapped particle fluxes. J. Geophys. Res. **71**, 1 (1966). https://doi.org/10.1029/JZ071i001p00001

D.R. Kenward, M.R. Lessard, B.A. Fritz, R. Varney, R.G. Michell, D. Hampton, Observations of ion upflow during pulsating aurora. J. Geophys. Res. Space Phys. (2019, submitted)

G.V. Khazanov, D.G. Sibeck, E. Zesta, Is diffuse aurora driven from above or below? Geophys. Res. Lett. **44**, 641–647 (2017). https://doi.org/10.1002/2016GL072063

S. Kirkwood, H. Opgenoorth, J.S. Murphree, Ionospheric conductivities, electric fields and currents associated with auroral substorms measured by the EISCAT radar. Planet. Space Sci. **36**(12), 1359–1380 (1988). https://doi.org/10.1016/0032-0633(88)90005-0

D.J. Knudsen, J.K. Burchill, T.G. Cameron, G.A. Enno, A. Howarth, A.W. Yau, The CASSIOPE/e-POP Suprathermal Electron Imager (SEI). Space Sci. Rev. **189**, 65 (2015)

D.J. Knudsen, J.K. Burchill, S.C. Buchert, A.I. Eriksson, R. Gill, J.-E. Wahlund, L. Åhlen, M. Smith, B. Moffat, Thermal ion imagers and Langmuir probes in the Swarm electric field instruments. J. Geophys. Res. Space Phys. **122**, 2655–2673 (2017). https://doi.org/10.1002/2016JA022571

M.J. Kosch, F. Honary, C.F. del Pozo, S.R. Marple, T. Hagfors, High-resolution maps of the characteristic energy of precipitating auroral particles. J. Geophys. Res. **106**(A12), 28925–28937 (2001). https://doi.org/10.1029/2001JA900107

S. Kurita, Y. Katoh, Y. Omura, V. Angelopoulos, C.M. Cully, O. Le Contel, H. Misawa, THEMIS observation of chorus elements without a gap at half the gyrofrequency. J. Geophys. Res. **117**, A11223 (2012). https://doi.org/10.1029/2012JA018076

B.S. Lanchester, M. Ashrafi, N. Ivchenko, Simultaneous imaging of aurora on small scale in OI (777.4 nm) and N2 1P to estimate energy and flux of precipitation. Ann. Geophys. **27**, 2881–2891 (2009). https://doi.org/10.5194/angeo-27-2881-2009

M.R. Lessard, in *A Review of Pulsating Aurora. Auroral Phenomenology and Magnetospheric Processes: Earth and Other Planets*. Geophys. Monogr. Ser., vol. 197 (2013), pp. 55–68. https://doi.org/10.1029/2011GM001187

W. Li, J. Bortnik, Y. Nishimura, R.M. Thorne, V. Angelopoulos, The origin of pulsating aurora: modulated whistler mode chorus waves, in *Auroral Phenomenology and Magnetospheric Processes: Earth and Other Planets*, ed. by A. Keiling, E. Donovan, F. Bagenal, T. Karlsson (2013). https://doi.org/10.1029/2011GM001164

J. Liang, V. Uritsky, E. Donovan, B. Ni, E. Spanswick, T. Trondsen, J. Bonnell, A. Roux, U. Auster, D. Larson, THEMIS observations of electron cyclotron harmonic emissions, ULF waves, and pulsating auroras. J. Geophys. Res. **115**, A10235 (2010). https://doi.org/10.1029/2009JA015148

J. Liang, B. Ni, E. Spanswick, M. Kubyshkina, E.F. Donovan, V.M. Uritsky, R.M. Thorne, V. Angelopoulos, Fast earthward flows, electron cyclotron harmonic waves, and diffuse auroras: conjunctive observations and a synthesized scenario. J. Geophys. Res. **116**, A12220 (2011). https://doi.org/10.1029/2011JA017094

J. Liang, E. Donovan, Y. Nishimura, B. Yang, E. Spanswick, K. Asamura, T. Sakanoi, D. Evans, R. Redmon, Low-energy ion precipitation structures associated with pulsating auroral patches. J. Geophys. Res. Space Phys. **120**, 5408–5431 (2015). https://doi.org/10.1002/2015JA021094

J. Liang, E. Donovan, B. Jackel, E. Spanswick, M. Gillies, On the 630 nm red-line pulsating aurora: red-line emission geospace observatory observations and model simulations. J. Geophys. Res. Space Phys. **121**(8), 7988–8012 (2016). https://doi.org/10.1002/2016JA022901

J. Liang, E. Donovan, A. Reimer, D. Hampton, S. Zou, R. Varney, Ionospheric electron heating associated with pulsating auroras: joint optical and PFISR observations. J. Geophys. Res. Space Phys. **123**, 4430–4456 (2018). https://doi.org/10.1029/2017JA025138

X. Liu, L. Chen, W. Gu, X.-J. Zhang, Electron cyclotron harmonic wave instability by loss cone distribution. J. Geophys. Res. **123**, 9035–9044 (2018). https://doi.org/10.1029/2018JA025925

A.T.Y. Lui, P. Perreault, S.I. Akasofu, C.D. Anger, The diffuse aurora. Planet. Space Sci. **21**(5), 857–858 (1973)

A.T.Y. Lui, C.D. Anger, W.J. Heikkila, J.R. Burrows, Simultaneous observations of particle precipitations and auroral. J. Geophys. Res. **82**(16), 2210–2226 (1977)

A.T.Y. Lui, L.L. Cogger, A. Howarth, A.W. Yau, First satellite imaging of auroral pulsations by the Fast Auroral Imager on e-POP. Geophys. Res. Lett. **42**, 6877–6882 (2015). https://doi.org/10.1002/2015GL065331

R. Lyons, Electron diffusion driven by magnetospheric electrostatic waves. J. Geophys. Res. **79**(4), 575–580 (1974). https://doi.org/10.1029/JA079i004p00575

J.E. Maggs, T.N. Davis, Measurements of the thicknesses of auroral structures. Planet. Space Sci. **16**(2), 205–209 (1968). https://doi.org/10.1016/0032-0633(68)90069-X

D.J. McEwan, C.N. Duncan, A campaign to study pulsating auroras. Can. J. Phys. **59**(8), 1029–1033 (1981). https://doi.org/10.1139/p81-135

D.J. McEwen, E. Yee, B.A. Whalen, A.W. Yau, Electron energy measurements in pulsating auroras. Can. J. Phys. **59**(8), 1106–1115 (1981). https://doi.org/10.1139/p81-146

D. McKay, N. Partamies, J. Vierinen, Pulsating aurora and cosmic noise absorption associated with growth-phase arcs. Ann. Geophys. **36**(1), 59–69 (2018). https://doi.org/10.5194/angeo-36-59-2018

D. McKay, T. Paavilainen, B. Gustavsson, A. Kvammen, N. Partamies, Lumikot: fast auroral transients during the growth phase of substorms. Geophys. Res. Lett. **46**, 7214–7221 (2019). https://doi.org/10.1029/2019GL082985

S.B. Mende, H.U. Frey, C.W. Carlson, J. McFadden, J. Gérard, B. Hubert et al., IMAGE and FAST observations of substorm recovery phase aurora. Geophys. Res. Lett. **29**(12), 1602 (2002). https://doi.org/10.1029/2001GL013027

S.B. Mende, S.E. Harris, H.U. Frey, V. Angelopoulos, C.T. Russell, E. Donovan, L.M. Peticolas, The THEMIS array of ground-based observatories for the study of auroral substorms. Space Sci. Rev. **141**(1–4), 357–387 (2008). https://doi.org/10.1007/s11214-008-9380-x

R.G. Michell, M. Samara, Ground magnetic field fluctuations associated with pulsating aurora. J. Geophys. Res. Space Phys. **120**, 9192–9201 (2015). https://doi.org/10.1002/2015JA021252

R.G. Michell, M.G. McHarg, M. Samara, D.L. Hampton, Spectral analysis of flickering aurora. J. Geophys. Res. **117**, A03321 (2012). https://doi.org/10.1029/2011JA016703

R.G. Michell, T. Grydeland, M. Samara, Characteristics of Poker Flat Incoherent Scatter Radar (PFISR) naturally enhanced ion-acoustic lines (NEIALs) in relation to auroral forms. Ann. Geophys. **32**, 1333–1347 (2014). https://doi.org/10.5194/angeo-32-1333-2014

Y. Miyoshi, Y. Katoh, T. Nishiyama, T. Sakanoi, K. Asamura, M. Hirahara, Time of flight analysis of pulsating aurora electrons, considering wave-particle interactions with propagating whistler mode waves. J. Geophys. Res. **115**, A10312 (2010). https://doi.org/10.1029/2009JA015127

Y. Miyoshi, S. Oyama, S. Saito, S. Kurita, H. Fujiwara, R. Kataoka, Y. Ebihara, C. Kletzing, G. Reeves, O. Santolik, M. Clilverd, C.J. Rodger, E. Turunen, F. Tsuchiya, Energetic electron precipitation associated with pulsating aurora: EISCAT and Van Allen Probe observations. J. Geophys. Res. **120**, 2754–2766 (2015a). https://doi.org/10.1002/2014JA020690

Y. Miyoshi, S. Saito, K. Seki, T. Nishiyama, R. Kataoka, K. Asamura, Y. Katoh, Y. Ebihara, T. Sakanoi, M. Hirahara, S. Oyama, S. Kurita, O. Santolik, Relation between fine structure of energy spectra for pulsating aurora electrons and frequency spectra of whistler mode chorus waves. J. Geophys. Res. **120**, 7728–7736 (2015b). https://doi.org/10.1002/2015JA021562

T. Motoba, Y. Ebihara, A. Kadokura, M.J. Engebretson, M.R. Lessard, A.T. Weatherwax, A.J. Gerrard, Fast-moving diffuse auroral patches: a new aspect of daytime Pc3 auroral pulsations. J. Geophys. Res. Space Phys. **122**, 1542–1554 (2017). https://doi.org/10.1002/2016JA023285

R. Nakamura, T. Oguti, Drifts of auroral structures and magnetospheric electric fields. J. Geophys. Res. **92**(A10), 11241–11247 (1987). https://doi.org/10.1029/JA092iA10p11241

P.T. Newell, Y.I. Feldstein, Y.I. Galperin, C.-I. Meng, Morphology of nightside precipitation. J. Geophys. Res. **101**(A5), 10737–10748 (1996). https://doi.org/10.1029/95JA03516

B. Ni, R.M. Thorne, Y.Y. Shprits, J. Bortnik, Resonant scattering of plasma sheet electrons by whistler-mode chorus: contribution to diffuse auroral precipitation. Geophys. Res. Lett. **35**, L11106 (2008). https://doi.org/10.1029/2008GL034032

B. Ni, R. Thorne, J. Liang, V. Angelopoulos, C. Cully, W. Li, X. Zhang, M. Hartinger, O. Le Contel, A. Roux, Global distribution of electrostatic electron cyclotron harmonic waves observed on THEMIS. Geophys. Res. Lett. **38**, L17105 (2011). https://doi.org/10.1029/2011GL048793

B. Ni, J. Liang, R.M. Thorne, V. Angelopoulos, R.B. Horne, M. Kubyshkina, E. Spanswick, E.F. Donovan, D. Lummerzheim, Efficient diffuse auroral electron scattering by electrostatic electron cyclotron harmonic waves in the outer magnetosphere: a detailed case study. J. Geophys. Res. **117**, A01218 (2012). https://doi.org/10.1029/2011JA017095

B. Ni, R.M. Thorne, X. Zhang, J. Bortnik, Z. Pu, L. Xie, Z.-J. Hu, D. Han, R. Shi, C. Zhou, X. Gu, Origins of the Earth's diffuse auroral precipitation. Space Sci. Rev. **200**(1–4), 205–259 (2016). https://doi.org/10.1007/s11214-016-0234-7

B. Ni, X. Gu, S. Fu, Z. Xiang, Y. Lou, A statistical survey of electrostatic electron cyclotron harmonic waves based on THEMIS FFF wave data. J. Geophys. Res. Space Phys. **122**, 3342–3353 (2017). https://doi.org/10.1002/2016JA023433

M.J. Nicolls, C.J. Heinselman, Three-dimensional measurements of traveling ionospheric disturbances with the poker flat incoherent scatter radar. Geophys. Res. Lett. **34**(21), L21104 (2007). https://doi.org/10.1029/2007GL031506

K. Nishi, K. Shiokawa, H. Spence, Magnetospheric source region of auroral finger-like structures observed by the RBSP-a satellite. J. Geophys. Res. Space Phys. **123**(9), 7513–7522 (2018a). https://doi.org/10.1029/2018JA025480

Y. Nishimura et al., Identifying the driver of pulsating aurora. Science **330**, 81 (2010)

Y. Nishimura et al., Estimation of magnetic field mapping accuracy using the pulsating aurora-chorus connection. Geophys. Res. Lett. **38**, L14110 (2011). https://doi.org/10.1029/2011GL048281

Y. Nishimura, J. Bortnik, W. Li, R.M. Thorne, B. Ni, L.R. Lyons et al., Structures of dayside whistler-mode waves deduced from conjugate diffuse aurora. J. Geophys. Res. Space Phys. **118**(2), 664–673 (2013). https://doi.org/10.1029/2012JA018242

Y. Nishimura, J. Bortnik, W. Li, J. Liang, R.M. Thorne, V. Angelopoulos, O. Le Contel, U. Auster, J.W. Bonnell, Chorus intensity modulation driven by time-varying field-aligned low-energy plasma. J. Geophys. Res. Space Phys. **120**, 7433–7446 (2015). https://doi.org/10.1002/2015JA021330

Y. Nishimura, J. Bortnik, W. Li, V. Angelopoulos, E.F. Donovan, E.L. Spanswick, Comment on "pulsating auroras produced by interactions of electrons and time domain structures" by Mozer et al. J. Geophys. Res. Space Phys. **123**, 2064–2070 (2018). https://doi.org/10.1002/2017JA024844

T. Nishiyama, T. Sakanoi, Y. Miyoshi, Y. Katoh, K. Asamura, S. Okano, M. Hirahara, The source region and its characteristic of pulsating aurora based on the Reimei observations. J. Geophys. Res. **116**, A03226 (2011a). https://doi.org/10.1029/2010JA015507

T. Nishiyama, T. Sakanoi, Y. Miyoshi, Y. Katoh, K. Asamura, S. Okano, M. Hirahara, The source region and its characteristic of pulsating aurora based on the Reimei observations. J. Geophys. Res. Space Phys. **116**, A03226 (2011b). https://doi.org/10.1029/2010JA015507

T. Nishiyama, T. Sakanoi, Y. Miyoshi, R. Kataoka, D. Hampton, Y. Katoh, K. Asamura, S. Okano, Fine scale structures of pulsating auroras in the early recovery phase of substorm using ground-based EMCCD camera. J. Geophys. Res. Space Phys. **117**(A10), A10229 (2012). https://doi.org/10.1029/2012JA017921

T. Nishiyama, T. Sakanoi, Y. Miyoshi, D.L. Hampton, Y. Katoh, R. Kataoka, S. Okano, Multiscale temporal variations of pulsating auroras: on-off pulsation and a few Hz modulation. J. Geophys. Res. Space Phys. **119**(5), 3514–3527 (2014). https://doi.org/10.1002/2014JA019818. https://agupubs.onlinelibrary.wiley.com/doi/abs/10.1002/2014JA019818

T. Nishiyama, Y. Miyoshi, Y. Katoh, T. Sakanoi, R. Kataoka, S. Okano, Substructures with luminosity modulation and horizontal oscillation in pulsating patch: principal component analysis application to pulsating aurora. J. Geophys. Res. Space Phys. **121**(3), 2360–2373 (2016). https://doi.org/10.1002/2015JA022288. https://agupubs.onlinelibrary.wiley.com/doi/abs/10.1002/2015JA022288

B.J. O'Brien, High-latitude geophysical studies with satellite Injun 3: 3. Precipitation of electrons into the atmosphere. J. Geophys. Res. **69**(1), 13–43 (1964). https://doi.org/10.1029/JZ069i001p00013

T. Oguti, K. Hayashi, Multiple correlation between auroral and magnetic pulsations: 2. Determination of electric currents and electric fields around a pulsating auroral patch. J. Geophys. Res. Space Phys. **89**(A9), 7467–7481 (1984). https://doi.org/10.1029/JA089iA09p07467

T. Oguti, S. Kokubun, K. Hayashi, K. Tsuruda, S. Machida, T. Kitamura, O. Saka, T. Watanabe, Statistics of pulsating auroras on the basis of all-sky TV data from five stations. I—Occurrence frequency. Can. J. Phys. **59**, 1150–1157 (1981a). https://doi.org/10.1139/p81-152

T. Oguti, S. Kokubun, K. Hayashi, K. Tsuruda, S. Machida, T. Kitamura, O. Saka, T. Watanabe, Latitudinally propagating on-off switching aurorae and associated geomagnetic pulsations: a case study of an event of February 20, 1980. Can. J. Phys. **59**, 1131–1136 (1981b)

Y. Omura, D. Nunn, Triggering process of whistler-mode chorus emissions in the magnetosphere. J. Geophys. Res. **116**, A05205 (2011). https://doi.org/10.1029/2010JA016280

Y. Omura, Y. Katoh, D. Summers, Theory and simulation of the generation of whistler-mode chorus. J. Geophys. Res. **113**, A04223 (2008). https://doi.org/10.1029/2007JA012622

Y. Omura, M. Hikishima, Y. Katoh, D. Summers, S. Yagitani, Nonlinear mechanisms of lower band and upper band VLF chorus emissions in the magnetosphere. J. Geophys. Res. **114**, A07217 (2009). https://doi.org/10.1029/2009JA014206

S. Oyama, K. Shiokawa, J. Kurihara, T.T. Tsuda, S. Nozawa, Y. Ogawa, B.J. Watkins, Lower-thermospheric wind fluctuations measured with an FPI during pulsating aurora at Tromsø, Norway. Ann. Geophys. **28**(10), 1847–1857 (2010). https://doi.org/10.5194/angeo-28-1847-2010

S. Oyama, K. Shiokawa, Y. Miyoshi, K. Hosokawa, B.J. Watkins, J. Kurihara, C.T. Fallen, Lower thermospheric wind variations in auroral patches during the substorm recovery phase. J. Geophys. Res. Space Phys. **121**(4), 3564–3577 (2016). https://doi.org/10.1002/2015JA022129

S. Oyama, A. Kero, C.J. Rodger, M.A. Clilverd, Y. Miyoshi, N. Paratamies, E. Turunen, T. Raita, P.T. Verronen, S. Saito, Energetic electron precipitation and auroral morphology at the substorm recovery phase. J. Geophys. Res. **122**, 6508–6527 (2017). https://doi.org/10.1002/2016JA023484

M. Ozaki et al., A direct link between chorus emissions and pulsating aurora on timescales from milliseconds to minutes: a case study at subauroral latitudes. J. Geophys. Res. Space Phys. **120**, 9617–9631 (2015). https://doi.org/10.1002/2015JA021381

M. Ozaki, K. Shiokawa, Y. Miyoshi, K. Hosokawa, S. Oyama, S. Yagitani, Y. Kasahara, Y. Kasaba, S. Matsuda, R. Kataoka, Y. Ebihara, Y. Ogawa, Y. Otsuka, S. Kurita, R.C. Moore, Y.-M. Tanaka Nos, T. Nagatsuma, M. Connors, N. Nishitani, Y. Katoh, M. Hikishima, A. Kumamoto, F. Tsuchiya, A. Kadokura, T. Nishiyama, T. Inoue, K. Imamura, A. Matsuoka, I. Shinohara, Microscopic observations of pulsating aurora associated with chorus element structures: coordinated arase satellite-pwing observations. Geophys. Res. Lett. **45**(22), 12,125–12,134 (2018). https://doi.org/10.1029/2018GL079812

M. Ozaki et al., Visualization of rapid electron precipitation via chorus element wave–particle interactions. Nat. Commun. **10**, 257 (2019). https://doi.org/10.1038/s41467-018-07996-z

N. Partamies, J.M. Weygand, L. Juusola, Statistical study of auroral omega bands. Ann. Geophys. **35**, 1069–1083 (2017b). https://doi.org/10.5194/angeo-35-1069-2017

N. Partamies, D. Whiter, A. Kadokura, K. Kauristie, H. Nesse Tyssøy, S. Massetti, P. Stauning, T. Raita, Occurrence and average behavior of pulsating aurora. J. Geophys. Res. Space Phys. **122**, 5606–5618 (2017a). https://doi.org/10.1002/2017JA024039

N. Partamies, K. Bolmgren, E. Heino, N. Ivchenko, J.E. Borovsky, H. Dahlgren, Patch size evolution during pulsating aurora. J. Geophys. Res. Space Phys. **124**, 4725–4738 (2019). https://doi.org/10.1029/2018JA026423

L.M. Peticolas, T.J. Hallinan, H.C. Stenbaek-Nielsen, J.W. Bonnell, C.W. Carlson, A study of black aurora from aircraft-based optical observations and plasma measurements on FAST. J. Geophys. Res. Space Phys. **107**(A8), 30 (2002). https://doi.org/10.1029/2001JA900157

M.H. Rees, Conjugate effects of atmospherically scattered auroral electrons. Radio Sci. **3**(7), 645–649 (1968). https://doi.org/10.1002/rds196837645

R. Robinson, New techniques and results from incoherent scatter radars. URSI Radio Sci. Bull. **2004**(311), 79–94 (2004). https://doi.org/10.23919/URSIRSB.2004.7909637

J.L. Roeder, H.C. Koons, A survey of electron cyclotron waves in the magnetosphere and the diffuse auroral electron precipitation. J. Geophys. Res. **94**, 2529–2541 (1989). https://doi.org/10.1029/JA094iA03p02529

O. Royrvik, T.N. Davis, Pulsating aurora: local and global morphology. J. Geophys. Res. **82**(29), 4720–4740 (1977)

S. Saito, Y. Miyoshi, K. Seki, Relativistic electron microbursts associated with whistler chorus rising tone elements: GEMSIS-RBW simulations. J. Geophys. Res. **117**, A10206 (2012). https://doi.org/10.1029/2012JA018020

S. Saito, Y. Miyoshi, K. Seki, Relativistic electron microbursts associated with whistler chorus rising tone elements: GEMSIS-RBW simulations. J. Geophys. Res. **117**, A10206 (2012). https://doi.org/10.1029/2012JA018020

M. Samara, R.G. Michell, Ground-based observations of diffuse auroral frequencies in the context of whistler mode chorus. J. Geophys. Res. **115**, A00F18 (2010). https://doi.org/10.1029/2009JA014852

M. Samara, R.G. Michell, K. Asamura, M. Hirahara, D.L. Hampton, H.C. Stenbaek-Nielsen, Ground-based observations of diffuse auroral structures in conjunction with Reimei measurements. Ann. Geophys. **28**(3), 873–881 (2010). https://doi.org/10.5194/angeo-28-873-2010

M. Samara, R.G. Michell, R.J. Redmon, Low- altitude satellite measurements of pulsating auroral electrons. J. Geophys. Res. Space Phys. **120**, 8111–8124 (2015). https://doi.org/10.1002/2015JA021292

M. Samara, R.G. Michell, G.V. Khazanov, First optical observations of interhemispheric electron reflections within pulsating aurora. Geophys. Res. Lett. **44**, 2618–2623 (2017). https://doi.org/10.1002/2017GL072794

I. Sandahl, The particle experiment on the substorm GEOS rockets. KGI Techn. Rep. 031 (1981)

I. Sandahl, Pitch angle scattering and particle precipitation in a pulsating aurora: an experimental study. KGI Rept. 185, Kiruna, Sweden (1985)

I. Sandahl, L. Eliasson, R. Lundin, Rocket observations of precipitating electrons over a pulsating aurora. Geophys. Res. Lett. **7**, 309–312 (1980). https://doi.org/10.1029/GL007i005p00309

L. Sangalli, N. Partamies, M. Syrjäsuo, C.-F. Enell, K. Kauristie, S. Mäkinen, Performance study of the new EMCCD-based all-sky cameras for auroral imaging. Int. J. Remote Sens. **32**(11), 2987–3003 (2011). https://doi.org/10.1080/01431161.2010.541505

O. Santolik, D.A. Gurnett, J.S. Pickett, M. Parrot, N. Cornilleau-Wehrlin, Spatio-temporal structure of storm-time chorus. J. Geophys. Res. **108**(A7), 1278 (2003). https://doi.org/10.1029/2002JA009791

O. Santolik, E. Macusova, I. Kolmasova, N. Cornilleau-Wehrlin, Y. de Conchy, Propagation of lower-band whistler-mode waves in the outer Van Allen belt: systematic analysis of 11 years of multi-component data from the cluster. Geophys. Res. Lett. **41**, 2729–2737 (2014). https://doi.org/10.1002/2014GL059815

K. Sato, M. Tsutsumi, T. Sato, T. Nakamura, A. Saito, Y. Tomikawa, K. Nishimura, M. Kohma, H. Yamagishi, T. Yamanouchi, Program of the Antarctic Syowa MST/IS radar (PANSY). J. Atmos. Sol.-Terr. Phys. **118A**, 2–15 (2014). https://doi.org/10.1016/j.jastp.2013.08.022

N. Sato, A. Kadokura, Y. Tanaka, T. Nishiyama, T. Hori, A.S. Yukimatu, Omega band pulsating auroras observed onboard THEMIS spacecraft and on the ground. J. Geophys. Res. Space Phys. **120**, 5524–5544 (2015). https://doi.org/10.1002/2015JA021382

N. Sato, A.S. Yukimatu, Y. Tanaka, T. Hori, Morphologies of omega band auroras. Earth Planets Space **69**, 103 (2017). https://doi.org/10.1186/s40623-017-0688-1

M.W.J. Scourfield, W.F. Innes, N.R. Parsons, Spatial coherency in pulsating aurora. Planet. Space Sci. **20**, 1843–1848 (1972)

J. Semeter, F. Kamalabadi, Determination of primary electron spectra from incoherent scatter radar measurements of the auroral e region. Radio Sci. **40**(2), 1–17 (2005). https://doi.org/10.1029/2004RS003042

J. Semeter, M. Mendillo, J. Baumgardner, Multispectral tomographic imaging of the midlatitude aurora. J. Geophys. Res. Space Phys. **104**(A11), 24565–24585 (1999). https://doi.org/10.1029/1999JA900305

V.A. Sergeev, E.M. Sazhina, N.A. Tsyganenko, J.A. Lundblad, F. Soraas, Pitch-angle scattering of energetic protons in the magnetotail current sheet as the dominant source of their isotropic precipitation into the nightside ionosphere. Planet. Space Sci. **31**(10), 1147–1155 (1983). https://doi.org/10.1016/0032-0633(83)90103-4

V.A. Sergeev, Y. Nishimura, M. Kubyshkina, V. Angelopoulos, R. Nakamura, H. Singer, Magnetospheric location of the equatorward pre-breakup arc. J. Geophys. Res. Space Phys. **117**(A1), A01212 (2012). https://doi.org/10.1029/2011JA017154

T. Sergienko, I. Sandahl, B. Gustavsson, L. Andersson, U. Brändström Å, Steen, a study of fine structure of diffuse aurora with ALIS-FAST measurements. **26** (2008). Retrieved from. www.ann-geophys.net/26/3185/2008/

K. Shiokawa, A. Nakajima, A. Ieda, K. Sakaguchi, R. Nomura, T. Aslaksen, M. Greffen, E. Donovan, Rayleigh-Taylor type instability in auroral patches. J. Geophys. Res. Space Phys. **115**(A2), A02211 (2009). https://doi.org/10.1029/2009JA014273

K. Shiokawa et al., Auroral fragmentation into patches. J. Geophys. Res. Space Phys. **119**, 8249–8261 (2014). https://doi.org/10.1002/2014JA020050

K. Shiokawa, Y. Katoh, Y. Hamaguchi, Y. Yamamoto, T. Adachi et al., Ground-based instruments of the pwing project to investigate dynamics of the inner magnetosphere at subauroral latitudes as a part of the erg-ground coordinated observation network. Earth Planets Space **69**(1), 160 (2017). https://doi.org/10.1186/s40623-017-0745-9

N. Sivadas, J. Semeter, Y. Nishimura, A. Kero, Simultaneous measurements of substorm-related electron energization in the ionosphere and the plasma sheet. J. Geophys. Res. Space Phys. **122**(10), 10,528–10,547 (2017). https://doi.org/10.1002/2017JA023995

N. Sivadas, J. Semeter, Y. Nishimura, S. Mrak, Optical signatures of the outer radiation belt boundary. Geophys. Res. Lett. **46**, 8588–8596 (2019). https://doi.org/10.1029/2019GL083908

M.J. Smith, D.A. Bryant, T. Edwards, Pulsations in auroral electrons and positive ions. J. Atmos. Terr. Phys. **42**, 167 (1980)

S.C. Solomon, P.B. Hays, V.J. Abreu, Tomographic inversion of satellite photometry. Appl. Opt. **23**(19), 3409 (1984). https://doi.org/10.1364/AO.23.003409

T. Sørensen, J. Bjordal, H. Trefall, G.J. Kvifte, H. Pettersen, Correlation between pulsations in auroral luminosity variations and X-rays. J. Atmos. Terr. Phys. **35**(5), 961–969 (1973). https://doi.org/10.1016/0021-9169(73)90075-5
E. Spanswick, E. Donovan, G. Baker, Pc5 modulation of high energy electron precipitation: particle interaction regions and scattering efficiency. Ann. Geophys. **23**, 1533–1542 (2005)
H.C. Stenbaek-Nielsen, Pulsating aurora—the importance of the ionosphere. Geophys. Res. Lett. **7**, 353–356 (1980). https://doi.org/10.1029/GL007i005p00353
H.C. Stenbaek-Nielsen, T.J. Hallinan, Pulsating auroras—evidence for noncollisional thermalization of precipitating electrons. J. Geophys. Res. **84**, 3257–3271 (1979). https://doi.org/10.1029/JA084iA07p03257
H.C. Stenbaek-Nielsen, T.J. Hallinan, L. Peticolas, Why do auroras look the way they do? Eos, Trans. Am. Geophys. Union **80**(17), 193 (1999). https://doi.org/10.1029/99EO00138
C. Størmer, *The Polar Aurora* (Clarendon, Oxford, 1955)
R.J. Strangeway, R.E. Ergun, Y.-J. Su, C.W. Carlson, R.C. Elphic, Factors controlling ionospheric outflows as observed at intermediate altitudes. J. Geophys. Res. **110**, A03221 (2005). https://doi.org/10.1029/2004JA010829
D. Summers, R. Tang, Y. Omura, D.-H. Lee, Parameter spaces for linear and nonlinear whistler-mode waves. Phys. Plasmas **20**, 072110 (2013). https://doi.org/10.1063/1.4816022
T. Takahashi et al., Depletion of mesospheric sodium during extended period of pulsating aurora. J. Geophys. Res. Space Phys. **122**, 1212–1220 (2017). https://doi.org/10.1002/2016JA023472
Y.-M. Tanaka, T. Aso, B. Gustavsson, K. Tanabe, Y. Ogawa, A. Kadokura, H. Miyaoka, T. Sergienko, U. Brändström, I. Sandahl, Feasibility study on generalized-aurora computed tomography. Ann. Geophys. **29**, 551–562 (2011). https://doi.org/10.5194/angeo-29-551-2011
Y. Tanaka, Y. Ogawa, A. Kadokura, N. Partamies, D. Whiter, C.-F. Enell, U. Brändström, T. Sergienko, B. Gustavsson, A. Kozlovsky, H. Miyaoka, A. Yoshikawa, Eastward-expanding auroral surges observed in the post-midnight sector during a multiple-onset substorm. Earth Planets Space **67**(1), 182 (2015). https://doi.org/10.1186/s40623-015-0350-8. ISSN 1880-5981
R.W. Thomas, H.C. Stenbaek-Nielsen, Recurrent propagating auroral forms in pulsating auroras. J. Atmos. Terr. Phys. **43**, 243 (1981)
R.M. Thorne, B. Ni, X. Tao, R.B. Horne, N.P. Meredith, Scattering by chorus waves as the dominant cause of diffuse auroral precipitation. Nature **467**, 943–946 (2010). https://doi.org/10.1038/nature09467
T.T. Tsuda et al., Decrease in sodium density observed during auroral particle precipitation over Tromsø, Norway. Geophys. Res. Lett. **40**, 4486–4490 (2013). https://doi.org/10.1002/grl.50897
K. Tsuruda, S. Machida, T. Oguti, S. Kokubun, K. Hayashi, T. Kitamura, O. Saka, T. Watanabe, Correlations between the very low frequency chorus and pulsating aurora observed by low-light-level television at $L = 4.4$. Can. J. Phys. **59**(8), 1042–1048 (1981). https://doi.org/10.1139/p81-137
E. Turunen, A. Kero, P.T. Verronen, Y. Miyoshi, S-I. Oyama, S. Saito, Mesospheric ozone destruction by high-energy electron precipitation associated with pulsating aurora. J. Geophys. Res. **121**, 11,155–11,879 (2016). https://doi.org/10.1002/2016JD025015
C.W. Unick, E. Donovan, M. Connors, B. Jackel, A dedicated H-beta meridian scanning photometer for proton aurora measurement. J. Geophys. Res. Space Phys. **122**(1), 753–764 (2017). https://doi.org/10.1002/2016JA022630
J.-E. Wahlund, H.J. Opgenoorth, I. Haggstrom, K.J. Winser, G.O.L. Jones, EISCAT observations of topside ionospheric ion outflows during auroral activity—revisited. J. Geophys. Res. **97**, 3019–3037 (1992)
M. Watanabe, A. Kadokura, N. Sato, T. Saemundsson, Absence of geomagnetic conjugacy in pulsating auroras. Geophys. Res. Lett. **34**, L15107 (2007). https://doi.org/10.1029/2007GL030469
J.M. Weygand, M.G. Kivelson, H.U. Frey, J.V. Rodriguez, V. Angelopoulos, R. Redmon, J. Barker-Ream Grocott, O. Amm, An interpretation of spacecraft and ground based observations of multiple omega band events. J. Atmos. Sol.-Terr. Phys. **133**, 185–204 (2015). https://doi.org/10.1016/j.jastp.2015.08.014
D.K. Whiter, B. Gustavsson, N. Partamies, L. Sangalli, A new automatic method for estimating the peak auroral emission height from all-sky camera images. Geosci. Instrum. Method. Data Syst. **2**, 131–144 (2013). https://doi.org/10.5194/gi-2-131-2013
C.R. Wilson, J.V. Olson, H.C. Stenbaek-Nielsen, High trace-velocity infrasound from pulsating auroras at Fairbanks, Alaska. Geophys. Res. Lett. **32**(14), n/a–n/a (2005). https://doi.org/10.1029/2005GL023188
Q. Wu, T.J. Rosenberg, High latitude pulsating aurora revisited. Geophys. Res. Lett. **19**(1), 69–72 (1992)
T. Yamamoto, On the temporal fluctuations of pulsating auroral luminosity. J. Geophys. Res. **93**(A2), 897–911 (1988). https://doi.org/10.1029/JA093iA02p00897
T. Yamamoto, T. Oguti, Recurrent fast motions of pulsating auroral patches: 1. A case study on optical and quantitative characteristics during a slightly active period. J. Geophys. Res. **87**(A9), 7603–7614 (1982). https://doi.org/10.1029/JA087iA09p07603

B. Yang, E. Donovan, J. Liang, J.M. Ruohoniemi, E. Spanswick, Using patchy pulsating aurora to remote sense magnetospheric convection. Geophys. Res. Lett. **42**, 5083–5089 (2015). https://doi.org/10.1002/2015GL064700

B. Yang, E. Donovan, J. Liang, E. Spanswick, A statistical study of the motion of pulsating aurora patches: using the THEMIS All-Sky Imager. Ann. Geophys. **35**, 217–225 (2017). https://doi.org/10.5194/angeo-35-217-2017

B. Yang, E. Spanswick, J. Liang, E. Grono, E. Donovan, Responses of different types of pulsating aurora in cosmic noise absorption. Geophys. Res. Lett. **46**, 5717–5724 (2019). https://doi.org/10.1029/2019GL083289

Y. Zhang, L.J. Paxton, D. Morrison, A.T.Y. Lui, H. Kil, B. Wolven, C.-I. Meng, A.B. Christensen, Undulations on the equatorward edge of the diffuse proton aurora: TIMED/GUVI observations. J. Geophys. Res. Space Phys. **110**(A8), A08211 (2005). https://doi.org/10.1029/2004JA010668

X. Zhang, V. Angelopoulos, B. Ni, R.M. Thorne, Predominance of ECH wave contribution to diffuse aurora in Earth's outer magnetosphere. J. Geophys. Res. **120**, 295–309 (2015). https://doi.org/10.1002/2014JA020455

Q. Zhou, F. Xiao, C. Yang, S. Liu, Y. He, D.N. Baker, H.E. Spence, G.D. Reeves, H.O. Funsten, Generation of lower and upper bands of electrostatic electron cyclotron harmonic waves in the Van Allen radiation belts. Geophys. Res. Lett. **44**, 5251–5258 (2017). https://doi.org/10.1002/2017GL073051

Space Sci Rev (2020) 216:46
https://doi.org/10.1007/s11214-020-00665-y

Physical Processes of Meso-Scale, Dynamic Auroral Forms

C. Forsyth[1] · V.A. Sergeev[2] · M.G. Henderson[3] · Y. Nishimura[4] · B. Gallardo-Lacourt[5,6]

Received: 29 April 2019 / Accepted: 28 March 2020 / Published online: 20 April 2020

Abstract Meso-scale auroral forms, such as poleward boundary intensifications, streamers, omega bands, beads and giant undulations, are manifestations of dynamic processes in the magnetosphere driven, to a large part, by plasma instabilities in the magnetotail. New observations from ground- and space-based instrumentation and theoretical treatments are giving us a clearer view of some of the physical processes behind these auroral forms. However, questions remain as to how some of these observations should be interpreted, given uncertainties in mapping auroral features to locations in the magnetotatil and due to the significant overlap in the results from a variety of models of different plasma instabilities. We provide an overview of recent results in the field and seek to clarify some of the remaining questions with regards to what drives some of the largest and most dynamic auroral forms.

Keywords Meso-scale aurora · Poleward boundary intensifications · Auroral streamers · Omega bands · Torches · Auroral beads · Giant undulations · Magnetic mapping · Magnetosphere ionosphere coupling · Plasma instabilities · Magnetotail

Auroral Physics
Edited by David Knudsen, Joe Borovsky, Tomas Karlsson, Ryuho Kataoka and Noora Partmies

✉ C. Forsyth

[1] UCL Mullard Space Science Laboratory, Dorking, Surrey, UK

[2] St. Petersburg State University, St. Petersburg, Russia

[3] Los Alamos National Laboratory, Los Alamos, NM, USA

[4] Department of Electrical and Computer Engineering and Center for Space Physics, Boston University, Boston, MA, USA

[5] Department of Physics and Astronomy, University of Calgary, Calgary, Canada

[6] NASA Goddard Space Flight Center, Greenbelt, MD, USA

1 Introduction

Meso-scale dynamic auroral forms, with scale sizes of 10s to 100s of kilometres, were reported as distinct features in some of the earliest schematic diagrams of the auroral oval. Through combining observations from orbiting and ground-based imagers with in-situ measurements of the magnetosphere and remote sensing observations of the ionosphere, we have begun to understand the physical processes at the source of these auroral forms and, in some cases, the associated acceleration of particles into the ionosphere. This is coupled with a growing body of theoretical and modelling work to help understand the key physical processes that can give rise to localised enhancements of the aurora. However, mapping specific auroral features into a dynamic plasma environment that encompasses millions of cubic kilometres is not straightforward, leaving the coupling of ground and space-based observations open to interpretation. Furthermore, different theoretical and modelling frameworks give overlapping results that fit to various aspects of the observations meaning that the physical processes behind meso-scale dynamic auroral forms are, in some cases, still to be resolved.

In this chapter, we review the current understanding of the physical processes associated with auroral beads, poleward boundary intensifications, streamers, omega bands and giant undulations. For each of these, we provide current definitions, a discussion of the physical drivers and sources of these features and some of the outstanding issues with regards to understanding the magnetosphere-ionosphere coupling of these meso-scale forms. We also discuss the role that magnetic mapping has in our understanding of the coupling of meso-scale auroral forms to magnetospheric drivers. We note that most of the auroral forms discussed have been linked, in some way, to the dynamics of substorm activity, although their appearance is not exclusively during substorm times. Here, we focus on the physical drivers of these auroral forms rather than their place in the phenomenology of auroral and geomagnetic activity. For a recent discussion from the magnetospheric viewpoint, we direct readers to reviews by Keiling et al. (2013) and Sitnov et al. (2019).

2 Poleward Boundary Intensifications

During undisturbed, geomagnetically quiet times, the nightside auroral oval predominantly consists of relatively diffuse auroral emission covering up to 10° of latitude. At times, however, the poleward edge of the nightside auroral oval can show rapid, latitudinally localised brightenings or the formation of new arcs along the poleward boundary that are now commonly refered to as Poleward Boundary Intensifications (PBIs). While observations of brightenings at the poleward edge of the auroral oval have been reported throughout studies of the meso-scale auroral morphology (e.g. Rostoker et al. 1987; de la Beaujardière et al. 1994; Henderson 1994), the terminology and investigation of PBIs as a distinct class of auroral structure largely follows on from Lyons et al. (1999).

There is some ambiguity in the literature regarding the definition of PBIs since the term is used as a general term for brightenings at or near the poleward edge of the auroral oval, with little or no distinction between different auroral morphologies. Zesta et al. (2002) suggested that three classes of PBIs could be observed: arcs that propagate across the auroral oval and are either (a) primarily north-south or (b) primarily east-west aligned arcs originating at or near the poleward boundary; or (c) arcs that remain at the poleward boundary that tend to be associated with eastward or westward moving swirls (examples of different PBIs are shown in Fig. 1). However, these distinctions are not often reported. Furthermore, auroral forms

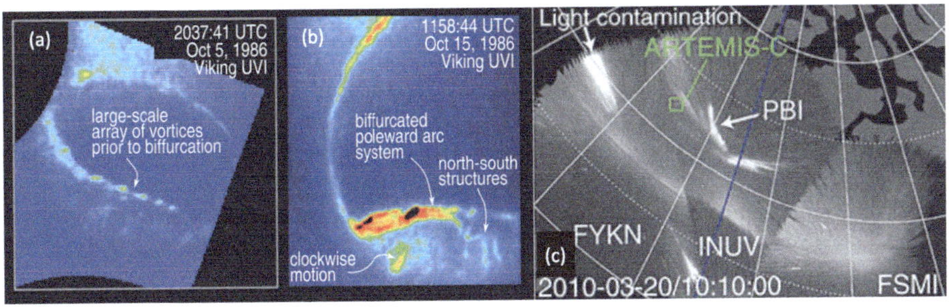

Fig. 1 Examples of poleward boundary intensifications seen in Polar (left, after Henderson 1994a), Viking (center, after Henderson et al. 1994) and THEMIS ASI data (right, after Nishimura et al. 2013c). Left: Spatially periodic bright spots (likely folds or spirals) appear on the poleward edge of the auroral oval prior to the bifurcation of the arc. Center: Example of a poleward boundary intensification associated with a bifurcation of arcs at the poleward boundary and associated equatorward ejection of streamers into the bulge. Right: Example of a quiet time PBI appearing at the poleward edge of a diffuse auroral oval

originating at the poleward boundary and extending equatorwards into the auroral oval have been described both as PBIs and auroral streamers (see Sect. 3). The point at which a PBI becomes a streamer is undefined, but, in general, streamers are considered to be auroral features that extend across the auroral oval, forming roughly north-south aligned arcs, and PBIs are transient auroral intensifications at the poleward edge of the auroral oval (Sandholt et al. 2002; Zesta et al. 2011). The lack of distinction between these different auroral forms in the literature means that there is a natural ambiguity in identifying their direct physical drivers. In the following, we concentrate on those observations at or close to the open-closed field-line boundary in order to consider the physical processes behind PBIs.

PBIs are reported at all levels of geomagnetic activity, from quiet or pre-substorm times (Lyons 2000; Kim et al. 2005; Nishimura et al. 2010b; Zou et al. 2010; Lyons et al. 2011; Zesta et al. 2011; Nishimura et al. 2013a), to during substorms (Sandholt et al. 2002; Voronkov et al. 2003; Pitkanen et al. 2013) and geomagnetic storm or storm-like events (Lyons 2000; Yue et al. 2013; Lyons et al. 2018). Global auroral imagers show enhancements of the poleward boundary after significant auroral activity which results in a double auroral oval (e.g. Henderson 1994). Much of the recent research in this field has concentrated on PBIs that extend equatorwards, rather than those that stay confined to the poleward boundary, and their potential role in the substorm cycle.

2.1 Drivers/Causes

Recent studies of PBIs and the physical processes that drive them have typically focussed on PBIs that result in auroral streamers by extending across the auroral oval. These are generally considered to be an auroral manifestation of localised reconnection in the magnetotail. Unlike the Near-Earth Neutral Line reconnection that occurs during substorms, which starts on closed field-lines and thus will result in brightenings within the auroral oval, the occurrence of an auroral enhancement at the poleward boundary of the aurora suggests an enhancement in reconnection occurring close to the distant neutral line. Evidence of bursts of reconnection around the time of PBIs has been provided by observations of fast plasma flows in the magnetotail plasmasheet, commonly known as bursty bulk flows or BBFs (Angelopoulos et al. 1992). BBFs are thought to be a signature of localised reconnection having happened further downtail than the observing spacecraft, with the enhanced reconnection electric field locally enhancing the convection of flux tubes through the plasma sheet. These signatures are often observed at times of PBI activity (Lyons et al. 1999; Zesta et al. 2000, 2006; Nishimura et al. 2010a,b; Zesta et al. 2011; Nishimura et al. 2013a,c; Pitkanen et al. 2013). However, given that the auroral oval is found on closed field-line and the poleward boundary of the auroral oval lies close to the open-closed field-line boundary, it is unlikely that observations of BBFs or reconnection signatures well within the magnetically closed plasmasheet map directly to PBIs. Notably, observations of BBFs inside of $\sim$20 R_E may be due to near-Earth or far-tail reconnection since the typical downtail distance of the near-Earth reconnection site is thought to be 20–30 R_E (Imber et al. 2011). Observations from the ARTEMIS mission at $\sim$60 R_E downtail show that fast plasma flows in close proximity to PBI activity (Nishimura et al. 2013c) are more likely to originate from a far-tail reconnection site. The crux of this may be the distinction between PBIs and auroral streamers combined with mapping effects: a deep-tail BBF may map to a region just inside the poleward edge of the auroral oval before it extends equatorwards and becomes an auroral streamer. This feature may be indistinguishable from any auroral features directly arising from the reconnection activity which created the BBF. Thus, while it is likely that PBIs are in some way associated with reconnection, whether they are also caused by propagating fast-plasma flows in the magnetotail remains unknown.

Observations of the plasma in the vicinity of a PBI have shown enhanced auroral precipitation, as expected for above bright aurora. Observations from the Cluster spacecraft at 3.5 R_E altitude enabled the time evolution of this particle population to be examined (Hull et al. 2010). These closely separated (in time) observations showed an injection of new plasmasheet plasma in concert with the PBI (see Fig. 2). This plasma initially showed a broadband energy distribution, associated with Alfvén wave acceleration, which merged with the existing plasmasheet population and evolved into a more mono-energetic population, generally associated with quasi-static electric potential drops. Similar multi-point observations by sounding rockets at lower altitude showed similar results (Mella et al. 2011; Lynch et al. 2012). Thus, the implication is that the initial formation of the PBI is driven by Alfvénically accelerated electrons, possibly arising from a reconnection site, which subsequently become quasi-statically accelerated electrons. Whether this is as a result of magnetosphere-ionosphere coupling and Alfvén waves 'digging out' an auroral cavity, or because the magnetospheric source of the aurora switches from an active reconnection site to the quasi-steady field-aligned currents supporting a BBF in the magnetotail remains an open question.

Fast plasma flows (up to 1000s of m/s) in the polar cap have been observed in both coherent and incoherent scatter radar measurements prior to the activation of a PBI (Lyons et al. 2011; Bristow et al. 2016). Figure 3 shows and example event from Nishimura et al. (2010b). These have been associated with the drift of polar cap patches across the polar cap originating from reconnection on the dayside (Nishimura et al. 2014; Zou et al. 2014). Statistics have shown that $\sim$50% of PBIs can be associated with an isolated fast equatorward flow in the polar cap, with a further 30% of PBIs associated with multiple polar cap flow channels (Zou et al. 2014). The isolated flow channels tend to appear in the polar cap a few minutes prior to the activation of the PBI and the speed of the observed flows can range from 1000s m/s in the F-region, to up to 400 m/s in the E-region, with E-region velocities being retarded to the ion acoustic speed. Calculations of the nightside reconnection rates at PBIs, based on the plasma flow across the OCB, show enhancements in reconnection localised to the PBI vicinity (Pitkänen et al. 2011). As such, the arrival of the fast flows at the polar cap boundary is a potential source of an enhanced electric field that drives a burst of reconnection, resulting in the PBI. However, reconnection in the far tail may also occur due to other changing conditions. Whether the reconnection event directly accelerates particles along the magnetic field to generate the auroral brightening or whether the subsequent fast magnetotail plasma flows is the source of the PBI remains unresolved.

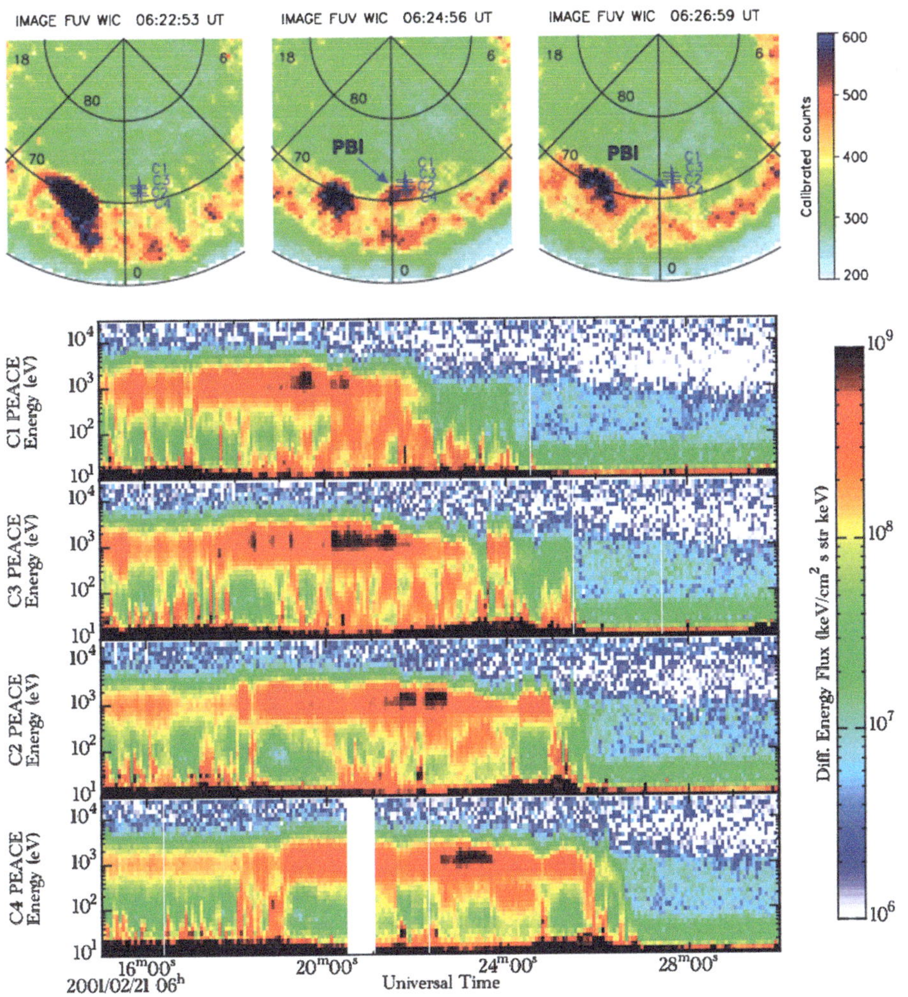

Fig. 2 Observations from IMAGE FUV-WIC and electron differential energy flux summed across all pitch angles from the Cluster PEACE instruments during a crossing of a PBI (adapted from Hull et al. 2010). Cluster 1 passed over the PBI region first (when there was no PBI), followed by Cluster 3, Cluster 2 and Cluster 4. The PEACE data shows the injection of new plasmasheet plasma (Cluster 3) next to a region of Alfvénically driven precipitation, which subsequently merges with the existing plasma and develops into a electrostatically accelerated population

One theoretical model of PBIs, based on the arrival of fast polar cap convection vortices at a conductivity boundary in the ionosphere through ionospheric polarization, suggests that the key aspect that results in the PBI is the change in ionospheric conductivity between the auroral oval and the polar cap (Ohtani et al. 2018; Ohtani and Yoshikawa 2016). In this model, the arrival of a pair of flow vortices surrounding an up-down field-aligned current pair transmitted by Alfvén waves at a gradient in the Pedersen and Hall conductivities results in enhanced field aligned currents at that boundary – currents that may be sufficiently large to result in the optical signature of a PBI (Fig. 4). Ohtani and Yoshikawa (2016) suggest that the finite conductivity of the auroral oval means that the auroral region is not shielded

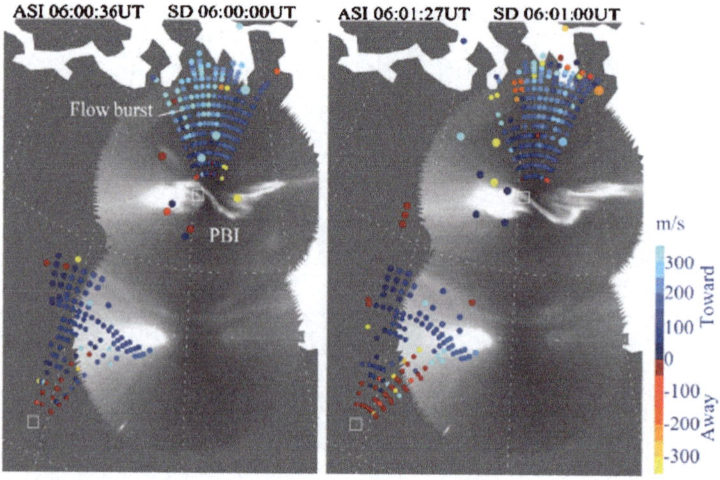

Fig. 3 Observations of ionospheric flow bursts seen in SuperDARN line-of-sight velocity data occurring in the polar cap, polewards of a PBI observed in the THEMIS ASI data (after Nishimura et al. 2010b)

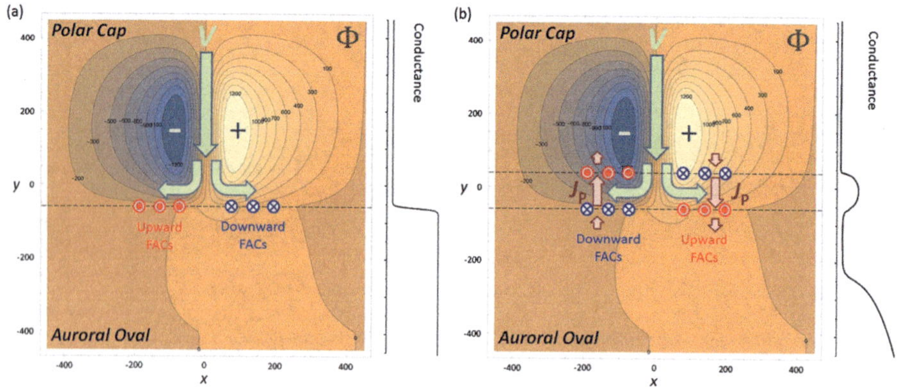

Fig. 4 From Ohtani et al. (2018): Schematic explanation of the deflection of the fast polar cap flow (light green arrow coming from the top) at (**a**) the step-like change of the ionospheric conductance and (**b**) the band of the enhanced conductance (as illustrated to the right of each panel). The electrostatic potential is shown by contours and colors, which was adopted from Fig. 5e of Ohtani and Yoshikawa (2016); although their model calculation was conducted for a step-like change of conductance, we adopted it here for both panels. The contours are squeezed out of the auroral oval (bottom half) to the polar cap (top half) because of electrostatic polarization at the poleward boundary. The vertically elongated area of dense potential contours in the middle of the polar cap corresponds to the flow channel. In panel (b) the intensities of the ionospheric Pedersen currents (J_P) inside and outside of the conductance band are schematically shown by the arrows in magenta. See Ohtani and Yoshikawa (2016) for more details

from the incoming flow and that a finite electric field along the conductivity boundary will permit field-lines to cross into the enhanced conductivity region without the need for reconnection (and thus that processes in the magnetosphere are being driven by the ionosphere). This would appear to be in contrast with observations of reconnection signatures in the magnetotail following PBIs, however the generation of Alfvén waves by reconnection may act to enhance the effects of ionospheric polarization.

2.2 Implications and Outstanding Issues

The evidence from in-situ plasma measurements of directly above a PBI and of fast plasma flows in the plasma sheet coupled with remote sensing observations of enhanced plasma flows (and hence enhanced electric fields) in the polar cap and around PBIs is strongly indicative that reconnection plays some role in the occurrence of certain forms of PBI. However, open questions remain as to the origins of PBIs and brightenings at the open-closed field-line boundary, for example: are they associated with reconnection and are brightenings that extend over significant portions of the nightside auroral oval or spatially periodic brightenings driven by reconnection or some other process? Furthermore, the question of the whether burst of reconnection at the dayside magnetopause remain cohesive in their passage through the magnetospheric lobes and lead to PBIs remains open. The modelling work by Ohtani et al. (2018), Ohtani and Yoshikawa (2016) seeks to address some of these issues by considering the interaction of plasma vortices with conductivity gradients in the ionosphere, it ignores any reconnection processes occurring in the magnetosphere. Observational studies have also shown the occurrence of dayside and nightside flow enhancements in concert with the onset of PBIs (Shi and Zesta 2014). However, the causal link between these events is yet to be conclusively demonstrated. Observations have shown that fast flows are observed typically 1–2 min before PBIs are identified (Zou et al. 2014) but often cannot discern whether the flow impinges on the OCB at the time of the PBI due to limitations in the ionospheric flow observations. Fast conjugate auroral and ionospheric observations will be key in resolving these ambiguities, since fast plasma flows need not pass through both lobes at the same speed or end up in the same location and thus PBIs without an associated plasma flow signature may be seen conjugate to a PBI with a plasma flow signature. Furthermore, the potential x-ray imaging of reconnection events and FTEs and their passage along the magnetopause in concert with auroral observations may provide new insights (Branduardi-Raymont et al. 2012; Branduardi-Raymont and Wang 2017; Sibeck et al. 2018), but will have to be coupled with advanced modelling work, such as that offered by the Vlasiator (Palmroth et al. 2013, 2018), that can predict the lifetime and cohesion of these structures as they pass from the dayside to nightside. If these processes are the result of reconnection, understanding the physical mechanisms as well as the global and local plasma conditions that lead to localized reconnection and dictate where the reconnection outflows and structures propagate to will be critical.

A major difficulty in unravelling the physical processes behind PBIs is in having spacecraft near or at the open-closed field-line boundary in the magnetotail at the time of these brightenings. To date, we are only aware of the study by Hull et al. (2010) using a low-altitude pass of the Cluster spacecraft to examine the changes in plasma at the poleward boundary of the auroral oval during a PBI. Observations from spacecraft in the magnetotail tend to be of plasma structures propagating through the closed plasmasheet rather which likely result from the same process that drives the PBI but are not, in themselves, the source. Furthermore, definitively determining the link between signatures in the far magnetotail and brightenings in the auroral oval requires a determination of the Alfvén velocity profile along the tail magnetic field lines such that the expected delay between in-situ measurements and their impact on the ionosphere can be accounted for.

While localised PBIs that result in auroral streamers have been widely studied, little attention has been given to PBIs that result in new arcs that remain at the poleward or the occurrence of trains of auroral swirls along this boundary. Ohtani et al. (2018) examined four events which extended longitudinally using a variety of ground- and space-based measurements and found that, in these events, the PBI was associated with an upward FAC region adjacent to a downward FAC region with a flow channel in between. They suggest that the Ohtani and Yoshikawa (2016) model can be adapted to reproduce the observed plasma conditions and thus that longitudinally extending PBIs are also the result of the arrival of fast plasma flows in the polar cap. However, the dynamics of the aurora during some of these events mean that the in-situ observations were not of the auroral features when they were at the poleward boundary but rather had already moved into the auroral oval. Also, a strong rotation of the structure during its mapping from equator to the ionosphere can participate in forming longitudinally extending PBIs (see Sect. 7). As such, further observations, particularly from the magnetospheric source are required to corroborate these results.

It should be noted that magnetic mapping is particularly challenging at the outer edge of the auroral oval. In this region, relatively small changes in latitude can map to vast distances downtail. As such, BBFs and fast plasma flows in the far magnetotail may map very close to the poleward boundary such as to appear as PBIs, and as such some PBIs may be driven by the physical processes associated with auroral streamers (as discussed in the following section). However, given the variety of forms of PBI, from those that extend across the oval to those that have a spatially periodic signature along the poleward boundary, it is unlikely that this explanation is universal.

3 Auroral Streamers

3.1 Definitions/Examples

As with many auroral features, streamers have been described by a number of different terms and have frequently appeared in schematics of the auroral region (e.g. Montbriand 1971; Akasofu 1976). Early studies often referred to North-South aligned auroral structures, easily distinguishable from standard east-west aligned auroral arcs by their orientation, however this alignment is no longer considered a key descriptor. Instead, a whole class of auroral features that extend across the auroral oval are now described as streamers.

Typically, auroral streamers are observed to extend out of a poleward boundary intensification (see Sect. 2) near the poleward edge of the oval or in the poleward-expanding bulge, and then propagate equatorward (Montbriand 1971; Henderson et al. 1994; Nakamura et al. 1993; Elphinstone et al. 1996; Henderson et al. 1998; Sergeev et al. 1999, 2000, 2004; Nakamura et al. 2001; Nishimura et al. 2010a; Henderson 2012). This dynamic motion away from their origin and (commonly) toward the equatorward boundary are now considered key aspects and implies that their magnetospheric source propagates from close to the open-closed field line boundary toward the inner magnetosphere. Figure 5 shows an example of several streamers observed by the Polar UVI instrument (adapted from Sergeev 2005). The propagation from the poleward to equator edge of aurora can be seen for the features labelled B, C, D and E. Their orientation may be north-south (structure A), or more inclined (structures B-E), sometimes being purely east-west in a particular local time sector (see, e.g., structure D between 20 and 24 h MLT). Frequently, E-W aligned streamers are an extension of N-S streamers that turn azimuthally following meso-scale convection pattern after approaching the equatorward boundary of the auroral oval (Henderson et al. 1994; Kauristie et al. 2003; Nishimura et al. 2010a). In the duskside convection cell, the azimuthal turning of streamers appear as a tracer of streamlines along the Harang flow shear with a clockwise loop opening to the west. Those streamers are called Harang aurora.

The identification and tracking of auroral streamers from ground or space-based imagers is non-trivial. In global auroral images, streamers are seen inside the poleward-expanding auroral bulge in the substorm expansion/recovery phases or during continuous geomagnetic

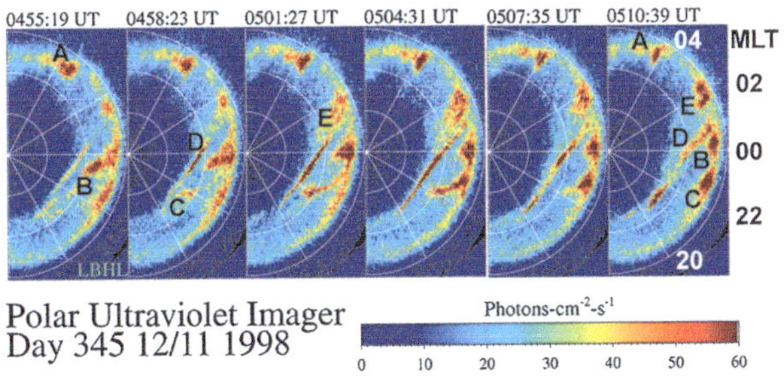

Fig. 5 Example of multiple auroral streamers. Development of four auroral streamers (B, C, D, E which eventually formed the bright patches B, C, E) and the eastward drift of bright patch A (created by previous streamer) as observed by Polar UVI imager during a steady convection event on December 11, 1998. The coordinates are AACGM – MLT (*after* Sergeev et al. 2005)

activity, in both cases in the so-called double auroral oval configuration (e.g. Rostoker et al. 1987; Henderson et al. 1994, 1998; Forsyth et al. 2008). However, the limited spatial and temporal resolution of global imagers (typically ∼100 km) means that the propagation may be missed, especially when the width of the auroral oval is small, and a streamer or streamers could be identified simply as a localised auroral enhancement. From ground-based imagers, N-S streamers can be seen without PBIs when PBIs are poleward of available imager coverage, and imagers may only detect the E-W segment of streamers when the N-S segment is outside coverage, which may be confused with more typical E-W aligned arcs.

Ionospheric convection signatures as well as field-aligned currents associated with the streamers have been examined using ground-based and in-situ measurements. The streamer itself is associated with a narrow region of intense (2–25 $\mu A/m^2$) upward field-aligned current, with a larger, more diffuse region of weaker downward field-aligned current observed to the dawnward side, both of which can be deconvolved from ground magnetometer data or observed by spacecraft (Amm et al. 1999; Sergeev et al. 2004; Nakamura et al. 2005; Forsyth et al. 2008). Enhanced equatorward convection electric fields resulting in plasma velocities of >1 km/s and a potential drop of up to 13 kV across the flow channel have also been observed in association with auroral streamers (Amm and Kauristie 2002; Sergeev et al. 2004; Nakamura et al. 2005; Amm et al. 2011; Pitkänen et al. 2011; Shi et al. 2012; Gallardo-Lacourt et al. 2014b). The equatorward flow is observed between the upward and downward field-aligned currents, with weaker poleward flows observed outside of the current system. Recent statistical study by Gallardo-Lacourt et al. (2014b) confirmed that in ∼90% of 135 studied streamers the SuperDARN radars detected an enhanced equatorward flow with an average width of 75 km offset by 58 km to the east from the narrow optical streamer.

The entire ionospheric field-aligned current pattern corresponds to the ∼1 hr MLT-wide equatorward flow channel with narrow downward and upward currents on its flanks. Observations by over-flying spacecraft have shown that a narrow region of monoenergetic electron precipitation is responsible for auroral signature of the streamer and coincides with the upward FAC on the dusk flank of the flow channel (Sergeev et al. 2004, see Fig. 6). Mapped into the equatorial magnetotail the streamer width is estimated to be 1.5–2.6R_E (Gallardo-Lacourt et al. 2014b) in keeping with the observed size of BBFs (e.g. Nakamura et al. 2004). This picture is valid for the poleward part of the streamer where, in addition, the return (poleward) flows are observed on flanks of the main flow channel in 10% of events (see also a case study by Pitkänen et al. 2011; Juusola et al. 2013; Aikio et al. 2018). When approaching the equatorward region where the streamer deforms, its properties change as described below.

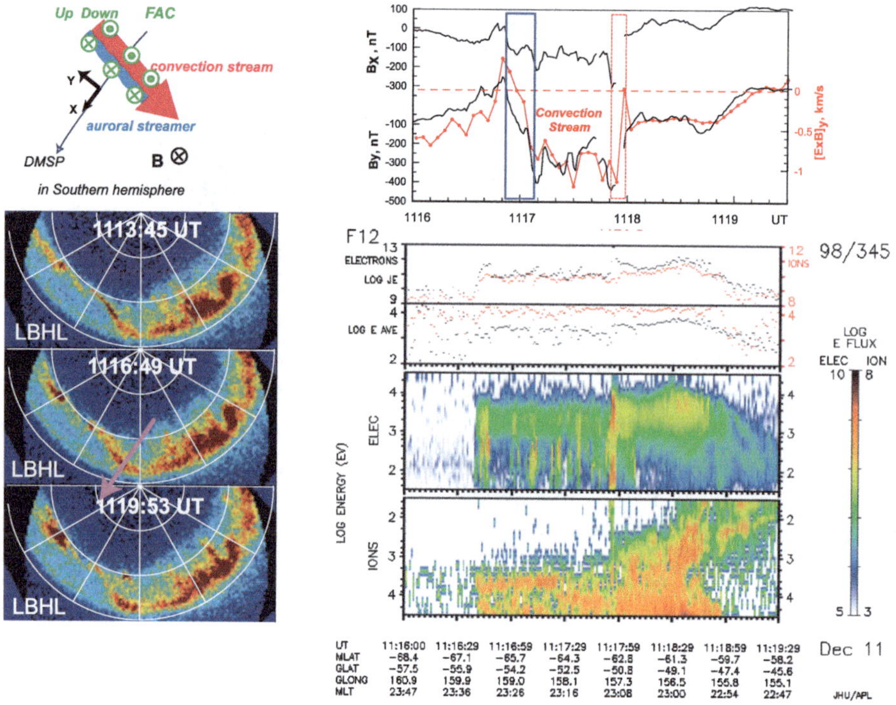

Fig. 6 Precipitation, convection and FAC patterns of auroral streamer observed by DMSP in the Southern hemisphere (adapted from Sergeev et al. 2004). Polar UVI observations in the northern hemisphere of the auroral streamer with conjugate DMSP traversal above the streamer (indicated by pink arrow) together with the schematic of observed patterns (as viewed from below in the southern hemisphere) are shown on the left panels. DMSP observations on right panels include particle spectrogram (bottom) together with horizontal magnetic field variations (black) and convection velocity changes (red line, corresponds to the flow component nearly aligned along the streamer). The entire pattern corresponds to the ∼1 hr MLT-wide equatorward flow channel with narrow downward (DFAC) and upward (UFAC) on its flank; narrow streamer precipitation coincides with the UFAC and located on the dusk flank of the flow channel

3.2 Drivers/Causes

Correlation has been established between appearance and location of auroral streamers and fast plasma jets (or BBFs) observed in-situ in the equatorial magnetotail (e.g., Henderson et al. 1998; Sergeev et al. 1999, 2000, 2004; Fairfield et al. 1999; Ieda et al. 2008; Nakamura et al. 2001; Zesta et al. 2000, 2002, 2006; Grocott et al. 2004 Forsyth et al. 2008; Pitkänen et al. 2011; Zou et al. 2010; Juusola et al. 2013). Various aspects of their relationships are illustrated in Fig. 7 (after Sergeev et al. 2000). In this event, during development of the streamers recorded by Polar UVI imager (image on the right), three magnetospheric spacecraft (Interball-Auroral, Geotail and LANL geostationary spacecraft) were magnetically conjugate to the streamer and registered different manifestations of flow channel development. Geotail (at $r \sim 26$ R_E, blue point on the sketch) detected the flow channel directly in the plasma sheet by observing two enhancements of earthward flow and BZ accompanied by plasma pressure decreases, which are basic signatures of flow channel. Interball-Auroral (at ~ 2 R_E altitude above the ionosphere, red point on sketch) recorded two intense time-of-flight dispersed injections of accelerated protons into the streamer-related magnetic flux tube, from which an inferred injection distance ($r \sim 40$ R_E in both cases) and injection time

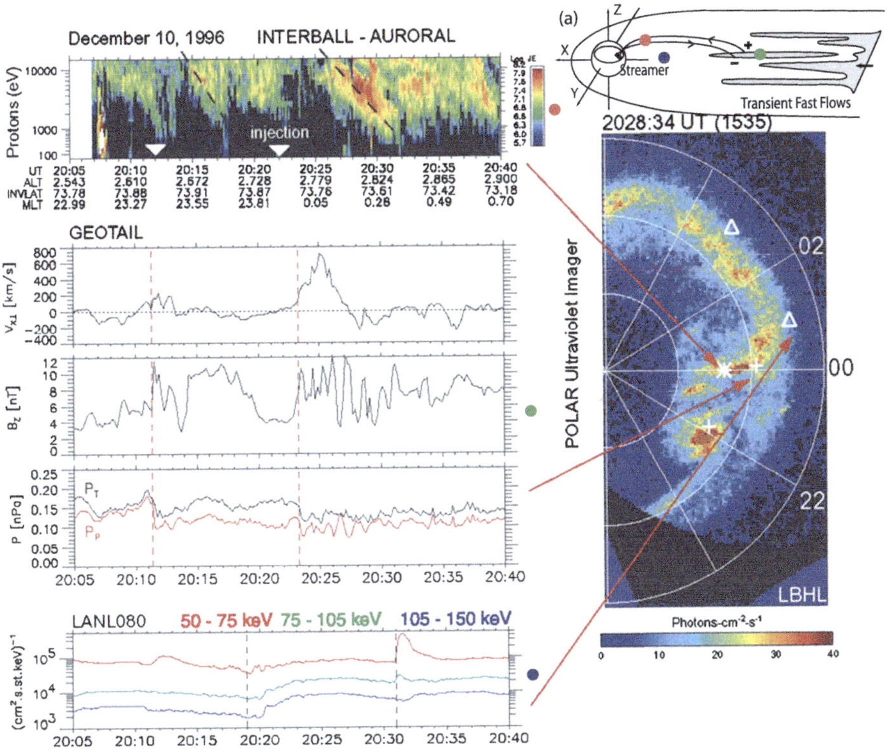

Fig. 7 Left: Magnetotail observations of 3 spacecraft, including: proton spectrogram at Interball (top); VxBz and plasma (and total) pressure observed by Geotail (center panel); energetic particle flux at LANL-080 geostationary spacecraft (bottom). Right panels show: a sketch of flow channels and their field-aligned currents with colored points indicating spacecraft locations (top); Polar UVI image showing the auroral streamer with mapped spacecraft locations indicated by asterisk (Interball), cross (Geotail) and triangle (LANL). (Adapted from Sergeev et al. 2000)

(white triangles in Fig. 11) are found, with both being in close agreement with commencements of Geotail flow channel events. At geosynchronous orbit, the LANL spacecraft (green point) observed two energetic electron flux increases with a ~8 min time delay after Geotail, suggesting that the flow channel injected energized plasma into the inner magnetosphere.

As discussed further in Sect. 7, mapping along the magnetic field lines using available magnetospheric models is rather uncertain, especially for the mapping in the meridional plane (latitude versus distance downtail). However magnetic mapping in azimuthal direction is less dependent on which model is used, and a distinct correspondence between Y-location of the flow channel and reconnection signatures in the plasma sheet and the magnetic local time of auroral streamer from global auroral images has been established (Nakamura et al. 2001). Typically, the mapped location of the magnetospheric signature is east of the auroral streamer, however Nakamura et al. (2001) showed that when the magnetospheric spacecraft crossed through the dusk flank of a BBF, there was better agreement between the streamer and mapped location. This helped to establish that streamers were mapped to the dusk flank of the flow channel (ΔMLT ~0 for positive values α_N) whereas the dawn flank crossings of the flow channel (negative values α_N) have ΔMLT ~1–1.5, implying that the entire width of streamer mapping is about 1 h MLT. This width is consistent with in-situ determinations of BBF cross-tail width of 2–4 R_E, obtained from statistical studies (Nakamura et al. 2004).

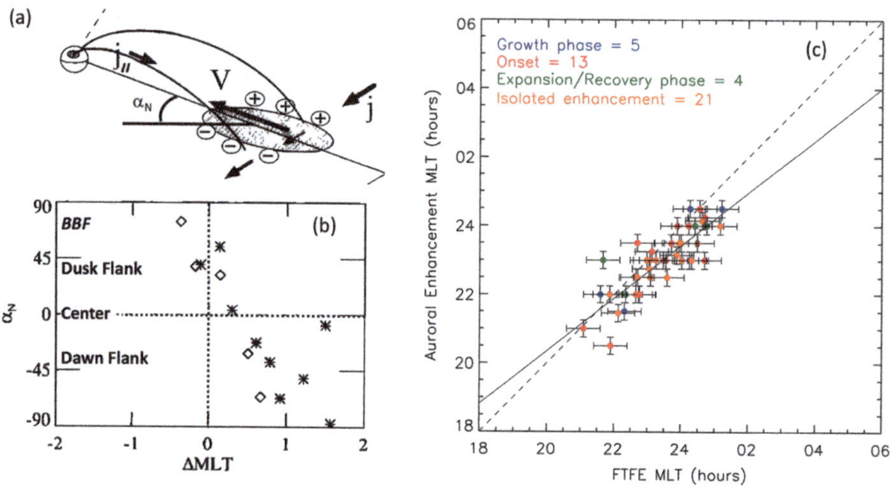

Fig. 8 (**a**) Scheme of fast flow channel (BBF), and, (**b**) – relationship of MLT difference between BBF ionospheric footpoint and streamer (DMLT = $MLT_{msph} - MLT_{isph}$) against the angle α_N characterizing which BBF side was crossed by spacecraft (both after Nakamura et al. 2001). Diamonds (or crosses) indicate the streamers (or localized brightenings) observed in the ionosphere. (**c**) MLT comparison of fast tailward flows (FTFE) footpoint and auroral brightenings from Case et al. (2017); red ellipse helps to see that ~1 h MLT-wide points cloud lay under the diagonal (shown by dashed black line) indicating a systematic dawnward shift of FTFE footpoints from their related optical effects. See text for discussion

Both conclusions (width and streamer location on the dusk side of the flow channel) are consistent with the characteristics of ionospheric pattern.

The well-developed auroral streamers are not as frequently observed in global UV images as the BBFs are observed in the plasma sheet. For example, Nakamura et al. (2001) found that more than half of the signatures of BBFs were localized auroral brightenings (marked by asterisks in Fig. 8b), rather than streamers (marked by diamonds). However, the spatial ordering of both streamers and localized brightenings with respect to the flow channel is the same, suggesting that auroral brightenings are also associated with the upward FAC region on the duskward flank of the BBF. Other studies have shown similar correspondence and MLT offsets between auroral enhancements and fast flows or reconnection signatures in the magnetotail (Ieda et al. 2001, 2008; Case et al. 2017), but did not explicitly examine auroral streamers. These results may reflect an observation bias, with global auroral imagers unable to resolve the detail of narrow auroral streamers, particularly when the auroral oval is relatively narrow.

Case studies of auroral streamers based on global auroral images are almost exclusively from intervals with a wide 'double oval' configuration, when the magnetic flux in the plasma sheet is large. In contrast, ground-based imaging with higher spatial resolution can detect streamers in a thin auroral oval during the substorm growth phase, where the oval latitudinal thickness can reduce down to a few degrees (Nishimura et al. 2011). It is also important to note that growth-phase and quiet-time streamers are dimmer by an order of magnitude and have little ground magnetic field perturbations and slower motion (c.f., Nishimura et al. 2013a,b). They may also only be apparent within a given wavelength range which can become washed out in white light images (Kepko et al. 2009). Their different properties may be related to their origins in the magnetotail, as well as to different energies of precipitating electrons and resulting Hall conductivity. Growth-phase and quiet-time streamers have been associated with Earthward flows at the lunar orbit, suggesting a distant neutral line origin

(Xing et al. 2010). The brightest streamers occur during the substorm expansion-phase and those streamers are associated with tailward flows at the lunar orbit, indicating near-Earth neutral line origin. BBFs originating closer to the Earth generally have lower entropy and involve larger acceleration, giving brighter streamers. However, the fact that non-streamer auroral signatures are reported conjugate to magnetotail reconnection signatures may suggest that auroral streamers are part of a wider class of reconnection flow related auroral brightenings.

In combination, all these results support the plasma sheet fast flow channel (BBF) as the origin of ionospheric manifestations of the auroral streamers. BBFs are natural products of magnetic reconnection pulses in the magnetotail, which are localized in the cross-tail direction. On the Earthward side of the reconnection X-line, the fast (initially Alfvénic) Earthward flow forms on the closed field lines connected to the ionosphere. These plasma tubes are underpopulated and carry the enhanced Bz compared to the ambient media (which are signatures of plasma bubbles), which causes their electric polarization shown in Fig. 7 and supports the interchange-type Earthward motion of the bubble, although it should be noted that under certain magnetotail configurations, in which there is a significant Bz maximum in the middle tail, the interchange instability can occur prior to the onset of reconnection (e.g., Pritchett 2005; Sitnov et al. 2013; Birn et al. 2018). A natural consequence of the polarized flow channel is the formation of Region 1-type field-aligned currents on its flanks, with upward (or downward) FAC delineating the dusk (or dawn) flank. Such plasma/FAC pattern is invariably reproduced in the RCM simulations (Yang et al. 2011) and MHD simulations of localized reconnection and plasma bubble motion (see, e.g., Birn et al. 2004; Birn and Hesse 2013) whose results are consistent with in situ spacecraft observations of BBF in the Earth's magnetotail (e.g. Sergeev et al. 2012; Wiltberger et al. 2015). The observed upward FAC region will require the acceleration of electrons along the magnetic field (Knight 1973), consistent with observed monoenergetic spectra in the auroral streamer (e.g., Fig. 6). It should be noted that not all streamers result from monoenergetic precipitation. Multi-wavelength observations of streamers have shown that some streamers appear as diffuse aurora in the 630 nm band, with the particle flux responsible for the auroral emission most likely coming from enhanced scattering of relatively cool plasma from the plasma sheet or an Alfvénic acceleration process (Kepko et al. 2009). This does not discount plasma sheet fast flows as a source of this precipitation. In fact, recent observations have shown that whistler-mode waves observed in fast plasma flows are the dominant scattering process and may enhance diffuse aurora (Wendel et al. 2019).

3.3 Implications and Outstanding Questions

Auroral streamers are a key indicator of magnetospheric dynamics and the coupling between the magnetosphere and the ionosphere. They are linked with Poleward Boundary Intensifications (Sect. 2) and omega band activity (Sect. 4) and their motion across the auroral oval is linked to the mapping and motion of readily identifiable plasma structures in the magnetosphere. However, questions remain with regards to streamer lifetimes and, in particular, what happens as their source flows enter the inner magnetosphere.

Observations of auroral streamers show that they can turn from N-S aligned arcs to E-W aligned arcs or rotate to follow convection around the Harang discontinuity as they approach the equatorward edge of the auroral oval, however the motion and evolution of the magnetospheric driver at this time remains poorly investigated even though some interesting material is already available. RCM and MHD simulations of flow channels approaching the inner magnetosphere show a prominent (about a factor 2) increase of plasma pressure ahead of the dipolarization front near its stopping point (Yang et al. 2011; Birn and Hesse 2013), which impacts on the pressure distribution across the magnetotail as well as the supporting field-aligned currents (Shiokawa et al. 1998). Supporting in-situ observations from the THEMIS spacecraft have been reported (Yao et al. 2013; Sergeev et al. 2014).

In the case of multiple streamers, Nishimura et al. (2011) observed intensifications of previous equatorward arcs when streamers approached close to but did not immediately reach those arcs, instead rotating towards the azimuthal direction. Such streamer rotation from roughly NS direction toward oval-aligned direction, reminiscent of the convection streamline, is very typical (Nishimura et al. 2010b). On occasion, this rotation has been associated with the reformation of convection pattern in this region as reflected in the change of equivalent current pattern (Kauristie et al. 2003). Furthermore, some streamers have been observed to split as they approach the equatorward edge of the auroral oval, which may be in indication that some form of ballooning instability is disrupting the plasma flow channel in the magnetosphere (Pritchett and Coroniti 2013; Pritchett et al. 2014).

While the link between streamers and fast plasma flows in the magnetotail has been extensively explored and reported, a number of outstanding issues remain. Firstly, individual flow bursts are associated with an ion-scale dipolarization front which carries a strong field-aligned current (Hwang et al. 2011; Liu et al. 2013; Sun et al. 2013; Zhou et al. 2013), however the corresponding narrow auroral feature is yet to be identified. Furthermore, the magnetospheric plasma flow is a transient feature, whereas streamers can be observed for 10 min or more. The streamer may be the footpoint of a channel of an under-populated fast flows, as seen in some simulations (e.g. Wiltberger et al. 2015), however how these flow channels persist and support sufficient field-aligned current to result in a bright auroral footprint, particularly in the flow braking region where the contrast between the plasma conditions in the flow and the surrounding plasma is diminished, remains an open question.

Some auroral streamers are seen disconnected from the poleward boundary of the aurora oval, as can be seen in Fig. 5 and 6. A number of explanations may be plausible. Firstly, if the fast plasma flow and streamer are the result of a burst reconnection in the magnetotail, the plasma flow will continue to exist after the reconnection ceases and thus the streamer would be disconnected from the poleward boundary. Alternatively, disconnected streamers may indicate that some streamers do not originate from PBIs but instead arise due to reconnection on closed field-lines within the plasma sheet. Furthermore, the strength of the FACs at the edge of the fast plasma flows, which is related to the brightness of the auroral signature, may vary with time or distance along the magnetotail. We note, in particular, that the auroral luminosity in the red-line aurora reported by Kepko et al. (2009) was weak and may be missed in white-light imagers, which predominantly observe green-line emissions. This raises questions as to the sources of streamers with respect to PBIs and the response of the auroral emission to the evolution of the fast plasma flows as they propagate through the magnetotail. Understanding these differences is critical in understanding the source of streamers and their potential implications in the wider dynamics of the magnetosphere.

The arrival of streamers at the equatorward edge of the auroral oval in the premidnight sector has been shown to be correlated with enhancements in subauroral (westward) convection events (Gallardo-Lacourt et al. 2017) which may play a role in the excitation of giant undulations (Sect. 6). The arrival of the streamer at the equatorward boundary also coincides with enhancements in the proton aurora at subauroral latitudes (Nishimura et al. 2014) and can excite omega band activity (Henderson et al. 2002; Sergeev et al. 2005; Henderson 2012; see Sect. 5). This indicates a prompt (on minutes scale) modification of inner magnetosphere convection by the intrusion of a magnetospheric flow channel. Furthermore, energetic particle injections at geosynchronous orbit have been observed in concert with streamers approaching the equatorward boundary of the auroral oval (Henderson et al. 1998; Sergeev et al. 2000). Lyons et al. (2002) reported similar results but describes the auroral activity as equatorward extending, which may also be described as streamers. These findings are consistent with the results of global RCM simulations by Yang et al. (2015) that show that flow bursts play an important role in the populating the ring current, providing as much as 20 to 60% of ion injections into the ring current during magnetic storms. While recent studies have shown links between injections in the inner magnetosphere and flow bursts further out (Runov et al. 2009; Gabrielse et al. 2012), other studies have shown that fast plasma flows are typically deflected around the Earth outside of geosynchronous orbit (Juusola et al. 2011; McPherron et al. 2011). Understanding how, when and why the magnetotail counterpart of streamers affects the inner magnetosphere, and how any feedback from the auroral enhancements impacts upon this, remains an open question.

4 Omega Band and Torches

4.1 Definitions

The omega band is a description of a dynamic diffuse auroral arc predominantly seen towards the end of substorm expansion phases or during substorm recovery phases. Although no longer commonly used, the term "band" was used by Akasofu and Kimball (1964) to describe an "arc" that has multiple folds in it and "drapery" as more active and complex displays of arcs containing rays, folds and loops. An omega band is a folded auroral arc in which the undulations of the poleward edge resemble a train of inverted capital greek letter omegas (Akasofu and Kimball 1964). These undulations are observed to move eastwards. The curved parts of the drifting inverted omegas are rooted to the equatorward regions of an expanded oval with the tongues of luminosity pointing polewards towards the more active expanded substorm bulge or double oval configuration. Along with the diffuse auroral emissions, the equatorward parts of the omega structures tend to merge together to form a wavy band of luminosity that dramatically contributes to the lower latitude component of the double oval distribution.

When a poleward-protruding tongue of luminosity (one of the legs of the omega) extends dramatically towards the poleward active arc system, the feature is referred to as an "auroral torch". These structures are most prominent at the western edge of omega band trains but can form anywhere along an already existing train. Dramatic examples of auroral torch structures can be seen in the DMSP imagery examined by Akasofu (1974) (see Fig. 9). "Brush-like" torches are wider than the "filamentary torch" and tend to appear off the equatorward auroral band, whereas the "filamentary torch" forms can connect all the way up into the high-latitude poleward arcs and are now more commonly referred to as auroral streamers (see Sect. 3).

High time resolution auroral imaging studies have revealed that auroral torch structures are "activity centers" for pulsating aurora (Oguti 1981; Tagirov 1993), with a complex "streaming type" of pulsation occurring within auroral torches inside a boundary enveloped by a non-pulsating peripheral region (see Fig. 10). The streaming striations have periods from 5–10 s and apparent speeds of 30–100 km/s. Equatorward of the streaming striations, diffuse patches are observed that display either quasi-periodic streaming and/or longitudinal "flooding motions" (patches brightening beyond their boundaries and enveloping neighboring patches). In addition, the intense energetic electron precipitation that accompanies omega band displays has distinctly different spectral characteristics from those found in

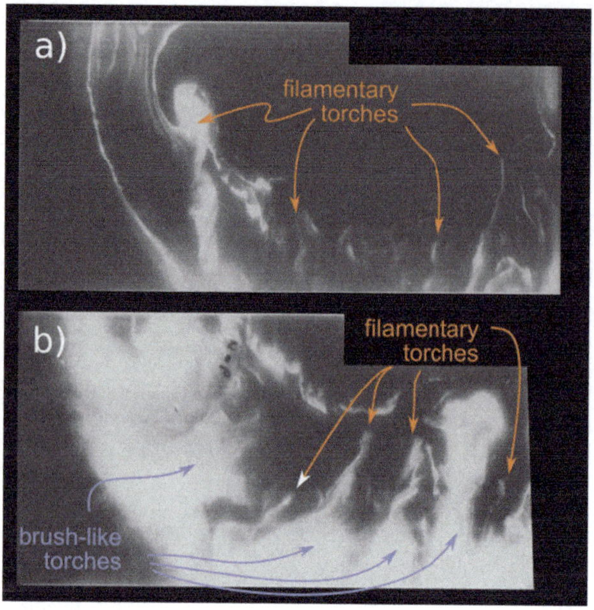

Fig. 9 A variety of auroral torches seen in DMSP imagery. The "filamentary torches" described by Akasofu were thought to protrude into the polar cap. Today, we would call these "auroral streamers" and they are known to originate at the poleward boundary arc system. (Adapted from Akasofu 1974)

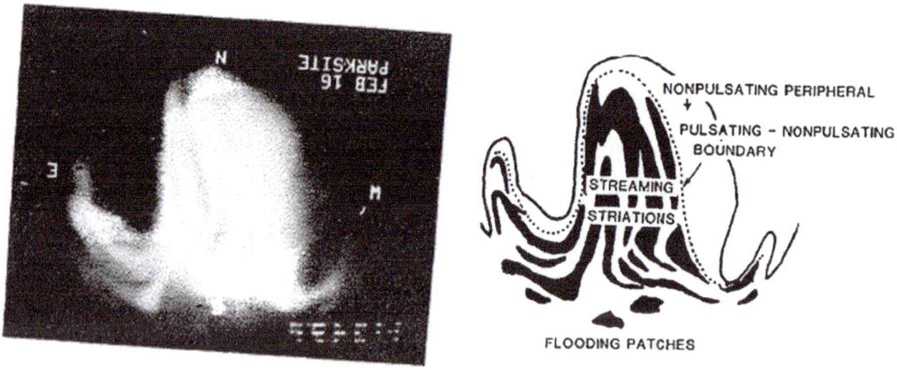

Fig. 10 An auroral torch is an activity center for pulsating aurora. (Adapted from Oguti 1981)

evening sector arcs or westward travelling surge type forms that are associated with substorm onset and expansion phase activity (e.g. Opgenoorth et al. 1994). The flux of low energy ($\sim<1$–2 keV) and high energy (>10–40 keV) electrons is higher above omega bands than for intense morning sector precipitation. Thus, the formation of torches results from the enhancement of auroral activity and may be related to the acceleration of particles in the magnetosphere and irregularities in the cold plasma (Oguti 1981).

On the ground, omega bands are associated with large-scale Ps6 magnetic pulsations that have periods of 10–30 min (Ps6 are a subclass of Pi3 pulsations) (Saito 1978). Opgenoorth et al. (1983) demonstrated that these pulsations are causally related to paired upward and downward field-aligned current (FAC) structures in which the bright tongues are associated with upward FAC and the dark voids are associated with downward FAC (also see Wild et al. 2000.) The enhanced electron fluxes also lead to low Pedersen/Hall conductivity ratios in this region and an overall lower altitude of the peak Hall conductivity layer.

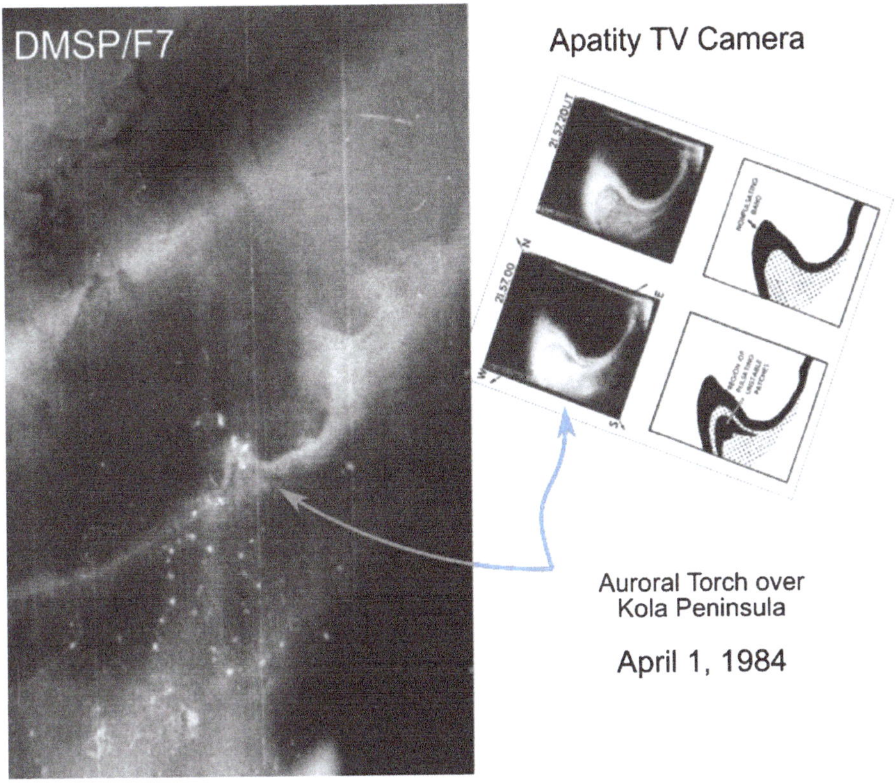

Fig. 11 High-time resolution ground-based imaging of an auroral troch-like structure observed over Apatity on April 1, 1984. Left: DMSP/F7 image showing torch/omega-band forms. Right: Apatity TV camera shows details of pulsating forms within the western-most torch. (Adapted from Tagirov 1993)

Omega bands have been typically thought to occur in the morning sector during the recovery phase of substorms (Akasofu 1974; Opgenoorth et al. 1994) or during periods with an active double oval that often develop out of substorms that remain continuously driven or from a reactivation of a quiescent high latitude arc system (i.e. Steady Magnetospheric Convection Type of activity). However, a recent statistical study that combined automated and manual detection algorithms shows that while omega band occurrence peaks in 02-03 MLT sector, approximately 10% of the observed omega bands appeared in the pre-midnight sectors. Furthermore, only two thirds of the observed omega bands appeared in the recovery phase, with the remaining third tending to be observed during substorm expansion phases (Partamies et al. 2017). Interestingly, omega bands tended to be associated with times of greater than average geomagnetic activity, with the IL index (AL index from the IMAGE magnetometer array) being -100 nT smaller than for all substorm, Dst reaching nearly -50 nT indicating storm-like activity, and Kp averaging 4.

4.2 Drivers/Causes

Mapping auroral omega bands observed in the ionosphere to their source locations in the magnetosphere is key to understanding their generation. In terms of contextual mapping, as distortions near the poleward regions of the diffuse auroral region and well away from the active high-latitude aurorae, they most likely have a near-Earth source region (Opgenoorth et al. 1994). In addition, several studies using empirical magnetic field models to map the structures to the equatorial plane (Min-B surface) concluded that omega bands map to regions quite close to the Earth ($<10\ R_E$) and thus that some type of instability in the inner magnetosphere is responsible for the development of the distortions (Pulkkinen et al. 1991; Tagirov 1993; Opgenoorth et al. 1994; Jorgensen et al. 1999). Nevertheless, several past studies incorrectly placed the omega band at the polar cap boundary rather than the edge of the diffuse aurora which led them to surmise that the distortions mapped to the open/closed boundary and the plasma sheet boundary layer on the nightside (e.g. Akasofu et el. 1971).

Due to the wave-like nature of omega bands combined with the pronounced eastward drift, many studies have invoked Kelvin-Helmholtz types of instabilities as an explanation (Rostoker and Samson 1984; Lyons and Walterscheid 1985; Buchert et al. 1990; Connors and Rostoker 1993; Wild et al. 2000). The growth rate of the KHI maximises for a velocity shear in the direction of the wave growth and perpendicular to the magnetic field (Southwood 1968; Walker 1981), resulting in large-scale, low frequency waves. Identifying the region where this occurs has proved trickier. Rostoker and Samson (1984) suggested that such conditions could arise between the low-latitude boundary layer and the surrounding plasma, potentially generating omega bands, the westward travelling surge and dayside Pc pulsations. Other mechanisms place the velocity shear at the inner edge of the plasma sheet (e.g. Connors and Rostoker 1993). Subsequent observations have shown that the edge of the diffuse aurora in omega bands corresponds to the boundary between the Region 1 and Region 2 current systems, mapping further downtail than the LLBL (Wild et al. 2011). Strong shear flows have been observed by low-altitude spacecraft in this region during the occurrence of omega bands (Liu et al. 2018), however these shear flows are a ubiquitous feature of the boundary between the R1 and R2 currents (Jiang et al. 2015; Archer et al. 2017) and as such, the occurrence of a shear flow does not appear to be sufficient to generate an omega band. This region is also generally stable to the KHI particularly when the coupling to the ionosphere is considered but may be unstable to the hybrid Kelvin-Helmholtz/Rayleigh Taylor Instability (Yamamoto 2009, 2011). However, none of these mechanisms, by themselves, appear to be consistent with the observation that omega band waveforms are often irregular in nature and it seems implausible that the KHI can produce the very pronounced torch-like structures that are commonly observed to extend far toward the poleward active auroras.

Up until the launch of the POLAR spacecraft, most platforms capable of observing omega bands were limited in what they could say about their development in a global context. Global auroral imagers either lacked spatial-resolution, time-resolution or sensitivity to observe the detailed formation of omega bands over large space and time scales (e.g., DMSP acquired only a single scanned image per orbit, Dynamics Explorer took $\sim$12 min to acquire an image, Viking was a snapshot style imager with a non-ideal orbit and launched during solar minimum.) And ground-based all-sky imagers had field of views that were too small for this purpose. The sensitivity and resolution of the POLAR auroral imagers (UVI and VIS) combined with the high apogee of the POLAR orbit (and operations throughout a solar maximum) offered the first opportunity to explore how omega bands form and evolve in relation to other types of activity especially over long time periods.

Using the POLAR VIS high-resolution imagery, Henderson et al. (2002) and Henderson (2009, 2012, 2016) have shown that auroral streamers ejected from the active poleward auroral arcs can evolve directly into auroral torch structures and contribute to the formation of auroral omega bands. Figure 12 (adapted from Henderson et al. (2002)) shows an example of this evolution, with a Poleward Boundary Intensification leading to an auroral streamer and ultimately a torch structure on the equatorward arc. Henderson et al. (2002)

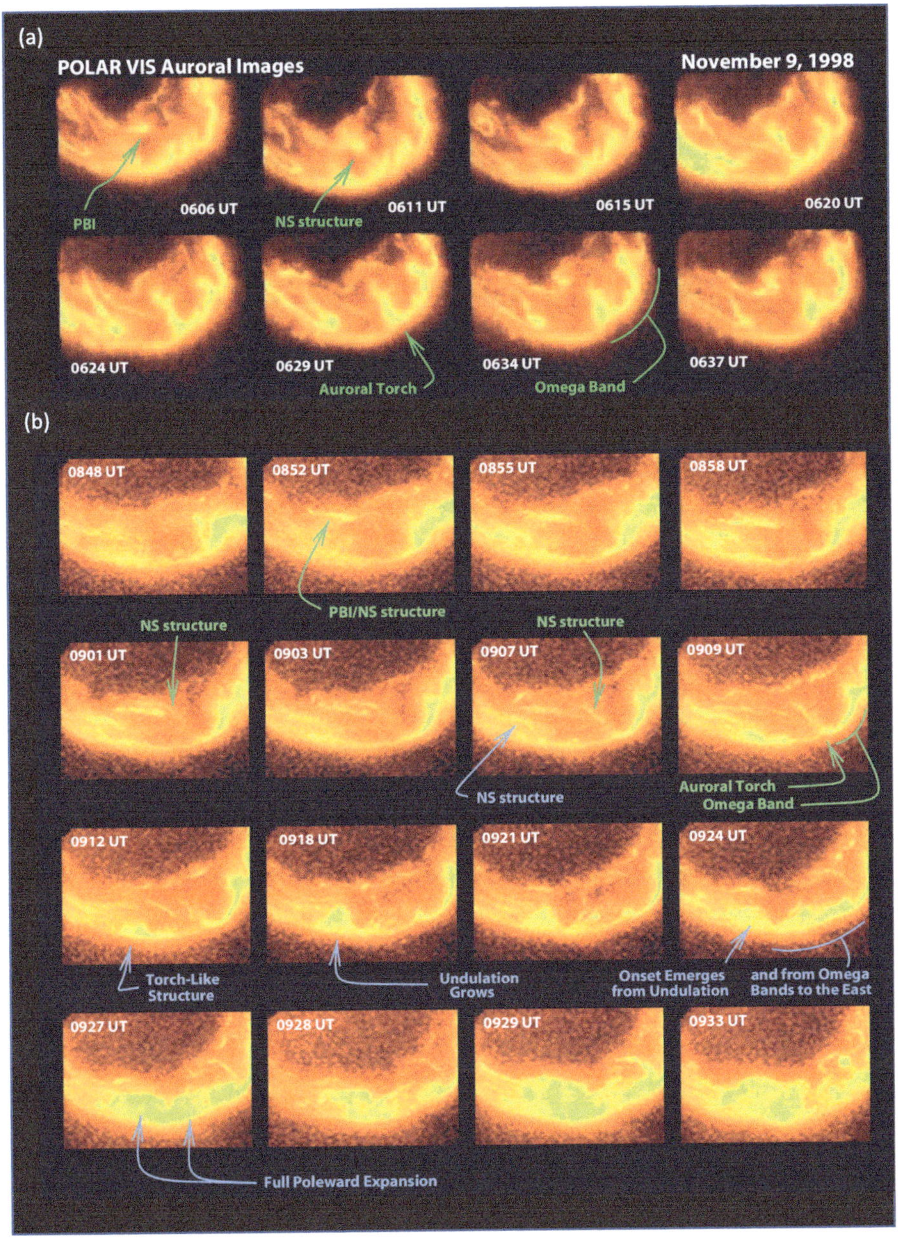

Fig. 12 Examples showing evolution of auroral streamers into torches and omega bands. (Adapted from Henderson et al. 2002)

put forth three scenarios by which the fast magnetospheric flows associated with streamers (see above) might result in the formation of omega bands: (1) a series of fast flows from the same region arriving at different times drift eastward creating a series of omega bands forming from the same location; (2) a group of fast flows from different locations arrive near-simultaneously at the auroral boundary and drift eastward creating a series of omega bands forming at different locations; (3) a single fast flow excites the growth of an azimuthal wave at the inner edge of the plasma sheet. Since BBFs already have an upward/downward FAC pair, this framework resolves issues with the FAC structuring and ground signatures and removes the need for the generated omega bands to be a periodic wave. In addition, multiple streamers can occur simultaneously and at different longitudes which explains how pre-existing omega-band structures can be interrupted by new torch-like forms (e.g. Fig. 13).

As the forms drift eastward, further evolution, growth and distortions of the torch-like disturbances occur. The arrival of the fast flow, which is typically a plasma-depleted flux tube (e.g. Chen and Wolf 1993; Sergeev et al. 1996) may be sufficient to create the inward energy density necessary for the generation of KH/RTI waves (Yamamoto 2009) and so these mechanisms may work together. However, magnetospheric observations of such a shear flow are yet to be reported. Observations from the magnetotail during the occurrence of omega bands have shown the occurrence of fast flows at the same time but different locations of omega bands, although fast flows are fairly ubiquitous during substorm activity (Weygand et al. 2015). Similarly, broadband, <1 keV electron populations have been observed close to the mapped location of omega bands, but the variation in electron fluxes did not align with the passage of the bands (Wild et al. 2011).

Figure 14b also shows the development of a torch-like structure on the western edge of the omega-band region, but this is seen to grow in intensity and expand poleward ultimately leading to the development of a new "embedded" auroral substorm onset. This type of activity had been noted much earlier by Oguti (1973) (but not in the context of streamers, torches and omega bands) who referred to such events as "contact breakups". It is likely that at least some of the streamer-induced breakups observed by Nishimura et al. (2010b) with the THEMIS ASI imagers are contact breakups of this sort.

4.3 Implications and Outstanding Issues

The proposed BBF generation mechanism for torches, omega bands and streamer-induced contact breakups provides a unifying framework for the complex auroral dynamics observed during sustained driving conditions that typically accompany driven substorm recovery phases and active double oval configurations (representative of SMC-like intervals). The mechanism resolves a number of previously puzzling features of omega bands including: why the bands are often so irregular, the nature of the paired up/down FAC structures and Ps6 pulsations, the fact that torches are activity centers for pulsating aurora, the energetic precipitation signatures. It also provides a potential mechanism by which the plasma in the inner magnetosphere may become unstable to KH-based instabilities that facilitate further excitation of distortions at the poleward edge of the diffuse aurora. However, several issues remain. Firstly, not all omega bands are observed to form following the arrival of streamers or magnetotail flows and the proportions of flow-driven bands is unclear. This also raises the questions as to what causes the non-flow driven bands. Furthermore, omega bands predominantly occur during periods of enhanced Kp while magnetotail fast flows are fairly ubiquitous during substorm so what plasma conditions prevent or allow the occurrence of omega bands. Also noted is that torches evolve over their lifetime, but it is unclear what controls this evolution and how this is linked to the evolution of magnetotail fast flows or depleted flux tubes in the magnetosphere.

The excitation of torches and omega bands in the westernmost auroral bulge seems to coincide with that region becoming more unstable to further distortions. This region (typically pre-midnight) is where auroral onsets (spontaneous or contact breakup types) tend to occur.

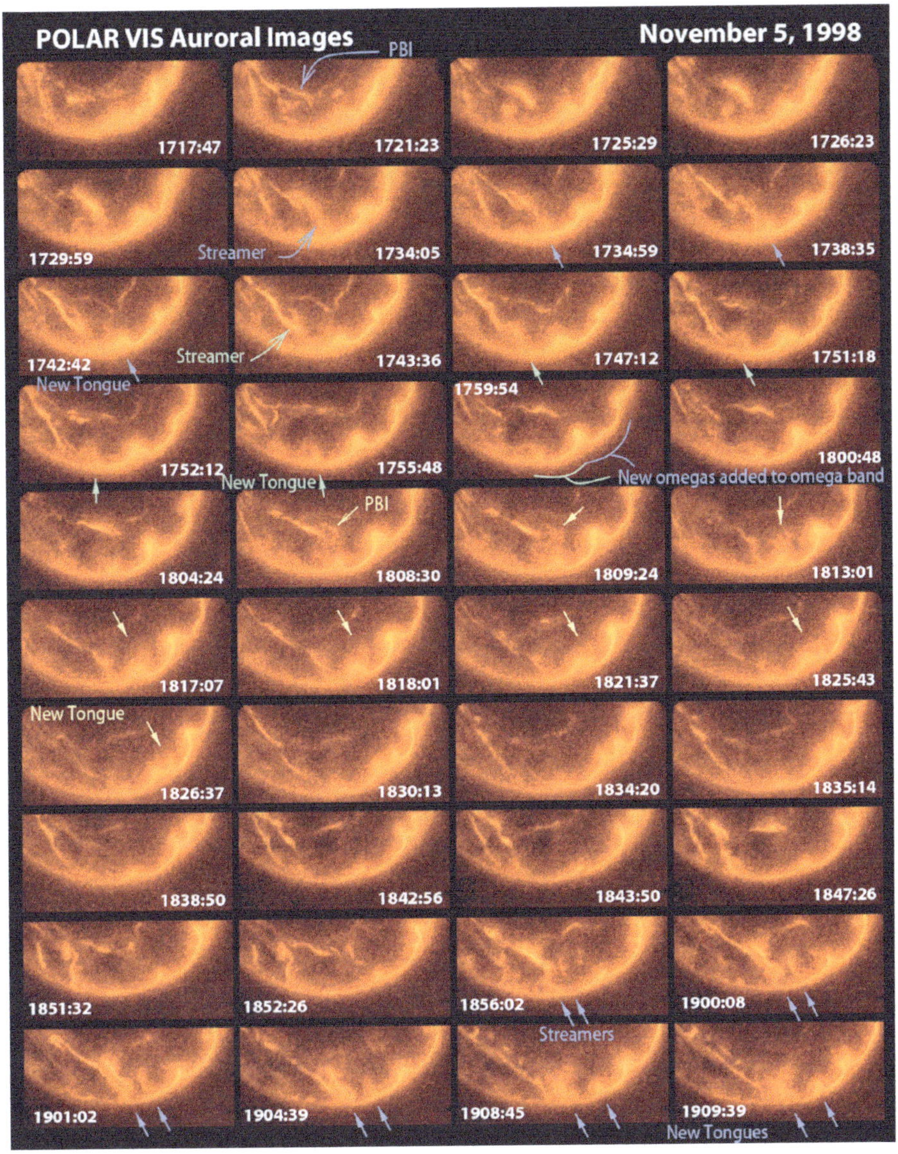

Fig. 13 Evolution of the northern auroral distribution over ~2 h time interval on Nov 5, 1998 acquired with the POLAR/VIS LR camera. This sequence clearly shows how streamers can evolve into torches and how pre-existing omega forms can be interrupted by new torches (see yellow annotation as an example). Note that additional evolution/growth occurred following the creation of each torch. Also note that not all instances of streamer -> torch formation has been annotated here

Thus, the occurrence of torches and omega bands may be indicative of changing plasma conditions in the magnetotail that increases the likelihood of different modes of variability. Understanding and isolating the different physical processes that arise throughout the magnetotail and how this arises from changing plasma conditions remains a key open question in magnetospheric physics and hence in the understanding of meso-scale auroral forms.

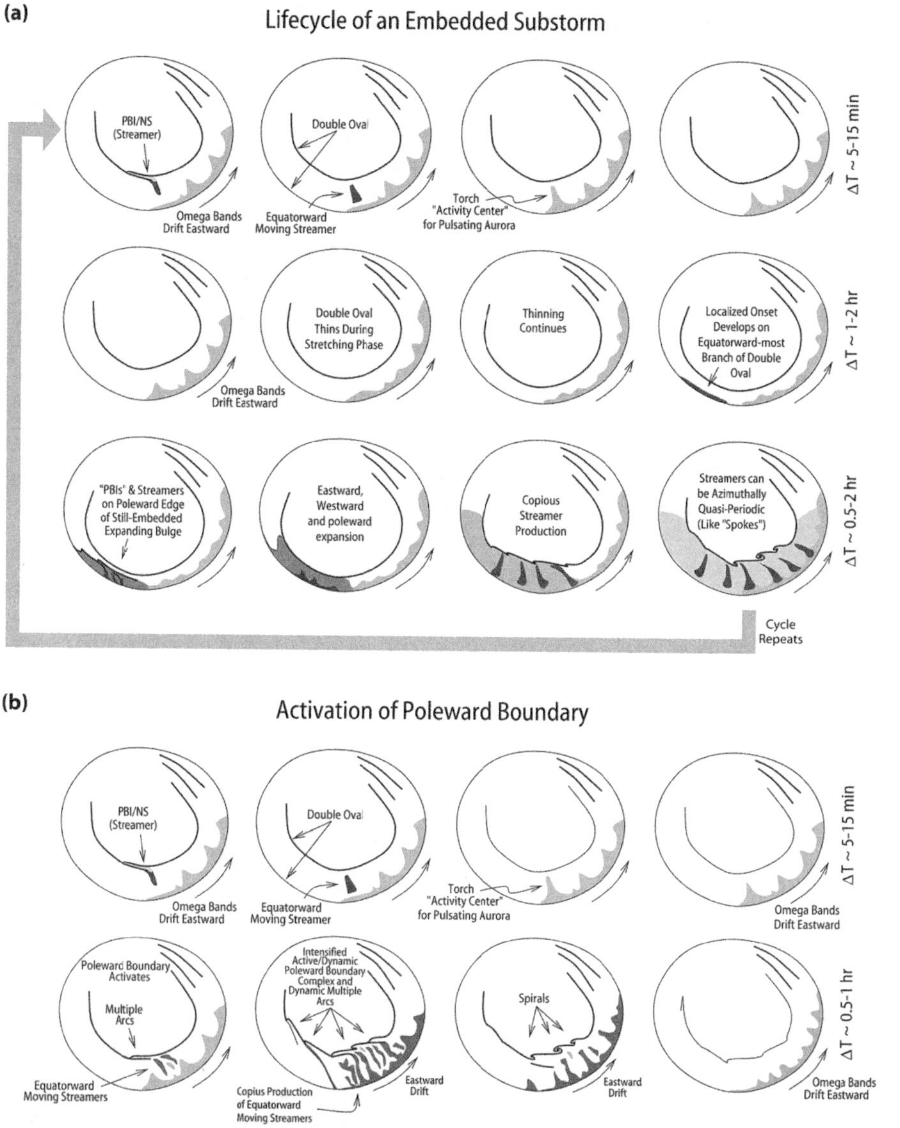

Fig. 14 A schematic showing the development of torches and omega bands. Top panel shows how this activity is related to the substorm cycle and the bottom panel shows an SMC-like interval of activity without the occurrence of new substorm onsets. From Henderson (2012)

5 Auroral Beads

5.1 Definitions

The term 'auroral beads' has become the commonly used term to describe spatially periodic auroral forms typically detected at or just before the substorm auroral onset. Originally described as 'azimuthally spaced auroral forms' (Henderson 1994; Elphinstone et al. 1995;

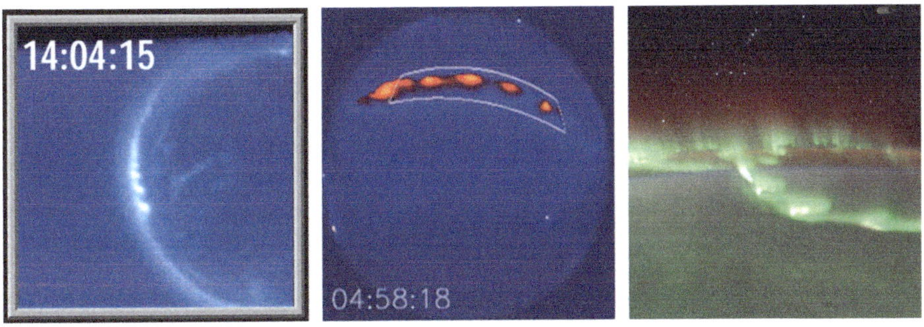

Fig. 15 Examples of auroral beads observed by the IMAGE FUV instrument (Henderson 2009), the THEMIS ASI at Gillam (Kalmoni et al. 2015), and from the International Space Station (from https://www.cosmos.esa.int/web/arrrgh/)

Voronkov et al. 1999, 2000) from data from the Viking global auroral imagers, the initial brightening along a preexisting or a newly formed arc at substorm onset is almost always observed in the form of a wavy auroral pattern (Kalmoni et al. 2017). The availability of data from networks of ground-based optical instruments has provided unprecedented observations of the these structures, which are now commonly referred to as auroral beads (Donovan et al. 2006; Liang et al. 2008; Rae et al. 2009; Sakaguchi et al. 2009). Unlike omega bands (see Sect. 4), auroral beads appear as are localised bright patches of aurora associated with an accelerated electron population (Motoba and Hirihara, 2016), as opposed to enhancements of diffuse aurora, that may or may not eventually take on the form of a series of auroral curls. Figure 15 shows a series of examples of auroral beads taken by the IMAGE FUV instrument, a THEMIS ASI and by astronauts on the International Space Station.

Early observations of auroral beads showed regularly spaced auroral bright points separated along a quiet arc by 130–585 km (Elphinstone et al. 1995). More recent statistical studies using ground-based data have shown wavelengths down to of 50–100 km (Liang et al. 2008; Nishimura et al. 2016) when tracking the brightest features in auroral keograms. However, quantitative analysis of the auroral power at all observable wavelengths has shown the presence of waves down to 1.6 km scales (Sakaguchi et al. 2009; Kataoka et al. 2011; Kalmoni et al. 2015), but that waves at the 100 km scale tend to have much larger power, thus are the most prominent. From both space and ground-based observations, the scales observed have been limited by the resolution of the instruments. These smallest observed wavelengths are smaller than the electron inertial lengths above the auroral region and that the larger wavelengths map to $\sim$1000–3000 km widths in the magnetotail.

One of the key features of auroral beads is the near simultaneous exponential growth in their brightness across a range of scales. The exponential growth rate of the total auroral brightness (Nishimura et al. 2016) and of the individual wavenumbers (Kalmoni et al. 2015) of the auroral beads is typically $\sim$0.04 s^{-1}, although different wavelengths grow at different rates and for different lengths of times. Growth rates above 0.1 s^{-1} have been derived from ASI data (Rae et al. 2010) and of 0.005 s^{-1} (Henderson 2009) have been reported from global auroral imagers. The differences in the growth rates from ground and space-based observations may result from differences in resolution; just as global imagers cannot resolve small scale ($\sim$10 km) waves, individual ground-based ASIs cannot examine larger scale ($>$500 km) waves, thus there are significant difficulties in examining the whole parameter space.

5.2 Drivers/Causes

The exponential growth of auroral beads, with similar rates to the growth in wave power observed by co-located ground magnetometers (Sakaguchi et al. 2009; Rae et al. 2012), and their spatially periodic form indicates that they arise from the linear phase of a plasma instability. Observations from conjugate auroral observatories has shown nearly coincident growth of auroral beads in both the northern and southern hemispheres, pointing to a common, magnetospheric source (Motoba et al. 2012) at a distance of $\sim 10\ R_E$ downtail. However, which instability remains an open question.

A number of instabilities have been implicated in the generation of auroral beads including shear flow ballooning (Viñas and Madden 1986; Voronkov et al. 1997), kinetic ballooning (Cheng and Lui 1998), kinetic ballooning interchange (Pritchett and Coroniti 2010) and cross-field current (Lui et al. 1991). Each have been suggested due to producing growth rates or wavenumbers similar to those observed for auroral beads for reasonable, but unobserved, conditions in the source region (Friedrich et al. 2001; Cheng and Zaharia 2004; Liang et al. 2008; Sakaguchi et al. 2009; Rae et al. 2010). However, there is sufficient variation in the plasma conditions close to the mapped source of the beads that the theoretical growth rates and wavenumbers can be made to match the observations. By examining the observed profile of the normalised growth rates against wavenumber, which showed a distinct peak for wavelengths of ~ 100 km, with the profiles predicted for the shear-flow ballooning and cross-field current instabilities, Kalmoni et al. (2015) found that while both provided comparable growth rates, only the shear-flow ballooning instability reproduced the observed growth rate vs wavenumber profile. From the same underlying data, Lui (2016) showed that the profile of the average, rather than normalised, growth rates could be reproduced by the cross-field current instability. A further statistical study examining the propagation speed vs wavelength of the dominant auroral beads and their growth rates vs duration also concluded that the cross-field current instability was unlikely to be the source of auroral beads, but that the kinetic ballooning interchange instability provided a good fit to the data (Nishimura et al. 2016). Results from these studies are summarised in Fig. 16.

Auroral observations provide information on the growth of the instability associated with auroral beads, but limited information about the plasma conditions and movement. Following on from observations of fast flows near auroral beads by Hosokawa et al. (2013), Gallardo-Lacourt et al. (2014a) used SuperDARN radars working in high-resolution mode to infer a clockwise flow shear around auroral beads, which indicated the presence of an upward field-aligned current in the center of the shear giving rise to the individual beading structures. These flows were qualititatively similar to those reported from simulations by Pritchett and Coroniti (2010, 2013), also associated with the kinetic ballooning interchange instability.

Understanding the mechanism by which particles are accelerated along the magnetic field to generate bright auroral beads provides constraints on the physical mechanisms at play. As has been examined at length (e.g. Paschmann et al. 2003), both broadband and mono-energetic electron populations can result in bright aurora, and the mechanisms by which the quiet arc from which auroral beads arise remain a mystery. As noted above, observations of the ionospheric plasma motions associate with auroral beads indicates that the auroral bright spots are associated with upward field-aligned currents. Given the known limitations of the magnetosphere-ionosphere system to support field-aligned currents (Knight 1973), the driving process or processes in the magnetotail must either accelerate electrons or provide the conditions for electron acceleration at some point along the magnetic field.

By comparing in-situ observations of the electron pitch-angle distribution taken by Reimei above auroral beads with optical emission at the spacecraft footpoint, Motoba and

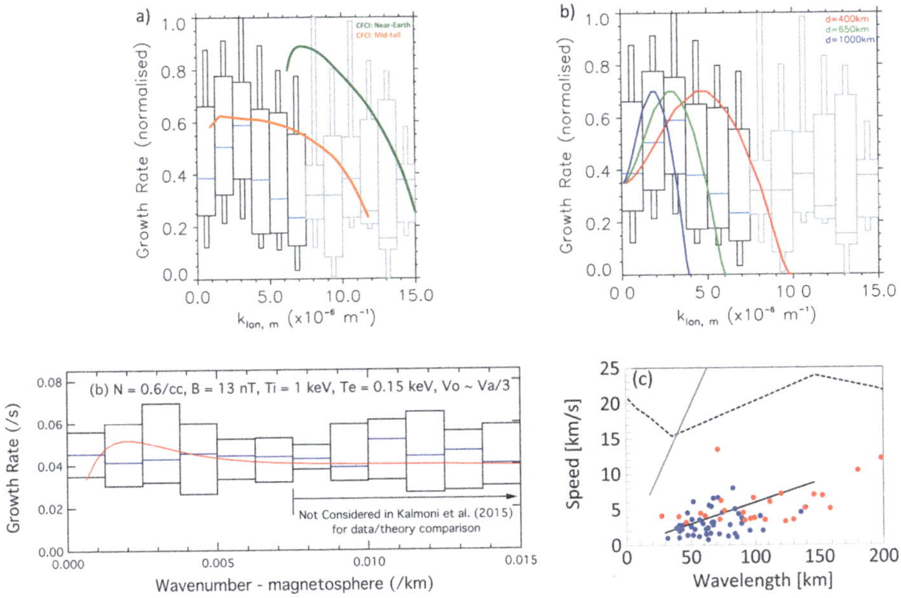

Fig. 16 Comparisons of auroral bead properties with those predicted for various instabilities. (Top) Normalised growth rates versus mapped longitudinal wave number overlaid by the expected results for (left) the cross-field current instability and (right) the shear-flow ballooning instability. Greyed out bars indicate those observations that are no statistically significant (from Kalmoni et al. 2015). (Bottom left) Growth rate versus mapped longitudinal wave number (from the same data) overlaid by an alternative calculation from the cross-field current instability (from Lui 2016). (Bottom right) Wavelength versus phases speed overlaid by the expected results for the current driver ion cyclotron instability (Perraut et al. 2000, grey line), the kinetic ballooning instability (Pritchett and Coroniti 2010, black line) and the cross-field current instability (Lui et al. 1991, dashed line) (from Nishimura et al. 2016)

Hirahara (2016) inferred that the optical emission was associated with a 3–7 keV mono-energetically accelerated electron population (see Fig. 17) associated with a quasi-static potential drop rather than from an Alfvén wave accelerated population. However, their observations do not show a clear horseshoe distribution as is typically associated with mono-energetic electron acceleration, but rather higher energies at pitch angles close to 90. Qualitatively similar distributions arise from simulations of inertial Alfvén waves (Watt et al. 2006). From detailed 2D analysis of the all-sky images of beads, Kalmoni et al. (2018) determined an observational dispersion relation from a set of auroral beads which had a remarkable match to that of waves on the kinetic branch of shear Alfvén waves with short perpendicular scales, as modeled by Watt et al. (2006), strongly suggesting a wave driven electron acceleration mechanism.

5.3 Outstanding Issues

While much progress has been made in understanding auroral beads, there is still no consensus as to which instability or instabilities are the drivers of these meso-scale auroral forms, although the various ballooning instabilities seem to be favoured. Largely, this has arisen due to qualitative rather than quantitative comparisons between observations and theory or simulations or comparisons based on different assumptions of the conditions in the source region. Given that the various instability theories give non-unique wave scale sizes, growth

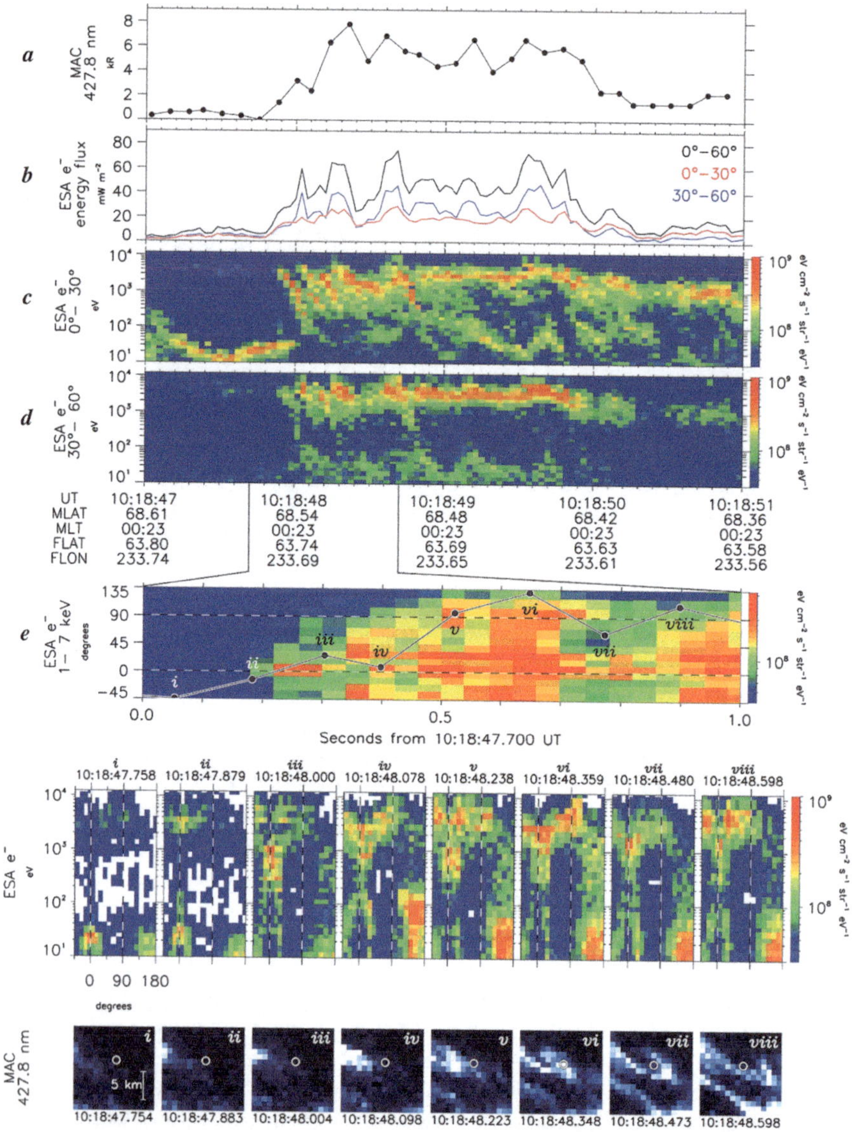

Fig. 17 Detailed comparisons between auroral and electron flux measurements on the same field line. (**a**) Temporal variation of the MAC 427.8 nm intensity at the Reimei footprint (see Sect. 2.2) for 10:18:47–10:18:51 UT. (**b**) Electron energy fluxes at pitch angles of 0–30° (red), 30–60° (blue), and 0–60° (black), respectively. (**c**) Electron energy spectrogram at 0–30°. (**d**) Electron energy spectrogram at 30–60°. (**e**) Electron pitch angle spectrogram averaged over energies of 1–7 keV for 1 s from 10:18:47.700 UT, together with the MAC 427.8 nm intensity variation. (**f**) Electron energy-pitch angle distribution at the nearest time to each of i to viii. (**g**) Zoomed-in MAC 427.8 nm images. Reproduced from Motoba and Hirahara (2016)

rates and propagation speeds for reasonable plasma conditions in the source region, it is unlikely that ground-based observations will be able to unequivocally identify the source mechanism. However, coordinated auroral and in-situ observations of the instability region will be challenging, given its predicted small size and the necessity to measure, for example, plasma gradients in three dimensions on short time-scales.

Auroral beads are uniquely discussed in terms of their association with substorm onset. Evidence suggests that most, if not all substorms, are preceded by auroral beads (Nishimura et al. 2016; Kalmoni et al. 2017). However, substorms are relatively easily identified in auroral data and can easily be searched for using automated routines (e.g. Forsyth et al. 2015). Conversely, auroral beads are relatively difficult to identify in such an automated manner and, as such, whether or not they appear at other times has largely been overlooked. Their appearance (or lack of appearance) at times not leading into substorm activity, and the plasma conditions at those times may provide useful insights into the conditions and physical processes that lead to large-scale dynamic changes in the magnetosphere.

6 Giant Undulations

6.1 Definition

Although the equatorward edge of the diffuse aurora is typically quite smooth, at times large-scale undulations can develop. These were first discovered by Lui et al. (1982) using Defense Meteorological Satellite Program (DMSP) imagery. They predominantly occur in the afternoon to evening sector and have amplitudes of 40–400 km and wavelengths of ∼200–900 km. When the amplitude of the undulation is comparable to the wavelength, they are classified as “Giant Undulations” (GUs). Some examples of GUs acquired with the DMSP/OLS imager are shown in Fig. 18 (from Henderson et al. 2010). Observations of GUs the TIMED/GUVI spacecraft-borne imager mostly occurred during geomagnetic storm times (Zhang et al. 2005), but observations from ground imagers across a solar cycle showed that more than half of the observed GU events occurred outside of storm times (Baishev et al. 2012).

GUs are quite different from omega bands. As discussed above, omega bands are a pre-midnight to morning sector phenomenon but, more importantly, the tongues of luminosity in omega bands point poleward, i.e. they are extensions into the auroral oval, while for GUs the tongues of luminosity point equatorward. In addition, omega bands are often quite irregular and occur near the poleward edge of the diffuse aurora while GUs are much more wave-like and appear on the equatorward edge of the diffuse aurora. To further illustrate the differences, Fig. 19 shows a composite image of DMSP/OLS and POLAR/VIS Earth camera data from a storm that occurred on Nov 9, 1998 (both images were taken at similar times.) The DMSP image shows a series of very well-defined GUs on the equatorward edge of the dusk-side diffuse aurora. Farther to the east, prominent auroral streamers can be seen which occurred copiously throughout this storm in association with torch and omega band formation (note that this is the same storm event that is shown in Fig. 12 above.)

Although DMSP photographs were incapable of detecting motion associated with the GUs, using successive orbital passes of the DMSP spacecraft Lui et al. (1982) were able to determine that GUs can persist for time periods of 0.5–3 hours, although subsequent statistical studies have shown typically shorter durations (Baishev et al. 2012). While early studies of undulations by Provadakes et al. (1989), Mendillo et al. (1989), Murphree and Johnson (1996) were not able to detect azimuthal motion of the forms, subsequent studies have consistently shown that the undulations propagate sunward on the duskside. Nishitani et al. (1994) reported a duskward phase speed of 540–650 m/s combined with a strong latitudinal velocity shear during a 22 min period of observation and Baishev et al. (2000) reported a

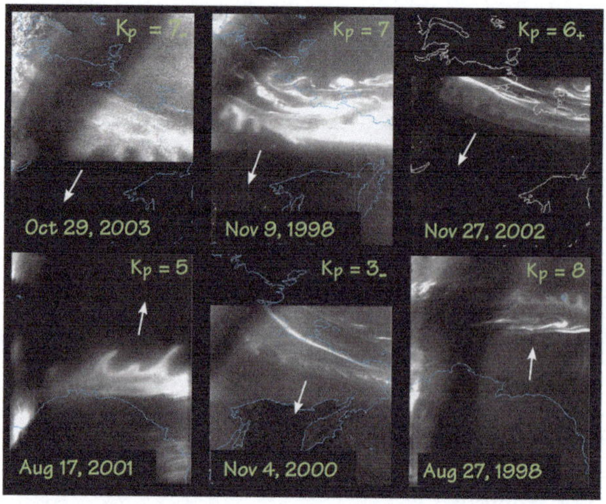

Fig. 18 An example of giant undulations of the equatorward edge of the diffuse aurora as seen in DMSP/OLS imagery. The value of the Kp index at the time of the undulations is shown. The arrows indicate the direction of the equator. Typically, giant undulations occur during high Kp intervals that are associated with high convection intervals. Adapted from Henderson et al. (2010)

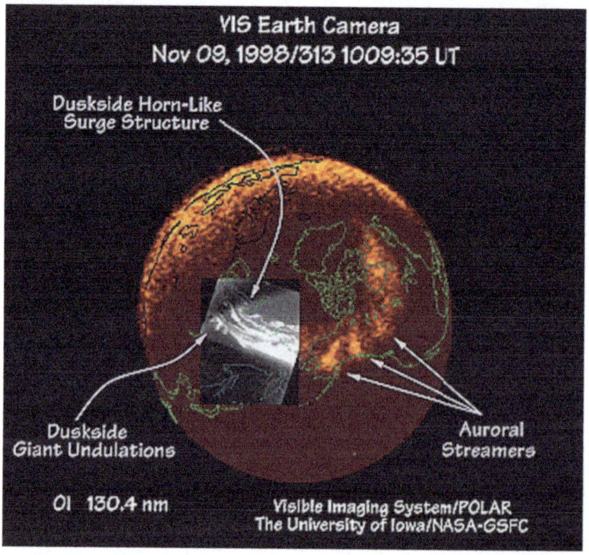

Fig. 19 A composite of a DMSP OLS image on top of a POLAR/VIS Earth Camera image showing where giant undulations tend to occur in relation to the large-scale auroral distribution. Omega bands are also typically seen to form in the region where the auroral streamers form. From Henderson et al. (2010)

duskward phase speed of ~700 m/s for undulations that developed following a substorm and persisted for more than an hour. Henderson et al. (2010) also reported a sunward phase speed of ~600 m/s also in association with strong SAPS observed by multiple DMSP crossings. Pc5 magnetic pulsations are also observed on the ground in association with the passage of giant undulations (Baishev et al. 2000; Henderson et al. 2010).

6.2 Drivers/Causes

Originally, Lui et al. (1982) suggested that undulations in the equatorward edge of the diffuse aurora may be due the Kelvin-Helmholtz Instability at the plasmapause. On the dusk-side, a velocity shear is expected between the corotating plasmasphere (tailward on the dusk-side) and the sunward plasma sheet flow. However, this hypothesis was called into question by

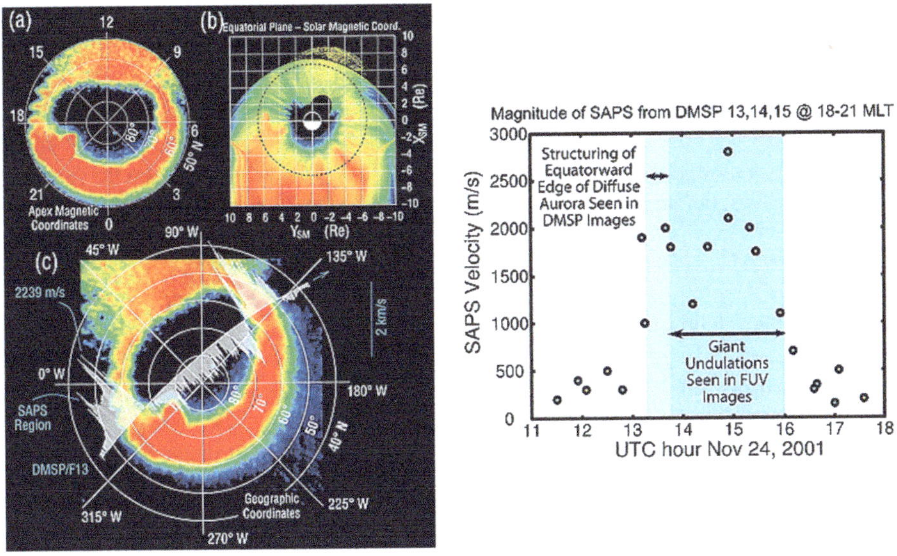

Fig. 20 Left panel: (**a**) Giants undulations seen in the IMAGE/FUV imagery (equatorward edge between 15–19 MLT). (**b**) Mapping to equatorial plane. (**c**) DMSP/F13 overflight shows GU region is associated with enhanced SAPS. Right panel: GUs occur when SAPS are high. From Henderson et al. (2010)

Kelley (1986) who showed, that the nominal shear at the plasmapause is much too small for such an instability to develop, based on the "shear flow ballooning instability" analysis of Viñas and Madden (1986). Instead, Kelley (1986) proposed that a much larger shear flow on the duskside associated with "subauroral ion drift" events (SAIDs) or "polarization jets" (PJs) was responsible. In more recent literature, these are referred to as SAPS (Foster and Burke 2002). Yamamoto et al. (1993, 1994) used to 2D particle simulation to suggest that a polarized arc sheet, arising from a quasi-statically driven auroral arc just equatorward of the diffuse proton aurora, would be unstable to the Kelvin-Helmholtz instability and could account for the observations of GUs. Using global auroral imaging results from POLAR and IMAGE, Henderson et al. (2010) studied the development of a GU event from start to finish and were able to confirm that the development of GUs is associated with strong SAPS (see Figs. 19 and 20).

Henderson et al. (2018) also showed an unusual breakup-like event that appears to have grown explosively in the dusk-to-midnight sector near a region of growing undulations (though not giant undulations). Following the onset, the equatorward edge of the diffuse aurora in this sector was violently disrupted in an unusual manner. They "speculate that the event may be due to an explosive disruption of the plasmapause—possibly due to an instability like the shear flow-ballooning instability driven by large shear flows between the plasmasheet and plasmasphere resulting from enhanced SAPS flows. These types of events might only occur during storms and may be responsible for structuring the plasmapause, feeding long-lived drainage plumes and possibly feeding into Steve-like subsuroral emissions."

Recent observations combining THEMIS ASI data and in-situ observations from the Van Allen probes suggest that auroral undulations are a manifestation of ULF waves in the plasmasphere (Motoba et al. 2015). To date, these observations appear to be the only reported observations of the in-situ plasma close to the source of auroral undulations. The ground observations were used to infer that the ULF waves associated with the undulations had an azimuthal wave number of $m \sim -103$, while the in-situ observations of the electric and magnetic fields revealed that the waves were a second harmonic poloidal wave existing just outside the plasmasphere. Motoba et al. (2015) reported, but did not show, an ionospheric flow of 300–450 m/s just equatorward of the undulations, hinting at a possible SAPS channel. The propagation of the undulations was 3–4 km/s, significantly faster than the reported ionospheric flows, but this may be as a result of the SuperDARN flows being determined by E-region echoes (as discussed above). However, although the ULF waves may provide the mechanism by which electrons with energies of 10s of keV are scattered into the ionosphere, particularly if they have a compressional component (Rae et al. 2018), the source of these waves and their relationship to SAPS was not resolved by this study.

6.3 Implications and Outstanding Questions

Giant undulations have, thus far, received relatively little attention in the literature. This may, in part, be to their relatively sparse occurrence (54 events in solar cycle 23 – Baishev et al. 2012) or due to the lack of corroborating in-situ observations in the magnetospheric source region. However, they are a clear, long-lived indicator of large-scale processes occurring in the inner magnetosphere, far from the plasmasheet dynamics that drive many other meso-scale auroral forms. As with other spatially period auroral forms, plasma instabilities in the source region are key to the growth of GUs. Both the shear-flow ballooning instability and Kelvin-Helmholtz instabilities remain potential candidates, but observations of a shear flow at near-ionospheric altitudes are insufficient to rule out either. Further robust modelling of this region, with a view to isolating the dominant instability, is required to provide indications of the required conditions to be measured by spacecraft and whether these plasma conditions can be measured by existing instrumentation. Such modelling must not only generate the undulation itself but, unlike auroral beads for which the instability can grow until the system breaks down, enable the large-scale waves to exist for long periods of time.

In-situ observations of the inner magnetosphere, in particular the conditions at and around the plasmapause, may provide further insights into the formation of GUs. The non-linear growth of instability responsible for GUs may lead to a layer of intermixed plasma near the plasmapause, which in turn may help to explain some observation of some puzzling subauroral structuring. Sustained activity of this sort may also feed long-lived drainage plumes observed by Borovsky et al. (2014). Finally, a firm grasp on the driving mechanism combined with observations of undulations may provide a new capability to remotely sense strong SAPS flows.

7 Magnetic Mapping

Whereas the short-time and small-scale auroral processes can significantly be controlled and/or modified by the ionospheric processes, it is common opinion that the magnetosphere is mainly responsible for generating and shaping the mesoscale auroral structures and large-scale auroral distributions. However, as discussed above, directly linking auroral features to plasma structures and dynamics in the magnetosphere is non-trivial, especially during active and dynamic intervals.

Directly mapping the magnetospheric source to bright aurorae by examining the particle distribution functions is challenging because the loss-cone size in the high-beta (source) region is so small as to be unresolved with current instrumentation. Similarly, the effect of magnetic focusing means that ion beams generated by auroral acceleration processes in the strong-B region above the ionosphere become increasingly narrow and similarly unresolvable by the time they reach the weakest magnetic field strength regions in the magnetotail. Thus, in order to learn about the magnetospheric roots of particular auroral forms we have to map these forms into the equatorial magnetosphere along the magnetic field lines using some magnetospheric model.

It is commonly accepted that most of auroral oval is mapped to the magnetotail plasma sheet, a large hot plasma reservoir captured on closed magnetic field lines. Most of its volume is occupied by the region of weak magnetic field in the plasma sheet center (say, where Bz is below 20–30 nT, a value comparable to the lobe magnetic field) where the plasma-β parameter is large. As magnetic field here is significantly modified by local electric currents generated by the hot plasma itself, and plasma motions are not fully dictated by the applied boundary conditions, this region plays an active role in the tail dynamics. This particularly manifests on the global scale as the energy storage/explosive dissipation phases in the magnetotail constituting the substorms, as well as a rich inhomogeneous structure of the active plasma sheet on medium-to-small scales, which exists at any activity level but highly enhanced during substorms.

One of most available mapping tools is still the mapping along the magnetic field lines using the magnetospheric magnetic field models. Currently three kinds of such models are available: statistical empirical models, adaptive models and global MHD models, each with advantages and drawbacks. In general, statistical models are relatively quick to run, but provide the average conditions and thus struggle with more dynamic times; adaptive models build on statistical models by providing further boundary conditions to be met, but are limited by the number of observations available; global MHD models can be slower to and more complex to run and, in the past, have not been accurate enough to represent magnetospheric magnetic field and its variations.

The most well-known and widely used models are statistical empirical magnetospheric models, constructed based on a huge historical archive of spacecraft observations. These models typically attempt to deconstruct the observed magnetic field into a series of defined current systems, the strength of which is then related to solar wind drivers and geomagnetic activity measures (see review by Tsyganenko 2013), although more recent models use empirical basis functions with no imposed current system configuration (e.g. Andreeva and Tsyganenko 2016; Tsyganenko and Andreeva 2016). Two major problems of these models for mapping purposes are known. Firstly, since these models describe average configurations, they fail to reproduce large dynamical changes or localised changes. For example, both the highly-stretched magnetotail and locally (and globally) dipolarised magnetotail, which develop during substorm growth and expansion phases significantly deviate from the average model. Statistically the average mapping errors from the nightside magnetosphere near GEO orbit to the ionosphere are estimated to be about 1–2° (Shevchenko et al. 2010). Another problem is that, because of their fitting method (minimizing the RMS deviation of modelled and observed B over the entire magnetosphere), these models do not particularly care about Bz values in the regions where Bz is small. Therefore, the details of Bz distribution and mapping are especially questionable in these models in the midtail region.

Adaptive modelling, which is a kind of data assimilation, combines in-situ magnetic field measurements with statistical models to optimize their input parameters. With good spatial coverage, this makes possible to reconstruct the concurrent configuration and follow its time-dependent behaviour in that region (Sitnov et al. 2008; Kubyshkina et al. 2011; Stephens et al. 2019) and generally improves the results compared to unoptimized versions of the models (Brito and Morley 2017). Quite expectedly, using more satellites in the optimization procedure yields more accurate results, thus the effectiveness of adaptive models

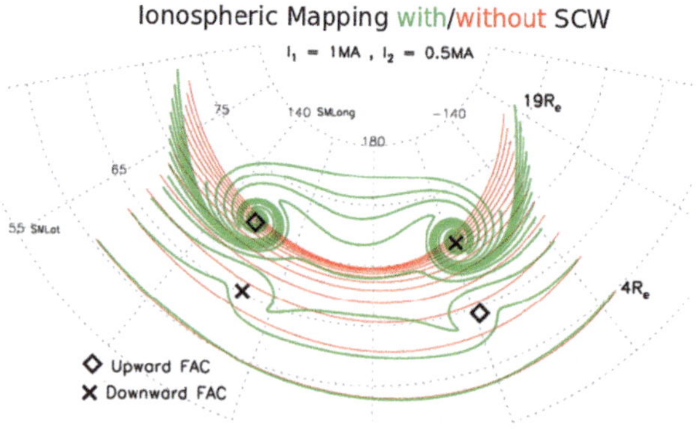

Fig. 21 Ionospheric mapping of equidistant equatorial contours (distances change by 1 R_E step between 4 R_E and 19 R_E, equatorial distances are labeled for selected circles). The red lines show neutral sheet mapping using the T89+IGRF model; the green lines show mapping of the same points using the SCW2+T89+IGRF model. Ionospheric locations of upward (downward) FACs are indicated by diamond (cross) symbols correspondingly (after Nikolaev et al. 2015)

is tied to the available observations. However, these models still rely on the underlying current systems prescribed within the model. While some prescriptions of these are fairly rigid (e.g. Kubyshkina et al. 2011), others expand the magnetic perturbations in terms of basis functions (e.g. Stephens et al. 2019). In either case, these models can only account for current systems that are included and may struggle to resolve small-scale current systems adequately.

Previously the global MHD models, which solve the physics equations under realistic time-varying solar wind conditions, were not accurate enough to represent magnetospheric magnetic field and its variations. However, recent benchmarking of all kinds of models indicate that predictions of current GMHD models (incorporating the drift physics in the inner magnetosphere) may approach and even outperform the predictions of best statistical models during disturbed conditions (Kubyshkina et al. 2019), so their further development may provide us a useful tool for future research.

The effects of large-scale field-aligned currents may be under-represented in the statistical magnetospheric models. Structured FACs of different scale (absent from these models) are formed in the plasma sheet and linked to the aurora during active conditions (e.g. Murphy et al. 2013; Forsyth et al. 2014; Kepko et al. 2015). FACs strongly affect magnetic mapping by changing both the shapes of field-lines and the latitude/longitude/distance range of projections, but so far this aspect is poorly studied. Nikolaev et al. (2015) numerically investigated the problem by applying the recently-developed two-loop SCW model whose parameters (intensity and longitudes of SCW currents) can be evaluated from mid-latitude magnetic variations. As illustrated by ionospheric mapping curves in Fig. 21, when the SCW is absent (red lines) the field line mapping of equidistant lines from the equatorial current sheet (tailward of 10 R_E) provides the smooth lines in the ionosphere resembling the auroral oval. The addition of a 1 MA (Region 1 current) SCW causes two major effects illustrated by green lines: (a) ionospheric projections leap poleward in the dipolarized sector of the SCW resembling the poleward expansion of auroras by as much as 8° latitude; (b) the overall deformation displays the pattern resembling an auroral bulge. The counter clockwise rotation pattern resembling the westward travelling surge is formed in the region around a major upward SCW current, it has a ~1 hr MLT azimuthal scale for the cited parameters. Similar deformations have been obtained in the MHD simulation of reconnection-related

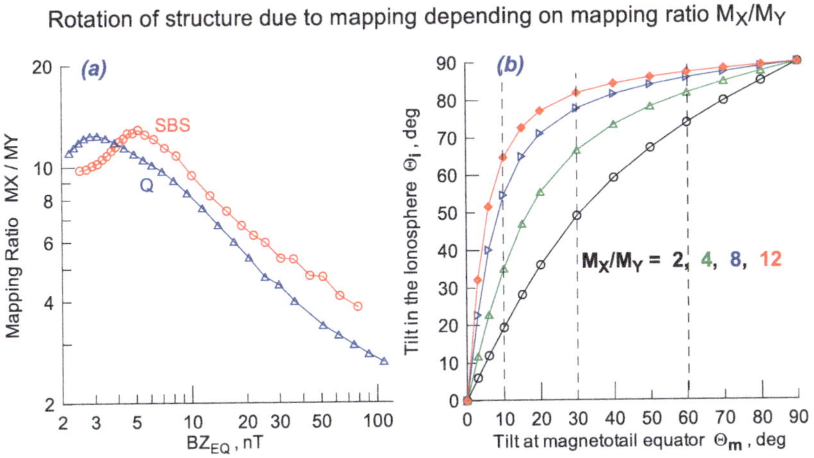

Fig. 22 Mapping ratio and rotation of the line segment during the mapping between equatorial magnetosphere and ionosphere (updated after Sergeev 2002). (**a**) Examples showing the Mx/My dependence on the equatorial BZ values along the tail axis obtained for samples of quiet and substorm conditions using recent Tsyganenko and Andreeva (2015) model. (**b**) Relationship between tilts of magnetospheric line segment and its ionospheric projection depending on Mx/My

dipolarization (Birn and Hesse 2013) and have been shown to significantly modify spacecraft footpoints such that they track variations in the aurora (Chu et al. 2014). Irrespective of the auroral acceleration mechanism, these modeling results suggest that auroral bulge-like shaped auroras (with poleward expansion and, not necessarily moving, WTS) will be seen as long as the 3-d configuration of SCW currents is persistent in the magnetotail. The expected close relationship between SCW current and poleward expansion magnitude still waits a quantitative statistical confirmation.

Another deformation, unrelated to field-aligned currents, is that that mapping factors M which describe the change of the magnetic flux tube size between the magnetosphere (index m) and ionosphere (i) are different for different directions across the flux tube (e.g., Kauffman et al. 1990). Taking Mx=Xm/Xi and My=Ym/Yi, where X and Y stands for the meridional and azimuthal directions across B, correspondingly, the ratio Mx/My is 2 for the dipole field, but significantly increases in the nightside current sheet region where B_Z is small, as shown on examples in Fig. 22a. One consequence is that a circular flux tube section in the tail current sheet is transformed into E-W aligned form in the ionosphere. Another consequence is a rotation of the line segment towards E-W alignment, which can be very significant if mapping is performed from the region having a small B_Z in the tail and thus a high Mx/My (Fig. 22b). For small Mx/My (close to a dipole field) the rotation is not large, but for large Mx/My (highly stretched fields) a nearly radial segment is rotated towards an azimuthal orientation in the ionosphere (say, by as much as 50° in the ionosphere for a 10° rotation in the magnetosphere if Mx/My$\sim$10). This could account for the dominant zonal orientation of auroral forms. Previously it was used to explain a frequent E-W alignment of PBIs/auroral streamers connected to the plasma sheet fast flow channels (Sergeev 2002; Zesta et al. 2006, see Fig. 23). It may also be invoked to understand the preferential oval-aligned orientation of the auroral arcs A critical magnetospheric quantity which controls the amount of rotation during the mapping is the magnitude of equatorial BZ in the plasma sheet (Sergeev 2002).

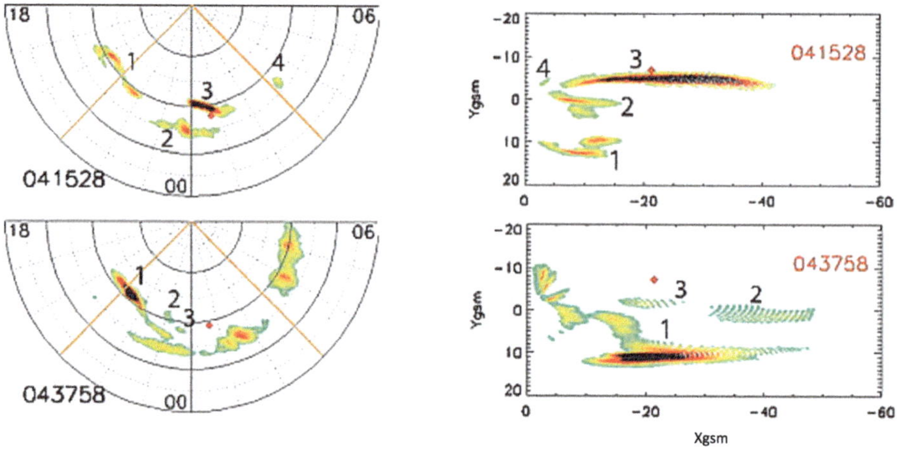

Fig. 23 IMAGE FUV data showing large-scale auroral structures, including PBIs, in magnetic coordinates and the projections of these auroral features along magnetic field lines to the center of the plasma sheet (right) (adapted from Zesta et al. 2006). The projected images show how apparently east-west aligned arcs may be produced from plasma structures aligned along the magnetotail

While magnetic modelling enables each point in the magnetotail to be traced to a unique point in each ionosphere, their limitations mean that they cannot be relied upon, particularly during active times. However, one can also apply different kinds of morphological mapping using tracers, which originate in specific places in the magnetotail but are observed in the ionosphere (see Paschmann et al. 2003). From them, the isotropy boundaries (IBs, the equatorward boundaries of extended isotropic loss-cone precipitation of energetic particles) are a most useful and well-known tool, providing useful low-altitude markers of equatorial B_Z values (where B_Z is roughly in the range 30 to 50 nT for 30 keV proton IB, and roughly ~5 nT for 30 keV electron IB). This method is useful in establishing the location of magnetic flux tubes containing specific auroras with respect to the high-beta plasma sheet region but requires in-situ observations of the low-altitude plasma in order to do so. Distinction from wave-induced precipitation mechanisms is often possible based on the energy dependence of isotropic loss-cone precipitation. A direct comparison between whistler-mode chorus activity in the magnetosphere and auroral brightness enabled Nishimura et al. (2010c) to exactly map the location of the THEMIS-A spacecraft when it was close to geosynchronous orbit, however such a mechanism requires that the source and resulting aurora can be readily identified. More recently, studies have shown that the utility of ULF wave Eigenfrequencies in magnetic mapping by comparing the Eigenfrequencies observed in-situ with the latitudinal Eigenfrequency profile determined by ground magnetometers (Rae et al. 2014) to determine the relative locations of mapped spacecraft footpoints to auroral features. Although the spatial resolution of the latitudinal Eigenfrequency profile is quite sparse, and the technique requires the identification of Eigenfrequencies on the ground and in space, the distribution of ground-based magnetometers in and around the auroral zone reduces the need for a low-altitude spacecraft overpass during times of interest.

8 Summary and Future Advancements

Since the turn of the century, there has been a major growth in both ground-based and magnetospheric observations, with numerous multi-spacecraft missions and multi-observatory

Table 1 Summary of mesoscale auroral forms, their size, where they are observed, their potential magnetotail drivers and locations of these drivers

Auroral form	Scale size	Auroral location	Potential driver	Driver location
Poleward boundary intensification	10s–100s km	Nightside, poleward edge of auroral oval	Fast flows arrivals from polar cap	Magnetotail lobes/deep tail plasma-sheet;
			Spontaneous or triggered in current sheet near or at OCB	
			Ionospheric plasma flows across conductivity gradient	
Auroral streamer	~10s km s across, stretching across the auroral oval	Nightside auroral oval, stretching from poleward to equatorward edge	Magnetotail fast plasma flow/depleted plasma channel within plasma-sheet	Magnetotail plasma-sheet
			Possible scattering of plasma-sheet particles by whistler-mode chorus	Magnetotail plasma-sheet
			Monoenergetic electron acceleration in upward FAC	Altitudes of a few R_E
Omega bands and torches	~200–500 km wide and stretching across the auroral oval	Dawn side, equatorward portion of expanded auroral oval	Kelvin-Helmholtz instability	Near-Earth magnetotail, ~7–12 R_E
			Hybrid Kelvin-Helmholtz/Rayleigh-Taylor instability	
			May results from injections associated with streamers	
Auroral beads	1–500 km, typically observed at ~100 km scales	Nightside, within auroral oval normally on pre-existing quiet arc	Ballooning instabilities	Near-Earth magnetotail, 10–12 R_E
			Cross-tail current instabilities	
Giant undulations	200–900 km	Equatorward edge of dusk auroral oval	Kelvin Helmholtz instability	Inner magnetosphere/ plasmapause
			Shear-flow ballooning instability	
			Magnetospheric source of SAPS	
Auroral bulge (for context but not covered in this review)	Up to 10° latitude and several hours of MLT	Activated part of nightside auroral oval	Dipolarized part of magnetic field configuration associated with intense reconnection and SCW	Tail plasma sheet earthward of near-Earth neutral line

programmes providing observations across multiple solar cycles. These datasets have already provided insights into the source and drivers of a range of meso-scale auroral forms, as laid out in this review and summarised in Table 1. However, multi-spacecraft missions, such as Cluster and THEMIS, have largely observed processes beyond 10 R_E, corresponding to the source region of auroral streamers, while low-altitude missions, such as FAST and Swarm, provide snapshots of plasma conditions just above the ionosphere but not always with auroral observations for context. The implication of the results reported in this review is that the transition region between stretched tail plasma sheet region and dipole-like the inner magnetosphere, ranging from the plasmapause out to $\sim$12 R_E is now a key region to observe in order to understand the mechanisms that drive many dynamic meso-scale auroral forms.

Many of the meso-scale auroral forms discussed here have been linked with a variety of plasma instabilities in the magnetotail: PBIs and streamers are linked to reconnection and its outflows; auroral beads are linked to ballooning or current instabilities; omega bands, torches and giant undulations are linked to Kelvin-Helmholtz-based or shear flow ballooning instabilities. In order to understand the underlying physics, it is imperative that we understand the scales in the magnetosphere on which these instabilities act in order to provide measurements on the necessary time and spatial scales. Furthermore, understanding the growth rates of the auroral signatures of these instabilities may help to constrain the plasma conditions required for their formation. Quantitative comparisons of the growth rates, scale sizes and plasma conditions will then reveal the underlying physical processes. Some of these necessary observations may already exist within the repositories of current and past missions, while as others may need new missions that can provide observations across multiple scales in the inner magnetosphere in conjunction with auroral and auroral plasma observations from low-Earth orbiting spacecraft. Ultimately, a global magnetospheric view will be needed to complement a global aurora view and make direct observations of plasma motions and instabilities.

Acknowledgements The authors thank the International Space Science Institute (ISSI) for supporting a workshop on Auroral Physics and the workshop convenors (D.J. Knudsen, J. Borovsky, T. Karlsson, R. Kataoka, N. Partamies, R. von Steiger) and participants for useful discussions around this topic. CF was supported by a NERC Independent Research Fellowship NE/N014480/1 and STFC Consolidated Grant ST/S000240/1. Y.N. was supported by NASA grant NNX17AL22G and 80NSSC18K0657, NSF grants AGS-1907698, AGS-1737823, and AFOSR FA9559-16-1-0364.

References

A.T. Aikio, H. Vanhamäki, A.B. Workayehu, I.I. Virtanen, K. Kauristie, L. Juusola et al., Swarm satellite and EISCAT radar observations of a plasma flow channel in the auroral oval near magnetic midnight. J. Geophys. Res. Space Phys. **123**, 5140–5158 (2018)

S.I. Akasofu, A study of auroral displays photographed from the DMSP-2 satellite and from the Alaska meridian chain of stations. Space Sci. Rev. **16**, 617 (1974). https://doi.org/10.1007/BF00182598

S.I. Akasofu, Recent progress in studies of DMSP auroral photographs. Space Sci. Rev. **19**, 169 (1976). https://doi.org/10.1007/BF00215692

S.-I. Akasofu, D.S. Kimball, The dynamics of the aurora, 1. Instabilities of the aurora. J. Atmos. Terr. Phys. **26**, 205 (1964)

S.-I. Akasofu, E.W. Hones Jr., M.D. Montgomery, S.J. Bame, S. Singer, Association of magnetotail phenomena with visible auroral features. J. Geophys. Res. **76**, 5985–6003 (1971)

O. Amm, K. Kauristie, Ionospheric signatures of bursty bulk flows. Surv. Geophys. **23**, 1–32 (2002)

O. Amm, A. Pajunpää, U. Brandström, Spatial distribution of conductances and currents associated with a north-south auroral form during a multiple-substorm period. Ann. Geophys. **17**, 1385–1396 (1999). https://doi.org/10.1007/s00585-999-1385-6

O. Amm, R. Nakamura, T. Takada, K. Kauristie, H.U. Frey, C.J. Owen, A. Aikio, R. Kuula, Observations of an auroral streamer in a double oval configuration. Ann. Geophys. **29**, 701–716 (2011). https://doi.org/10.5194/angeo-29-701-2011

V.A. Andreeva, N.A. Tsyganenko, Reconstructing the magnetosphere from data using radial basis functions. J. Geophys. Res. Space Phys. **121**, 2249–2263 (2016). https://doi.org/10.1002/2015JA022242

V. Angelopoulos, W. Baumjohann, C.F. Kennel, F.V. Coroniti, M.G. Kivelson, R. Pellat, R.J. Walker, H. Lühr, G. Paschmann, Bursty bulk flows in the inner central plasma sheet. J. Geophys. Res. **97**(A4), 4027–4039 (1992). https://doi.org/10.1029/91JA02701

W.E. Archer, D.J. Knudsen, J.K. Burchill, B. Jackel, E. Donovan, M. Connors, L. Juusola, Birkeland current boundary flows. J. Geophys. Res. Space Phys. **122**, 4617–4627 (2017). https://doi.org/10.1002/2016JA023789

D.G. Baishev, E.S. Barkova, S.I. Solovyev, K. Yumoto, M.J. Engebretson, A.V. Koustov, Formation of large-scale giant undulations at the equatorial boundary of diffuse aurora and Pc5 magnetic pulsations during the January 14, 1999 magnetic storm, in *Proceedings of the Fifth International Conference on Substorms*, ed. by A. Wilson. Eur. Space Agency Spec. Publ., vol. ESA SP-443 (2000), pp. 427–430

D.G. Baishev, E.S. Barkova, K. Yumoto, Optical observations of large-scale undulations in the 23rd cycle of solar activity. Geomagn. Aeron. **52**(2), 197–203 (2012). https://doi.org/10.1134/s0016793212020028

J. Birn, M. Hesse, The substorm current wedge in MHD simulations. J. Geophys. Res. Space Phys. **118**, 3364–3376 (2013). https://doi.org/10.1002/jgra.50187

J. Birn, J. Raeder, Y.L. Wang, R.A. Wolf, M. Hesse, On the propagation of bubbles in the geomagnetic tail. Ann. Geophys. **22**, 1773–1786 (2004). https://doi.org/10.5194/angeo-22-1773-2004

J. Birn, V.G. Merkin, M.I. Sitnov, A. Otto, MHD stability of magnetotail configurations with a Bz hump. J. Geophys. Res. Space Phys. **123**, 3477–3492 (2018). https://doi.org/10.1029/2018JA025290

J.E. Borovsky, D.T. Welling, M.F. Thomsen, M.H. Denton, Long-lived plasmaspheric drainage plumes: where does the plasma come from? J. Geophys. Res. Space Phys. **119**, 6496–6520 (2014). https://doi.org/10.1002/2014JA020228

G. Branduardi-Raymont, C. Wang, A novel and global way to explore solar-terrestrial relationships: SMILE. Cospar Inf. Bull. **200**, 32–39 (2017)

G. Branduardi-Raymont, S.F. Sembay, J.P. Eastwood et al., Exp. Astron. **33**, 403 (2012). https://doi.org/10.1007/s10686-011-9239-0

W.A. Bristow, D.L. Hampton, A. Otto, High-spatial-resolution velocity measurements derived using Local Divergence-Free Fitting of SuperDARN observations. J. Geophys. Res. Space Phys. **121**(2), 1349–1361 (2016)

T.V. Brito, S.K. Morley, Improving empirical magnetic field models by fitting to in situ data using an optimized parameter approach. Space Weather **15**, 1628–1648 (2017). https://doi.org/10.1002/2017SW001702

S. Buchert, G. Haerendel, W. Baumjohann, A model for the electric fields and currents during a strong Ps 6 pulsation event. J. Geophys. Res. **95**(A4), 3733–3743 (1990). https://doi.org/10.1029/JA095iA04p03733

N.A. Case, A. Grocott, S.E. Milan, T. Nagai, J.P. Reistad, An analysis of magnetic reconnection events and their associated auroral enhancements. J. Geophys. Res. Space Phys. **122**, 2922–2935 (2017). https://doi.org/10.1002/2016JA023586

C.X. Chen, R.A. Wolf, Interpretation of high-speed flows in the plasma sheet. J. Geophys. Res. **98**(A12), 21409–21419 (1993). https://doi.org/10.1029/93JA02080

C.Z. Cheng, A.T.Y. Lui, Kinetic ballooning instability for substorm onset and current disruption observed by AMPTE/CCE. Geophys. Res. Lett. **25**, 4091–4094 (1998). https://doi.org/10.1029/1998GL900093

C.Z. Cheng, S. Zaharia, MHD ballooning instability in the plasma sheet. Geophys. Res. Lett. **31**, L06809 (2004). https://doi.org/10.1029/2003GL018823

X. Chu et al., Development and validation of inversion technique for substorm current wedge using ground magnetic field data. J. Geophys. Res. Space Phys. **119**, 1909–1924 (2014). https://doi.org/10.1002/2013JA019185

M. Connors, G. Rostoker, Source mechanisms for morning auroral features. Geophys. Res. Lett. **20**(15), 1535–1538 (1993). https://doi.org/10.1029/93gl01594

O. de la Beaujardière, L.R. Lyons, J.M. Ruohoniemi, E. Friis-Christensen, C. Danielsen, F.J. Rich, P.T. and-Newell, Quiet-time intensifications along the poleward auroral boundary near midnight. J. Geophys. Res. **99**(A1), 287–298 (1994). https://doi.org/10.1029/93JA01947

E.F. Donovan, S. Mende, B. Jackel, H. Frey, M. Syrjäsuo, I. Voronkov, T. Trondsen, L. Peticolas, V. Angelopoulos, S. Harris, M. Greffen, M. Connors, The THEMIS all-sky imaging array—system design and initial results from the prototype imager. J. Atmos. Terr. Phys. **68**, 1472–1487 (2006)

R.D. Elphinstone, D.J. Hearn, L.L. Cogger, J.S. Murphree, H. Singer, V. Sergeev, K. Mursula, D.M. Klumpar, G.D. Reeves, M. Johnson, S. Ohtani, T.A. Potemra, I. Sandahl, E. Nielsen, M. Persson, H. Opgenoorth, P.T. Newell, Y.I. Feldstein, Observations in the vicinity of substorm onset: implications for the substorm process. J. Geophys. Res. **100**, 7937–7969 (1995). https://doi.org/10.1029/94JA02938

R.D. Elphinstone, J.S. Murphree, L.L. Cogger, What is a global auroral substorm? Rev. Geophys. **34**, 169–232 (1996)

D.H. Fairfield et al., Earthward flow bursts in the inner magnetotail and their relation to auroral brightening, AKR intensifications, geosynchronous particle injections and magnetic activity. J. Geophys. Res. **104**, 355 (1999)

C. Forsyth, M. Lester, S.W.H. Cowley, I. Dandouras, A.N. Fazakerley, R.C. Fear, H.U. Frey, A. Grocott, A. Kadokura, E. Lucek, H. Rème, S.E. Milan, J. Watermann, Observed tail current systems associated with bursty bulk flows and auroral streamers during a period of multiple substorms. Ann. Geophys. **26**, 167–184 (2008). https://doi.org/10.5194/angeo-26-167-2008

C. Forsyth, A.N. Fazakerley, I.J. Rae, C.E.J. Watt, K. Murphy, J.A. Wild, T. Karlsson, R. Mutel, C.J. Owen, R. Ergun, A. Masson, M. Berthomier, E. Donovan, H.U. Frey, J. Matzka, C. Stolle, Y. Zhang, In situ spatiotemporal measurements of the detailed azimuthal substructure of the substorm current wedge. J. Geophys. Res. Space Phys. **119**, 927–946 (2014). https://doi.org/10.1002/2013JA019302

C. Forsyth, I.J. Rae, J.C. Coxon, M.P. Freeman, C.M. Jackman, J. Gjerloev, A.N. Fazakerley, A new technique for determining Substorm Onsets and Phases from Indices of the Electrojet (SOPHIE). J. Geophys. Res. Space Phys. **120**, 10,592–10,606 (2015). https://doi.org/10.1002/2015JA021343

J.C. Foster, W.J. Burke, SAPS: a new categorization for sub-auroral electric fields. Eos Trans. AGU **83**(36), 393–394 (2002). https://doi.org/10.1029/2002EO000289

E. Friedrich, J.C. Samson, I. Voronkov, Ground-based observations and plasma instabilities in auroral substorms, wave-like formations observed in aurora during substorm onset. Phys. Plasmas **8**(4), 1104–1110 (2001). https://doi.org/10.1063/1.1355678

C. Gabrielse, V. Angelopoulos, A. Runov, D.L. Turner, The effects of transient, localized electric fields on equatorial electron acceleration and transport toward the inner magnetosphere. J. Geophys. Res. **117**, A10213 (2012). https://doi.org/10.1029/2012JA017873

B. Gallardo-Lacourt, Y. Nishimura, L.R. Lyons, J.M. Ruohoniemi, E. Donovan, V. Angelopoulos, K.A. McWilliams, N. Nishitani, Ionospheric flow structures associated with auroral beading at substorm auroral onset. J. Geophys. Res. Space Phys. **119**, 9150–9159 (2014a). https://doi.org/10.1002/2014JA020298

B. Gallardo-Lacourt, Y. Nishimura, L.R. Lyons, S. Zou, V. Angelopoulos, E. Donovan, et al., Coordinated SuperDARN THEMIS ASI observations of mesoscale flow bursts associated with auroral streamers. J. Geophys. Res. Space Phys. **119**, 142–150 (2014b). https://doi.org/10.1002/2013JA019245

B. Gallardo-Lacourt, Y. Nishimura, L.R. Lyons, E.V. Mishin, J.M. Ruohoniemi, E.F. Donovan, V. Angelopoulos, N. Nishitani, Influence of auroral streamers on rapid evolution of ionospheric SAPS flows. J. Geophys. Res. Space Phys. **122**, 12,406–12,420 (2017). https://doi.org/10.1002/2017JA024198

A. Grocott, T.K. Yeoman, R. Nakamura, S.W.H. Cowley, H.U. Frey, H. Rème, B. Klecker, Multi-instrument observations of the ionospheric counterpart of a bursty bulk flow in the near-Earth plasma sheet. Ann. Geophys. **22**, 1–15 (2004)

M.G. Henderson, Implications of Viking Imager Results for Substorm Models. PhD Thesis, Department of Physics and Astronomy, University of Calgary (1994)

M.G. Henderson, Observational evidence for an inside-out substorm onset scenario. Ann. Geophys. **27**, 2129–2140 (2009). https://doi.org/10.5194/angeo-27-2129-2009

M.G. Henderson, Recurrent embedded substorms during the 19 October 1998 GEM storm. J. Geophys. Res. Space Phys. **121**, 7847–7859 (2016). https://doi.org/10.1002/2015JA022014

M.G. Henderson, G.D. Reeves, J.S. Murphree, The activation of the dusk-side and the formation of north-south aligned structures during substorms, in *Proc. Second International Conference on Substorms (ICS-2)*, ed. by J.R. Kan, J.D. Craven, S.-I. Akasofu (1994), p. 37. Geophysical Institure, University of Alaska Fairbanks

M.G. Henderson, G.D. Reeves, J.S. Muphree, Are north-south aligned auroral structures an ionospheric manifestation of bursty bulk flows? Geophys. Res. Lett. **25**, 3737–3740 (1998)

M.G. Henderson, L. Kepko, H.E. Spence, M. Connors, J.B. Sigwarth, L.A. Frank, H.J. Singer, K. Yumoto, The evolution of north-south aligned auroral forms into auroral torch structures: the generation of omega bands and Ps6 pulsations via flow bursts, in *Proc. Sixth Int. Conf. on Substorms*, ed. by R.M. Winglee (University of Washington, Seattle, 2002), pp. 169–174. http://permalink.lanl.gov/object/tr?what=info:lanl-repo/lareport/LA-UR-02–lanl-repo/lareport/LA-UR-3972

M.G. Henderson, E.F. Donovan, J.C. Foster, I.R. Mann, T.J. Immel, S.B. Mende, J.B. Sigwarth, Start-to-end global imaging of a sunward propagating, SAPS-associated giant undulation event. J. Geophys. Res. **115**, A04210 (2010). https://doi.org/10.1029/2009JA014106

M.G. Henderson, Auroral substorms, poleward boundary activations, auroral streamers, omega bands, and onset precursor activity, in *Auroral Phenomenology and Magnetospheric Processes: Earth and Other Planets*, ed. by A. Keiling et al. Geophysical Monograph Series, vol. 197, (American Geophysical Union, Washington, 2012), pp. 39–54

M.G. Henderson, S.K. Morley, L.E. Kepko, SAPS-associated explosive brightening on the duskside: a new type of onset-like disturbance. J. Geophys. Res. Space Phys. **123**, 197–210 (2018). https://doi.org/10.1002/2017JA024472

K. Hosokawa, S.E. Milan, M. Lester, A. Kadokura, N. Sato, G. Bjornsson, Large flow shears around auroral beads at substorm onset. Geophys. Res. Lett. **40**, 4987–4991 (2013). https://doi.org/10.1002/grl.50958

A.J. Hull, M. Wilber, C.C. Chaston, J.W. Bonnell, J.P. McFadden, F.S. Mozer, M. Fillingim, M.L. Goldstein, Time development of field-aligned currents, potential drops, and plasma associated with an auroral poleward boundary intensification. J. Geophys. Res. Space Phys. **115**, A06211 (2010). https://doi.org/10.1029/2009JA014651

K.-J. Hwang, M.L. Goldstein, E. Lee, J.S. Pickett, Cluster observations of multiple dipolarization fronts. J. Geophys. Res. **116**, A00I32 (2011). https://doi.org/10.1029/2010JA015742

A. Ieda, D.H. Fairfield, T. Mukai, Y. Saito, S. Kokubun, K. Liou, C.-I. Meng, G.K. Parks, M.J. Brittnacher, Plasmoid ejection and auroral brightenings. J. Geophys. Res. **106**(A3), 3845–3857 (2001). https://doi.org/10.1029/1999JA000451

A. Ieda et al., Longitudinal association between magnetotail reconnection and auroral breakup based on Geotail and Polar observations. J. Geophys. Res. **113**, A08207 (2008). https://doi.org/10.1029/2008JA013127

S.M. Imber, J.A. Slavin, H.U. Auster, V. Angelopoulos, A THEMIS survey of flux ropes and traveling compression regions: location of the near-Earth reconnection site during solar minimum. J. Geophys. Res. **116**, A02201 (2011). https://doi.org/10.1029/2010JA016026

F. Jiang, M.G. Kivelson, R.J. Strangeway, K.K. Khurana, R. Walker, Ionospheric flow shear associated with the preexisting auroral arc: a statistical study from the FAST spacecraft data. J. Geophys. Res. Space Phys. **120**, 5194–5213 (2015). https://doi.org/10.1002/2013JA019255

A.M. Jorgensen, H.E. Spence, T.J. Hughes, D. McDiarmid, A study of omega bands and Ps6 pulsations on the ground, at low altitude and at geostationary orbit. J. Geophys. Res. **104**(A7), 14705–14715 (1999). https://doi.org/10.1029/1998JA900100

L. Juusola, N. Østgaard, E. Tanskanen, Statistics of plasma sheet convection. J. Geophys. Res. Space Phys. **116**, A08201 (2011). https://doi.org/10.1029/2011JA016479

L. Juusola, M. Kubyshkina, R. Nakamura, T. Pitkänen, O. Amm, K. Kauristie, N. Partamies, H. Rème, K. Snekvik, D. Whiter, Ionospheric signatures of a plasma sheet rebound flow during a substorm onset. J. Geophys. Res. Space Phys. **118**, 350–363 (2013). https://doi.org/10.1029/2012JA018132

N.M.E. Kalmoni, I.J. Rae, C.E.J. Watt, K.R. Murphy, C. Forsyth, C.J. Owen, Statistical characterization of the growth and spatial scales of the substorm onset arc. J. Geophys. Res. Space Phys. **120**, 8503–8516 (2015). https://doi.org/10.1002/2015JA021470

N.M.E. Kalmoni, I.J. Rae, K.R. Murphy, C. Forsyth, C.E.J. Watt, C.J. Owen, Statistical azimuthal structuring of the substorm onset arc: implications for the onset mechanism. Geophys. Res. Lett. (2017). https://doi.org/10.1002/2016gl071826

N.M.E. Kalmoni, I.J. Rae, C.E.J. Watt, K.R. Murphy, M. Samara, R.G. Michell et al., A diagnosis of the plasma waves responsible for the explosive energy release of substorm onset. Nat. Commun. **9**(1) (2018). https://doi.org/10.1038/s41467-018-07086-0

R. Kataoka, Y. Miyoshi, T. Sakanoi, A. Yaegashi, K. Shiokawa, Y. Ebihara, Turbulent microstructures and formation of folds in auroral breakup arc. J. Geophys. Res. **116**, A00K02 (2011). https://doi.org/10.1029/2010JA016334

R.L. Kauffman, D.J. Larson, C. Lu, Mapping and distortions of auroral structures in the quiet magnetosphere. J. Geophys. Res. Space Phys. **95**(A6), 7973–7994 (1990). https://doi.org/10.1029/JA095iA06p07973

K. Kauristie, V.A. Sergeev, O. Amm, M.V. Kubyshkina, J. Jussila, E. Donovan, K. Liou, Bursty bulk flow intrusion to the inner plasma sheet as inferred from auroral observations. J. Geophys. Res. **108**(A1), 1040 (2003). https://doi.org/10.1029/2002JA009371

A. Keiling, K. Shiokawa, V. Uritsky, V. Sergeev, E. Zesta, L. Kepko, N. Ostgaard, Auroral signatures of the dynamic plasma sheet, in *Auroral Phenomenology and Magnetospheric Processes: Earth and Other*

Planets, ed. by A. Keiling. Geophysical Monograph Series, vol. 197 (2013), pp. 317–335. American Geophysical Union

M.C. Kelley, Intense sheared flow as the origin of large-scale undulations of the edge of the diffuse aurora. J. Geophys. Res. **91**(A3), 3225–3230 (1986). https://doi.org/10.1029/JA091iA03p03225

L. Kepko, E. Spanswick, V. Angelopoulos, E. Donovan, J. McFadden, K.-H. Glassmeier, J. Raeder, H.J. Singer, Equatorward moving auroral signatures of a flow burst observed prior to auroral onset. Geophys. Res. Lett. **36**, L24104 (2009). https://doi.org/10.1029/2009GL041476

L. Kepko, R.L. McPherron, O. Amm, S. Apatenkov, W. Baumjohann, J. Birn, M. Lester, R. Nakamura, T.I. Pulkkinen, V. Sergeev, Substorm current wedge revisited. Space Sci. Rev. **190**, 1–46 (2015). https://doi.org/10.1007/s11214-014-0124-9

K.H. Kim, K. Takahashi, D.H. Lee, P.R. Sutcliffe, K. Yumoto, Pi2 pulsations associated with poleward boundary intensifications during the absence of substorms. J. Geophys. Res. Space Phys. **110**(A1) (2005). https://doi.org/10.1029/2004JA010780

S. Knight, Parallel electric fields. Planet. Space Sci. **21**(5), 741–750 (1973). https://doi.org/10.1016/0032-0633(73)90093-7

M. Kubyshkina, V. Sergeev, N. Tsyganenko, V. Angelopoulos, A. Runov, E. Donovan, H. Singer, U. Auster, W. Baumjohann, Time-dependent magnetospheric configuration and breakup mapping during a substorm. J. Geophys. Res. **116**, A00I27 (2011). https://doi.org/10.1029/2010JA015882

M. Kubyshkina, V.A. Sergeev, N.A. Tsyganenko, Y. Zheng, Testing efficiency of empirical, adaptive, and global MHD magnetospheric models to represent the geomagnetic field in a variety of conditions. Space Weather **17**(5), 672–686 (2019). https://doi.org/10.1029/2019SW002157

J. Liang, E.F. Donovan, W.W. Liu, B. Jackel, M. Syrjäsuo, S.B. Mende, H.U. Frey, V. Angelopoulos, a. Connors, Intensification of preexisting auroral arc at substorm expansion phase onset: wave-like disruption during the first tens of seconds. Geophys. Res. Lett. **35**, L17S19 (2008). https://doi.org/10.1029/2008GL033666

J. Liu, V. Angelopoulos, A. Runov, X.-Z. Zhou, On the current sheets surrounding dipolarizing flux bundles in the magnetotail: the case for wedgelets. J. Geophys. Res. Space Phys. **118**, 2000–2020 (2013). https://doi.org/10.1002/jgra.50092

J. Liu, L.R. Lyons, W.E. Archer, B. Gallardo-Lacourt, Y. Nishimura, Y. Zou, et al., Flow shears at the poleward boundary of omega bands observed during conjunctions of Swarm and THEMIS ASI. Geophys. Res. Lett. **45**, 1218–1227 (2018). https://doi.org/10.1002/2017GL076485

A.T.Y. Lui, Cross-field current instability for auroral bead formation in breakup arcs. Geophys. Res. Lett. **43**, 6087–6095 (2016). https://doi.org/10.1002/2016GL069892

A.T.Y. Lui, C.-I. Meng, S. Ismail, Large amplitude undulations on the equatorward boundary of the diffuse aurora. J. Geophys. Res. **87**(A4), 2385 (1982). https://doi.org/10.1029/ja087ia04p02385

A.T.Y. Lui, C.-L. Chang, A. Mankofsky, H.-K. Wong, D. Winske, A cross-field current instability for substorm expansions. J. Geophys. Res. **96**, 11,389–11,401 (1991). https://doi.org/10.1029/91JA00892

K.A. Lynch, D. Hampton, M. Mella, B.Z. Zhang, H. Dahlgren, M. Disbrow, P.M. Kintner, M. Lessard, E. Lundberg, H.C. Stenbaek-Nielsen, Structure and dynamics of the nightside poleward boundary: sounding rocket and ground-based observations of auroral electron precipitation in a rayed curtain. J. Geophys. Res. Space Phys. **117** (2012). https://doi.org/10.1029/2012JA017691

L.R. Lyons, Geomagnetic disturbances: characteristics of, distinction between types, and relations to interplanetary conditions. J. Atmos. Sol.-Terr. Phys. **62**(12), 1087–1114 (2000)

L.R. Lyons, R.L. Walterscheid, Generation of auroral omega bands by shear instability of the neutral winds. J. Geophys. Res. **90**, 12321–12329 (1985)

L.R. Lyons, T. Nagai, G.T. Blanchard, J.C. Samson, T. Yamamoto, T. Mukai, et al., Association between Geotail plasma flows and auroral poleward boundary intensifications observed by CANOPUS photometers. J. Geophys. Res. **104**(A3), 4485–4500 (1999). https://doi.org/10.1029/1998JA900140

L.R. Lyons, E. Zesta, Y. Xu, E.R. Sánchez, J.C. Samson, G.D. Reeves, J.M. Ruohoniemi, J.B. Sigwarth, Auroral poleward boundary intensifications and tail bursty flows: a manifestation of a large-scale ULF oscillation? J. Geophys. Res. **107**(A11), 1352 (2002). https://doi.org/10.1029/2001JA000242

L.R. Lyons, Y. Nishimura, H.J. Kim, E. Donovan, V. Angelopoulos, G. Sofko, M. Nicolls, C. Heinselman, J.M. Ruohoniemi, N. Nishitani, Possible connection of polar cap flows to pre- and post-substorm onset PBIs and streamers. J. Geophys. Res. Space Phys. **116** (2011). https://doi.org/10.1029/2011JA016850

L.R. Lyons et al., Driving of strong nightside reconnection and geomagnetic activity by polar cap flows: application to CME shocks and possibly other situations. J. Atmos. Sol.-Terr. Phys. **177**, 73–83 (2018)

R.L. McPherron, T.-S. Hsu, J. Kissinger, X. Chu, V. Angelopoulos, Characteristics of plasma flows at the inner edge of the plasma sheet. J. Geophys. Res. **116**, A00I33 (2011). https://doi.org/10.1029/2010JA015923

M.R. Mella, K.A. Lynch, D.L. Hampton, H. Dahlgren, P.M. Kintner, M. Lessard, D. Lummerzheim, E.T. Lundberg, M.J. Nicolls, H.C. Stenbaek-Nielsen, Sounding rocket study of two sequential auroral poleward boundary intensifications. J. Geophys. Res. Space Phys. **116**, A00K18 (2011)
M. Mendillo, J. Baumgardner, J. Providakes, Ground based imaging of detached arcs, ripples in the diffuse aurora, and patches of 6300Å emission. J. Geophys. Res. **94**, 5367–5381 (1989)
L.E. Montbriand, The proton aurora substorm, in *The Radiating Atmosphere*, ed. by B.M. McCormac (D. Reidel, Norwell, 1971), pp. 366–373
T. Motoba, M. Hirahara, High-resolution auroral acceleration signatures within a highly dynamic onset arc. Geophys. Res. Lett. **43**, 1793–1801 (2016). https://doi.org/10.1002/2015GL067580
T. Motoba, K. Hosokawa, A. Kadokura, N. Sato, Magnetic conjugacy of northern and southern auroral beads. Geophys. Res. Lett. **39**, L08108 (2012). https://doi.org/10.1029/2012GL051599
T. Motoba, K. Takahashi, A. Ukhorskiy, M. Gkioulidou, D.G. Mitchell, L.J. Lanzerotti, G.I. Korotova, E.F. Donovan, J.R. Wygant, C.A. Kletzing, W.S. Kurth, J.B. Blake, Link between premidnight second harmonic poloidal waves and auroral undulations: conjugate observations with a Van Allen Probe spacecraft and a THEMIS all-sky imager. J. Geophys. Res. Space Phys. **120**, 1814–1831 (2015). https://doi.org/10.1002/2014JA020863
J.S. Murphree, M.L. Johnson, Clues to plasma processes based on Freja UV observations. Adv. Space Res. **18**, 95–105 (1996)
K.R. Murphy, I.R. Mann, I.J. Rae, C.L. Waters, H.U. Frey, A. Kale, H.J. Singer, B.J. Anderson, H. andKorth, The detailed spatial structure of field-aligned currents comprising the substorm current wedge. J. Geophys. Res. Space Phys. **118**, 7714–7727 (2013). https://doi.org/10.1002/2013JA018979
R. Nakamura, T. Oguti, T. Yamamoto, S. Kokubun, Equatorward and poleward expansion of the auroras during auroral substorms. J. Geophys. Res. **98**(A4), 5743–5759 (1993). https://doi.org/10.1029/92JA02230
R. Nakamura, W. Baumjohann, R. Schödel, M. Brittnacher, V.A. Sergeev, M. Kubyshkina, T. Mukai, K. Liou, Earthward flow bursts, auroral streamers, and small expansions. J. Geophys. Res. **106**, 10,791–10,802 (2001)
R. Nakamura et al., Spatial scale of high-speed flows in the plasma sheet observed by cluster. Geophys. Res. Lett. **31**, L09804 (2004). https://doi.org/10.1029/2004GL019558
R. Nakamura, O. Amm, H. Laakso, N.C. Draper, M. Lester, A. Grocott, B. Klecker, I.W. McCrea, A. Balogh, H. Rème, M. André, Localized fast flow disturbance observed in the plasma sheet and in the ionosphere. Ann. Geophys. **23**, 553–566 (2005). https://doi.org/10.5194/angeo-23-553-2005
A.V. Nikolaev, V.A. Sergeev, N.A. Tsyganenko, M.V. Kubyshkina, H. Opgenoorth, H. Singer, V. Angelopoulos, A quantitative study of magnetospheric magnetic field line deformation by a two-loop substorm current wedge. Ann. Geophys. **33**, 505–517 (2015). https://doi.org/10.5194/angeo-33-505-2015
Y. Nishimura, L. Lyons, S. Zou, V. Angelopoulos, S. Mende, Substorm triggering by new plasma intrusion: THEMIS all-sky imager observations. J. Geophys. Res. **115**, A07222 (2010a). https://doi.org/10.1029/2009JA015166
Y. Nishimura et al., Preonset time sequence of auroral substorms: coordinated observations by all-sky imagers, satellites, and radars. J. Geophys. Res. **115**, A00I08 (2010b). https://doi.org/10.1029/2010JA015832
Y. Nishimura, J. Bortnik, W. Li, R.M. Thorne, L.R. Lyons, V. Angelopoulos, S.B. Mende, J.W. Bonnell, O. Le Contel, C. Cully, R. Ergun, U. Auster, Identifying the driver of pulsating aurora. Science **330**, 6000 (2010c). https://doi.org/10.1126/science.119318
Y. Nishimura, L.R. Lyons, V. Angelopoulos, T. Kikuchi, S. Zou, S.B. Mende, Relations between multiple auroral streamers, pre-onset thin arc formation, and substorm auroral onset. J. Geophys. Res. **116**, A09214 (2011). https://doi.org/10.1029/2011JA016768
Y. Nishimura, L.R. Lyons, X. Xing, V. Angelopoulos, E.F. Donovan, S.B. Mende, J.W. Bonnell, U. Auster, Identifying the magnetotail source region leading to preonset poleward boundary intensifications. J. Geophys. Res. Space Phys. **118**, 4335–4340 (2013a). https://doi.org/10.1002/jgra.50407
Y. Nishimura, L.R. Lyons, K. Shiokawa, V. Angelopoulos, E.F. Donovan, S.B. Mende, Substorm onset and expansion phase intensification precursors seen in polar cap patches and arcs. J. Geophys. Res. Space Phys. **118**, 2034–2042 (2013b). https://doi.org/10.1002/jgra.50279
Y. Nishimura, L.R. Lyons, X. Xing, V. Angelopoulos, E.F. Donovan, S.B. Mende, J.W. Bonnell, U. Auster, Tail reconnection region versus auroral activity inferred from conjugate ARTEMIS plasma sheet flow and auroral observations. J. Geophys. Res. Space Phys. **118**, 5758–5766 (2013c). https://doi.org/10.1002/jgra.50549
Y. Nishimura et al., Day-night coupling by a localized flow channel visualized by polar cap patch propagation. Geophys. Res. Lett. **41**(11), 3701–3709 (2014)
Y. Nishimura, J. Yang, P.L. Pritchett, F.V. Coroniti, E.F. Donovan, L.R. Lyons, R.A. Wolf, V. Angelopoulos, S.B. Mende, Statistical properties of substorm auroral onset beads/rays. J. Geophys. Res. Space Phys. **121**, 8661–8676 (2016). https://doi.org/10.1002/2016JA022801

N. Nishitani, G. Hough, M.W.J. Scourfield, Spatial and temporal characteristics of giant undulations. Geophys. Res. Lett. **21**, 2673–2676 (1994)

T. Oguti, Hydrogen emission and electron aurora at the onset of the auroral breakup. J. Geophys. Res. **78**(31), 7543–7547 (1973). https://doi.org/10.1029/JA078i031p07543

T. Oguti, Tv observations of auroral arcs, in *Physics of Auroral Arc Formation*, ed. by e.S. Akasofu, J. Kan (1981). https://doi.org/10.1029/GM025p0031

S. Ohtani, A. Yoshikawa, The initiation of the poleward boundary intensification of auroral emission by fast polar cap flows: a new interpretation based on ionospheric polarizarion. J. Geophys. Res. Space Phys. **121**, 10910–10928 (2016). https://doi.org/10.1002/2016JA023143

S. Ohtani, T. Motoba, J.W. Gjerloev, J.M. Ruohoniemi, E.F. Donovan, A. Yoshikawa, Longitudinal development of Poleward Boundary Intensifications (PBIs) of auroral emission. J. Geophys. Res. Space Phys. **123**(11), 9005–9021 (2018)

H.J. Opgenoorth, J. Oksman, K.U. Kaila, E. Nielsen, W. Baumjohann, Characteristics of eastward drifting omega bands in the morning sector of the auroral oval. J. Geophys. Res. **88**(A11), 9171–9185 (1983). https://doi.org/10.1029/JA088iA11p09171

H.J. Opgenoorth, M.A.L. Persson, T.I. Pulkkinen, R.J. Pellinen, Recovery phase of magnetospheric substorms and its association with morning-sector aurora. J. Geophys. Res. **99**(A3), 4115–4129 (1994). https://doi.org/10.1029/93JA01502

M. Palmroth, I. Honkonen, A. Sandroos, Preliminary testing of global hybrid-Vlasov simulation: magnetosheath and cusps under northward interplanetary magnetic field. J. Atmos. Sol.-Terr. Phys. **99**, 41–46 (2013). https://doi.org/10.1016/j.jastp.2012.09.013

M. Palmroth, U. Ganse, Y. Pfau-Kempf, M. Battarbee, L. Turc, T. Brito et al., Vlasov methods in space physics and astrophysics. Living Rev. Comput. Astrophys. **4**(1) (2018). https://doi.org/10.1007/s41115-018-0003-2

N. Partamies, J.M. Weygand, L. Juusola, Statistical study of auroral omega bands. Ann. Geophys. **35**, 1069–1083 (2017). https://doi.org/10.5194/angeo-35-1069-2017

G. Paschmann, S. Haaland, R. Treumann (eds.), *Auroral Plasma Physics* (2003). Springer Netherlands. https://doi.org/10.1007/978-94-007-1086-3.

S. Perraut et al., Current-driven electromagnetic ion cyclotron instability at substorm onset. J. Geophys. Res. **105**, 21,097–21,107 (2000). https://doi.org/10.1029/2000JA900059

T. Pitkänen, A.T. Aikio, O. Amm, K. Kauristie, H. Nilsson, K.U. Kaila, EISCAT-cluster observations of quiet-time near-Earth magnetotail fast flows and their signatures in the ionosphere. Ann. Geophys. **29**, 299–319 (2011). https://doi.org/10.5194/angeo-29-299-2011

T. Pitkanen, A.T. Aikio, L. Juusola, Observations of polar cap flow channel and plasma sheet flow bursts during substorm expansion. J. Geophys. Res. Space Phys. **118**(2), 774–784 (2013)

P.L. Pritchett, Externally driven magnetic reconnection in the presence of a normal magnetic field. J. Geophys. Res. **110**, A05209 (2005). https://doi.org/10.1029/2004JA010948

P.L. Pritchett, F.V. Coroniti, A kinetic ballooning/interchange instability in the magnetotail. J. Geophys. Res. **115**, A06301 (2010). https://doi.org/10.1029/2009JA014752

P.L. Pritchett, F.V. Coroniti, Structure and consequences of the kinetic ballooning/interchange instability in the magnetotail. J. Geophys. Res. Space Phys. **118**, 146–159 (2013). https://doi.org/10.1029/2012JA018143

P.L. Pritchett, F.V. Coroniti, Y. Nishimura, The kinetic ballooning/interchange instability as a source of dipolarization fronts and auroral streamers. J. Geophys. Res. Space Phys. **119**(6), 4723–4739 (2014)

J.F. Provadakes, M.C. Kelley, W.E. Swartz, M. Mendillo, Radar and optical measurements of ionospheric processes associated with intense subauroral electric fields. J. Geophys. Res. **94**, 5350–5366 (1989)

T.I. Pulkkinen, R.J. Pellinen, H.E.J. Koskinen, H.J. Opgenoorth, J.S. Murphree, V. Petrov, A. Zaitzev, E. Friis-Christensen, Auroral signatures of substorm recovery phase: a case study, in *In Magnetospheric Substorms*, ed. by J.R. Kan et al.. Geophys. Monogr. Ser., vol. 64 (AGU, Washington, 1991), pp. 333–341

I.J. Rae et al., Near-Earth initiation of a terrestrial substorm. J. Geophys. Res. **114**, A07220 (2009). https://doi.org/10.1029/2008JA013771

I.J. Rae, C.E.J. Watt, I.R. Mann, K.R. Murphy, J.C. Samson, K. Kabin, V. Angelopoulos, Optical characterization of the growth and spatial structure of a substorm onset arc. J. Geophys. Res. **115**, A10222 (2010). https://doi.org/10.1029/2010JA015376

I.J. Rae, C.E.J. Watt, K.R. Murphy, H.U. Frey, L.G. Ozeke, D.K. Milling, I.R. Mann, The correlation of ULF waves and auroral intensity before, during and after substorm expansion phase onset. J. Geophys. Res. **117**, A08213 (2012). https://doi.org/10.1029/2012JA017534

I.J. Rae, K.R. Murphy, C.E.J. Watt, G. Rostoker, R. Rankin, I.R. Mann, C.R. Hodgson, H.U. Frey, A.W. Degeling, C. Forsyth, Field line resonances as a trigger and a tracer for substorm onset. J. Geophys. Res. Space Phys. **119**, 5343–5363 (2014). https://doi.org/10.1002/2013JA018889

I.J. Rae, K.R. Murphy, C.E.J. Watt, A.J. Halford, I.R. Mann, L.G. Ozeke, et al., The role of localized compressional ultra-low frequency waves in energetic electron precipitation. J. Geophys. Res. Space Phys. **123**, 1900–1914 (2018). https://doi.org/10.1002/2017JA024674

G. Rostoker, J.C. Samson, Can substorm expansive phase effects and low frequency Pc magnetic pulsations be attributed to the same source mechanism? Geophys. Res. Lett. **11**(3), 271–274 (1984). https://doi.org/10.1029/gl011i003p00271

G. Rostoker, A.T.Y. Lui, C.D. Anger, J.S. Murphree, North-south structures in the midnight sector auroras as viewed by the Viking imager. Geophys. Res. Lett. **14**, 407–410 (1987)

A. Runov, V. Angelopoulos, M.I. Sitnov, V.A. Sergeev, J. Bonnell, J.P. McFadden, D. Larson, K.-H. Glassmeier, U. Auster, THEMIS observations of an earthward-propagating dipolarization front. Geophys. Res. Lett. **36**, L14106 (2009). https://doi.org/10.1029/2009GL038980

T. Saito, Space Sci. Rev. **21**, 427 (1978). https://doi.org/10.1007/BF00173068

K. Sakaguchi, K. Shiokawa, A. Ieda, R. Nomura, A. Nakajima, M. Greffen, E. Donovan, I.R. Mann, H. Kim, M. Lessard, Fine structures and dynamics in auroral initial brightening at substorm onsets. Ann. Geophys. **27**(2), 623–630 (2009). https://doi.org/10.5194/angeo-27-623-2009

P.E. Sandholt, C.J. Farrugia, M. Lester, S. Cowley, S. Milan, W.F. Denig, B. Lybekk, E. Trondsen, V. Vorobjev, Multistage substorm expansion: Auroral dynamics in relation to plasma sheet particle injection, precipitation, and plasma convection. J. Geophys. Res. Space Phys. **107**(A11) (2002). https://doi.org/10.1029/2001JA900116

V.A. Sergeev, Ionospheric signatures of magnetospheric particle acceleration in substorms—how to decode them? in *Proceedings of the 6th International Conference on Substorms*, ed. by R.M. Winglee (Univ. of Washington, Seattle, 2002), pp. 39–46

V.A. Sergeev, *Bursty Bulk Flows and Their Ionospheric Footprints, Multiscale Processes in the Earth's Magnetosphere: From Interball to Cluster* (2005), pp. 289–297. Springer Netherlands

V.A. Sergeev, V. Angelopoulos, J.T. Gosling, C.A. Cattell, C.T. Russell, Detection of localized, plasma-depleted flux tubes or bubbles in the midtail plasma sheet. J. Geophys. Res. **101**(A5), 10817–10826 (1996). https://doi.org/10.1029/96JA00460

V.A. Sergeev, K. Liou, C.-I. Meng, P.T. Newell, M. Brittnacher, G. Parks, G.D. Reeves, Development of auroral streamers in association with localized impulsive injections to the inner magnetotail. Geophys. Res. Lett. **26**, 417–420 (1999)

V.A. Sergeev, J.-A. Sauvaud, D. Popescu, R.A. Kovrazhkin, K. Liou, P.T. Newell, M. Brittnacher, G. Parks, R. Nakamura, T. Mukai, Multiple-spacecraft observation of a narrow transient plasma jet in the Earth's plasma sheet. Geophys. Res. Lett. **27**, 851–854 (2000)

V.A. Sergeev, K. Liou, P.T. Newell, S.I. Ohtani, M.R. Hairston, F. Rich, Auroral streamers: characteristics of associated precipitation, convection and field-aligned currents. Ann. Geophys. **22**, 537–548 (2004)

V.A. Sergeev, K. Liou, S.-I. Ohtani, M. Hairston, M. Thomsen, Narrow plasma streams as a candidate to populate the inner magnertosphere, in *The Inner Magnetosphere: Physics and Modeling*, ed. by T.I. Pulkkinen, N.A. Tsyganenko, R.H.W. Friedel. Geophys. Monogr., vol. 155 (2005), pp. 55–60. https://doi.org/10.1029/155GM07

V.A. Sergeev, V. Angelopoulos, R. Nakamura, Recent advances in understanding substorm dynamics. Geophys. Res. Lett. **39**, L05101 (2012). https://doi.org/10.1029/2012GL050859

V.A. Sergeev, I.A. Chernyaev, V. Angelopoulos, A.V. Runov, R. Nakamura, Stopping flow bursts and their role in the generation of the substorm current wedge. Geophys. Res. Lett. **41**, 1106–1112 (2014). https://doi.org/10.1002/2014GL059309

I.G. Shevchenko, V. Sergeev, M. Kubyshkina, V. Angelopoulos, K.H. Glassmeier, H.J. Singer, Estimation of magnetosphere-ionosphere mapping accuracy using isotropy boundary and THEMIS observations. J. Geophys. Res. **115**, A11206 (2010). https://doi.org/10.1029/2010JA015354

Y. Shi, E. Zesta, Global-scale ionospheric flow and aurora precursors of auroral substorms: coordinated SuperDARN and IMAGE/WIC observations. J. Geophys. Res. Space Phys. **119**, 4860–4871 (2014). https://doi.org/10.1002/2013JA019175

Y. Shi, E. Zesta, L.R. Lyons, J. Yang, A. Boudouridis, Y.S. Ge, J.M. Ruohoniemi, S. Mende, Two-dimensional ionospheric flow pattern associated with auroral streamers. J. Geophys. Res. **117**, A02208 (2012). https://doi.org/10.1029/2011JA017110

K. Shiokawa, G. Haerendel, W. Baumjohann, Azimuthal pressure gradient as driving force of substorm currents. Geophys. Res. Lett. **25**(7), 959–962 (1998). https://doi.org/10.1029/98gl00540

D.G. Sibeck, R. Allen, H. Aryan et al., Space Sci. Rev. **214**, 79 (2018). https://doi.org/10.1007/s11214-018-0504-7

M.I. Sitnov, N.A. Tsyganenko, A.Y. Ukhorskiy, P.C. Brandt, Dynamical data-based modeling of the storm-time geomagnetic field with enhanced spatial resolution. J. Geophys. Res. **113**, A07218 (2008). https://doi.org/10.1029/2007JA013003

M.I. Sitnov, N. Buzulukova, M. Swisdak, V.G. Merkin, T.E. Moore, Spontaneous formation of dipolarization fronts and reconnection onset in the magnetotail. Geophys. Res. Lett. **40**(1), 22–27 (2013). https://doi.org/10.1029/2012GL054701

M. Sitnov, J. Birn, B. Ferdousi et al., Space Sci. Rev. **215**, 31 (2019). https://doi.org/10.1007/s11214-019-0599-5

D.J. Southwood, The hydromagnetic stability of the magnetospheric boundary. Planet. Space Sci. **16**, 587 (1968)

G.K. Stephens, M.I. Sitnov, H. Korth, N.A. Tsyganenko, S. Ohtani, M. Gkioulidou, A.Y. Ukhorskiy, Global empirical picture of magnetospheric substorms inferred from multimission magnetometer data. J. Geophys. Res. Space Phys. **124**, 1085–1110 (2019). https://doi.org/10.1029/2018JA025843

W.J. Sun, S.Y. Fu, G.K. Parks, J. Liu, Z.H. Yao, Q.Q. Shi, Q.-G. Zong, S.Y. Huang, Z.Y. Pu, T. Xiao, Field-aligned currents associated with dipolarization fronts. Geophys. Res. Lett. **40**, 4503–4508 (2013). https://doi.org/10.1002/grl.50902

V.R. Tagirov, Auroral torch structures: results of optical observations. J. Atmos. Sol.-Terr. Phys. **55**(14), 1775–1787 (1993). https://doi.org/10.1016/0021-9169(93)90144-n

N.A. Tsyganenko, Data-based modelling of the Earth's dynamic magnetosphere: a review. Ann. Geophys. **31**, 1745–1772 (2013). https://doi.org/10.5194/angeo-31-1745-2013

N.A. Tsyganenko, V.A. Andreeva, A forecasting model of the magnetosphere driven by an optimal solar wind coupling function. J. Geophys. Res. Space Phys. **120**, 8401–8425 (2015). https://doi.org/10.1002/2015JA021641

N.A. Tsyganenko, V.A. Andreeva, An empirical RBF model of the magnetosphere parameterized by interplanetary and ground-based drivers. J. Geophys. Res. Space Phys. **121**, 10,786–10,802 (2016). https://doi.org/10.1002/2016JA023217

A.F. Viñas, T.R. Madden, Shear flow-ballooning instability as a possible mechanism for hydromagnetic fluctuations. J. Geophys. Res. **91**(A2), 1519–1528 (1986). https://doi.org/10.1029/JA091iA02p01519

I. Voronkov, R. Rankin, P. Frycz, V.T. Tikhonchuk, J.C. Samson, Coupling of shear flow and pressure gradient instabilities. J. Geophys. Res. **102**(A5), 9639–9650 (1997). https://doi.org/10.1029/97JA00386

I. Voronkov, E. Friedrich, J.C. Samson, Dynamics of the substorm growth phase as observed using CANOPUS and SuperDARN instruments. J. Geophys. Res. **104**(A12), 28491–28505 (1999). https://doi.org/10.1029/1999JA900362

I.O. Voronkov, E.F. Donovan, B.J. Jackel, J.C. Samson, Large-scale vortex dynamics in the evening and midnight auroral zone: observations and simulations. J. Geophys. Res. **105**, 18505–18518 (2000)

I.O. Voronkov, E.F. Donovan, J.C. Samson, Observations of the phases of the substorm. J. Geophys. Res. Space Phys. **108**(A2) (2003). https://doi.org/10.1029/2002JA009314

A.D.M. Walker, The Kelvin-Helmholtz instability in the low-latitude boundary layer. Planet. Space Sci. **29**, 1119 (1981)

C.E.J. Watt, R. Rankin, I.J. Rae, D.M. Wright, Inertial Alfvén waves and acceleration of electrons in nonuniform magnetic fields. Geophys. Res. Lett. **33**, L02106 (2006). https://doi.org/10.1029/2005GL024779

D.E. Wendel, G.V. Khazanov, A.K. Tripathi, R.P. Singhal, E. Zesta, Source of the bursty bulk flow diffuse aurora: electrostatic cyclotron harmonic and whistler waves in the coupling of bursty bulk flows to auroral precipitation. J. Geophys. Res. Space Phys. **124**, 6669–6690 (2019). https://doi.org/10.1029/2019JA026606

J.M. Weygand, M.G. Kivelson, H.U. Frey, J.V. Rodriguez, V. Angelopoulos, R. Redmon et al., An interpretation of spacecraft and ground based observations of multiple omega band events. J. Atmos. Sol.-Terr. Phys. **133**, 185–204 (2015). https://doi.org/10.1016/j.jastp.2015.08.014

J.A. Wild, T.K. Yeoman, P. Eglitis, H.J. Opgenoorth, Multi-instrument observations of the electric and magnetic field structure of omega bands. Ann. Geophys. **18**, 99–110 (2000). https://doi.org/10.1007/s00585-000-0099-6

J.A. Wild, E.E. Woodfield, E. Donovan, R.C. Fear, A. Grocott, M. Lester et al., Midnight sector observations of auroral omega bands. J. Geophys. Res. Space Phys. **116**(A5) (2011). https://doi.org/10.1029/2010ja015874

M. Wiltberger, V. Merkin, J.G. Lyon, S. Ohtani, High-resolution global magnetohydrodynamic simulation of bursty bulk flows. J. Geophys. Res. Space Phys. **120** (2015). https://doi.org/10.1002/2015JA021080

X. Xing, L. Lyons, Y. Nishimura, V. Angelopoulos, D. Larson, C. Carlson, J. Bonnell, U. Auster, Substorm onset by new plasma intrusion: THEMIS spacecraft observations. J. Geophys. Res. **115**, A10246 (2010). https://doi.org/10.1029/2010JA015528

T. Yamamoto, Hybrid Kelvin-Helmholtz/Rayleigh-Taylor instability in the plasma sheet. J. Geophys. Res. **114**, A06207 (2009). https://doi.org/10.1029/2008JA013760

T. Yamamoto, A numerical simulation for the omega band formation. J. Geophys. Res. Space Phys. **116**, A2 (2011)

T. Yamamoto, K. Makita, C.-I. Meng, A particle simulation of "giant" undulations on the evening diffuse auroral boundary. J. Geophys. Res. **98**(A4), 5785–5800 (1993). https://doi.org/10.1029/92JA01434

T. Yamamoto, M. Ozaki, S. Inoue, K. Makita, C.-I. Meng, Convective generation of "giant" undulations on the evening diffuse auroral boundary. J. Geophys. Res. **99**(A10), 19499–19512 (1994). https://doi.org/10.1029/94JA00273

J. Yang, F.R. Toffoletto, R.A. Wolf, S. Sazykin, RCM-E simulation of ion acceleration during an idealized plasma sheet bubble injection. J. Geophys. Res. **116**, A05207 (2011). https://doi.org/10.1029/2010JA016346

J. Yang, F.R. Toffoletto, R.A. Wolf, S. Sazykin, On the contribution of plasma sheet bubbles to the storm time ring current. J. Geophys. Res. Space Phys. **120**, 7416–7432 (2015). https://doi.org/10.1002/2015JA021398

Z.H. Yao et al., Conjugate observations of flow diversion in the magnetotail and auroral arc extension in the ionosphere. J. Geophys. Res. Space Phys. **118**, 4811–4816 (2013). https://doi.org/10.1002/jgra.50419

C. Yue, Y. Nishimura, L.R. Lyons, V. Angelopoulos, E.F. Donovan, Q.Q. Shi, Z.H. Yao, J.W. Bonnell, Coordinated THEMIS spacecraft and all-sky imager observations of interplanetary shock effects on plasma sheet flow bursts, poleward boundary intensifications, and streamers. J. Geophys. Res. Space Phys. **118**(6), 3346–3356 (2013)

E. Zesta et al., Ionospheric convection signatures of tail fast flows during substorms and Poleward Boundary Intensifications (PBI). Geophys. Res. Lett. **38** (2011). https://doi.org/10.1029/2011GL046758

E. Zesta, L.R. Lyons, E. Donovan, The auroral signature of earthward flow burst observed in the magnetotail. Geophys. Res. Lett. **27**(20), 3241–3244 (2000). https://doi.org/10.1029/2000GL000027

E. Zesta, E. Donovan, L. Lyons, G. Enno, J.S. Murphree, L. Cogger, Two-dimensional structure of auroral poleward boundary intensifications. J. Geophys. Res. **107**(A11), 1350 (2002). https://doi.org/10.1029/2001JA000260

E. Zesta, L. Lyons, C.-P. Wang, E. Donovan, H. Frey, T. Nagai, Auroral poleward boundary intensifications (PBIs): their two-dimensional structure and associated dynamics in the plasma sheet. J. Geophys. Res. **111**, A05201 (2006). https://doi.org/10.1029/2004JA010640

Y. Zhang, L.J. Paxton, D. Morrison, A.T.Y. Lui, H. Kil, B. Wolven, C.-I. Meng, A.B. Christensen, Undulations on the equatorward edge of the diffuse proton aurora: TIMED/GUVI observations. J. Geophys. Res. **110**, A08211 (2005). https://doi.org/10.1029/2004JA010668

M. Zhou et al., Cluster observations of kinetic structures and electron acceleration within a dynamic plasma bubble. J. Geophys. Res. Space Phys. **118**, 674–684 (2013). https://doi.org/10.1029/2012JA018323

S. Zou et al., Identification of substorm onset location and preonset sequence using Reimei, THEMIS GBO, PFISR, and Geotail. J. Geophys. Res. Space Phys. **115** (2010). https://doi.org/10.1029/2010JA015520

Y. Zou, Y. Nishimura, L.R. Lyons, E.F. Donovan, J.M. Ruohoniemi, N. Nishitani, K.A. McWilliams, Statistical relationships between enhanced polar cap flows and PBIs. J. Geophys. Res. Space Phys. **119**(1), 151–162 (2014)

Space Sci Rev (2021) 217:10
https://doi.org/10.1007/s11214-020-00776-6

Proton Aurora and Optical Emissions in the Subauroral Region

B. Gallardo-Lacourt[1,2] · H.U. Frey[3] · C. Martinis[4]

Received: 5 November 2019 / Accepted: 27 November 2020 / Published online: 6 January 2021

Abstract Optical structures located equatorward of the main auroral oval often exhibit different morphologies and dynamics than structures at higher latitudes. In some cases, questions arise regarding the formation mechanisms of these photon-emitting phenomena. New developments in space and ground-based instruments have enabled us to acquire a clearer view of the processes playing a role in the formation of subauroral structures. In addition, the discovery of new optical structures helps us improve our understanding of the latitudinal and altitudinal coupling that takes place in the subauroral region. However, several questions remain unanswered, requiring the development of new instruments and analysis techniques. We discuss optical phenomena in the subauroral region, summarize observational results, present conclusions about their origin, and pose a number of open questions that warrant further investigation of proton aurora, detached subauroral arcs and spots, stable auroral red (SAR) arcs, and STEVE (Strong Thermal Emission Velocity Enhancement).

Keywords Optical structures · Subauroral region

1 Introduction

Proton aurora and the structures observed at subauroral latitudes occur in a region of magnetic field transition. In this region, Earth's magnetic field changes from a stretched topology closer to the auroral oval, into more dipolar field lines towards the equator. Optical structures in this region are diverse in morphology, propagation, dynamics, and lifetimes. In addition,

Auroral Physics
Edited by David Knudsen, Joe Borovsky, Tomas Karlsson, Ryuho Kataoka and Noora Partmies

✉ B. Gallardo-Lacourt

1 NASA Goddard Space Flight Center, Greenbelt, MD, USA

2 Universities of Space Research Association, Columbia, MD, USA

3 Space Sciences Laboratory, University of California, Berkeley, CA, USA

4 Center of Space Physics, Boston University, Boston, MA, USA

some of these structures exhibit different generation mechanisms than aurorae; specifically, some of these subauroral optical structures are not caused by the excitation from energetic particle precipitation as is the case with auroral arcs (Vegard 1932).

In this chapter, we review the recent advances made on proton aurora, detached subauroral arcs and spots, Subauroral Red (SAR) arcs, and the newly discovered phenomenon named STEVE (Strong Thermal Emission Velocity Enhancement). For each of these topics we provide a description of their main features, drivers and formation mechanisms. At the end of the chapter, we highlight some of the open questions that still remain for this region.

2 Proton Aurora

There are two common uses of the term "proton aurora". In some contexts, it is used to describe any aurora that is generated by energetic protons. As energetic electrons and protons interact with the atomic and molecular oxygen and nitrogen of Earth's upper atmosphere, ionization and excitation occur. Spectroscopic observations of photons generated by this process do not provide information weather the initial atomic excitation—and subsequent emission—was stimulated by energetic electrons or protons. Accordingly, the term "proton aurora" refers to excitation produced by energetic proton precipitation.

When spectroscopic data are used to determine the mean energy and flux of energetic electrons, it is almost always assumed that the relative contribution from protons is negligible (Germany et al. 1994; Galand and Lummerzheim 2004). One exception is the use of IMAGE-FUV data where the proton contribution could be determined and a proper correction to the electron contribution could be made (Frey et al. 2003). Another possibility is the use of a specific model and its comparison to measurement comparisons (Galand and Lummerzheim 2004). So, in this case additional measurements for instance by low-altitude satellites with particle detectors are required to determine the relative contribution of energetic protons to the overall aurora generation.

The other common use of the term "proton aurora" refers strictly to emission that comes from precipitating energetic protons themselves. Bare protons cannot emit any photons but during precipitation they can undergo charge exchange collisions with atmospheric constituents, briefly become neutral hydrogen atoms, and are then able to emit hydrogen-specific photons. The brightest and most prominent emissions are the Lyman-α at 121.57 nm in the ultraviolet and the Balmer-α and Balmer-β lines at 656.3 and 486.1 nm in the visible, respectively. The process of charge exchange decouples the hydrogen atoms from the guiding magnetic field, at which point they will travel in straight lines (unless neutral-neutral collisions become important at low altitude) until their electron is stripped off and they become positively charged protons again (Fig. 1). This leads to a spatial broadening of any existing small-scale structure in the proton precipitation. The resulting proton aurora does not show fine structure and appears relatively homogeneous on spatial scales of $\sim$100 km (Fig. 2). Fang et al. (2004) found that the main dispersion region for a fine proton beam is located in the altitude range of around 250–450 km, where the first few charge exchange collisions typically occur. As the energetic protons move at high speed, the resulting hydrogen atoms emit photons with a specific Doppler shift and any emission is shifted towards blue if seen from the ground or towards red if seen from space (see e.g. Chamberlain 1961). High-resolution spectroscopic measurements then allow estimates of the proton energy (Galand and Chakrabarti 2006). High-resolution spectroscopic separation of the Doppler-shifted Lyman-α from the much brighter cold emission of the geocorona was the operational principle of the IMAGE Spectrographic Imager SI-12 proton aurora channel which allowed for

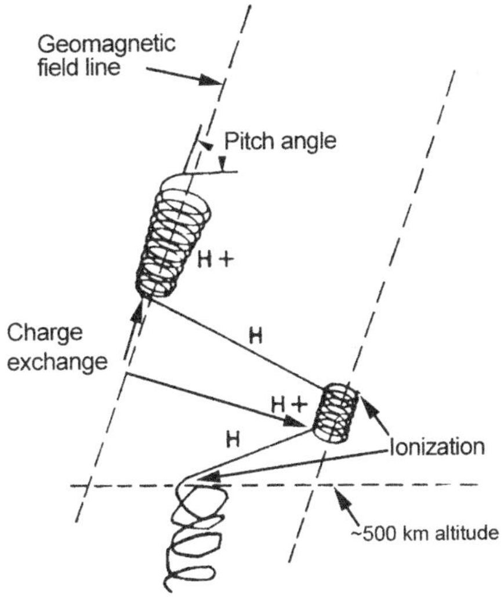

Fig. 1 Schematic of the charge exchange process for energetic protons which become neutral hydrogen, can emit a photon, leave the original magnetic field line, and get ionized again at some distance from the original travel direction. (From Egeland and Burke 2019)

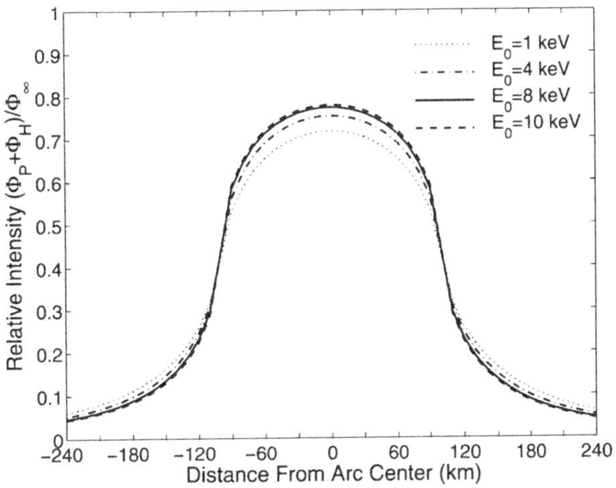

Fig. 2 Relative intensity of the total downward H+ and H fluxes at 300 km altitude with respect to the homogeneous input at the top. The incident proton fluxes have a Maxwellian distribution with the characteristic energy of E0 = 1, 4, 8, and 10 keV. The semi-width of the proton arc is W = 100 km. (From Fang et al. 2004)

a substantial reduction of the background and a clear selection of red-shifted signal coming from protons with more than 1 keV energy (Mende et al. 2000). In the remainder of this paper we will use this second interpretation of the term "proton aurora" to clearly identify signal that can only originate from energetic proton precipitation.

The charge exchange and photon emission can generally be described as:

$$p^+ + X \quad \rightarrow H^* + X^- \quad \rightarrow H + h\nu + X^- \tag{1}$$

where p^+ is the energetic proton, X is an atmospheric constituent (mostly N_2 but also N, O, O_2), H* is the excited hydrogen atom, X^- the ionized atmospheric constituent, H the hydrogen atom in its ground state, and hν the hydrogen photon. The still energetic hydrogen atom can ionize again (Fig. 1) and continue as an energetic proton. The process repeats itself until the energetic proton achieves thermodynamic equilibrium with the surrounding medium. Additionally, the ionized constituent can emit auroral photons (for instance, the N_2

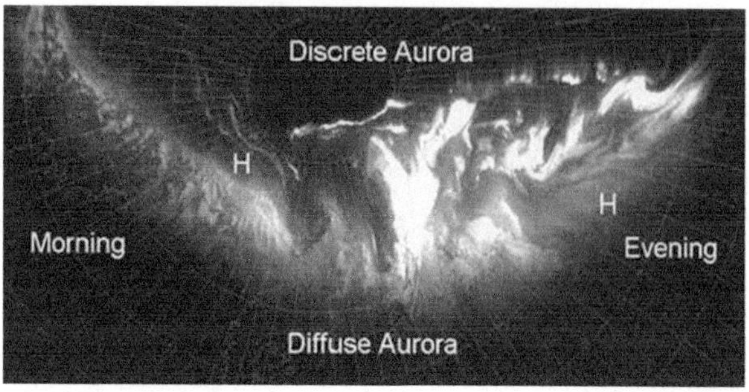

Fig. 3 A synoptic image of nightside aurora showing the relative location of the hydrogen emissions – marked with H – versus the mostly electron generated oval, in both the morning (left) and evening (right). On the evening side the Balmer radiations are equatorward of the electron aurora, while on the morning side (left) the opposite is the case. (From Egeland and Burke 2019)

1st negative emission) which do not carry information about the generation partner and will generally be misidentified as "electron aurora".

First identifications of hydrogen Balmer lines were reported in the 1930/40 s (see Chamberlain 1961). Significant effort was then put into the identification of emission lines and the source of energetic protons (Egeland and Burke 2019). With improved ground instrumentation in the 1950/1960 s, particularly space-based particle detectors and imagers, the distribution and importance of proton precipitation in certain regions was recognized. One important difference between electron generated aurora and proton aurora is the reversal of the relative magnetic latitude distribution between both types of aurora (Fig. 3). Before magnetic midnight the proton aurora oval spreads equatorward, and peaks in intensity equatorward of the electron-generated green line. After magnetic midnight the hydrogen emissions are mainly observed poleward of the electron oval (i.e., MLat $> 70°$). However, the latitude distributions also depend on the energies of precipitating protons. Today proton aurora observations allow the assessment of the source region of energetic protons. They also can be used to identify the associated magnetospheric processes and to evaluate the ionospheric perturbations induced by energetic protons. The source regions include the cusp, the low-latitude boundary layer, the mantle, and the plasma sheet, including their dayside extensions (Galand and Chakrabarti 2006).

Secondary electrons produced within the proton beam contribute to "electron auroral" emissions. Since they are less energetic than the secondary electrons produced in electron aurora, they have a different spectral signature. In addition, for a given energy flux, protons are usually more efficient at ionizing than electrons and yield larger values of the Pedersen ionospheric conductance (Galand and Lummerzheim 2004). This difference between proton and electron aurora can lead to misinterpretation when brightness ratios are used to derive ionospheric conductances with parameterizations that are based on electron aurora. At the equatorward edge of the afternoon oval, where protons are a significant energy source, the traditional auroral analysis (Rees and Luckey 1974) using two lines to determine electron energy is limited. In regions of >4 keV electron precipitation, the presence of proton precipitation (even modest $\sim 10\%$), yields a large underestimation of both the electron mean energy and the energy flux. Overall, the presence of proton precipitation yields a poor estimation of the electron mean energy. In proton-dominated aurora, the Pedersen and Hall conductances are often underestimated. However, in location where the protons are not dominant and the electron precipitation is not too energetic, it is legitimate to estimate the particle characteristics and ionospheric conductances from the FUV brightnesses assuming pure electron precipitation. This is true in particular for the region around midnight (1900–0400 MLT), at a magnetic latitude of 65–67° (Galand and Lummerzheim 2004).

The general view that proton aurora is located equatorward of electron aurora in the dusk-side of the auroral oval was later slightly modified as it was shown that the location depends on the level of geomagnetic activity (Zou et al. 2012). A statistical study of the absolute and relative locations of the equatorward boundaries of the proton and diffuse electron auroras used meridian scanning photometers to observe 630.0 nm emission from "electron aurora" and 486.1 nm ("H-β") emissions of proton aurora together with particle data from satellites. The statistical studies showed that the quiet time dusk-premidnight proton aurora extends slightly equatorward of the electron aurora. This is reversed on the dawnside. In contrast, the electron aurora during moderate and active intervals lies equatorward of the proton aurora throughout the nightside region (20–03 MLT). There is a dawn-dusk offset in the auroral oval location, with the proton aurora shifting toward premidnight and the electron aurora toward postmidnight. The penetration of the electron boundary equatorward of the proton precipitation across this MLT range had not been previously identified. Zou et al. (2012) attributed this to the fact that the equatorward part of the diffuse electron aurora is produced by lower energy electrons to which the 630.0 nm emission is more responsive and so marks a more realistic location of the electron equatorward boundary than do the 557.7 nm and UV emissions used in previous intercomparisons of these boundaries.

On the nightside, the bright proton aurora forms a band of diffuse aurora near the equatorward boundary of the electron auroral oval. Donovan et al. (2012) demonstrated that the precipitation is due to strong pitch angle diffusion that is thought to be the result of non-adiabatic motion of sub-keV to tens of keV central plasma sheet (CPS) protons as they traverse the tightly curved magnetic field topology in the vicinity of the neutral sheet. These authors also provided an overview of the relationship between the spatiotemporal evolution of the proton aurora and magnetospheric dynamics. They focused on the equatorward boundary of the proton aurora, the latitude of which has been shown to be strongly correlated with magnetic field line stretching in the inner CPS, and they provided the first-ever identification of the position of the ion isotropy boundary relative to equatorial magnetospheric spacecraft.

The bright proton aurora is generally understood to be the projection of the CPS onto the ionosphere. Particles in the CPS have sufficient particle energy to cause bright proton aurora luminosity after strong pitch angle scattering (presumably due to field line curvature). This region is often interpreted as the transition region between dipolar and tail-like magnetic topologies. Spanswick et al. (2017) presented statistics of proton aurora luminosity computed from THEMIS electrostatic analyzer measurements for various radial distances in the magnetotail. These results were compared to ground observations of proton auroral luminosity and used to derive a statistical source region of the bright proton aurora. They found that near magnetic midnight, the bright proton aurora rarely, if ever, maps to a location beyond 10 RE. They also showed that once inside of 6 RE, the predicted proton auroral luminosities become large relative to those typically observed. This likely indicates that 6 RE is typically inside of the ion scattering (isotropy) boundary. Spanswick et al. (2017) further asserted that this implies the bright proton aurora typically maps to the region between 6 RE and 10 RE in the tail.

Optical auroral measurements have repeatedly revealed the existence of a transitional stage between a quiescent preexisting arc and its significant auroral expansion in a substorm onset. Such a transitional stage of substorm onset (TSSO) is characterized by a gradual intensification and the emergence of auroral beads, along the preexisting arc. Prior studies on TSSO were limited to electron auroras which are dominant in optical luminosity. Liang et al. (2018) used 12 substorm events to investigate the proton auroral features during the late growth phase and the TSSO. They confirm the previous notion that the onset electron auroral arc is usually located at the poleward "shoulder" of the main proton auroral band. Furthermore, while the electron auroral arc typically intensifies by a few times or even an order of magnitude during the TSSO, the concurrent proton aurora at the same location as the electron auroral arc shows much less noticeable variations. The proton auroral variations averaged over the arc band, as well as that integrated over the entire latitudinal range, are mostly within 10% of their mean late-growth-phase levels during the TSSO. Substantial intensifications of proton auroras occur after the poleward expansion of electron auroras. Even considering the spatial spreading of proton auroras, they estimated that the variation of ion precipitation fluxes on top of the ionosphere would be typically $<30\%$ during the TSSO. Liang et al. (2018) interpreted it as non-significant ion energization or large-scale magnetic field reconfiguration during the TSSO. Instead, it is likely that the underlying mechanism of the TSSO might be some instability wave mode. The instability wavelength is comparable to the ion gyroradius, so that the ions are demagnetized, suppressing ion flux variations with the instability. Enhancing upward electric fields in the auroral acceleration region during the TSSO may also play a partial role in weakening the ion precipitation flux variation.

3 Detached Subauroral Arcs

3.1 Afternoon Detached Auroral Arcs

Afternoon detached auroral arcs are equatorward to the main auroral oval, often connected to it near noon (Immel et al. 2002). Their properties and conditions are summarized in Table 1 of (Frey 2007). They occur in the general magnetic local time (MLT) region of 12–18 hours after a change of either the Interplanetary Magnetic Field (IMF) Bz or B_y from negative to positive (Burch et al. 2002). Early after their discovery it was speculated that these arcs are "leftovers" of a preexisting arc when the auroral oval either contracted (Moshupi et al. 1979) or moved poleward in response to the B_z or B_y change, respectively (Burch et al. 2002). Early observations of detached arcs by the ISIS-2 photometers (Anger et al. 1978) detected only precipitating electrons, while later observations by IMAGE-FUV detected strong proton precipitation that could account for the entire "electron aurora" signal (Frey 2007). It is unclear if the electron and proton arcs are inherently different, or if the early photometers were not sensitive enough to detect the missing proton component. It should be noted that the positions of electron and ion precipitation boundaries in the dusk sector are nearly coincident for all levels of magnetic activity. However, the latitudinal distribution of energy fluxes indicates that the positions of electron and ion precipitation maxima are spatially separated. Maximum energy fluxes of ions are observed at the equatorial precipitation boundary, while those of electrons at the poleward one (Vorobjev and Yagodkina 2014).

Burch et al. (2002) suggested that detached afternoon arcs are generated by a well-established sequence of events. During stronger geomagnetic activity, which is often related to extended periods of southward IMF, ions are injected at the night side, drift around the dusk side towards the dayside, and precipitate into a proton arc along the equatorward part of the main afternoon auroral oval. That afternoon oval is mostly created by low energy electrons and has a general diffuse aurora appearance. A polarity change of the IMF Bz or B_y causes a contraction and poleward motion of the diffuse auroral oval. However, the ring current proton precipitation remains unchanged and leaves the proton arc mostly unaffected; thus, the gap between the oval and the detached arc develops (Fig. 4).

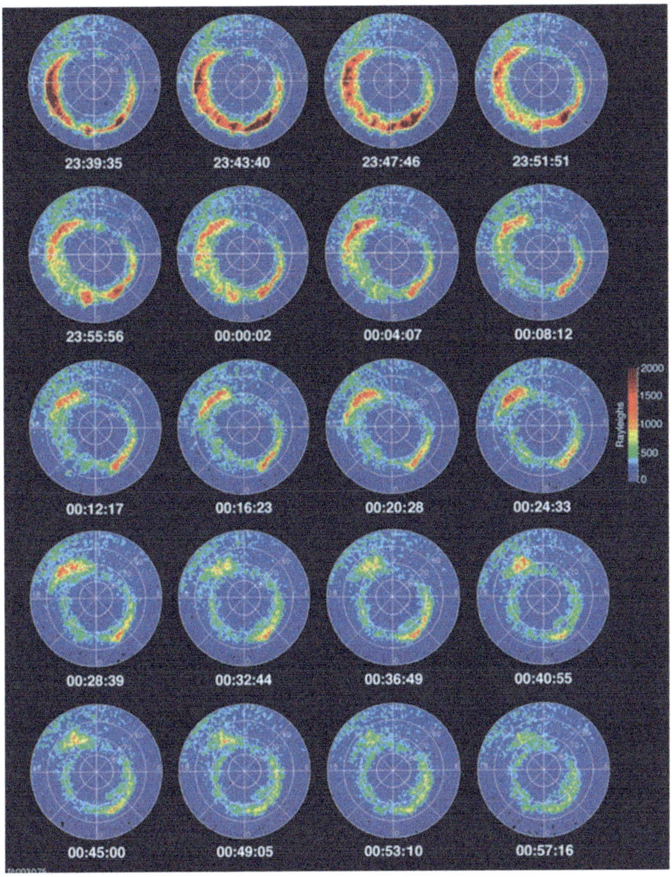

Fig. 4 Proton aurora images on 9–10 November 2000 demonstrating the separation of a detached afternoon arc from the auroral oval after a strong magnetic storm and IMF rotation to northward direction. Times shown are in universal time. Pixels were mapped into geomagnetic latitude and local time coordinates. Noon is at the top. The false color indicates the emission intensity in Rayleighs. (From Burch et al. 2002)

Spasojevic et al. (2004) present a case study in which coincident observations of IMAGE in the far ultraviolet (FUV) and extreme ultraviolet (EUV) establish a causal relationship between a detached afternoon proton arc, and a plasmaspheric plume that extended from a contracted plasmasphere towards the dayside magnetopause. During a geomagnetic disturbance, the plasmasphere contracted and a distinct plume formed because of enhanced sunward convection and corotation flows across the dayside (Fig. 5). After a polarity change of Bz the convection decreased, and corotation dominated the flow inside the magnetosphere. This caused the plasmaspheric plume to further corotate into the afternoon sector where a detached proton arc had developed. Magnetic mapping using the IGRF plus Tsyganenko-96 magnetic field model (Tsyganenko 1995) established the magnetic connection between the two phenomena. It was early on speculated that the proton precipitation may be due to pitch angle scattering of energetic protons by electromagnetic ion cyclotron (EMIC) waves, which may be preferentially generated within the plume since the presence of cold, dense ions lowers the threshold for the EMIC instability (Spasojevic et al. 2005; Spasojevic and Fuselier 2009).

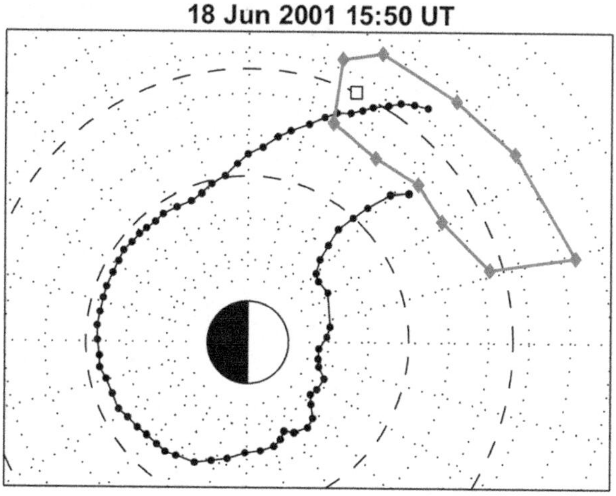

Fig. 5 Contour plot of the plasmapause (black circles) and the mapped proton arc (gray diamonds) in the geomagnetic equatorial plane. The coincidence of both features creates the link between the plasmaspheric plume and the afternoon arc (reproduced from Spasojevic et al. 2004)

Electromagnetic ion cyclotron (EMIC) waves are generated in the equatorial region of the plasmasphere-magnetosphere by internal wave-particle interaction with ring current ions (Fraser et al. 2006; Ebihara and Khazanov 2016). In ground observations, they are observed as Pc 1-2 (0.1–5 Hz) waves, and one group of waves exhibits a spectral fine structure that has been classically explained by bouncing packet field-aligned propagation. An unstructured class of Pc 1-2 waves, including intervals of pulsations with diminishing period, lacks a fine structure pattern and is the dominant emission observed in the middle and outer magnetosphere by satellites. The cold/cool magnetospheric plasma has a profound effect on the generation and spectral properties of EMIC waves (de Soria-Santacruz et al. 2013). The waves are often observed at geostationary orbit within the outer magnetospheric extension of plasmaspheric plasma plumes as seen by the IMAGE spacecraft in the EUV imager instrument and are associated with subauroral proton arcs seen by the FUV imager. This provides evidence in support of a ring current loss mechanism induced by pitch angle scattering of protons by EMIC waves (Jordanova 2020). Characteristic frequencies introduced into the cold/cool plasma by heavy O+ and He+ ions determine the EMIC wave spectra, and these may be used in plasma diagnostic studies.

Jordanova et al. (2007), confirmed the connection between EMIC waves and afternoon detached arcs with calculations using a kinetic physics-based model coupled with a dynamic plasmasphere model. These authors calculated the growth rate of EMIC waves self-consistently with the evolving ring current H+, O+, and He+ ion distributions. The aforementioned study found that EMIC waves are preferentially excited, and proton precipitation maximizes, within regions of spatial overlap of energetic ring current protons and dayside plasmaspheric plumes and along steep density gradients at the plasmapause. The model matches very well the temporal and spatial evolution of FUV observations of the detached proton arc (Pierrard et al. 2009).

The connection between particle precipitation and EMIC waves was also seen on the ground as geomagnetic Pc1 pulsations and Intervals of Pulsations with Diminishing Periods (IPDP) (Yahnin and Yahnina 2007; Yahnin et al. 2009). Particle precipitation signatures of ion-cyclotron interaction were described from low-altitude satellite measurements of the energetic proton fluxes as well as from observations of the proton aurora. As a result, localized proton precipitation patterns were shown as an elongated region of the precipitation, presumably mapped onto the plasmaspheric plume. Clear evidence of the pitch-angle scattering associated with the ion-cyclotron wave activity was found near the equatorial plane in the region conjugated with the localized proton precipitation at low altitude. Thus, the revealed precipitation patterns determined the location of the region of intense pitch-angle scattering of energetic protons. The proton arcs tend to appear at lower latitudes at later magnetic local times (MLTs). This agrees with the fact that the IPDP occurrence exhibits a similar behavior and that the IPDP end frequency tends to increase with increasing MLT (Yahnin et al. 2009). The spatial-temporal correlation of IPDP with proton aurora arcs confirms the expectation that the proton arcs are the ionospheric image of the region where the ion cyclotron instability develops in the equatorial magnetosphere. In the case of IPDP, the instability develops when drifting proton clouds resulting from particle injections in the night sector contact the plasmaspheric plume onto which the proton arcs map.

Ground-based observations of aurora and geomagnetic pulsations at subauroral latitude, using an all-sky imager and an induction magnetometer revealed isolated auroral arcs separated equatorward from the main auroral oval (Sakaguchi et al. 2008). The arcs were observed in both pre- and post-midnight sectors during the late recovery phase of geomagnetic storms. They found that as isolated arcs moved equatorward (poleward), the frequencies of the simultaneous Pc 1 pulsations increased (decreased). Using the Tsyganenko-02 magnetic field model (Tsyganenko 2002), the observed Pc 1 frequencies were almost the same as the frequencies of He+ EMIC waves at the equatorial plane connected to observed isolated arcs. These results also imply that the dynamics and instabilities in the inner magnetosphere can be monitored from the ground as low latitude auroral emissions away from the ordinary auroral oval.

The characteristics of precipitating ring current (RC) ions/electrons and precipitating radiation belt electrons associated with wave-particle interactions in the plasmaspheric plume in the main phase of a geomagnetic storm was investigated by Yuan et al. (2010, 2013). With observations of the NOAA-16 satellite, within the anisotropic zone, the peak of precipitating RC electron flux was equatorward to that of precipitating RC proton flux in a plasmaspheric plume recognized by the IMAGE and LANL-91/94 satellites. An enhancement of precipitating flux for >3 MeV electrons was simultaneously observed by NOAA-16 with the increase of precipitating RC proton flux within the anisotropic zone. Theoretical calculations of pitch angle diffusion coefficients for RC protons and for radiation belt electrons caused by EMIC waves demonstrated that precipitating flux enhancements of RC protons and >3 MeV radiation belt electrons are a result of EMIC wave-particle interactions in the plasmaspheric plume. These results suggest that EMIC waves in the plasmaspheric plume can scatter not only RC ions but also radiation belt electrons into the loss cone, thus explaining the existence of both proton and detached electron arcs.

3.2 Subauroral Spots

There are two types of subauroral spots, those with pure electron precipitation and those with pure proton precipitation. Their size is of the order of ~100–300 km diameter, they can last for many hours, and they generally corotate with 70-95% of the full corotation speed. General properties and conditions are summarized by Frey (2007). The very first study of such patches in the evening region, coined Evening Corotating Patches (ECP), concluded that they are relatively rare, occur during weak geomagnetic activity, and are created by high-energy electrons (Moshupi et al. 1977). Their location and coincident plasma measurements led to the conclusion that they had to be created by particles originating at the plasmapause. The electron cyclotron wave resonance was suggested as a potential generation mechanism.

ECP also showed up in ground-based all-sky camera observations in many different wavelength bands (Kubota et al. 2003). The patches were clearly visible in emissions that

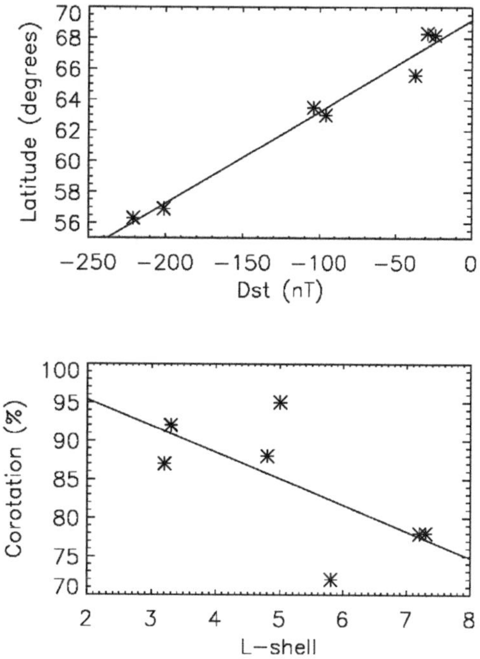

Fig. 6 (Top) Relationship between the latitude of SAMPS observation and the minimum Dst of the previous geomagnetic storm. The correlation is 0.98. (Bottom) Relationship between the corotation speed of SAMPS and the L-shell of the location. The correlation coefficient is −0.68. (From Frey et al. 2004)

are generated by higher energy electrons, but they were absent in low-energy electron aurora and proton aurora. Simultaneous particle measurements by a low-altitude satellite confirmed the absence of proton precipitation, and the electron energy was greater than 10 keV. Earlier observations from a different study placed the patches at plasmapause latitude (Moshupi et al. 1979). The interaction of cold plasmaspheric plasma at plasmapause undulations with drifting higher energy electrons, could account for the scattering of electrons into the loss cone and cause their precipitation into the ionosphere (Krall and Huba 2016).

The second category of subauroral spots created by pure proton precipitation was first described as Subauroral Morning Proton Spots (SAMPS), which were seen in the proton aurora channel of the IMAGE-FUV Spectrographic Imager (Frey et al. 2004). In the aforementioned study, SAMPS were observed in the Wideband Imaging Camera that is primarily sensitive to emissions from electron precipitation; however, energetic protons will also excite higher energy levels of atmospheric constituents (N_2 and O) that will then appear as electron aurora. Two passes of low-altitude spacecraft through the same flux tubes confirmed the pure proton precipitation and the lack of higher energy electrons.

The long observational periods from the high-altitude IMAGE orbit together with the long-lasting duration of these spots allowed for a detailed investigation into their occurrence and their dynamics. Frey et al. (2004) found subauroral spots to appear in the recovery phase of geomagnetic storms. The spots could be seen for several hours moving towards later local times with a little bit of lack behind true corotation speed. Different spots appeared at different geomagnetic latitudes. The authors found a good relationship between the minimum Dst of the prior geomagnetic storm, together with a strong relationship between their geomagnetic latitude (or L-shell) and their corotation speed (Fig. 6).

Coincident observations of the plasmasphere by IMAGE-EUV confirmed the relationship between SAMPS spots and spatial modulations of the plasmapause in two instances. In these two cases, the auroral spots magnetically mapped once to a notch and once to a shoulder in

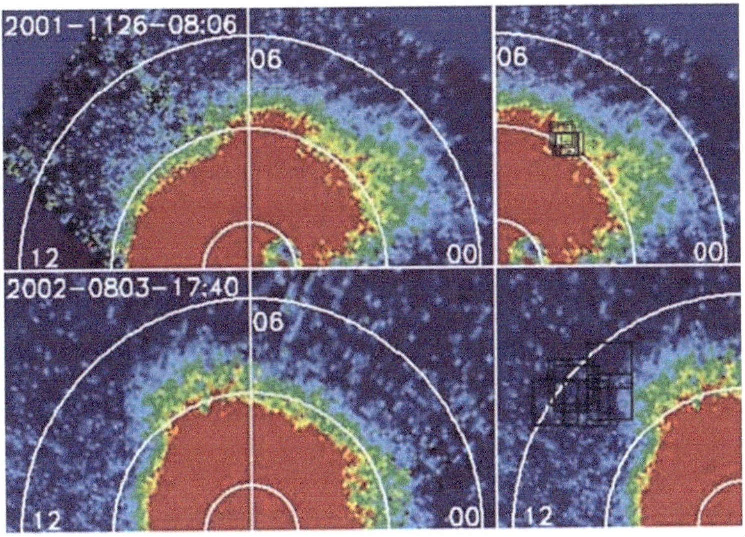

Fig. 7 (Top) EUV image of the plasmasphere at 08:06 on 26 November 2001, as mapped to the geomagnetic equator. The dynamic range was exaggerated to show the notch in the plasmapause at L = 3 and MLT = 4 hours. The left part shows the EUV image with the noon meridian to the left. The right part shows the location of SAMPS at ±5 min around the EUV center time mapped to the magnetic equatorial plane. (Bottom) The same for 3 August 2002 at 1740 UT when SAMPS were observed around 1000 MLT. The size of the squares in the two right panels represents the diameter of the magnetic flux tubes originating in the spots with the larger area for flux tubes originating at higher latitude and mapping to larger distances. (From Frey et al. 2004)

the plasmapause. Figure 7 shows these two cases with the SAMPS locations mapped into the EUV images. The speeds of the spot and the plasmaspheric shoulder were 88% and 94% of the corotation speed, respectively.

The above mentioned connection between the locations of spots and of modulations in the plasmapause suggested that wave-particle interactions can cause the precipitation of energetic protons. After strong geomagnetic storms, the originally eroded plasmasphere expands again. The cold plasma corotates and interacts with enhanced fluxes of trapped ring current ions. The energy limit for the interaction between ring current ions and EMIC waves is lowered, allowing a larger portion of the ion distribution function to be scattered into the loss cone.

The hypothesis that this proton aurora is produced by energetic proton precipitation after the interaction of ring current particles with electromagnetic ion cyclotron (EMIC) waves in the equatorial plane of the magnetosphere was further confirmed by ground-based observations of Pc1 pulsations (Yahnin et al. 2007). Geomagnetic pulsations Pc1, which are a ground signature of EMIC waves, were found when the proton spot is nearly conjugated with the ground station equipped with a pulsation magnetometer. Moreover, there is a good agreement between the appearance/disappearance of the spot and the beginning/end of the Pc1 train. Thus, the subauroral proton spots are images on the ionospheric "screen" of magnetospheric regions where the ion cyclotron instability develops leading to an intense scattering of energetic protons into the loss cone.

Proton aurora patches similar to SAMPS were also observed in the same evening local time region as the ECP (Zhang et al. 2005). However, simultaneous low altitude satellite particle measurements exhibited the absence of any electron precipitation and showed

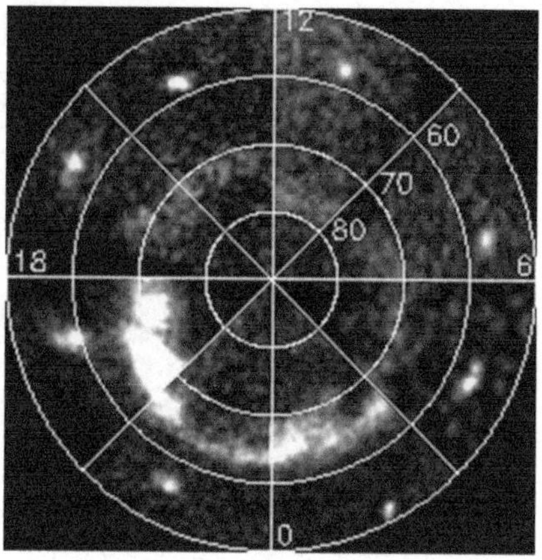

Fig. 8 Collage of "pie-sections" from eight different original images by IMAGE-SI demonstrating the occurrence of localized proton spots at all local times. The eight original raw images were collected on different dates in 2000–2002

pure proton spectra with average energies of 10–20 keV. Evening proton spots appeared at lower magnetic latitude of 45°–55° during the main phase of strong storms, while the electron spots generally appeared at higher latitude of 58°–65° after minor to moderate storms (Moshupi et al. 1979). This difference may not be important as bigger storms erode the plasmasphere stronger, but it is still unclear what causes in one case the wave-particle interaction of EMIC to scatter ions, and in other cases the electron cyclotron waves to scatter electrons. Are different processes responsible for the ion and electron scattering, or are there just two aspects of the same process at work? Observations by the IMAGE Spectrographic Imager (Mende et al. 2000) revealed more cases of pure proton spots occuring at all local times and not just in the morning and evening (Fig. 8).

In a later study, Sakaguchi et al. (2012) presented ground-based observations using all-sky imager and search coil magnetometer networks, which provide the large-scale distribution and motion of the EMIC wave-particle interaction regions. They observed several spots of isolated proton auroras coincident with Pc1/EMIC waves at subauroral latitudes during the expansion phase of a storm-time substorm. The isolated auroras were distributed over 4-hours MLT preceding midnight. The POES-17 satellite confirmed enhancements of 30-keV proton precipitation over the isolated auroras. The equatorward motion of the auroras and frequency drift of the wave were consistent with the plasmasphere eroding due to a polar cap potential enhancement modeled by a numerical simulation (Sakaguchi et al. 2016). Similar results were reported by Søraas et al. (2013) who determined proton energies of up to 100 keV going into a proton spot, which mapped to the equatorial plasmapause.

Observational evidence of rapid luminous modulation of isolated proton aurora (IPA) spots correlated with Pc1 waves simultaneously observed from the ground, which are equivalent to the electromagnetic ion cyclotron (EMIC) waves in the magnetosphere as reported by Ozaki et al. (2018). The fastest luminous modulation of IPA was observed in the 1 Hz frequency range, which was twice the frequency of the related Pc1 waves. The time lag between variations of Pc1 wave power and the IPA luminosity suggests that the source regions of IPA are distributed near the magnetic equator, suggesting an EMIC wave-energetic (a few tens of keV) proton or relativistic (MeV or sub-MeV) electron interaction. The generation mechanism of this 1 Hz luminous modulation remains an open issue.

Yahnin et al. (2013) compared the projection of "proton aurora spots" with the location of the plasmapause. Their plasmapause formation model is based on the quasi-interchange instability mechanism and was used to determine the plasmapause location. This comparison showed that often the proton aurora spot source is located near the plasmapause location or in the cold plasma gradient region just inside the plasmapause. In some events, the proton aurora spots mapped well outside the plasmapause. They suggested that in the latter case the ion cyclotron instability develops when westward drifting energetic protons interact with cold plasma that was earlier detached from the plasmasphere.

In a different study, two Pc1 bands were registered in the event when two proton auroral spots were simultaneously observed at different latitudes (Yahnina and Yahnin 2012). In this case, the Pc1 intensity distribution maximum at lower frequencies was related to a proton auroral spot at a higher latitude and vice versa. Such a spatial correlation between Pc1 pulsations and proton auroral spots, together with the previously established time correlation between these phenomena, demonstrates that subauroral proton spots delineate the region of ion cyclotron instability in the equatorial magnetosphere at the level of the ionosphere.

The relationship between evening corotating patches and processes in the plasmasphere and ring current were recently investigated through aurora and simultaneous particle observations by DMSP/SSUSI (Yao et al. 2019). The patches corotated with the Earth and stayed at a constant magnetic latitude. The particle measurements confirmed that energetic protons with energies above 20 keV without any significant electron contribution were responsible for the patches. The authors argued that the energetic ions producing ECP originate from the ring current, which is assumed to be filled with hot ions during the recovery phase of magnetic storms.

4 Stable Auroral Red (SAR) Arcs

In addition to the bright and variable optical emissions observed in the auroral oval, another zone of visible Magnetosphere-Ionosphere (M-I) coupling is the sub-auroral ionosphere-thermosphere system. Here, one of the optical structures typically observed is known as a SAR (Stable Auroral Red) arc. Its location and occurrence are directly linked to the ionospheric trough/plasmapause region– a region of reduced electron density. The poleward wall of the trough is caused by ionospheric plasma produced by the precipitation of low energy ions and electrons. The equatorward wall of the trough is the gradient caused by normal post-sunset decay ionospheric plasma. During geomagnetic quiet times, the trough is at approximately 60° magnetic latitude. This corresponds to an L shell value of 4, with L being the distance at the magnetic equator measured in Earth's radii. The magnetospheric region interior to this boundary is the *plasmasphere*—a region filled by upward flowing cold ionospheric plasma. Within the magnetosphere, this important field-line boundary is called the *plasmapause*, and it corresponds, approximately, to the separation of co-rotating plasma from plasma undergoing magnetospheric convection. During geomagnetic storms the ionospheric trough and the plasmapause move to lower latitudes—most often to about $L = 3$. Auroral features are enhanced and the whole auroral oval moves equatorward. The SAR arcs intensify and can be observed at $L = 3$. Figure 9 shows a schematic representation of the inner magnetosphere adapted from Kozyra et al. (1987). The two rings labeled 'Conjugate SAR Arcs' are the ionospheric foot-points of the overlapping plasmasphere and ring current regions. A SAR arc is almost always subvisual, typically 10–20 times fainter than visible aurora. Initial observations showed that brightness levels remain virtually constant for hours, the emission was described as "spectrally pure" red (630.0 nm-only photons) from atomic oxygen excited to the ^{1}D (~2 eV) state by ambient ionospheric electrons with $T_e > 3000$ K, and was a narrow emission feature (50–200 km) in N-S width, but with great E-W extent (1000's km) (Kozyra et al. 1997). This increased electron temperature is produced by heat

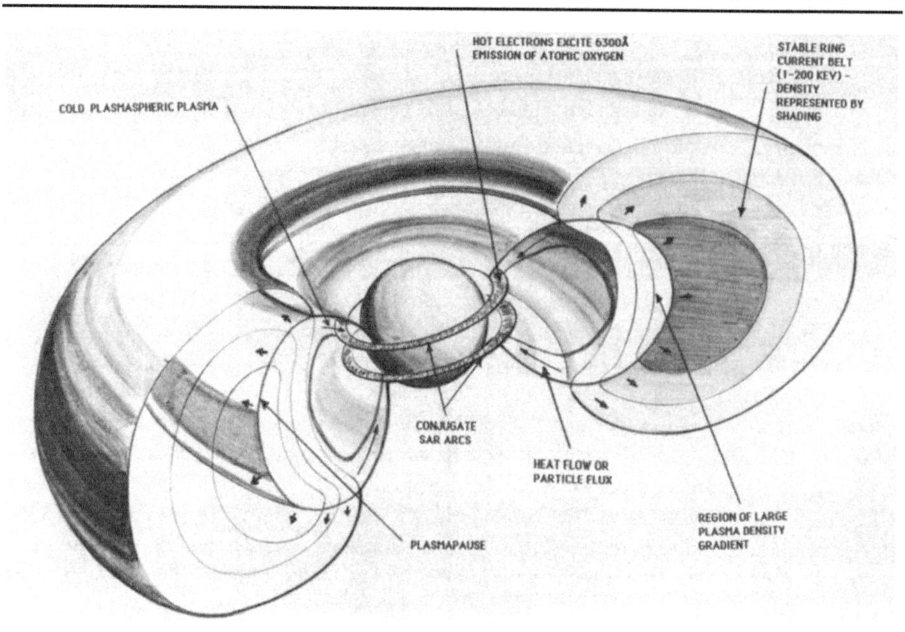

Fig. 9 Schematic representation of the inner magnetosphere showing the overlap between the ring current and the plasmasphere. SAR arc emissions occur at the foot of the field lines that thread this overlapping region (Kozyra et al. 1987)

conduction or low energy precipitating electron flux from the ring current (Rees and Roble 1975).

Using 10 years of SAR arc intensities measured by a network of scanning photometers during 1978–1988, Lobzin and Pavlov (1999) presented a statistical study correlating these observations with different solar and geomagnetic activity indices. They showed that the electron temperature in the SAR arc region does not strongly depend on the electron density and is strongly coupled to the additional heating of the electron gas due to interaction of the ring current ions and plasmaspheric electrons. As a result, variations of electron temperature in the SAR arc region with characteristic time scales from several minutes to several hours could be a manifestation of time variations of ring current parameters.

Takagi et al. (2018) performed a statistical analysis of SAR arc detachments from the main oval based on 11 years of all-sky camera observations obtained at subauroral latitudes (at Athabasca, $L = 4.3$). The authors showed that the yearly variation of SAR arc detachments has a better correlation with the geomagnetic Ap index than the solar F10.7 index. SAR arcs tend to occur pre-midnight, and at the beginning of a substorm recovery phase. They concluded that the SAR arcs detach from the main oval as the main auroral oval returns to higher latitudes at the beginning of the recovery phase. An average SYM-H index of ~ -20 nT was observed, indicating that the observed detachments could be related to weak geomagnetic storm conditions or non-storm time substorms.

Seasonal studies have been described by several groups using ground-based data from a northern hemisphere site (Fok et al. 1993; Kozyra et al. 1997). Thus, the seasonal patterns they addressed were acquired sequentially and, during different geomagnetic storms. Their statistical findings were that emission levels were lowest in summer months. Only recently, SAR arcs were observed simultaneously in both hemispheres and their characteristics were investigated (Martinis et al. 2019; Hong et al. 2020).

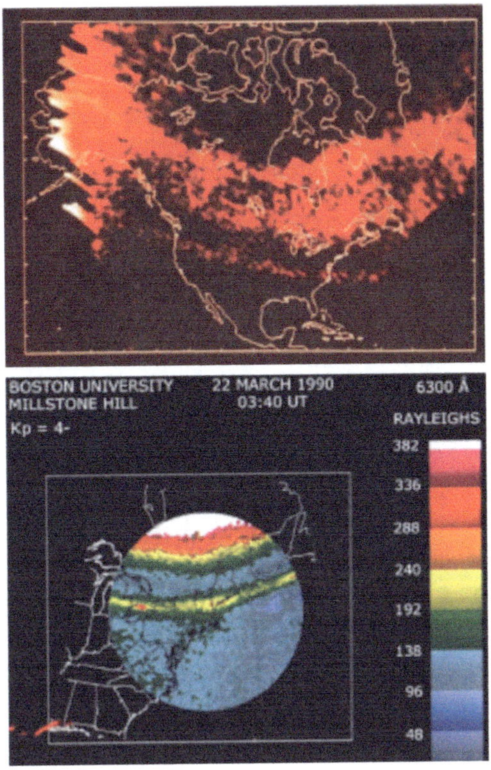

Fig. 10 (Top) Dynamic Explorer 2 satellite observations of the northern hemisphere showing the red aurora at higher latitudes and a relatively thin arc equatorward it. This thin arc represents a SAR arc. (bottom) Ground-based all-sky image from Millstone Hill Imager showing a bright auroral display and a uniform east-west elongated arc at subauroral latitudes (adapted from Mendillo et al. 2013)

The following sections summarize past and present observations of SAR arcs, energy sources for their formation, and examples of SAR arcs that diverge from their classical definition.

4.1 Observations: Past and Present

The first observations of a SAR arc from space occurred with the Dynamic Explorer (DE) polar orbiting spacecraft. Figure 10 (top) shows a SAR arc over North America. The arc is the southernmost band of emission separated from the more northerly aurora (Craven et al. 1982).

Optical ground-based observations of SAR arcs were performed using scanning photometers and all-sky imagers. Figure 10 (bottom) shows a SAR arc, equatorward from the diffuse aurora, observed with the Millstone Hill all-sky imager (Mendillo et al. 2013). A relatively uniform and stable structure was typically observed. Radar measurements have revealed characteristic ionospheric parameters such as electron temperature (Te) and electron density (Ne) within a SAR arc. Figure 11 (left) shows results from the Millstone Hill incoherent scatter radar. Near 44° magnetic latitude decreased Ne and elevated Te in the altitude range from ~300 to ~600 km were measured. In addition, a strong westward drift was observed, consistent with a SubAuroral polarization stream (SAPS) co-located with a SAR arc (Spiro et al. 1979; Foster and Burke 2002). These results indicate that SAR arcs and SAPS might be related phenomena occurring simultaneously.

Previous studies have reported in-situ satellite observations of SAR arcs and SAPS, i.e., increased Te, decreased Ne, and elevated westward plasma flow (Rees and Roble 1975;

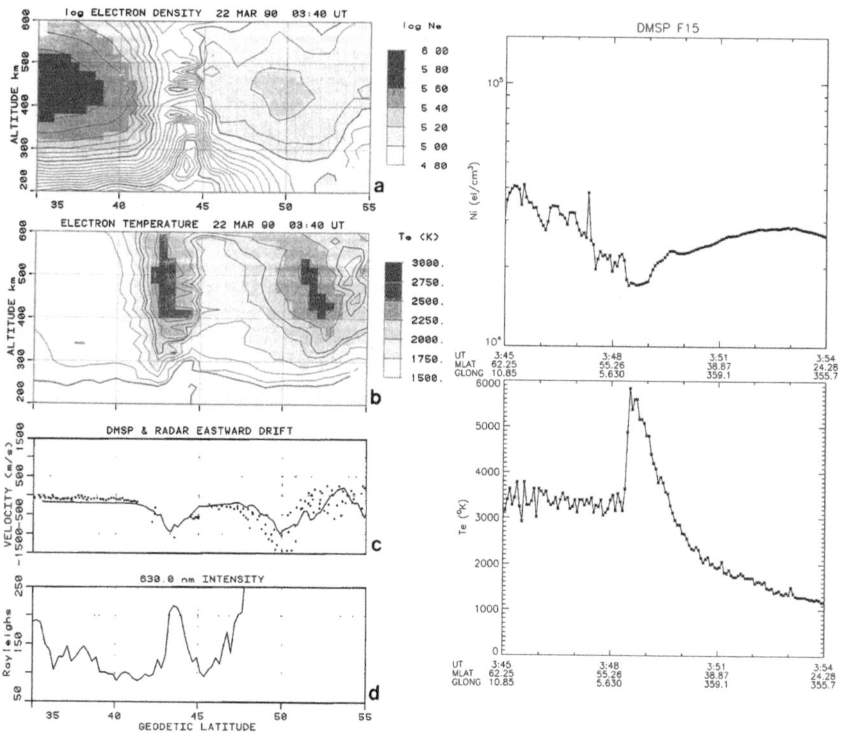

Fig. 11 (Left) From top to bottom: (**a**) electron density height profile using the Millstone Hill ISR; (**b**) electron temperature, showing enhanced values near 43°; (**c**) Westward plasma flows obtained from radar and DMSP data; (**d**) 630.00 nm brightness showing a SAR arc (from Foster et al. 1994). (Right) DMSP F15 data showing ion density (top) and electron temperature Te (bottom). SAR arc signatures are visible as the spike in Te associated with a minimum in ion density (adapted from Mendillo et al. 2016a)

Kozyra et al. 1997). For example, Fig. 11 (right) shows data from DMSP, orbiting at ~840 km, during the geomagnetic storm of 12 September 2013. The left panel shows Ne and Te measurements (a, b) and drifts (c) from Millstone Hill ISR. Airglow intensity from a collocated all-sky imager (ASI) is shown at the bottom (d). These plots clearly show the reduced Ne (a), increased Te (b), and westward SAPS flow enhancement (c) collocated with the higher brightness below 45° latitude indicating the presence of a SAR arc (d).

Evidence of wide longitudinal occurrence and long lifetime was reported by the DE-1 spacecraft when a SAR arc was observed in four subsequent, consecutive orbits spanning 28 hours (Craven et al. 1982). Recent observations (Mendillo et al. 2013) using multiple ground-based all-sky imagers have shown that SAR arcs extend along the entire globe, as depicted in Fig. 9. Three ASIs, located in North America, Europe and New Zealand detected a SAR arc during the 27 September 2011 storm as shown in Fig. 12. This study represents the first time that the same SAR arc has been seen from ground-based images from multiple continents in both hemispheres. The center of the SAR arc in each image was determined along the N-S meridian to create the local time versus latitude pattern for that night. Those locations were then mapped to the magnetospheric equatorial plane and their positions versus Local Time (LT) and L value appear at the bottom of Fig. 12. There is a clear trend of the SAR arc/plasmapause moving earthward from dusk to dawn (O'Brien and Moldwin 2003), similar to the morphology observed in SAPS (Foster and Vo 2002). To merge the

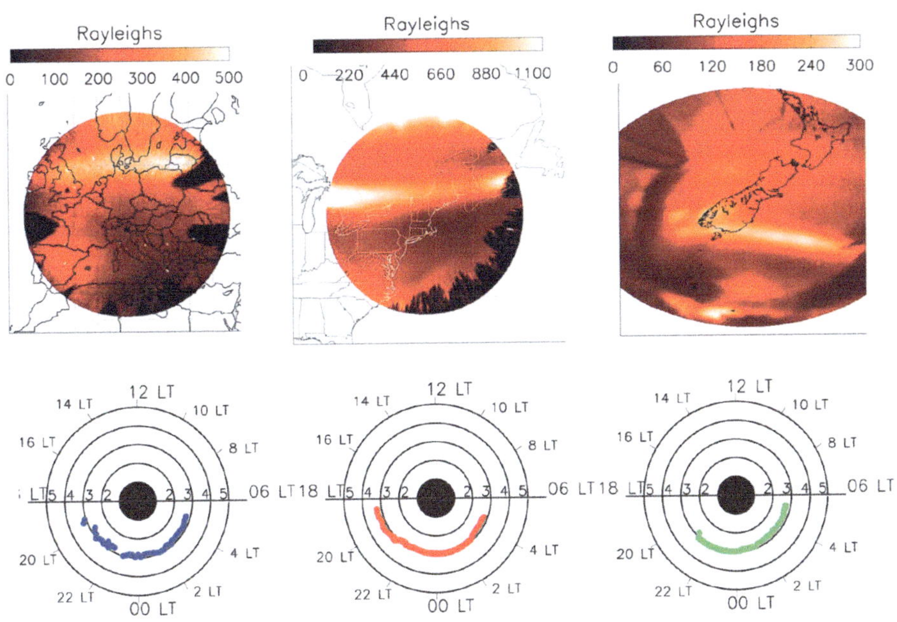

Fig. 12 (Top) Sample images of a SAR arc at Asiago, Millstone Hill, and Mount John during the storm of 26–27 September 2011. (Bottom) Equatorial plane representation of the location of the arc (adapted from Mendillo et al. 2013)

three-site LT observations into a single composite result, the positions from each station were re-mapped into a UT and L value format (Fig. 13). The total data set spans darkness over ~24 h of UT storm time. When the end of coverage at one station overlaps the onset of coverage at the next station, the dusk-dawn asymmetry in plasmapause location can be seen at specific UT periods. The clearest example of this occurs between 08:00 and 10:00 UT on 27 September when the dusk plasmapause is at $L = 3$ (from Mount John data) while the dawn plasmapause is at $L = 2.7$ (from Millstone Hill data).

Magnetically Conjugate observations of SAR arcs using satellite data have been reported by Reed and Blamont (1974) and LaValle and Elliott (1972). The Reed and Blamont's study was a brief report on the SAR arc of 28–30 September 1967 observed by the photometer onboard the Ogo-4 spacecraft. The Ogo-4 data in the southern hemisphere located the SAR arc within a few tenths of the L-value recorded in Idaho ($L \sim 3$). LaValle and Elliott (1972) analyzed three SAR arcs from the satellite OV1-10 that was launched in 1966 in a polar orbit. One of the SAR arcs was observed as a conjugate phenomenon on 16 February 1967. Brightness values did not differ much in both hemispheres: ~ 500 R in the northern hemisphere as compared with ~ 650 R in the southern hemisphere. Ground based studies in both hemispheres were presented by Roach and Roach (1963). They reported SAR arc observations in North America and in New Zealand, but at different times, i.e., not simultaneous conjugate observations.

SAR arc studies using two ground-based data sets linked by the same geomagnetic field line, i.e., *conjugate-point studies*, allow seasonal patterns and other hemispheric characteristics to be obtained *simultaneously*, not sequentially, as presented above. Martinis et al. (2019) showed results from two magnetic conjugate stations, Millstone Hill (USA) and Rothera (Antarctica), that are part of a network of all-sky imagers built and operated worldwide by Boston University (Martinis et al. 2018). Previous works addressing 'seasonal

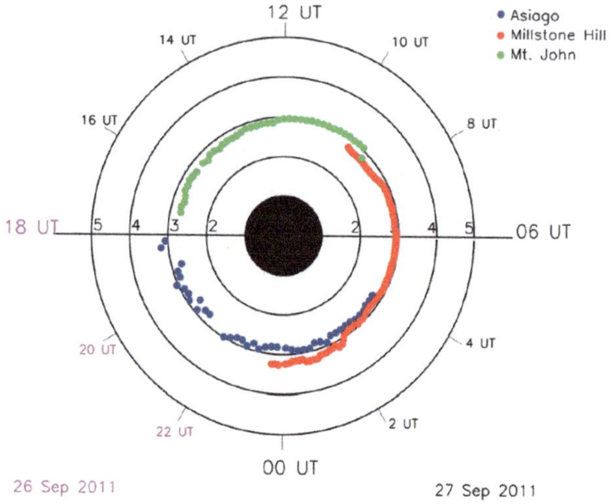

Fig. 13 Location of the SAR arc of 26–27 September 2001 in UT obtained by merging the data from the local time patterns in Fig. 12. (From Mendillo et al. 2013)

asymmetries' relied on measurements at a single site during different seasons (implying different background conditions) and different storm conditions. The work by Martinis et al. (2019) showed for the first time simultaneous inter-hemispheric characteristics of SAR arcs during the 1 June 2013 geomagnetic storm.

Figure 14 shows unwrapped all-sky images at Millstone Hill (top) and the corresponding Rothera images (bottom). The circle in each image represents the zenith location and the asterisk indicates the conjugate location of the imager in the opposite hemisphere. The arcs are observed at the same magnetic latitudes in both hemispheres, i.e., they are magnetically conjugate. The middle panel shows the arc overhead at Millstone Hill and the arc observed at Rothera falls right on top of the conjugate location of Millstone Hill. Both arcs seem to have similar width, although one of the main differences is that the poleward edge of the arc at Rothera is sharper and less diffuse than the one observed at Millstone Hill. This effect is not related to viewing geometry or differences in the two optical systems. Figure 15 shows brightness values (background subtracted) for the SAR arcs at Millstone Hill and Rothera for all the images taken on 1 June 2013. The ~30-minute data gap in the Rothera curve represents a brief period when clouds affected the calibration at this site. A ±25% uncertainty in brightness is shown in each data point. The SAR arc at Rothera is clearly brighter than the one observed at Millstone Hill, at times by a factor of 2–3. Both arcs show simultaneous brightness variations during the night. Martinis et al. (2019) argued that the observed changes were related to changes in solar wind/magnetospheric activity, as measured by different solar wind parameters, geomagnetic indices and particle flux data. The main conclusions in the study showed that: (a) the SAR arc observed at one site is seen at its magnetically conjugate location; (b) the SAR arc at Rothera (winter) is brighter than the one observed at Millstone Hill (summer), sometimes by a factor of 2–3 and similar changes in brightness are observed at both sites throughout the night; (c) both arcs show the same equatorward motion and have similar latitudinal width, although the poleward edges of the arcs at Rothera are sharper.

A recent study by Hong et al. (2020) used the Millstone Hill ASI and a conjugate ASI at the King Sejong Station, close to the Rothera ASI, to compare SAR arcs during the St. Patrick's Day storm in 2015. They determined that ionospheric irregularities occurred only near the diffuse aurora and not in the SAR arc and found that the Millstone Hill arc showed a broader latitudinal extent.

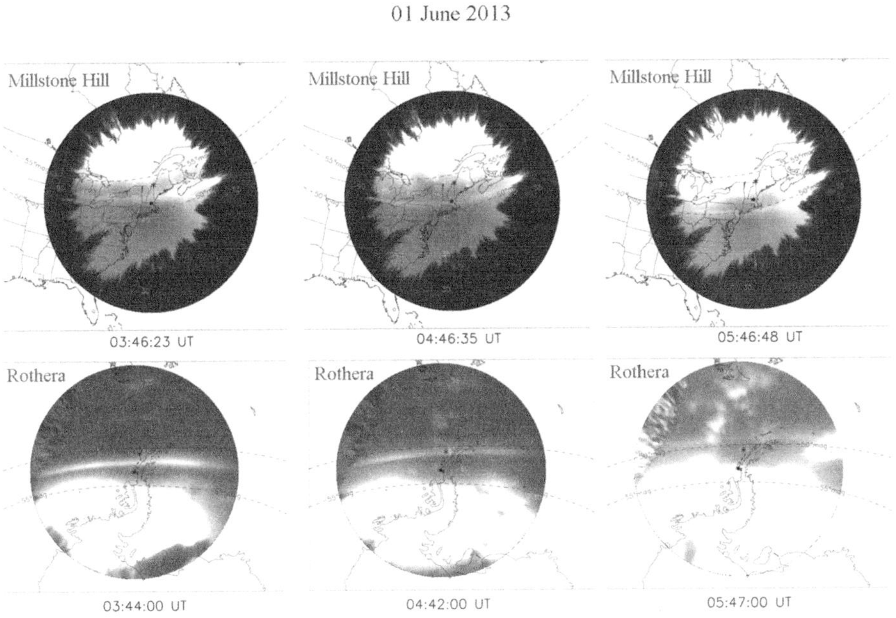

Fig. 14 (Top) Simultaneous images from Millstone Hill all-sky imager (ASI) and (bottom) Rothera ASI at three different times. The black circles indicate the zenith location and asterisks show their respective conjugate points. (From Martinis et al. 2019)

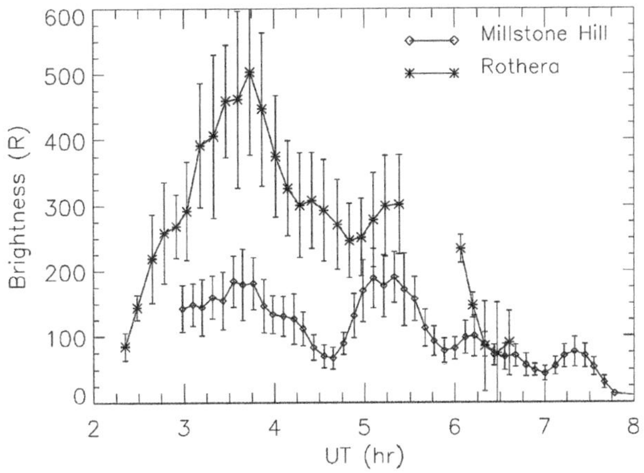

Fig. 15 (Millstone Hill (diamonds) and Rothera (asterisk) SAR arc brightness (above background airglow levels). The SAR arcs curves at both sites follow a similar trend, with Rothera values being larger than Millstone Hill values. Error bars represent the uncertainty in the determination of the brightness values. (From Martinis et al. 2019)

4.2 Energy Source of SAR Arcs

There are no conclusive observations or theories explaining how the energy required to feed SAR arcs is generated. All the candidate energy source mechanisms require that a fraction of storm energy be transported from the inner magnetosphere (at $L \sim 3$) along magnetic field lines down to F-region heights via heat conduction or soft electron precipitation (Rees and Roble 1975). According to Fok et al. (1991, 1993), this heat conduction is understood as the downward transport (into the ionosphere along field-lines) of energy produced by the interaction of energetic ring current ions with plasmaspheric electrons.

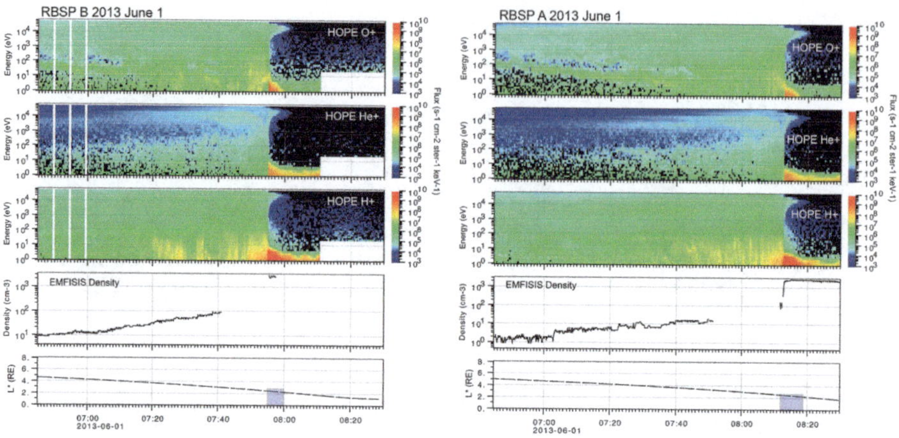

Fig. 16 HOPE and EMFSIS data for VAP B (left) and VAP A (right). The location of the plasmapause coincides with the increase in fluxes in the three ion species between ~08:00UT and 08:20UT. Both probes sample the field lines that connect to the subauroral ionosphere where a SAR arc is being measured by the Millstone Hill ASI (Showed as the shaded blue region)

Cole (1965) described the connection between SAR arc emissions and collisional interaction between ring current ions and electrons in the plasmasphere. The amount of energy transferred to the thermal electrons is a function of the temperature and density of the plasmaspheric thermal electrons and the magnitude, composition and energy distribution of the ring current. Using satellite data and ground-based scanning photometers Kozyra et al. (1993) explained the occurrence of SAR arcs during the different phases of a storm. For SAR arcs occurring during the recovery phase, it was observed that O^+ was the dominant energy source for SAR arcs; while during the main phase of the storm, medium energy H^+ was significantly enhanced. Column electron heating rates were computed using a well-established model (Rees and Roble 1975) that takes into account the flux and energy of ring current ions and their interaction with plasmaspheric thermal electrons at a given temperature and density. Calculations show that the maximum energy losses depend on the interaction between ring current ions of particular energy ranges with thermal plasmaspheric electron populations. For a SAR arc's typical electron density (Ne) and temperature (Te), the peak in energy loss for H^+ and He^+ occurs at $E < 10$ keV, and for O^+ it occurs between 30 keV and 50 keV. These average maximum energy loss rates are shifted toward lower energy ions if the plasmaspheric Ne and Te are smaller (Kozyra et al. 1993). Modelling results show that after the interaction between the high energy ions and the plasmaspheric population an enhanced flux of very low energy ions (~10's eV) is observed (Fok et al. 1993).

Ongoing studies are combining Van Allen Probes data with ground-based all-sky imagers detecting SAR arcs, trying to find out if these eV ion populations are observed near L-shell magnetic lines connecting with the sub-auroral ionosphere. Low energy ions measured by the Helium, Oxygen, Proton, and Electron (HOPE) instrument have been observed to increase near the magnetic equator on magnetic field lines that connect to locations where SAR arcs are observed. In Fig. 16 we show low-energy ion fluxes measured by HOPE, as well as cold density profiles derived from upper hybrid wave measurements provided by the Electric and Magnetic Field Instrument Suite and Integrated Science (EMFISIS). The blue-shaded regions mark the mapped locations of the ground-based observed SAR arc, for each probe (B is leading A). It can be seen that the arc maps to the plasmapause region and coincides with enhanced low energy ion fluxes. Figure 17 shows pitch-angle information

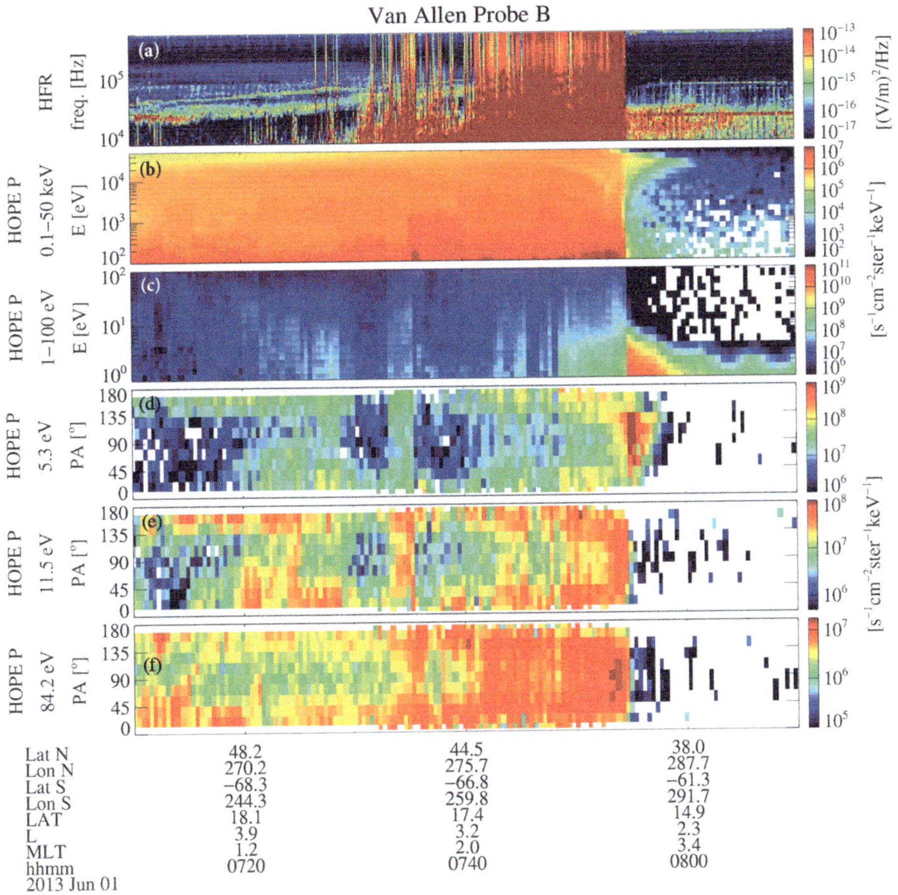

Fig. 17 The H^+ ion distributions observed by Van Allen Probe B. (**a**) The high frequency wave electric spectrogram observed by HFR instrument; the energy spectrograms of spin-averaged proton fluxes observed by HOPE at energies (**b**) from 500 eV to 50 keV and (**c**) from 1 eV to 100 eV; the pitch angle spectrograms of proton fluxes observed by HOPE at energies of (**d**) 5.3 eV, (**e**) 11.5 eV, and (**f**) 84.2 eV. The spacecraft travelled from the plasma trough to the plasmasphere at ~07:49 UT. The pitch angle distribution of protons evolved from mostly field-aligned to pancake-like during the plasmapause crossing, indicated by the appearance of trapped protons during ~07:40–08:00 UT

for H^+ ions at several energies. When VAP-B crosses the magnetic field lines linked to the SAR arc predominantly trapped particles were observed, i.e., no evidence of ion outflows, as shown in the 11.5 eV and 82.4 eV panels between ~07:30 and 08:00 UT. We also looked at the wave data for this event. During 07:30–8:00 UT, the EMFISIS data (not shown) indicate that no EMIC (electromagnetic ion cyclotron) waves were detected by either VAP A or B. VAP EFW (electric fields and waves) data published by Thaller et al. (2015) indicate persistent SAPS presence in the magnetosphere during this period.

Other generation mechanisms include interactions between thermal electrons and low frequency EMIC waves (Cornwall et al. 1971) and Landau damping of kinetic Alfven waves (Hasegawa and Mima 1978). Direct observations of ion cyclotron waves in the SAR arc region ($L < 4$) are not common. Anderson et al. (1992) studied the Pc 1-2 waves (0.1–4 Hz) in the L shell range 3.5–9 near the equatorial plane. They found that for L< 5 Pc 1-2 waves occurred 1–2% of the time, while this percentage increased to 10–20% on the dayside at $L > 7$. This result was also obtained with the CRRES satellite. Iyemori et al. (1994) studied the relationship between Te enhancements in the sub-auroral ionosphere and Pc 1 waves measured near the plasmapause using DE 2 spacecraft. After examining thousands of orbits, only 22 events with large amplitude waves were found. Some of these events were accompanied by Te enhancements. Lanzerotti et al. (1978) examined two intervals of irregular magnetic activity during SAR arc observations on December 17–18, 1971. By assuming a wave amplitude of 10 nT the authors concluded that it was plausible that sufficient wave energy might be available to produce the SAR arc. Kozyra et al. (1987) showed a case where intense ELF hiss was observed at a field line in the magnetosphere that connected to an optically observed SAR arc. Recently, Wang et al. (2019) performed test-particle simulations at $L = 5$ to demonstrate the possible cold (1–10 eV) electron heating by EMIC waves with wave amplitudes of a few nT through nonlinear Landau resonance. The importance of plasma waves in SAR arc formation still remains poorly understood.

The role of large perpendicular electric fields in the ionosphere on the formation of SAR arcs has also been considered. At sub-auroral latitudes in the F-region it is fairly common to observe east-west elongated structures characterized by large northward electric fields of up to $\sim$100 mV/m, or a few mV/m in the equatorial plane, confined to narrow bands inside the main ionospheric trough in the evening local time sector. Depending on their characteristics they can be SubAuroral Polarization Stream (SAPS) (Foster and Burke 2002; Foster and Vo 2002; Clausen et al. 2012), SubAuroral Ion Drift (SAID) (Spiro et al. 1979; Mishin 2013), or Polarization Jet (PJ) structures (Galperin 2002). Sazykin et al. (2002) suggested that the presence of strong localized SAPS-like electric fields is potentially capable of powering weak ($\sim$100 Rayleigh) SAR arcs. SAID events are supersonic westward plasma drifts on the equatorward edge of the diffuse aurora in the evening and nighttime sector. Their optical F-region signatures are weak 630.0 nm red arcs collocated with regions of fast convection. These weak arcs resemble SAR arcs observed during the recovery phase of magnetic storms, but have lower intensities, shorter lifetimes, and occur without a significant heat flux from the magnetosphere. Ion-neutral collisional heating and ion composition changes during PJ events may be an additional source of 630.0 nm emission. Ion-neutral frictional heating in the F-region may be capable of providing energy enough to excite weak emissions even without the heat flux from the magnetosphere postulated as the source of SAR arcs. As a test to distinguish PJ-related SAR arcs from ring current-induced arcs, Sazykin et al. (2002) suggested that some enhancement of the highly forbidden 5200 NI spectral line is expected in PJ-related arcs due to optical relaxation of $N(^2D)$.

It is possible that more than one mechanism is playing a role in the generation of energy to feed a SAR arc. For example, Jordanova et al. (1999) considered interactions between ring current protons and EMIC waves of the He^+ wave branch. The maximum wave gain occurred in the post-noon to midnight local time sector, just inside the plasmapause. Landau damping of these waves could represent an additional heat source for the elevated electron temperature observed at F region heights during the storm period. Total magnetospheric heating, resulting from both Coulomb collisions and wave-particle interactions will thus maximize near dusk, in agreement with previously reported westward intensifications of SAR arcs emission (Kozyra et al. 1993; Mendillo et al. 2016b).

The relative contribution of the above-mentioned heating processes to the electron temperature enhancement at sub-auroral latitudes and SAR arcs excitation should be quantitatively investigated in future studies.

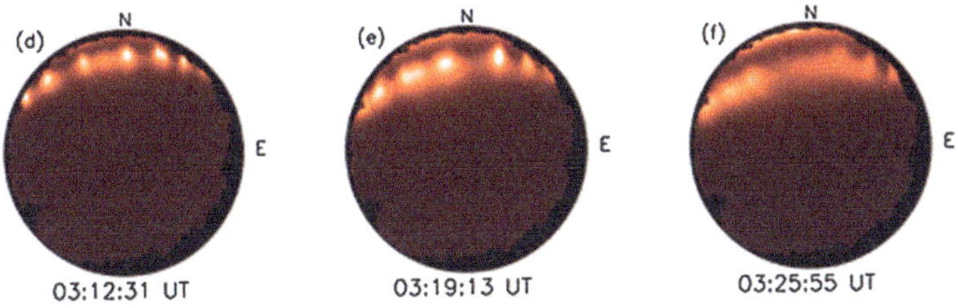

Fig. 18 SAR arc with embedded structures observed on the night of 2 October 2013 at Millstone Hill. After ~30 minutes, the beaded structures begin to dissipate and a 'regular' uniform arc is left (adapted from Mendillo et al. 2016a)

4.3 Departure from 'Stable' and 'Homogenous' SAR Arcs

Recent studies have shown that SAR arcs are far from being 'stable' structures and not all of them agree with the "classic" definition of a single east-west, geomagnetic latitude-aligned, uniform emission feature. Mendillo et al. (2016a) presented observations of a stable auroral red (SAR) arc observed from the Millstone Hill observatory that contained embedded sub-structures. Patches of emission occurred during a ~30 min period in the pre-midnight sector that coincided with the main phase onset time of a moderate geomagnetic storm. Later in the evening, the pattern evolved into a classic SAR arc of uniform brightness east to west. Figure 18 shows these 'beaded' structures superimposed on the uniform arc. Several brightness peaks with average longitude spacing of ~4° can be identified. Concurrent images in the oxygen green line (5577 Å) showed similar patches of emission but without a background arc. Observations of total electron content (TEC) from GPS satellites showed increases in TEC and phase fluctuations when the ray paths crossed the optical features. This case seems to represent an example of two emission mechanisms within a single SAR arc: a dominant level of emission due to thermal excitation, augmented by small bursts of soft energetic particles that excite additional 6300 Å radiation and 5577 Å. These observations led Mendillo et al. (2016a) to hypothesize that the inner edge of the plasma sheet (source of the soft precipitation that creates diffuse aurora) and the ring current-plasmapause interaction region (source of the downward heat flux that creates the SAR arc) occupied the same flux tube of 0.2 R_E extent in the geomagnetic equatorial plane during the ~30 min period when the beaded structures were observed.

Other morphological departures from single and 'stable arcs' are shown in Fig. 19. The left panel shows an image with two SAR arcs, representing the presence of two boundaries in the plasmasphere. Evidence of this kind of structuring seems to be possible as seen in Nishimura et al. (2019) who showed the presence of two separated ionospheric troughs and electron temperature peaks when looking at DMSP 17 data (see their Fig. 2a). The middle panel shows an arc with finger-like structures extending to the north, evidence of irregular and dynamic poleward boundaries. In the right panel, several small-scale patches of enhanced brightness are observed along the arc. The cases shown in Fig. 19 indicate that the assumption of a uniform Magnetosphere-Ionosphere interaction region needs to be reconsidered to take into account the non-uniform ionospheric response as a result of associated complex plasmapause/ionospheric trough boundaries.

5 STEVE

The discovery of a new phenomenon located several degrees equatorward from the auroral oval is an excellent example illustrating that new developments are still occurring in auroral

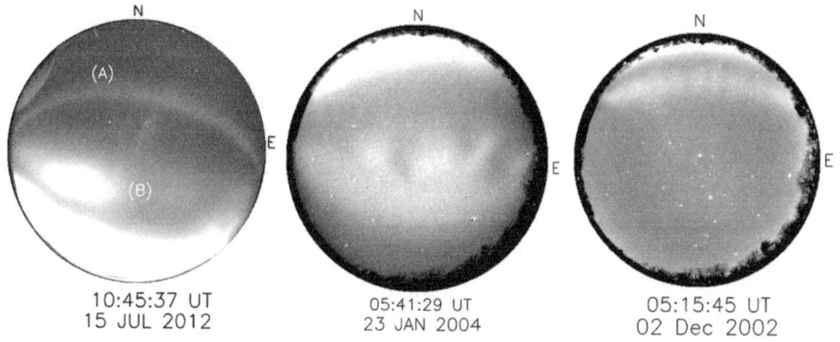

Fig. 19 Examples of lack of uniformity observed during three different SAR arc events. (Left): A double SAR arc observed at Mount John in New Zealand; the arc closer to the diffuse aurora is brighter than that located equatorward. (Middle): Northward finger-like structures accompany the SAR arc observed at Millstone Hill. (Right): Several small-scale structures are seen within the SAR arc. (Figure adapted from Mendillo et al. 2016b)

and subauroral research. Amateur night sky watchers and auroral photographers have documented this upper atmospheric phenomenon for decades, but the scientific community only recently began analyzing it in further detail. The story of this discovery has been recently published by Gallardo-Lacourt et al. (2019).

Based on auroral photographers' observations, the optical structure appears as a narrow luminous mauve structure across the night sky over thousands of kilometers in the east-west direction. Due to the location of the emission (equatorward of the auroral oval), the phenomenon was initially incorrectly named 'proton arc'. Observers later changed the name to 'Steve', a denomination without physical implications. As described below, based on the observed characteristics of this phenomenon, the name 'Steve' was subsequently changed to Strong Thermal Emission Velocity Enhancement, or STEVE.

In 2018, MacDonald et al. reported a STEVE event for the first time in the scientific literature. One of the primary tasks in this study was to identify STEVE in the scientific optical dataset with the aid of amateur auroral photographer data. As observed in red-line All-Sky Imager (ASI) data, STEVE appeared adjacent to, and equatorward of, the aurora oval (i.e., in the subauroral region). During this event, the structure extended thousands of kilometers in the east-west direction but had a width of only tens of kilometers in latitude. Figure 20 shows the STEVE event observed on July 25, 2016 reported by MacDonald et al. (2018). The event lasted for about an hour while propagating from east to west, covering the entire longitudinal extent of the ASI.

In addition to STEVE, MacDonald et al. (2018) reported a green feature resembling a picket fence that is sometimes observed together with STEVE. This green structure had smaller longitudinal and latitudinal dimensions than STEVE.

At the same time of the STEVE observation, the Swarm-A satellite crossed the location of the emission corroborating that the structure was located in the subauroral region. In addition, Swarm-A revealed that the observed luminosity was collocated with a hot ($\sim$6000K electron temperature) stream of fast moving ($>$6 km/sec) plasma less than 50 km in north-south extent. In addition, Swarm-A measured a small magnetic field perturbation consistent with a small downward field-aligned current (FAC). This result suggested that STEVE may not be generated like auroral arcs, which are associated with upward FACs (downward or precipitating electrons) (Kamide and Akasofu 1976). Based on these observed characteristics, scientists converted "Steve" into the backronym of Strong Thermal Emission Velocity

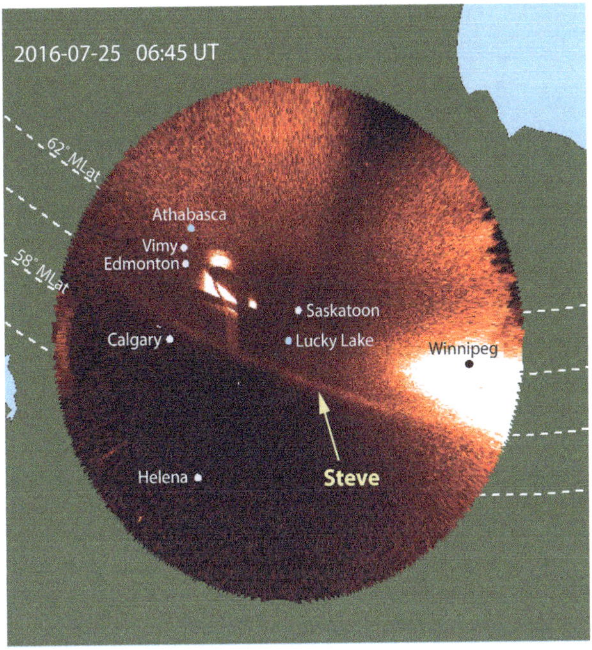

Fig. 20 STEVE event reported by MacDonald et al. (2018) observed in REGO at Lucky Lake on July 25, 2016

Enhancement, or STEVE. As shown in Fig. 21, STEVE's maximum luminosity (Panel a) is collocated with fast moving ions (Panel b), extremely high electron temperature (Panel c), and the plasma trough (Panel d). The in-situ observations shown in Fig. 21 are consistent with a Subauroral Ion Drift (SAID); Archer et al. (2019a) recently corroborated the relationship between STEVE and extreme SAIDs which are characterized by faster moving ions, higher electron temperatures, and lower electron densities than regular SAIDs.

This initial case study by MacDonald et al. (2018) suggested STEVE to be fundamentally different from traditional aurora. Several studies over the next two years investigated STEVE with as many different instruments as possible in order to further our understanding of this phenomena.

5.1 STEVE'S Generation Mechanisms

In an attempt to understand the generation mechanism behind STEVE's formation, Gallardo-Lacourt et al. (2018a) analyzed data from the Polar Orbiting Environmental Satellite (POES)-17 for one STEVE event that occurred on March 28, 2008 and was identified by the Time History of Events and Macroscale Interactions during Substorm (THEMIS) ASI. Figure 22 shows energy flux for ions (left) and electrons (right) for this event. The trapped particle population (90° view angle) and precipitating particles (0° view angle) are in black and blue curves, respectively (except for panel j that only shows 0° and 30° view angle). The green vertical line indicates the time when the magnetic footprint of POES-17 crosses STEVE. For protons, there is no clear evidence of precipitating particles at the time of STEVE. The situation is similar for high energy electrons (>30 keV); however, for low energy particles (50 eV–1 keV) the detector captured a significant increase in the precipitating electron flux (several orders of magnitude) at the time of STEVE. Nevertheless, the authors reported that the integrated energy flux for this event is at least two orders of magnitude

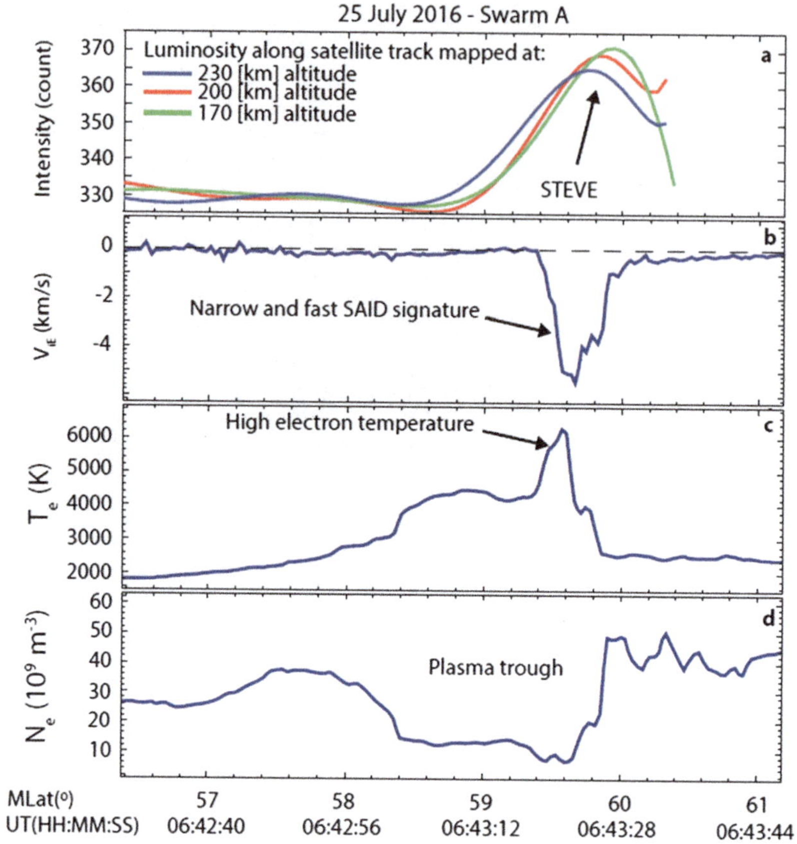

Fig. 21 Swarm-A satellite crossing STEVE on July 25, 2016. Adapted from MacDonald et al. (2018)

smaller than the average energy fluxes reported for visible auroras (Zhang and Paxton 2008; Newell et al. 2010), concluding that these low precipitating energy fluxes cannot be responsible for the luminosity observed in STEVE.

This result suggests that STEVE is not produced by particle precipitation similar to the aurora but generated locally in the ionosphere.

A more recent study by Nishimura et al. (2019) investigated the magnetospheric drivers of several STEVE events and the "picket fence" originally identified by MacDonald et al. (2018). Although the mauve emission observed in STEVE could be explained by F-region thermal emission (like in SAR arcs), this mechanism could not explain the green features observed in the picket fence. The authors analyzed the processes involved in the generation of both phenomena by using a combination of non-traditional data from citizen scientists' photographs, ground and space-based imagers, and satellite data. The study showed for the first-time conjugate features for STEVE in the two hemispheres by using citizen scientist data in the northern hemisphere and the first STEVE observation measured by the Special Sensor Ultraviolet Spectrographic Imager (SSUSI) instrument on board of DMSP F-17 satellite in the southern hemisphere. The reported absence of proton precipitation for these STEVE events supported previous findings and confirmed that STEVE does not correspond to proton precipitation or a proton arc. The authors also reported that the picket fence struc-

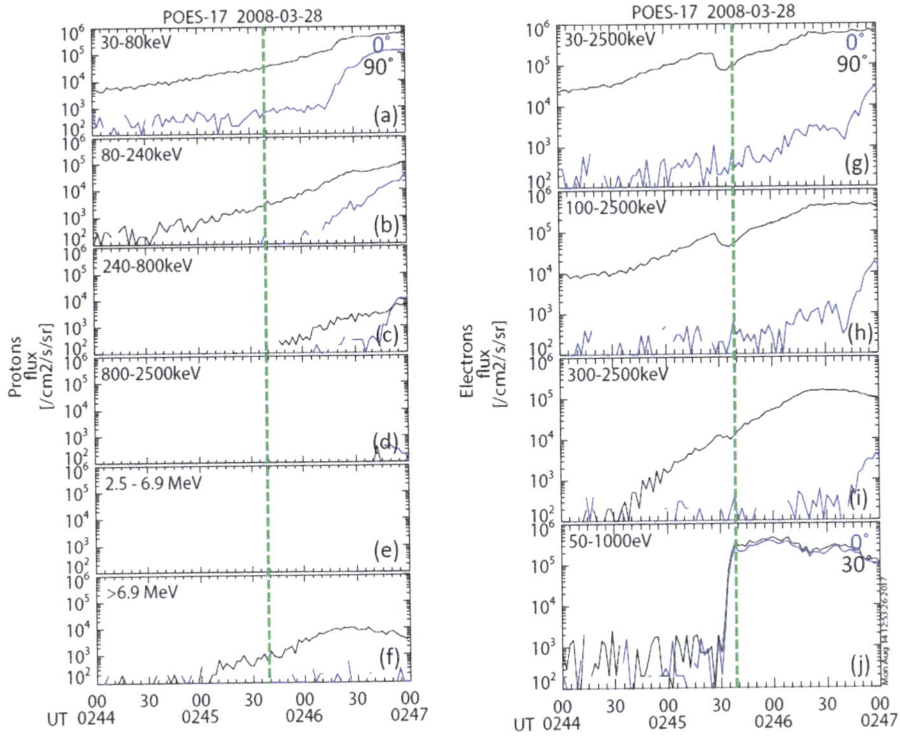

Fig. 22 POES-17 satellite particle data. Proton (left) and electron (right) flux. (From Gallardo-Lacourt et al. 2018a)

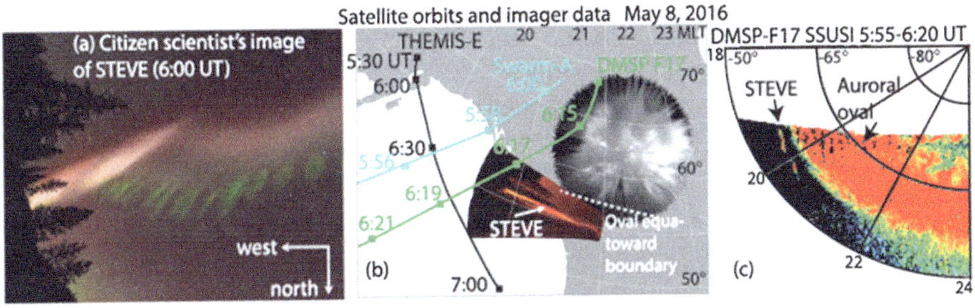

Fig. 23 STEVE event observed in the northern and southern hemisphere together with a picket fence occurring on 8 May 2016. (**a**) Citizen scientist observation at the West Coast of North America. (**b**) REGO ASI observed STEVE at the Lucky Lake station, while THEMIS ASI measures auroral activity at higher latitude. (**c**) SSUSI instrument on board of DMSP F-17 detected the STEVE arc in the southern hemisphere. Figure adapted from Fig. 1 by Nishimura et al. 2019

ture was associated with electron precipitation of the order of 10 keV and weak upward Field-aligned current (FAC). Figure 23 has been adapted from Nishimura et al. (2019) and shows a STEVE event and picket fence occurring on 8 May 2016.

According to Nishimura et al. (2019), electron precipitation was only observed for events where the picket fence was present; in absence of the green-rayed structure no electron

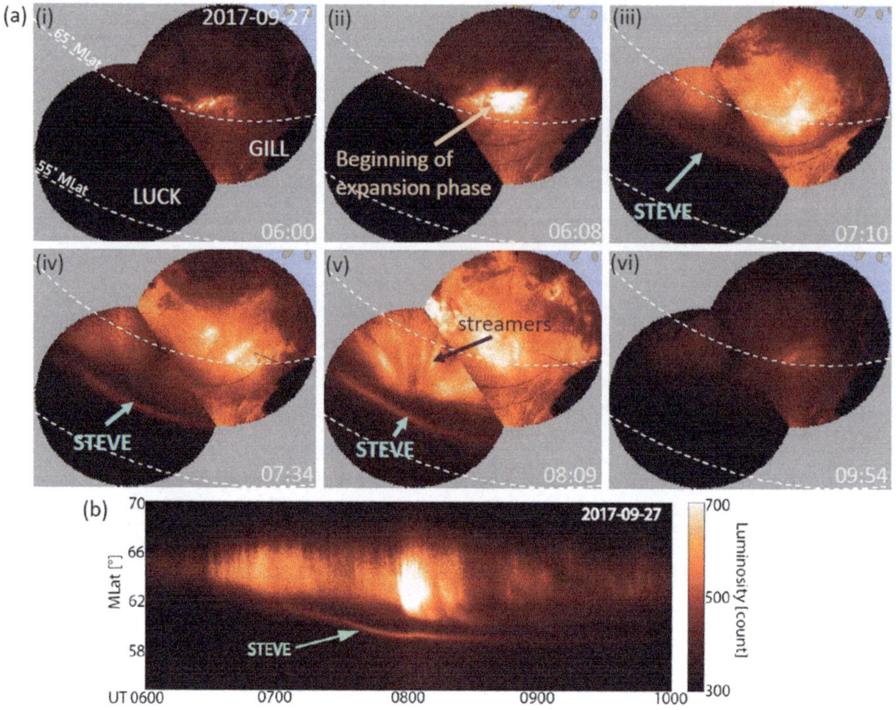

Fig. 24 STEVE event captured with REGO ASI at Lucky Lake (LUCK) on September 27, 2017. (**a**) STEVE is a narrow structure located equatorward of the auroral oval. In this event, STEVE lasted about 2 hours. (**b**) Keogram with STEVE event showing the structure's equatorward displacement. (From Gallardo-Lacourt et al. 2018b)

precipitation was detected. These observations support previous findings establishing that STEVE is locally produced in the ionosphere, but also suggest that the picket fence structure is formed by electron precipitation, which is an auroral-like process. In addition, the authors reported observations of waves and strong electric fields that could be responsible for producing heating in the F region ionosphere, as well as a structured electron boundary that may drive precipitation for the picket fence structure.

5.2 STEVE's Morphology and Geomagnetic Conditions

Gallardo-Lacourt et al. (2018b) reported a statistical analysis of STEVE's properties for 28 STEVE events identified in two optical data bases (THEMIS and REGO ASI) from December 2007 to December 2017. They reported that all the events occurred between 22-01 MLT and on average STEVE lasted for about one hour and its latitudinal width was ~20 km. STEVE also showed a latitudinal displacement (equatorward for most of the events) of about 50 km on average during its entire duration. STEVE's longitudinal extent was ~2145 km across the North American sector, limited by the ASI field-of view. Figure 24 shows an example of the STEVE events analyzed by Gallardo-Lacourt et al. (2018a, 2018b). STEVE lasted about 2 hours and covered the entire FOV of the ASI. The keogram (panel b) illustrates the equatorward motion that was typically observed for STEVE events.

Gallardo-Lacourt et al. (2018b) analyzed magnetosphere and solar wind data to investigate the geomagnetic conditions under which STEVE develops. The Dst index for the events

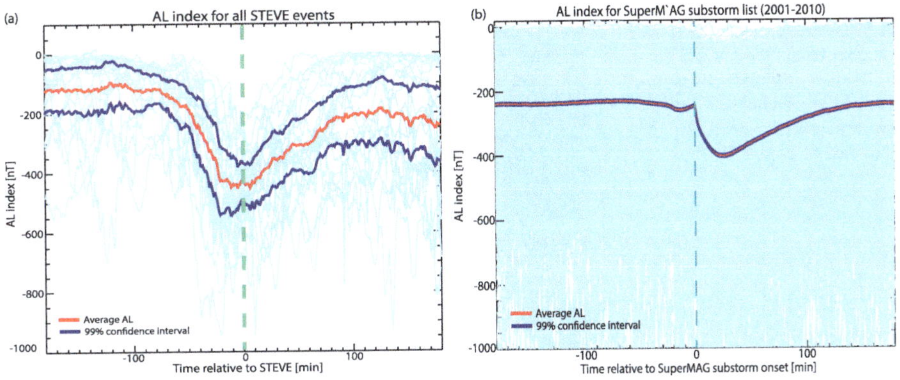

Fig. 25 AL-index for 28 STEVE events (left) and for SuperMAG substorms (right). Adapted from Gallardo-Lacourt et al. (2018b)

was relatively moderate at ~20 nT and Kp increased from 2 to 3.5 about 12 hours before STEVE was observed. In addition, the average solar wind speed and dynamic pressure were ~550 km/s and ~3 nPa, respectively.

From a superposed epoch analysis of AL index, the STEVE events reported in the study occurred about 1 hour after substorm onset at the end of a prolonged expansion phase. Figure 25 (adapted from Gallardo-Lacourt et al. 2018b) shows the AL index for the STEVE events (left panel) and the average AL index for the 16k substorms from the SuperMAG database. This result suggested that for STEVE to be observed, large amount of plasma may be injected from the magnetotail towards the inner magnetosphere. It is possible these injections effectively increase the ring current pressure and further enhance the subauroral westward flows, an effect that is consistent with the Swarm observations from MacDonald et al. (2018) and Archer et al. (2019a).

Recently, Archer et al. (2019b) analyzed photographers' data to estimate the altitude at which STEVE and the picket fence are produced. They used triangulation to analyze one event in which both structures were observed. They concluded that the picket fence was located from roughly 95- to 150-km altitude and STEVE to range from roughly 130- to 270-km altitude. During this event, they suggested that STEVE and the picket fence were aligned with each other along similar magnetic field lines. Figure 26 shows (a) the estimated altitudes for STEVE and the picket fence and (b) the median brightness for the red, green and blue channels, for the photograph in panel a. The authors highlighted that these qualitative observations should be interpreted with caution as the photographs have not been color calibrated for scientific purposes.

5.3 STEVE's Spectrum

Recent observations using an imaging spectrograph from the recently deployed Transition Region Explorer (TREx) array showed the spectrum of the picket fence and STEVE for the first time (Gillies et al. 2019). The spectrograph is designed to capture nighttime aeronomical phenomena with emissions between 400–800 nm, with a resolution of 0.4 nm.

Figure 27 shows the picket fence spectrum. The blue curve corresponds to the luminosity of the picket fence and the green (red) line corresponds to the background luminosity poleward (equatorward) of the picket fence structure, as shown in the inset. The location of the picket fence was obtained by fitting a Gaussian curve to the luminosity profile and the

Fig. 26 (**a**) Estimated altitude for STEVE event and picket fence on 16 September 2017. (**b**) Median brightness from red, blue and green channels for each row of pixels in panel a. Adapted from Archer et al. (2019b)

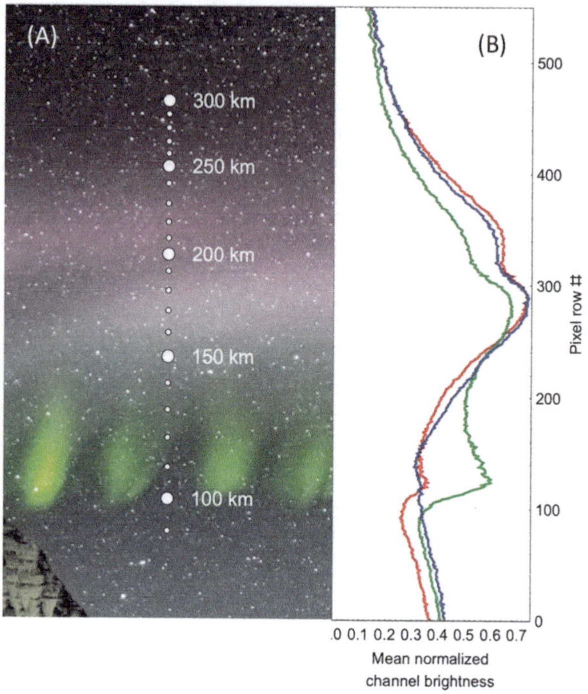

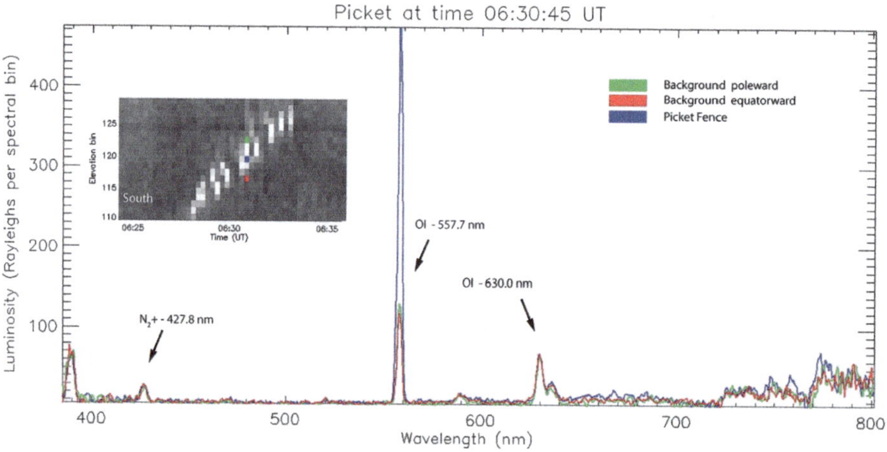

Fig. 27 Spectrum of picket fence structure obtained using the Transition Region Explorer (TREx) imaging spectrograph. Inset shows the locations where the respective luminosities were extracted. Adapted from Gillies et al. (2019)

background was defined by extracting the luminosity at 3σ poleward and equatorward. Figure 27 shows that the blue line (picket fence) clearly differs from the background at 557.7 nm which corresponds to the green line emission. The authors concluded that the picket fence is dominated by OI 557.7 emission with intensities higher than the normal level of airglow, indicating that the picket fence is likely produced by electron precipitation.

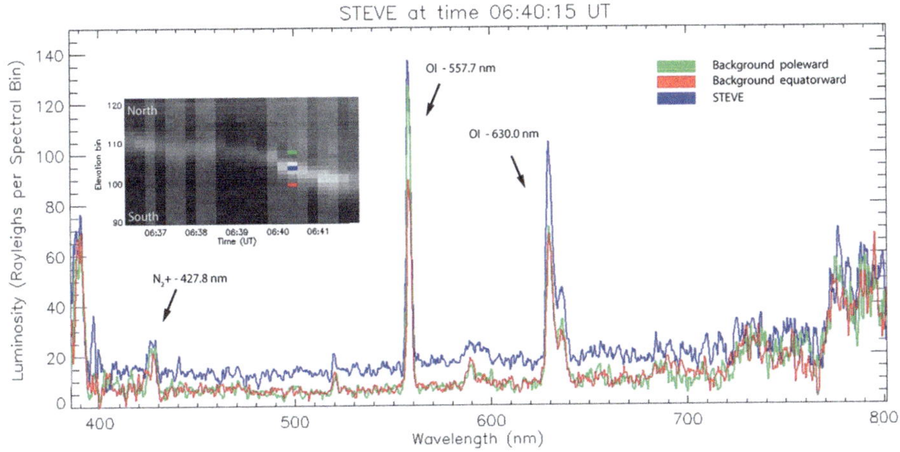

Fig. 28 STEVE spectrum obtained with the TREX imaging spectrograph. Adapted from Gillies et al. (2019)

The Gillies et al. (2019) study also presented a STEVE spectrum. The methodology to obtain STEVE and the background luminosity was similar to the one applied to the picket fence described above. The blue line in Fig. 28 shows STEVE's spectrum. The green and red lines indicate the background luminosities poleward and equatorward, respectively. STEVE spectrum shows an elevated emission continuum at all wavelengths, with slightly higher intensity at ~630.0 nm emission. The total luminosity of STEVE was approximately 6 kR higher than the background. The authors concluded that STEVE's spectrum contains two major contributing components: the OI red-line doublets (630.0 and 636.4 nm) and an overall enhancement of a continuous spectrum spanning between ~400 and 730 nm. The 557.7-nm green-line emission within STEVE only contributes a minor portion to the total STEVE brightness and is comparable to the night airglow intensity. Finally, Gillies et al. (2019) asserted that the observations essentially exclude STEVE from being an auroral phenomenon, consistent with previous studies (Gallardo-Lacourt et al. 2018a; Nishimura et al. 2019).

6 Outstanding Issues

Although much progress has been made to understand the structures at subauroral latitudes, there are several unresolved issues. For proton aurora for example, understanding the mechanism that produces proton scattering in the magnetosphere will give us important information about the interaction between the central plasma sheet and the inner magnetosphere. Some of these important open questions are for example: What are the differences between the electron and proton aurora spots? Which particular wave-particle interaction processes lead to the precipitation of protons?

In the subauroral region, several questions remain unanswered. Regarding SAR arcs, the mechanism responsible for the generation of energy that produces SAR arcs is not fully understood. The role that SAPS play on the formation of SAR arcs is still a topic under investigation. In addition, more studies are necessary to know how often, and under what conditions, sub-structures within SAR arcs develop. With respect to STEVE, although its spectrum has been analyzed, we do not understand the underlying chemistry playing a role in its formation at the predicted altitudes. In addition, the physical mechanisms responsible for the optical signature of STEVE have not yet been identified and the relationship between STEVE and the picket fence phenomenon is still unknown. Lastly, even though STEVE and SAR are observed at similar latitudes, it is not known if or how STEVE can be related to SAR arcs.

In general, the overarching question regarding all subauroral phenomena relates to their potential commonality and their specific differences. Although the subauroral structures discussed in this chapter share a similar location with respect to the auroral oval, it is not known whether they are generated by the same wave-particle processes. The development of new ground-based observatories, together with simulation studies, would help advance our understanding of proton aurora and the optical structures in the subauroral regions. In addition, new satellite missions, both planned and under development, will bring a clearer understanding and complete view of the highly coupled system at these latitudes.

Acknowledgements B. Gallardo-Lacourt wants to thank Mei-Ching Fok for her invaluable thoughts and discussions related to this topic. C. Martinis thanks Wen Li and Qianli Ma for valuable inputs when discussing VAP data. H.U. Frey was supported by the NSF award AGS-1004736 and NASA's Explorers Program through contracts NNG12FA45C and NNG12FA42I. B. Gallardo-Lacourt was supported by a NASA postdoctoral program fellowship. C. Martinis was supported by NASA grant 80NSSC19K0546 issued through the Heliophysics Guest Investigator Program and by NSF Aeronomy and Office of Polar Program award #1246423'. We acknowledge the Van Allen probes data from the EMFISIS instrument obtained from http://emfisis.physics.uiowa.edu/Flight/, and data from the HOPE instrument obtained from http://www.rbsp-ect.lanl.gov/data_pub/.

References

B.J. Anderson, R.E. Erlandson, L.J. Zanetti, A statistical study of Pc 1-2 magnetic pulsations in the equatorial magnetosphere, 1. Equatorial occurrence distributions. J. Geophys. Res. **97**, 3075 (1992)

C.D. Anger, M.C. Moshupi, D.D. Wallis, J.S. Murphree, L.H. Brace, G.G. Shepherd, Detached auroral arcs in the trough region. J. Geophys. Res. **83**(A6), 2683–2689 (1978). https://doi.org/10.1029/JA083iA06p02683

W.E. Archer, B. Gallardo-Lacourt, G.W. Perry, J.-P. St.-Maurice, S.C. Buchert, E.F. Donovan, Steve: the optical signature of intense subauroral ion drifts. Geophys. Res. Lett. **46**, 6279–6286 (2019a). https://doi.org/10.1029/2019GL082687

W.E. Archer, J.-P. St.- Maurice, B. Gallardo-Lacourt, G.W. Perry, C.M. Cully, E. Donovan et al., The vertical distribution of the optical emissions of a Steve and picket fence event. Geophys. Res. Lett. **46**, 10719–10725 (2019b). https://doi.org/10.1029/2019GL084473

J.L. Burch, W.S. Lewis, T.J. Immel, P.C. Anderson, H.U. Frey, S.A. Fuselier, J.-C. Gerard, S.B. Mende, D.G. Mitchell, M.F. Thomsen, Interplanetary magnetic field control of afternoon-sector detached proton auroral arcs. J. Geophys. Res. **107**, SMP 17 (2002). https://doi.org/10.1029/2001JA007554

J.W. Chamberlain, *Physics of the Aurora and Airglow* (Academic Press, New York, 1961). https://doi.org/10.1029/SP041

L.B.N. Clausen et al., Large-scale observations of a subauroral polarization stream by midlatitude SuperDARN radars: Instantaneous longitudinal velocity variations. J. Geophys. Res. **117** (2012). https://doi.org/10.1029/2011JA017232

K.D. Cole, Stable auroral red arc, sinks for energy of Dst main phase. J. Geophys. Res. **70**, 1689 (1965)

J.M. Cornwall, F.V. Coroniti, R.M. Thorne, Unified theory of SAR-arc formation at the plasmapause. J. Geophys. Res. **76**, 4428 (1971)

J.D. Craven, L.A. Frank, K.L. Ackerson, Global observations of a SAR ARC. Geophys. Res. Lett. **9**, 961–964 (1982). https://doi.org/10.1029/GL009i009p00961

M. de Soria-Santacruz, M. Spasojevic, L. Chen, EMIC waves growth and guiding in the presence of cold plasma density irregularities. Geophys. Res. Lett. **40**, 1940–1944 (2013). https://doi.org/10.1002/grl.50484

E. Donovan, E. Spanswick, J. Liang, J. Grant, B. Jackel, M. Greffen, Magnetospheric dynamics and the proton aurora, in *Auroral Phenomenology and Magnetospheric Processes: Earth and Other Planets*. Geophysical Monograph Series, vol. 197 (2012). https://doi.org/10.1029/2012GM001241

Y. Ebihara, G.V. Khazanov, Ring current, in *Space Weather Fundamentals*, ed. by G.V. Khazanov (CRC Press, Boca Raton, 2016), pp. 149–172. https://doi.org/10.1201/9781315368474

A. Egeland, W.J. Burke, Auroral hydrogen emissions: a historical survey. Hist. Geo- Space Sci. **10**, 201–213 (2019). https://doi.org/10.5194/hgss-10-201-2019

X. Fang, M.W. Liemohn, J.U. Kozyra, S.C. Solomon, Quantification of the spreading effect of auroral proton precipitation. J. Geophys. Res. **109**, A04309 (2004). https://doi.org/10.1029/2003JA010119

M.-C. Fok, J.U. Kozyra, L.H. Brace, Solar cycle variation in the subauroral electron temperature enhancement: comparison of AE-C and DE 2 satellite observations. J. Geophys. Res. **96**(A2), 1861–1866 (1991). https://doi.org/10.1029/90JA02377

M.C. Fok, J.U. Kozyra, A.F. Nagy, C.E. Rasmussen, G.V. Khazanov, Decay of equatorial ring current ions and associated aeronomical consequences. J. Geophys. Res. **98**(A11), 19381–19393 (1993). https://doi.org/10.1029/93JA01848

J.C. Foster, W.J. Burke, SAPS: a new categorization for sub-auroral electric fields. Eos Trans. AGU **83**(36), 393 (2002). https://doi.org/10.1029/2002EO000289

J.C. Foster, H.B. Vo, Average characteristics and activity dependence of the subauroral polarization stream. J. Geophys. Res. **107**(A12), 1475 (2002). https://doi.org/10.1029/2002JA009409

J.C. Foster, M.J. Buonsanto, M. Mendillo, D. Nottingham, F. Rich, W. Denig, Coordinated stable auroral red arc observations: relationship to plasma convection. J. Geophys. Res. **99**(A6), 11429–11439 (1994). https://doi.org/10.1029/93JA03140

B.J. Fraser, T.M. Loto'aniu, H.J. Singer, K. Takahashi, P.J. Chi, R.E. Denton, R.L. Lysak, Electromagnetic ion cyclotron waves in the magnetosphere, in *Magnetospheric ULF Waves: Synthesis and New Directions*, Chapman Conference on Magnetospheric ULF Waves. Geophysical Monograph Book Series, vol. 169 (2006), pp. 195–212. https://doi.org/10.1029/169GM13

H.U. Frey, Localized aurora beyond the auroral oval. Rev. Geophys. **45**, RG1003 (2007). https://doi.org/10.1029/2005RG000174

H.U. Frey, S.B. Mende, T.J. Immel, J.-C. Gerard, B. Hubert, S. Habraken, J. Spann, G.R. Gladstone, D.V. Bisikalo, V.I. Shematovich, Summary of quantitative interpretation of IMAGE far ultraviolet auroral data. Space Sci. Rev. **109**, 255–283 (2003). https://doi.org/10.1023/B:SPAC.0000007521.39348.a5

H.U. Frey, G. Haerendel, S.B. Mende, W.T. Forrester, T.J. Immel, N. Ostgaard, Sub-Auroral Morning Proton Spots (SAMPS) as a result of plasmapause-ring-current interaction. J. Geophys. Res. **109**(A10), A10304 (2004). https://doi.org/10.1029/2004JA010516

M. Galand, S. Chakrabarti, Proton aurora observed from the ground. J. Atmos. Sol.-Terr. Phys. **68**, 1488–1501 (2006). https://doi.org/10.1016/j.jastp.2005.04.013

M. Galand, D. Lummerzheim, Contribution of proton precipitation to space-based auroral FUV observations. J. Geophys. Res. **109**, A03307 (2004). https://doi.org/10.1029/2003JA010321

B. Gallardo-Lacourt, J. Liang, Y. Nishimura, E. Donovan, On the origin of STEVE: particle precipitation or ionospheric skyglow? Geophys. Res. Lett. **45**, 7968–7973 (2018a). https://doi.org/10.1029/2018GL078509

B. Gallardo-Lacourt, Y. Nishimura, E. Donovan, D.M. Gillies, G.W. Perry, W.E. Archer et al., A statistical analysis of STEVE. J. Geophys. Res. Space Phys. **123**, 9893–9905 (2018b). https://doi.org/10.1029/2018JA025368

B. Gallardo-Lacourt, G.W. Perry, W.E. Archer, E. Donovan, How did we miss this? An upper atmospheric discovery named STEVE. Eos **100** (2019). https://doi.org/10.1029/2019EO117351

Y.I. Galperin, Polarization jet: characteristics and a model. Ann. Geophys. **20**(3), 391–404 (2002)

G.A. Germany, M.R. Torr, D.G. Torr, P.G. Richards, Use of FUV auroral emissions as diagnostic indicators. J. Geophys. Res. **99**, 383–388 (1994). https://doi.org/10.1029/93JA02357

D.M. Gillies, E. Donovan, D. Hampton, J. Liang, M. Connors, Y. Nishimura et al., First observations from the TREx spectrograph: the optical spectrum of STEVE and the picket fence phenomena. Geophys. Res. Lett. **46**, 7207–7213 (2019). https://doi.org/10.1029/2019GL083272

A. Hasegawa, K. Mima, Anomalous transport produced by kinetic Alfven wave turbulence. J. Geophys. Res. **83**, 1117 (1978)

J. Hong, J.-H. Kim, J.-K. Chung, Y.H. Kim, H. Kam, J. Park, M. Mendillo, Simultaneous observations of SAR arc and its ionospheric response at subauroral conjugate points ($L \simeq 2.5$) during the St. Patrick's Day Storm in 2015. J. Geophys. Res. Space Phys. **125** (2020). https://doi.org/10.1029/2019JA027321

B. Hultqvist, H. Borg, P. Christophersen, W. Riedler, Observations of magnetic field-aligned anisotropy for 1 and 6 keV positive ions in the upper atmosphere. Planet. Space Sci. **19**, 279 (1971)

T.J. Immel, S.B. Mende, H.U. Frey, L.M. Peticolas, C.W. Carlson, J.-C. Gerard, B. Hubert, S.A. Fuselier, J.L. Burch, Precipitation of auroral protons in detached arcs. Geophys. Res. Lett. **29** (2002). https://doi.org/10.1029/2001GL013847

T. Iyemori et al., Localized injection of large-amplitude Pc 1 waves and electron temperature enhancement near the plasmapause observed in DE 2 in the upper atmosphere. J. Geophys. Res. **99**, 6187 (1994)
V.K. Jordanova, Ring current decay in the Quest for Space Weather Prediction (2020), pp. 181–223. https://doi.org/10.1016/B978-0-12-815571-4.00006-8
V.K. Jordanova, R.B. Torbert, R.M. Thorne, H.L. Collin, J.L. Roeder, J.C. Foster, Ring current activity during the early Bz < 0 phase of the January 1997 magnetic cloud. J. Geophys. Res. **104**(A11), 24895–24914 (1999). https://doi.org/10.1029/1999JA900339
V.K. Jordanova, M. Spasojevic, M.F. Thomsen, Modeling the electromagnetic ion cyclotron wave-induced formation of detached subauroral proton arcs. J. Geophys. Res. **112**, A08209 (2007). https://doi.org/10.1029/2006JA012215
Y. Kamide, S.-I. Akasofu, The location of the field-aligned currents with respect to discrete auroral arcs. J. Geophys. Res. **81**(22), 3999–4003 (1976). https://doi.org/10.1029/JA081i022p03999
J.U. Kozyra, E.G. Shelly, R.H. Comfort, L.H. Brace, T.E. Cravens, A.F. Nagy, The role of ring current O^+ in the formation of stable red arcs. J. Geophys. Res. **92**, 7487 (1987)
J.U. Kozyra, M.O. Chandler, D.C. Hamilton, W.K. Peterson, D.M. Klumpar, D.W. Slater, M.J. Buonsanto, H.C. Carlson, The role of ring current nose events in producing stable auroral red arc intensifications during the main phase: observations during the September 19-24, 1984 equinox transition study. J. Geophys. Res. **98**, 9267 (1993)
J.U. Kozyra, A.F. Nagy, D.W. Slater, High-altitude energy source(s) for stable auroral red arcs. Rev. Geophys. **35**, 155–190 (1997)
J. Krall, J.D. Huba, Plasmasphere, in *Space weather fundamentals*, ed. by G.V. Khazanov (CRC Press, Boca Raton, 2016), pp. 185–198. https://doi.org/10.1201/9781315368474
M. Kubota, T. Nagatsuma, Y. Murayama, Evening co-rotating patches: a new type of aurora observed by high sensitivity all-sky cameras in Alaska. Geophys. Res. Lett. **30** (2003). https://doi.org/10.1029/2002GL016652
L.J. Lanzerotti, A. Hasegawa, C.G. Maclennan, Hydromagnetic waves as a cause of a SAR arc event. Planet. Space Sci. **26**, 777 (1978)
S.R. LaValle, D.D. Elliott, Observations of SAR arcs from OV1-10. J. Geophys. Res. **77**(10) (1972)
J. Liang, E. Donovan, D. Gillies, E. Spanswick, M. Connors, Proton auroras during the transitional stage of substorm onset. Earth Planets Space **70**, 126 (2018). https://doi.org/10.1186/s40623-018-0899-0
V. Lobzin, A. Pavlov, Correlations between SAR arc intensity and solar and geomagnetic activity. Ann. Geophys. **17**, 770–781 (1999)
L. Lyons, Generation of large-scale regions of auroral currents, electric potentials, and precipitation by the divergence of the convection electric field. J. Geophys. Res. **85**(A1), 17–24 (1980). https://doi.org/10.1029/JA085iA01p00017
L.R. Lyons, Discrete aurora as the direct result of an inferred high-altitude generating potential distribution. J. Geophys. Res. **86**(A1), 1–8 (1981). https://doi.org/10.1029/JA086iA01p00001
MacDonald et al., New science in plain sight: Citizen scientists lead to the discovery of optical structure in the upper atmosphere. Sci. Adv. **4**(3) (2018). https://doi.org/10.1126/sciadv.aaq0030
C. Martinis, J. Baumgardner, J. Wroten, M. Mendillo, All-sky-imaging capabilities for ionospheric space weather research using geomagnetic conjugate point observing sites. Adv. Space Res. (2018). https://doi.org/10.1016/j.asr.2017.07.021
C. Martinis, J. Baumgardner, M. Mendillo, M.J. Taylor, T. Moffat-Griffin, J. Wroten, C. Sullivan, R. Macinnis, B. Alford, Y. Nishimura, First ground-based conjugate observations of Stable Auroral Red (SAR) Arcs, submitted. J. Geophys. Res. Space Phys. **124**(6), 4658–4671 (2019). https://doi.org/10.1029/2018JA026017
S.B. Mende, H. Heetderks, H.U. Frey, J.M. Stock, M. Lampton, S.P. Geller, R. Abiad, O.H.W. Siegmund, S. Habraken, E. Renotte, C. Jamar, P. Rochus, J.-C. Gerard, R. Sigler, H. Lauche, Far ultraviolet imaging from the IMAGE spacecraft: 3. Spectral imaging of Lyman alpha and OI 135.6 nm. Space Sci. Rev. **91**, 287–318 (2000). https://doi.org/10.1023/A:1005292301251
M. Mendillo, J. Baumgardner, J. Wroten, C. Martinis, S. Smith, K.-D. Merenda, T. Fritz, M. Hairston, R. Heelis, C. Barbieri, Imaging magnetospheric boundaries at ionospheric heights. J. Geophys. Res. Space Phys. **118**, 7294–7305 (2013). https://doi.org/10.1002/2013JA019267
M. Mendillo, R. Finan, J. Baumgardner, J. Wroten, C. Martinis, M. Casillas, A stable auroral red (SAR) arc with multiple emission features. J. Geophys. Res. Space Phys. **121**, 10,564–10,577 (2016a). https://doi.org/10.1002/2016JA023258
M. Mendillo, J. Baumgardner, J. Wroten, SAR arcs we have seen: evidence for variability in stable auroral red arcs. J. Geophys. Res. Space Phys. **121**, 245–262 (2016b). https://doi.org/10.1002/2015JA021722
E.V. Mishin, Interaction of substorm injections with the subauroral geospace: 1. Multispacecraft observations of SAID. J. Geophys. Res. **118**(9), 5782–5796 (2013). https://doi.org/10.1002/jgra.50548

M.C. Moshupi, L.L. Cogger, D.D. Wallis, J.S. Murphree, C.D. Anger, Auroral patches in the vicinity of the plasmapause. Geophys. Res. Lett. **4**, 37–40 (1977)
M.C. Moshupi, C.D. Anger, J.S. Murphree, D.D. Wallis, J.H. Whitteker, L.H. Brace, Characteristics of trough region auroral patches and detached arcs observed by ISIS 2. J. Geophys. Res. **84**, 1333–1346 (1979)
P.T. Newell, A.R. Lee, K. Liou, S.-I. Ohtani, T. Sotirelis, S. Wing, Substorm cycle dependence of various types of aurora. J. Geophys. Res. **115**, A09226 (2010). https://doi.org/10.1029/2010JA015331
Y. Nishimura, B. Gallardo-Lacourt, Y. Zou, E. Mishin, D.J. Knudsen, E.F. Donovan et al., Magnetospheric signatures of STEVE: implications for the magnetospheric energy source and interhemispheric conjugacy. Geophys. Res. Lett. **46**, 5637–5644 (2019). https://doi.org/10.1029/2019GL082460
T.P. O'Brien, M.B. Moldwin, Empirical plasmapause models from magnetic indices. Geophys. Res. Lett. **30**, 1152 (2003). https://doi.org/10.1029/2002GL016007
M. Ozaki, K. Shiokawa, Y. Miyoshi, R. Kataoka, M. Connors, T. Inoue, S. Yagitani, Y. Ebihara, C.-W. Jun, R. Nomura, K. Sakaguchi, Y. Otsuka, H.A. Uchida, I. Schofield, D.W. Danskin, Discovery of 1 Hz range modulation of isolated proton aurora at subauroral latitudes. Geophys. Res. Lett. **45**, 1209–1217 (2018). https://doi.org/10.1002/2017GL076486
V. Pierrard, J. Goldstein, N. Andre, V.K. Jordanova, Recent progress in physics-based models of the plasmasphere. Space Sci. Rev. **145**, 193–229 (2009). https://doi.org/10.1007/s11214-008-9480-7
E.I. Reed, J.E. Blamont, Observations of the conjugate SAR arcs of September 28-30, 1967. J. Geophys. Res. **79**(16), 2524–2525 (1974)
M.H. Rees, D. Luckey, Auroral electron energy derived from ratio of spectroscopic emissions 1. Model computations. J. Geophys. Res. **79**(34), 5181–5186 (1974). https://doi.org/10.1029/JA079i034p05181
M.H. Rees, R.G. Roble, Observations and theory of the formation of stable auroral red arcs. Rev. Geophys. **13**, 201 (1975)
F. Roach, J. Roach, Stable 6300 Å auroral arcs in mid-latitudes. Planet. Space Sci. **11**, 523–545 (1963)
K. Sakaguchi, K. Shiokawa, Y. Miyoshi, Y. Otsuka, T. Ogawa, K. Asamura, M. Connors, Simultaneous appearance of isolated auroral arcs and Pc 1 geomagnetic pulsations at subauroral latitudes. J. Geophys. Res. **113**, A05201 (2008). https://doi.org/10.1029/2007JA012888
K. Sakaguchi, Y. Miyoshi, E. Spanswick, E. Donovan, I.R. Mann, V. Jordanova, K. Shiokawa, M. Connors, J.C. Green, Visualization of ion cyclotron wave and particle interactions in the inner magnetosphere via THEMIS-ASI observations. J. Geophys. Res. **117**, A10204 (2012). https://doi.org/10.1029/2012JA018180
K. Sakaguchi, K. Shiokawa, Y. Miyoshi, M. Connors, Isolated proton auroras and Pc1/EMIC waves at subauroral latitudes, in *Auroral Dynamics and Space Weather*, ed. by Y. Zhang, L.J. Paxton. Geophysical Monograph, vol. 215 (Wiley, New York, 2016)
S. Sazykin, B.G. Fejer, Y.I. Galperin, L.V. Zinin, S.A. Grigoriev, M. Mendillo, Polarization jet events and excitation of weak SAR arcs. Geophys. Res. Lett. **29**(12), 1586 (2002). https://doi.org/10.1029/2001GL014388
F. Søraas, K.M. Laundal, M. Usanova, Coincident particle and optical observations of nightside subauroral proton precipitation. J. Geophys. Res. Space Phys. **118**, 1112–1122 (2013). https://doi.org/10.1002/jgra.50172
E. Spanswick, E. Donovan, L. Kepko, V. Angelopoulos, The magnetospheric source region of the bright proton aurora. Geophys. Res. Lett. **44**, 10,094–10,099 (2017). https://doi.org/10.1002/2017GL074956
M. Spasojevic, S.A. Fuselier, Temporal evolution of proton precipitation associated with the plasmaspheric plume. J. Geophys. Res. Space Phys. **114** (2009). https://doi.org/10.1029/2009JA014530
M. Spasojevic, H.U. Frey, M.F. Thomsen, S.A. Fuselier, S.P. Gary, B.R. Sandel, U.S. Inan, The link between a detached subauroral proton arc and a plasmaspheric plume. Geophys. Res. Lett. **31** (2004). https://doi.org/10.1029/2003GL018389
M. Spasojevic, M.F. Thomsen, P.J. Chi, B.R. Sandel, Afternoon subauroral proton precipitation resulting from ring current-plasmasphere interaction, in *Inner Magnetosphere Interactions: New Perspectives from Imaging*. Geophysical Monograph Series, vol. 159 (AGU, Washington, 2005)
R.W. Spiro, R.A. Heelis, W.B. Hanson, Rapid sub-auroral ion drifts observed by Atmospheric Explorer C. Geophys. Res. Lett. **6**, 657 (1979)
Y. Takagi, K. Shiokawa, Y. Otsuka, M. Connors, I. Schofield, Statistical analysis of SAR arc detachment from the main oval based on 11-year, all-sky imaging observation at Athabasca, Canada. Geophys. Res. Lett. **45**, 11,539–11,546 (2018). https://doi.org/10.1029/2018GL079615
S.A. Thaller, J.R. Wygant, L. Dai, A.W. Breneman, K. Kersten, C.A. Cattell, S.R. Bounds, Van Allen probes investigation of the large-scale duskward electric field and its role in ring current formation and plasmasphere erosion in the 1 June 2013 storm. J. Geophys. Res. Space Phys. **120**(6), 4531–4543 (2015). https://doi.org/10.1002/2014JA020875
N.A. Tsyganenko, Modeling the Earth's magnetospheric magnetic field confined within a realistic magnetopause. J. Geophys. Res. **100**(A4), 5599–5612 (1995). https://doi.org/10.1029/94JA03193

N.A. Tsyganenko, A model of the magnetosphere with a dawn-dusk asymmetry. 1. Mathematical structure. J. Geophys. Res. **107**(A8), SMP 12 (2002). https://doi.org/10.1029/2001JA000219

L. Vegard, Results of the investigations of the auroral spectrum during the years 1921–1926. Geofys. Publik. **9**, 1–71 (1932)

V.G. Vorobjev, O.I. Yagodkina, Comparative characteristics of ion and electron precipitation in the dawn and dusk sectors. Geomagn. Aeron. **54**, 50–58 (2014). https://doi.org/10.1134/S0016793214010162

B. Wang, P. Li, J. Huang, B. Zhang, Nonlinear Landau resonance between EMIC waves and cold electrons in the inner magnetosphere. Phys. Plasmas **26**(4), 042903 (2019)

A.G. Yahnin, T.A. Yahnina, Energetic proton precipitation related to ion-cyclotron waves. J. Atmos. Sol.-Terr. Phys. **69**, 1690–1706 (2007). https://doi.org/10.1016/j.jastp.2007.02.010

A.G. Yahnin, T.A. Yahnina, H.U. Frey, Subauroral proton spots visualize the Pc1 source. J. Geophys. Res. Space Phys. **112**, A10223 (2007). https://doi.org/10.1029/2007JA012501

A.G. Yahnin, T.A. Yahnina, H.U. Frey, T. Bosinger, J. Manninen, Proton aurora related to intervals of pulsations of diminishing periods. J. Geophys. Res. Space Phys. **114**, A12215 (2009). https://doi.org/10.1029/2009JA014670

A.G. Yahnin, T.A. Yahnina, H. Frey, V. Pierrard, Sub-oval proton aurora spots: mapping relatively to the plasmapause. J. Atmos. Sol.-Terr. Phys. **99**, 61–66 (2013). https://doi.org/10.1016/j.jastp.2012.09.018

T.A. Yahnina, A.G. Yahnin, Comparing the latitudinal distribution of the Pc1 intensity and the position of subauroral proton spots. Geomagn. Aeron. **52**, 624–628 (2012). https://doi.org/10.1134/S0016793212040159

J. Yao, Y. Cai, Y. Ma, S. Zhou, Evening corotating patches (ECP) observed by DMSP/SSUSI. J. Atmos. Sol.-Terr. Phys. **186**, 82–87 (2019). https://doi.org/10.1016/j.jastp.2019.02.008

Z.G. Yuan, X.H. Deng, X. Lin, Y. Pang, M. Zhou, P.M.E. Décréau, J.G. Trotignon, E. Lucek, H.U. Frey, J.F. Wang, Link between EMIC waves in a plasmaspheric plume and a detached sub-auroral proton arc with observations of Cluster and IMAGE satellites. Geophys. Res. Lett. **37** (2010). https://doi.org/10.1029/2010GL042711

Z. Yuan, M. Li, Y. Xiong, H. Li, M. Zhou, D. Wang, S. Huang, X. Deng, J. Wang, Simultaneous observations of precipitating radiation belt electrons and ring current ions associated with the plasmaspheric plume. J. Geophys. Res. Space Phys. **118**, 4391–4399 (2013). https://doi.org/10.1002/jgra.50432

Y. Zhang, L.J. Paxton, An empirical Kp-dependent global auroral model based on TIMED/GUVI data. J. Atmos. Sol.-Terr. Phys. **70**, 1231–1242 (2008)

Y. Zhang, L.J. Paxton, D. Morrison, B. Wolven, H. Kil, S. Wing, Nightside detached auroras due to precipitating protons/ions during intense magnetic storms. J. Geophys. Res. **110** (2005). https://doi.org/10.1029/2004JA010498

Y. Zou, Y. Nishimura, L.R. Lyons, E.F. Donovan, A statistical study of the relative locations of electron and proton auroral boundaries inferred from meridian scanning photometer observations. J. Geophys. Res. **117**, A06206 (2012). https://doi.org/10.1029/2011JA017357

Space Sci Rev (2019) 215:51
https://doi.org/10.1007/s11214-019-0617-7

Dayside Aurora

Harald U. Frey[1] · Desheng Han[2] · Ryuho Kataoka[3,4] · Marc R. Lessard[5] · Stephen E. Milan[6,7] · Yukitoshi Nishimura[8] · Robert J. Strangeway[9] · Ying Zou[10,11]

Received: 14 May 2019 / Accepted: 19 October 2019 / Published online: 12 November 2019

Abstract Dayside aurora is related to processes in the dayside magnetosphere and especially at the dayside magnetopause. A number of dayside aurora phenomena are driven by reconnection between the solar wind interplanetary magnetic field and the Earth's internal magnetic field at the magnetopause. We summarize the properties and origin of aurora at the cusp foot point, High Latitude Dayside Aurora (HiLDA), Poleward Moving Auroral Forms (PMAFs), aurora related to traveling convection vortices (TCV), and throat aurora. Furthermore we discuss dayside diffuse aurora, morning side diffuse aurora spots, and shock aurora.

Auroral Physics
Edited by David Knudsen, Joe Borovsky, Tomas Karlsson, Ryuho Kataoka and Noora Partmies

✉ H.U. Frey
hfrey@ssl.berkeley.edu

R. Kataoka
kataoka.ryuho@nips.ac.jp

1 Space Sciences Laboratory, University of California, Berkeley, CA 94720, USA

2 State Key Laboratory of Marine Geology, School of Ocean and Earth Science, Tongji University, Shanghai, China

3 National Institute of Polar Research, 10-3 Midori-cho, Tachikawa, Tokyo 190-8518, Japan

4 Department of Polar Science, SOKENDAI, 10-3 Midori-cho, Tachikawa, Tokyo 190-8518, Japan

5 Space Science Center, University of New Hampshire, Durham, NH, USA

6 Department of Physics and Astronomy, University of Leicester, Leicester, UK

7 Birkeland Centre for Space Sciences, University of Bergen, Bergen, Norway

8 Department of Electrical and Computer Engineering and Center for Space Physics, Boston University, 725, Commonwealth Avenue, Boston, MA, USA

9 Department of Earth, Planetary, and Space Sciences, UCLA, Los Angeles, CA, USA

10 Department of Astronomy and Center for Space Physics, Boston University, Boston, MA, USA

11 Cooperative Programs for the Advancement of Earth System Science, University Corporation for Atmospheric Research, Boulder, CO, USA

Keywords Auroral phenomena · Dayside aurora · Cusp · Magnetic reconnection · MI-coupling · Energetic particles · Field-aligned currents and current systems · Energetic particle precipitation

1 Introduction

Aurora is generally considered a nightside phenomenon, primarily because that's where people can easily see it. However, it has been recognized already before the dawn of the space age, that aurora occurs at all local times around the auroral oval, even at local noon (Feldstein 1973). Nightside aurora is related to processes in the magnetotail. Dayside aurora on the other hand is related to processes in the dayside region of the magnetosphere, especially the dayside magnetopause. Therefore it is of particular interest as processes in the dayside magnetosphere are directly related to the solar wind–magnetosphere interaction.

Dayside aurora is difficult to observe from the ground. Because of the offset between the geographic and geomagnetic poles and the distribution of land and oceans there are only very few places on the ground that can actually observe the sky at local magnetic noon, most noticeably on Svalbard and in the Antarctic interior. Furthermore optical observations require darkness which only exists for 1–3 months in local winter. Therefore, dayside aurora optical observations have been rather limited and mostly been done from space where observations in the ultra-violet can separate aurora from the dayglow background.

Many of the localized dayside auroral phenomena have been reviewed in Sandholt et al. (2002a), Frey (2007). All of them occur under specific solar wind conditions and most of them are directly related to magnetic reconnection between the solar wind interplanetary magnetic field (IMF) and the Earth magnetic field at the dayside magnetopause. A summary of the preferred solar wind conditions and the aurora properties are given in Table 1 of Frey (2007) and will not be repeated here. In this paper we will describe a few more dayside auroral phenomena and review the progress in our understanding since 2007.

The vast majority of our discussion will be related to observations in the northern hemisphere. At a few places we will refer to a similar or different behavior in the southern hemisphere.

2 Magnetopause Reconnection

Many dayside auroras are a consequence of magnetic reconnection occurring between the interplanetary magnetic field (IMF) and the terrestrial magnetic field. Reconnection is thought to occur most efficiently where there is a large magnetic shear across the magnetopause. If reconnection occurs where the magnetic shear angle is strictly 180°, then it is referred to as antiparallel reconnection; if the shear angle is less than this, then it is termed component reconnection. The terrestrial field presented to the incoming solar wind is directed northwards at low latitudes and swings around to be directed southwards at high latitudes, pointing towards the northern cusp magnetopause indentation and away from the southern cusp. The orientation of the IMF in the Y–Z plane with respect to the Z-axis, that is its clock angle, θ, is largely preserved as the solar wind traverses the magnetosheath, while the X component is modified by the variable sheath speed so that the sheath magnetic field drapes over the magnetopause. The location and extent on the magnetopause at which reconnection occurs is then largely determined by the clock angle of the IMF and the orientation of the field within. When the IMF is directed southwards ($\theta \approx 180°$) the magnetic shear is greatest across a large swath of the low latitude magnetopause, where the terrestrial magnetic field lines are closed. For northwards IMF ($\theta \approx 0°$) the magnetic shear is greatest at the high latitude magnetopause, tailward of the cusp openings, where the terrestrial field lines tend to be open and comprise the magnetotail lobes (see Fig. 1). At intermediate clock angles, the two fields are strictly antiparallel in two regions. For $B_Y > 0$, one region starts at the northern cusp and extends southeastwards or northeastwards around the dusk flank if $B_Z < 0$ or $B_Z > 0$; the other extends dawnwards from the southern cusp opening, northwestwards or southwestwards for $B_Z < 0$ or $B_Z > 0$. The dawn and dusk sense of these regions are reversed for $B_Y < 0$. Except for $\theta = 180°$, the antiparallel regions avoid the subsolar magnetopause, such that reconnection at dawn and dusk occurs independently, with ramifications for ionospheric convection patterns (e.g., Coleman et al. 2001). However, there is also considerable evidence that reconnection does occur in the vicinity of the subsolar point when $|\theta| \gtrsim 90°$. It has been proposed, then, that in this clock angle range antiparallel reconnection occurs at the high latitude flanks and component reconnection occurs across a tilted region passing between the antiparallel regions through the subsolar point (e.g., Fuselier et al. 2011). On average the IMF has a Parker spiral configuration with $B_z = 0$, and purely northward and purely southward IMF configurations are rare. Hence Fig. 2(b) and 2(c) are representative of the most common reconnection geometries.

The rate of reconnection, and the steadiness or impulsiveness of reconnection, are then controlled by conditions at the reconnection site. For instance, where the magnetosheath flow is super-Alfvénic steady-state reconnection cannot occur and the interaction is necessarily impulsive. The evolution of newly-opened field lines across the magnetopause, and hence the motion of their footprints in the ionosphere, is controlled by magnetic tension forces and the magnetosheath flow (e.g., Cooling et al. 2001; Fear et al. 2007) and/or by the conductivity (load) of the ionosphere in the hemisphere to which the field line is connected. Many of the influences on the occurrence of reconnection are poorly understood, and are difficult to investigate in situ with spacecraft due to the vastness and sporadic motions of the magnetopause. Much, however, can be gleaned from a study of the convective ionospheric motions excited by reconnection and the associated auroras created by particles injected from the magnetosheath on newly-opened field lines. Understanding magnetopause reconnection is a major motivation for studying the dayside auroras.

3 Ionospheric Convection

Magnetic reconnection at the magnetopause and in the magnetotail drives a circulation of plasma within the magnetosphere, which determines the general structure of the magnetosphere and the morphology and dynamics of the auroras. This circulation is most easily monitored in the ionosphere, using e.g. the Super Dual Auroral Radar Network (SuperDARN) (Chisham et al. 2007). The behaviour of this circulation depends mainly on the north/south orientation of the IMF (see Fig. 1).

When the IMF is directed southwards, reconnection occurs across a broad region of the dayside magnetopause, opening previously closed field lines (Fig. 1a). This leads to an antisunwards ionospheric convection in the dayside polar cap, but a sunwards (equatorwards) motion of the dayside polar cap boundary (PCB) and the dayside auroras as the open magnetic flux content of the magnetosphere increases (Cowley and Lockwood 1992). This represents the loading due to the expansion of the polar cap and lobe pressure increase and is often referred to as substorm growth phase. Periods of southwards IMF are usually followed by the onset of magnetic reconnection in the magnetotail, substorm expansion phase

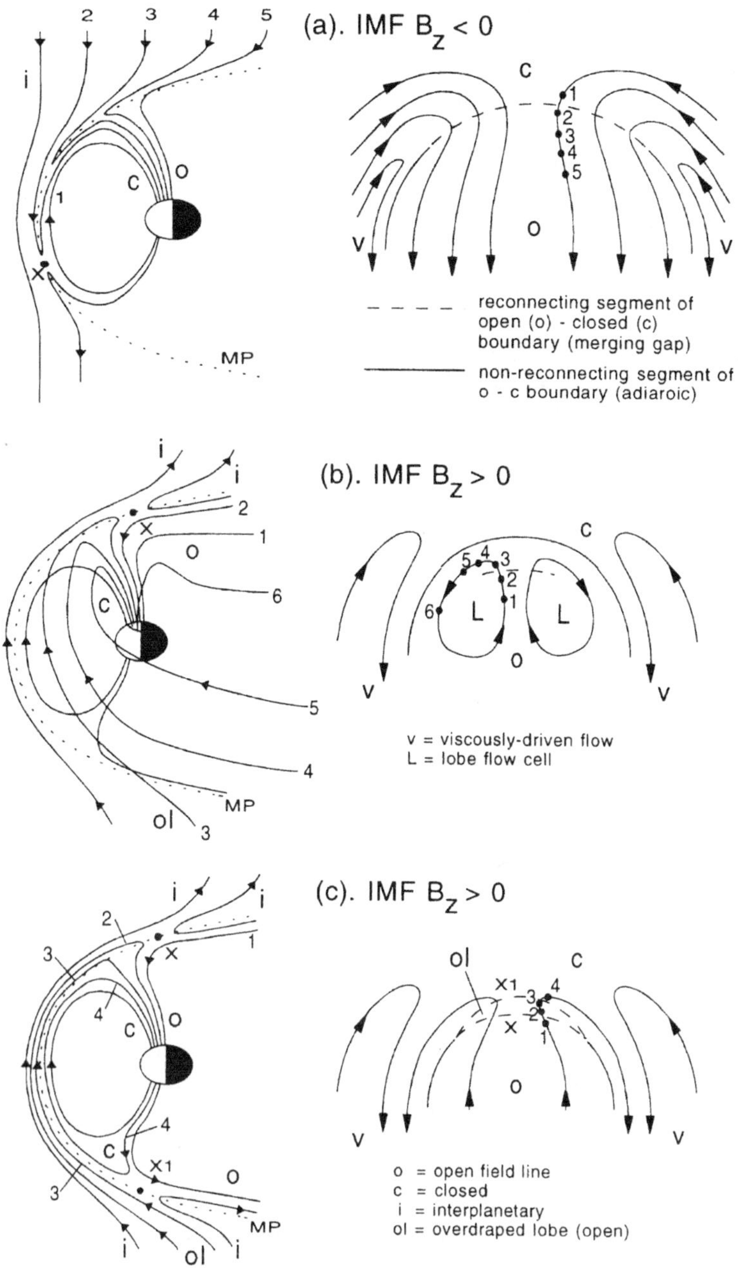

Fig. 1 Schematic illustrations of (left) the evolution of reconnected field lines in the magnetosphere, as seen from the dusk flank and (right) the corresponding flow in the northern hemisphere ionosphere for magnetopause reconnection during (**a**) southward IMF and (**b** and **c**) northward IMF, with $B_y \approx 0$. Field lines and regions that are open, closed, interplanetary and overdraped lobe are labelled, respectively, o, c, i, and ol. From Lockwood (1998)

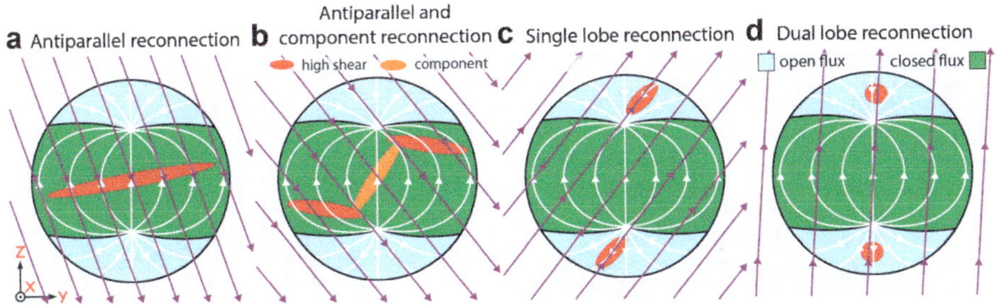

Fig. 2 Four panels showing possible reconnection geometries for different orientations of the IMF. (**a**) Clock angle near 180°, where antiparallel reconnection occurs across the dayside magnetopause at low latitudes. (**b**) Clock angle near 135°, where antiparallel reconnection occurs away from the equatorial plane and component reconnection occurs near the subsolar point. (**c**) Clock angle near 45°, where antiparallel reconnection occurs tailwards of the cusps, but with different IMF field lines in the two hemispheres. (**d**) Clock angle near 0°, where antiparallel reconnection occurs tailwards of the cusps, but with the same IMF field line in the two hemispheres

onset, to reduce the open flux content of the magnetosphere, and a cyclical expansion and contraction of the polar cap can ensue (Milan et al. 2003, 2007), leading to polewards and equatorwards motions of the dayside auroras. Within the dayside polar cap, the B_Y component of the IMF leads to a dawn-dusk asymmetry in the flows, with a dawnward (duskward) sense to the flows for $B_Y > 0$ ($B_Y < 0$) in the northern hemisphere (and opposite in the southern hemisphere).

For northwards IMF, reconnection occurs with the open field lines of the lobes, producing sunwards convection in the dayside polar cap. Single-lobe reconnection, when reconnection occurs independently in each hemisphere (Fig. 1b), leads to twin-reverse convection cells contained entirely within the polar cap; the relative size of the two cells depends on the polarity of IMF B_y, producing a larger pre-noon (post-noon) cell for $B_Y > 0$ in the northern (southern) hemisphere. Dual-lobe reconnection (Fig. 1c) closes open flux and sunward convection occurs across the dayside polar cap boundary as the polar cap contracts (e.g., Imber et al. 2006).

The transmission of stress from the magnetopause and magnetosphere to the ionosphere to cause convection is accomplished by electrical currents that flow along magnetic field lines into and out of the ionosphere. Upwards electrical currents are carried by downwards-precipitating electrons, often producing auroras. The large-scale auroral morphology is determined by the configuration of these currents (e.g. Milan et al. 2017), and the dayside auroras reviewed in this paper sit within this overall context.

4 Source Regions for Auroral Particles

The ionospheric regions where dayside aurora occurs map along magnetic field lines to the dayside region of the magnetosphere. Many years of particle measurements were used to infer the magnetospheric source region from the general energy spectral properties of precipitating electrons (Newell et al. 2005) and ions (Newell et al. 2009). Figure 3 shows the distribution in magnetic latitude and local time for the different inferred source regions for varying solar wind conditions. As Newell et al. (2005) point out these maps are averages and on a particular day or time a specific auroral precipitation region can show up at a different location. But these maps provide an excellent basis for the identification of the most likely

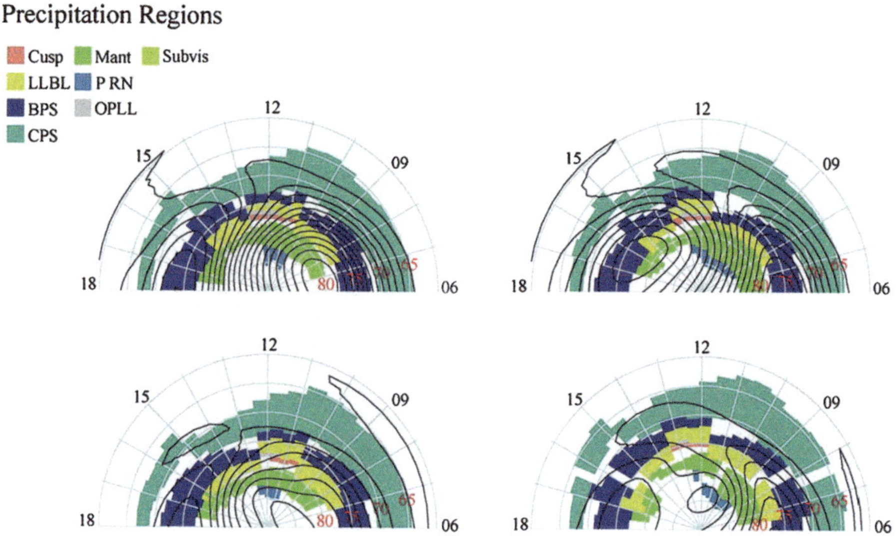

Fig. 3 A map of the dayside precipitation regions binned by interplanetary magnetic field (IMF). Top left, $B_z < -1$ nT, $B_y < -3$ nT; top right, $B_z < -1$ nT, $B_y > 3$ nT. Bottom left, $B_z > 1$ nT, $B_y < -3$ nT. Bottom right, $B_z > 1$ nT, $B_y > 3$ nT. ("OPLL" refers to open low-latitude boundary layer; "P RN" refers to polar rain; "LLBL" refers to Low Latitude Boundary Layer; "BPS" refers to Boundary Plasma Sheet; "CPS" refers to Central Plasma Sheet). From Newell et al. (2005)

source region of auroral particles if in situ measurements are not available (Newell et al. 2004). Furthermore, the investigations determine the average probability of electron acceleration and energy flux coming from these inferred source regions. The highest probability for electron acceleration events comes from the boundary plasma sheet (BPS), followed by the cusp and the Low Latitude Boundary Layer (LLBL) just equatorward of the cusp.

Electron acceleration in the cusp is surprisingly low. Even though acceleration events are often observed in the cusp, their typical energy is usually less than a few hundred electronvolts. Electron acceleration events are quite common in the LLBL, their average energy can exceed many hundreds of electronvolts. Newell et al. (2005) speculate that the higher energy makes the related aurora easily visible from ground-based optical observation sites.

As a general result Newell et al. (2005) find that the regions lying further equatorward, and particularly on closed field lines, have accelerated electrons with higher associated energy fluxes and thus show aurora more frequently. The post-noon LLBL and, especially, the post-noon BPS have the highest number of events and the largest associated energy fluxes. The peak energy of precipitation also rises moving equatorward, with the exception that the events occurring within the cusp and open LLBL have even lower average energy than those in the mantle.

5 Aurora Related to Dayside Reconnection

The majority of dayside aurora are related to dayside reconnection (see Sect. 2). In this section we will review those phenomena that are created by or related to dayside reconnection while later sections will summarize phenomena that are related to other processes.

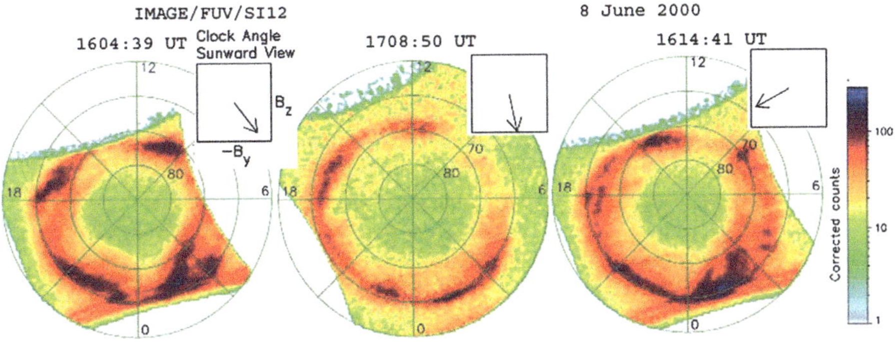

Fig. 4 Three images during a southward IMF interval showing that the cusp footpoint moves as the IMF B_Y changes. The clock angles are shown as viewed from the Earth. The cusp is located on the dawnside (duskside) of the auroral oval when the IMF B_Y is negative (positive). From Fuselier et al. (2003)

5.1 Cusp Aurora During Southward IMF

As discussed in Sect. 3, during southwards IMF, the opening of closed field lines at the magnetopause leads to poleward convection across the dayside polar cap boundary (PCB). Equatorwards of the PCB, auroral emission is dominated by diffuse green line aurora at 557.7 nm (see also Sect. 7), produced by trapped, hot magnetospheric plasma. Polewards of the PCB, the auroras are dominated by red line 630.0 nm emissions produced by precipitation of cool magnetosheath plasma, injected along newly-opened field lines (e.g., Lockwood et al. 1993; Lockwood 1997).

The time-of-flight of magnetosheath ions from the magnetopause to the ionosphere depends on their energy, leading to a latitudinal energy dispersion on the poleward-convecting field lines (e.g., Reiff et al. 1977; Woch and Lundin 1992; Trattner et al. 2002). Discontinuities in the energy-dispersion, leading to a "staircase" appearance are linked to non-steady reconnection at the magnetopause (e.g., Newell and Meng 1991; Lockwood et al. 1998). Such signatures have been observed in conjunction with poleward convection and quasi-period poleward-moving auroral forms (PMAFs) (e.g., Yeoman et al. 1997; Lockwood et al. 2001a). A similar association has been drawn between pulsed bipolar magnetic field signatures at the magnetopause, known as flux transfer events (FTEs) and also interpreted as a result of transient reconnection, and pulsed ionospheric flows and radar auroras (e.g., Neudegg et al. 2000; Wild et al. 2001). The study of PMAFs has since focused on understanding the role of transient reconnection in solar wind-magnetosphere coupling, which occurs even during periods of steady southwards IMF (see Sect. 5.4 for more details).

In global space-based views the cusp aurora during southward IMF appears as a brighter latitudinally thin and longitudinally extended region (Frey et al. 2003b) that is merged with the poleward part of the dimmer dayside diffuse aurora (Fig. 4). The local time location changes with the IMF B_y as the reconnection site at the low latitude dayside magnetopause moves in longitude. Magnetic mapping confirmed this connection (Frey 2007). The relative displacement in magnetic latitude in response to changes in the magnitude of the IMF B_z will be discussed in Sect. 5.2.

5.2 Cusp Aurora During Northward IMF

During northward IMF the reconnection site at the dayside magnetopause moves to the high latitude lobes (see Fig. 1). The cusp foot point moves to higher latitudes and gets clearly separated from the dayside auroral oval. As a result, global images show a clear cusp spot (e.g., Milan et al. 2000b) which has been extensively studied using IMAGE FUV proton aurora images (see e.g. Frey et al. 2002, 2003c; Fuselier et al. 2002; Frey 2007).

Earlier studies from the ground already revealed many of the general properties of cusp aurora and its dependence on the IMF B_z, B_y and solar wind dynamic pressure (see Sandholt et al. 2002a for event studies and references within). Even though ground-based observations can often detect small-scale features of the cusp aurora it also has to be realized that the limited coverage occasionally can lead to wrong conclusions about the disappearance of aurora which in reality just moved outside of the observation region. Furthermore, these observations are limited to just a few weeks of dark skies around local winter solstice. The main properties of cusp aurora during northward IMF and in the northern hemisphere can be summarized as follows (Sandholt et al. 2002a):

1. Strong emission both in 557.7 nm and 630.0 nm
2. Occurrence generally within the 0900-1500 MLT and 75°–80° MLAT region
3. Postnoon and prenoon occurrence depending on positive/negative IMF B_y
4. B_y dependent east-west motion of brightening events
5. Short term variability of brightness with stepwise poleward expansions followed by equatorward retreat
6. Decreasing brightness towards background when the IMF rotates from northward to a radial orientation

The proton aurora channel of the IMAGE FUV instrument allowed for extended studies of cusp aurora dynamics on the one hand because of the nonexistent background even during full solar illumination of the cusp foot point. On the other hand, the long orbital period of the satellite allowed for extended times of uninterrupted observations. The early studies established the strong relationship between the solar wind dynamic pressure and the cusp aurora brightness and confirmed particle measurements that ion precipitation can provide up to 60% of the total energy flux into the cusp (Frey et al. 2002; Østgaard et al. 2005).

Magnetic field line mapping with modern geomagnetic field models is much more reliable on the dayside than on the nightside because the dayside magnetosphere is compressed by the solar wind and much more stable than on the nightside where field lines can get substantially stretched. That allowed in several cases the mapping either from a spacecraft down to the cusp foot point, or from the cusp aurora out into the magnetosphere and to the magnetopause. When the Cluster spacecraft crossed the magnetopause and observed diverging proton jets from antiparallel reconnection at the high-latitude magnetopause the proton aurora imager observed a bright aurora spot (Phan et al. 2003). The field line mapping connected the satellite at the magnetopause during the jet observations with the bright cusp aurora spot and provided the direct evidence for their relationship.

Long periods of steady northward IMF provided unprecedented opportunities for the investigation of the steadiness of reconnection and cusp proton aurora generation (Frey et al. 2003c). Figure 5 shows examples of mapped proton aurora observations during one of these occasions where the inserts of the IMF orientation track the prenoon and postnoon motion of the proton aurora spot remarkable well. The IMAGE SI-12 channel had an imaging cadence of one image every two minutes. Therefore, nothing can be said about shorter period fluctuations. However, the continued presence of the proton aurora spot for many hours allowed (Frey et al. 2003c) to conclude that reconnection for northward IMF is a directly driven process that never stops on a global scale but is continuous and quasi-steady for as long as the driving solar wind conditions do not change substantially.

The mapping of the cusp aurora was used to estimate the length of the reconnection line at the magnetopause. During southward IMF a much longer X-line of 10–25 R_e was

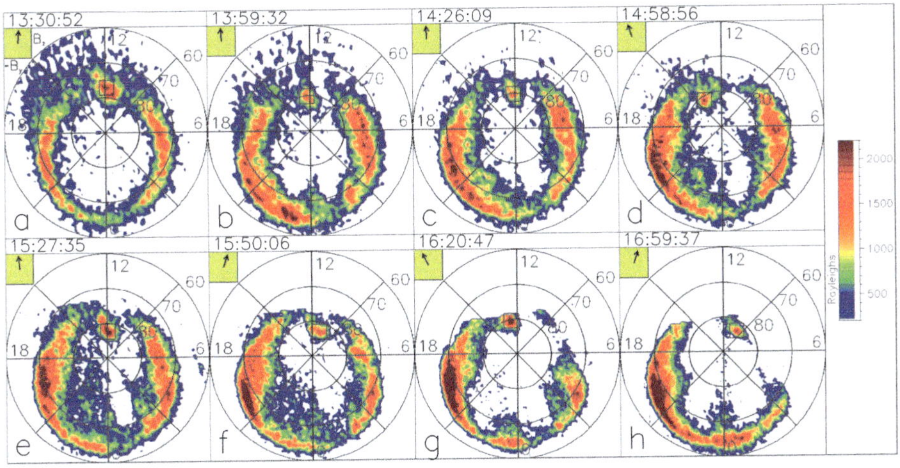

Fig. 5 Snapshots of the proton aurora oval and spot on 18 March 2002 showing the continuous presence of the proton aurora spot. The Spectrographic Imager channel SI-12 on board the IMAGE spacecraft observes Doppler-shifted Lyman-α emission from precipitating protons with energy of several keV. The images are shown in a geomagnetic grid of latitudes and local time with the noon meridian at the top and morning 06:00 h pointing to the right. The solar-wind magnetic field in the y–z GSM plane is shown in the upper left insert, with north ($B_z > 0$) pointing up and east ($B_y > 0$) pointing to the left. The black square in each panel covers the 500×500 km^2 area around the spot. The dayside proton aurora spot is seen uninterruptedly over ~4 hours. The spot appears on the dayside at ~80° latitude. Its location in magnetic local time (MLT) is correlated with the y-component of the solar-wind magnetic field, being in the pre-noon (post-noon) sector for negative (positive) B_y. From Frey et al. (2003c)

determined compared to the much shorter estimate of 3–5 R_e for northward IMF (Fuselier et al. 2002). The longer X-line for southward IMF explains the larger size and brightness of the cusp aurora during southward IMF compared to northward conditions and is consistent with measurements of much larger ion number fluxes (Frey 2007).

One remarkable similarity between cusp aurora during northward and southward IMF is their MLT location dependence on the IMF B_y. Positive and negative IMF B_y move the cusp foot point, and the corresponding reconnection site at the magnetopause, to later and earlier local time, respectively. During northward IMF the dependence is:

$$\mathrm{MLT} = 11.8\ \text{hours} + 0.127 * B_y \tag{1}$$

While during southward IMF a dependence of

$$\mathrm{MLT} = 12.6\ \text{hours} + 0.120 * B_y \tag{2}$$

Was determined (Frey et al. 2002, 2003b).

The cusp footpoint moves in magnetic latitude during southward IMF. Observation with the TIMED-GUVI instrument provided a linear relationship of 1.1° change in magnetic latitude for each Nano-Tesla of southward IMF (Zhang et al. 2005). For northward IMF no such relationship could be found indicating that the reconnection site under these conditions must be quite fixed at the high latitude lobe while during southward IMF the reconnection region can move much more freely along the dayside magnetopause (Frey et al. 2002). These results from aurora observations are in very good agreement with cusp modeling and observations which found the following relationships for the Cusp equatorward boundary

(Ceb; Wing et al. 2005):

$$\text{Ceb} = (0.81 \pm 0.05)\text{IMF}B_z + (77.7 \pm 0.2)^\circ \text{ for southward IMF and} \tag{3}$$

$$\text{Ceb} = (0.04 \pm 0.06)\text{IMF}B_z + (78.1 \pm 0.2)^\circ \text{ for northward IMF} \tag{4}$$

Other relationships of the equatorward and poleward cusp boundaries with respect to the IMF B_z and other solar wind coupling functions can be found in Newell et al. (2006), Johnsen and Lorentzen (2012).

The operation of the Polar UVI in the south and IMAGE FUV instrument in the north allowed for observations to confirm the simultaneous presence of the cusp aurora in both hemispheres. Reconnection during northward IMF should occur at both high latitude lobes and the cusp spots should be visible. The cusp spots were found at different local times (1000–1200 MLT in the north and 1200–1700 in the south) which is to be expected as a response to opposite IMF B_y control (Østgaard et al. 2005). The difference in latitude between the hemispheres was explained with the dipole tilt angle that shifted the reconnection site in the south to higher latitudes compared to the north.

5.3 High Latitude Dayside Aurora (HiLDA)

Few reports had been published about an aurora at dayside high latitudes before observations by the IMAGE FUV instrument allowed for a more detailed and systematic study of this phenomenon (Frey et al. 2003a). A very localized aurora could be seen in UV images of an otherwise empty polar cap which could last for many hours during otherwise quiet geomagnetic conditions. The absence of a signature in the proton aurora and a few in situ measurements by low altitude satellites (Fig. 6) confirmed their generation by pure high energy electron precipitation (Frey et al. 2003a). Field lines from this High Latitude Dayside Aurora (HiLDA) mapped very far into the tail while field lines from a simultaneously observed northward IMF cusp footprint (see Sect. 5.2) mapped to the high latitude magnetopause confirming the occurrence of reconnection but pointing to potentially lobe reconnection for the HiLDA spot. Very distinct external conditions for the HiLDA occurrence were northward IMF B_z, a strong dominating positive IMF B_y, very low solar wind density, and northern summer (Frey et al. 2004).

Model calculations of the ionospheric convection with these external conditions put the cusp aurora at the foot point of a downward field-aligned current in the center of a convection cell at the dayside. The HiLDA spot is in the center of the dominating clockwise convection cell with upward field-aligned current. These observational and modeling results then led to an explanation of the HiLDA occurrence as the result of a multi-step process. The northward B_y dominated IMF drives high latitude reconnection with a downward current into the cusp. Ionospheric convection drives the upward current over the HiLDA and both currents are connected by Pedersen currents in the high-conductivity sunlit summer ionosphere. The drop of the solar wind density before the HiLDA occurrence reduces the number of available current carriers (electrons) in the upward field-aligned current leg and the system sets up an accelerating potential in order to drive more of the available electrons into the loss cone to keep the current going. In astrophysics this process is called 'current starvation' (Kuijpers et al. 2015).

Conjugate observations in both hemispheres were used to study the asymmetric response of geospace to a strong IMF B_y driving during a geomagnetic storm (Østgaard et al. 2018). The strong IMF combined with a large dipole tilt led not only to large asymmetries in the

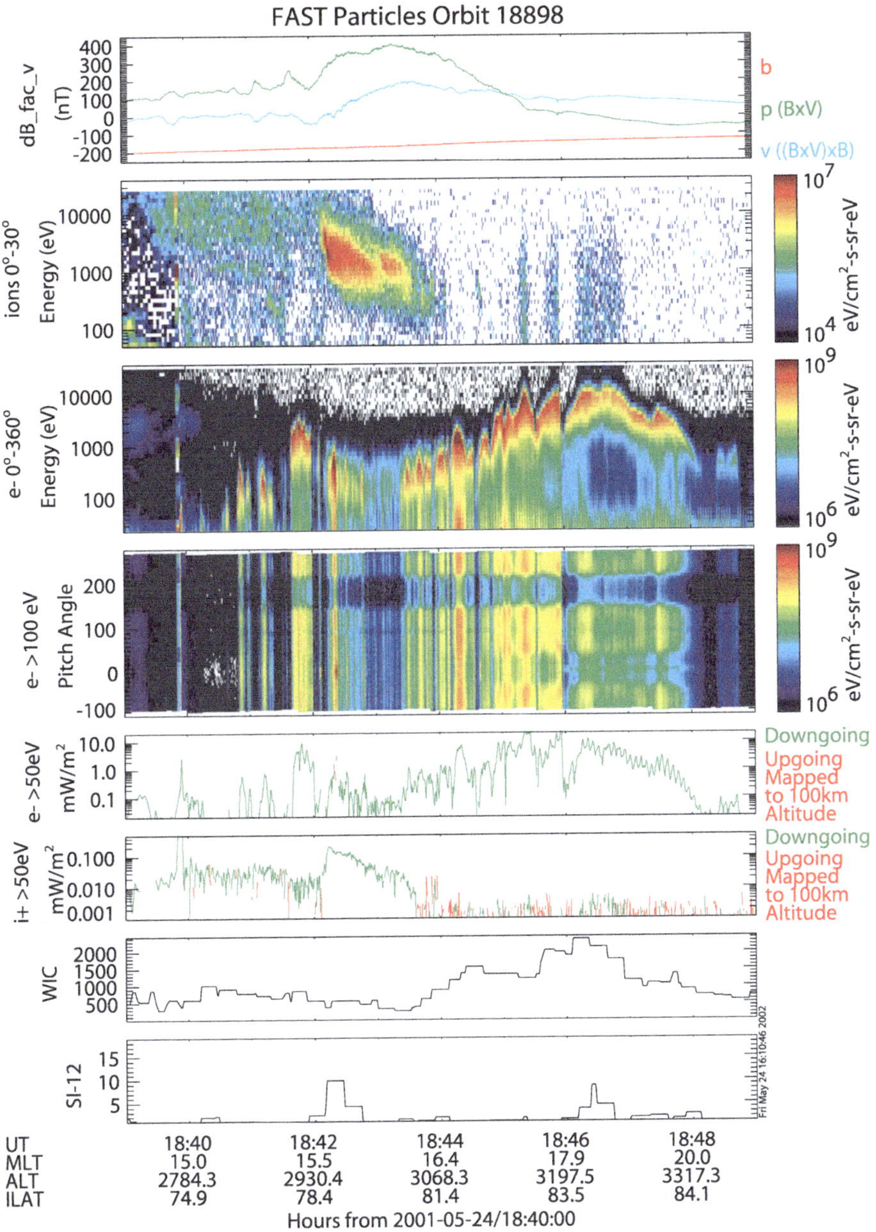

Fig. 6 FAST measurements and FUV observations of a HiLDA spot. The background IGRF magnetic field was subtracted from the FAST magnetic field measurements and the top panel shows the three components of the remaining magnetic field disturbance along the background field (red), perpendicular to the background field and the spacecraft velocity vector (green) and along the spacecraft track (blue). The green line represents the magnetic field disturbance that is created by field-aligned currents with a positive (negative) slope for a downward (upward) current. The next panel is the ion energy spectrogram in the loss cone. Then follow the electron energy spectrogram and the pitch angle distribution. The next two panels summarize the energy fluxes of electrons and ions, mapped to 100 km altitude. The last two panels are the count rates in the six consecutive electron and proton aurora images along the footprint of FAST (reproduced from Frey et al. 2003a)

dawn and dusk auroral ovals but also to the occurrence of the HiLDA spot only in the northern hemisphere, but not in the south. These observations were interpreted as a strong signature that lobe reconnection only occurred in the northern summer hemisphere.

Based on the distinct external conditions for HiLDA in the northern hemisphere Frey (2007) predicted a similar occurrence during summer in the southern hemisphere with negative IMF B_y and positive B_x. These predictions were recently confirmed by Carter et al. (2018) who used a large database of auroral observations and field-aligned current measurements. They found bright HiLDA spots only during southern summer with the predicted IMF orientation confirming the IMF-magnetosphere electrodynamic coupling that creates this particular localized aurora.

5.4 Poleward Moving Auroral Forms (PMAFs)

Dayside discrete aurora often shows a repetitive poleward motion of auroral arcs. It was first documented in the 1970's (Vorobjev et al. 1975; Horwitz and Akasofu 1977), and was recognized as an ionospheric signature of transient reconnection along the dayside magnetopause. This type of aurora, later referred to as poleward moving auroral forms (PMAFs) (Fasel et al. 1993), is characterized by brightening near the equatorward boundary of dayside discrete aurora (equatorward boundary intensification or EBI), followed by poleward motion. This sequence manifests an ionospheric signature of flux transfer events (FTEs, see below for more details) (Sandholt et al. 1986).

Early observations of PMAFs were made with meridian scanning photometers (e.g., Sandholt and Newell 1992), showing that these features were embedded within red line auroras and hence were probably associated with newly-opened field lines. Milan et al. (1999a, 1999b) demonstrated a close association between PMAFs and poleward bursts of convective flow. In addition, a sawtooth motion of the PCB was observed during its general equatorward progression, which suggested periodic erosion of closed flux associated with each burst of reconnection. Still to be resolved, however, is why individual PMAFs are latitudinally separated by regions devoid of auroras: each new region of open flux should be spatially contiguous, and hence no "gap" is expected in the auroral emission. Lockwood et al. (2001a) speculated that the auroral forms were produced by field-aligned currents at the boundaries between adjacent regions of newly-opened flux, associated with convection shears driven by differential azimuthal tension forces associated with IMF B_Y; however, no subsequent studies have followed up on this suggestion.

Figure 7 shows a typical event that presents a series of PMAFs after an IMF southward turning. Figure 8 schematically illustrates the PMAF sequence and the relation to large-scale precipitation and convection. Each PMAF starts with an EBI and then propagates poleward. The equatorward boundary of the 630.0 nm wavelength aurora overall moves equatorward as expected from enhanced magnetopause reconnection under the southward IMF. Note that the equatorward motion is not monotonic but stepwise, where the boundary rapidly shifts equatorward by $\sim 0.5^\circ$ magnetic latitude (MLAT) with each EBI initiation and then slows down or slightly retreats poleward (Horwitz and Akasofu 1977; Pudovkin et al. 1992). This boundary motion indicates a transient enhancement and then reduction of magnetopause reconnection associated with each PMAF.

Each PMAF is oriented approximately azimuthally ($\sim$500 km east-west and $\sim$50 km north-south). As Figs. 7c–h show, the equatorward boundary motion associated with the EBIs is also localized, indicating that the enhanced magnetopause reconnection is localized azimuthally. PMAFs have a lifetime and recurrence period of $<\sim$10 min (Fasel 1995) and a poleward propagation speed of $\sim$1 km/s (Oksavik et al. 2005). PMAFs can also extend over

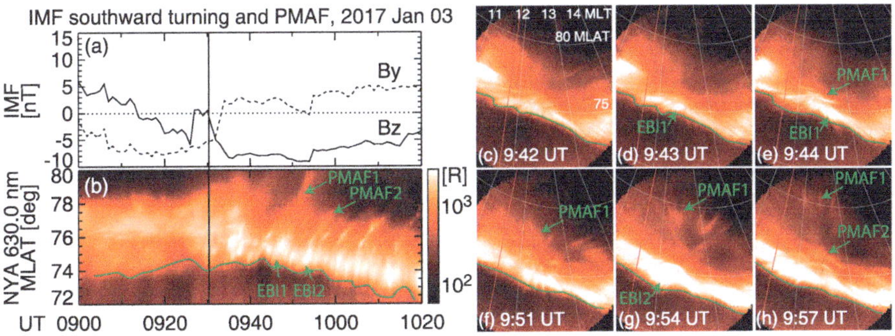

Fig. 7 A series of PMAFs associated with IMF southward turning, measured by the Ny-Alesund (NYA) all-sky imager at Svalbard on 3 January 2017. (**a**) IMF B_y and B_z, (**b**) 630.0 nm wavelength keogram, and (**c–h**) the sequence of two of the PMAFs. The green curve traces the 1 kR contour as a proxy of the open-closed boundary. The red line in Panels (**c–h**) shows the magnetic noon

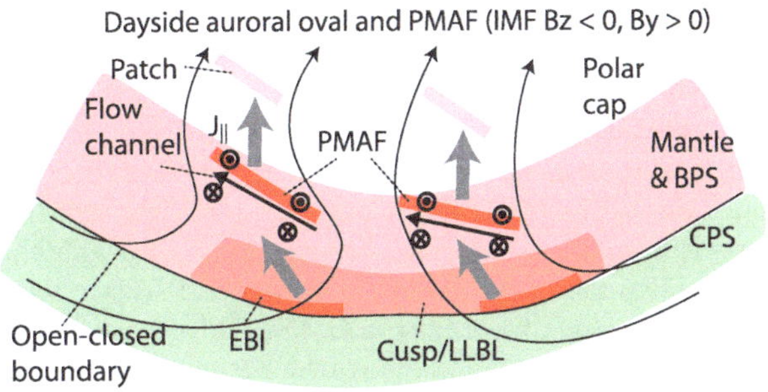

Fig. 8 Schematic illustration of PMAF and related processes in the dayside auroral oval under a typical IMF condition for PMAF occurrence ($B_Z < 0$ and $B_Y > 0$). The gray lines depict poleward propagation of the EBI-PMAF-patch sequence

much larger east-west size (>7 h MLT extent) and the azimuthal size has been suggested to be controlled by the solar wind speed (Milan et al. 2016).

PMAFs are generally a phenomenon under southward IMF. A typical example is shown in Fig. 7. Initially the dayside discrete aurora stays at higher latitudes without PMAFs under northward IMF, and when an IMF southward turning occurs, the pre-existing discrete aurora fades and a new discrete auroral activity emerges equatorward of the pre-existing aurora and a series of PMAFs occurs as the discrete aurora overall moves equatorward (Sandholt and Farrugia 2002). However, while more than half ($\sim$60%) of PMAFs occurs under southward IMF, it is nevertheless not uncommon to have PMAFs under northward IMF (Fasel 1995; Drury et al. 2003; Xing et al. 2012). The PMAF occurrence and location are rather strongly controlled by IMF B_y, where PMAFs are more frequent under positive IMF B_y in the northern hemisphere at post-noon (negative B_y at pre-noon in the southern hemisphere) (Sandholt et al. 2004). PMAFs under negative IMF B_y occur at pre-noon, and the occurrence rate near noon is smaller than pre- and post-noon (midday gap) (Fasel 1995; Sandholt et al. 2004). PMAFs are most often found at $|B_y|/|B_z| \geq 1 (B_z < 0)$, while aurora under $|B_y|/|B_z| < 1 (B_z < 0)$ is characterized as a quasi-steady longitudinally extended band at lower intensity with fewer PMAFs. A majority of PMAFs rebrighten during poleward propagation (Fasel 1995). The rebrightening time scale (a few minutes) is comparable to the Alfvén transit time and is considered to be due to a sequence of Alfvén waves launched from the reconnection region.

PMAFs under positive IMF B_y propagate dawnward (Fig. 8), and PMAFs under negative IMF B_y propagate duskward. The IMF B_y dependence is consistent with the release of magnetic stress of the reconnected field lines between the IMF and dayside magnetosphere. The azimuthal motion of PMAFs can be considerably large, giving rise to PMAFs moving ~3–5 h MLT away from noon (Wang et al. 2016a).

PMAFs do not necessarily occur only at a single location in the dayside auroral oval but can occur at more than one location. During time intervals of IMF $|B_y|/|B_z| \geq 1 (B_z < 0)$, PMAFs are found both at pre-noon and post-noon with a gap at noon (see Fig. 8 for $B_y > 0$), while PMAFs in different regions are not synchronized (Sandholt et al. 2003a; Maynard et al. 2006). This structure is interpreted as activation of magnetopause reconnection at multiple sites.

It is important to identify whether PMAFs are triggered by solar wind parameter changes or occur spontaneously. Both types of events have been reported. PMAFs can occur in association with IMF orientation or solar wind dynamic pressure changes (Sandholt et al. 2003b; Maynard et al. 2006; Mende et al. 2009). PMAFs can also occur under quasi-steady solar wind (Sandholt and Farrugia 2002; Sandholt et al. 2003b). Until recently, there was no systematic study of how often PMAFs are triggered or occur spontaneously. In addition, solar wind observations far from the subsolar bow shock have been used in past studies, which are problematic for estimating actual solar wind conditions reaching the bow shock. The solar wind could evolve during the propagation or the solar wind may contain smaller-scale structures that may miss the satellite location but reach the bow shock. To address these issues, Wang et al. (2016b) used the THEMIS outer probes-all sky imager conjunctions and found that 70% of PMAFs are preceded by IMF orientation changes, indicating that a majority of PMAFs are triggered by solar wind parameter changes. The IMF at the THEMIS outer probes are markedly different from OMNI data; the IMF southward turnings seen at THEMIS do not always exist in OMNI. This means that IMF evolution or localized structures are important for accurately understanding dayside magnetospheric responses.

PMAFs are part of a dynamical and localized magnetosphere-ionosphere coupling system that involves a localized flow channel and current system (Fig. 8). Localized field-aligned currents (FACs) have been detected in association with PMAFs and flow channels (Sandholt and Newell 1992), and the FACs and horizontal currents create poleward moving magnetic field perturbations on the ground (Milan et al. 2000a). PMAFs lie along the flank of the flows where the velocity shear is clockwise when viewing from above (Oksavik et al. 2004, 2005), as expected from the relation between FACs and flow shears.

Precipitating particle measurements show that PMAFs are associated with enhanced electron precipitation of hundreds of eV (Oksavik et al. 2005; Lorentzen et al. 2010). Electron precipitation creates substantially enhanced density, electron heating and ion upflow in the F-region ionosphere, while ion temperature responses depend on the location relative to the cusp (Lockwood et al. 2000; Skjaveland et al. 2011). Joule heating is also enhanced during PMAFs and is estimated to consume a tenth of the solar wind energy that has entered the magnetopause reconnection region (Pudovkin et al. 1992).

As mentioned above, PMAFs occur in association with flows of enhanced velocities ($>\sim 1$ km/s) that are superimposed on the cusp convection (Thorolfsson et al. 2000; Neudegg et al. 2001). These flows are important as they reflect the momentum transport aspect of FTEs. Depending on their exact characteristics, and the parameter in which they are identified, the flows are named as flow/convection channels, flow bursts, pulsed ionospheric flows, or poleward moving radar auroral forms (PMRAFs) (see Wild et al. 2001; Davies et al. 2000). The differently named flows are very much related, and in some cases describe the same phenomena (Davies et al. 2000). Moen et al. (2012) have summarized four representative types of flow channels.

Similarly to the PMAF optical properties, the enhanced flows generally occur in the pre-noon (post-noon) sector when the IMF By is positive (negative) (Provan et al. 1999). They are not limited to southward IMF conditions but can occur during prolonged northward IMF intervals (Provan et al. 2005). They occur repetitively on time scales of minutes to a few tens of minutes and the repetition rate may be associated with fluctuations in the solar wind (Prikryl et al. 2002; Rae et al. 2004). While statistically the flow occurrence rate appears to decrease with increasing repetition period (McWilliams et al. 2000a), the distribution is fitted well with a power-law-like power spectrum, implying no actual preferred timescale (Abel and Freeman 2002). The flows vary broadly in size from tens to thousands of km (Provan and Yeoman 1999; Oksavik et al. 2005) and factors controlling the size are not clearly understood. The flows contribute significantly to plasma transport with a potential drop of a few to tens of kV, which is often a few tenths of the total transpolar potential drop (Lockwood et al. 1990; Glassmeier and Stellmacher 1996).

Observations also reveal a new category of cusp flows that are directed oppositely to the background convection, termed as reversed flow events (RFEs) (Rinne et al. 2007). While they were initially interpreted as return flows of the increased anti-sunward flows, they are later noticed not to always follow the propagation of PMAFs but associate with Birkeland current arcs moving with the cusp/cleft boundary (Moen et al. 2008). The exact driver has not yet been clearly resolved.

The magnetospheric driver of PMAFs is identified as FTEs, products of time varying magnetopause reconnection (Haerendel et al. 1978; Russell and Elphic 1979), or solar wind dynamic pressure pulses. The former is a more common driver than the latter (Xing et al. 2012). After formation, FTEs propagate anti-sunward away from the X-line where they can be identified through their characteristic bipolar magnetic field perturbations normal to the magnetopause and mixed plasma populations of magnetospheric and magnetosheath origin (Paschmann et al. 1982; Farrugia et al. 1988). Southwood (1987) predicts that FTEs connect to the ionosphere via a pair of FACs and that the upward FAC illuminates as auroral transients. The FACs also transfer momentum of FTEs down to the ionosphere, inducing the flows of enhanced velocities that have been discussed above. This model is supported by numerous space-ground coordinated observations (Moen et al. 2012; Carlson 2012).

A key question regarding transient reconnection is the length of the reconnection X-line on the magnetopause, the extent of the PCB over which the associated PMAF extends, and hence the contribution to the total reconnection voltage (e.g., Lockwood et al. 1995). Many early studies of PMAFs were undertaken with ground-based cameras with a limited field-of-view, which lead to an underestimate of the open magnetic flux in each FTE ($<1\%$ of the pre-existing polar cap flux) and hence the contribution of FTEs to convection. Space-based auroral imagery of PMAFs revealed that coherent auroral features could occupy up to 7 hours of MLT around the boundary and contain up to 10% of the polar cap flux, sufficient to represent the total voltage associated with the growing polar cap (e.g., Milan et al. 2000a). Moreover, these features began their poleward progression first near noon and then at later local times, suggesting a reconnection X-line that grew from the subsolar point around the flank of the magnetopause, a suggestion first proposed by Lockwood et al. (1995). More recent studies have confirmed that such FTEs are a common occurrence, and have allowed the magnetopause structure of FTEs to be explored (e.g., Fear et al. 2017). There are still, however, significant questions regarding why reconnection is pulsed, the exact relationship between FTEs at the magnetopause and their auroral signatures in the ionosphere, and what controls the local time extent of the reconnection X-line (e.g., Milan et al. 2016).

Case studies show that FTEs and the enhanced flows are one-to-one related (Wild et al. 2003; Marchaudon et al. 2004) and that the shape, east-west extent, and drift velocity of

FTEs are consistent with those of the flows (Lockwood et al. 2001a,b). The motion of FTEs agrees with the motion of PMAFs (Zhang et al. 2010). A fortuitous event even captures Alfvén waves carrying FACs down the reconnected flux tubes to the enhanced flows (Farrugia et al. 2004). Statistically, 77% of FTEs are associated with the enhanced flows, and 64% of the enhanced flows are associated with FTEs (Neudegg et al. 2000).

PMAFs and related ionospheric enhanced flows have been advantageous in uncovering the macroscopic properties of FTEs that are not easily accessible from the spacecraft point measurements alone. For example, Lockwood et al. (2000) and Wild et al. (2001) have estimated the spatial extent of FTEs and found their observed FTEs span >4 h in local time. McWilliams et al. (2000b) observed that the source of FTEs, i.e., the region of active reconnection, shifts azimuthally over time. Fear et al. (2017) estimated the magnetic flux transferred by individual FTEs to be tens of MWb, which is ∼10% of the flux over the entire polar cap.

PMAFs are not only related to the reconnection at the magnetopause but also to the plasma convection pattern as can be determined from radar measurements (Tremsina et al. 2017). The IMF B_y can shift the convection pattern asymmetrically towards dawn or dusk and will thus determine the direction of motion of PMAFs. Part of the asymmetry between PMAFs during northward and southward IMF B_z may arise from the difficulty in distinguishing PMAFs and other type of auroral emissions with a north-south drift.

A number of in-situ measurements have provided important information about PMAFs, their evolution and effects. Oksavik et al. (2005) used a wide range of ground- and space-based data, including three low-altitude polar-orbiting spacecraft, to highlight a train of narrow flow channels drifting into the polar cap, related to a PMAF event. Lorentzen et al. (2010) used data from the Investigation of Cusp Irregularities 2 rocket to show that PMAFs continuously evolve to patches.

The Sounding of the Cusp Ion Fountain Energization Region (SCIFER) sounding rocket reached an apogee of nearly 1500 km. Kintner et al. (1996) (also see Arnoldy et al. 1996b and Lorentzen et al. 1996) provide an abundance of data explaining the physics of ionospheric plasma fountain formation in association with a PMAF. Related work is provided by Lund et al. (2012), who used SCIFER2 data to show electron heating signatures at altitudes of nearly 1500 km, above PMAF activity below.

The Rocket Experiment for Neutral Upwelling 2 (RENU2) rocket was launched on Dec 13, 2015 at 07:34 UT (Lessard et al. 2019). The solar wind speed averaged 520 km/s with B_z 1.5 nT and B_y 3.5 nT, producing the desired PMAF activity. The payload reached its apogee of 447 km as it crossed over Svalbard. Data acquired during the flight include electron distribution functions (Kenward et al. 2019), ion distribution functions and upflow/downflow (Godbole et al. 2019), neutral populations (Clemmons et al. 2019; Fritz et al. 2019) and photometers (Hecht et al. 2019), as well as electric and magnetic fields and a comprehensive set of ground instrumentation.

RENU2 data highlight two important processes. One has to do with the basic PMAF structure. Although in a typical keogram, PMAF activity appears as the poleward motion of a broad, isolated arc, RENU2 observations show that this aurora actually consists of clusters of highly-dynamic, multiple thin arcs, near 100 m in thickness, driven by soft electron precipitation (Kenward et al. 2019). With an apparent poleward motion that is faster than measured convection speeds, the implication is that the passage of a sequence of PMAFs delivers a series of transient, quasi-periodic energy fluxes to any given flux tube.

In addition, timescales for electron, ion and neutral particle dynamics range from a few seconds to tens of minutes. Timescales between PMAF passages are typically less than the time required to heat the neutral particles (10 min) (Carlson et al. 2012). Therefore, each

PMAF encounters a region having an initial condition that includes the recent history of the region—that is, hysteresis in the system is important. Perhaps more importantly, the differences in timescales tend to drive altitudinal and latitudinal neutral density structuring, including up to altitudes near 500 km or higher.

Although PMAFs gradually fade as they propagate away from the cusp, the magnetic flux tube continues to propagate into the polar cap. Chapter (XX paper on polar cap patches of this volume XX) describes more details on how PMAFs connect to polar cap patches and nightside auroral activity.

5.5 TCV Aurora

Traveling Convection Vortices (TCVs) are ionospheric structures having vortical flows around pairs of field-aligned currents in opposite hemispheres, typically taken to be signatures of dayside reconnection that propagate anti-sunward at high-latitudes. Murr et al. (2002) describe typical properties of TCVs, noting that "they form near noon at latitudes typically near 73° magnetic latitude and propagate antisunward with velocities of 5–10 km/s. They have scale sizes of 1000–3000 km and are observed most frequently a few hours prenoon and postnoon with the majority of events observed in the prenoon sector."

Initial studies of TCVs and related phenomena attempted to understand ground-based magnetometer signatures associated with a so-called flux transfer event (or FTE, which had been identified as evidence for sporadic reconnection on the dayside Russell and Elphic 1978; Berchem and Russell 1984), Lanzerotti et al. (1986) present observations showing two such events, but without explicitly calling them TCVs. The basis for event identification in that study included the use of a simple model showing the expected topology and ground magnetometer signatures as described by Lee and Fu (1985).

Interhemispheric FTE signatures were considered by Lanzerotti et al. (1990), who examined magnetic field data acquired at high-latitude, near-conjugate stations and identified numerous events that could be interpreted as having signatures of field-aligned currents above the stations. These authors introduced the term "Magnetic Impulse Events" (MIE) for these events. Related work by Friis-Christensen et al. (1988) also examined magnetometer signatures of such events, though these authors used the term "Traveling Convection Vortices" (TCVs), which has become the more common term.

Auroral observations associated with TCVs (or MIEs) have been observed in 630.0 and 427.8 nm all-sky images (Mende et al. 1990, 2001; Kataoka et al. 2001), though with the 630.0 emission line being dominant. A typical example of ground-based observations of a TCV aurora event is given in Fig. 9 from Kim et al. (2017b). Weatherwax et al. (1999) showed riometer signatures associated with 427.8 nm emissions. Murr et al. (2002) also reported a 427.8 nm auroral arc, extending ∼600 km in length. Kataoka et al. (2003) reported transient enhancements in UV auroral intensity, extending ∼500 km in length and width, related to the passage of a TCV current system in POLAR-UVI images. Finally, Ebihara et al. (2008) report observations of proton aurora during TCVs, typically ∼300–500 km in length and ∼150–200 km in width that lasted for ∼1–2 minutes.

A signature of proton aurora is consistent with various observations of ion cyclotron waves occurring with TCVs (Arnoldy et al. 1996a; Sakaguchi et al. 2007; Yahnin et al. 2007; Yahnina et al. 2008), where the waves may have precipitated the protons. Engebretson et al. (2013) showed evidence of ion cyclotron wave signatures in the ground during a TCV that transited Svalbard. The waves were associated with proton precipitation and were deemed to have been generated on closed field lines (see, also, Posch et al. 2013 and Kim et al. 2017a).

Finally, coupling of the vortical to the thermosphere has been shown for one TCV event, over Svalbard by Kim et al. (2017b), who also show ion cyclotron wave occurrences at the

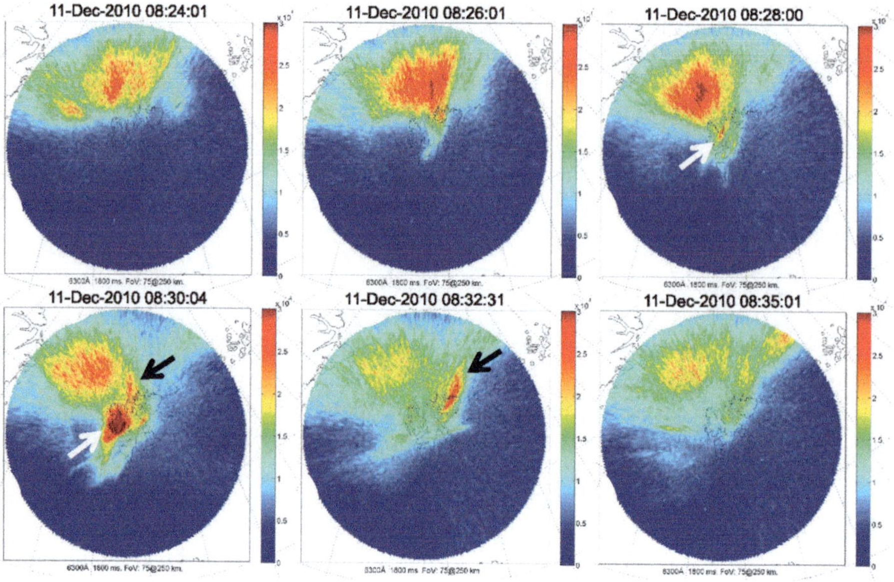

Fig. 9 All-sky camera images at 630.0 nm acquired during a TCV event. The emissions occurring poleward of the station before the event are thought to be cusp precipitation. The emissions associated with the TCVs are the brief, bright spots. Adapted from Kim et al. (2017b)

approximate time of the TCV arrival, implying that they were generated by a temperature anisotropy resulting from a compression on the dayside magnetosphere. The possible mechanisms to cause the magnetopause compression include localized pressure pulses (Glassmeier and Heppner 1992), hot flow anomalies (Sitar et al. 1998; Sibeck et al. 1999), and foreshock cavities (Murr and Hughes 2003).

5.6 Throat Aurora

Throat aurora is a particular auroral form observed around the dayside ionospheric convection throat region, which was first defined during an extensive study on dayside diffuse aurora (Han et al. 2015). Throat auroras typically exhibit as north-south aligned discrete auroral forms extending from the equatorward edge of the east-west-aligned auroral oval toward low latitude as shown in Fig. 10. The main observational properties of throat aurora have been investigated (Han et al. 2015, 2016, 2017a, 2018; Chen et al. 2017). It has been confirmed that throat auroras are caused by precipitation of particles originated from the magnetosheath (Han et al. 2016). Recently, Han et al. (2018) provide apparent one-to-one correspondences between throat auroras observed on the ground and magnetopause transients observed by MMS satellites near the subsolar magnetopause, which have been suggested to be the direct evidence for throat auroras being the ionospheric signature of magnetopause indentations. In addition, it was found that the spatial scale of the magnetopause indentation can be as large as $\sim 2.0 \times 3.0\ R_e$ after mapping a throat aurora to the geomagnetic equatorial plane, and the daily occurrence rate is higher than $\sim 50\%$ (Han et al. 2017a). These observational results indicate that throat aurora may reflect an important process that commonly occurs on the subsolar magnetopause.

On considering how the throat auroras are generated, Han et al. (2017a) pointed out that stripy diffuse aurora connecting to the discrete aurora oval seems to be a necessary

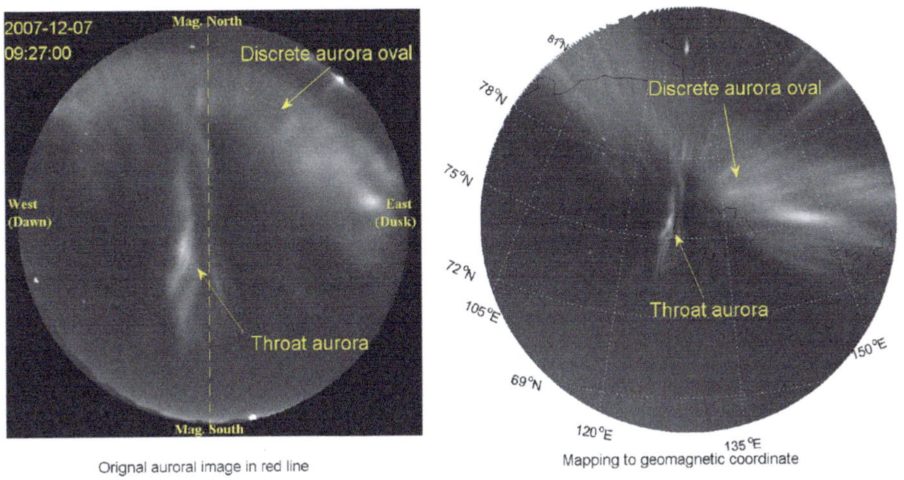

Fig. 10 A typical throat aurora observed at Yellow River Station by all-sky camera on 7 December 2007 shown (left) in original observation and (right) after mapping to geomagnetic coordinates (from Han et al. 2018). According to previous studies (e.g., Lockwood 1997), the equatorward edge of the discrete aurora oval near local noon can be regarded as the open-closed field line boundary

condition for the occurrence of throat aurora. It was also noticed that orientations of throat aurora are convection-aligned (Han et al. 2017a). Because diffuse auroras are caused by precipitation of particles from the central plasma sheet that is inside the magnetosphere and the convection is also a magnetosphere internal process, these observational results indicate that the occurrence of throat aurora is affected by factors from inside the magnetosphere. At the same time, Han et al. (2017a) found that the occurrence rate of throat aurora shows a clear dependence on the IMF cone angle ($\arccos[|B_x|/B_{total}]$). This suggests that some outside factors, such as magnetosheath high-speed jets, might be a driver for producing the throat auroras (Han et al. 2017a; Plaschke et al. 2018). It is puzzling why the generation of throat aurora depends on factors either inside or outside of the magnetosphere.

Recently, Han (2019) proposed a conceptual model for throat aurora. The model suggests that precipitation of a stripy diffuse aurora can lead to ionospheric conductivity enhancement and thus produce a polarization electric field in the ionosphere. This electric field maps to the magnetopause along closed field lines, and may affect the magnetopause reconnection to develop an inward motion and result in a throat aurora. This model can explain why the occurrence of throat aurora depends on either inside or outside factors, because it has been confirmed that two types of diffuse aurora exist near magnetic local noon, which are related with inside and outside factors, respectively (Han et al. 2017b). Furthermore, using coordinated observations from THEMIS and all-sky camera observations at South Pole station, Wang et al. (2018a) showed that magnetosheath high-speed jets were well associated the localized diffuse auroral brightening observed at South Pole station near local noon. This provides support to the model of Han (2019). According to this model, the outside factors may not drive throat aurora directly, but produce diffuse aurora first, change the ionospheric conductivity, set up a polarization electric field, and some feedback could then affect the magnetopause reconnection and result in throat aurora. Some observational support for the model was presented in Han (2019).

Very recent results provide evidence for throat aurora being associated with magnetopause reconnection (Han et al. 2019). Throat auroras have been suggested to be related to indentations on the subsolar magnetopause. However, the indentation generation process and the resulting ionospheric responses have remained unknown. An EISCAT Svalbard Radar experiment enabled the authors for the first time to observe the temporal and spatial evolution of flow reversals, Joule heating, and ion upflows associated with throat aurora. The high-resolution data allowed to discriminate that the flow bursts and Joule heating were concurrent and co-located, but were always observed on the west side of the associated throat auroras, reflecting that the upward/downward field-aligned currents associated with throat aurora are always to the east/west, respectively. These results are consistent with the geometry of the flux transfer event model (Southwood 1987) and provide strong evidence for throat aurora being associated with magnetopause reconnection.

6 Shock Aurora

Shock aurora is a subset of dayside aurora associated with the sudden enhancements of solar wind dynamic pressure that occurs at an interplanetary shock. The processes responsible for the shock aurora are complicated, as the Interplanetary Magnetic Field (IMF) is also highly variable. Consequently, some of the signatures of shock aurora are associated with changes in the reconnection topology. Indeed Zhou et al. (2017) list four different processes that could generate shock aurora. These are:

1. Enhanced pitch angle diffusion associated with direct compression of the magnetospheric magnetic field (Zhou and Tsurutani 1999; Zhou et al. 2003, 2009);
2. Fast mode (compressional) waves launched by the shock and propagating through the magnetosphere, generating traveling convection vortices (TVCs) and the associated field-aligned currents that generate aurora (Zhou et al. 2010; Zhang et al. 2002);
3. Alfvén waves launched through magnetic reconnection, with these waves supporting parallel electric fields, and hence particle acceleration through kinetic effects, i.e., electron thermal pressure (Chaston et al. 2005; Zhou et al. 2009) and
4. Flow and magnetic field shears that generate field-aligned currents that in turn require parallel electrostatic potentials, much like the discrete aurora on the nightside.

We note that many of these processes are discussed in other sections of this chapter, and indeed many of the processes are not unique to shock aurora. Haerendel (2011), in discussing different auroral generators, includes both Alfvén wave generated aurora (process 3 above) and quasi-static potentials (process 4). In this Chapter TCVs are discussed in Sect. 5.5 as a consequence of reconnection imposed flows, while Sect. 7 discusses diffuse aurora and the association with pitch angle scattering. The difference here is that there is a direct association of the arrival of an interplanetary shock and the resultant aurora.

The most directly shock-related aurora is that associated with enhanced wave-induced particle precipitation. As discussed by Zhou and Tsurutani (1999) and Zhou et al. (2017), the effect of the shock-induced magnetic field compression is to enhance the perpendicular energy of the near equatorial particle distribution in the magnetosphere. This in turn results in wave instabilities that isotropize the distribution, increasing the energy flux in the loss cone. This in turn leads to diffuse aurora.

Fast-mode wave-related TCVs are also primarily associated with solar wind dynamic pressure enhancements. In this case the fast-mode waves convert to shear-mode waves, as these are the only waves that carry field-aligned current. This allows the vortical flow around the fast-mode wave to be transmitted to the ionosphere. As with nightside aurora, the associated field-aligned currents could have energetic (100 eV to several keV) electron precipitation associated with them either through quasi-static electric fields (Knight 1973), or through Alfvén waves (Chaston et al. 2005). In the latter case the Alfvén waves can carry a field-aligned electric field through either kinetic (electron pressure) or inertial (electron inertia) effects (see Lysak and Song 2003 for a discussion of how non-ideal-MHD Alfvén waves can have a parallel electric field associated with them).

Simultaneous observations of the electron and proton aurora during a solar wind pressure pulse showed the slightly different behavior to the shock arrival (Meurant et al. 2003). It was found that the electron and the proton precipitation both start in the post noon sector and expand concurrently, but the expansion into the nightside starts sooner for the protons than for the electrons. In situ electron measurements in the afternoon sector indicate that the shock has a significant effect on the electron spectral characteristics. It is suggested that the various Alfvén frequencies generated by the shock account for the two different speeds of propagation of the disturbance. Other observations showed that the propagation from the noon to the night sector mainly occurs through the afternoon region for proton precipitation and the morning sector for electron aurora, as expected from the azimuthal drift of newly injected plasma (Meurant et al. 2004). The asymmetry of the precipitation distribution around the noon-midnight axis is more pronounced during negative B_z periods, when activity is the most important. It was suggested that adiabatic compression and plasma waves play an important role on the locations of electron and proton precipitation in the dayside. It was also shown that some shocks are able to trigger substorms when they hit an unstable magnetosphere (Meurant et al. 2005).

Figure 11 shows the complexity of the various processes associated with shock aurora and the resultant precipitating electron signatures. The left hand side of the figure shows an electron energy-time spectrogram as the FAST spacecraft passes from low to high latitudes, leaving the plasma-sheet (as evidenced by several keV electrons) into the polar cap. For this pass the spacecraft passes through a region of intense moderate energy (<500 eV) electrons. Magnetometer data (not shown) show highly variable FAC structure throughout this region. The right-hand side of the figure shows a 2-D phase space density contour plot for the time indicated by the vertical white line in the left-hand panel. The distribution, which has been clipped at 10^{-10} s^3 cm^{-3} km^{-3}, shows a clear loss-cone on the left-hand side. These particles, with speeds $>10^4$ km/s (>300 eV), are mainly of magnetosheath origin, although they may have been further energized by waves at higher altitudes. The lower energy electrons show bi-directional fluxes. This is indicative of local wave acceleration.

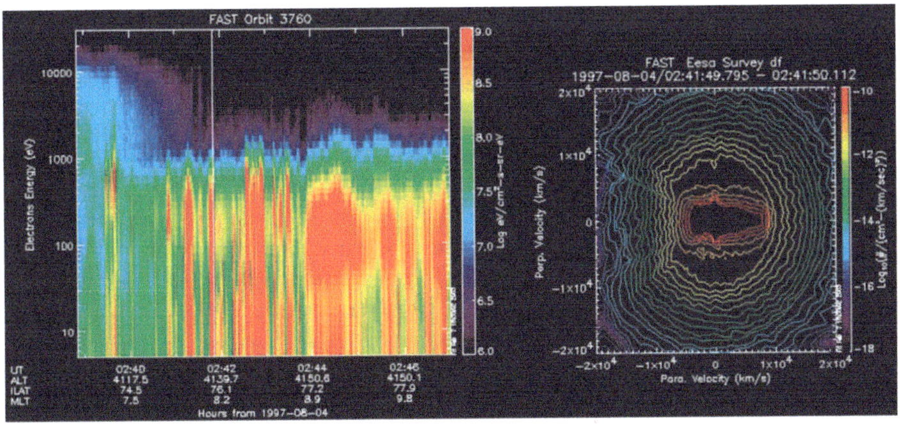

Fig. 11 Example of shock aurora observed by the FAST spacecraft (after Zhou et al. 2009)

In summary, shock aurora is a highly complex phenomenon. Some of the signatures are directly attributable to the shock passage (diffuse aurora from wave-induced precipitation, TCVs from compressional fast-mode waves), while others are reminiscent of processes associated with nightside aurora, but in this case connected to the dayside magnetopause, through either reconnection or boundary layer motion through Kelvin-Helmholtz waves. The shock aurora in this case is either directly related to FACs, or parallel acceleration by kinetic or inertial Alfvén waves.

7 Dayside Diffuse Aurora

The equatorward portion of the dayside auroral oval is dominated by diffuse aurora including pulsating aurora (Lorentzen et al. 1996; Sandholt et al. 2002b). The spectral intensity is primarily at the 557.7 nm wavelength ($I_{630.0}/I_{557.7} < 1$), and the emission is driven by precipitating electrons of a few keV energy scattered by waves in the dayside plasma sheet through wave-particle interaction. Ni et al. (2016) provide an extensive review of the wave-particle interaction and driving processes of diffuse auroral precipitation.

Dayside diffuse aurora is more intense at pre-noon than post-noon, and is a common structure particularly at magnetically quiet times, although it is more intense at active times (Newell et al. 2009; Han et al. 2015). This is consistent with the fact that whistler-mode waves in the magnetosphere occur more frequently on the dayside than nightside even with low amplitudes (Li et al. 2011) and that plasma sheet electron fluxes are higher at pre-noon because electron populations decrease as they experience magnetic drifts from the nightside through dawn.

Despite the diffuse appearance, a variety of structures can be found in the diffuse auroral region:

1. Pulsating aurora occurs in the poleward portion of the diffuse aurora region and is associated with energy-dispersed electron precipitation (Lorentzen et al. 1996). Dayside pulsating aurora intensifies by extending from the nightside during substorms (Royrvik and Davis 1977).
2. Particularly near noon, the diffuse aurora shows patches, bands, stripes and irregular forms of some tens of km size. Examples are shown in Fig. 12. The diffuse auroral structures drift at the large-scale convection speed, and diffuse auroral stripes are suggested to occur in the dayside convection throat (Ebihara et al. 2007; Han et al. 2015). Some of them have been shown to connect to nearly north-south aligned discrete aurora called throat aurora (Han et al. 2016) (see Sect. 5.6).
3. Faster propagation of patches beyond convection speed ($\sim$20 km/s) has also been found, in association with Pc3 ULF pulsations (Motoba et al. 2017).
4. Black aurora is also embedded in the structure-less diffuse auroral background (Han et al. 2015).

Satellite-ground conjugate observations have shown that the diffuse auroral patches correspond to low-energy ($<\sim$50 eV) plasma density structures in the dayside outer magnetosphere at $L > \sim 6\ R_e$ (Nishimura et al. 2013). Whistler-mode waves are intensified in the high-density regions, while plasma sheet electrons of >1 keV do not show similar structures (Nishimura et al. 2013). Growth rates of whistler-mode waves are enhanced in the high-density regions due to lower resonant energy, and the waves increase pitch angle scattering of plasma sheet electrons into the loss cone. This process explains the formation of the localized patches of diffuse aurora and is consistent with the motion of patches being at

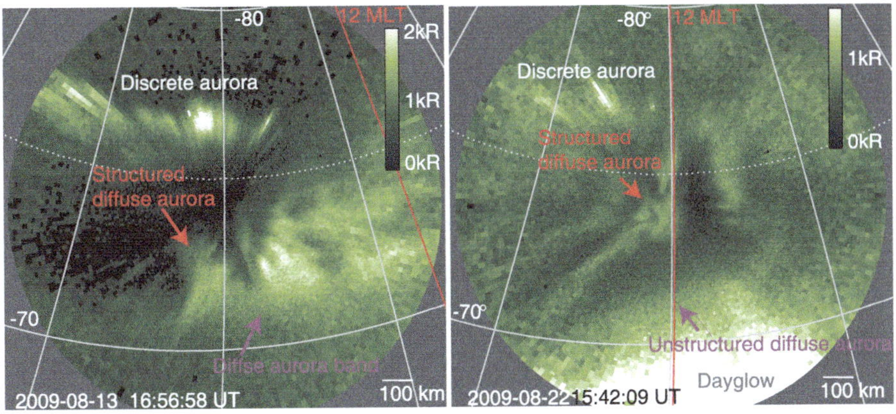

Fig. 12 Examples of dayside diffuse aurora. Diffuse aurora occurs equatorward of discrete aurora and shows different types of structures (Nishimura et al. 2013)

large-scale convection speed rather than magnetic drift speed of energetic electrons. Numerical calculations have also demonstrated that enhanced waves can drive sufficient electron precipitation flux for the measured level of diffuse auroral intensity (Ni et al. 2014).

Dayside diffuse aurora also shows rapid temporal variations. Solar wind dynamic pressure enhancements associated with interplanetary shocks create abrupt intensifications of diffuse aurora due to large-scale magnetosphere compression (Zhou et al. 2003; Nishimura et al. 2016) (see also Sect. 6). Recently, diffuse aurora has also been found to brighten associated with IMF discontinuities, some of which create compressions and rarefactions through interaction with the bow shock (foreshock transients). The 2D structure and propagation of diffuse auroral features were able to reveal a few R_e azimuthal size of localized compression and its azimuthal propagation at $\sim$100 km/s (Wang et al. 2018b). Dayside diffuse aurora also present localized modulation without appreciable changes in the solar wind. Those are particularly seen under B_x-dominant IMF, where magnetosheath high-speed jets (HSJs) are created and compress the magnetopause locally (a few R_e azimuthal size) (Wang et al. 2018a).

8 Morning Side Diffuse Aurora Spots

A most likely diffuse spot of aurora was often seen in X-ray images by the Polar spacecraft PIXIE instrument (Imhof et al. 1995), while it was not seen in better spatial resolution observations by Polar-UVI (Torr et al. 1995) or VIS (Frank et al. 1995). The "traditional" UV aurora instruments UVI and VIS are sensitive to emissions that are generated by precipitating electrons up to $\approx$25 keV, PIXIE was specifically sensitive to emissions from high energy electrons (>10 keV). Delayed about 30 minutes relative to substorm onsets, a localized maximum of X-ray emission was observed at 5–9 hours magnetic local time (Østgaard et al. 1999) where no corresponding UV emissions were seen (Fig. 13). By identifying the location of the injection region and determining the substorm onset time it was speculated that this maximum most probably is caused by electrons injected in the midnight sector and then drifting (i.e., gradient and curvature drift) into a region in the dawnside magnetosphere where some mechanism effectively scatters the electrons into the loss cone. Later more de-

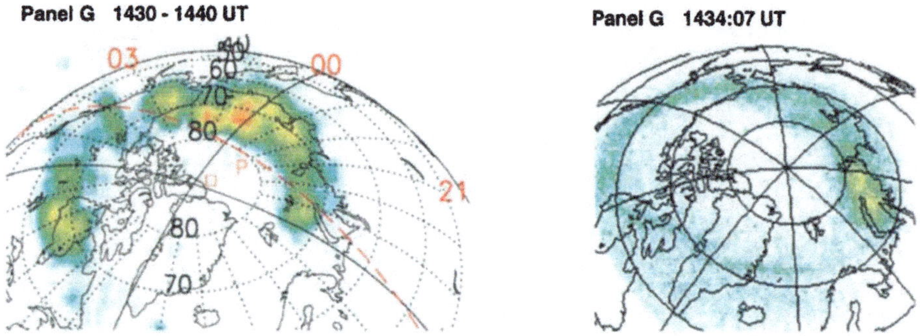

Fig. 13 Example of a diffuse morning side spot emission in X-rays (left) but nothing is seen in UV emissions (right). Adapted from Østgaard et al. (1999)

tailed investigations including ground-based and in situ measurements from high- and low-altitude satellites provided more information about the global characteristics of such events (Østgaard et al. 2000a). It could be confirmed that the time delay between substorm onset around midnight and the appearance of the spots on the morning side is similar to the drift time of energetic electrons. The electron spectra measured in the early stage of the localized morning maximum of X-ray emission strongly indicate that the scattering of >2–10 keV electrons by wave-particle interaction into the loss cone is the main mechanism for this precipitation (Østgaard et al. 2000b). Such high energy electrons (up to 200 keV) have already been found in this general local time region much earlier (McDiarmid et al. 1975). It was also found that a high energy ($\approx$30–100 keV) component of the precipitating electron population contributes $\approx 30\%$ of the total electron energy flux in that region (Cummer et al. 2000).

9 Conclusions

Most dayside aurora phenomena are driven by reconnection between the solar wind and the terrestrial magnetic field at the dayside magnetopause. Studying dayside aurora can thus provide a lot of insight into the solar wind–magnetosphere interaction. This review summarizes the properties, dynamics, and driving solar wind conditions for dayside aurora phenomena. Even though past space missions and ground-based observations have provided a lot of information about these processes there are still a number of open questions. For instance, what is the time scale between the arrival of solar wind structures at the magnetopause and the visible response in the dayside aurora dynamics or appearance? Is there a quantitative relationship between the magnitude and orientation of the IMF and the response in dayside auroral forms? Why do PMAFs occur during northward and southward IMF? Does steady or pulsed reconnection display a similar temporal behavior in cusp aurora brightness and appearance? Can we infer lobe reconnection properties from observations of HiLDA? Can the extent of the reconnection X-line be inferred from PMAF observations? Is the proposed model for throat aurora correct or does need to be adjusted with more detailed observations? Why does shock aurora appear to be related to the shock passage and dayside processes but also to nightside processes? Unfortunately, there are no immediate plans for the launch of satellites that could provide long-term and continuous observations of dayside aurora and we can only hope that in-situ measurements and model calculations with appropriate inputs will enhance our knowledge about dayside aurora and the interactions at the magnetopause.

Acknowledgements HUF was supported by NSF award AGS-1004736 and NASA's Explorers Program through contracts NNG12FA45C and NNG12FA42I. The work of D. Han is supported by the National Key R & D Program of China (2018YFC1407303) and National Natural Science Foundation of China (NSFC) (41774174). SEM was supported by the Science and Technology Facilities Council (STFC), UK, grant no. ST/N000749/1. The work at the Birkeland Centre for Space Centre, University of Bergen, Norway, was supported by the Research Council of Norway/CoE under contract 223252/F50.

Publisher's Note Springer Nature remains neutral with regard to jurisdictional claims in published maps and institutional affiliations.

References

G.A. Abel, M.P. Freeman, A statistical analysis of ionospheric velocity and magnetic field power spectra at the time of pulsed ionospheric flows. J. Geophys. Res. **107**, 1470 (2002). https://doi.org/10.1029/2002JA009402

R.L. Arnoldy, M.J. Engebretson, J.L. Alford, R.E. Erlandson, B.J. Anderson, Magnetic impulse events and associated Pc 1 bursts at dayside high latitudes. J. Geophys. Res. **101**, 7793 (1996a). https://doi.org/10.1029/95JA03378

R.L. Arnoldy, K.A. Lynch, P.M. Kintner, J. Bonnell, T.E. Moore, C.J. Pollock, SCIFER—Structure of the Cleft Ion Fountain at 1400 km altitude. Geophys. Res. Lett. **23**, 1869–1872 (1996b). https://doi.org/10.1029/96GL00475

J. Berchem, C.T. Russell, Flux transfer events on the magnetopause—spatial distribution and controlling factors. J. Geophys. Res. **89**, 6689 (1984). https://doi.org/10.1029/JA089iA08p06689

H.C. Carlson, Sharpening our thinking about polar cap ionospheric patch morphology, research, and mitigation techniques. Radio Sci. **47**, RS0L21 (2012). https://doi.org/10.1029/2011RS004946

H.C. Carlson, T. Spain, A. Aruliah, A. Skjaeveland, J. Moen, First principles physics of cusp/polar cap thermospheric disturbances. Geophys. Res. Lett. **39**, L19103 (2012). https://doi.org/10.1029/2012GL053034

J.A. Carter, S.E. Milan, A.R. Fogg, L.J. Paxton, B.J. Anderson, The association of high-latitude dayside aurora with NBZ field-aligned currents. J. Geophys. Res. (2018). https://doi.org/10.1029/2017JA025082

C.C. Chaston, L.M. Peticolas, C.W. Carlson, J.P. McFadden, F. Mozer, M. Wilber, G.K. Parks, A. Hull, R.E. Ergun, R.J. Strangeway, M. Andre, Y. Khotyaintsev, M.L. Goldstein, M. Acuna, E.J. Lund, H. Reme, I. Dandouras, A.N. Fazakerley, A. Balogh, Energy deposition by Alfvén waves into the dayside auroral oval: Cluster and FAST observations. J. Geophys. Res. **110**, A02211 (2005). https://doi.org/10.1029/2004JA010483

X.C. Chen, D.S. Han, D.A. Lorentzen, K. Oksavik, J.I. Moen, L.J. Baddeley, Dynamic properties of throat aurora revealed by simultaneous ground and satellite observations. J. Geophys. Res. **122**, 3469–3486 (2017). https://doi.org/10.1002/2016JA023033

G. Chisham, M. Lester, S.E. Milan, M.P. Freeman, W.A. Bristow, A. Grocott, K.A. McWilliams, J.M. Ruohoniemi, T.K. Yeoman, P.L. Dyson, R.A. Greenwald, T. Kikuchi, M. Pinnock, J.P.S. Rash, N. Sato, G.J. Sofko, J.-P. Villain, A.D.M. Walker, A decade of the Super Dual Auroral Radar Net-work (SuperDARN): scientific achievements, new techniques and future directions. Surv. Geophys. **28**, 33–109 (2007). https://doi.org/10.1007/s10712-007-9017-8

J.H. Clemmons, R.L. Walterscheid, D.G. Brinkman, J.H. Hecht, M.R. Lessard, B.A. Fritz, D.R. Kenward, L.B.N. Clausen, D.L. Hysell, The neutral atmosphere during the RENU2 investigation. Geophys. Res. Lett. (2019, submitted)

I.J. Coleman, G. Chisham, M. Pinnock, M.P. Freeman, An ionospheric convection signature of antiparallel reconnection. J. Geophys. Res. **106**, 28,995–29,007 (2001)

B.M.A. Cooling, C.J. Owen, S.J. Schwartz, Role of the magnetosheath flow in determining the motion of open flux tubes. J. Geophys. Res. **106**, 18,763–18,776 (2001)

S.W.H. Cowley, M. Lockwood, Excitation and decay of solar wind-driven flows in the magnetosphere-ionosphere system. Ann. Geophys. **10**, 103–115 (1992)

S.A. Cummer, R.R. Vondrak, N. Østgaard, J. Stadsnes, J. Bjordal, D.L. Chenette, M.J. Brittnacher, G.K. Parks, J.B. Sigwarth, L.A. Frank, Global multispectral auroral imaging of an isolated substorm. Geophys. Res. Lett. **27**, 637–640 (2000). https://doi.org/10.1029/1999GL003678

J.A. Davies, T.K. Yeoman, I.J. Rae, S.E. Milan, M. Lester, M. Lockwood, A. McWilliams, Ground-based observations of the auroral zone and polar cap ionospheric responses to dayside transient reconnection. Ann. Geophys. **20**, 781–794 (2000). https://doi.org/10.5194/angeo-20-781-2002

E.E. Drury, S.B. Mende, H.U. Frey, J.H. Doolittle, Southern Hemisphere poleward moving auroral forms. J. Geophys. Res. **108**, 1114 (2003). https://doi.org/10.1029/2001JA007536

Y. Ebihara, Y.M. Tanaka, S. Takasaki, A.T. Weatherwax, M. Taguchi, Quasi-stationary auroral patches observed at the South Pole Station. J. Geophys. Res. **112**, A01201 (2007). https://doi.org/10.1029/2006JA012087

Y. Ebihara, R. Kataoka, A.T. Weatherwax, M. Yamauchi, J. Geophys. Res. **113**, A07209 (2008). https://doi.org/10.1029/2008JA013099

M.J. Engebretson et al., Multi-instrument observations from Svalbard of a traveling convection vortex, electromagnetic ion cyclotron wave burst, and proton precipitation associated with a bow shock instability. J. Geophys. Res. **118**, 2975 (2013). https://doi.org/10.1002/jgra.50291

C.J. Farrugia, R.P. Rijnbeek, M.A. Saunders, D.J. Southwood, D.J. Rodgers, M.F. Smith, L.J.C. Woolliscroft, A multi-instrument study of flux transfer event structure. J. Geophys. Res. **93**, 14,465–14,477 (1988). https://doi.org/10.1029/JA093iA12p14465

C.J. Farrugia et al., Pulsed flows at the high-altitude cusp poleward boundary, and associated ionospheric convection and particle signatures, during a Cluster–FAST–SuperDARN- Søndrestrøm conjunction under a southwest IMF. Ann. Geophys. **22**, 2891–2905 (2004). https://doi.org/10.5194/angeo-22-2891-2004

G.J. Fasel, Dayside poleward moving auroral forms: a statistical study. J. Geophys. Res. **100**(A7), 11891–11905 (1995). https://doi.org/10.1029/95JA00854

G.J. Fasel, L.C. Lee, R.W. Smith, A mechanism for the multiple brightenings of dayside poleward-moving auroral forms. Geophys. Res. Lett. **20**(20), 2247–2250 (1993). https://doi.org/10.1029/93GL02487

R.C. Fear et al., Motion of flux transfer events: a test of the Cooling model. Ann. Geophys. **25**, 1669–1690 (2007). www.ann-geophys.net/25/1669/2007/

R.C. Fear, L. Trenchi, J.C. Coxon, S.E. Milan, How much flux does a flux transfer event transfer? J. Geophys. Res. **122**, 12,310–12,327 (2017). https://doi.org/10.1002/2017JA024730

Y.I. Feldstein, Auroral oval. J. Geophys. Res. **78**, 1210 (1973)

L.A. Frank, J.B. Sigwarth, J.D. Craven, J.P. Cravens, J.S. Dolan, M.R. Dvorsky, P.K. Hardebeck, J.D. Harvey, D.W. Muller, The Visible Imaging System (VIS) for the polar spacecraft. Space Sci. Rev. **71**, 297–328 (1995). https://doi.org/10.1007/BF00751334

H.U. Frey, Localized aurora beyond the auroral oval. Rev. Geophys. **45** (2007). https://doi.org/10.1029/2005RG000174

H.U. Frey, S.B. Mende, T.J. Immel, S.A. Fuselier, E.S. Claflin, J.-C. Gérard, B. Hubert, Proton aurora in the cusp. J. Geophys. Res. **107**, 1091 (2002). https://doi.org/10.1029/2001JA900161

H.U. Frey, T.J. Immel, G. Lu, J. Bonnell, S.A. Fuselier, S.B. Mende, B. Hubert, N. Østgaard, G. Le, Properties of localized, high latitude, dayside aurora. J. Geophys. Res. **108**, 8008 (2003a). https://doi.org/10.1029/2002JA009332

H.U. Frey, S.B. Mende, S.A. Fuselier, T.J. Immel, N. Østgaard, Proton aurora in the cusp during southward IMF. J. Geophys. Res. **108**, 1277 (2003b). https://doi.org/10.1029/2003JA009861

H.U. Frey, T.D. Phan, S.A. Fuselier, S.B. Mende, Continuous magnetic reconnection at Earth's magnetopause. Nature **426**, 533–537 (2003c)

H.U. Frey, N. Østgaard, T.J. Immel, H. Korth, S.B. Mende, Seasonal dependence of localized, High Latitude Dayside Aurora (HiLDA). J. Geophys. Res. **109**, A04303 (2004). https://doi.org/10.1029/2003JA010293

E. Friis-Christensen, M.A. McHenry, C.R. Clauer, S. Vennerstrøm, Ionospheric traveling convection vortices observed near the polar cleft—a triggered response to sudden changes in the solar wind. Geophys. Res. Lett. **15**, 253 (1988). https://doi.org/10.1029/GL015i003p00253

B.A. Fritz, M.R. Lessard, K.F. Dymond, D.L. Kenward, K.A. Lynch, J.H. Hecht, J.H. Clemmons, RENU2 UV PMT observations in the cusp. Geophys. Res. Lett. (2019). https://doi.org/10.1029/2019GL082314

S.A. Fuselier, H.U. Frey, K.J. Trattner, S.B. Mende, J.L. Burch, Cusp aurora dependence on IMF B_Z. J. Geophys. Res. **107** (2002). https://doi.org/10.1029/2001JA900165

S.A. Fuselier, S.B. Mende, T.E. Moore, H.U. Frey, S.B. Petrinec, E.S. Claflin, M.R. Collier, Cusp dynamics and ionospheric outflow. Space Sci. Rev. **109**, 285–312 (2003). https://doi.org/10.1023/B:SPAC.0000007522.71147.b3

S.A. Fuselier, K.J. Trattner, S.M. Petrinec, Antiparallel and component reconnection at the dayside magnetopause. J. Geophys. Res. **116**, A10227 (2011). https://doi.org/10.1029/2011JA016888

K.-H. Glassmeier, C. Heppner, Traveling magnetospheric convection twin vortices: another case study, global characteristics, and a model. J. Geophys. Res. **97**(A4), 3977 (1992). https://doi.org/10.1029/91JA02464

K.H. Glassmeier, M. Stellmacher, Mapping flux transfer events to the ionosphere. Adv. Space Res. **18**(8), 151 (1996)

N.H. Godbole et al., RENU2 observations of ion upflow and downflow. Geophys. Res. Lett. (2019, submitted)

G. Haerendel, Six auroral generators: a review. J. Geophys. Res. **116**, A00K05 (2011). https://doi.org/10.1029/2010JA016425. Printed **117**(A1) (2012)

G. Haerendel, G. Paschmann, N. Sckopke, H. Rosenbauer, P.C. Hedgecock, The frontside boundary layer of the magnetosphere and the problem of reconnection. J. Geophys. Res. **83**(A7), 3195 (1978). https://doi.org/10.1029/JA083iA07p03195

D.S. Han, Ionospheric polarization electric field guiding magnetopause reconnection: a conceptual model of throat aurora. Sci. China Earth Sci. **62** (2019). https://doi.org/10.1007/s11430-019-9358-8

D. Han, X.-C. Chen, J.-J. Liu, Q. Qiu, K. Keika, Z.-J. Hu, J.-M. Liu, H.-Q. Hu, H.-G. Yang, An extensive survey of dayside diffuse aurora based on optical observations at Yellow River Station. J. Geophys. Res. **120**(9), 7447 (2015). https://doi.org/10.1002/2015ja021699

D.S. Han, Y. Nishimura, L.R. Lyons, H.Q. Hu, H.G. Yang, Throat aurora: the ionospheric signature of magnetosheath particles penetrating into the magnetosphere. Geophys. Res. Lett. **43**, 1819–1827 (2016). https://doi.org/10.1002/2016GL068181

D.S. Han, H. Hietala, X.C. Chen, Y. Nishimura, L.R. Lyons, J.J. Liu, H.Q. Hu, H.G. Yang, Observational properties of dayside throat aurora and implications on the possible generation mechanisms. J. Geophys. Res. **122**, 1853 (2017a). https://doi.org/10.1002/2016ja023394

D.S. Han, J.-X. Li et al., Coordinated observations of two types of diffuse auroras near magnetic local noon by Magnetospheric Multiscale mission and ground all-sky camera. Geophys. Res. Lett. **44**(16), 8130 (2017b). https://doi.org/10.1002/2017gl074447

D.-S. Han, J.J. Liu, X.C. Chen, T. Xu, B. Li, Z.J. Hu, H.Q. Hu, H.G. Yang, S.A. Fuselier, C.J. Pollock, Direct evidence for throat aurora being the ionospheric signature of magnetopause transient and reflecting localized magnetopause indentations. J. Geophys. Res. (2018). https://doi.org/10.1002/2017ja024945

D.S. Han, T. Xu, Y. Jin, K. Oksavik, X.C. Chen, J.J. Liu et al., Observational evidence for throat aurora being associated with magnetopause reconnection. Geophys. Res. Lett. **46**, 7113–7120 (2019). https://doi.org/10.1029/2019GL083593

J.H. Hecht, J.H. Clemmons, M.R. Lessard, D.L. Kenward, B.F. Sadler, B.A. Fritz, J.S. Evans, K.A. Lynch, A new technique for estimating the lifetime of bursts of electron precipitation from sounding rocket measurementsb. Geophys. Res. Lett. (2019). https://doi.org/10.1029/2019GL082894

J.L. Horwitz, S.I. Akasofu, The response of the dayside aurora to sharp northward and southward transitions of the interplanetary magnetic field and to magnetospheric substorms. J. Geophys. Res. **82**(19), 2723 (1977). https://doi.org/10.1029/JA082i019p02723

S.M. Imber, S.E. Milan, B. Hubert, Ionospheric flow and auroral signatures of dual lobe re-connection. Ann. Geophys. **24**, 3115–3129 (2006)

W.L. Imhof, K.A. Spear, J.W. Hamilton, B.R. Higgins, M.J. Murphy, J.G. Pronko, R.R. Vondrak, D.L. McKenzie, C.J. Rice, D.J. Gorney, D.A. Roux, R.L. Williams, J.A. Stein, J. Bjordal, J. Stadsnes, K. Njoten, T.J. Rosenberg, L. Lutz, D. Detrick, The Polar Ionospheric X-Ray Imaging Experiment (PIXIE). Space Sci. Rev. **71**, 385–408 (1995). https://doi.org/10.1007/BF00751336

M.G. Johnsen, D.A. Lorentzen, A statistical analysis of the optical dayside open/closed field line boundary. J. Geophys. Res. **117**, A02218 (2012). https://doi.org/10.1029/2011JA016984

R. Kataoka, H. Fukunishi, L.J. Lanzerotti, C.G. Maclennan, H.U. Frey, S.B. Mende, J.H. Doolittle, T.J. Rosenberg, A.T. Weatherwax, Magnetic impulse event: a detailed case study of extended ground and space observations. J. Geophys. Res. **106**(A11), 25873 (2001). https://doi.org/10.1029/2000JA000314

R. Kataoka, H. Fukunishi, K. Hosokawa, H. Fujiwara, A.S. Yukimatu, N. Sato, Y.-K. Tung, Transient production of F-region irregularities associated with TCV passage. Ann. Geophys. **21**, 1531 (2003). https://doi.org/10.5194/angeo-21-1531-2003

D.L. Kenward, M.R. Lessard, B.A. Fritz, I.J. Cohen, Characterization of soft electron precipitation in the cusp region during poleward moving auroral form event. Geophys. Res. Lett. (2019, submitted)

H. Kim et al., Conjugate observations of electromagnetic ion cyclotron waves associated with traveling convection vortex events. J. Geophys. Res. **122**, 7336 (2017a). https://doi.org/10.1002/2017JA024108

H. Kim et al., Simultaneous observations of traveling convection vortices: ionosphere-thermosphere coupling. J. Geophys. Res. **122**, 4943 (2017b). https://doi.org/10.1002/2017JA023904

P.M. Kintner et al., The SCIFER experiment. Geophys. Res. Lett. **23**, 1865 (1996). https://doi.org/10.1029/96GL01260

S. Knight, Parallel electric fields. Planet. Space Sci. **21**, 741–750 (1973)

J. Kuijpers, H.U. Frey, L. Fletcher, Electric current circuits in astrophysics. Space Sci. Rev. **188**, 3–57 (2015). https://doi.org/10.1007/s11214-014-0041-y

L.J. Lanzerotti, L.C. Lee, C.G. Maclennan, A. Wolfe, L.V. Medford, Possible evidence of flux transfer events in the polar ionosphere. Geophys. Res. Lett. **13**, 1089 (1986). https://doi.org/10.1029/GL013i011p01089

L.J. Lanzerotti, A. Wolfe, N. Trivedi, C.G. Maclennan, L.V. Medford, Magnetic impulse events at high latitudes—magnetopause and boundary layer plasma processes. J. Geophys. Res. **95**, 97 (1990). https://doi.org/10.1029/JA095iA01p00097

L.C. Lee, Z.F. Fu, A theory of magnetic flux transfer at the Earth's magnetopause. Geophys. Res. Lett. **12**, 105 (1985). https://doi.org/10.1029/GL012i002p00105

M.R. Lessard et al., Overview of the Rocket Experiment for Neutral Upwelling sounding Rocket 2 (RENU2). Geophys. Res. Lett. (2019). https://doi.org/10.1029/2018GL081885

W. Li, J. Bortnik, R.M. Thorne, V. Angelopoulos, Global distribution of wave amplitudes and wave normal angles of chorus waves using THEMIS wave observations. J. Geophys. Res. **116**, A12205 (2011). https://doi.org/10.1029/2011JA017035

M. Lockwood, Relationship of dayside auroral precipitations to the open-closed separatrix and the pattern of convective flow. J. Geophys. Res. **102**(17), 17,475–17,487 (1997)

M. Lockwood, Identifying the open-closed field line boundary, in *Polar Cap Boundary Phenomena*, ed. by J. Moen, A. Egeland, M. Lockwood. NATO ASI Series, vol. 509 (Kluwer Academic Publishers, Dordrecht, 1998), pp. 73–90

M. Lockwood, S.W.H. Cowley, P.E. Sandholt, R.P. Lepping, The ionospheric signatures of flux transfer events and solar wind dynamic pressure changes. J. Geophys. Res. **95**, 17,113 (1990). https://doi.org/10.1029/JA095iA10p17113

M. Lockwood, H.C. Carlson, P.E. Sandholt, The implications of the altitude of transient 630 nm dayside auroral emission. J. Geophys. Res. **98**, 15,571–15,587 (1993)

M. Lockwood, S.W.H. Cowley, M.F. Smith, R.P. Rijnbeek, R.C. Elphic, The contribution of flux transfer events to convection. Geophys. Res. Lett. **22**, 1185–1188 (1995)

M. Lockwood, C.J. Davis, T.G. Onsager, J.A. Scudder, Modelling signatures of pulsed magnetopause reconnection in cusp ion dispersion signatures seen at middle altitudes. Geophys. Res. Lett. **25**, 591–594 (1998)

M. Lockwood, I.W. McCrea, S.E. Milan, J. Moen, J.C. Cerisier, A. Thorolfsson, Plasma structure within poleward-moving cusp-cleft auroral transients: EISCAT Svalbard radar observations and an explanation in terms of large local time extent of events. Ann. Geophys. **18**, 1027 (2000)

M. Lockwood, S.E. Milan, T. Onsager, C.H. Perry, J.A. Scudder, C.T. Russell, M. Brittnacher, Cusp ion steps, field-aligned currents and poleward moving auroral forms. J. Geophys. Res. **106**, 29,555–29,569 (2001a)

M. Lockwood, H. Opgenoorth et al., Coordinated Cluster, ground-based instrumentation and low-altitude satellite observations of transient poleward-moving events in the ionosphere and in the tail lobe. Ann. Geophys. **19**, 1589–1612 (2001b). https://doi.org/10.5194/angeo-19-1589-2001

D.A. Lorentzen, C.S. Deehr, J.I. Minow, R.W. Smith, H.C. Stenbaek-Nielsen, F. Sigernes, SCIFER-dayside auroral signatures of magnetospheric energetic electrons. Geophys. Res. Lett. **23**, 1885 (1996). https://doi.org/10.1029/96GL00593

D.A. Lorentzen, J. Moen, K. Oksavik, F. Sigernes, Y. Saito, M.G. Johnsen, In situ measurement of a newly created polar cap patch. J. Geophys. Res. **115**, A12323 (2010). https://doi.org/10.1029/2010JA015710

E.J. Lund et al., Electron temperature in the cusp as measured with the SCIFER-2 sounding rocket. J. Geophys. Res. **117**, A06326 (2012). https://doi.org/10.1029/2011JA017404

R.L. Lysak, Y. Song, Kinetic theory of the Alfvén wave acceleration of auroral electrons. J. Geophys. Res. **108**(A4), 8005 (2003). https://doi.org/10.1029/2002JA009406

A. Marchaudon, J.-C. Cerisier, J.-M. Bosqued, M.W. Dunlop, J.A. Wild, P.M.E. Decreau, M. Forster, D. Fontaine, H. Laakso, Transient plasma injections in the dayside magnetosphere: one-to-one correlated observations by Cluster and SuperDARN. Ann. Geophys. **22**, 141–158 (2004). https://doi.org/10.5194/angeo-22-141-2004

N.C. Maynard et al., Characteristics of merging at the magnetopause inferred from dayside 557.7 nm all-sky images: IMF drivers of poleward moving auroral forms. Ann. Geophys. **24**, 3071 (2006). https://doi.org/10.5194/angeo-24-3071-2006

I.B. McDiarmid, J.R. Burrows, E.E. Budzinski, Average characteristics of magnetospheric electrons (140 eV to 200 keV) at 1400 kilometers. J. Geophys. Res. **80**, 73–79 (1975)

K.A. McWilliams, T.K. Yeoman, S.W.H. Cowley, Two-dimensional electric field measurements in the ionospheric footprint of a flux transfer event. Ann. Geophys. **18**, 1584–1598 (2000a). https://doi.org/10.1007/s00585-001-1584-2

K.A. McWilliams, T.K. Yeoman, G. Provan, A statistical survey of dayside pulsed ionospheric flows as seen by the CUTLASS Finland HF radar. Ann. Geophys. **18**, 445–453 (2000b). https://doi.org/10.1007/s00585-000-0445-8

S.B. Mende, R.L. Rairden, L.J. Lanzerotti, C.G. Maclennan, Magnetic impulses and associated optical signatures in the dayside aurora. Geophys. Res. Lett. **17**, 131 (1990). https://doi.org/10.1029/GL017i002p00131

S.B. Mende, H.U. Frey, J.H. Doolittle, L. Lanzerotti, C.G. Maclennan, Dayside optical and magnetic correlation events. J. Geophys. Res. **106**, 24637 (2001). https://doi.org/10.1029/2001JA900065

S.B. Mende, H.U. Frey, J. McFadden, C.W. Carlson, V. Angelopoulos, K.H. Glassmeier, D.G. Sibeck, A. Weatherwax, Coordinated observation of the dayside magnetospheric entry and exit of the THEMIS satellites with ground-based auroral imaging in Antarctica. J. Geophys. Res. **114**, A00C23 (2009). https://doi.org/10.1029/2008JA013496

M. Meurant, J.-C. Gerard, B. Hubert, V. Coumans, C. Blockx, N. Østgaard, S.B. Mende, Dynamics of global scale electron and proton precipitation induced by a solar wind pressure pulse. Geophys. Res. Lett. **30**, 2032 (2003). https://doi.org/10.1029/2003GL018017

M. Meurant, J.-C. Gerard, C. Blockx, B. Hubert, V. Coumans, Propagation of electron and proton shock-induced aurora and the role of the interplanetary magnetic field and solar wind. J. Geophys. Res. **109**, A10210 (2004). https://doi.org/10.1029/2004JA010453

M. Meurant, J.-C. Gerard, C. Blockx, V. Coumans, B. Hubert, M. Connors, L.R. Lyons, E. Donovan, Comparison of intense nightside shock-induced precipitation and substorm activity. J. Geophys. Res. **110**, A07228 (2005). https://doi.org/10.1029/2004JA010916

S.E. Milan, M. Lester, S.W.H. Cowley, J. Moen, P.E. Sandholt, C.J. Owen, Meridian-scanning photometer, coherent HF radar, and magnetometer observations of the cusp: a case study. Ann. Geophys. **17**, 159–172 (1999a)

S.E. Milan, T.K. Yeoman, M. Lester, J. Moen, P.E. Sandholt, Post-noon two-minute period pulsating aurora and their relationship to the dayside convection pattern. Ann. Geophys. **17**(7), 877 (1999b). https://doi.org/10.1007/s00585-999-0877-8

S.E. Milan, M. Lester, S.W.H. Cowley, M. Brittnacher, Convection and auroral response to a southward turning of the IMF: polar UVI, CUTLASS, and IMAGE signatures of transient magnetic flux transfer at the magnetopause. J. Geophys. Res. **105**(A7), 15741 (2000a). https://doi.org/10.1029/2000JA900022

S.E. Milan, M. Lester, S.W.H. Cowley, M. Brittnacher, Dayside convection and auroral morphology during an interval of northward interplanetary magnetic field. Ann. Geophys. **18**, 436–444 (2000b)

S.E. Milan, M. Lester, S.W.H. Cowley, K. Oksavik, M. Brittnacher, R.A. Greenwald, G. Sofko, J.-P. Villain, Variations in polar cap area during two substorm cycles. Ann. Geophys. **21**, 1121–1140 (2003)

S.E. Milan, G. Provan, B. Hubert, Magnetic flux transport in the Dungey cycle: a survey of dayside and nightside reconnection rates. J. Geophys. Res. **112**, A01209 (2007). https://doi.org/10.1029/2006JA011642

S.E. Milan, S.M. Imber, J.A. Carter, M.T. Walach, B. Hubert, What controls the local time extent of flux transfer events? J. Geophys. Res. **121**, 1391 (2016). https://doi.org/10.1002/2015JA022012

S.E. Milan, L.B.N. Clausen, J.C. Coxon, J.A. Carter, M.-T. Walach, K. Laundal, N. Østgaard, P. Tenfjord, J. Reistad, K. Snekvik, H. Korth, B.J. Anderson, Overview of solar wind-magnetosphere-ionosphere-atmosphere coupling and the generation of magnetospheric currents. Space Sci. Rev. **206**, 547–573 (2017). https://doi.org/10.1007/s11214-017-0333-0

J. Moen, Y. Rinne, H.C. Carlson, K. Oksavik, R. Fujii, H. Opgenoorth, On the relationship between thin Birkeland current arcs and reversed flow channels in the winter cusp/cleft ionosphere. J. Geophys. Res. **113**, A09220 (2008). https://doi.org/10.1029/2008JA013061

J. Moen, H. Carlson, Y. Rinne, A. Skjaveland, Multi-scale features of solar terrestrial coupling in the cusp ionosphere. J. Atmos. Sol.-Terr. Phys. **87**, 11 (2012). https://doi.org/10.1016/j.jastp.2011.07.002

T. Motoba, Y. Ebihara, A. Kadokura, M.J. Engebretson, M.R. Lessard, A.T. Weatherwax, A.J. Gerrard, Fast-moving diffuse auroral patches: a new aspect of daytime Pc3 auroral pulsations. J. Geophys. Res. **122**, 1542 (2017). https://doi.org/10.1002/2016JA023285

D.L. Murr, W.J. Hughes, Solar wind drivers of Traveling Convection Vortices. Geophys. Res. Lett. **30**(7), 1354 (2003). https://doi.org/10.1029/2002GL015498

D.L. Murr, W.J. Hughes, A.S. Rodger, E. Zesta, H.U. Frey, A.T. Weatherwax, Conjugate observations of traveling convection vortices: the field-aligned current system. J. Geophys. Res. **107**, 1306 (2002). https://doi.org/10.1029/2002JA009456

D.A. Neudegg et al., A survey of magnetopause FTEs and associated flow bursts in the polar ionosphere. Ann. Geophys. **18**, 416–435 (2000). https://doi.org/10.1007/s00585-000-0416-0

D.A. Neudegg et al., The UV aurora and ionospheric flows during flux transfer events. Ann. Geophys. **19**, 179–188 (2001). https://doi.org/10.5194/angeo-19-179-2001

P.T. Newell, C.-I. Meng, Ion acceleration at the equatorward edge of the cusp: low altitude observations of patchy merging. Geophys. Res. Lett. **18**, 1829–1832 (1991)

P.T. Newell, J.M. Ruohoniemi, C.-I. Meng, Maps of precipitation by source region, binned by IMF, with inertial convection streamlines. J. Geophys. Res. **109**, A10206 (2004). https://doi.org/10.1029/2004JA010499

P.T. Newell, S. Wing, C.I. Meng, Spectral properties and source regions of dayside electron acceleration events. J. Geophys. Res. **110**, A11205 (2005). https://doi.org/10.1029/2005JA011264

P.T. Newell, T. Sotirelis, K. Liou, C.-I. Meng, F.J. Rich, Cusp latitude and the optimal solar wind coupling function. J. Geophys. Res. **111**, A09207 (2006). https://doi.org/10.1029/2006JA011731

P.T. Newell, T. Sotirelis, S. Wing, Diffuse, monoenergetic, and broadband aurora: the global precipitation budget. J. Geophys. Res. **114**, A09207 (2009). https://doi.org/10.1029/2009JA014326

B. Ni, J. Bortnik, Y. Nishimura, R.M. Thorne, W. Li, V. Angelopoulos, Chorus wave scattering responsible for the Earth's dayside diffuse auroral precipitation: a detailed case study. J. Geophys. Res. **119**, 897 (2014). https://doi.org/10.1002/2013JA019507

B. Ni, R.M. Thorne, X. Zhang, J. Bortnik, Z. Pu, L. Xie, Origins of the Earth's diffuse auroral precipitation. Space Sci. Rev. **200**, 205 (2016). https://doi.org/10.1007/s11214-016-0234-7

Y. Nishimura, J. Bortnik, W. Li, R.M. Thorne, B. Ni, L.R. Lyons, Structures of dayside whistler-mode waves deduced from conjugate diffuse aurora. J. Geophys. Res. **118**, 664 (2013). https://doi.org/10.1029/2012JA018242

Y. Nishimura, T. Kikuchi, Y. Ebihara, A. Yoshikawa, S. Imajo, W. Li, H. Utada, Evolution of the current system during solar wind pressure pulses based on aurora and magnetometer observations. Earth Planets Space **68**(1), 144 (2016)

K. Oksavik, J. Moen, H.C. Carlson, High-resolution observations of the small-scale flow pattern associated with a poleward moving auroral form in the cusp. Geophys. Res. Lett. **31**, L11807 (2004). https://doi.org/10.1029/2005GL019838

K. Oksavik, J. Moen, H.C. Carlson, R.A. Greenwald, S.E. Milan, M. Lester, R.J. Barnes, Multi-instrument mapping of the small-scale flow dynamics related to a cusp auroral transient. Ann. Geophys. **23**, 2657 (2005)

N. Østgaard, J. Stadsnes, J. Bjordal, R.R. Vondrak, S.A. Cummer, D.L. Chenette, G.K. Parks, M.J. Brittnacher, D.L. McKenzie, Global-scale electron precipitation features seen in UV and X rays during substorms. J. Geophys. Res. **104**, 10191–10204 (1999). https://doi.org/10.1029/1999JA900004

N. Østgaard, J. Stadsnes, J. Bjordal, G.A. Germany, R.R. Vondrak, G.K. Parks, S.A. Cummer, D.L. Chenette, J.G. Pronko, Auroral electron distributions derived from combined UV and X-ray emissions. J. Geophys. Res. **106**, 26081–26090 (2000a). https://doi.org/10.1029/2001JA000031

N. Østgaard, J. Stadsnes, J. Bjordal, R.R. Vondrak, S.A. Cummer, D.L. Chenette, M. Schulz, J.G. Pronko, Cause of the localized maximum of X-ray emission in the morning sector: a comparison with electron measurements. J. Geophys. Res. **105**, 20869–20884 (2000b). https://doi.org/10.1029/1999JA000354

N. Østgaard, S.B. Mende, H.U. Frey, J.B. Sigwarth, Simultaneous imaging of the reconnection spot in the opposite hemispheres during northward IMF. Geophys. Res. Lett. **32**, L21104 (2005). https://doi.org/10.1029/2005GL024491

N. Østgaard et al., The asymmetric geospace as displayed during the geomagnetic storm on 17 August 2001. Ann. Geophys. **36**, 1577 (2018). https://doi.org/10.5194/angeo-36-1577-2018

G. Paschmann, G. Haerendel, I. Papamastorakis, N. Sckopke, S.J. Bame, J.T. Gosling, C.T. Russell, Plasma and magnetic field characteristics of magnetic flux transfer events. J. Geophys. Res. **87**, 2159 (1982). https://doi.org/10.1029/JA087iA04p02159

T.D. Phan et al., Simultaneous Cluster and IMAGE observations of cusp reconnection and auroral proton spot for northward IMF. Geophys. Res. Lett. **30**(10), 1509 (2003). https://doi.org/10.1029/2003GL016885

F. Plaschke, H. Hietala, M. Archer, X. Blanco-Cano, P. Kajdic, T. Karlsson, H.L. Sun, N. Omidi, M. Palmroth, V. Roytershteyn, Jets downstream of collisionless shocks. Space Sci. Rev. **214**(5), 81 (2018)

J.L. Posch, M.J. Engebretson, A.J. Witte, D.L. Murr, M.R. Lessard, M.G. Johnsen, H.J. Singer, M.D. Hartinger, Simultaneous traveling convection vortex events and Pc1 wave bursts at cusp latitudes observed in Arctic Canada and Svalbard. J. Geophys. Res. Space Phys. **118**, 6352 (2013). https://doi.org/10.1002/jgra.50604

P. Prikryl, G. Provan, K.A. McWilliams, T.K. Yeoman, Ionospheric cusp flows pulsed by solar wind Alfvén waves. Ann. Geophys. **20**, 161–174 (2002). https://doi.org/10.5194/angeo-20-161-2002

G. Provan, T.K. Yeoman, Statistical observations of the MLT, latitude and size of pulsed ionospheric flows with the CUTLASS Finland radar. Ann. Geophys. **17**, 855–867 (1999). https://doi.org/10.1007/s00585-999-0855-1

G. Provan, T.K. Yeoman, S.W.H. Cowley, The influence of the IMF B_y component on the locations of pulsed flows in the dayside ionosphere as observed by HF radar. Geophys. Res. Lett. **26**, 521 (1999)

G. Provan, M. Lester, A. Grocott, S.W.H. Cowley, Pulsed flows observed during an interval of prolonged northward IMF. Ann. Geophys. **23**, 1207–1225 (2005). https://doi.org/10.5194/angeo-23-1207-2005

M.I. Pudovkin, S.A. Zaitseva, P.E. Sandholt, A. Egeland, Dynamics of aurorae in the cusp region and characteristics of magnetic reconnection at the magnetopause. Planet. Space Sci. **40**, 879 (1992)

I.J. Rae, F.R. Fenrich, M. Lester, K.A. McWilliams, J.D. Scudder, Solar wind modulation of cusp particle signatures and their associated ionospheric flows. J. Geophys. Res. **109**, A03223, 10/1029/2003JA010188 (2004)

P.H. Reiff, T.W. Hill, J.L. Burch, Solar wind plasma injection at the dayside magnetospheric cusp. J. Geophys. Res. **82**, 479–487 (1977)

Y. Rinne, J. Moen, K. Oksavik, H.C. Carlson, Reversed flow events in the winter cusp ionosphere observed by the European Incoherent Scatter (EISCAT) Svalbard radar. J. Geophys. Res. **112**, A10313 (2007). https://doi.org/10.1029/2007JA012366

O. Royrvik, T.N. Davis, Pulsating aurora: local and global morphology. J. Geophys. Res. **82**(29), 4720 (1977). https://doi.org/10.1029/JA082i029p04720

C.T. Russell, R.C. Elphic, Initial ISEE magnetometer results—magnetopause observations. Space Sci. Rev. **22**, 681 (1978). https://doi.org/10.1007/BF00212619

C.T. Russell, R.C. Elphic, ISEE observations of flux transfer events at the dayside magnetopause. Geophys. Res. Lett. **6**, 33 (1979). https://doi.org/10.1029/GL006i001p00033

K. Sakaguchi, K. Shiokawa, A. Ieda, Y. Miyoshi, Y. Otsuka, T. Ogawa, M. Connors, E.F. Donovan, F.J. Rich, Simultaneous ground and satellite observations of an isolated proton arc at subauroral latitudes. J. Geophys. Res. **112**, A04202 (2007). https://doi.org/10.1029/2006JA012135

P.E. Sandholt, C.J. Farrugia, Monitoring magnetosheath-magnetosphere interconnection topography from the aurora. Ann. Geophys. **20**, 629 (2002)

P.E. Sandholt, P.T. Newell, Ground and satellite observations of an auroral event at the cusp/cleft equatorward boundary. J. Geophys. Res. **97**(A6), 8685 (1992). https://doi.org/10.1029/91JA02995

P.E. Sandholt, C.S. Deehr, A. Egeland, B. Lybekk, R. Viereck, G.J. Romick, Signatures in the dayside aurora of plasma transfer from the magnetosheath. J. Geophys. Res. **91**(A9), 10063 (1986). https://doi.org/10.1029/JA091iA09p10063

P.E. Sandholt, H.C. Carlson, A. Egeland, *Dayside and Polar Cap Aurora*. Astrophysics and Space Science Library, vol. 270 (Kluwer Academic Publishers, Dordrecht/Boston/London, 2002a)

P.E. Sandholt, W.F. Denig, C.J. Farrugia, B. Lybekk, E. Trondsen, Auroral structure at the cusp equatorward boundary: relationship with the electron edge of low-latitude boundary layer precipitation. J. Geophys. Res. **107**(A9), 1235 (2002b). https://doi.org/10.1029/2001JA005081

P.E. Sandholt, C.J. Farrugia, W.F. Denig, S.W.H. Cowley, M. Lester, Spontaneous and driven cusp dynamics: optical aurora, particle precipitation, and plasma convection. Planet. Space Sci. **51**(12), 797 (2003a). https://doi.org/10.1016/S0032-0633(03)00114-4

P.E. Sandholt, J. Moen, C.J. Farrugia, S.W.H. Cowley, M. Lester, S.E. Milan, C. Valladares, W.F. Denig, S. Eriksson, Multi-site observations of the association between aurora and plasma convection in the cusp/polar cap during a southeastward ($B_y \sim \|B_z\|$) IMF orientation. Ann. Geophys. **21**, 539 (2003b). https://doi.org/10.5194/angeo-21-539-2003

P.E. Sandholt, C.J. Farrugia, W.F. Denig, Dayside aurora and the role of IMF $|B_y/B_z|$: detailed morphology and response to magnetopause reconnection. Ann. Geophys. **22**, 613 (2004). https://doi.org/10.5194/angeo-22-613-2004

D.G. Sibeck et al., Comprehensive study of the magnetospheric response to a hot flow anomaly. J. Geophys. Res. **104**(A3), 4577 (1999). https://doi.org/10.1029/1998JA900021

R.J. Sitar, J.B. Baker, C.R. Clauer, A.J. Ridley, J.A. Cumnock, V.O. Papitashvili, J. Spann, M.J. Brittnacher, G.K. Parks, Multi-instrument analysis of the ionospheric signatures of a hot flow anomaly occurring on July 24, 1996. J. Geophys. Res. **103**(A10), 23357 (1998). https://doi.org/10.1029/98JA01916

A. Skjaveland, J. Moen, H.C. Carlson, On the relationship between flux transfer events, temperature enhancements, and ion upflow events in the cusp ionosphere. J. Geophys. Res. **116**, A10305 (2011). https://doi.org/10.1029/2011JA016480

D.J. Southwood, The ionospheric signature of flux transfer events. J. Geophys. Res. **92**(A4), 3207 (1987). https://doi.org/10.1029/JA092iA04p03207

A. Thorolfsson, J.C. Cerisier, M. Lockwood, P.E. Sandholt, C. Senior, M. Lester, Simultaneous optical and radar signatures of poleward moving auroral forms. Ann. Geophys. **18**, 1054 (2000). https://doi.org/10.1007/s00585-000-1054-2

M.R. Torr, D.G. Torr, M. Zukic, R.B. Johnson, J. Ajello, P. Banks, K. Clark, K. Cole, C. Keffer, G. Parks, B. Tsurutani, J. Spann, A far ultraviolet imager for the International Solar-Terrestrial Physics Mission. Space Sci. Rev. **71**, 329 (1995). https://doi.org/10.1007/BF00751335

K.J. Trattner, S.A. Fuselier, W.K. Peterson, M. Boehm, D. Klumpar, C.W. Carlson, T.K. Yeoman, Temporal versus spatial interpretation of cusp ion structures observed by two spacecraft. J. Geophys. Res. **107**(A10), 1287 (2002). https://doi.org/10.1029/2001JA000181

E.A. Tremsina, S.B. Mende, H.U. Frey, Identifying the evolution of Southern Hemisphere poleward moving auroral forms (PMAFs) in the context of plasma convection and magnetic reconnection. J. Geophys. Res. Space Phys. **122**, 4037–4050 (2017). https://doi.org/10.1002/2016JA023426

V.G. Vorobjev, G. Gustafsson, G.V. Starkov, Y.I. Feldstein, N.F. Shevnina, Dynamics of day and night aurora during substorms. Planet. Space Sci. **23**, 269 (1975). https://doi.org/10.1016/0032-0633(75)90132-4

B. Wang, Y. Nishimura, L.R. Lyons, Y. Zou, H.C. Carlson, H.U. Frey, S.B. Mende, Analysis of close conjunctions between dayside polar cap airglow patches and flow channels by all-sky imager and DMSP. Earth Planets Space **68**(1), 150 (2016a). https://doi.org/10.1186/s40623-016-0524-z

B. Wang, Y. Nishimura, Y. Zou, L.R. Lyons, V. Angelopoulos, H. Frey, S. Mende, Investigation of triggering of poleward moving auroral forms using satellite-imager coordinated observations. J. Geophys. Res. **121**, 10,929 (2016b). https://doi.org/10.1002/2016JA023128

B. Wang, Y. Nishimura, H. Hietala, L. Lyons, V. Angelopoulos, F. Plaschke, Impacts of magnetosheath high-speed jets on the magnetosphere and ionosphere measured by optical imaging and satellite observations. J. Geophys. Res. **123**, 4879 (2018a). https://doi.org/10.1029/2017JA024954

B. Wang, Y. Nishimura, H. Hietala, X.C. Shen, Q. Shi, H. Zhang, Dayside magnetospheric and ionospheric responses to a foreshock transient on June 25, 2008: 2.2-D evolution based on dayside auroral imaging. J. Geophys. Res. **123**, 6347 (2018b). https://doi.org/10.1029/2017JA024846

A.T. Weatherwax, H.B. Vo, T.J. Rosenberg, S.B. Mende, H.U. Frey, L.J. Lanzerotti, C.G. Maclennan, A dayside ionospheric absorption perturbation in response to a large deformation of the magnetopause. Geophys. Res. Lett. **26**, 517 (1999). https://doi.org/10.1029/1999GL900017

J.A. Wild et al., First simultaneous observations of flux transfer events at the high-latitude magnetopause by the Cluster spacecraft and pulsed radar signatures in the conjugate ionosphere by the CUTLASS and EISCAT radars. Ann. Geophys. **19**, 1491–1508 (2001). https://doi.org/10.5194/angeo-19-1491-2001

J.A. Wild, S.E. Milan, S.W.H. Cowley, M.W. Dunlop, C.J. Owen, J.M. Bosqued, H. Reme, Coordinated interhemispheric SuperDARN radar observations of the ionospheric response to flux transfer events observed by the Cluster spacecraft at the high-latitude magnetopause. Ann. Geophys. **21**, 1807 (2003). https://doi.org/10.5194/angeo-21-1807-2003

S. Wing, P.T. Newell, C.-I. Meng, Cusp modeling and observations at low latitude. Surv. Geophys. **26**, 341–367 (2005). https://doi.org/10.1007/s10712-005-1886-0

J. Woch, R. Lundin, Magnetosheath plasma precipitation in the polar cusp and its control by the interplanetary magnetic field. J. Geophys. Res. **97**, 1421–1430 (1992)

Z.Y. Xing et al., Poleward moving auroral forms (PMAFs) observed at the Yellow River Station: a statistical study of its dependence on the solar wind conditions. J. Atmos. Sol.-Terr. Phys. **86**, 25 (2012). https://doi.org/10.1016/j.astp.2012.06.004

A.G. Yahnin, T.A. Yahnina, H.U. Frey, Subauroral proton spots visualize the Pc1 source. J. Geophys. Res. **112**, A10223 (2007). https://doi.org/10.1029/2007JA012501

T.A. Yahnina, H.U. Frey, T. Bösinger, A.G. Yahnin, Evidence for subauroral proton flashes on the dayside as the result of the ion cyclotron interaction. J. Geophys. Res. **113**, A07209 (2008). https://doi.org/10.1029/2008JA013099

T.K. Yeoman, M. Lester, S.W.H. Cowley, S.E. Milan, J. Moen, P.E. Sandholt, Simultaneous observations of the cusp in optical, DMSP and HF radar data. Geophys. Res. Lett. **24**, 2251–2254 (1997)

Y. Zhang, L.J. Paxton, T.J. Immel, H.U. Frey, S.B. Mende, Sudden solar wind dynamic pressure enhancements and dayside detached auroras: IMAGE and DMSP observations. J. Geophys. Res. **108**(A4), 8001 (2002). https://doi.org/10.1029/2002JA009355

Y. Zhang, C.-I. Meng, L.J. Paxton, D. Morrison, B. Wolven, H. Kil, P. Newell, S. Wing, A.B. Christensen, Far-ultraviolet signature of polar cusp during southward IMF B_Z observed by TIMED/Global Ultraviolet Imager and DMSP. J. Geophys. Res. **110**, A01218 (2005). https://doi.org/10.1029/2004JA010707

Q.H. Zhang et al., Simultaneous observations of reconnection pulses at Cluster and their effects on the cusp aurora observed at the Chinese Yellow River Station. J. Geophys. Res. **115**, A10237 (2010). https://doi.org/10.1029/2010JA015526

X.-Y. Zhou, B.T. Tsurutani, Rapid intensification and propagation of the dayside aurora: large scale interplanetary pressure pulses (fast shocks). Geophys. Res. Lett. **26**, 1097–1100 (1999)

X.Y. Zhou, R.J. Strangeway, P.C. Anderson, D.G. Sibeck, B.T. Tsurutani, G. Haerendel, H.U. Frey, J.K. Arballo, Shock aurora: FAST and DMSP observations. J. Geophys. Res. **108**(A4), 8019 (2003). https://doi.org/10.1029/2002JA009701

X.-Y. Zhou, K. Fukui, H.C. Carlson, J.I. Moen, R.J. Strangeway, Shock aurora: ground-based imager observations. J. Geophys. Res. **114**, A12216 (2009). https://doi.org/10.1029/2009JA014186

X.-Y. Zhou, D.G. Sibeck, O. Amm, J.M. Ruohoniemi, Interplanetary shock generated ionospheric traveling convection vortex near local noon. Adv. Geosci. **21**, 367–378 (2010)

X.-Y. Zhou, G. Haerendel, J.I. Moen, E. Trondsen, L. Clausen, R.J. Strangeway, B. Lybekk, D.A. Lorentzen, Shock aurora: field-aligned discrete structures moving along the dawnside oval. J. Geophys. Res. **122**, 3145–3162 (2017). https://doi.org/10.1002/2016JA022666

Space Science Reviews (2021) 217:60
https://doi.org/10.1007/s11214-021-00830-x

On the Relationship Between Shear Alfvén Waves, Auroral Electron Acceleration, and Field Line Resonances

R. Rankin[1] · D.M. Gillies[2] · A.W. Degeling[3]

Received: 5 August 2020 / Accepted: 6 May 2021 / Published online: 14 June 2021

Abstract
This article describes the relationship between shear Alfvén waves and auroral electron acceleration, with an emphasis on long-period standing waves that correlate with redline auroral arcs in the Earth's magnetosphere. Discrete auroral arcs were correlated with high-latitude field line resonances in the early 1990's. The past decade has seen advances in all-sky camera technology improve the detection and categorization of "FLR arcs" and establish them as a distinct population. We review observations of redline arcs and discuss estimates of wave amplitudes, wavelengths perpendicular to the geomagnetic field, and saturation times obtained within the framework of two-fluid theories. The two-fluid theory explains the spatial and temporal evolution of FLR optical signatures, but the estimated parallel electric field strengths are insufficient to accelerate electrons and produce 6300 Å auroral emissions. A kinetic theory of FLRs is necessary since electron bounce motion in long-wavelength standing waves affects the ac conductivity and hence the strength of parallel electric fields. In the kinetic theory, the current-voltage relation comprises a conductivity kernel that is a function of the wave frequency, field line length, electron thermal speed, and the number of electron trajectories nearly parallel to geomagnetic field lines close to the ionosphere. The ensuing nonlocal relationship between wave parallel currents and parallel electric fields provides a feasible explanation of the correlation between long-period field line resonances and redline arcs in the terrestrial magnetosphere. The mirror force and particle trapping in the wave fields of shear Alfvén waves are demonstrated to be important aspects of the kinetics of FLRs.

Keywords Field line resonance · Shear Alfvén waves · Auroral arc · Auroral electron acceleration · Magnetosphere-ionosphere coupling

Auroral Physics
Edited by David Knudsen, Joe Borovsky, Tomas Karlsson, Ryuho Kataoka and Noora Partmies

✉ A.W. Degeling

1 Department of Physics, University of Alberta, Edmonton, Alberta, Canada

2 Department of Physics and Astronomy, University of Calgary, Calgary, Alberta, Canada

3 Institute of Space Science, Shandong University, Weihai, China

1 Introduction

We shall review theory and observations of magnetospheric field line resonances (FLRs) in Earth's magnetosphere. The relationship between ultra-low-frequency (ULF) standing waves and the associated periodic modulations in auroral luminosity in the wavelength range of 6300 Å redline emissions (associated with precipitating electron energies of less than 1 keV) is examined. The material presented here is not intended as a general overview of ULF pulsations, which is well covered elsewhere.

The optical signature of field-line resonances (FLRs) holds interest in the attempt to determine the mechanism behind the generation of auroral arcs. Despite decades of research spanning nearly 70 years, a definitive mechanism has remained elusive. The relationship between MHD waves, particle precipitation, and auroral arc formation has a well-established and solid foundation (Berger et al. 1970; Rankin et al. 1999a; Milan et al. 2001). As detailed in the review by Karlsson and colleagues (this issue), quasi-static arcs, sometimes referred to as inverted-V arcs (which are not predominantly redline arcs), have received much greater attention. This section's relevance to discrete arc formation is motivated by the marked similarity between the electric field and field-aligned current morphologies generated by FLR arcs and long east-west aligned auroral arcs (Karlsson et al. 2020). The absence of a comprehensive database of FLR arc observations may be the most significant difficulty in determining how they relate to other arc types and for adapting arc generation theories to account for the observed periodicity. Recently, the situation has improved (Gillies et al. 2018).

Standing Alfvén wave FLRs occur over a relatively broad range of latitude spanning the auroral zone (with frequencies in the Pc4-5 range of the ULF band, between 2 and 22 mHz), and are correlated with spatial and temporal modulations of a particular morphology. The east-west extent of redline discrete auroral arcs and FLRs is up to several thousand km (Samson et al. 2003). We focus on this subset of discrete redline auroral arcs and place them in the context of electron acceleration by higher-frequency ULF and VLF (very-low-frequency) band waves. It is believed that the major source of long-period toroidal FLRs is near-monochromatic compressional Alfvén waves, for example, excited either by magnetopause buffeting along the dayside magnetopause (Kepko and Spence 2003; Kessel et al. 2004; Kim et al. 2002), or by Kelvin-Helmholtz-type shear flow instability along the magnetopause flanks (Chen and Hasegawa 1974; Mills et al. 1999; Rae et al. 2005; Walker 1981). These waves propagate across geomagnetic field lines where they resonantly excite shear Alfvén waves. These standing shear Alfvén waves are localized to the L-shells upon which they are excited and can grow to large amplitudes. As opposed to the compressional wave, FLRs carry field-aligned currents (FACs) that can exceed 10 μAm^{-2} near the ionosphere, which may result in electron acceleration. The auroral morphology of FLRs is well explained by their localization in latitude, extended range of longitude, and poleward propagation (Samson et al. 2003). However, a fully self-consistent current-voltage relation for FLRs is not yet available as it is complicated by the bounce motion of electrons in the background geomagnetic field topology and inductive wave fields that cause charged particle acceleration and potentially auroral emissions at 6300 Å and sometimes 5577 Å (Rankin et al. 1999b).

Samson et al. (1971) were the first to identify ground magnetic signatures of ULF field line resonances at high latitudes over northern Canada. The waves are classified as toroidal mode shear Alfvén waves because of their relatively low azimuthal wavenumbers, roughly speaking $m \lesssim 8$. A characteristic of FLRs is that the horizontal component of the excited wave magnetic field observed on the ground as a function of latitude reverses at magnetic latitudes of roughly 69° and 60°, respectively. A theoretical interpretation of the observations made by Samson et al. (1971) was provided by Southwood (1974), who inferred that the source of ULF wave excitation is most likely compressional Alfvén waves excited on the flanks of the magnetopause. These low-m compressional waves propagate earthward and tailward onto relatively low L-shells where they excite toroidal Alfvén waves earthward of the turning points of the compressional waves (Rankin et al. 1993).

FLRs have been studied in-situ using the Cluster (Chaston et al. 2005) and THEMIS spacecraft (Sarris et al. 2009, 2010; Takahashi et al. 2015). Their detection by ground-based magnetometers is common and is used to monitor cold plasma mass density dynamics (Kabin et al. 2007; Waters et al. 1996, 2007). When the wavelength of the FLR perpendicular to the geomagnetic field is on the order of the electron inertial scale in the ionosphere or the ion-acoustic (or thermal ion) gyro-radius in the equatorial magnetosphere, ideal MHD breaks down. Waves of this scale are sometimes called inertial or kinetic, cf. Chen and Hasegawa (1974). They provide a natural pathway to significant electrodynamic energy and particle flows along geomagnetic field lines (Hasegawa 1976; Samson et al. 2003; Streltsov and Lotko 1999). After reviewing the basic ideal MHD theory of FLRs, we will explore the relationship between small perpendicular scale Alfvén waves and redline arcs.

Theories and simulations have been used for decades to study the optimal development of MHD FLRs (Kivelson and Southwood 1985; Singer et al. 1981; Southwood 1974; Southwood and Hughes 1983; Streltsov and Lotko 1999; Degeling et al. 2010, 2018; Lu et al. 2003a; Rankin et al. 1993, 2004, 2005a; Rickard and Wright 1995). Nevertheless, there are few self-consistent models of wave-particle interactions for inertial and kinetic FLRs (Damiano et al. 2003; Rankin et al. 2007; Nakamura 2000 and Damiano et al. 2018). Recent work on kinetic scale FLRs in the inner magnetosphere suggest that the occurrence of these waves peaks at and may be related to substorm onset (e.g. Chaston et al. 2014).

It is in fact incorrect to refer to FLRs as being "inertial" or "kinetic," since they experience both warm and cold plasma regimes along geomagnetic field lines. Small wavelength modulated arcs, called "FLR-arcs" here, are a non-ideal MHD phenomenon. They are possibly intensified by the ionospheric feedback instability, or nonlinear feedback processes, and their sources have been suggested to include the ionosphere (Pokhotelov et al. 2002).

This paper is organized as follows. Section 2 reviews recent observations of redline optical signatures of field line resonances based on a statistical study of 6300 Å auroral emissions. In Sect. 3 the characteristics of ground magnetometer observations of FLRs are explained using the ideal MHD approximation. Section 4 discusses FLRs from the point of view of two-fluid theory, which is used to introduce wave dispersion, parallel electric fields, and nonlinear behaviour. Local and nonlocal kinetic theory of magnetospheric field lines is discussed in Sect. 5. In Sect. 6, kinetic simulations of field line resonances are detailed.

2 Redline Auroral Arcs and Field Line Resonances

Though FLR theory and modelling reproduces many of the characteristics of periodic arcs (as discussed in Sect. 3), important issues remain to be addressed. Kinetic theory and simulations of long-period FLRs demonstrate that parallel electric fields and dissipation of these waves may accelerate electrons to energies greater than 1 keV (Damiano et al. 2019), especially in the presence of deep density cavities (Samson et al. 2003). This has also been found in the hybrid MHD-kinetic simulations of Damiano and Johnson (2012), which show substantial mono-energetic electron energization (in excess of a keV), even at MHD scales. FLR arcs are distinguished from quiet, discrete "inverted-V" arcs which have a peak precipitation energy of 5–10 keV and occur more frequently in the post midnight sector, as reviewed by (Karlsson et al. 2020).

The lack of a systematic database of observations of FLR arcs into the 1990's presented a significant challenge for determining how they relate to other types of arcs and for adapting arc generation theories to account for the periodicity seen. Early observations of FLRs and their relation to auroral arcs were predominantly represented by individual case studies. Samson et al. (1996) demonstrated the existence of a distinct link between discrete auroral arcs and the magnetic field-aligned current (FAC) morphology observed both in auroral arcs and in magnetic FLRs. In the Samson et al. (1996) study, an MSP (Meridian Scanning Photometer) located in Gillam, Manitoba was utilized to observe a poleward moving periodic arc system. By comparing arc scale size and power spectra of both the optical and co-located magnetometer traces, they were able to conclude that a magnetic FLR could produce a discrete auroral arc. They also found a link between FLR arcs and intensifications of substorms in the high-latitude ionosphere by observing the topology of a large-scale auroral vortex structure. The results were also consistent with the periodicities of FLRs related to substorm intensifications (Samson et al. 1992).

Though redline periodic arcs or "FLR arcs" have been suggested to be a new type of auroral arc since the 1990s, the improved observational capabilities of modern all-sky cameras have only recently verified the suggestion (Gillies et al. 2018). Individual observations of poleward moving periodic auroral arcs recorded by ground-based cameras are indicative of this. FLR arcs may now be classified as a natural component of the aurora, but thousands of observations of discrete arcs by all-sky cameras indicate that they are a minority arc species. Observational campaigns continue to utilize optical observations of discrete periodic FLR arcs and other arc types to further our understanding of their generation mechanisms.

Milan et al. (2001) used MSP observations to model the poleward moving auroral forms based on the behaviour predicted by MHD theory. They used models to reproduce the field aligned current pattern typical of the auroral arc as well as the magnitudes observed in situ. Lotko et al. (1998) discuss the frequency of periodic auroral arcs observed by the CANOPUS all-sky camera located in Gillam, Manitoba, although examples of the auroral forms were not included in the paper. A more recent study (Baddeley et al. 2017) utilized data from the DMSP satellites to directly measure precipitating particle populations inside FLR arcs and the corresponding ionospheric ion temperature enhancements in-between propagating arcs. They found that although the currents associated with these auroral arcs were smaller than those commonly found (or modelled) in other studies (Milan et al. 2001; Rankin et al. 2005a) the particle energies observed were sufficient to induce aurora at both the 6300 Å and 5577 Å emission wavelengths.

The deployment of large-scale ground-based observing networks such as THEMIS and REGO has enabled optical observations of FLR arcs and their occurrence rates to be examined statistically (Gillies et al. 2018). To illustrate the challenges in the identification of FLR arcs, consider Fig. 1 (top), which shows an example of an FLR auroral feature observed by the THEMIS ASI network of "white light" imagers (Mende et al. 2008).

In addition to capturing auroras of all energies and wavelengths, the white light ASI measures undesirable phenomena such as clouds. The top panel of Fig. 1 shows dynamic features including clouds, pulsating auroral patches, and FLR auroral arcs over the course of an hour. In this image, the optical signature of redline 6300 Å FLR arcs, which can be clearly seen in the bottom panel, is lost in the brighter and more dynamic auroral activity. The lower panel of Fig. 1 presents REGO (Redline Emission Geospace Observatory) observations of the same interval, with periodic poleward moving redline arcs more clearly visible. These periodic features also show evidence of phase mixing discussed in Sect. 3.2. The REGO observatory has enabled a much larger database of FLR arc observations, and as this database continues to grow, it will be useful for statistical studies of this new type of discrete arc phenomenon. The REGO ASI network was deployed across the Canadian/Alaskan

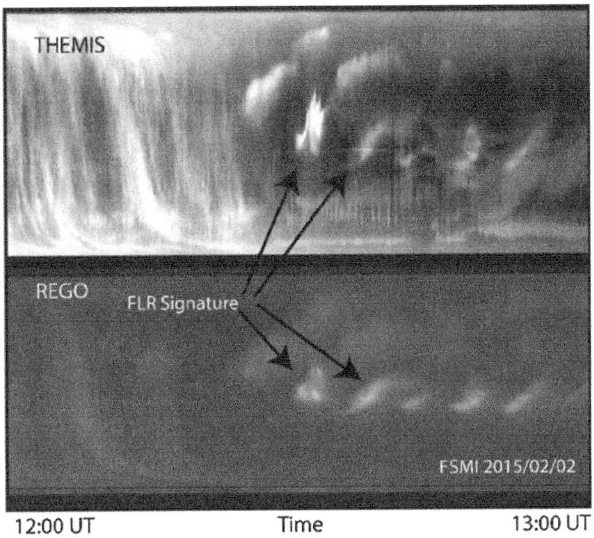

Fig. 1 FLR auroral arcs seen in ASI over Fort Smith, NWT on February 2nd, 2015. (top) THEMIS ASI keogram. (bottom) REGO (630.0 nm) ASI keogram. From Gillies et al. (2018)

sector in 2014. REGO cameras have a similar, slightly larger projected field of view than the THEMIS cameras, but are filtered to observe only the 6300 Å auroral emission associated with precipitating electrons with energies less than 1 keV (Kosch et al. 2001). Much of the auroral activity (pulsating aurora, dynamic arcs) associated with high-energy precipitation seen in the top panel of Fig. 1 is filtered to reveal the redline aurora generated predominately by lower-energy precipitation (Fig. 1 bottom). Sampling the redline aurora generated by sub-keV precipitation allows one to identify FLR arcs. Gillies et al. (2018) used the REGO observing network to catalog FLRs using keograms like Fig. 1. With the aid of these keograms, they found 400 intervals containing a FLR arc lasting at least three periods. The REGO database of FLR arcs represents an increase by two orders of magnitude in sampling of this type of arc morphology, enabling for the first time large-scale statistical studies of 6300 Å redline arcs.

In Gillies et al. (2018) the REGO database of occurrence statistics of 6300 Å FLR auroral arcs was compared to "standard" (non-periodic) discrete redline auroral arcs. The REGO results indicate a marked difference between FLR arcs and non-periodic arcs (Fig. 2). Discrete non-periodic redline auroral arcs are found to be most common just prior to magnetic midnight in accordance with a previous study of THEMIS white light auroral arcs (Gillies et al. 2014). The origin of these non-periodic arcs is currently under speculation. While possible that they are the redline component of inverted-V auroral arcs, establishing this link was beyond the scope of the Gillies et al. (2018) study which focused primarily on aurora at the 630 nm wavelength and contained no in situ observations of electron precipitation or observations at other wavelengths (such as the green 557.7 nm emission). Investigation into the nature of the non-periodic redline arcs remains a subject of future work. In contrast, the MLT distributions of redline periodic arcs show peak occurrences just after dusk and just before dawn. From a ground perspective, the magnetic signature of FLRs is a well-established and well-understood process. FLRs usually manifest as Pc4-5-range waves characteristic of the oscillations of closed geomagnetic field lines. The magnitudes of these oscillations are typically highest at auroral latitudes (Samson et al. 1971). At these auroral latitudes, magnetic FLRs exhibit a distinct dawn-dusk occurrence asymmetry (Rae et al. 2007), with a tendency towards anti-sunward propagation and more frequent occurrence under northward IMF con-

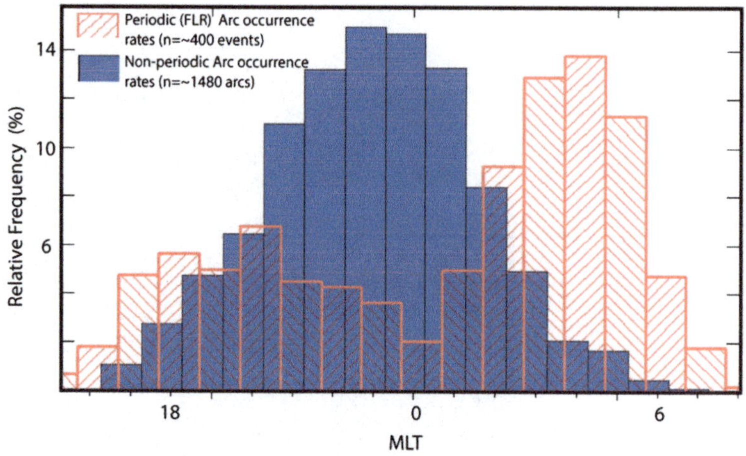

Fig. 2 Histogram of relative magnetic local time (MLT) occurrence rates of discrete, non-periodic 630.0 nm auroral arcs (purple bins) and field-line resonance arcs (red bins). Non periodic arcs data set contains images obtained between January and March 2015. FLR arcs observed between April 2014 and March 2016 (Gillies et al. 2018)

ditions (Baker et al. 2003; Ziesolleck and Mcdiarmid 1994). The occurrence statistics of FLR arcs observed by Gillies et al. (2018) are consistent with the statistical distribution of Pc5 wave observations of the ground magnetometers (Takahashi et al. 2015), which is an indication that the two phenomena are related.

Analysis of data from riometer networks (Spanswick et al. 2005) suggests that electromagnetic fields of FLRs are also effective at interacting with the high-energy electron population, predominantly in the dawn sector. On the other hand, there is little evidence of modulated precipitation associated with FLRs in the dusk sector. Spanswick et al. (2005) attributed the asymmetry to how particles gain access to the region in which FLRs are present. They proposed that the dawn sector had sufficient particle flux in the appropriate range of energy, and that pitch angle scattering induced by the FLR triggered the associated modulation in precipitation. They concluded that a sufficient flux of particles, in addition to the presence of an FLR, are necessary conditions for modulated high-energy electron precipitation. They further suggested the enhanced high-energy electron flux is due to substorm injection electrons that then gradient-curvature drift into the morning sector. These conditions are not met on the dusk flank where modulated high-energy precipitation is not expected to occur.

The presence of redline periodic auroral arcs on the dusk flank seen by Gillies et al. (2018) indicates that conditions are met for significant wave-particle interactions at relatively low electron energies, hundreds of eV, and that there is sufficient observational capacity to observe the phenomenon on both flanks. In Korth et al. (1999) it was postulated that there is either (1) sufficient particle access for low-energy electrons, making any FLR-associated pitch-angle scattering a viable generation mechanism (similar to high-energy electron precipitation modulation), or (2) a different interaction mechanism is responsible for the optical signatures such as parallel acceleration. As more FLR-arcs are identified, the auroral scientific community is closer to being able to accurately model how redline periodic arcs form, and the magnetospheric-ionospheric conditions needed.

3 Ideal MHD Theory of Magnetospheric Field Line Resonances

Sections 3.1 and 3.2 provide an overview of the features of toroidal mode field resonances under the approximation of ideal MHD, first in straight field geometry and then in dipole geometry. This follows the historical development of the subject.

3.1 Box-Model Approximation

Analysis of cold plasma ideal MHD field line resonances in straight magnetic field lines is often used to elucidate physical properties of the waves. The simplest model consists of straight geomagnetic field lines embedded in a cold plasma bounded by partially conducting sheets at each polar ionosphere (Southwood 1974). The methodology involves linearizing the ideal MHD equations in Cartesian geometry (x, y, z, where z is parallel to the magnetic field direction, and x is analogous to L-shell) to obtain an equation for the x-component of the fluid displacement near the field line resonance position, x_r, on the L-shell where $\omega^2 - k_z^2 v_A^2 \sim 0$. The triad ($\omega$, k_z, v_A) is defined by the wave frequency, wavenumber along the ambient magnetic field, and the Alfvén speed on the resonant magnetic field, respectively, with v_A taken to be a function of x. The equation for the fluid displacement in a meridional plane has been considered by many authors and takes the form (Samson and Rankin 1994)

$$\frac{d^2\xi_x}{dx^2} + \frac{1}{x - x_r + i\gamma}\frac{d\xi_x}{dx} - k_y^2\xi_x = 0 \tag{1}$$

$$\gamma = \frac{\mathrm{Im}\{\varepsilon(x_r)\}}{\frac{d}{dx}\mathrm{Re}\{\varepsilon(x_r)\}} \tag{2}$$

where k_y and γ in these expressions represent the azimuthal wavenumber and ionospheric losses, respectively, and $\mathrm{Re}\{\varepsilon(x_r)\} = \frac{\omega^2}{v_A^2(x_r)} - [\mathrm{Re}\{k_z\}]^2$. The solution for the fluid displacement near the resonance location is given by

$$\xi_x = \xi_0 \ln\big[k_y(x - x_r + i\gamma)\big]. \tag{3}$$

The equation for the fluid displacement must also be supplemented with boundary conditions at the ionosphere where $\boldsymbol{E}$ and $\boldsymbol{B}$ are related by $B \times \hat{z} = \pm\mu_0\Sigma_P E$, where Σ_P is the height-integrated Pedersen conductivity. With Maxwell's equations and the boundary conditions at the ionosphere, the electric field can be found as $E_x(z,t) = 2E_0\cos(k_z z + \phi)e^{-i\omega t}$. The complex wavenumber k_z is such that $\mathrm{Im}(k_z) = -\ln|R|/l$, $\mathrm{Re}(k_z) = \theta + n\pi/l = m\pi/l$, where l is the length of a geomagnetic field line and the complex phase ϕ and reflectivity $R = |R|e^{i\theta}$ at the ionosphere are defined by (Hughes and Southwood 1976),

$$|R| = \frac{\Sigma_A - \Sigma_P}{\Sigma_A + \Sigma_P} \tag{4}$$

$$\phi = -\frac{i}{2}\ln|R| - \frac{\theta}{2} - n\pi \tag{5}$$

with $\Sigma_A = 1/\mu_0 v_A$ representing the Alfvén wave conductivity. Equation (3) for ξ_x gives the spatial variation of the electric field E_x around the resonance,

$$E_x(x) \sim \frac{ik_y E_0}{(\mathrm{Re}\,\varepsilon - k_y^2)(x - x_r + i\gamma)} \tag{6}$$

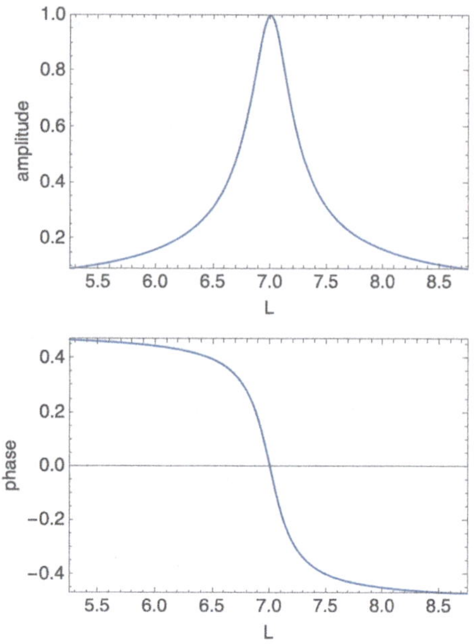

Fig. 3 FLR amplitude and phase versus L. Azimuthal wavenumber $m = 8$, resonant field line $L_0 = 7$, reflection coefficient $|R| = 0.8$. Amplitude is in arbitrary units and wave phase is in radians. Note the 180-degree phase change across the resonance

which lead to expressions for the variation of the amplitude and phase across the resonant field line. The amplitude of the field is expressed as follows:

$$\frac{|E_x|}{E_0} \sim \frac{1}{k_y\sqrt{(x - x_r)^2 + \gamma^2}} \tag{7}$$

where in the expression $\frac{\omega^2}{v_A^2(x)} - k_z^2 \ll k_y^2$ is assumed, $k_y = m/LR_E$, and m is the azimuthal wavenumber. An expression for the loss term γ in a dipole-like magnetosphere with $B_0 \propto x^{-3}$ can be found under the same condition on k_y,

$$\gamma = -\ln|R|\left(\frac{1}{v_A}\frac{dv_A}{dx}\right)^{-1} = -\frac{x_r}{3\pi}\ln|R| \tag{8}$$

Here, γ is the half-width-half-maximum of the electric field E_x, i.e., $|x_h - x_r| \sim \gamma$. The amplitude and phase are plotted in Fig. 3 for $L = 7$ and ionospheric reflectivity $|R| = 0.8$. The full-width half-maximum of the FLR in the equatorial plane is $\sim R_E/3$. The phase change with latitude in Fig. 3 has been confirmed by observation and is characteristic of field line resonances (Rae et al. 2007; Samson et al. 2003; Waters et al. 2007).

3.2 Dipole Magnetic Field Geometry

The field-aligned current (FAC) model of Sect. 3.1 provides useful information on FLR morphology at the ionosphere. However, magnetic topology must be realistic for quantitative analyses.

Here, we present an approximate analytic theory of FLRs in dipolar geometry (Cummings et al. 1969; Obana et al. 2008; Southwood and Hughes 1983; Taylor and Walker 1984; Voronkov et al. 1997).

The linearized ideal MHD equations in curvilinear coordinates for a cold one-fluid magneto plasma can be solved as an eigenvalue problem given boundary conditions and an ionospheric conductance. Equation (9) defines the azimuthal magnetic field component B_3, with subscripts 1, 2 and 3 representing the field-aligned, cross-field (meridional) and azimuthal directions, respectively:

$$\frac{\partial^2}{\partial t^2}(h_3 B_3) = \frac{1}{h_2^2}\frac{\partial}{\partial x_1}\left[\frac{v_A^2}{h_3^2}\frac{\partial}{\partial x_1}(h_3 B_3)\right]. \tag{9}$$

The dipole scale factors are defined by $h_3(s) = (1 - s^2)^{3/2} L R_E$, $h_2(s) = h_3 L R_E/(1 + 3s^2)^{1/2}$, and $h_1(s) = h_2 h_3 L R_E$, where $s = \cos\theta$, L is the McIlwain parameter, R_E is Earth's radius, and the dipole field-aligned coordinate $x_1(s) = s L^{-2} R_E^{-2}(1 - s^2)^{-2}$. It is convenient to substitute $h_3 B_3 = \mathrm{Re}\{b(x_2)e^{i(\omega t - m\phi)} S_N(x_1)\}$ into Eq. (9), with $b(x_2)$ describing the variation of the toroidal wave magnetic field amplitude in the equatorial plane, and $S_N(x_1)$ the eigenfuction expressed as a function of x_1. The eigenfunction and eigenfrequency can be solved for analytically for a dipole magnetic field and a density profile that varies as r^{-6}. Then, $S_N(s) = A\sin(f_N s)$, where A is a normalization constant, $f_N(s) = \pi s_m/2$, $s_m = \sqrt{1 - L^{-1}}$, $\omega = f_N v_{A0}/L R_E$, and v_{A0} is the Alfvén speed $v_A(s) = v_{A0}(L)\sqrt{1 + 3s^2}/(1 - s^2)$ at $s = 0$ (the equator). The wave frequency of the toroidal mode is dominated by the Alfvén speed in the equatorial plane and therefore the assumptions made are not overly restrictive.

With knowledge of the eigenfunction and eigenfrequency of a freely evolving shear Alfvén wave, a model describing wave growth and phase mixing to the perpendicular scale γ calculated in Sect. 3.1 can be formulated. To excite a field line resonance, a constant frequency forcing term $R_D \sin(\omega_0 t - m\phi)$ is applied uniformly across geomagnetic field lines. The driven solution for the toroidal mode magnetic field B_3 on a field line labeled by L can then be calculated (Rankin et al. 1999a; Voronkov et al. 1997; Wang et al. 2018),

$$h_3 B_3 = \frac{1}{2} R_D \omega_0 t S_N(s) \cos(\omega t - m\phi)\,\mathrm{Sinc}\left(\frac{\Delta\omega t}{2}\right) \tag{10}$$

where $\Delta\omega = \omega - \omega_0$ is the frequency mismatch between the toroidal mode eigenfrequency ω and the constant driver frequency ω_0. Note that for $\Delta\omega = 0$ the wave amplitude grows linearly with time, $B_3 \propto \omega_0 t$ on the resonant field line and more slowly away from the resonant field line, resulting in a phenomenon called phase mixing. Heyvaerts and Priest (1983) presented the first full MHD study of this process, while Mann et al. (1995) explored the temporal evolution of FLR widths driven by fast cavity modes, finding a relationship between the phase mixing length and associated timescale of adjacent resonant waves. Phase mixing recognizes that for a wave limited to a small range of latitude there is a redistribution of energy among the Fourier components, leading to a delta-function asymptotically in time. In the presence of dissipation, the limiting scale length perpendicular to geomagnetic field lines is determined by ionospheric conductivity.

Taking the time derivative of B_3 and substituting it into the azimuthal component of Faraday's law gives an expression for the meridional electric field E_2:

$$h_2 E_2 = -\frac{1}{2} R_D \frac{v_A^2}{h_3^2} \omega_0 t \frac{\partial S_N(s)}{\partial x_1} \sin(\omega t - m\phi)\,\mathrm{Sinc}\left(\frac{\Delta\omega t}{2}\right). \tag{11}$$

A compressional component of the wave magnetic field also exists to satisfy the field-aligned component of Faraday's law. This field component is not calculated here because it is not used in the discussion that follows. The magnetic field components of the wave are in phase

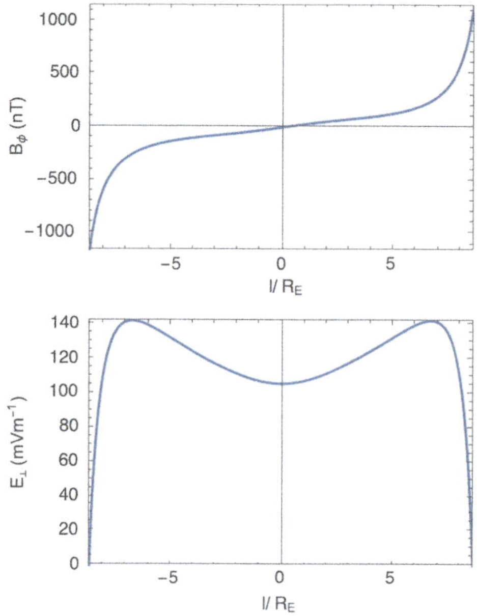

Fig. 4 Solutions to the MHD equations for shear Alfvén waves in a dipole magnetic field. The top panel shows the azimuthal magnetic field; the bottom panel shows the perpendicular (radial in the equatorial plane) electric field with a ninety-degree phase difference between them. The horizontal axis shows distance along an $L = 7$ geomagnetic field line measured from the equator where $l = 0$. The wave frequency is 10.5 mHz

as opposed to being in anti-phase as indicated in Zong et al. (2017) and Wang et al. (2018). In these articles an error in plotting the fields was made but the test-particle calculations in the papers used the correct phasing of the fields. Figure 4 illustrates the electric and magnetic field at $L = 7$ as a function of distance along the field, measured from the equator, for an equatorial electric field strength of 100 $\mathrm{mVm^{-1}}$. The field-aligned current $j_\parallel$ on the resonant field line at the width defined in Eq. (8) is a few $\mu\mathrm{Am}^{-2}$ when scaled to the ionosphere.

To examine the range of L-shell spanned by B_3 at time t it is convenient to express the Alfvén speed in the form $v_A(L) = v_{A0}L_0^3/L^3$, where v_{A0} is the Alfvén speed at the equator on the resonant L-shell for the driver wave of frequency ω_0, $\Delta\omega = \omega_0(L - L_0)/l_\omega$ is a frequency mismatch parameter, and l_ω is a characteristic length scale for the variation of the Alfvén speed; $l_\omega = R_E(\frac{1}{v_A}\frac{dv_A}{dL})^{-1}$ (Rankin et al. 1994). The resonant Alfvén wave localizes on the resonant field line where $\omega_0 = \omega(L_0)$, and phase mixes to increasingly narrower scales as a function of time according to $\Delta L(t) \sim 2\pi l_\omega R_E^{-1}\Delta L_0/\omega_0 t$, where this expression follows from Eq. (11).

The phase mixing time t_{mix} for ideal MHD waves in a dipole magnetic field is obtained by comparing $\Delta L(t)$ with Eq. (8):

$$t_{mix} \sim \frac{\pi}{2}\frac{\Delta L_0}{\ln|R|}\tau_{saw} \tag{12}$$

Here, ΔL_0 is the initial FWHM of the resonance expressed in terms of L, and the wave period is τ_{saw}. The phase mixing time t_{mix} evaluated with $|R| = 0.8$, $\Delta L_0 = L_0/5$, and $L_0 = 7$, is on the order of 10 wave periods and the resonance width of the electric field and magnetic field in the ionosphere is $\sim$ 18 km. This scale is comparable to latitudinal widths of Pc5 waves with periods of hundreds of seconds (Motoba et al. 2002; Walker et al. 1992). Figure 5 shows phase mixing for a wave of different amplitude that is resonant at $L_0 = 5.7$. In the REGO keogram of Fig. 1, a similar evolution can be seen, although the observed wave eventually fades.

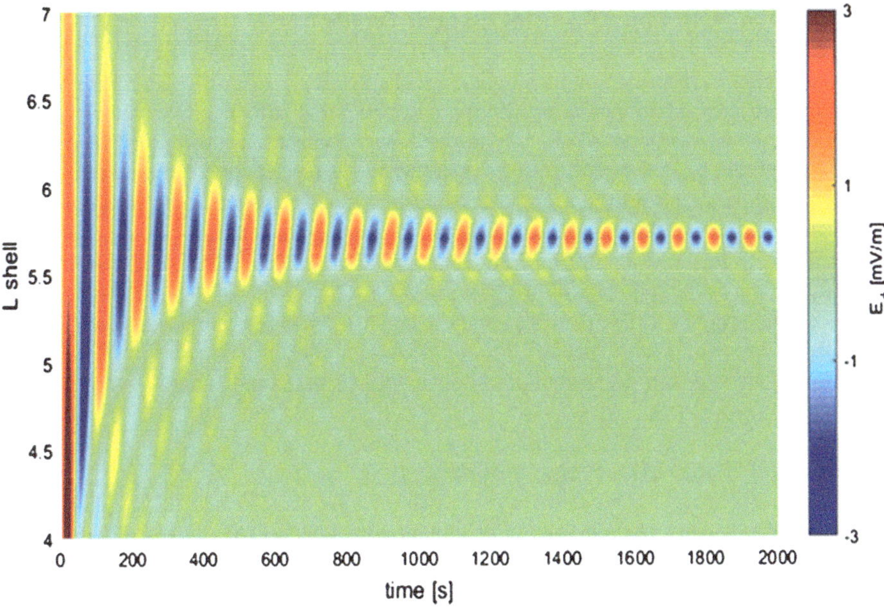

Fig. 5 Phase mixing of perpendicular electric field $E_\perp = E_\phi$ as a function of L-shell *and* time. The resonant field line where $\Delta\omega = 0$ occurs at $L = 7$. The resonance frequency is approximately 10 mHz and the electric field amplitude is in mVm^{-1}

Figure 1 of Rankin et al. (2004) shows a more striking example of phase mixing in a higher latitude FLR-arc. The luminosity occurs when the FAC is upward (away from the ionosphere) and it is proportional to the gradient of the wave across the background magnetic field. Thus, the FAC has an effective width less than B_3. The FAC can be calculated from Ampere's law using the observed spatial scale of the current perpendicular to the magnetic field. It is mainly carried by electrons and is intensified by phase mixing provided the amplitude dissipation rate of the wave is much smaller than the wave frequency. Phase mixing will narrow the wave to a scale where ideal MHD breaks down if wave dissipation is small enough. The limiting width of the FLR is then dependent on the electron skin depth c/ω_{pe} near the ionosphere, and the scales c_s/Ω_{ci} and v_{Ti}/Ω_{ci} near the equatorial magnetosphere.

Estimation of wave frequencies and profiles of electric and magnetic fields is more challenging when the magnetic topology is non-dipolar (Singer et al. 1981). For specific case studies of the midnight region of the magnetotail, stretched magnetic field topologies have been determined on the basis of MSP observations (Wanliss et al. 2000) and stochastic diffusion (Zelenyi et al. 1990). Although the approach is relatively simple, reasonable agreement between observed and modeled FLR eigenfrequencies in these topologies are achieved (Wanliss et al. 2002). More advanced FLR theory for non-dipolar magnetic fields has also been developed (Cheng and Zaharia 2003; Rankin et al. 2006; Waters et al. 2013; Degeling et al. 2010) and compared favourably with ground-based and THEMIS spacecraft observations (Sarris et al. 2010).

Parallel electric fields are formed by a toroidal Alfvén wave at the inertial and ion-acoustic length scales and are of interest because they may precipitate electrons that form FLR-arcs in the high latitude ionosphere. The properties of parallel electric fields in toroidal shear Alfvén waves are explained in Sects. 4–6.

4 Wave Dispersion and Parallel Electric Fields

The ideal MHD wave theory of FLRs explains characteristic features of modulated long period discrete arcs, including their latitude and azimuthal extent, phase velocities, and phase variation with latitude. The FACs in these waves can reach upward of 10 μAm^{-2}, which makes them a good candidate for explaining the observed optical signature of FLR-arcs. The long periods (up to ten minutes) of these waves in comparison with the bounce time of the thermal electron population is expected to invalidate Ohm's Law, i.e., such that $\sigma(\omega) \neq j_{\parallel}/E_{\parallel}$ (Samson et al. 2003; Tikhonchuk and Rankin 2000, 2002). For example, use of Ohm's law in two-fluid theory of long period waves yields small parallel electric field amplitudes on the order of a few μV m^{-1}, which is much smaller than required value (Rankin et al. 1999b; Wei et al. 1994). A brief consideration of parallel electric fields in FLRs is presented first, followed by a discussion of kinetic theory (Sect. 5) and simulations of toroidal mode standing waves (Sect. 6).

4.1 Two-Fluid Theory of Field Line Resonances

The development of parallel electric fields along geomagnetic field lines is influenced by two parameters, namely, the density of electrons, n_e, and the temperature, T_e. It has been analyzed using several two-fluid theories that differ in their approximations (Bhattacharjee et al. 1999; Frycz et al. 1998; Rankin et al. 1999a; Streltsov and Lotko 1995). These models are based on assumptions that reduce an intrinsically kinetic process to a macroscopic one. However, a comprehensive theory of plasma dynamics in FLRs must ultimately account for changes in particle distribution functions that support the waves. A variety of approaches have been proposed to explain the generation of parallel electric fields, for example, by considering anomalous resistivity associated with microturbulence (Lotko et al. 1998; Streltsov and Lotko 1999), mirror force contributions (Nakamura 2000; Stark et al. 2011; Damiano and Johnson 2012), and nonlinear electron inertial effects (Wright et al. 2002). The latter was extended to a kinetic description in Wright and Hood (2003). Using kinetic theory, Nakamura (2000) shows that the $E_{\parallel}$ associated with kinetic and inertial Alfvén wave regimes can be exceeded by the contribution provided by the mirror force for low frequency waves, describing this limit as the "mirror kinetic Alfvén wave". Stark et al. (2011) examines how the different terms in the generalized Ohm's law compare in the generation of $E_{\parallel}$, including mirror force contributions.

The set of two-fluid nonlinear equations developed by Rankin et al. (1999a) describes coupling between toroidal, Alfvén, and slow mode waves via the action of the wave ponderomotive force. Similar ideas have also been developed by Boehm et al. (1990), and Allan et al. (1990, 1993). The discussion that follows is motivated by the author's obvious familiarity with the Rankin et al. (1999a) model, and does not suggest criticism of other models (Lotko et al. 1998; Streltsov and Lotko 1995, 2008; and the references mentioned above). Where appropriate, key results from other two-fluid models are discussed. The linearized form of the equations used in Rankin et al. (2004) reproduce the eigenmode solutions discussed in Sect. 3 and can be solved numerically (Lu et al. 2003a, 2003b; Rankin et al. 2005a, 2005b). Solutions to the equations are qualitatively consistent with those obtained from other models mentioned, specifically Streltsov and Lotko (1996, 1997), and Streltsov (1999). The model equations are equivalent to those used by Chaston et al. (2003). The equations are summarized below:

$$\frac{1}{v_A^2}\frac{d\nabla^2\phi}{dt} + (\boldsymbol{b}\cdot\nabla)\nabla^2 A = \nabla\cdot\left(\frac{\mu_0}{B_0}\boldsymbol{b}\times\nabla P\right) \tag{13}$$

$$\frac{\partial^2 A}{\partial t^2}+(\boldsymbol{b}\cdot\nabla)\frac{\partial\phi}{\partial t}=\lambda_e^2\frac{d}{dt}\left(\frac{\partial\nabla^2 A}{\partial t}\right)-\lambda_e^2(\boldsymbol{b}\cdot\nabla)v_{Te}^2(\boldsymbol{b}\cdot\nabla)\nabla^2 A \tag{14}$$

$$\rho\frac{dV_\parallel}{dt}=-(\boldsymbol{b}\cdot\nabla)P+\frac{\rho}{B_0^2}\nabla A\cdot\left(\frac{d\nabla\phi}{dt}\right)+\frac{1}{B_0}\nabla A\times\nabla P \tag{15}$$

$$\frac{1}{\rho}\frac{d\rho}{dt}=\frac{1}{B_0}\frac{\partial B_\parallel}{\partial t}-(\boldsymbol{b}\cdot\nabla)V_\parallel. \tag{16}$$

Equations (13)–(15) are defined in terms of the magnetic vector potential A along the geomagnetic field (with direction defined by the unit vector $\boldsymbol{b}$), the electric scalar potential ϕ, the parallel ion velocity $V_\parallel$, the plasma density ρ and magnetic field of the compressional wave $B_\parallel$. Equations (15)–(16) describe excitation of ion acoustic waves by the ponderomotive force of shear Alfvén waves (the second term on the right-hand side of Eq. (15)). The equation for the compressional component of the wave field is given by

$$\nabla B_\parallel=-\frac{\mu_0}{B_0}\nabla P-\boldsymbol{b}\times\frac{1}{v_A^2}\frac{d\nabla^2\phi}{dt}. \tag{17}$$

Note that in Eqs. (13)–(17) $\frac{d}{dt}=\frac{\partial}{\partial t}+(\boldsymbol{b}\times\nabla\phi/B_0)\cdot\nabla$ is used and $v_{Te}^2=T_e/m_e$ is defined without the factor of two used in Sect. 3. The equations require two small parameters (Frycz et al. 1998): the ratio of the inhomogeneity scale of the toroidal wave amplitude perpendicular to the magnetic field ($L_\perp$) to the scale along the magnetic field line, ($L_\parallel$) and the ratio of the toroidal mode frequency (ω), to the ion cyclotron frequency (ω_{ci}). The shear Alfvén wave dynamics is described by two scalar functions that define the magnetic field, $\boldsymbol{B}=\boldsymbol{B}_0+\nabla_\perp A\times\boldsymbol{b}$, and the electric field, $\boldsymbol{E}=-\nabla\Phi-\boldsymbol{b}\partial A/\partial t$, where $\nabla_\perp$ is the gradient across the magnetic field. Linearizing Eqs. (13)–(17), we obtain the zeroth-order solutions of Sect. 3.2. It is easily seen that $E_\parallel=-\frac{\partial A}{\partial t}-(\boldsymbol{b}\cdot\nabla)\phi$ is small compared to $E_\perp$ and so the eigenmodes $S_N(l)$ for the latter are assumed to be those shown in Fig. 4; $E_\parallel$ can then be expressed in terms of the toroidal mode magnetic field B_3

$$E_\parallel\propto-\omega_0\left(\lambda_e^{eq}\right)^2B_0^{eq}\frac{\partial B_3}{\partial x_2}G_\parallel(s)\exp\left[i(m\phi-\omega_0 t)\right] \tag{18}$$

$$G_\parallel(s)=\frac{\rho_0^{eq}}{\rho_0(s)}\frac{h_1^{eq}}{h_1(l)}\left[\left(\frac{v_{Te}^2}{v_A^2}-1\right)S_N(l)-\frac{v_A^2}{\omega_0^2}h_2^2\frac{\partial S_N}{\partial l}\frac{\partial}{\partial l}\left(\frac{v_{Te}^2}{v_A^2h_2^2}\right)\right] \tag{19}$$

$$B_3\propto\left(\frac{h_3^{eq}}{h_3}\right)S_N(l). \tag{20}$$

In Eqs. (18)–(20), $dl=h_1dx_1$ and "eq" denotes equatorial value. For profiles of the electron density and temperature as a function of $s=\cos\theta$ we take $n_e(s)=a_0+a_1/(1-s^2)^3+a_2\exp[-L(1-s^2)+1.05]$ and $T_e(s)=[a_0+a_1/(1-s^2)^3]^{-1}$. Here, units of density and temperature are in cm^{-3} and eV, respectively and the numerical values of parameters are $a_0=1.77$, $a_1=0.23$, and $a_2=1.7\times10^5$. Profiles of the Alfvén speed and thermal speed are shown in Fig. 6, with the bottom panel displaying these quantities overlaid on a logarithmic scale. The thermal speed drops below the Alfvén speed at $\sim 2R_E$ off the equator (along $\boldsymbol{B}_0$), thus delineating the warm plasma and inertial regimes.

For a wave with a frequency of 10.5 mHz that resonates at $L=7$, the perpendicular and parallel electric field components computed using Eqs. (18)–(20) and Eq. (11) are shown in Fig. 7. The top panel shows the real and imaginary (quadrature) components of the perpendicular electric field. The increasing quadrature component is due to the presence of

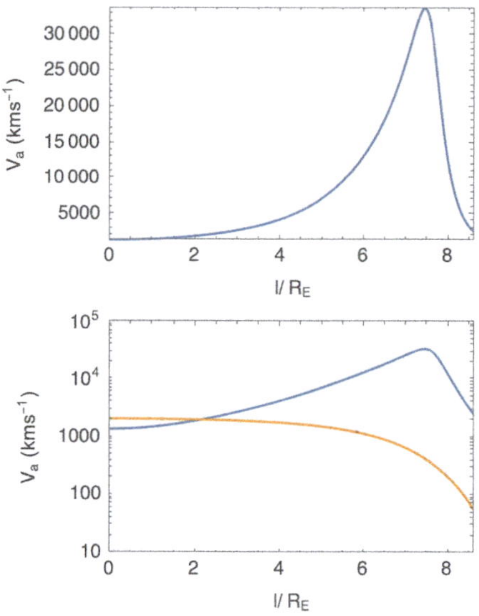

Fig. 6 Top panel: Alfvén speed as a function of distance along a $L = 7$ dipole field line measured from the equator in units of Earth's radius. Bottom panel: Alfvén speed overlaid with the electron thermal speed. The profiles are described in the text

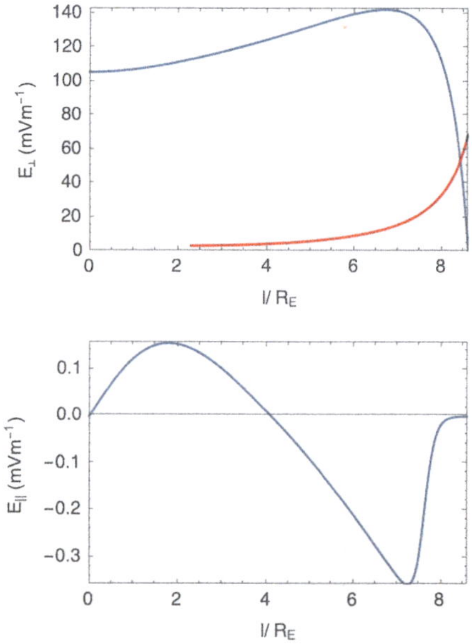

Fig. 7 Top panel: perpendicular electric field (blue line) and its quadrature component (red line) as a function of distance along a dipole field line $L = 7$ measured from the equator. Bottom panel: parallel electric field. Wave period 10.5 mHz, ionospheric damping time 32.75 minutes

ionospheric damping, as indicated in the figure caption. The ratio of the maximum of the parallel to perpendicular electric field components is about 1/400, which is about 40–50 times smaller than that of a 3 Hz propagating wave in the same setting of plasma parameters and spatial scale. The wave parallel potential drop $\Phi(s) = \int_{s_0}^{s} E_\parallel \frac{h_3}{h_2 ds}$ ($E_\parallel$ is negative when directed upward) in Fig. 7 is approximately 100 V, which is insufficient for producing auroral emissions even though the value of $E_\perp = 140\ \text{Vm}^{-1}$ is large and the wave frequency ~ 10 mHz is at the upper range for Pc5 waves. The small parallel electric field in the two-fluid models of long-period FLRs requires examination since it excludes electron bounce

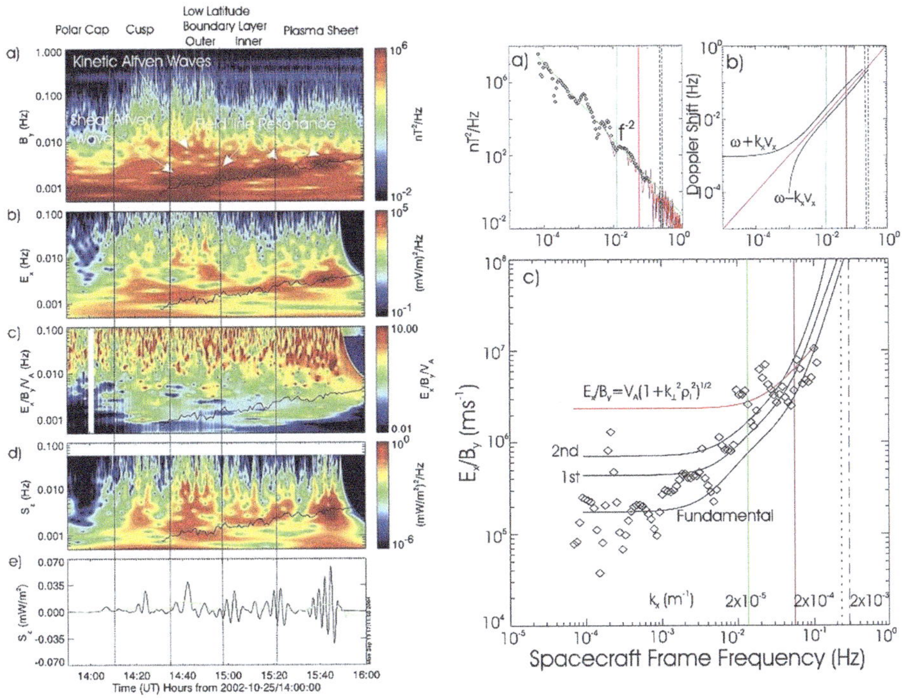

Fig. 8 Cluster-1 wavelet spectra. Black lines in panels a-d in the left column shows the frequency of the FLR with (**a**) wavelet power spectra of azimuthal magnetic field (**b**) wavelet spectra of radial electric field (**c**) E_X/B_Y ratio normalized by local Alfvén speed (**d**) and (**e**) field-aligned wave Poynting flux (**e**) field-aligned wave Poynting flux. In the right column the red curve shows the kinetic Alfvén wave dispersion based on local parameters at Cluster-1 while the black curves show the FLR case for fundamental and harmonic eigenmodes

motion, which can have timescales comparable to the wave period. Section 5.2 discusses bounce motion.

The linear eigenmode solutions are useful for understanding properties of FLRs but do not capture all of the physics. For example, the focusing of wave energy in FLRs because of the opposing dispersive characteristics of inertial and kinetic Alfvén waves, as discussed by Streltsov and Lotko (1996, 1997) leads to non-radiative resonance layers with large $j_{\|}$ (and hence $E_{\|}$). Two-fluid simulation of dispersive FLRs including ion gyroradius effects were presented in Streltsov et al. (1998) and illustrate, among other things, how ion gyroradius effects can act to reduce $E_{\|}$.

The linear eigenmode solutions, however, *have* proved useful in the analysis of satellite data. Figure 8 is a reproduction of Figs. 4 and 6 from Chaston et al. (2005) showing ULF waves seen on the FAST and Cluster-1 satellites when in magnetic conjunction. Data obtained during the pass through the polar cusp region onto closed field lines show FLRs with scales down to the local ion gyro-radii and electron inertial length. The source region for the FLRs was attributed to surface waves associated with tailward flows on the low latitude boundary layer. The right panel of Fig. 8 shows the dispersion relation is described by a standing wave paradigm, with the wavelength along the field primarily the fundamental mode. The wave Poynting flux and the field-aligned electrons observed during the event at FAST satellite altitudes are sufficient to produce visible aurora.

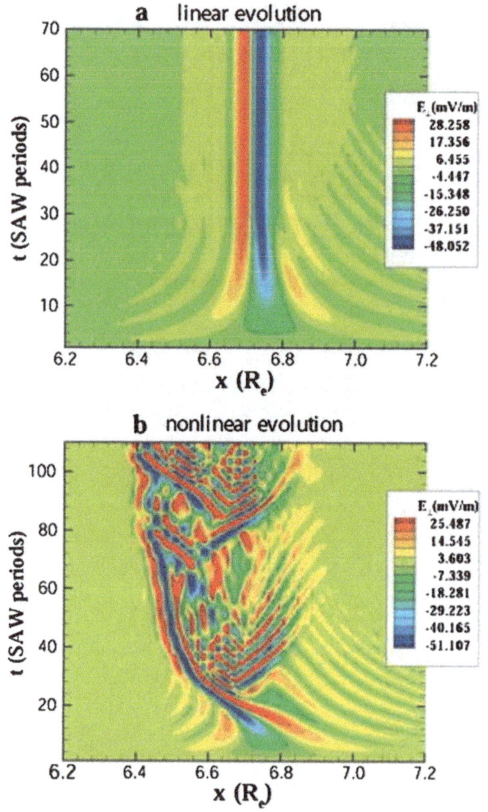

Fig. 9 Temporal evolution of the perpendicular electric field $E_\perp$ in the equatorial plane in the (**a**) linear and (**b**) nonlinear evolution of FLRs. The spatial structuring in the lower panel is a consequence of time-varying density variations induced by the ponderomotive force of the toroidal mode. Coordinate x is equivalent to L-shell

4.2 Nonlinear Effects in Two-Fluid Field Line Resonances

Section 4.1 demonstrates that perpendicular wave dynamics and parallel ion motion can be analyzed using two-fluid theory. It is interesting to discuss nonlinear solutions to Eqs. (13)–(17) and examine their influence on the strength of parallel electric fields. Figure 9 reproduces Fig. 1 from Lu et al. (2003b) and illustrates how the results when nonlinear dynamics and dispersion are taken account of differ from those of ideal MHD. These processes produce an Earthward displacement of the FLR and a very broad, highly structured waveform. The nonlinear spatial structuring in Fig. 9b is due to the dependence of the dispersive saturation time and width on the dispersion parameter δ and the Alfvén speed gradient scale l_ω. Expressions for these parameters are defined by Eqs. (21) and (22).

$$t_{disp} = \frac{1}{2}\tau_{saw}^2\left(\frac{l_\omega^2}{|\delta|}\right)^{\frac{1}{3}}, \quad l_{disp} \sim \left(l_\omega|\delta|\right)^{\frac{1}{3}} \tag{21}$$

$$\delta = \frac{(h_2^{eq})^2}{N}\int_0^{l_{max}}\frac{dl}{h_1}\left[\frac{\rho_i^2 v_A^2}{\omega^2}\left(\frac{\partial S_N}{\partial l}\right)^2 + \frac{v_{Te}^2}{\omega^2}\frac{\partial S_N}{\partial l}\frac{\partial(\lambda_e^2 S_N)}{\partial l} - \lambda_e^2 S_N^2\right] \tag{22}$$

In Eq. (22), $N = 2\int_0^{l_{max}} dl S_N^2 h_2/h_3$, and $\delta = \frac{3\rho_i^2}{4} - \lambda_e^2(1 - v_{Te}^2/v_A^2)$ for straight magnetic field lines (Rankin et al. 1999a). Equation (21) shows that when density fluctuations steepen the Alfvén speed profile, the local gradient scale l_ω also steepens and dynamic structures form on time and spatial scales t_{sat} and l_{sat}, respectively. They differ from the ideal MHD estimates Eqs. (8) and (12). The cross-field structure in Fig. 9b is confined to a density cavity as the ponderomotive force drives parallel ion motion along geomagnetic field lines to the equator. The decrease in density near the ionosphere enhances $j_{\|}$ at low altitude and strengthens parallel electric fields.

There continues to be debate about the role of the ionosphere in the evolution of shear Alfvén waves. When the ionospheric Pedersen conductance varies in response to currents, ionization via soft electron precipitation, and heating by global convection electric fields and wave dynamics are naturally more varied (Lysak 1991; Pokhotelov et al. 2002; Streltsov and Lotko 2008; Streltsov 2008; Streltsov and Mishin 2018a,b; Sydorenko and Rankin 2017). The characteristic electron-ion recombination time in the ionosphere makes a quasi-static approximation applicable (Lysak 1991). The ionospheric density can be expressed as $S_e + S_c = \alpha n_e^2$, where α is the recombination rate, S_c is an ionization source due to cosmic radiation, and S_e is the ionization rate due to precipitating electrons. The root-mean-square Pedersen conductance is defined by,

$$\Sigma_p(t) = \left(\Sigma_{p0} + Q^2\varepsilon\right)^{\frac{1}{2}}, \quad Q = \frac{1260K_e}{(16 + K_e^2)} \tag{23}$$

where Σ_{p0} is the initial value of the conductivity, Q is a coefficient with units of mho-m/W$^{1/2}$, K_e is the energy of precipitating electrons in keV, and ε is the precipitation energy flux of electrons in Wm^{-2}. The latter is defined by $\varepsilon = -j_{\|}K_e/e$ for upward-directed FAC's and is zero otherwise. Equation (2) in Prakash et al. (2003) provides an estimate of the damping rate of the FLR by the Pedersen current in the ionosphere,

$$\Gamma = \frac{1}{\mu_0 \Sigma_p N}\left(S_N^2 \frac{h_2}{h_3}\right)_{l_{max}} \tag{24}$$

Based on Eq. (24), the resonance width at the equator due to ionospheric dissipation follows:

$$l_{\Sigma_p} = l_\omega \frac{\Gamma}{\omega} \tag{25}$$

Equation (25) can be regarded as a more accurate formula than Eq. (8) for FLR-arc widths in a dissipation-dominated ionosphere. Together with Eq. (21), different regimes of FLR dynamics are possible depending on characteristic widths achieved: $l_{FLR} = \max(l_{\Sigma_p}, l_{disp})$. These equatorial widths can be mapped along field lines from the equator according to:

$$l_{Ion} = l_{FLR} \frac{h_2(l)}{h_2^{eq}} \tag{26}$$

If the scale l_{Σ_p} is reached first, nonlinear ionospheric feedback effects are triggered. The condition for this nonlinear feedback is that the dissipative FLR-arc width should be larger than the dispersive width, which translates to $\Gamma > \omega(\delta/l_\omega^2)^{1/3}$.

Statistically, low conductivity has been identified as preferential to auroral arc formation (Newell et al. 1996) with the suggestion that feedback from the ionosphere is one of the necessary conditions. Prakash et al. (2003) show that electron precipitation and low conductivity intensifies the wave FAC and results in large-amplitude spatially structured current oscillations for Pedersen conductivities in the range $0.5 \lesssim \Sigma_p \lesssim 1$ mho, cf. Figs. 1 and 2 in Prakash et al. (2003). This behaviour is due to periodic transitions between dissipation-dominant and dispersion-dominant regimes of FAC dynamics. They also find that feedback nonlinearity does not affect FAC dynamics for large conductance because the dissipative width is always smaller than the dispersive width. It is interesting to note that in nonlinear feedback regimes, the spatially structured current oscillations can continue to evolve over several wave periods even when the wave source is turned off. This partially self-sustaining evolution of a nonlinear dispersive and dissipative FLR is distinct from the ionospheric feedback instability, but it may make it difficult to distinguish between the two. The FAC in nonlinear feedback follows from Ampere's Law; it decreases ionospheric damping, activating a feedback response from the ionosphere.

Cohen et al. (2013) observed wave-like structures near the edges of a discrete arc in the return current region and attributed them to the ionospheric feedback instability. The plasma at the altitude of the ACES-High sounding (~ 345 km) rockets had a temperature of fractions of an eV with a density enhancement attributed to the wave ponderomotive force. The observations were simulated with the model of Streltsov and Lotko (2008), which captured aspects of the observed wave dynamics and density enhancement. However, intense small-scale current filamentation occurred throughout the pair of upward-downward FACs in the model and the density enhancement was significantly smaller than in the ACES-High observations. A single downward current system was simulated using a model similar to Streltsov and Karlsson (2008) was used to launch a single downward FAC as opposed to an opposing pair (Cohen et al. 2013). The results from this simulation, shown in the right panel of Fig. 10, agree better with the observations from ACES-High in that they show evidence of small-scale electromagnetic waves toward the equatorward and poleward sides of the downward FAC region (top-left panel of Fig. 10). The current system appears to broaden with time in the simulation, raising the question of whether single FAC current filamentation is fundamentally different from the paired current system simulated by Streltsov and Lotko (2008). Although the results shown in Fig. 10 are for waves characteristic of the IAR, they make the point that FLR and IAR wave dynamics share common characteristics.

In the original formulation, the ionospheric feedback instability has an energy source provided by the magnetospheric convection electric field (Lysak 1991; Pokhotelov et al. 2002) Recently, the viability of the classical linear feedback instability has been questioned for waves excited in the ionospheric resonator. It is suggested by Sydorenko and Rankin (2017) that velocity shear due to the variation of the ion mobility with altitude in the ionosphere stabilizes the instability. These authors question the validity of the height-integrated conductivity assumption that is used in studies of the IFI. Using a numerical model, they demonstrate that in a finite-thickness realistic ionosphere, velocity shear stabilizes the instability. Watanabe and Maeyama (2018) questioned these results by constructing a simplified model describing a much longer wavelength (FLR-type) system. In their linear eigenmode analyses, a model magnetosphere with straight magnetic field lines is connected to a finite thickness model ionosphere with a stationary (analytic) ion velocity shear. Unstable modes are found, which is not unexpected because the wavelength is much larger than the thickness of the ionosphere, opposite of what Sydorenko and Rankin (2017) considered.

The results of Sydorenko and Rankin (2017) have also been questioned based on the point of view that the results are wrong because they contradict the large literature on the IFI (Streltsov and Mishin 2018a). All previously published work, however, uses a height-integrated conductivity approximation. Regardless of whether the IFI or nonlinear feedback process is more efficient, FACs carried by Alfvén waves are predicted to be intensified by magnetosphere-ionosphere coupling, providing an obvious potential link to associated auroral luminosity. The debate around the viability of the IFI would clearly benefit from further study

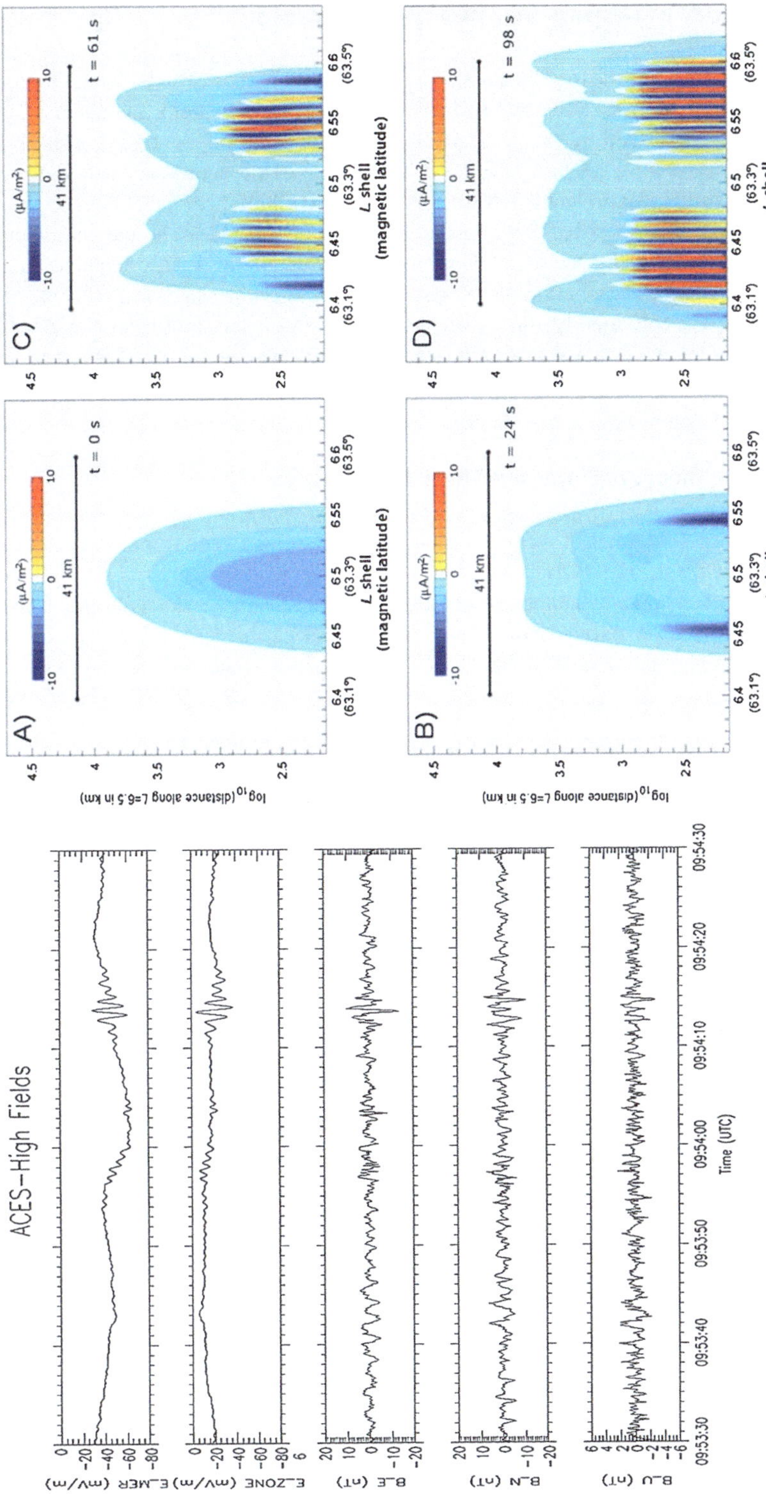

Fig. 10 Left panel: ACES-High small-scale electromagnetic waves at roughly 09:53:58 UTC and 09:54:15 UTC at the equatorward and poleward boundaries of the downward current region. Right panel: Simulated downward current system detailed in Cohen et al. (2013)

5 Kinetic Theory of Dispersive Shear Alfvén waves

Sections 3 and 4 illustrate the significant strengths and weaknesses of ideal MHD theory and two-fluid theory applied to Shear Alfvén Waves and especially, field line resonances. Wave dispersion is described by two-fluid theory, and interesting linear and nonlinear behaviour manifests that have relevance to the interpretation of observations. Although the development of parallel electric fields is also captured by two-fluid theory, the interpretation of observations is to a large extent made in terms of the FACs produced by the waves, cf., Figs. 9 and 10. When the characteristic electron bounce time along the geomagnetic field lines is comparable to the wave period, the electron dynamics becomes nonlocal with respect to the fields of the wave. Evidently, for high-frequency waves, such as those associated with the IAR, non-local effects are not as significant, while they are critical for FLRs. It is instructive to examine differences between local and nonlocal behaviour of electron dynamics using plasma kinetic theory.

5.1 Local Kinetic Theory of Dispersive Shear Alfvén Waves

The inertial and kinetic regimes of shear Alfvén waves can be understood from consideration of the local kinetic dispersion relation for an infinite homogeneous magnetized plasma. For waves of frequency ω and wavenumber $\boldsymbol{k}$ the analysis proceeds under the assumption that the polarization current is carried mainly by ions, while electrons carry the wave parallel current. The $\omega \ll \omega_{ci}$ small parallel wavelength limit is given in Lysak and Lotko (1996),

$$\frac{\omega^2}{k_\parallel^2 v_A^2} \sim \frac{1}{\frac{v_A^2}{c^2} + \frac{[1-\Gamma_0(\mu_i)]}{\mu_i}} + \frac{k_\perp^2 \rho_s^2}{\Gamma_0(\mu_e)[1+\zeta_e Z(\zeta_e)] + k_\parallel^2 \lambda_D^2} \tag{27}$$

where $\Gamma_0(\mu_{e,i}) = e^{-\mu_{e,i}} I_0(\mu_{e,i})$, I_0 is the modified Bessel function of the first kind, $Z(\xi_e)$ is the electron plasma dispersion function, $\mu_{e,i} = k_\perp^2 \rho_{e,i}^2$, $\rho_s^2 = (T_e/T_i)\rho_i^2 = \frac{c_s^2}{\omega_{ci}^2} = \frac{c^2 \lambda_D}{v_A^2}$, $\zeta_e = \frac{\omega}{k_\parallel v_{Te}}$, and $v_{Te}^2 = \frac{2T_e}{m_e}$. For $v_A^2 \ll c^2$, and $k_\parallel^2 \lambda_D^2 \ll 1$, the kinetic dispersion relation simplifies to,

$$\frac{\omega^2}{k_\parallel^2 v_A^2} = \frac{\mu_i}{1-\Gamma_0(\mu_i)} + \frac{k_\perp^2 \rho_s^2}{\Gamma_0(\mu_e)[1+\zeta_e Z(\zeta_e)]} \tag{28}$$

For a wave of given frequency ω, Eq. (28) can be applied pointwise along geomagnetic field lines provided $\lambda_\parallel \ll 2l_{max}$, $\omega \ll \omega_{ci}$, and $k_\parallel^2 v_{Ti}^2/2 \ll \omega_{ci}^2$, where $2l_{max}$ is the length of a field line, $\lambda_\parallel$ is the parallel wavelength, and $v_{Ti}^2 = 2T_i/m_i$. In the cold plasma limit, $T_{e,i} \to 0$, the kinetic dispersion relation reduces to $\omega^2 = k_\parallel^2 v_A^2/(1 + k_\perp^2 \lambda_e^2)$, while for hot electrons and in the limit of small ion gyroradius, the kinetic dispersion relation reduces to $\omega^2 = k_\parallel^2 v_A^2 [1 + k_\perp^2(\rho_s^2 + \frac{3}{4}\rho_i^2)]$. It is informative to compare these kinetic limits with the two-fluid dispersion relation. A commonly used two-fluid dispersion relation is (Streltsov and Lotko 2008),

$$\frac{\omega^2}{k_\parallel^2 v_A^2} \sim \frac{1 + k_\perp^2 \rho_s^2}{1 + k_\perp^2 \lambda_e^2} \tag{29}$$

which differs somewhat from different limits of the kinetic dispersion relation. Jones and Su (2008) utilized the two-fluid dispersion relation Eq. (15) to analyze spacecraft observations near Io, concluding that it implied parallel electric fields must be very small near this moon. However, the kinetic dispersion relation applied under similar plasma conditions suggests otherwise; cf., Fig. 1 of Watt and Rankin (2009a);. The discrepancy comes from the fact that in the two-fluid theory the complex amplitude and phase of $E_\perp$ and $E_\parallel$ vary continuously as the ratio v_{Te}/v_A is varied, with the imaginary part becoming comparable to the real part for conditions near Io. Even so, the two-fluid theory, in tandem with test-particle simulation, reproduces accelerated wave signatures produced by inertial Alfvén waves lasting seconds (Kletzing et al. 2003; Kletzing and Hu 2001).

For the case of a warm electron plasma with cold ions and weak damping, Nakamura (2000) shows that

$$\frac{E_\parallel}{E_\perp} = k_\perp k_\parallel \frac{v_{Te}^2}{2v_A^2} \lambda_e^2 \frac{1}{1+\xi_e Z(\xi_e)} \tag{30}$$

and the linear kinetic dispersion relation of Lysak and Lotko (1996) can be simplified to

$$1 + \frac{\omega^2 - k_\perp^2 v_A^2}{k_\perp^2 k_\parallel^2 \lambda_e^2 v_{Te}^2} Z'(\xi_e) = 0 \tag{31}$$

which is found to be in good agreement with the kinetic simulation of Watt and Rankin (2007b).

In Eqs. (30)–(31) the cold plasma (inertial) and the hot electron limits are distinct regimes and in a formal sense should not be combined in the manner of Chaston et al. (2003). Neither can they be applied to field line resonances, which simultaneously thread warm plasma sheet and cold ionosphere plasmas. There is not much advantage over two-fluid theory in using Eqs. (30)–(31) in test-particle simulations involving inertial waves (Kletzing and Hu 2001). A different approach is gyrofluid simulation methodology incorporating electron inertia, electron pressure gradients, and finite ion gyroradius effects. Su et al. (2004) combined a linear gyrofluid model with electron test-particle simulations to reproduce features of Alfvénic electron acceleration observed by low-altitude satellites. Damiano et al. (2015) used a 2D ion gyrofluid electron-kinetic model to simulate kinetic Alfvén wave propagation from an equatorial source region in the magnetotail to the ionospheric boundary of the simulation domain. They found a faster travel time of the wave to the ionosphere as $k_\perp \rho i$ increases, and a decrease in both the FAC (Fig. 11a) and electron energization (Fig. 11c) due to ion temperature effects. As illustrated in Fig. 11b, the electron distribution in the simulation is elongated in the parallel direction with a slight peak to positive $v_{//}$ (owing to the negative current), which is consistent with observations (Janhunen et al. 2005; Wygant et al. 2002).

Numerical models to solve the nonlinear electron drift-kinetic equation have also been developed (Watt and Rankin 2007a). These models reproduce observations by NASA's POLAR and FAST spacecraft of dispersive Alfvén waves as they encounter the plasma sheet boundary layer (Wygant et al. 2002). Simulations show that bulk-heating and beam-like electron signatures are observed by FAST at conjugate altitudes below POLAR. The simulated POLAR observations exhibit strong particle trapping in the wave parallel potential that suppresses Landau damping before the waves can reach the location where electrons escape the wave to form the beam observed by FAST (Watt and Rankin 2012, 2009b).

Figure 12 shows the observed and simulated electron distribution functions for the POLAR satellite. Results are time-averaged at a cadence of 13 s to match the particle detector on POLAR. The left panels show simulated electron distribution functions following the injection of an electric field pulse $E_\perp = k_\perp \phi_0(t)$, with $\phi_0^{max} = 100$ V in the top-left panel

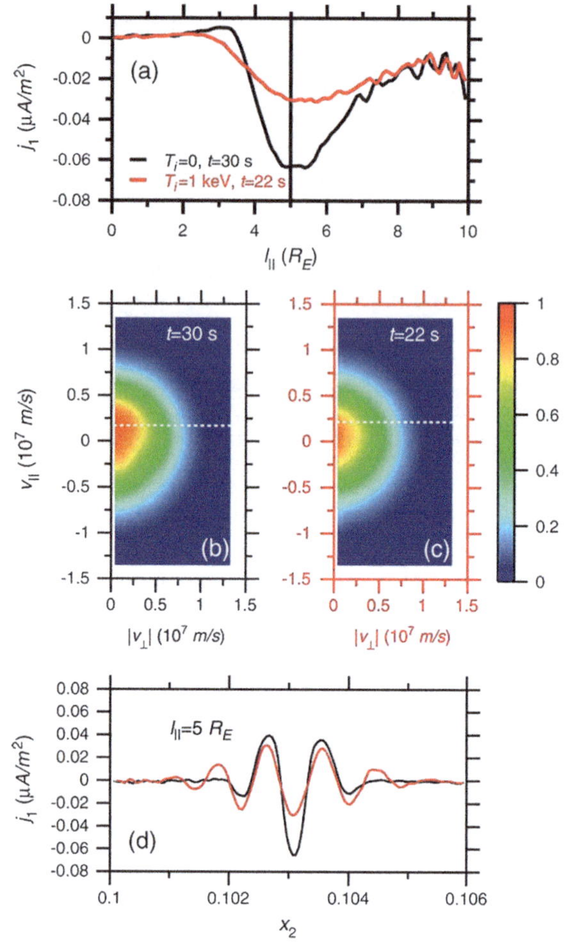

Fig. 11 Panel (**a**) Parallel current density along the field line measured from the equator for $T_i = 0$ (black) and $T_i = 1$ keV (red) at indicated times. (**b**) normalized electron velocity distribution function at the location indicated by the vertical black line in panel (**a**) for $T_i = 0$ and $t = 30$ s. (**c**) same as (**b**) but $T_i = 1$ keV and $t = 22$ s. White broken lines in panels **b** and **c** indicate parallel wave phase velocities. (**d**) Profile of parallel current density at times indicated in panel **a**

and $\phi_0^{max} = 400$ V in the bottom-left panel; for each the wave Poynting flux is $\sim 7\ \mu\text{Wm}^{-2}$. The distribution function elongates as the wave amplitude is increased, as in Damiano et al. (2015) which is supportive of the idea that auroral brightening seen by POLAR is causally related to kinetic scale Alfvén waves observed in the vicinity of the satellite. Other simultaneous observations of waves in the PSBL by FAST and POLAR make the case that auroral brightening can be ascribed to dispersive Alfvén waves (Dombeck et al. 2005).

5.2 Nonlocal Kinetic Theory of Dispersive Shear Alfvén Waves

The narrative in Sect. 5.1 states that dispersive Alfvén waves trap or reflect electrons, accelerating them to form an electron beam that propagates downward, causing aurora. Local kinetic theory captures the essence of this Alfvénic acceleration process although the actual auroral signature is not calculated. A similar argument cannot be made for field line resonances because according to Eq. (30), $E_{\|}$ becomes very small in the long wavelength limit. Nonlinear effects discussed in Sect. 4.2 can enhance $E_{\|}$, but nonlocal electron dynamics has a larger effect. The kinetic theory developed by Rankin et al. (1999a) and Tikhonchuk and Rankin (2002) for FLRs with frequencies in the mHz range predicts significantly larger

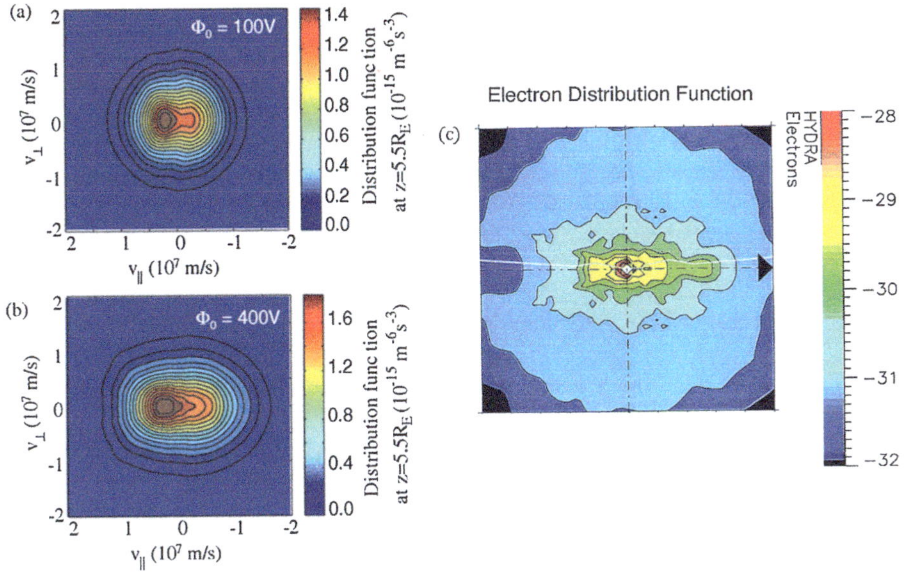

Fig. 12 Left column: simulated electron distributions for two amplitudes of injected kinetic Alfvén waves. Right column: electron distribution observed by the POLAR satellite showing a beam-like feature during a period of kinetic Alfvén waves

parallel electric fields than two-fluid theory. It is found that magnetic mirroring of electrons significantly reduces the field line conductivity and can result in parallel fields up to two orders of magnitude larger than obtained from the two-fluid treatment (Samson et al. 2003; Tikhonchuk and Rankin 2002).

The details of the nonlocal theory of FLRs are complicated and here we only sketch the main ideas. In FLRs the bounce motion of electrons is affected by the wave parallel current dictated by Ampere's law. Electron trajectories perturbed by the wave form different trapped populations that can be ordered by energy and pitch angle. Electrons on open, semi-open, or closed orbits have a significant influence on the formation of parallel electric fields in FLRs (Tikhonchuk and Rankin 2000, 2002). Their dynamics can be described under the drift-kinetic approximation, as discussed in Hazeltine and Waelbroeck (1998),

$$(\partial_t \pm v_\parallel \partial_l)\delta f_e = \pm e E_\parallel v_\parallel \partial_w f_0 \tag{32}$$

$$f_0(w) = \frac{n_h}{(2\pi)^{3/2} v_h^3} \exp\left(-\frac{w}{T_h}\right) + \frac{n_c}{(2\pi)^{3/2} v_c^3} \exp\left(-\frac{w}{T_c}\right) \tag{33}$$

Here, the perturbed distribution of electrons is expressed as $\delta f_e = f_e - f_0$, and $v_\parallel$ is the parallel component of the electron velocity with plus and minus signs corresponding to electrons travelling along the magnetic field and in the opposite direction, respectively. For convenience, a background ambient Maxwellian electron distribution f_0 is assumed, which is a combination of two Maxwellians with hot and cold electron temperatures T_h and T_c, respectively. The corresponding thermal speeds are expressed as $v_{h,c} = \sqrt{T_{h,c}/m_e}$. The electron energy $w = w_\perp + w_\parallel$ defines the parallel energy $w_\parallel = \sqrt{(2/m_e)(w - \mu B_0 + e\Phi_0)}$, where $\mu = w_\perp / B_0$ is the electron magnetic moment and Φ_0 (not the quantity in Sect. 5.1) is the quasi-static electric potential supporting the electron density and temperature profiles n_e and T_e, respectively.

Rankin et al. (1999a) compute wave eigenfunctions and frequencies using the methodology of Sect. 3, and then relate the wave FAC to the parallel component of Ampere's law, $j_\parallel \sim \mu_0^{-1}\nabla_\perp^2 \partial A/\partial t$, where A is the parallel component of the magnetic vector potential. Mathematically, $j_\parallel(s) = B_0(s)/B_0^I \int_0^s \sigma(s,s')E_\parallel(s')ds'$, where $s = \cos(\theta)$, B_0^I is the magnetic field strength at the ionosphere, and $\sigma(s,s')$ is a nonlocal conductivity that defines the contribution of the parallel electric field at position s' to the current $j_\parallel(s)$ at position s on the geomagnetic field line. The nonlocal conductivity $\sigma(s,s')$ and $E_\parallel$ are defined by an integral equation which can be solved if $j_\parallel$ is specified (Samson et al. 2003). This can be justified because the inhomogeneity of the shear Alfvén wave azimuthal magnetic field in the meridional direction acts as a current generator. Electrons distributed on geomagnetic field lines serve as an inductive load for this current with a conductivity kernel relating $j_\parallel$ to $E_\parallel$.

5.2.1 Nonlocal Kinetic Theory: Homogeneous Bounded Plasma

The simplest case of a bounded homogeneous plasma with constant density, temperature, and magnetic field can be evaluated analytically by prescribing the parallel current $j_\parallel(l) = j_{max}\sin(\pi l/2l_{max})$, where $2l_{max}$ is the length of the system. This mathematics is straightforward in the warm plasma limit for which $V_A \ll V_{Te}$ (the warm plasma limit of two-fluid theory). Then, the leading-order expression for the kinetic conductivity is identical to the two-fluid expression obtained in the electron thermal pressure limit. The parallel potential $\Phi_\parallel = -\int_0^{l_{max}} E_\parallel dl$ at constant density, ambient magnetic field strength, and temperature, gives the maximum current density in terms of the parallel electric field,

$$j_{max} \sim -i\frac{e^2 n_e}{m_e\omega}\left(\frac{\omega l_{max}}{V_{Te}}\right)^2 E_{max} \tag{34}$$

where $E_\parallel = E_{max}\sin(\pi l/2l_{max})$ was used. Equation (34) illustrates that in a homogeneous plasma with $V_A \ll V_{Te}$, the ac conductivity is proportional to the square of the ratio of the transit time of a thermal electron to the wave period (Rankin et al. 1999a). Consequently, the parallel electric field E_{max} is very small for long-period FLRs unless $\omega l_{max} \ll V_{Te}$, which is not the case for FLR frequencies on the order of a few mHz. On the other hand, if the parallel electric field is confined to a region $\Delta l \ll l_{max}$, Eq. (34) predicts that a large reduction in the ac conductivity results. This situation is not realized in homogeneous plasma, but Eq. (34) should be approximately satisfied if l_{max} is substituted by the length of field line over which $V_A < V_{Te}$. In the bottom panel of Fig. 6, which is based on two-fluid theory, $l_{max} \sim 2R_E$.

5.2.2 Nonlocal Kinetic Theory: Inhomogeneous Bounded Plasma

The uniform case, Eq. (34), excludes the mirror force and the rapid increase in the density of plasma near the ionosphere. On dipole-like magnetic field lines the spatial structure of $E_\parallel$ becomes critically dependent on the profile of the FAC near the foot of the field line where $j_\parallel$ is close to its maximum, cf. Eq. (34). $E_\parallel$ and the net potential drop are then confined to the region of field line where the static potential Φ_0 (see Eq. (35)) begins to increase rapidly, causing a large reduction in the field line conductivity. There is (so far) no general current-voltage expression for FLRs equivalent to the Knight relation (Knight 1973) because of the dependence of the ac conductivity on details of the density and temperature profile on geomagnetic field lines.

The reduction in conductivity predicted by the kinetic theory of FLRs detailed in Tikhonchuk and Rankin (2000, 2002) who exploit that ambient plasma is cold and dense near the ionosphere and warm and tenuous at altitudes above $\sim 4\text{–}5R_E$. They find that the auroral plasma modified by the FLR consists of a bulk cold electron component and a population that has interacted with the wave nonlocally on timescales comparable to the wave period. Density and temperature profiles for the admixture of H^+ and O^+ they consider are obtained by integrating Eq. (33) for f_0 over velocity:

$$n_e = n_{e0}\exp\left(\frac{e\Phi_0}{T_{e0}}\right) + n_{ec}\exp\left(\frac{e\Phi_0}{T_{ec}}\right) \tag{35}$$

$$T_e = (T_{e0} - T_{ec})\frac{n_{eo}}{n_e}\exp\left(\frac{e\Phi_0}{T_{e0}}\right) + T_{ec}\frac{n_{ec}}{n_e}\exp\left(\frac{e\Phi_0}{T_{ec}}\right) \tag{36}$$

Equations (35)–(36) can be solved for Φ_0 by assuming $n_{ec} \ll n_{e0}$, making use of quasi-neutrality $n_e = n_i$, and setting $n_i = n_H + n_O\exp[-\frac{(l_{max}-|l|)}{h_O}]$. Here, n_H and n_O are constants related to densities of H^+ and O^+, and h_O equals the oxygen scale height of the ionosphere. Equations (35)–(36) define an effective electron temperature T_e that varies smoothly from a value T_{e0} near the equator to T_{ec} at the ionosphere provided the O^+ density is very small at the equator where $n_{e0} \sim n_H$ and $\Phi_0(0) = 0$. The solution to Eq. (32), with f_0 given by Eq. (33), gives a nonlocal relationship between the FAC and parallel electric field of the FLR,

$$j_\parallel(l_{max}) = -i\frac{e^2 n_{eq}\omega}{T_e}\int_0^{l_{max}} E_\parallel(l)c(l)dl \tag{37}$$

$$c(l) = \int_0^l dl'\frac{B_0^{max}}{B_0}\left[1 - \sqrt{1 - \frac{B_0}{B_0^{max}}\exp\left(-\frac{eB_0}{T_e}\frac{\Phi_0^{max} - \Phi_0}{B_0^{max} - B_0}\right)}\right]\exp\left(\frac{e\Phi_0}{T_e}\right) \tag{38}$$

The function $c(l)$ takes account of the variation of the plasma density and ambient magnetic field. It depends on B_0^{max} and Φ_0^{max}, the values of the geomagnetic field and quasi-static potential at the foot of the field line, respectively. Equation (38) is valid for any monotonic density profile under the condition $\frac{\Phi_0'}{B_0'} > (\Phi_0^{max} - \Phi_0)/(B_0^{max} - B_0)$.

It is interesting that the integrand in Eq. (37), which depends on the un-trapped electron population, has the same form as the Knight relation (Knight 1973). Damiano and Johnson (2013) also found a degree of consistency with the Knight relation owing to the low frequency nature of the long period standing waves. Significant differences from the Knight relation are that $j_\parallel$ depends on $E_\parallel$ through Eq. (37), which involves spatial integration accounting for all trajectories of an electron (of energy w) that enable it to move from position l' to l along a field line. Electrons at l' can only reach position l provided $\mu \le (w + q\Phi_0(l))/B_0(l)$. When the maximum possible μ is realized for given w, the conductivity is dominated by thermal electrons that have the largest bounce time $\omega_b \sim V_{Te}/l_{max}$. This is the scenario defined by Eq. (34) for a homogeneous plasma, which is valid for electron orbits satisfying $v_\parallel \lesssim \omega l_{max}$.

The key aspect of low frequency standing waves in a dipole-like magnetic field is that the number of electrons satisfying $v_\parallel \lesssim \omega l_{max}$ is significantly diminished by the mirror force. This arises because $\omega \ll v_{Te}/l_{max}$ and thus the current at the ionosphere where $j_\parallel \sim j_{max}$ must be carried by the remaining relatively small number of electrons that move nearly parallel to the magnetic field. The large deficiency in the number of such current carriers carrying the FAC dictated by Ampere's Law means that the conductivity is much smaller than in Eq. (34). The field line conductivity is reduced and l_{max} in Eq. (34) is effectively substituted by a length scale $\Delta l \ll l_{max}$. The scale Δl of the wave potential turns out to

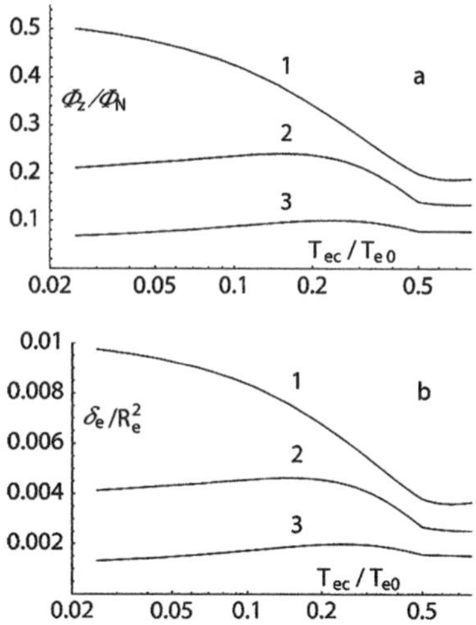

Fig. 13 Dependence of (**a**) the wave parallel potential Φ_z and (**b**) the wave dispersion coefficient δ_e on the ratio of the cold (ionosphere) to hot (magnetosphere electron temperatures T_{ec} and T_{e0}, respectively). Curves 1, 2, and 3 correspond to oxygen density scale heights $h_O = 0.15R_E$, $0.25R_E$, and $0.4R_E$, respectively. Profiles of the density and temperature are defined by Eqs. (35)–(36). Other parameters are defined in the text

be on the order of $1R_E$ for the range of density scale heights shown in Fig. 13, which is a parametrically determined solution for the wave potential at the ionosphere. Here, the scale height of the ionosphere $h_O \sim T_{ec}/m_O g$, where m_O is the mass of O^+ (see Fig. 13). The altitude range over which the wave potential is significant is generally less than about $1R_E$, and is centered on the position $l_\parallel$ measured from the equator where $E_\parallel = E_{max}$,

$$l_\parallel \sim l_{max} - h_O \ln \frac{n_O}{n_H} \tag{39}$$

From Eq. (39), the location of the parallel electric field maximum is quite sensitive to the density near the ionosphere and the scale height h_O. The wave parallel potential is found to be less sensitive to these parameters and is therefore a more useful estimate of the expected energy gain of electrons. It is estimated using $\Phi_\parallel = -E_{max}\Delta l$, where Δl is the extent of the potential drop, and

$$j_{max} \sim -iC\left(\frac{e^2 n_H \omega l_{max}}{T_{e0}}\right)E_{max} \tag{40}$$

$$C \sim 1 + 3\frac{h_O}{R_E}\frac{\Phi_0^{max}}{T_{e0}} + 1.6\left(\frac{h_O}{R_E}\right)^{\frac{5}{2}}\frac{\Phi_0^{max}}{T_{ec}}\ln\frac{T_{e0}}{T_{ec}} \tag{41}$$

where C is a coefficient related to the maximum of the static potential, the hot and cold electron temperatures, and the scale height of the ionosphere, h_O. Equations (40)–(41) are valid provided $0.025 < \frac{T_{ec}}{T_{e0}} < 1$ and $0.1 < \frac{h_O}{R_E} < 0.5$. The density scale height h_O is treated as a free parameter in the analyses and depends on the cold electron temperature. Note that in Eq. (41), $\Phi_0^{max} \sim T_{ec}\ln(n_O/n_{ec})$ and $n_{ec} \sim n_O \exp(-l_{max}/l)$.

In Fig. 13, the wave parallel potential $\Phi_\parallel$ and the kinetic expression for the wave dispersion coefficient δ_e (cf., Eq. (22)) are illustrated as a function of the ratio of the cold (ionosphere) to hot (magnetosphere) electron temperatures T_{ec} and T_{e0}, respectively. Curves 1, 2, and 3 correspond to oxygen density scale heights $h_O = 0.15R_E$, $0.25R_E$, and $0.4R_E$, respectively. The wave potential is normalized by the potential $\Phi_N = j_{max}T_{e0}/e^2 n_{e0}\omega l_{max}$, where n_{e0} is the density at the equator, ω is the wave frequency, and l_{max} is the half-length of the field line. Taking $j_{max} \sim 1\ \mu\mathrm{Am}^{-2}$ and a hot (magnetosphere) electron temperature

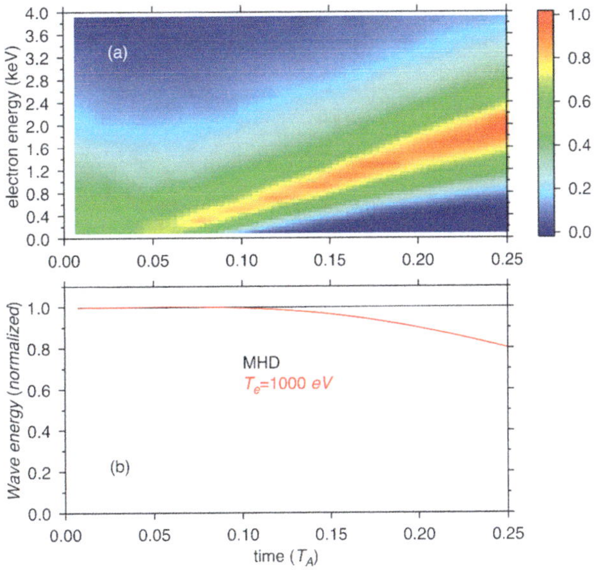

Fig. 14 Energy of the accelerated electron population in the region of the FLR upward parallel current maximum as a function of time. (**b**) Comparison of wave energy in MHD and hybrid fluid-kinetic systems for the same FLR wave perturbation as a function of time

of a few hundred eV the wave potential drop can by several hundred eV and can reach 1–2 keV at higher current densities. While this approach provides estimates of the parallel potential drop, at high current densities the electron dynamics become nonlinear, and kinetic simulations are needed to provide a fully self-consistent electron description; see Sect. 6.

6 Kinetic Simulations of Field Line Resonances

We conclude our consideration of kinetic processes by reviewing kinetic simulations of field line resonances. Simulations of electron energization in field line resonances have been conducted using a 2 D hybrid MHD-kinetic electron model in dipolar coordinates (Damiano et al. 2003, 2007) and its gyrofluid extension (Damiano et al. 2015). These models have been used to investigate Alfvén wave electron interactions in both large perpendicular scale field line resonances (see Damiano et al. 2007; Damiano and Johnson 2012; Damiano and Wright 2008) in the auroral zone and kinetic scale field line resonances in the inner magnetosphere (Damiano et al. 2018). Kinetic scale FLRs associated with substorm onset, although not necessarily long period resonances, are characterized by parallel elongated electron distributions, associated with electron trapping in these waves (e.g. Chaston et al. 2015; Hull et al. 2016; Damiano et al. 2018). It has been reported that these kinetic scale FLRs can also lead to substantial stochastic ion heating (Chaston et al. 2016) and ionospheric oxygen outflow facilitated by the soft electron precipitation (Chaston et al. 2015).

Figure 14 reproduces Fig. 4 from Damiano and Johnson (2012) and shows results of a 2D kinetic simulation of a large perpendicular scale fundamental mode FLR centered at $L = 10$. There is electron energization to the keV energies generally associated with the generation of auroral emissions. The top panel shows electron energy at the position of peak parallel current at the ionospheric boundary of the simulation as a function of time. The parallel current is carried by electrons that aided by the parallel electric field, overcome mirror force resistance along the field line. The electron energy peaks at around 2 keV when the parallel current reaches its maximum magnitude. The bottom panel of Fig. 14 compares

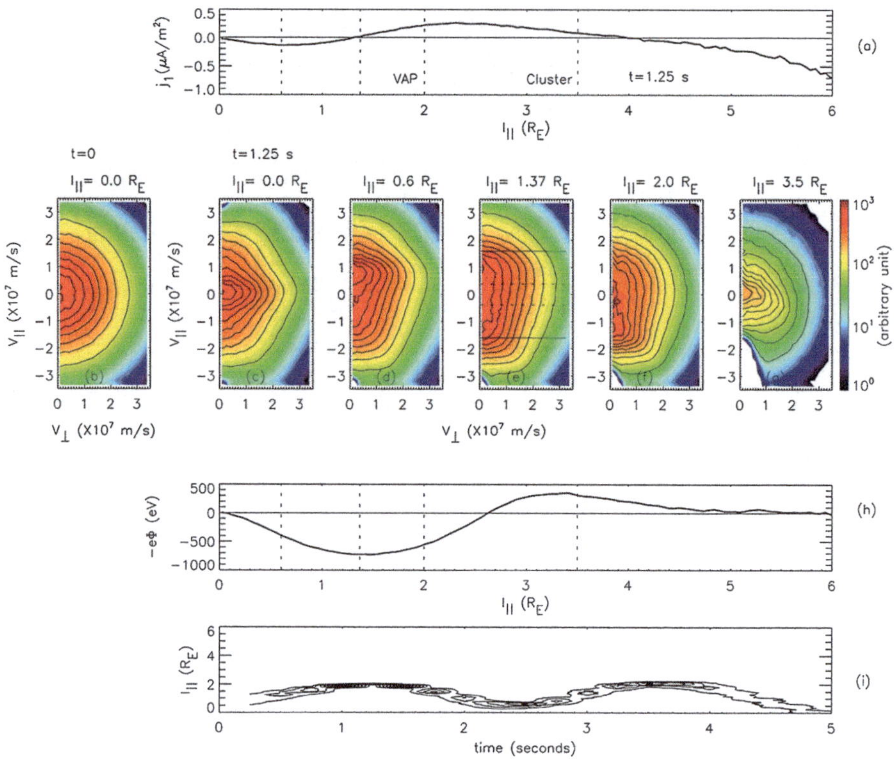

Fig. 15 Top panel: Parallel current density as a function of distance along the field line ($l_{\|}$ is measured from the equator) at $t = 1.25$ s (1/4 period). (**b**) Maxwellian distribution at $t = 0$. (**c**)–(**g**) Electron distribution functions at $t = 1.25$ s and indicated values of $l_{\|}$ (vertical dashed lines in panels **a** and **h**). Parallel dashed lines panel **e** illustrates phase velocity ($\pm v_{ph}$) while parallel solid lines indicate the theoretical trapping width ($\pm v_{tr}$). (**h**) Effective potential $\Phi = -\int E_{\|} dl_{\|}$ at $t = 1.25$ s. Note that $v_{\perp}$ is the gyro-averaged perpendicular velocity and $v_{\|} = v_1$. (i) Evolution of an ensemble of trapped electrons from the core of the distribution evident in panel **f**. The structure is associated with the interval at which particle data are saved in the simulation (Damiano et al. 2018)

the Alfvén wave energy in the MHD limit and for the hybrid MHD-kinetic system with a background Maxwellian electron distribution temperature of 1 keV. The decrease in wave energy for the hybrid system (reaching 20% during the first 1/4 Alfvén period) is the result of the electron energization noted in the top panel. This illustrates that an undriven large-scale FLR would be effectively damped by electron energization alone over a few Alfvén cycles, which is a similar level of wave energy dissipation associated with Joule heating by ionospheric Pedersen currents.

Figure 15 reproduces Fig. 4 in Damiano et al. (2018) and shows results from a simulation of a kinetic-scale FLR centered on an $L = 6$ field line. The top panel shows the field-aligned profile of the wave parallel current for an $m = 5$ standing wave after the system has evolved for 1/4 of an Alfvén wave period. Panels (c) to (g) illustrate corresponding distribution functions at indicated positions along the field line. The elongated beam-like structures in the cores of the distributions illustrated in panels (d)–(f) are due to electron trapping (Wygant et al. 2002; Watt and Rankin 2009b; Chaston et al. 2015; Damiano et al. 2016) in the wave potential well (panel h). Panel e, corresponding to a parallel current node, exhibits a roughly symmetric plateau around $v_{\|} = 0$.

The plateaus of the distributions in panels (d) and (f) have preferential maxima (the core of the distribution is angled) opposite in sign to that of the parallel current (panel a) illustrating that the skew in the plateau is associated with carrying the parallel current. Further up the field line from the equatorial region, large loss cones at 180° (panels f + g) are associated with significant soft electron precipitation driven by the wave parallel electric field. This precipitation is important for magnetosphere-ionosphere coupling as it can drive significant ionospheric oxygen outflow (Chaston et al. 2015). These modeled features of the electron distribution functions in KFLRs are in good qualitative agreement with satellite observations (Chaston et al. 2014; Hull et al. 2016; Damiano et al. 2018). The bottom panel shows the evolution of an ensemble of trapped electrons from the distribution illustrated in panel (f). These electrons remain closely confined to the region of the parallel current node over the duration of the simulation. They eventually would become de-trapped when the wave effective potential changes sign.

7 Conclusions

Naturally occurring ultralow frequency (ULF) waves in the magnetosphere have been studied for about five decades. Much understanding of their properties has been achieved through theory and observations, but until now there has been no rigorous theory about whether they can directly produce aurorae. Observationally, the results of Gillies et al. (2018) confirm the existence of a class of modulated discrete arcs that follow a pattern of spatiotemporal evolution consistent with field line resonances. The nonlocal kinetic theory of FLRs discussed in this review suggests that precipitation under periodic redline arcs FLRs is produced directly by the field aligned currents and parallel electric fields carried by these waves. It has also been demonstrated that dispersive shear Alfvén waves can produce accelerated electron distributions at both high and low latitude. At high latitudes in the inertial regime of wave dispersion the mirror force and density cavities can enhance $E_{\|}$ sufficiently to produce auroral particle acceleration that matches observed particle and wave signatures from spacecraft. In long period FLRs the mirror force and quasi-static potential exclude all but electrons with very small magnetic moments, depleting the current-carrying population sufficiently that large (mV/m) parallel electric fields are generated.

The historical ideal MHD approach applied to long period FLRs has stood the test of time at explaining characteristic features of FLRs routinely observed by ground-based magnetometer networks and all-sky cameras. Two-fluid theories of shear Alfvén waves have improved our understanding of the different regimes of parallel electric field generation in warm and cold plasma. For high frequency (Hz) waves in warm plasma, two-fluid theories predict parallel electric fields that are large enough to accelerate electrons to observed energies. Paradoxically, the two-fluid theory has the weakness that in this situation the waves are very strongly damped. Kinetic theory solves this problem by demonstrating that particle trapping reduces Landau damping rates in warm plasma and enables the trapped population to be carried along the magnetic field, gaining energy in the process. These trapped electrons carried by the wave form a beam at lower altitude in the inertial regime. For propagating high frequency Alfvén waves in cold plasma typical of the ionosphere, the mirror force is intimately involved in the formation of parallel electric fields, which are easily strong enough to accelerate electrons from hundreds of eV to over 1 keV without appreciable Landau damping.

Linear kinetic theory of FLRs contradicts the results of two-fluid theories in predicting that most of the potential drop in a low frequency standing wave occurs close to the ionosphere at a location where the plasma density starts to increase rapidly (cf. Eq. (39)). The width of the parallel potential is proportional to the scale height $h_O \sim T_c/m_O g$, with a current-voltage relation defined by Eq. (40), which assumes $\Phi_\parallel = -E_{max}\Delta l$ and $\Delta l \gtrsim h_O$. The maximum of the parallel electric field E_{max} is related to the field-aligned current at the ionosphere, which can be on the order of several to tens of μAm^{-2}. The depletion of current-carrying electrons at low altitude that occurs through the action of the mirror force has been confirmed using gyrofluid-kinetic simulations (Damiano and Johnson 2013; Damiano and Wright 2008). Particle trapping in field line resonances observed by Cluster have been confirmed in realistic simulations by Damiano et al. (2018) and in straight magnetic field line geometry (Rankin et al. 2007).

The various space and timescales controlling amplitudes, widths, and the temporal evolution of FLRs summarized by Eqs. (8) & (12), (26)–(27), and (29)–(31) apply in situations where dissipation, wave dispersion, or wave dispersion and nonlinearity dominate the evolution of FLRs. A corollary to this statement is that when the dissipation is time-dependent, for example due to conductivity modification by fluxes of precipitating electrons, FACs and parallel electric fields are enhanced through nonlinear feedback with the ionosphere. This feedback mechanism is not an instability and is a distinct mechanism from the ionospheric feedback instability. As more FLR redline arcs are identified the auroral scientific community is coming closer to being able to accurately model how this arc type forms, and the magnetospheric and ionospheric conditions that are needed.

Acknowledgements The authors wish to thank the International Space Science Institute (ISSI), Bern, for organization of the reviews and for support of the team meeting. R. Rankin and D.M. Gillies acknowledge financial support from NSERC and the Canadian Space Agency. A.W. Degeling is supported by the National Natural Science Foundation of China (41774172).

References

W. Allan, The ponderomotive force of standing Alfvén waves in a dipolar magnetosphere. J. Geophys. Res. **98**, 1409–1417 (1993)

W. Allan, J.R. Manuel, E.M. Poultier, Does the ponderomotive force move mass in the magnetosphere? Geophys. Res. Lett. **17**, 917–920 (1990)

L.J. Baddeley, D.A. Lorentzen, N. Partamies, M. Denig, V.A. Pilipenko, K. Oksavik et al., Equatorward propagating auroral arcs driven by ULF wave activity: multipoint ground- and space-based observations in the dusk sector auroral oval. J. Geophys. Res. Space Phys. **122**(5), 5591–5605 (2017). https://doi.org/10.1002/2016JA023427

G.J. Baker, E.F. Donovan, B.J. Jackel, A comprehensive survey of auroral latitude Pc5 pulsation characteristics. J. Geophys. Res. Space Phys. **108**(A10), 1384 (2003). https://doi.org/10.1029/2002JA009801

M.J. Berger, S.M. Seltzer, K. Maeda, Energy deposition by auroral electrons in the atmosphere. J. Atmos. Terr. Phys. **32**(6), 1015–1045 (1970). https://doi.org/10.1016/0021-9169(70)90115-7

A. Bhattacharjee, C.A. Kletzing, Z.W. Ma, C.S. Ng, N.F. Otani, X. Wang, Four-field model for dispersive field-line resonances: effects of coupling between shear-Alfvén and slow modes. Geophys. Res. Lett. **26**(21), 3281–3284 (1999). https://doi.org/10.1029/1999GL003628

M.H. Boehm, C.W. Carlson, J.P. McFadden, J.H. Clemmons, F.S. Mozer, High-resolution sound rocket observations of large amplitude Alfvén waves. J. Geophys. Res. **95**, 12,157–12,171 (1990)

C.C. Chaston, J.W. Bonnell, C.W. Carlson, J.P. McFadden, R.J. Strangeway, R.E. Ergun, Kinetic effects in the acceleration of auroral electrons in small scale Alfvén waves: a FAST case study. Geophys. Res. Lett. **30**(6), 1289 (2003). https://doi.org/10.1029/2002GL015777

C.C. Chaston, L.M. Peticolas, C.W. Carlson, J.P. McFadden, F. Mozer, M. Wilber et al., Energy deposition by Alfvén waves into the dayside auroral oval: cluster and FAST observations. J. Geophys. Res. Space Phys. **110**(A2), A02211 (2005). https://doi.org/10.1029/2004JA010483

C.C. Chaston, J.W. Bonnell, J.R. Wygant, F. Mozer, S.D. Bale, K. Kersten et al., Observations of kinetic scale field line resonances. Geophys. Res. Lett. **41**, 209–215 (2014). https://doi.org/10.1002/2013GL058507

C.C. Chaston, J.W. Bonnell, J.R. Wygant, C.A. Kletzing, G.D. Reeves, A. Gerrard et al., Extreme ionospheric ion energization and electron heating in Alfvén waves in the storm time inner magnetosphere. Geophys. Res. Lett. **42**, 10,531–10,540 (2015). https://doi.org/10.1002/2015GL066674

C.C. Chaston, J.W. Bonnell, G.D. Reeves, R.M. Skoug, Driving ionospheric outflows and magnetospheric O^+ energy density with Alfvén waves. Geophys. Res. Lett. **43**, 4825–4833 (2016). https://doi.org/10.1002/2016GL069008

L. Chen, A. Hasegawa, Theory of long-period magnetic pulsations. 1. Steady-state excitation of field line resonance. J. Geophys. Res. **79**(7), 1024–1032 (1974). https://doi.org/10.1029/JA079i007p01024

C.Z. Cheng, S. Zaharia, Field line resonances in quiet and disturbed time three-dimensional magnetospheres. J. Geophys. Res. **108**(A1), 1001 (2003). https://doi.org/10.1029/2002JA009471

I.J. Cohen, M.R. Lessard, S.R. Kaeppler, S.R. Bounds, C.A. Kletzing, A.V. Streltsov et al., Auroral Current and Electrodynamics Structure (ACES) observations of ionospheric feedback in the Alfvén resonator and model responses. J. Geophys. Res. Space Phys. **118**(6), 3288–3296 (2013). https://doi.org/10.1002/jgra.50348

W. Cummings, R.J. O'Sullivan, P. Coleman, Standing Alfvén waves in magnetosphere. J. Geophys. Res. **74**(3), 778 (1969). https://doi.org/10.1029/JA074i003p00778

P.A. Damiano, J.R. Johnson, Electron acceleration in a geomagnetic field line resonance. Geophys. Res. Lett. **39**, L02102 (2012). https://doi.org/10.1029/2011GL050264

P.A. Damiano, J.R. Johnson, Mirror force induced wave dispersion in Alfvén waves. Phys. Plasmas **20**(6), 062901 (2013). https://doi.org/10.1063/1.4810788

P.A. Damiano, A.N. Wright, Electron thermal effects in standing shear Alfvén waves. J. Geophys. Res. Space Phys. **113**(A9), A09219 (2008). https://doi.org/10.1029/2008JA013087

P.A. Damiano, R.D. Sydora, J.C. Samson, Hybrid magnetohydrodynamic-kinetic model of standing shear Alfvén waves. J. Plasma Phys. **69**, 277–304 (2003). https://doi.org/10.1017/S0022377803002216

P.A. Damiano, A.N. Wright, R.D. Sydora, J.C. Samson, Energy dissipation via electron energization in standing shear Alfvén waves. Phys. Plasmas **14**, 062904 (2007). https://doi.org/10.1063/1.2744226

P.A. Damiano, J.R. Johnson, C.C. Chaston, Ion temperature effects on magnetotail Alfvén wave propagation and electron energization. J. Geophys. Res. Space Phys. **120**(7), 5623–5632 (2015). https://doi.org/10.1002/2015JA021074

P.A. Damiano, J.R. Johnson, C.C. Chaston, Ion gyroradius effects on particle trapping in kinetic Alfvén waves along auroral field lines. J. Geophys. Res. Space Phys. **121**, 10,831–10,844 (2016). https://doi.org/10.1002/2016JA022566

P.A. Damiano, C.C. Chaston, A.J. Hull, J.R. Johnson, Electron distributions in kinetic scale field line resonances: a comparison of simulations and observations. Geophys. Res. Lett. **45**, 5826–5835 (2018). https://doi.org/10.1029/2018GL077748

P.A. Damiano, E.-H. Kim, J.R. Johnson, P. Porazik, Electron energization by parallel electric fields in poloidal standing waves. J. Geophys. Res. **124**, 6691–6700 (2019). https://doi.org/10.1029/2019JA026849

A.W. Degeling, R. Rankin, K. Kabin, I.J. Rae, F.R. Fenrich, Modeling ULF waves in a compressed dipole magnetic field. J. Geophys. Res. Space Phys. **115**, A10212 (2010). https://doi.org/10.1029/2010JA015410

A.W. Degeling, I.J. Rae, C.E.J. Watt, Q.Q. Shi, R. Rankin, Q.-G. Zong, Control of ULF wave accessibility to the inner magnetosphere by the convection of plasma density. J. Geophys. Res. Space Phys. **123**(2), 1086–1099 (2018). https://doi.org/10.1002/2017JA024874

J. Dombeck, C. Cattell, J.R. Wygant, A. Keiling, J. Scudder, Alfvén waves and Poynting flux observed simultaneously by Polar and FAST in the plasma sheet boundary layer. J. Geophys. Res. Space Phys. **110**(A12), A12S90 (2005). https://doi.org/10.1029/2005JA011269

P. Frycz, R. Rankin, J.C. Samson, V.T. Tikhonchuk, Nonlinear field line resonances: dispersive effects. Phys. Plasmas **5**(10), 3565–3574 (1998). https://doi.org/10.1063/1.873075

D.M. Gillies, D.J. Knudsen, E.F. Donovan, E.L. Spanswick, C. Hansen, D. Keating, S. Erion, A survey of quiet auroral arc orientation and the effects of the interplanetary magnetic field. J. Geophys. Res. Space Phys. **119**(4), 2550–2562 (2014). https://doi.org/10.1002/2013JA019469

D.M. Gillies, D. Knudsen, R. Rankin, S. Milan, E. Donovan, A statistical survey of the 630.0-nm optical signature of periodic auroral arcs resulting from magnetospheric field line resonances. Geophys. Res. Lett. **45**(10), 4648–4655 (2018). https://doi.org/10.1029/2018GL077491

A. Hasegawa, Particle acceleration by MHD surface-wave and formation of aurora. J. Geophys. Res. Space Phys. **81**(28), 5083–5090 (1976). https://doi.org/10.1029/JA081i028p05083

R.D. Hazeltine, F.L. Waelbroeck, *The Framework Of Plasma Physics* (1998). https://doi.org/10.1201/9780429502804. Avalon Publishing

J. Heyvaerts, E.R. Priest, Coronal heating by phase mixed shear Alfvén waves. Astron. Astrophys. **117**, 220 (1983)

W. Hughes, D. Southwood, Screening of micropulsation signals by atmosphere and ionosphere. J. Geophys. Res. Space Phys. **81**(19), 3234–3240 (1976). https://doi.org/10.1029/JA081i019p03234

A.J. Hull, C.C. Chaston, H.U. Frey, M.O. Fillingim, M.L. Goldstein, J.W. Bonnell, F.S. Mozer, The "Alfvénic surge" at substorm onset/expansion and the formation of "Inverted Vs": Cluster and IMAGE observations. J. Geophys. Res. Space Phys. **121**, 3978–4004 (2016). https://doi.org/10.1002/2015JA022000

P. Janhunen, A. Olsson, N.A. Tsyganenko, C.T. Russell, H. Laakso, L.G. Blomberg, Statistics of a parallel Poynting vector in the auroral zone as a function of altitude using Polar EFI and MFE data and Astrid-2 EMMA data. Ann. Geophys. **23**(5), 1797–1806 (2005). https://hal.archives-ouvertes.fr/hal-00317830

S.T. Jones, Y.-J. Su, Role of dispersive Alfvén waves in generating parallel electric fields along the Io-Jupiter fluxtube. J. Geophys. Res. Space Phys. **113**(A12), A12205 (2008). https://doi.org/10.1029/2008JA013512

K. Kabin, R. Rankin, C.L. Waters, R. Marchand, E.F. Donovan, J.C. Samson, Different eigenproblem models for field line resonances in cold plasma: effect on magnetospheric density estimates. Planet. Space Sci. **55**(6), 820–828 (2007). https://doi.org/10.1016/j.pss.2006.03.014

T. Karlsson, L. Andersson, D.M. Gillies, K. Lynch, O. Marghitu, N. Partamies, N. Sivadas, J. Wu, Space Sci. Rev. **216**(16) (2020). https://doi.org/10.1007/s11214-020-0641-7

L. Kepko, H.E. Spence, Observations of discrete, global magnetospheric oscillations directly driven by solar wind density variations. J. Geophys. Res. **108**(A6), 1257 (2003). https://doi.org/10.1029/2002JA009676

R. Kessel, I. Mann, S. Fung, D. Milling, N. O'Connell, Correlation of Pc5 wave power inside and outside the magnetosphere during high speed streams. Ann. Geophys. **22**(2), 629–641 (2004)

K.-H. Kim, C.A. Cattell, D.-H. Lee, K. Takahashi, K. Yumoto, K. Shiokawa, M. Andre, Magnetospheric responses to sudden and quasiperiodic solar wind variations. J. Geophys. Res. **107**(A11), 1406 (2002). https://doi.org/10.1029/2002JA009342

M. Kivelson, D. Southwood, Resonant Ulf waves – a new interpretation. Geophys. Res. Lett. **12**(1), 49–52 (1985). https://doi.org/10.1029/GL012i001p00049

C.A. Kletzing, S. Hu, Alfvén wave generated electron time dispersion. Geophys. Res. Lett. **28**(4), 693–696 (2001). https://doi.org/10.1029/2000GL012179

C.A. Kletzing, S.R. Bounds, J. Martin-Hiner, W. Gekelman, C. Mitchell, Measurements of the shear Alfvén wave dispersion for finite perpendicular wave number. Phys. Rev. Lett. **90**(3), 035004 (2003). https://doi.org/10.1103/PhysRevLett.90.035004

S. Knight, Parallel electric-fields. Planet. Space Sci. **21**(5), 741–750 (1973). https://doi.org/10.1016/0032-0633(73)90093-7

H. Korth, M.F. Thomsen, J.E. Borovsky, D.J. McComas, Plasma sheet access to geosynchronous orbit. J. Geophys. Res. Space Phys. **104**(A11), 25047–25061 (1999). https://doi.org/10.1029/1999JA900292

M.J. Kosch, F. Honary, C.F. del Pozo, S.R. Marple, T. Hagfors, High-resolution maps of the characteristic energy of precipitating auroral particles. J. Geophys. Res. Space Phys. **106**(A12), 28925–28937 (2001). https://doi.org/10.1029/2001JA900107

W. Lotko, A.V. Streltsov, C.W. Carlson, Discrete auroral arc, electrostatic shock and suprathermal electrons powered by dispersive, anomalously resistive field line resonance. Geophys. Res. Lett. **25**(24), 4449–4452 (1998). https://doi.org/10.1029/1998GL900200

J.Y. Lu, R. Rankin, R. Marchand, V.T. Tikhonchuk, J. Wanliss, Finite element modeling of nonlinear dispersive field line resonances: trapped shear Alfvén waves inside field-aligned density structures. J. Geophys. Res. Space Phys. **108**(A11), 1394 (2003a). https://doi.org/10.1029/2003JA010035

J.Y. Lu, R. Rankin, R. Marchand, V.T. Tikhonchuk, Nonlinear acceleration of dispersive effects in field line resonances. Geophys. Res. Lett. **30**(10), 1540 (2003b). https://doi.org/10.1029/2003GL016929

R. Lysak, Feedback instability of the ionospheric resonant cavity. J. Geophys. Res. Space Phys. **96**(A2), 1553–1568 (1991). https://doi.org/10.1029/90JA02154

R.L. Lysak, W. Lotko, On the kinetic dispersion relation for shear Alfvén waves. J. Geophys. Res. Space Phys. **101**(A3), 5085–5094 (1996). https://doi.org/10.1029/95JA03712

I.R. Mann, A.N. Wright, P.S. Cally, Coupling of magnetospheric cavity modes to field line resonances: a study of resonance widths. J. Geophys. Res. **100**, 19,441–19,456 (1995)

S.B. Mende, S.E. Harris, H.U. Frey, V. Angelopoulos, C.T. Russell, E. Donovan et al., The THEMIS array of ground-based observatories for the study of auroral substorms. Space Sci. Rev. **141**(1–4), 357–387 (2008). https://doi.org/10.1007/s11214-008-9380-x

S.E. Milan, N. Sato, M. Ejiri, J. Moen, Auroral forms and the field-aligned current structure associated with field line resonances. J. Geophys. Res. Space Phys. **106**(A11), 25825–25833 (2001). https://doi.org/10.1029/2001JA900077

K. Mills, A. Wright, I. Mann, Kelvin-Helmholtz driven modes of the magnetosphere. Phys. Plasmas **6**(10), 4070–4087 (1999)

T. Motoba, T. Kikuchi, H. Luhr, H. Tachihara, T.I. Kitamura, K. Hayashi, T. Okuzawa, Global Pc5 caused by a DP 2-type ionospheric current system. J. Geophys. Res. Space Phys. **107**(A2), 1032 (2002). https://doi.org/10.1029/2001JA900156

T.K. Nakamura, Parallel electric field of a mirror kinetic Alfvén wave. J. Geophys. Res. Space Phys. **105**(A5), 10729–10737 (2000). https://doi.org/10.1029/1999JA900494

P.T. Newell, K.M. Lyons, C.-I. Meng, A large survey of electron acceleration events. J. Geophys. Res., Atmos. **101**, 2599–2614 (1996). https://doi.org/10.1029/95ja03147

Y. Obana, F.W. Menk, M.D. Sciffer, C.L. Waters, Quarter-wave modes of standing Alfvén waves detected by cross-phase analysis. J. Geophys. Res. Space Phys. **113**(A8), A08203 (2008). https://doi.org/10.1029/2007JA012917

D. Pokhotelov, W. Lotko, A.V. Streltsov, Harmonic structure of field line eigenmodes generated by ionospheric feedback instability. J. Geophys. Res. Space Phys. **107**(A11), 1363 (2002). https://doi.org/10.1029/2001JA000134

M. Prakash, R. Rankin, V.T. Tikhonchuk, Precipitation and nonlinear effects in geomagnetic field line resonances. J. Geophys. Res. Space Phys. **108**(A4), 8014 (2003). https://doi.org/10.1029/2002JA009383

I.J. Rae, E.F. Donovan, I.R. Mann, F.R. Fenrich, C.E.J. Watt, D.K. Milling, A. Balogh, Evolution and characteristics of global Pc5 ULF waves during a high solar wind speed interval. J. Geophys. Res. **110**, A12211 (2005). https://doi.org/10.1029/2005JA011007

I.J. Rae, I.R. Mann, Z.C. Dent, D.K. Milling, E.F. Donovan, E. Spanswick, Multiple field line resonances: optical, magnetic and absorption signatures. Planet. Space Sci. **55**(6), 701–713 (2007). https://doi.org/10.1016/j.pss.2006.02.009

R. Rankin, J. Samson, P. Frycz, Simulations of driven field line resonances in the Earths magnetosphere. J. Geophys. Res. Space Phys. **98**(A12), 21341–21352 (1993). https://doi.org/10.1029/93JA02083

R. Rankin, P. Frycz, V. Tikhonchuk, J. Samson, Nonlinear standing shear Alfvén waves in the Earths magnetosphere. J. Geophys. Res. Space Phys. **99**(A11), 21291–21301 (1994). https://doi.org/10.1029/94JA01629

R. Rankin, J.C. Samson, V.T. Tikhonchuk, I. Voronkov, Auroral density fluctuations on dispersive field line resonances. J. Geophys. Res. Space Phys. **104**(A3), 4399–4410 (1999a). https://doi.org/10.1029/1998JA900106

R. Rankin, J.C. Samson, V.T. Tikhonchuk, Parallel electric fields in dispersive shear Alfvén waves in the dipolar magnetosphere. Geophys. Res. Lett. **26**(24), 3601–3604 (1999b). https://doi.org/10.1029/1999GL010715

R. Rankin, J.Y. Lu, R. Marchand, E.F. Donovan, Spatiotemporal characteristics of ultra-low frequency dispersive scale shear Alfvén waves in the Earth's magnetosphere. Phys. Plasmas **11**(4), 1268–1276 (2004). https://doi.org/10.1063/1.1647138

R. Rankin, K. Kabin, J.Y. Lu, I.R. Mann, R. Marchand, I.J. Rae, V.T. Tikhonchuk, E.F. Donovan, Magnetospheric field-line resonances: ground-based observations and modeling. J. Geophys. Res. **110**, A10S09 (2005a). https://doi.org/10.1029/2004JA010919

R. Rankin, R. Marchand, J.Y. Lu, K. Kabin, V.T. Tikhonchuk, Theory of dispersive shear Alfvén wave focusing in Earth's magnetosphere. Geophys. Res. Lett. **32**(5), L05102 (2005b). https://doi.org/10.1029/2004GL021831

R. Rankin, K. Kabin, R. Marchand, Alfvenic field line resonances in arbitrary magnetic field topology. Adv. Space Res. **38**, 1720–1729 (2006). https://doi.org/10.1016/j.asr.2005.09.034

R. Rankin, C.E.J. Watt, J.C. Samson, Self-consistent wave-particle interactions in dispersive scale long-period field-line-resonances. Geophys. Res. Lett. **34**(23), L23103 (2007). https://doi.org/10.1029/2007GL031317

G. Rickard, A. Wright, Ulf pulsations in a magnetospheric wave-guide – comparison of real and simulated satellite data. J. Geophys. Res. Space Phys. **100**(A3), 3531–3537 (1995). https://doi.org/10.1029/94JA02935

J.C. Samson, R. Rankin, The coupling of solar wind energy to MHD cavity modes, waveguide modes, and field line resonances in the Earth's magnetosphere, in *Solar Wind Sources of Magnetospheric Ultra-Low-Frequency Waves*, ed. by M.J. Engebretson, K. Takahashi, M. Scholer (1994). https://doi.org/10.1029/GM081p0253

J. Samson, J. Jacobs, G. Rostoker, Latitude-dependent characteristics of long-period geomagnetic micropulsations. J. Geophys. Res. **76**(16), 3675 (1971). https://doi.org/10.1029/JA076i016p03675

J. Samson, D. Wallis, T. Hughes, F. Creutzberg, J. Ruohoniemi, R. Greenwald, Substorm intensifications and field line resonances in the nightside magnetosphere. J. Geophys. Res. Space Phys. **97**(A6), 8495–8518 (1992). https://doi.org/10.1029/91JA03156

J.C. Samson, L.L. Cogger, Q. Pao, Observations of field line resonances, auroral arcs, and auroral vortex structures. J. Geophys. Res. Space Phys. **101**(A8), 17373–17383 (1996). https://doi.org/10.1029/96JA01086

J.C. Samson, R. Rankin, V.T. Tikhonchuk, Optical signatures of auroral arcs produced by field line resonances: comparison with satellite observations and modeling. Ann. Geophys. **21**(4), 933–945 (2003). https://doi.org/10.5194/angeo-21-933-2003

T.E. Sarris, W. Liu, K. Kabin, X. Li, S.R. Elkington, R. Ergun et al., Characterization of ULF pulsations by THEMIS. Geophys. Res. Lett. **36**, L04104 (2009). https://doi.org/10.1029/2008GL036732

T.E. Sarris, W. Liu, X. Li, K. Kabin, E.R. Talaat, R. Rankin, V. Angelopoulos, J. Bonnell, K.-H. Glassmeier, THEMIS observations of the spatial extent and pressure-pulse excitation of field line resonances. Geophys. Res. Lett. **37**, L15104 (2010). https://doi.org/10.1029/2010GL044125

H. Singer, D. Southwood, R. Walker, M. Kivelson, Alfvén-wave resonances in a realistic magnetospheric magnetic-field geometry. J. Geophys. Res. Space Phys. **86**(A6), 4589–4596 (1981). https://doi.org/10.1029/JA086iA06p04589

D. Southwood, Some features of field line resonances in magnetosphere. Planet. Space Sci. **22**(3), 483–491 (1974). https://doi.org/10.1016/0032-0633(74)90078-6

D. Southwood, W. Hughes, Theory of hydromagnetic-waves in the magnetosphere. Space Sci. Rev. **35**(4), 301–366 (1983)

E. Spanswick, E. Donovan, G. Baker, Pc5 modulation of high energy electron precipitation: particle interaction regions and scattering efficiency. Ann. Geophys. **23**(5), 1533–1542 (2005). https://doi.org/10.5194/angeo-23-1533-2005

C.R. Stark, A.P. Cran-McGreehin, A.N. Wright, Contributions to the magnetospheric parallel electric field. J. Geophys. Res. **116**, A07216 (2011). https://doi.org/10.1029/2010JA016270

A.V. Streltsov, Dispersive width of the Alfvénic field line resonance. J. Geophys. Res. Space Phys. **104**, 22,657–22,666 (1999)

A.V. Streltsov, Effects of ionospheric heating on feedback-unstable electromagnetic waves. J. Geophys. Res. Space Phys. **113**(A9), A09211 (2008). https://doi.org/10.1029/2008JA013199

A.V. Streltsov, T. Karlsson, Small-scale, localized electromagnetic waves observed by Cluster: Result of magnetosphere-ionosphere interactions. Geophys. Res. Lett. **35**, L22107 (2008). https://doi.org/10.1029/2008GL035956

A. Streltsov, W. Lotko, Dispersive field line resonances on auroral field lines. J. Geophys. Res. Space Phys. **100**(A10), 19457–19472 (1995). https://doi.org/10.1029/95JA01553

A. Streltsov, W. Lotko, The fine structure of dispersive, nonradiative field line resonance layers. J. Geophys. Res. **101**(A3), 5343–5358 (1996)

A.V. Streltsov, W. Lotko, Dispersive, nonradiative field line resonances in a dipolar magnetic field geometry. J. Geophys. Res. **102**, 27,121–27,135 (1997)

A.V. Streltsov, W. Lotko, Small-scale, "electrostatic" auroral structures and Alfvén waves. J. Geophys. Res. Space Phys. **104**(A3), 4411–4426 (1999). https://doi.org/10.1029/1998JA900153

A.V. Streltsov, W. Lotko, Coupling between density structures, electromagnetic waves and ionospheric feedback in the auroral zone. J. Geophys. Res. Space Phys. **113**(A5), A05212 (2008). https://doi.org/10.1029/2007JA012594

A. Streltsov, E. Mishin, On the existence of ionosoheric feedback instability in the Earth's magnetosphere-ionosphere system. J. Geophys. Res. Space Phys. **123**(11), 8951–8957 (2018a). https://doi.org/10.1029/2018JA025942

A. Streltsov, E. Mishin, Ultralow frequency electrodynamics of magnetosphere-ionosphere interactions near the plasmapause during substorms. J. Geophys. Res. Space Phys. **123**(9), 7441–7451 (2018b). https://doi.org/10.1029/2018JA025899

A.V. Streltsov, W. Lotko, J.R. Johnson, C.Z. Cheng, Small-scale, dispersive field line resonances in the hot magnetospheric plasma. J. Geophys. Res. **103**, 26,559–26,572 (1998). https://doi.org/10.1029/98JA02679

Y.J. Su, S.T. Jones, R.E. Ergun, S.E. Parker, Modeling of field-aligned electron bursts by dispersive Alfvén waves in the dayside auroral region. J. Geophys. Res. Space Phys. **109**(A11), A11201 (2004). https://doi.org/10.1029/2003JA010344

D. Sydorenko, R. Rankin, The stabilizing effect of collision-induced velocity shear on the ionospheric feedback instability in Earth's magnetosphere. Geophys. Res. Lett. **44**(13), 6534–6542 (2017). https://doi.org/10.1002/2017GL073415

K. Takahashi, M.D. Hartinger, V. Angelopoulos, K.-H. Glassmeier, A statistical study of fundamental toroidal mode standing Alfvén waves using THEMIS ion bulk velocity data. J. Geophys. Res. Space Phys. **120**(8), 6474–6495 (2015). https://doi.org/10.1002/2015JA021207

J. Taylor, A. Walker, Accurate approximate formulas for toroidal standing hydromagnetic oscillations in a dipolar geomagnetic-field. Planet. Space Sci. **32**(9), 1119–1124 (1984). https://doi.org/10.1016/0032-0633(84)90138-7

V.T. Tikhonchuk, R. Rankin, Electron kinetic effects in standing shear Alfvén waves in the dipolar magnetosphere. Phys. Plasmas **7**(6), 2630–2645 (2000). https://doi.org/10.1063/1.874105

V.T. Tikhonchuk, R. Rankin, Parallel potential driven by a kinetic Alfvén wave on geomagnetic field lines. J. Geophys. Res. Space Phys. **107**(A7), 1104 (2002). https://doi.org/10.1029/2001JA000231

I. Voronkov, R. Rankin, V.T. Tikhonchuk, J.C. Samson, Nonlinear shear Alfvén resonances in a dipolar magnetic field. J. Geophys. Res. Space Phys. **102**(A12), 27137–27143 (1997). https://doi.org/10.1029/97JA02533

A.D.M. Walker, The Kelvin-Helmholtz instability in the low-latitude boundary-layer. Planet. Space Sci. **29**(10), 1119–1133 (1981)

A. Walker, J. Ruohoniemi, K. Baker, R. Greenwald, J. Samson, Spatial and temporal behavior of Ulf pulsations observed by the Goose Bay Hf Radar. J. Geophys. Res. Space Phys. **97**(A8), 12187–12202 (1992). https://doi.org/10.1029/92JA00329

C. Wang, R. Rankin, Y. Wang, Q.-G. Zong, X. Zhou, K. Takahashi et al., Poloidal mode wave-particle interactions inferred from Van Allen probes and CARISMA ground-based observations. J. Geophys. Res. Space Phys. **123**(6), 4652–4667 (2018). https://doi.org/10.1029/2017JA025123

J.A. Wanliss, J.C. Samson, E. Friedrich, On the use of photometer data to map dynamics of the magnetotail current sheet during substorm growth phase. J. Geophys. Res. **105**, 27,673 (2000)

J.A. Wanliss, R. Rankin, J.C. Samson, V.T. Tikhonchuk, Field line resonances in a stretched magnetotail: CANOPUS optical and magnetometer observations. J. Geophys. Res. **107**, A7 (2002). https://doi.org/10.1029/2001ja000257

T.-H. Watanabe, S. Maeyama, Unstable eigenmodes of the feedback instability with collision-induced velocity shear. Geophys. Res. Lett. **45**(19), 10043–10049 (2018). https://doi.org/10.1029/2018GL079715

C.L. Waters, J.C. Samson, E.F. Donovan, Variation of plasmatrough density derived from magnetospheric field line resonances. J. Geophys. Res. Space Phys. **101**(A11), 24737–24745 (1996). https://doi.org/10.1029/96JA01083

C.L. Waters, K. Kabin, R. Rankin, E. Donovan, J.C. Samson, Effects of the magnetic field model and wave polarisation on the estimation of proton number densities in the magnetosphere using field line resonances. Planet. Space Sci. **55**(6), 809–819 (2007). https://doi.org/10.1016/j.pss.2006.04.041

C.L. Waters, R.L. Lysak, M.D. Sciffer, On the coupling of fast and shear Alfvén wave modes by the ionospheric Hall conductance. Earth Planets Space **65**, 385–396 (2013)

C.E.J. Watt, R. Rankin, Electron acceleration due to inertial Alfvén waves in a non-Maxwellian plasma. J. Geophys. Res. Space Phys. **112**(A4), A04214 (2007a). https://doi.org/10.1029/2006JA011907

C.E.J. Watt, R. Rankin, Parallel electric fields associated with inertial Alfvén waves. Planet. Space Sci. **55**(6), 714–721 (2007b). https://doi.org/10.1016/j.pss.2006.04.042

C.E.J. Watt, R. Rankin, Comment on "Role of dispersive Alfvén waves in generating parallel electric fields along the Io-Jupiter fluxtube" by S.T. Jones and Y.-J. Su. J. Geophys. Res. Space Phys. **114**, A04212 (2009a). https://doi.org/10.1029/2009JA014083

C.E.J. Watt, R. Rankin, Electron trapping in shear Alfvén waves that power the aurora. Phys. Rev. Lett. **102**(4), 045002 (2009b). https://doi.org/10.1103/PhysRevLett.102.045002

C.E.J. Watt, R. Rankin, Alfvén wave acceleration of auroral electrons in warm magnetospheric plasma, in *Auroral Phenomenology and Magnetospheric Processes: Earth and Other Planets*, vol. 197, ed. by A. Keiling, E. Donovan, F. Bagenal, T. Karlsson (Amer. Geophysical Union, Washington, 2012), pp. 251–260. https://doi.org/10.1029/2011GM001171

C.Q. Wei, J.C. Samson, R. Rankin, P. Frycz, Electron inertial effects on geomagnetic field line resonances. J. Geophys. Res. **99**, 11,265–11,276 (1994)

A.N. Wright, A.W. Hood, Field-aligned electron acceleration in Alfvén waves. J. Geophys. Res. **108**(A3), 1135 (2003). https://doi.org/10.1029/2002JA009551

A.N. Wright, W. Allan, M.S. Ruderman, R.C. Elphic, The dynamics of current carriers in standing Alfvén waves: parallel electric fields in the auroral acceleration region. J. Geophys. Res. **107**(A7), 1120 (2002). https://doi.org/10.1029/2001JA900168

J.R. Wygant, A. Keiling, C.A. Cattell, R.L. Lysak, M. Temerin, F.S. Mozer et al., Evidence for kinetic Alfvén waves and parallel electron energization at 4-6 R-E altitudes in the plasma sheet boundary layer. J. Geophys. Res. Space Phys. **107**(A8), 1201 (2002). https://doi.org/10.1029/2001JA900113

L. Zelenyi, A. Galeev, C.F. Kennel, Ion precipitation from the innerplasma sheet due to stochastic diffusion. J. Geophys. Res. **95**, 3871 (1990)

C. Ziesolleck, D. Mcdiarmid, Auroral latitude Pc5 field line resonances – quantized frequencies, spatial characteristics, and diurnal-variation. J. Geophys. Res. Space Phys. **99**(A4), 5817–5830 (1994). https://doi.org/10.1029/93JA02903

Q. Zong, R. Rankin, X. Zhou, The interaction of ultra-low-frequency pc3-5 waves with charged particles in Earth's magnetosphere. Rev. Mod. Plasma Phys. **1**(1), 10 (2017). https://doi.org/10.1007/s41614-017-0011-4

Space Sci Rev (2020) 216:15
https://doi.org/10.1007/s11214-020-0637-3

Aurora in the Polar Cap: A Review

Keisuke Hosokawa[1] · Anita Kullen[2] · Steve Milan[3] · Jade Reidy[4,5] · Ying Zou[6] · Harald U. Frey[7] · Romain Maggiolo[8] · Robert Fear[4]

Received: 30 April 2019 / Accepted: 23 January 2020 / Published online: 5 February 2020

Abstract This paper reviews our current understanding of auroral features that appear poleward of the main auroral oval within the polar cap, especially those that are known as Sun-aligned arcs, transpolar arcs, or theta auroras. They tend to appear predominantly during periods of quiet geomagnetic activity or northwards directed interplanetary magnetic field (IMF). We also introduce polar rain aurora which has been considered as a phenomenon on open field lines. We describe the morphology of such auroras, their development and

Auroral Physics
Edited by David Knudsen, Joe Borovsky, Tomas Karlsson, Ryuho Kataoka and Noora Partmies

✉ K. Hosokawa
keisuke.hosokawa@uec.ac.jp

A. Kullen
kullen@kth.se

S. Milan
steve.milan@leicester.ac.uk

J. Reidy
jr10g11@soton.ac.uk

Y. Zou
yz0025@uah.edu

H.U. Frey
hfrey@ssl.berkeley.edu

R. Maggiolo
romain.maggiolo@aeronomie.be

R. Fear
R.C.Fear@soton.ac.uk

1 Dept. of Comm. Engineering and Informatics, Univ. of Electro-Communications, Tokyo, Japan

2 Space and Plasma Physics, School of Electrical Engineering and Computer Science, KTH Royal Institute of Technology, Stockholm, Sweden

3 Dept. of Physics and Astronomy, University of Leicester, Leicester, UK

4 Dept. of Physics and Astronomy, University of Southampton, Southampton, UK

dynamics in response to solar wind-magnetosphere coupling processes, and the models that have been developed to explain them.

Keywords Polar cap aurora · Transpolar arc · Sun-aligned arc · Polar rain aurora · Ionospheric convection · Magnetosphere · Reconnection

1 Introduction

1.1 Transpolar Arc or Theta Aurora

The first report of auroral arcs appearing poleward of the main oval stem from the early 19th century (Mawson 1916, 1925). In the 1970s polar cap arcs were examined more systematically with the help of ground-based cameras. These observational studies revealed that polar cap arcs are typically thin, Sun-aligned auroral arcs that occur nearly exclusively during quiet times (e.g., Ismail et al. 1977; Lassen and Danielsen 1978). The first global images of the auroral oval from high-altitude became available in the early 1970s. High-altitude space-based observations have the advantage that the whole polar cap can be seen at once and long duration observations of the same feature are possible. High altitude observations have to be done in the ultraviolet (UV) because observations in visible light would severely suffer from the large brightness difference between the dark nightside and the bright dayside. High-altitude observations started with a scanning photometer onboard the ISIS-II satellite in the early 1970s. However, the global structure of polar arcs was not understood until the 1980s when the Spin-scan Auroral Imaging (SAI) UV photometer onboard DE-1 (Dynamics Explorer 1: Frank et al. 1981) and later-on a UV imager onboard the Viking satellite produced instant images with a (near) global view of the oval. The SAI UV photometer onboard DE-1 was also the instrument that for the first time provided observations of small-scale polar cap arcs, which will be introduced in Sect. 1.2, and allowed an investigation of the auroral distribution, current systems, and convection electric fields (Frank et al. 1982). When the first global auroral images from DE-1 were taken, it became clear that much larger auroral structures may appear in the quiet-time polar cap. Frank et al. (1982, 1986) published DE-1 images where 1000–1500 km-long Sun-aligned auroral arcs span the entire polar cap and connect the nightside to the dayside oval (see Fig. 1, left plot). Due to the resulting theta-shape of the global auroral pattern, Frank et al. (1982) suggested the name "theta aurora" for this phenomenon. Further progress was made in the 1990s–2000s with the Polar UVI (Ultraviolet Imager: Torr et al. 1995) and IMAGE (Imager for Magnetopause-to-Aurora Global Exploration) FUV (Far-Ultraviolet) instruments (Mende et al. 2000) that provided true 2-dimensional imaging in different wavelengths, which allowed the estimation of the mean precipitation electron energy and the distinction between electron and proton aurora, respectively. Due to the high temporal resolution of Polar and IMAGE (producing UV images at a 2-min cadence) the temporal evolution of large-scale polar arcs could be studied in much more detail than with previous instruments. A large number of observations of large-scale transpolar arcs reported in this section were obtained by these two instruments.

[5] British Antarctic Survey, Natural Environment Research Council, Cambridge, UK

[6] Dept. of Space Science, University of Alabama in Huntsville, Huntsville, USA

[7] Space Sciences Laboratory, University of California Berkeley, Berkeley, CA, USA

[8] Belgian Institute for Space Aeronomy, Brussels, Belgium

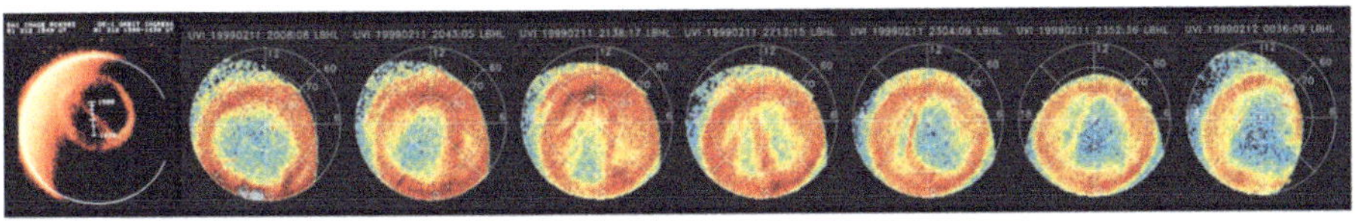

Fig. 1 (left) DE-1 image of a transpolar arc forming a true theta aurora in the northern polar cap (from Frank et al. 1986); (right) Time sequence of Polar UV images in CGlat coordinates showing a TPA that forms at dawn and moves over the entire polar cap until it reaches the dusk oval side. Note, a smaller duskside oval-aligned TPA appears simultaneously (from Kullen et al. 2002)

Nowadays, the expression 'transpolar arc' (TPA) is more commonly used for that type of large-scale polar arc than 'theta aurora'. By definition, a true theta aurora spans the polar cap from midnight to noon. However, this is not always the case during the arc's entire lifetime. Often, a TPA starts to form at the dawn or duskside oval before it moves poleward and becomes a true theta aurora (Cumnock et al. 1997). Thus, the oval resembles the Greek letter theta only during part of a TPA's several-hour-long lifetime (average lifetime of 2 hours according to Kullen et al. 2002). In rare cases TPAs move even over the entire polar cap until they reach the opposite side of the oval (Huang et al. 1989; Cumnock et al. 2002; Kullen et al. 2002; Cumnock 2005). An example of such a dawn-dusk moving TPA is shown in Fig. 1, right plot. However, most TPAs never become true theta auroras; those which form close to the dawn- or duskside oval often remain there. Such TPAs are also referred to as oval-aligned arcs (Murphree and Cogger 1981; Elphinstone et al. 1990; Kullen et al. 2002). TPAs are in general rare events, they are visible on auroral images during only 10–16% of the time (Kullen et al. 2002). Those TPAs which show a considerable dawn-dusk motion are even less common (15% of all TPAs according to Kullen et al. (2002), and 24% according to Fear and Milan (2012a)).

On rare occasions (7% of all TPAs according to Kullen et al. 2002), a TPA starts to form at the nightside oval and stretches from there into the empty polar cap. Such nightside-originating arcs have occasionally been reported (Craven et al. 1986; McEwen and Zhang 2000; Kullen et al. 2002; Milan et al. 2005; Goudarzi et al. 2008; Fear and Milan 2012b). Nightside-originating TPAs often evolve at the end of substorm recovery from a bulge at the poleward boundary of a typically thick oval and stretch towards noon. At the time of formation such TPAs are often more patchy than the usually thinner Sun-aligned TPA. Recovery oval bulges that reach far into the polar cap do not always develop into TPAs that stretch all the way to noon but collapse after a while (Kullen et al. 2002; Goudarzi et al. 2008).

There exist numerous reports about TPAs that deviate in formation and motion from the most typical TPA evolution. For example, Eriksson et al. (2005) showed a TPA case where the dayside part is split in two branches. Sometimes, oval-aligned arcs appear simultaneously on both oval sides, causing an oval shape that resembles the so-called horse-collar oval (Hones et al. 1989)—the region between oval-aligned arc and oval is in such cases typically filled with small-scale arcs, see Sect. 1.2. On other occasions, one TPA moves poleward while the TPA on the other oval side remains static (see Fig. 1, right plot). That two TPAs appear simultaneously, is not uncommon, this happens during about 30% of all TPA events (Kullen et al. 2002). Although TPAs are generally Sun-aligned, occasionally their tailward end becomes (during part of their lifetime) strongly hook-shaped (Ismail and Meng 1982; Murphree et al. 1982; Gusev and Troshichev 1986). This should not be confused with bending arcs (Kullen et al. 2002). The bending arcs are typically faint polar cap arcs where the dayside tip of the arc splits from the oval and moves successively antisunward, while the tailward arc end remains attached to the oval, until the entire arc is strongly curved before disappearing within 10 s of minutes (Kullen et al. 2015; Carter et al. 2015).

Sometimes, the polar cap is filled with multiple polar cap arcs that evolve in a seemingly unordered manner, both from the dawnside, duskside as well as the nightside oval boundaries (Newell et al. 1999; Kullen et al. 2002). Observations in the last decade from UV imagers on low-orbiting satellites (they have a better resolution but not a full global view on the oval) revealed that multiple arc events are far more common than previously anticipated. During extended periods of northward IMF the entire polar cap may in rare cases even disappear completely (Zhang et al. 2009). Zhang et al. (2016) showed that multiple arcs are rather 'cusp-aligned' than sun-aligned, as their dayside ends are connected to the cusp, one of these cusp-aligned events (an example of such cusp-aligned multiple arcs will be shown later in Fig. 13a in 2.3). Possible explanations for that behavior are discussed in Sect. 2.3. Multiple ion beams appearing several Earth radii above the polar cap (Maggiolo et al. 2006, 2011) have been shown to be the high-altitude signatures of multiple polar cap arcs (Maggiolo et al. 2012). These are described in more detail in Sect. 3.4.

The disadvantage of most global auroral imagers is their poor spatial resolution (e.g. 1 pixel corresponds to 35–50 km in Polar UVI) and low luminosity sensitivity, making it difficult to identify small-scale Sun-aligned arcs or any TPA fine structure. With help of simultaneous ground-based cameras (Feldstein et al. 1995) and particle detectors (Makita et al. 1991; Cumnock et al. 2009) it was shown that a TPA appearing often as one 50–300 km wide structure on a satellite image from DE-1, Polar, or IMAGE, consists in reality of multiple thin auroral arcs. Due to the lower luminosity sensitivity of global imagers, another problem occurs: even when a TPA seems to be far inside the polar cap on a global image (i.e., the arc is clearly separated by a dark region from the main oval on the image) it is often still attached to the oval by (in the image) subvisual particle precipitation regions, as pointed out by Newell et al. (2009) (e.g. Fig. 2, right plot). However, at least those TPAs that start to form as oval-aligned arcs, but then move over the entire polar cap to the opposite oval side will at some point become completely separated from the oval side they came from (see Fig. 2). Thus Newell et al.'s (2009) categorization into oval-aligned TPAs (that are attached to the oval) and true theta aurora does not make much sense for TPAs that have such an evolution. A TPA like the ones shown in Figs. 1 and 2 would at the beginning of its lifetime belong to the category 'oval aligned TPA' while when reaching the noon-midnight meridian the same TPA would be classified a 'true theta aurora'.

As discussed in detail in the excellent review by Zhu et al. (1997), studying polar cap arcs from ground-based cameras (where the large-scale auroral pattern remains unclear due to the limited field of view), and space based global auroral imagers (where small-scale Sun-aligned arcs remain subvisual due to low resolution and high luminosity threshold), resulted in sometimes contradicting results regarding particle characteristics and source regions of polar cap arcs. Zhu et al. (1997) suggested that part of the confusion appears because small-scale Sun-aligned arcs and TPA probably have different source regions. For small-scale Sun-aligned arcs it was for a long time unclear whether they appear on open or closed field lines (e.g. Bonnell et al. 1999), however more recent results point again to an open field line source region (see Sect. 3.3 and 5.2 for further details). The studies from the 1980s and 1990s focusing on large-scale TPAs all pointed to the plasma sheet or its boundary layer as the origin, due to slightly lower particle energies, but otherwise with similar particle characteristics as auroral arcs of the main oval (e.g., Frank et al. 1986; Peterson and Shelley 1984).

Further indication that TPAs map to the plasma sheet is the fact that these seem to lie on closed field lines. The most direct evidence stems from Fear et al. (2014), who showed

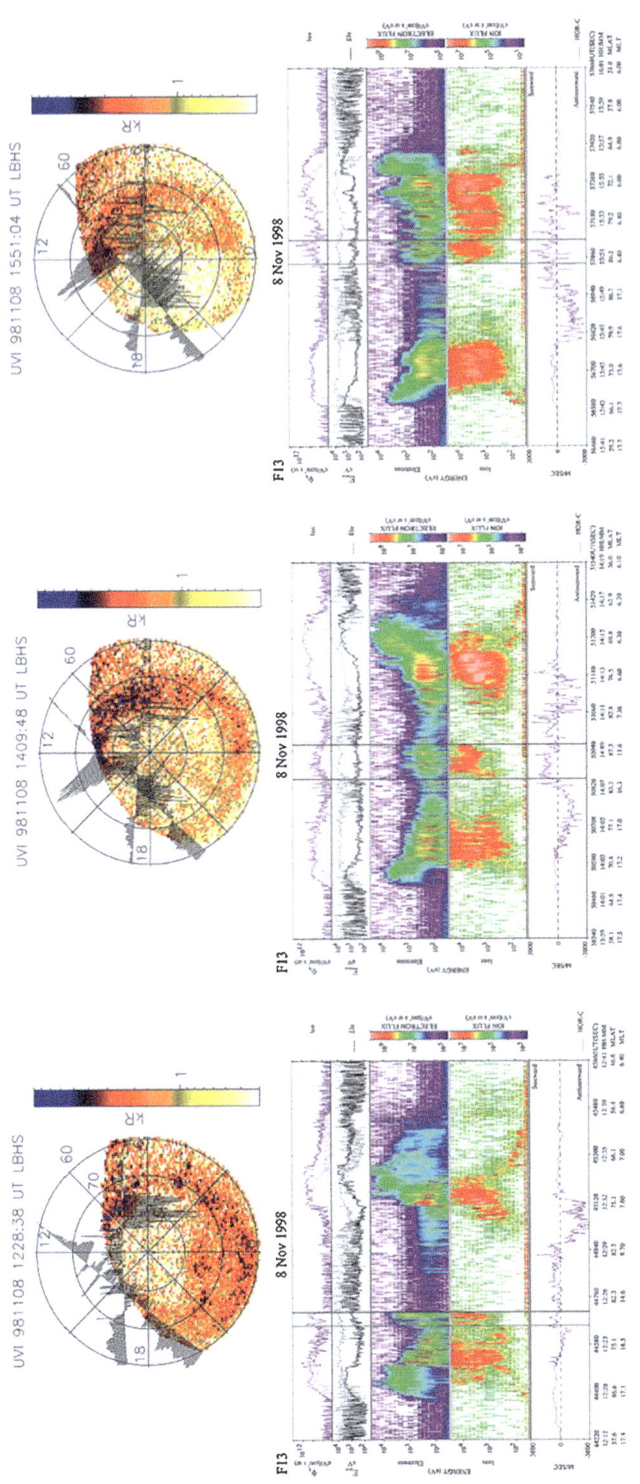

Fig. 2 DE-1 images and DMSP particle data from a TPA that forms at the dusk oval and moves subsequently over the entire polar cap until it reaches the dawn oval side. On the left image, particle data show that the TPA is still attached to the dusk side, on the right image it has merged with the dawn side oval (figures from Cumnock et al. 2002)

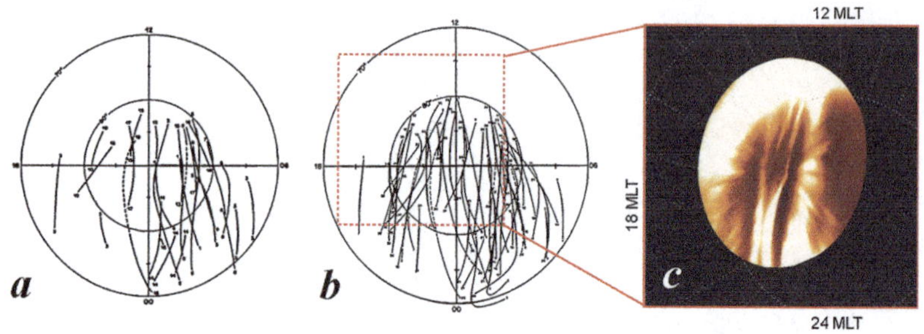

Fig. 3 (**a**) Collection of small-scale Sun-aligned arcs (taken from Berkey et al. 1976). (**b**) Similar statistical collection of the arcs (taken from Ismail et al. 1977). (**c**) Example of ground-based all-sky image of small-scale Sun-aligned arcs obtained by an all-sky cooled CCD imager at Resolute Bay

observations of double-loss cone electrons in the magnetotail mapping to a TPA. Another indication for closed field lines is the observed TPA conjugacy. There exist many reports that TPAs are a conjugate phenomenon appearing simultaneously in the northern and southern hemispheres on opposite oval sides and moving in opposite directions (e.g., Mizera et al. 1987; Obara et al. 1988; Huang et al. 1989; Craven et al. 1991; Cumnock et al. 2006; Carter et al. 2017; Xing et al. 2018). However, Østgaard et al. (2003) showed that during certain IMF B_x conditions, a clear TPA appears only in one hemisphere (see Sect. 3.6 for more details). That TPAs appear within sunward plasma flow in the polar cap was taken as further evidence for a plasma sheet source (e.g., Meng 1981; Peterson and Shelley 1984; Frank et al. 1986; Frank and Craven 1988; Huang et al. 1987, 1989). However, as Liou et al. (2005) pointed out, these studies were all based on satellite measurements from the dayside polar cap. Their own observations, and those of Nielsen et al. (1990) showing anti-sunward flows along the nightside part of TPAs let them suggest that (at least the nightside part of a) TPA may map to the LLBL (Low-Latitude Boundary Layer) at the flanks. A large part of this review is dedicated to the still ongoing discussion about the origin of TPAs. For a more general survey of the magnetosphere's behavior during periods of northward IMF, we refer the interested reader to a recent review by Fear (2019).

1.2 Sun-Aligned Arcs and Horse-Collar Aurora

As mentioned in the previous section, larger and brighter arcs in the polar cap region are traditionally called 'theta aurora' or 'transpolar arcs'. Such large-scale arcs have been observed by several space-based optical instruments onboard DE-1, Polar, and IMAGE. In contrast, small and relatively dim (subvisible in most cases) arcs have been observed by ground-based optical instruments. The morphology of such small-scale arcs was studied intensively in the 70's and early 80's by using ground-based intensified photometers located in the polar cap (e.g., Lassen 1972; Ismail et al. 1977) and since then they have been known as 'Sun-aligned arcs'. It was also indicated that such small-scale sub-visual Sun-aligned arcs are present in the polar cap more often than is implied by spaced-based data (Weber and Buchau 1981). Figures 3a and 3b show collections of Sun-aligned arcs detected by one of the earlier studies (Berkey et al. 1976; Ismail et al. 1977). One end of the arcs is directed to the Sun, which is the reason why small-scale arcs in the polar cap are called 'Sun-aligned arcs.' In the 70's or 80's, people already knew that the small-scale Sun-aligned arcs are indeed a northward B_z phenomenon. Since then, Sun-aligned arcs have been studied extensively by combining space-based and ground-based instruments and such earlier observations up to the late 90's

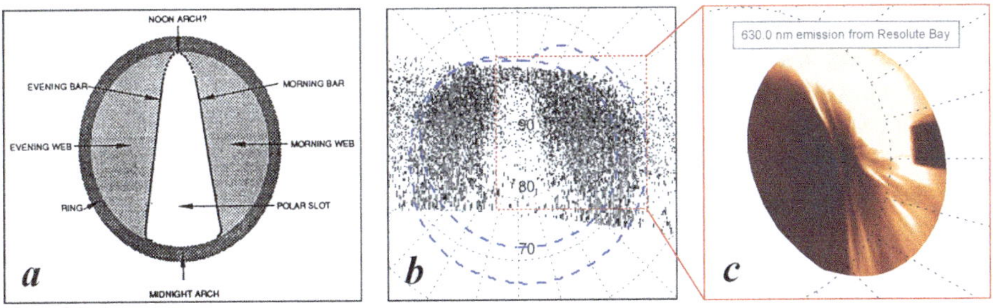

Fig. 4 (**a**) Schematic illustration of horse-collar aurora during northward IMF condition (taken from Hones et al. 1989). (**b**) Space based optical observations from DMSP SUSSI instrument on January 6, 2013. (**c**) Simultaneous ground-based 630.0 nm optical observations from an all-sky imager at Resolute Bay, Canada

are summarized in the review of Zhu et al. (1997). After Zhu et al. (1997), highly-sensitive all-sky imagers have been employed to observe weak Sun-aligned arcs (e.g., Hosokawa et al. 2011). An example of the 630.0 nm red line emission is shown in Fig. 3c. A few thin arcs traversing the field-of-view from the bottom to the top are the manifestations of Sun-aligned arcs. In addition to such a large-scale structure of the arcs, small radially aligned features are embedded within the arcs, which are internal substructures of the arcs representing their extension in the direction of altitude. These substructures are not associated with the Sun-aligned elongation of the arcs.

During periods of prolonged northward IMF conditions, the polar cap boundaries on the duskside and dawnside move poleward to form a large-scale configuration known as "horse-collar aurora" (Hones et al. 1989). Figure 4a is a schematic illustration from Hones et al. (1989), in which the light grey region is a region of soft particle precipitation called 'web'. Figure 4b is a space-based optical observation from the SUSSI (Special Sensor Ultraviolet Spectrographic Imager) instrument onboard the Defense Meteorological Satellite Program (DMSP) satellite. The shape is similar to the cartoon shown in Fig. 4a. If we look at the dawnside part of the collar (Fig. 4c), by using highly-sensitive ground-based all-sky camera, the morning web of the horse collar aurora is composed of a number of small-scale Sun-aligned arcs. That is, small-scale Sun-aligned arcs and horse collar aurora geometry are closely associated with each other, which is consistent with the fact that the most frequently occurring polar cap arcs are the arcs confined either in the morning sector or the evening sector of the polar cap (e.g., Zhu et al. 1996).

1.3 Polar Rain Aurora

Aurora and particle precipitation occurring in the polar cap are naturally associated with each other. Weak, uniform, electron precipitation, known as polar rain, has been found to be present across the entire polar cap both, during southward and northward IMF conditions (Winningham and Heikkila 1974). This kind of precipitation originates in the solar wind and enters the ionosphere via open field lines (Baker et al. 1986). Hardy et al. (1982) reported polar cap arcs embedded within polar rain, which led them to suggest the arcs themselves were occurring on open field lines.

Structured, localized precipitation, known as polar showers, interrupts the soft electron precipitation during northward IMF (Hardy 1984; Hardy et al. 1986). Shinohara and Kokubun (1996) classified two different types of polar showers using one year of particle data from a Defense Meteorological Satellite Program (DMSP) spacecraft. The first ('type A') is associated with ion flux; the second ('type B') is not. The type A polar showers were found to be more intense than type B. Furthermore, type A showed no dependence on the IMF B_x or B_y components whereas type B demonstrated a statistically similar dependence on the IMF condition to typical polar rain. For example, the type B polar showers occurred predominantly in the northern hemisphere for negative IMF B_x and in the southern hemisphere for positive IMF B_x, as does polar rain (Yeager and Frank 1976; Meng and Kroehl 1977). This led Shinohara and Kokubun (1996) to suggest that the properties of the type A polar showers are similar to plasma sheet boundary layer plasma and hence occur on closed field lines, whereas the type B polar showers have similar properties to the solar wind and hence occur on open field lines. Type A polar showers are suggested to be linked with large-scale theta aurora (e.g. Frank et al. 1982), discussed in Sect. 1.1. However, as discussed above, theta auroras are known to have a dependence on the IMF B_y component whereas the type A polar showers were found to be independent of IMF B_y and B_x. This apparent discrepancy can be attributed to Shinohara and Kokubun (1996) using the instantaneous IMF conditions instead of accounting for the time for the IMF B_y to propagate into the magnetotail. This propagation time was discussed by Fear and Milan (2012a), who found a 1–3 hour time delay (dependent on solar wind conditions) in the correlation between IMF B_y and the position where transpolar arcs form in the polar cap.

One of the key questions surrounding polar cap arcs occurring on open field lines is whether they result from accelerated polar rain, and if so, what is the acceleration mechanism? Carlson and Cowley (2005) addressed this question by assuming the source region of polar cap arcs to be polar rain and then estimated several electrodynamic parameters using kinetic theory. They found good agreement between their estimates and observations of polar cap arcs from Carlson et al. (1988). Their analysis indicates that polar cap arcs are produced by field-aligned acceleration of polar rain electrons. Carlson and Cowley (2005) state that any modest mechanism that drives shear flows across open field lines could cause this acceleration, for example at the boundaries between the 'lobe' cells and 'merging' cells. The authors state that they expect polar cap aurora caused by accelerated polar rain to be present whenever the IMF is northward, which is approximately half the time.

It has been suggested that polar cap arcs occurring on open field lines, consistent with accelerated polar rain, can be identified by electron-only particle precipitation. This was discussed by Newell et al. (2009) in a review of polar cap precipitation and aurora. They argue that weaker Sun-aligned arcs are distinct from the large-scale transpolar arcs that occur on closed field lines (discussed in Sect. 1.2). Reidy et al. (2017) observed a Sun-aligned arc associated with electron-only precipitation, using particle data and conjugate UV images of the aurora from a DMSP spacecraft. Several aspects of this arc were consistent with observed dependencies for polar rain, namely it occurred on the duskside of the northern hemisphere during negative B_x and positive B_y conditions. These properties are similar to the 'type B' polar showers found by Shinohara and Kokubun (1996). Reidy et al. (2017) concluded, using these observations combined with the electron-only precipitation, that this arc was consistent with formation on open field lines. Figure 5, taken from the Reidy et al. (2017) survey, shows an observation of polar cap arcs associated with different types of particle precipitation occurring in the same auroral UV image. The corresponding particle data (right) shows the precipitation measured as the spacecraft crossed the polar cap. The red lines in both the auroral image and particle spectrogram indicate when the spacecraft sampled a polar cap arc associated with ion and electron precipitation, consistent with the type A polar showers; the orange lines show the spacecraft measuring a polar cap arc associated with electron-only particle precipitation, consistent with the type-B polar showers. These 'electron-only' arcs were found to be consistent with accelerated polar rain, as proposed by Carlson and

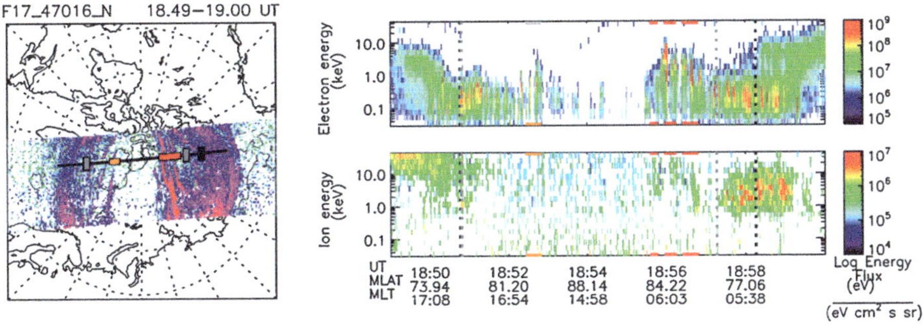

Fig. 5 UV image from the SSUSI instrument (left) with the corresponding particle measurements from SSJ/5 (left) onboard DMSP F17 taken between 18:49–19:00 UT on 15 December 2015 (taken from Reidy et al. 2017). The black line in the SSUSI image represents the footprint of the DMSP spacecraft. The orange and red lines on the footprint and in the particle spectrogram correspond to electron-only and electron and ion signatures within the polar cap respectively, these features were identified from the particle spectrogram and can be seen to coincide with polar cap arcs in the SSUSI images

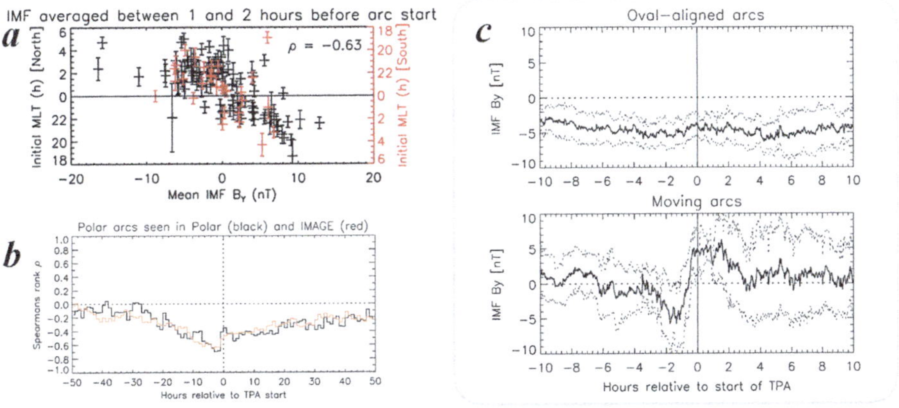

Fig. 6 (**a**) IMF B_y 1–2 h before TPA formation versus MLT location of TPA intersection with the nightside oval (Fear and Milan 2012a), (**b**) correlation between initial arc location and hourly average IMF B_y up to ±50 h from TPA formation for Kullen et al.'s (2002) (black curve) and Fear and Milan's (2012a) (red curve) TPA datasets, (**c**) Superposed epoch analysis plot of IMF B_y up to ±10 hours from TPA formation, centered at TPA start time. The sign of IMF B_y is reversed for all arcs that form on the dusk oval side (Kullen et al. 2015)

Cowley (2005). However, Reidy et al. (2018) did speculate, in this case, that the lack of ion precipitation could be due to instrument sensitivity. Further work is needed to fully understand these polar-rain-accelerated arcs and what exactly is causing them.

1.4 IMF Dependence of TPAs

It has been known for several decades that both small-scale Sun-aligned arcs (Valladares et al. 1994) and TPAs are a predominantly northward IMF phenomenon, while IMF B_y controls their location and motion (e.g., Gussenhoven 1982; Frank et al. 1986; Elphinstone et al. 1990). Most duskside TPAs appear during duskward IMF conditions, most dawnside TPAs during dawnward IMF in the northern hemisphere (Elphinstone et al. 1990; Kullen

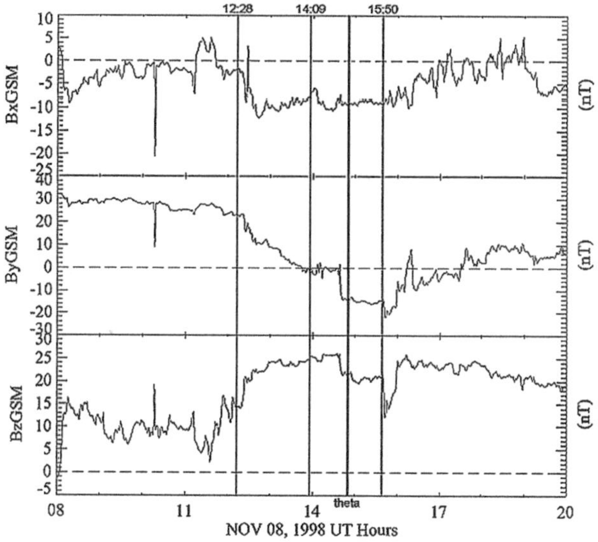

Fig. 7 IMF conditions for the dawn-duskward moving TPA shown in Fig. 2. The vertical bars mark the points of time for which the three images are shown in Fig. 2. The vertical lines have been shifted 15 min to account for the time delay between the Wind satellite at $X = +75\ R_E$ and $X = 0\ R_E$ (figure from Cumnock et al. 2002)

et al. 2002). Fear and Milan (2012a) showed that the statistical dependence of TPA location and formation on IMF B_y is even more exact. There is a clear correlation between IMF magnitude and MLT location of the initial nightside oval intersection when a TPA starts to form. As can be seen in Fig. 6a, this is only statistically correct. Some (especially dawnside) TPAs form on the 'wrong' oval side for a certain IMF B_y value, often in connection with the simultaneous appearance of another TPA on the 'correct' oval side (Hones et al. 1989; Elphinstone et al. 1990; Kullen et al. 2002). Much confusion about the IMF dependence of TPAs stems from the fact that we do not know the exact time delay between IMF conditions and their effect on TPA location. In many studies with varying IMF B_y and B_z, the time of TPA formation was compared to concurrent IMF conditions, which resulted in partially incorrect conclusions. Fear and Milan (2012a) and Kullen et al. (2015) showed with independent TPA datasets, that the time delay is at least 1–2 hours. As shown in Fig. 6b, the correlation between IMF B_y and initial TPA location is high even for B_y values several hours before the TPA forms. The time delay cannot be determined more exactly, as most TPAs (visible in global images) appear during times with strong IMF. Kullen et al. (2002) showed that TPAs form preferably during solar wind conditions with a high magnetic energy flux $v \times B^2$ and strong northward IMF (Kullen et al. 2002), probably due to the reported correlation between magnetic energy flux and TPA luminosity in the dayside polar cap (Kullen et al. 2008). Such solar wind conditions appear typically during CMEs or similar events, where the IMF magnitude is large, and the IMF components remain more or less constant during several hours. This explains the nearly constant average IMF B_y values during the hours around oval-aligned TPAs, as seen in Polar UV statistics (Kullen et al. 2015) and shown in Fig. 6c (top).

A connection between the poleward motion of TPAs, and an IMF B_y sign reversal was first discovered by Cumnock et al. (1997). A duskside TPA in the northern hemisphere will move over the polar cap towards dawn after IMF B_y has changed sign from duskward to dawnward, see the TPA example shown in Fig. 2 and corresponding IMF data in Fig. 7. For an opposite sign change, a dawnside TPA is expected to move towards dusk. Cumnock (2005) showed with a statistical study that one clear IMF B_y sign change during constantly northward IMF is a sufficient condition for a poleward motion of a TPA to occur, as long as the solar wind magnetic energy flux is high enough (Kullen et al. 2008). The different examples discussed in Kullen et al. (2002) and Cumnock (2005) show that TPAs that move over the entire polar cap to the other oval side require an IMF B_y sign reversal, while for TPAs reaching only the center of the polar cap, an IMF B_y magnitude decrease is sufficient. Cumnock's (2005) observations that 16 of the 19 TPAs in her study had already formed before an IMF B_y sign change took place, shows, that an IMF B_y sign change does not necessarily trigger the TPA formation itself (as originally assumed by, e.g. Chang et al. (1998) or Kullen (2000)) but only its subsequent motion in direction of IMF B_y. This can also be seen from Fig. 6c (bottom), where an IMF B_y sign change appears on average 40 min before a moving TPA forms, which is clearly shorter than the at least 1–2 hours time delay required for IMF B_y to affect the TPA location at formation (Fear and Milan 2012a, Kullen et al. 2015).

So far, the IMF B_y effect has been discussed only for TPAs in the northern hemisphere. From the observational reports of conjugate TPAs appearing in the northern and southern hemisphere on opposite oval sides and with opposite motion (see Sect. 1.1 for references), it can be inferred that IMF B_y has an opposite effect on TPA location and motion in the southern hemisphere. However, the interhemispheric TPA conjugacy may be prevented by a strong IMF B_x component. Several statistical studies show a (weak) correlation between IMF B_x and the occurrence of high-latitude aurora in general. TPAs (Kullen et al. 2002), small-scale polar arcs (Lassen and Danielsen 1978), and also localized High-Latitude Dayside Aurora (HiLDA) (Frey 2007 and references therein), are in the northern hemisphere more commonly observed during negative than during positive IMF B_x. The conjugacy issue and its connection to IMF B_x is discussed in more detail in Chap. 3.6 in connection with a discussion of non-conjugate TPA cases observed by Østgaard et al. (2003).

2 Proposed Generation Models

2.1 IMF B_y Switching for TPA

The reason why a TPA forms has been discussed extensively in the last decades. As described in detail in Kullen's (2012) review about large-scale TPAs, there exist many different models trying to explain their cause. Most of these models are conceptual, and are based on observations of IMF control of TPAs. The IMF dependence of TPA formation and motion is already described in Sect. 1.4. In the present chapter, the different TPA models are presented. Many TPA models focus on the topological connection between TPAs and the tail plasma sheet, as the plasma sheet or its boundary region is generally believed to be the source region of TPAs (see Sect. 1.1 and 5.2 for more details). Meng (1981) proposed that an oval-aligned TPA appears on the poleward boundary of an expanded closed field line region. Makita et al. (1991) suggested among other ideas, that the poleward expansion of the closed field line region is caused by a tail configuration where the entire plasma sheet is twisted around the x-axis, such that one plasma sheet flank appears on very high latitudes and the other on much lower. This is shown in Fig. 8b (see also Sect. 2.4 where this model is discussed again in connection with small-scale Sun-aligned arcs). Such a tail twist appears indeed for nonzero IMF B_y (see Fig. 8a), as first proposed by Cowley (1981) and later on confirmed both observationally (Kaymaz et al. 1994; Owen et al. 1995; Tsyganenko and Fairfield 2004) and with MHD (Magnetohydrodynamics) simulations (Walker et al. 1999; Kullen and Janhunen 2004). Tsyganenko et al.'s (2004) observational study as well as both MHD simulations showed that the degree of IMF B_y-induced tail twisting is much stronger

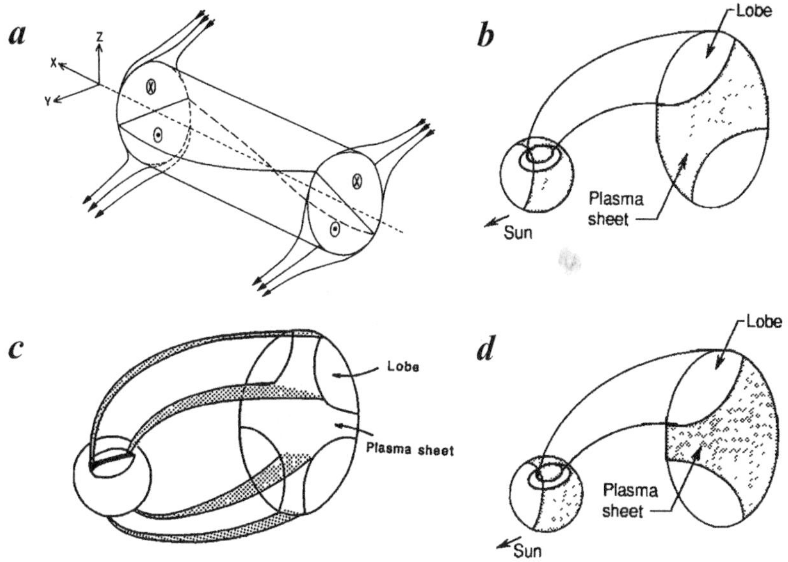

Fig. 8 (**a**) Tail twist for duskward IMF (Cowley 1981) (**b**) oval-aligned TPA at dusk due to twist of tail plasma sheet for duskward IMF (Makita et al. 1991), (**c**) bifurcated tail plasma sheet mapping to a TPA in the center of the polar cap, and a conjugate TPA in the other hemisphere (Obara et al. 1988), (**d**) possible relations of polar cap arc formation with plasma sheet configuration (taken from Makita et al. 1991)

during northward than during southward IMF, which would explain why a considerable poleward expansion of the closed field line region (and thus the possibility of an oval-aligned TPA to form at its poleward boundary) appears only during northward IMF. With this simple model, it can be easily explained why oval-aligned TPAs appear typically at the dawnside oval during dawnward IMF and at the duskside oval during duskward IMF (see Sect. 1.4 for references), as well as why a conjugate TPA typically appears simultaneously in the southern hemisphere on the opposite oval side (see Sect. 1.1 for references about conjugate TPAs). Not explained with this model is, how exactly an oval-aligned TPA during fairly constant IMF B_y may become separated from the main oval. There, the tail reconnection during IMF northward non-substorm intervals (TRINNI) model by Milan et al. (2005) offers a possible explanation (see Sect. 2.2).

In the case that an oval-aligned TPA moves poleward and develops into a true theta aurora, the tail topology becomes more complicated. As already suggested by Frank et al. (1982, 1986) a TPA in the center of the polar cap should map to a plasma sheet filament that reaches from the main plasma sheet high into the lobe, such that the plasma sheet becomes bifurcated. A conjugate TPA will appear in the southern hemisphere that will map to a plasma sheet filament in the southern lobe (see Fig. 8c taken from Obara et al. 1988). As mentioned already in Sect. 1.1, such plasma filaments inside the lobe during TPA events have indeed been observed by e.g., Huang et al. (1989) and Fear et al. (2014).

As discussed in Sect. 1.4, a TPA moving over the entire polar cap is typically connected to an IMF B_y sign change. Thus, it can be expected that a B_y sign change from dawnward to duskward causes an originally clockwise-twisted plasma sheet to transform in an intermediate state into a bifurcated plasma sheet and finally into a new equilibrium state with an anti-clockwise twist for the new B_y direction. Kullen (2000) proposed that an IMF B_y sign change may cause such a tail reconfiguration, assuming that the tail will not change

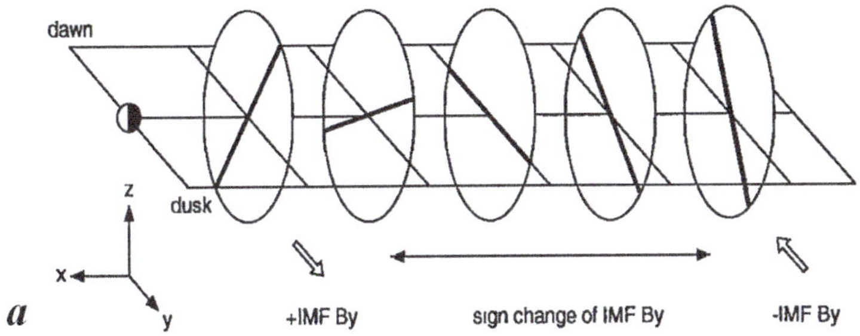

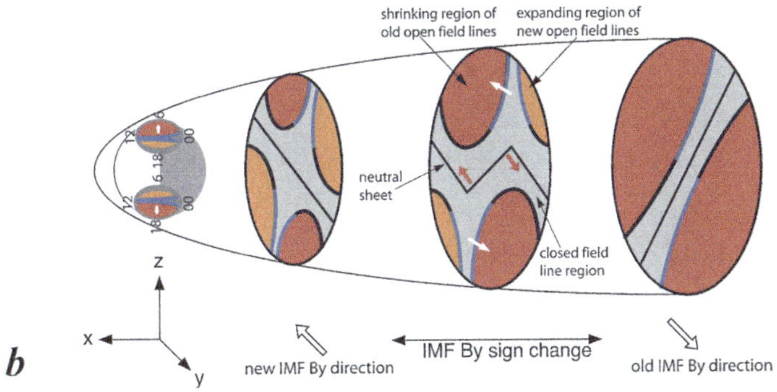

Fig. 9 (**a**) Simple tail twist rotation model for an IMF B_y sign change by Kullen (2000), (**b**) tail topology after an IMF B_y sign change during northward IMF, as it appears in MHD simulations (from Kullen 2012)

its B_y-induced twist at once from e.g. clockwise to anti-clockwise, but with a time delay until far tail regions are affected. Thus, in an intermediate state, near-Earth and far tail regions are oppositely twisted (see Fig. 9a). By modifying the T89 model (Tsyganenko 1989) in such a way, Kullen (2000) showed, the closed field line region becomes indeed bifurcated in the near-Earth tail and maps to a finger of closed field lines in the polar cap. The closed field line finger moves from dawn to dusk while the tail rotation propagates tailward, as is expected for a TPA occurring for an IMF B_y sign change from negative to positive. MHD simulations with several different codes confirm in principle this scenario (Slinker et al. 2001; Maynard et al. 2003; Kullen and Janhunen 2004; Tanaka et al. 2004; Naehr and Toffoletto 2004). However, while the plasma sheet in Kullen (2000) was forced to stay straight while rotating (i.e. no bifurcation of the plasma sheet occurs), MHD results show that the reconfiguration does not only propagate tailward but also inward towards the tail center. As a result, the plasma sheet gets in an intermediate kinked (weakly bifurcated) state (see black line, indicating the neutral sheet in Fig. 9b). The MHD results are discussed in more detail in Sect. 4.1.

2.2 TRINNI for TPA

As discussed in Sect. 1.1., some large-scale TPAs appear to grow into the polar cap from the nightside auroral oval. The growth and evolution of one such event was studied in global auroral imagery (IMAGE) and global convection mapping (Super Dual Auroral Radar Network: SuperDARN, Chisham et al. 2007) of the northern hemisphere by Milan et al. (2005), a subset of the observations being presented in Fig. 10. Three snapshots of auroral imagery

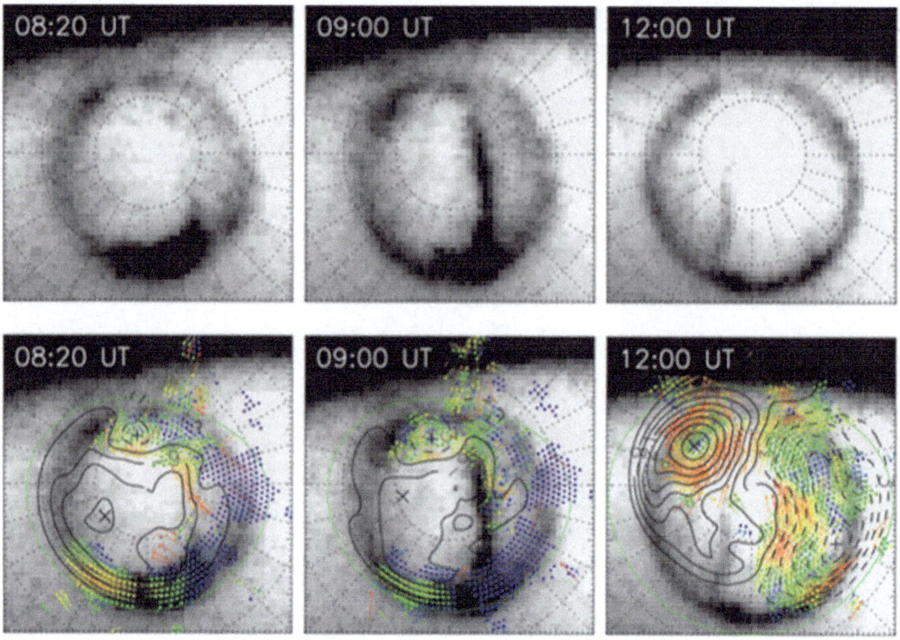

Fig. 10 (Upper panels) Three auroral images captured by the Far Ultraviolet Wideband Imaging Camera (FUV-WIC) on the IMAGE spacecraft on 19 Jan 2002. (Lower panels) The same auroral images, with convection measurements by SuperDARN superimposed (after Milan et al. 2005)

from the Far Ultraviolet Wideband Imaging Camera (FUV-WIC) on the IMAGE spacecraft are shown from 08:20, 09:00, and 12:00 UT on 19 Jan 2002. They identified a modest brightening across the midnight sector of the auroral oval prior to the event and the development of a poleward intrusion into the polar cap of an auroral arc at the post-midnight edge of this brightening (left panel), followed by the subsequent growth of a TPA into the dawn-side portion of the polar cap (middle panel).

The convection observations showed that simultaneous to, and coincident with, the initial auroral brightening, occurring during a period of IMF $B_z > 0$ and $B_y < 0$, fast westward ionospheric flows developed across the midnight sector, with much slower eastward flows to the east. Milan et al. (2005) noted that this distorted twin-cell return flow pattern was highly reminiscent of fast nightside flows that had been reported by Grocott et al. (2003, 2004) to occur during periods of prolonged northward IMF, and to have an east-west sense of flow that was determined by the B_y component of the IMF, and which was opposite in the northern and southern hemispheres. Such flow events have come to be known as tail reconnection during IMF northward non-substorm intervals (TRINNIs).

Grocott et al. (2003) had proposed a model for the formation of TRINNIs that invoked the induction of a B_y component in the magnetotail lobe field by magnetic tension forces associated with magnetopause reconnection occurring during periods of northward but B_y-dominated IMF; reconnection of these lobe field lines by tail reconnection then leads to newly-closed field lines that are tilted out of meridional planes, leading to asymmetrical sunward return flows, with fast eastward or westward flow across the midnight meridian. The observations of Milan et al. (2005) suggested that the TPA initially formed close to the location where TRINNI return flows diverged eastward and westward. This led them

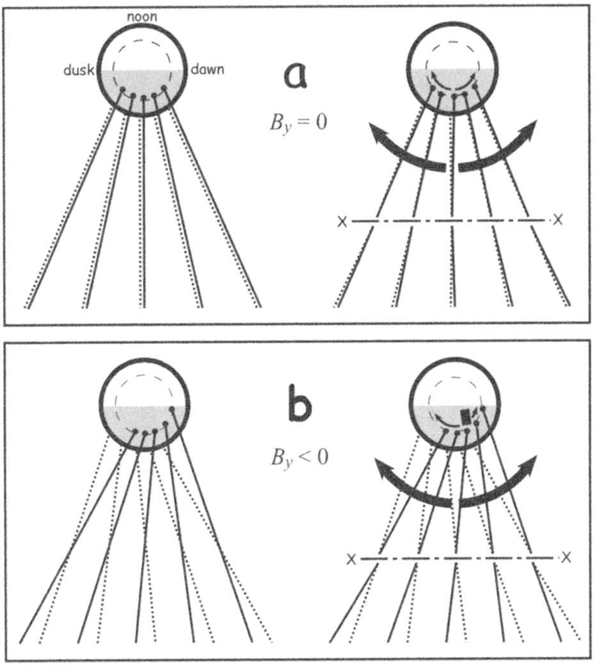

Fig. 11 The model proposed by Milan et al. (2005) for the relationship between TRINNI occurrence and TPA formation. (**a**) With no B_y component in the tail, magnetotail field lines closed by reconnection are contained in meridian planes and can convect sunward. (**b**) With a B_y component, closed field lines are bent out of meridian planes, creating asymmetric return flows in the ionosphere as they convect sunwards, where the fast azimuthal flows across midnight are identified as the TRINNI; field lines crossing the equatorial plane near midnight have no easy sunward route and build up in the tail before protruding into the open polar cap

to speculate (see Fig. 11) that the TPA was formed by field lines which had undergone reconnection in the midnight sector of the tail and which consequently straddled the midnight meridian with footprints that were located in the dusk sector in one hemisphere and the dawn sector in the opposite hemisphere. They suggested that these field lines were unable to evolve back to the dayside in the convection return flow, accumulated in the midnight sector, and eventually protruded into the polar cap. The TPA was then formed by precipitation of trapped particles on these closed field lines.

The B_y component of the magnetic field in the lobe that leads to the formation of TRINNIs, and by implication this class of TPAs, is created by reconnection at the magnetopause under conditions of significant IMF B_y. It might not be expected that field lines of one sense of B_y or another immediately impact the reconnection in the magnetotail, and promptly influence the east-west sense of the TRINNI flows. Fear and Milan (2012a) studied the time delay between periods of IMF of one east-west sense or the other and the control of the dawn-or-dusk sector formation of TPAs. They found that a delay of around 4 hours best explained their observations (see red curve in Fig. 6b), consistent with a reasonable estimate of the propagation time of newly-opened field lines to the tail reconnection site. A more recent study (Kullen et al. 2015) confirmed the overall trends found by Fear and Milan (2012a) but found the correlation peaked on a slightly shorter timescale (around 2 hours), see black curve in Fig. 6b. However, both studies are compatible with the TRINNI formation mechanism being responsible for at least a significant proportion of TPAs.

2.3 Instability for TPA

An alternative model to explain TPAs that form at the nightside oval and stretch from there sunward into the polar cap has been proposed by Rezhenov (1995) and in a slightly different variant by Golovchanskaya et al. (2006). Instead of invoking tail reconnection to explain the

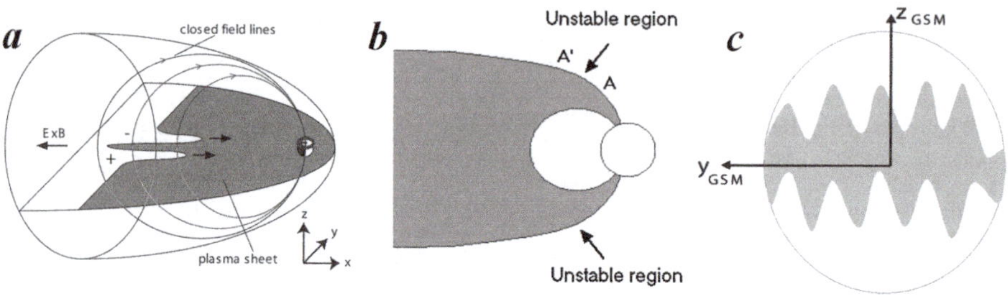

Fig. 12 Interchange instability in strongly curved B-field region causing plasma filaments stretching into the lobes according to (**a**) Rezhenov (1995) and (**b** and **c**) Golovchanskaya et al. (2006). The filaments are assumed to map to nightside originating TPAs

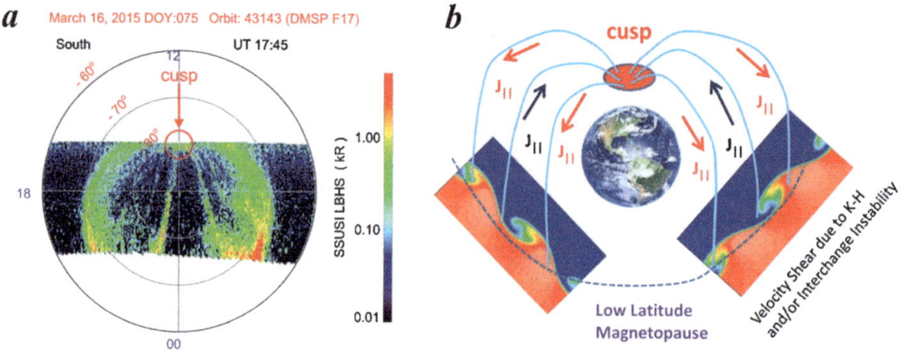

Fig. 13 (**a**) Multiple cusp-aligned arcs seen in SSUSI LBHS images from the DMSP F17 satellite; (**b**) Model by Zhang et al. (2016) explaining the occurrence of multiple cusp-aligned arcs as a result of the Kevin-Helmholtz instability occurring during northward IMF along the dayside magnetopause (Both Figures are taken from Zhang et al. 2016)

sunward growth of the TPA as in the conceptual model by Milan et al. (2005) described in Sect. 2.2., these authors showed with theoretical calculations that TPAs forming at the nightside oval may appear as a result of the ballooning instability acting on highly curved magnetic field lines. The instability causes the growth of plasma sheet filaments at the plasma sheet boundary into the lobes that would map to a sunward growing TPA. Note, this type of TPA model does not require a considerable IMF B_y component, as is the case in e.g., Kullen's TPA model (Kullen and Janhunen 2004) but needs solar wind conditions causing a strongly curved magnetic field, such as northward IMF, high solar wind speed and/or substorm recovery. Rezhenov (1995) suggested such an instability would appear at the tailward end of a short plasma sheet during northward IMF (resulting in a tailward-growing plasma sheet filament that would map to a sunward growing TPA, see Fig. 12a) while Golovchanskaya et al. (2006) located the instability at the strongly curved near-Earth part of the tail magnetic field lines (Fig. 12b), causing plasma sheet filaments to grow into the tail lobes. According to Golovchanskaya et al. (2006) this could also explain the emergence of multiple plasma sheet filaments (Fig. 12c), causing multiple nightside originating TPAs, as has been observed by e.g. Huang et al. (1989).

Zhang et al. (2016) suggested the Kelvin Helmholtz (KH) instability (which appears predominantly during northward IMF) and/or the interchange instability as a possible source of multiple polar arcs (see Fig. 13a). In their model, the instability appears along the low-

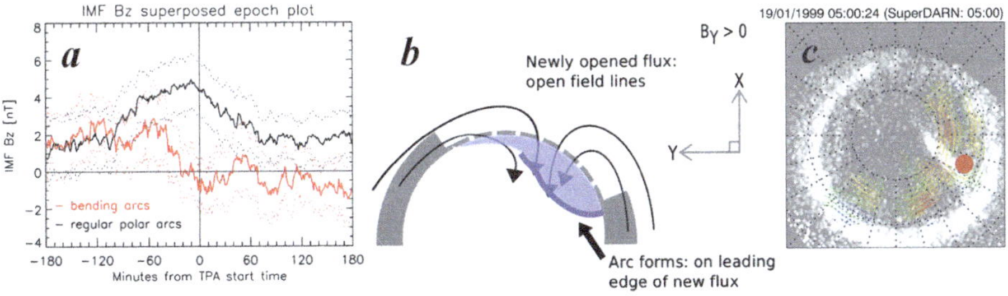

Fig. 14 (**a**) Average IMF B_z up to 3 hours before and after formation of bending arcs (red) and regular TPAs (black); (**b**) bending arc model; (**c**) Polar UV image of bending arc overlaid by 10-min averaged SuperDARN plasma flows (figures from Kullen et al. 2015 and Carter et al. 2015)

latitude dayside magnetopause (see Fig. 13b for a schematic). Then the instability-generated waves lead to strong shear flows, which in turn may generate multiple field-aligned currents that map into the polar cap. Yamamoto and Ozaki (2005) suggested that this process is in principle possible by running a numerical model which produces dayside and polar cap arcs by FACs originating from the LLBL that is modulated by a KH-instability. As shown in Fig. 13b, plasma velocity shears exist on both sides of the open-closed field line boundary, thus some arcs may appear on open and other arcs on closed field lines. This would be one possible explanation for the open-closed field line issue of TPAs (see also 1.4 and 5.2). Zhang et al.'s (2016) model may explain the observed cusp-alignment of multiple polar arcs (see Fig. 13b) as well. However, the cusp-alignment of TPAs can also be explained in a different way. Fear et al. (2015) argued that large-scale TPAs should end up intersecting the cusp as their motion is controlled by the ionospheric/magnetospheric flow. Yet, another explanation has been proposed by Zhang et al. (2009) for a rare case where the entire polar cap is completely filled with auroral arcs. As this appears only during long-lasting strongly northward IMF, they suggested that dual lobe reconnection may be responsible for the nearly entire closure of open flux in the polar cap.

2.4 Plasma Sheet Thickening for Small-Scale Sun-Aligned Arcs

In Sect. 1.2, it was documented that small-scale Sun-aligned arcs and so-called horse-collar aurora are closely related geometrically. Specifically, the morning/evening web regions in Fig. 4a are filled with subvisible small-scale auroral arcs. One of the mechanisms for the formation of horse-collar aurora geometry is auroral oval expansion on the dusk and dawn sides during northward IMF conditions, which was first proposed by Meng (1981). Such poleward expansion of the oval corresponds to either the thickening or tilting of the central plasma sheet. These two models were later illustrated by Makita et al. (1991) as shown in Fig. 8b (plasma sheet tilting) and Fig. 8d (plasma sheet thickening). As shown in Fig. 8d, the central plasma sheet becomes thicker on the duskside which could lead to the poleward expansion of the oval as the evening web in the schematic illustration of horse-collar aurora by Hones et al. (1989) shown in Fig. 4a. The thickening of the plasma sheet may occur on the dawn and dusk sides simultaneously. In such a case, poleward expansion of the oval can happen on both sides and complete horse-collar aurora will appear.

2.5 Dayside Reconnection for TPA

Several studies in the 1990s proposed that a jump in the location of the reconnection site arising from an abrupt change in the IMF, either from northward to southward IMF or due to a change in the B_y component, could somehow lead to the development of a new polar cap region equatorward of the old polar cap, causing a TPA to appear between the two regions of open field lines (Newell and Meng 1995; Newell et al. 1997; Chang et al. 1998). However, Fear and Milan (2012a) noted that the Chang et al. (1998) mechanism was incompatible with the fact that all points on the dayside magnetopause map to the open-closed field line boundary in the ionosphere. Furthermore, a correlation between an IMF B_z sign change and TPA formation could not be confirmed by observations (Kullen et al. 2002; Fear and Milan 2012a) except for the subgroup of TPAs called bending arcs (Kullen et al. 2002, 2015). For a description of bending arcs, see Sect. 1.1. The IMF conditions during which bending arcs occur deviate strongly from regular TPAs. As shown in Fig. 14a (red curve), most bending arcs form when IMF B_z is close to zero (or changes sign to weakly negative values). They appear at the dusk oval side for negative IMF B_y and at dawn for positive B_y (Kullen et al. 2002, 2015). Carter et al. (2015) proposed that the bending of the arc is probably caused by strong anti-sunward polar cap plasma flows in connection with dayside reconnection during $|B_y| \gg |B_z|$ conditions (Fig. 14b and 14c). They suggest that bending arcs have, despite some apparent differences in shape, location, recurrence frequency and lifetime, a similar origin as poleward moving auroral forms (PMAFs). PMAFs are small polar cap arcs that split off the dayside oval and move poleward within a few minutes until they disappear (Fasel 1995; Milan et al. 2000), and are generally believed to be caused by flux transfer events (Frey 2007, and references therein). A detailed discussion of PMAFs can be found in Frey et al. (2019). In Kullen et al. (2015) several of Carter et al.'s (2015) ideas could be confirmed for a larger dataset. The connection to in situ FTE signatures, however, remains to be shown.

3 Recent Observations

3.1 Arc Characteristics Seen in Ground-Based Optical Observations

Polar cap arcs have been routinely observed in ground-based all-sky imagers, for example imagers located at Longyearbyen, Svalbard (78.2° N, 15.7° E), Ny-Ålesund, Svalbard (78.95° N, 11.9° E), and Resolute Bay, Canada (74.7° N, 265.1° E). The imagers are equipped with multiple optical filters and the most extensively utilized emissions for polar cap arc observations are at 557.7 and 630.0 nm wavelengths. The imagers have enabled comprehensive investigations of the motion of polar cap arcs. Hosokawa et al. (2011) carried out a statistical analysis using a large dataset by employing an automated algorithm for detecting polar cap arcs. They examined the motion of the Sun-aligned arcs on the duskside, midnight and dawnside separately. Figure 15 shows the distribution of the duskward velocity (on the vertical axis) of the duskside arcs (left), midnight arcs (middle) and dawnside arcs (right) as a function of the IMF clock-angle on the horizontal axis. The grey scale demonstrates the density of data points in percentage on a linear scale, self-normalized. Most of the arcs, especially those seen in the midnight sector, move in the direction of IMF B_y, which is consistent with the results of Valladares et al. (1994). However, many arcs move just poleward irrespective of the sign of IMF B_y. Such B_y-independent arcs are mostly seen either in the dusk side or dawn side. In particular, those seen on the dawnside have already been reported by a series of papers of Shiokawa et al. (1995, 1996, and 1997). They indicated that poleward moving Sun-aligned arcs are very common on the dawnside. The spatial structure is mostly Sun-aligned, but the temporal variations of these arcs are much more dynamic as compared to the B_y-dependent slowly moving arcs. The origin of this type of Sun-aligned

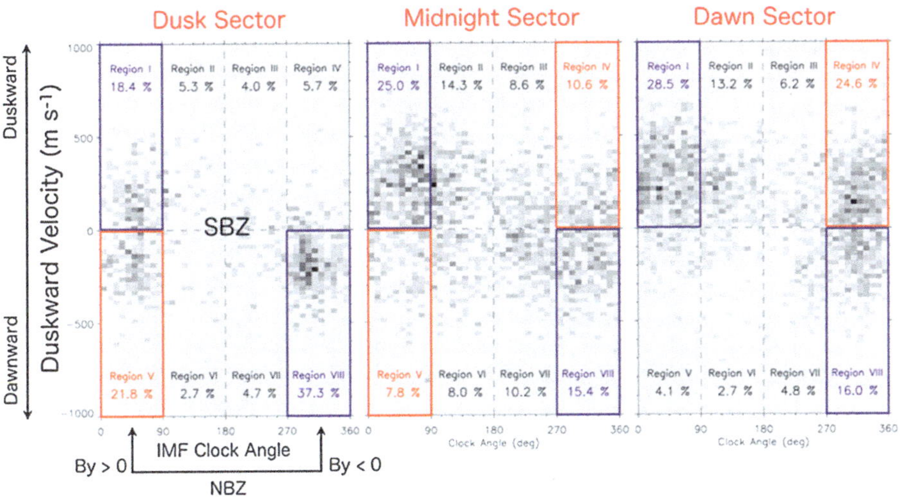

Fig. 15 Statistical results on the dependence of the duskward velocity of polar cap arcs on the clock angle of IMF (taken from Hosokawa et al. 2011)

arc is still unknown, but Shiokawa and colleagues speculated, at that time, that these are associated with some boundary phenomena like Kelvin-Helmholtz instability, for example.

Most ground-based auroral observations of polar cap arcs are based on all-sky imager observations, as discussed above, but another type of observation can be made by examining auroral emissions on a very fine spatial scale. The Auroral Structure and Kinetics (ASK) instrument is a multi-spectral high-resolution imager located on Svalbard (Ashrafi 2007; Dahlgren et al. 2016). Due to its high-latitude location, ASK is perfectly situated for the study of polar cap aurora. ASK is capable of imaging auroral emission in three narrow (and separated) wavelength bands, in a field of view that corresponds to 55 km at 100 km altitude, and can resolve auroral structure down to spatial scales of tens of meters and temporal scales of tens of milliseconds. This capability has only been applied to a small number of events so far; the Reidy et al. (2017) multi-instrument study (discussed above) used ASK to conclude that a polar cap protrusion over Svalbard was consistent with formation on closed field lines. By comparing the ratio of two prompt emission wavelengths measured by ASK, they determined the energy of the precipitation causing the arc, which is usually only possible from space. The precipitation energy was found to be of the order of 5 keV, suggesting the electrons must have been accelerated. Coupled with the highly dynamic auroral structure seen by ASK, this observation demonstrates that this polar cap protrusion was occurring on closed field lines.

3.2 Small-Scale Electrodynamics in the Vicinity of Arcs

Measurements of small or meso-scale convection systems in the vicinity of polar cap arcs were carried out by using data from low-altitude satellite and incoherent scatter radars in 80's to early 90's (Burke et al. 1982; Frank et al. 1986; Robinson et al. 1987; Carlson et al. 1988; Valladares and Carlson 1991). One consensus from these series of measurements is that polar cap arcs are always associated with strong shear in the background convection (i.e., $\mathbf{E} \times \mathbf{B}$ drift), which corresponds to the electric field structure converging towards the arc. Such converging electric field structure can drive horizontal Pedersen currents also flowing towards the arc which are closed by the upward field-aligned currents within the arcs.

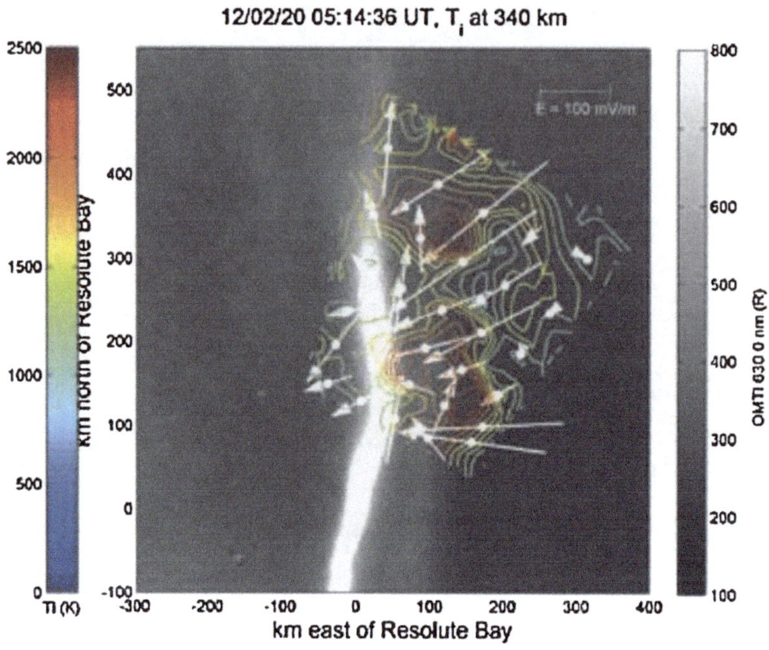

Fig. 16 Simultaneous radar and optical observations of a polar cap arc at Resolute Bay, Canada (taken from Dahlgren et al. 2014)

In these earlier studies, however, due to the limitation in the temporal resolution of the radar measurements, it was difficult to visualize the convection pattern in a two-dimensional fashion. Recent simultaneous ground-based imager and radar (both coherent and incoherent scatter) measurements have been of great value in revealing the electrodynamics of polar cap arcs. Especially, in the last decade, a phased-array incoherent scatter radar at Resolute Bay, Canada has been used for observing 3D electrodynamics in the vicinity of polar cap arcs in combination with all-sky imagers situated in the same place. In addition, several coherent high-frequency (HF) radars of SuperDARN were deployed in the polar cap latitudes which can be used for simultaneous observations of Sun-aligned arcs with all-sky imagers.

Koustov et al. (2008, 2012) found that in comparison to the background polar cap, polar cap arcs are associated with decreased and increased power of radar echoes coming back from the E- and F-regions, respectively. The spectrum of these echoes is broadened. Plasma flow is enhanced on the dawn- and duskside of the arc and across the arc the flow exhibits a sharp reversal from sunward to anti-sunward direction. Since the plasma motion is governed by the local $\mathbf{E} \times \mathbf{B}$-drift, the radar's measurements of plasma flow represent the distribution of converging electric fields in the vicinity of the arc. Figure 16 shows one of the other examples from simultaneous radio and optical observations of a Sun-aligned arc by the incoherent scatter radar and all-sky imager at Resolute Bay (Dahlgren et al. 2014). They observed strong electric field directed towards the arc, which is consistent with the converging electric field reported in the earlier studies. They also indicated that the ion temperature was significantly enhanced in the region adjacent to the arc through the Joule heating effect of the electric field; thus, the magnitude of the flow (and corresponding electric field) was large only in a localized region near the arc.

By examining the similar incoherent scatter radar's electron density measurements, Perry et al. (2015) noted that the leading edge of the arc is associated with a meridionally extended plasma density depletion in the F-region ionosphere, located partially within a downward field-aligned current region. Samara and Michell (2013) found that the altitude profile of the ionization produced by polar cap arcs is consistent with that expected for precipitating electrons accelerated by Alfvén waves. Kozlovsky et al. (2007) studied morning-side arcs and found that each arc was associated with four specific FAC regions: the upward FAC region containing the optical arc, the return downward FAC region poleward of the arc; and a secondary weaker arc equatorward of the main arc with a pair of FACs similar to the main arc. Zhang et al. (1999) identified a strong horizontal electrojet flowing along the arcs as carried by the electric field drift of the low-energy electrons within the arc.

3.3 Optical and in-Situ Particle Signatures at Low-Altitude

Over the last 15 years, four SSUSI instruments have been deployed on DMSP spacecraft F16-F19 (Paxton et al. 2002). When over the polar regions, the SSUSI instruments scan anti-sunward along their orbit, building up an image of a swath of the auroral region over 20 minutes. These instruments provide an excellent opportunity for near-simultaneous, inter-hemispheric observation of polar cap arcs. The DMSP satellites have also carried sensors to measure auroral precipitation since 1973 (Hardy 1984). Hardy et al. (1982) found examples of polar cap arcs embedded within polar rain, suggesting to them that the source region of these arcs was likely to be on open field lines. More recently, a renewed interest in these in-situ measurements have been reported, particularly since the SSUSI instruments were placed on DMSP satellites.

Reidy et al. (2017) combined SSUSI images with multiple other instruments, to demonstrate that polar cap arcs occurring on both open and closed field lines could occur simultaneously. The SSUSI observations showed a feature protruding into the duskside of the polar cap in both hemispheres, which lasted several hours. This feature was termed a 'failed transpolar arc' as it did not cross the entire polar cap but was consistent with a closed field line formation mechanism. The arc 'failed' due to a southward turning of the IMF. A clear Sun-aligned arc stretching across the polar cap was observed on the dawnside of the northern hemisphere in the same SSUSI image. This arc was consistent with formation on open field lines (see also Sect. 1.3 and 5.2). Carter et al. (2017) and Xing et al. (2018) presented separate SSUSI observations of polar cap arcs occurring simultaneously in both hemispheres that were found to be consistent with formation on closed field lines. In the case of Carter et al. (2017), they showed two transpolar arcs occurring in both hemispheres simultaneously. This observation demonstrated the complex configuration of the Earth's magnetotail during northward IMF, such that two 'tongues' of closed flux could co-exist in the otherwise open polar cap. By surveying one month of SSUSI images with the corresponding particle data, Reidy et al. (2018) found examples of polar cap arcs consistent with open and closed field lines. These studies demonstrate that low-altitude UV imagers can provide near-simultaneous measurements of both hemispheres and are capable measuring polar cap arcs occurring on both closed and open field lines.

Low altitude in-situ particle data, such as that from DMSP, provide an excellent insight into the topology of the polar cap. Newell et al. (2009) used these data to suggest that there are three different formation mechanisms for polar cap arcs: two which occur on closed fields and one on open. Fear et al. (2014), Carter et al. (2017) and Xing et al. (2018) presented DMSP particle observations, as well as observations from multiple ground-based and spacecraft instruments, of transpolar arcs occurring on closed field lines. The DMSP particle detector measured ion and electron precipitation associated with each of these arcs. The

DMSP data presented by Fear et al. (2014) also showed electron-only signatures on either side of the transpolar arc that did not correspond to arcs observed by the IMAGE FUV imager, suggesting the possible simultaneous presence of open field line polar cap arcs, as subsequently shown by Reidy et al. (2017). Cumnock et al. (2009) used DMSP particle data to examine the structure within large scale PCAs, finding a small number of them to be composed of multiple distinct thin arcs, each associated with ion precipitation. Additionally, Reidy et al. (2017) presented an observation of a Sun-aligned arc seen in the SSUSI images that was associated with electron-only precipitation in the corresponding DMSP particle data. This arc was found to be consistent with open field lines formed by accelerated polar rain. Also discussed above, the Reidy et al. (2018) survey found examples of arcs occurring on both open and closed field lines using corresponding auroral and particle data from DMSP satellites. These studies demonstrate the importance of in-situ particle data and show that these low-altitude observations provide information about the topology of the polar cap that is not evident from images alone.

3.4 Fine-Scale Particle Characteristics at High-Altitude

In the 1990s the Polar spacecraft and the ISEE (International Sun-Earth Explorer) spacecraft provided the first in-situ observations at high-altitude above polar cap arcs. These high-altitude observations were used to investigate the magnetic field line topology of polar cap arcs and the magnetospheric region to which they are connected (see Zhu et al. 1997 and references therein). Since their launch in 2000 into a polar orbit, the 4 Cluster satellites have sampled the lobe region above the polar caps at high altitude, providing new insight on the configuration of the high latitude magnetosphere during periods of low geomagnetic activity. They revealed the presence of upflowing ionospheric ion beams with typical inverted-V structures in time-energy spectrograms above the polar cap during periods of low geomagnetic activity typically at altitudes between $\sim$5 R_E and $\sim$9 R_E (Maggiolo et al. 2006). These polar cap ion beams (PCIBs) are accelerated by quasi-static electric field structures located above the polar ionosphere during periods of northward IMF. In a latter statistical study, Maggiolo et al. (2011) showed that PCIBs display similar properties as small-scale Sun-aligned arcs, like their dependency on IMF or their width and orientation in the Sun-Earth direction. Later the association between PCIBs and small-scale Sun-aligned arcs was definitely proven during a conjunction between Cluster and TIMED (Thermosphere Ionosphere Mesosphere Energetics Dynamics) satellites (see Fig. 17 taken from Maggiolo et al. (2012) for detail).

Contrary to low altitude observations (see Sect. 3.3) which are restricted to the observation of precipitating particles, high altitude observations sample the particle and fields inside or above the acceleration region. PCIBs are generated by quasi-static electric field structures which accelerate ions upward and electron downward, triggering the optical emission of the associated polar cap arcs. Interestingly, Cluster observations at high altitude reveals that the electric field parallel to the magnetic field extends to higher altitudes above the polar caps (up to $\sim$5 R_E) compared to similar structures in the auroral zone. Furthermore, the total field aligned potential drop in the polar cap region is typically of a few hundreds of Volts, one order of magnitude lower than the potential drop associated with inverted-Vs in the auroral zone. As discussed in Sect. 1.3, particle observations at high altitude by the Cluster spacecraft revealed that PCIBs are associated with the precipitation of polar rain-like electrons and often with the simultaneous presence of plasma sheet like electrons which leads to questions about the magnetic field line topology.

In-situ observations at high altitude benefit from the low spacecraft velocity and the large spatial extension of the magnetic flux tubes connected to polar cap arcs. This allows a much

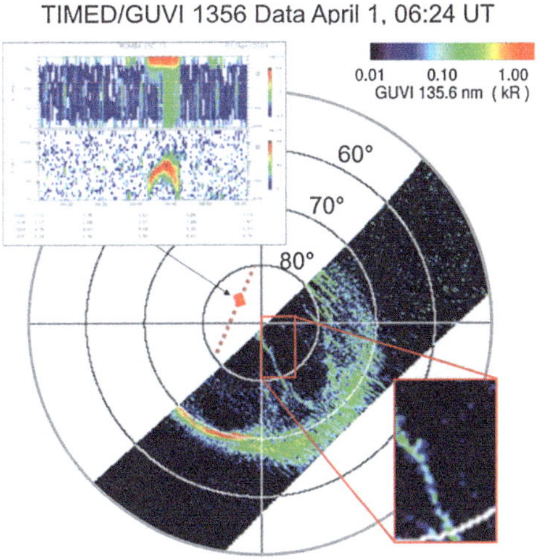

Fig. 17 Image of the northern polar cap from the GUVI experiment onboard the TIMED spacecraft. The red dotted line shows the projected orbit of the Cluster spacecraft and the red square the projected location of the inverted-V detected by Cluster. Bottom right: detail of the polar cap arc detected by GUVI. Top left: Cluster HIA observation of the ion inverted-V associated with the polar cap. Top: ion pitch angle distribution from HIA for all ions. Bottom: time-energy spectrogram from HIA for all ions. See Maggiolo et al. (2012) for more details

better sampling of the particles and fields. For instance, the return current region associated with PCIBs and thin Sun-aligned polar cap arcs, whose width is only of a few km at ionospheric altitude, has been clearly identified with the Cluster spacecraft (Teste et al. 2007). Carried by upflowing low-energy electrons accelerated by a quasi-static electric field (typically ~50 V), the return current has the same intensity as the upward current inside which points to a local closure of the current system. High-altitude observations thus have significant advantages. However, as they are made far from the ionospheric level where polar cap arc emissions occur, it is not always obvious to make a one-to-one correspondence between them and polar cap arcs. That is why they clearly benefit from conjugate observations made either by imagers or low-altitude spacecraft.

3.5 Relationship with Large-Scale Convection

The large-scale convection pattern can be observed with the SuperDARN (Chisham et al. 2007). Studies have been undertaken which determine the general morphology of convection during periods of northwards IMF, and there are studies which focus specifically on convection at times when TPAs are present: for instance, Milan et al. (2005), presented in Fig. 9, and Goudarzi et al. (2008), presented in Fig. 18. During northward IMF conditions, three main features are observed: (1) a cessation of the typical twin-cell convection pattern associated with the Dungey cycle during periods of southwards IMF, (2) sporadic but weak bursts of nightside reconnection which lead to enhanced nightside convection, often with a significant eastward or westward component to the flow, known as TRINNIs (see Sect. 2.2), (3) lobe reconnection at the magnetopause which leads to sunwards convection and the formation of reverse lobe cells in the noon sector polar cap. Features (1) and (3) are well-known from early studies of the average morphology of convection for different orientations of the IMF (e.g. Ruohoniemi and Greenwald 1996). Feature (2), first studied in-depth by Grocott et al. (2003, 2004), was less-well understood due to the sporadic nature of the flows; moreover, as TRINNIs do not have a significant auroral or geomagnetic signature associated with them (Milan et al. 2005), they were not discovered until regular measurements of convection were available.

It is found that the convection during TPA events is no different from the general northward IMF morphology, but that TPA formation and motion can be understood in terms of this

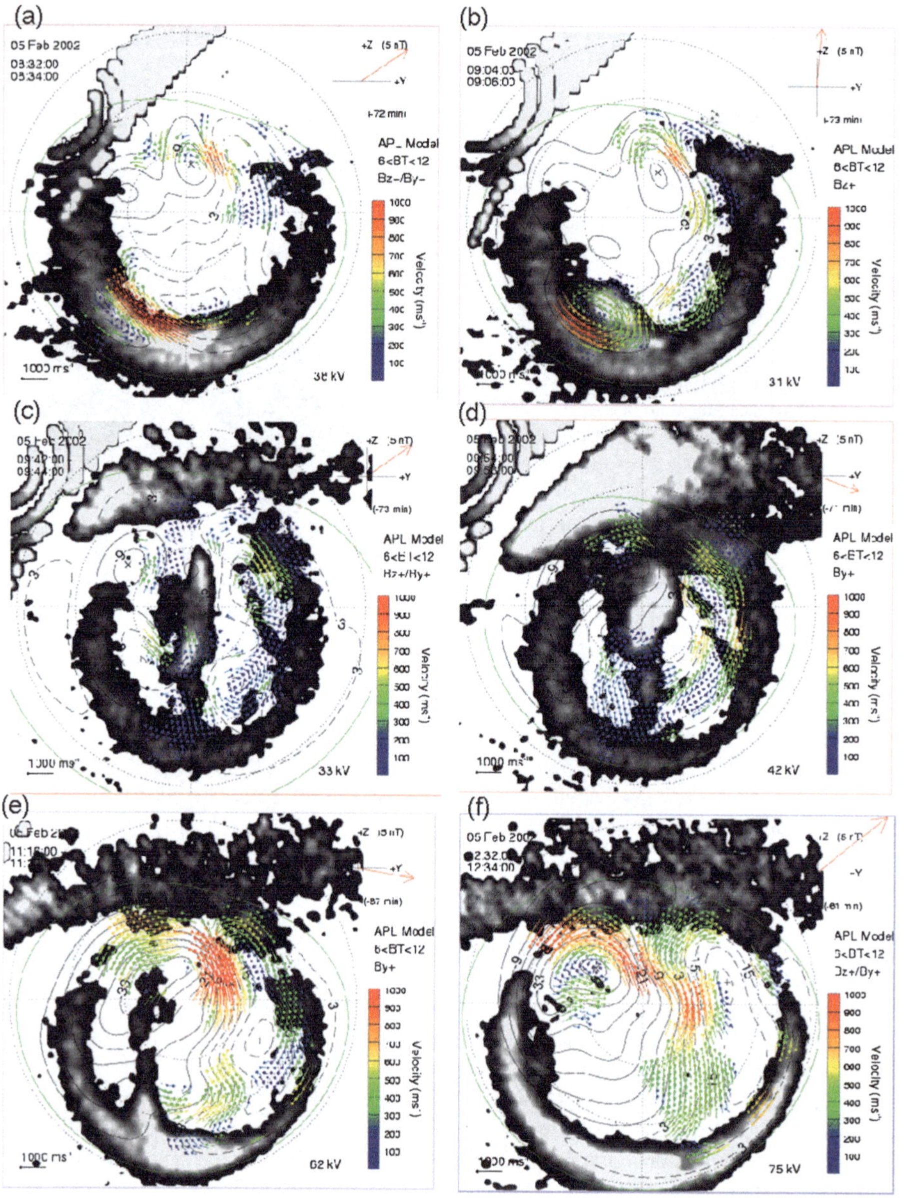

Fig. 18 A sequence Far Ultraviolet Wideband Imaging Camera (FUV-WIC) images from 5 Feb 2002, with superimposed SuperDARN convection maps. (From Goudarzi et al. 2008)

morphology. Specifically, under certain circumstances, which are presently not understood, the occurrence of a TRINNI can lead to the protrusion of a tongue of closed magnetic flux into the otherwise open polar cap (see Sect. 2.2), with the TPA being associated with plasma sheet-like precipitation on these closed field lines (Milan et al. 2005; Fear and Milan 2012b; Fear et al. 2014). Once formed, the closed field lines which constitute the TPA are entrained within the general polar cap convection of the surrounding open field lines, whatever the source of that convection may be. For instance, "lobe stirring" produced by lobe reconnection can cause dawnward and duskward motion of the TPA (Milan et al. 2005; Carter et al. 2017); twin-cell convection associated with southwards IMF can cause the TPA to be pushed towards the nightside (Milan et al. 2005; Reidy et al. 2017); or twin-cell convection with a significant east-west asymmetry associated with IMF B_y can lead to dawnward or duskward motions (Goudarzi et al. 2008).

3.6 Interhemispheric Symmetry/Asymmetry

So far there have only been two opportunities for true simultaneous optical observations of polar cap aurora from space, by the DE-1 and Viking satellites in the 1980s (Craven et al. 1991), and by Polar and IMAGE in the 2000s (Østgaard et al. 2003). As discussed in Sect. 1.1., earlier particle measurements in both hemispheres hinted at some conjugacy between transpolar arcs and a similar but opposite motion towards dawn/dusk depending on the sign of the IMF B_y component. This symmetry was used as an argument for magnetic mapping from a single magnetospheric source, and thus the occurrence of transpolar arcs on closed field lines (Craven et al. 1991). But true interhemispheric observations and the determination of the degree of conjugacy was only possible with space-based imagers.

During a period in 1986 simultaneous images by Viking in the sunlit northern hemisphere and by DE-1 in the dark southern hemisphere revealed the simultaneous presence of transpolar arcs in both hemispheres (Craven et al. 1991). Mapping of the arcs between the hemispheres showed that they were slightly rotated with respect to the noon-midnight meridian and that they were slightly offset in the dawn-dusk direction. These measurements also confirmed their occurrence during northward IMF and a dawnward motion in the northern hemisphere during positive B_y and an opposite motion in the southern hemisphere. Overall there has been only a 20-minute overlap of the observations and a number of open questions remained unanswered like topological variations of the magnetotail in response to changes in the IMF orientation and the related direction of motion of the arcs.

The ultraviolet cameras on IMAGE and Polar provided a new opportunity for interhemispheric aurora observations. In two cases a transpolar arc could only be observed in one hemisphere, but not the other. These two cases were discussed in 3 different papers (Østgaard et al. 2003, 2007; Frey 2007). In one particular case (see Fig. 19) also several passes by DMSP satellites allowed for the measurement of precipitating electrons and ions and plasma drift. The northern hemisphere images showed a clear transpolar arc with an intensity similar to the brightness of the auroral oval. In the southern hemisphere however, there was no sign of any aurora inside the polar cap. The particle measurements in the northern hemisphere show that the arc was produced by 0.4–1 keV electrons while simultaneous 1–10 keV ions indicated that the arc was on closed field lines with surrounding open field line regions on both sides (Fig. 19d). Contrary, in the southern hemisphere no indication of any energetic precipitation could be detected but rather polar rain features that are signatures of open field lines. Plasma drift measurements in the northern hemisphere showed two narrow regions of sunward flows adjacent to an anti-sunward flow at the location of the transpolar arc. The additional presence of a bright proton aurora cusp spot (Frey et al. 2019) further supported the overall picture of a 4-cell convection pattern during northward IMF with the arc located between the cells (Newell et al. 2009). A similar convection pattern was observed in the southern hemisphere but as it was not related to any visible aurora or particle precipitation it appeared that the theta aurora was a non-conjugate phenomenon (Østgaard et al. 2003).

The authors discussed several reasons for the absence of theta aurora in one hemisphere, including conductivity differences in the sunlit and dark ionospheres (Shue et al. 2001;

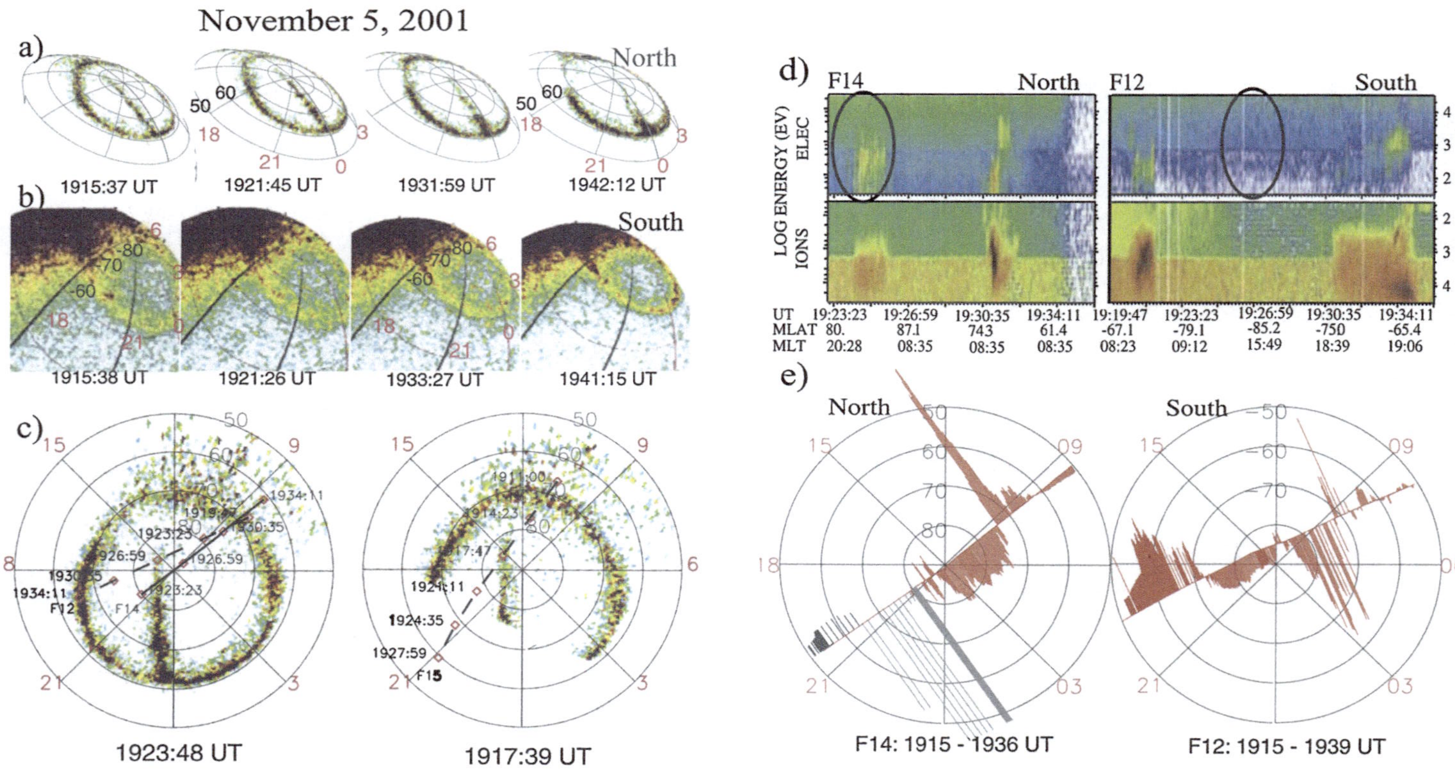

Fig. 19 Interhemispheric observations of a non-conjugate transpolar arc on November 5, 2001 (from Østgaard et al. 2003). (**a**) The IMAGE SI13 observed the northern hemisphere (NH). (**b**) The Polar VIS Earth camera observed the southern hemisphere (SH). (**c**) Trajectories of nearly simultaneous DMSP passes (solid: NH, dashed: SH) mapped onto simultaneous SI13 images from the NH. (**d**) Spectrograms from DMSP, electrons and ions, where the ovals signify the expected location of the theta auroral precipitation (if any). (**e**) Cross track plasma drift data from DMSP. Grey color indicate data where the amplitude is not correct

Wiltberger et al. 2009) and the dipole tilt angle (Kullen 2012). In the end they found it most reasonable that the IMF B_x component determines in which hemisphere lobe reconnection is most efficient, in the northern hemisphere for negative B_x and in the southern hemisphere for positive B_x (Stubbs et al. 2005; Laundal et al. 2018). This is in agreement with statistical studies about the B_x effect on high-latitude aurora (see Chap. 1.4 for references). The resulting 4-cell convection pattern can then drive flow shears with sufficient strength to generate strong electric fields just in one hemisphere which then accelerate electrons to create aurora (Østgaard et al. 2003, 2005, 2007).

The interhemispheric observational results can now also be modeled with magnetohydrodynamic simulations (Watanabe and Hairston 2016). A specific field-aligned current system is set up in connection to the formation of drifting theta aurora in response to the B_y direction during northward IMF. When the IMF B_y changes from positive to negative a transpolar arc in the northern hemisphere moves dawnward. In the southern hemisphere the arc moves into the opposite direction. The polarity of the field-aligned current system reverses accordingly.

The combination of IMAGE in the southern hemisphere and one of the DMSP satellites in the northern hemisphere allowed for some study of the dynamics of transpolar arcs in both hemispheres (Carter et al. 2017). The dawnside arc in the northern hemisphere traversed to the duskside responding to a negative to positive change of the IMF B_y. At the same time, the duskside arc in the southern hemisphere traversed the polar cap to the dawnside. As the arcs are connected to closed field lines a complex magnetic field line topology must develop protruding into the otherwise open polar cap. These observations supported the hypothesis that transpolar arcs form on closed field lines and are entrained within the ionospheric convection. The IMAGE/Polar combination of interhemispheric observations was also used to determine asymmetries in the polar cap boundary location (Laundal et al. 2010). It was shown that the northern and southern auroral ovals encircle the same amount of magnetic flux and that the aurora poleward boundaries thus mark the open/closed field line boundary. Rapid magnetic flux closure in the magnetotail however can lead to strong asymmetries between the hemispheres as a response to differences in the ionospheric convection. The different amount of solar illumination is the most logical reason for the differences in convection.

3.7 Ionospheric Scintillation Associated with Polar Cap Arcs

Data from ground-based imagers, which are introduced in 3.1, have been compared with Global Navigation Satellite Systems (GNSS) to study how polar cap arcs ionize and structure the ionospheric plasma and affect the performance of the Global Positioning System (GPS) in the polar regions. Jayachandran et al. (2009, 2012) found that the ionospheric total electron content (TEC) can vary by $\sim$3 TEC units during the transit of the arc, the amplitude of the variation decreasing as arcs move poleward from the source regions. Jayachandran et al. (2017) detected amplitude and phase scintillations, which are rapid random fluctuations of the amplitude and phase of trans-ionospheric radio signals, in close association with polar cap arcs, indicating the occurrence of kilometer scale electron density fluctuations, i.e., irregularities, in the ionosphere. On the other hand, van der Meeren et al. (2016) used DMSP SSUSI images to study the presence of irregularities in the ionosphere and scintillation caused by polar cap arcs. They used multiple other data sources and concluded, in this case, that the polar cap arc was only a weak source of irregularity.

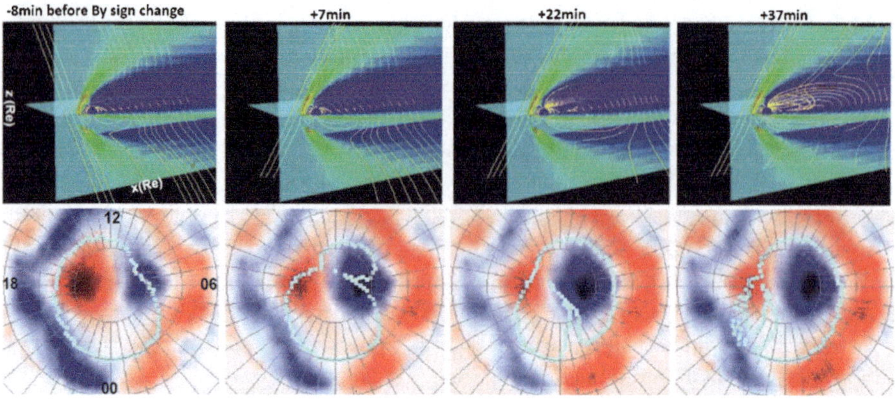

Fig. 20 MHD simulations of an IMF B_y sign reversal during northward IMF by Kullen and Janhunen (2004). The top row shows IMF and tail B-field in a 3-D view of all B-field lines crossing the x-axis. The bottom row shows the northern hemisphere in a polar view with the sun at the top. The blue and red colors mark the current densities, overlaid in light blue is the open-closed field line boundary

4 Recent Modeling Results

4.1 MHD Simulation for B_y-Switching

As discussed in Sect. 1.4, a TPA that moves over the entire polar cap from one oval side to the other, appears typically in connection with an IMF B_y sign change during predominantly northward IMF conditions (e.g., Cumnock 2005). Simulating such IMF conditions with an MHD model results in conjugate TPAs that move in each hemisphere in opposite directions over the model polar cap, as expected from observations. This is shown in Fig. 20, containing a 3-D view of IMF and magnetosphere B-field lines from a simulation by Kullen and Janhunen (2004) in the first row, and a polar view on the northern hemisphere in the second row. Four points of time are shown before and after the sign change has reached the dayside magnetopause. The plots show that the open-closed field line boundary in the model ionosphere (light-blue curve in bottom plot) becomes bifurcated after the B_y sign change, and that this finger of closed field lines (which is assumed to be the location of a TPA) propagates over the entire polar cap while the tail B-field undergoes a major tail reconfiguration.

The good agreement between IMF B_y sign-change simulation results of several different MHD codes allows us to draw conclusions about the connection between TPA motion and changes in the magnetotail topology. Slinker et al. (2001) used for their simulations the Fedder and Lyons model, Kullen and Janhunen (2004) the GUMICS-3 code, Naehr and Toffoletto (2004) the BATSRUS model, and Tanaka et al. (2004) an earlier version of REPPU code. The results from the different MHD codes are compared in Fig. 21. It contains ionospheric and magnetotail plots from these four different MHD codes about 20–30 minutes after an IMF B_y sign change reached the dayside magnetopause. The top row shows the open-closed field line boundary in the northern ionosphere, the bottom row the plasma pressure distribution (in the right plot only the open-closed field line boundary) at tail cross sections in the mid tail. In each MHD simulation the plasma sheet (closed field line region in right plot) has become strongly kinked (bifurcated). The bifurcated plasma sheet part maps to a finger of closed field lines that stretches into the polar cap (the assumed location of a TPA), as is easily seen in the middle plots where the black squares along the TPA are mapped back to the plasma sheet in the tail cross section. Note, also Maynard et al. (2003)

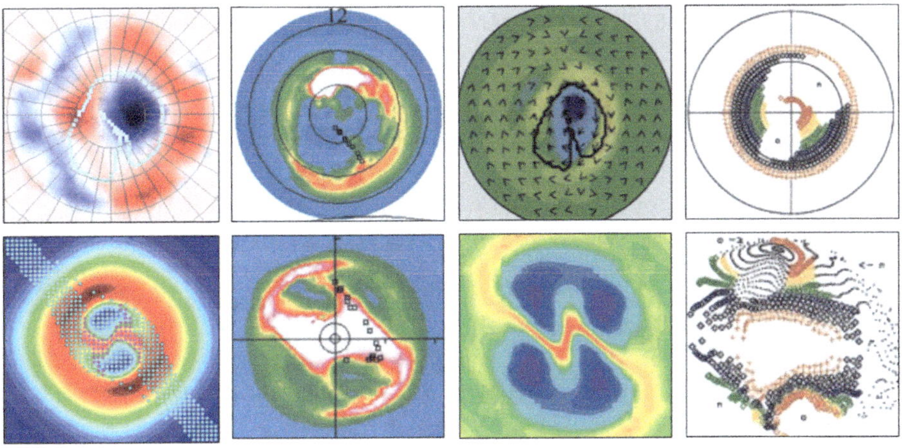

Fig. 21 MHD simulations of a B_y sign reversal during northward IMF by Kullen and Janhunen (2004) (left), Tanaka et al. (2004) (middle left), Naehr and Toffoletto (2004) (middle right), and Slinker et al. (2001) (right column) 20 min, 24 min, 26 min, and 30 min after the sign reversal. The top row shows the open-closed field line boundary in the northern ionosphere (polar view with Sun at the top) (from left to right) as a blue curve, black squares, and a black curve. The background colors mark current density (left and right plots) and plasma pressure isocontours (middle plot). The bottom row shows the tail cross section at (from left to right) $x = -30\ R_E$, $-15\ R_E$, and $-24\ R_E$ with color coded plasma pressure isocontour lines. In the left plot the open (blue dots) and closed (pink dots) field lines are overlaid over the plasma pressure, in the right plot only the closed field line region is shown (plasma pressure plots not shown in Slinker et al. 2001)

studied a B_y sign change run (not shown here), however in the IRM model they used, only small bulges appeared at the open closed field line boundary, which the authors assumed to represent small-scale arcs.

How such a deformed plasma sheet shape evolves after an IMF B_y sign change, is examined in detail in Kullen and Janhunen (2004) and in Naehr and Toffoletto (2004). The results have been summarized in the schematic of Fig. 9b. As mentioned already in Sect. 2.1., the original (IMF B_y induced) plasma sheet twist does not change immediately in the entire tail after a sign reversal. The plasma sheet flanks and the near-Earth tail are affected first. Thus, in an intermediate state, the plasma sheet in near-Earth and far tail are oppositely twisted. Looking at a tail cross section in the mid tail, the flanks' parts are already twisted in the new direction while the central part keeps its old twist, resulting in the strongly kinked (bifurcated) shape, shown in Fig. 21 for the different models. As time passes, the bifurcated part propagates through the lobe from one flank to the other. It maps to the closed field line strip (TPA) that moves over the entire polar cap, until the entire tail is twisted in the opposite direction and the model TPA disappears.

Studying the reconfiguration of the tail lobes during the sign change shows that a new region of open field lines appears (for a sign change from dusk- to dawnward) in the northern hemisphere at the dusk flank (marked in orange in the schematic of Fig. 9b) which expands at the expense of the old open flux region at the dusk lobe (marked in red). In the southern lobe, new and old regions appear on the opposite sides. The formation of the new lobe can be understood by considering the redistribution of open field lines through high-latitude lobe reconnection during northward IMF. In general, old lobe field lines are transformed through reconnection with IMF field lines into new open field lines that are first draped over the dayside magnetopause until they eventually move along the flanks tailward. For dusk-northward IMF, a (at the high-latitude lobe reconnected) newly oriented open field line will have the same dawn-dusk tilt as the surrounding IMF, and thus propagates from the dayside magnetopause towards the dawnside flank and further tailward (e.g., Fig. 1 in Crooker and Rich 1993). After an IMF B_y sign change, a new (in the northern hemisphere created) open field lines will no longer move along the dawn flank but along the dusk flank to the tail (as is expected for the new, dawnward B_y direction). Thus, the duskside grows at the expense of the dawnside part of the northern lobe.

The different MHD studies could clarify several other IMF effects on TPAs as well. An IMF B_y sign change during southward IMF does not cause a TPA (Maynard et al. 2003), which is in agreement with observations. This is probably due to the smaller plasma sheet twist and a shorter tail (i.e. shorter region of closed tail field lines) during southward IMF (Kullen and Janhunen 2004). Slinker et al. (2001) showed by comparing observations with simulations that an IMF B_y sign change during weak and strong northward IMF results in both cases in a TPA that moves over the entire polar cap, but in the latter case the polar cap is considerably contracted. Their results demonstrate that this is connected to a considerable thickening off the plasmas sheet, especially at its flanks (see also Sect. 4.2 about MHD simulations of the horse collar oval). Naehr and Toffoletto (2004) simulated two B_y sign changes separated by one hour, resulting in the formation of a second TPA at the opposite oval side. Their results showed that the first TPA did not disappear but moved slightly back to the oval side it came from, thus offering a possible explanation for the often observed simultaneous existence of two TPAs on opposite oval sides (see Sect. 1.1 and Fig. 1).

There are a number of qualitative differences between observations and the simulation results, such as a too short and thick TPA, too short TPA lifetime, a too short tail etc. However, these discrepancies may disappear with better resolution in the model tail and more advanced modeling of the ionosphere (see, e.g. the promising mapping results from Tanaka et al. 2017 presented in Sect. 4.2). A more fundamental question is, whether a plasma sheet filament (that stretches into the lobes and maps to a TPA) can form without local tail reconnection or not. According to Milan et al.'s (2005) conceptual TPA model, such reconnection is required (see Sect. 2.2). On the other hand, Tanaka et al.'s (2004) detailed examination of the magnetic field line convection after the B_y sign change revealed, that a return flow blockage in the convection cells after an IMF B_y sign change may be responsible for the pileup of closed flux at the location of the bifurcated plasma sheet. Whether this pile-up of closed flux is created by reconnection or is just a result of redistributing field lines is not addressed by Tanaka et al. (2004) and remains to be shown (note that, just recently, a newer version of REPPU code was applied to the theta aurora case and the results are introduced in Tanaka et al. 2018).

Further investigations (both simulation-based and observational) should be performed to examine this matter in more detail. Note also, that all TPA simulations discussed in this section require an IMF sign change for a TPA to appear and move over the polar cap. As discussed earlier (in Sect. 1.4) a sign change is not an observational necessity for a TPA to occur (e.g., Fear and Milan 2012a). This is indicated in more recent MHD simulation runs for northward IMF with a constant IMF B_y component as well. In Maynard et al.'s (2012) simulations a plasma sheet bifurcation occurs close to the flanks, in Tanaka et al.'s (2017) simulations polar cap arcs occur close to the dawn and dusk oval. The results of the latter model are discussed in more detail in the next section.

4.2 MHD Simulation of Sun-Aligned Arcs

As introduced in Sect. 4.1, several global MHD simulations have been carried out to understand the formation process of large-scale transpolar arcs. In contrast, it has been difficult to

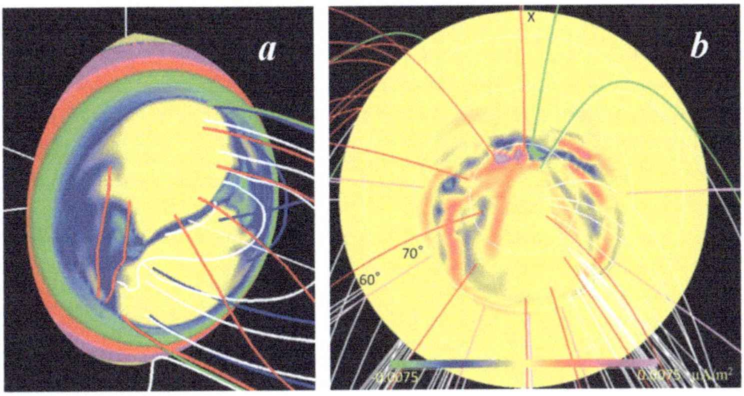

Fig. 22 (**a**) Cross section of pressure distribution (colors), last closed magnetic file lines, and the nightside separator. The radial distance of the cross section is 22 R_E. (**b**) Equivalent ionospheric projection of the FAC and magnetic field lines. The size of the sphere is 3 R_E. This figure is looked down from just over the North Pole. The FAC is shown by colors (− upward, + downward). The 60°, 70°, and 80° northern latitudes are back projected from the ionosphere along the dipole field (taken from Tanaka et al. 2017)

reproduce small-scale Sun-aligned arcs or horse-collar aurora in MHD simulations because of insufficient spatial resolution. Recently, Tanaka et al. (2017) reproduced signatures of small-scale Sun-aligned arcs on both the dawn and dusk sides of the polar cap (i.e., horse-collar aurora) by improving the mesh resolution in the ionospheric side of the simulation. Figure 22a shows a cut of the tail and the blue structure in the pressure is the plasma sheet. Tanaka and colleagues did not consider any big changes in the IMF B_y. However, significant thickening of the plasma sheet is seen on the duskside and dawnside. Figure 22b is the field-aligned current distribution in the ionosphere, and the horse collar structure is clearly reproduced as a bundle of small-scale Sun-aligned arcs. Tanaka et al. (2017) noted that the origin of the shear flow producing FAC in the region of Sun-aligned arcs is still unclear, but at least they demonstrated that Sun-aligned arcs in the horse collar aurora system are caused by this kind of thickening of the plasma sheet on closed field lines.

5 Unclarified Issues

5.1 Origin of the Dawn-Dusk Motion of Arcs

As introduced in 1.4, the dawn-dusk motion of large-scale TPA is known to be controlled to a large extent by the B_y component of the IMF (e.g., Frank et al. 1986; Rairden and Mende 1989; Kullen et al. 2002; Milan et al. 2005; Carter et al. 2017). Not only large-scale arcs, but also small-scale Sun-aligned arcs in the polar cap region are known to move in the dawn or dusk direction depending on the sign of the upstream IMF B_y, as for example demonstrated by Hosokawa et al. (2011) and introduced in 3.1. These studies indicated that in the northern hemisphere polar cap arcs move in the direction of IMF B_y. That is, when the IMF B_y is positive arcs move duskward while they move dawnward when the IMF B_y is negative. The speed of the arcs ranges from 50 to 150 m/s in the dawn-dusk direction.

The entrainment of the TPA within the convection pattern is exemplified by the examples reported by Milan et al. (2005) and Goudarzi et al. (2008), shown in Figs. 10 and 18, respectively. In Fig. 10, both before and after the TPA has formed, flows in the noon sector polar cap suggest the presence of reverse lobe cells associated with lobe reconnection (left and middle panels). In the middle panel, the prenoon lobe cell is focused on the sunward tip of the TPA, and the redistribution of open/closed flux within the polar cap associated with this convection is thought to give rise to the dawn-dusk motion of the TPA (middle and right panels). Furthermore, following a southward turning of the IMF, the development of typical Dungey cycle twin-cell convection flow is observed (right panel), and the addition of new open flux to the dayside polar cap by low latitude reconnection increases the size of the polar cap (seen as an equatorward expansion of the auroral oval) and pushes the TPA towards the nightside. The control of the motion of TPA, in particular in the dawn-dusk motion, by the B_y and B_z components of IMF is schematically depicted in Fig. 8 of Milan et al. (2005). This model can also explain the B_y-controlled motion of small-scale Sun-aligned arcs (Hosokawa et al. 2011). In Fig. 18 the sequence of images shows that following the formation of the TPA, a southward turning of the IMF with significant $B_y > 0$ leads to twin-cell convection with most convection flow into the dawn-sector polar cap, causing a duskward motion of the TPA (panels (e) and (f)).

5.2 Open/Closed Issue

Conflicting conclusions on whether polar cap arcs occur on open field lines (e.g. Hardy et al. 1982) or closed field lines (e.g. Frank et al. 1982) have been proposed since the 1980s. Despite a number of studies dedicated to this issue, the magnetic field line configuration is still debated. The topology of the magnetic field lines connected to polar cap arcs can be inferred from the in-situ observation of particles. The detection of plasma sheet-like ion and electron populations points toward a closed field line topology while the detection of polar rain-like electrons and the absence of ion precipitation points toward an open field line topology. Another means of inferring the magnetic field line topology is to investigate the conjugacy of polar cap arcs. Conjugate observation of polar cap arcs in both hemispheres fits with a closed field line topology while non-conjugate polar cap arcs rather suggest an open field line topology. The ionospheric convection within and around polar cap arcs can also provide information on the magnetic field topology.

Over the last 10 years or so, new analysis methods and data sets have suggested that polar cap arcs forming on both topologies may occur. That is, there is more than one formation mechanism of polar cap arcs. Hosokawa et al. (2011) found, in a statistical study using ground-based images, two distinct types of polar cap arcs based on their motion. They found the motion of some of the arcs to be dependent on the sign of the IMF B_y component, whilst others moved completely independently from it. They suggested that the motion of the B_y-dependent arcs is caused by flux transfer due to lobe reconnection suggesting that the plasma source for these arcs is on or adjacent to open field lines. As the B_y-independent arcs are seemingly unaffected by lobe reconnection, Hosokawa et al. (2011) argue that they are formed on closed field lines. However, Reidy et al. (2017) suggested that the motion pattern observed by Hosokawa et al. (2011) could alternatively be explained by polar cap convection as outlined by Milan et al. (2005) (a closed field line mechanism discussed in Sect. 2.2). Newell et al. (2009), based on DMSP observations of particle precipitation at low altitudes, suggested that there are three different ‘types’ of polar cap arcs, two of which occur on closed field lines and the other on open field lines. The first type discussed by Newell et al. (2009), lacks ion precipitation and thus likely occurs on open magnetic field lines, formed by accelerated polar rain (discussed in Sect. 1.3). Such arcs likely correspond to the type B polar showers as defined by Shinohara and Kokubun (1996) and to the Sun-aligned arcs discussed in Reidy et al. (2017, 2018) (see also Sect. 1.3). Such polar cap arcs compatible with an open field line topology are associated with weak emissions. Of the two closed field line arcs (both of which are associated with plasma sheet-like ion and electron precipitation, which are consistent with the type A polar showers defined by Shinohara and Kokubun 1996), one is suggested to form on the contracted field lines of the auroral oval (and hence not actually occurring within the polar cap), and the other is completely detached from it and considered to be a more classical 'TPA' (as discussed in Sect. 1.1). However, these categorizations can be problematic when considering TPA's which form on one oval side before moving poleward and becoming 'detached' (as discussed in Sect. 1.1 and shown in Fig. 2). It is therefore necessary to look at more than one time instance to determine the nature of the arc, for example by using measurements from the multiple DMSP spacecraft.

The two possibilities (open and closed field lines) are not necessarily mutually exclusive as illustrated by the simultaneous observation of polar cap arcs on open and closed magnetic field lines by Reidy et al. (2017) (see Sect. 3.3). Due to the low altitude orbit of DMSP, the SSUSI observations could be considered to be an intermediary between ground-based and spacecraft observations. Conflicting reports suggesting polar cap arcs occur around 40% of the time in ground-based data (Valladares et al. 1994) but only 10% in UV images from high latitude spacecraft (Kullen et al. 2002) have led to some confusion. It is most likely that these differences are due to ground-based studies being able to measure weaker Sun-aligned arcs (consistent with open field lines) that are below sensitivity thresholds of UV imagers on board high-altitude spacecraft, which predominantly observe the larger-scale theta aurora (consistent with closed field lines). Reidy et al. (2018) observed polar cap arcs 20% of the time when surveying the SSUSI images over 4 months in 2015. This value is between the previous estimates discussed above and perhaps suggests that the SSUSI images could bridge the gap between the contradicting ground-based versus high altitude spacecraft observations.

In addition, some polar cap arcs do not neatly fit in the closed or open field line category. For instance, the second category of polar cap arcs discussed in Newell et al. (2009), is associated with ion and electron precipitation, about the auroral oval in particle observations, and is commonly associated with multiple convection reversals in their vicinity. These observational facts are not compatible with a closed or open magnetic field line topology. Similarly, apparently contradictory observations were reported by the Cluster satellites at higher altitude. A statistical study by Maggiolo et al. (2011) of field aligned acceleration regions above the polar caps showed that a plasma sheet-like ion population coexists with polar rain like electrons for 40% of the events. This is not consistent with a closed or an open field line topology. For the other 60% of the events, no ions could be detected, consistent with an open magnetic field line topology. However, it appears that the isotropic ion density was highly variable, from typical plasma sheet densities down to very low density without a sharp delimitation. Consequently, the distinction between events accompanied or not by plasma sheet like ions is dependent on the sensitivity of the ion detector on board Cluster. It raises questions about the impact of instrument sensitivity when using particle observations to determine the magnetic field line topology as also pointed out by Reidy et al. (2018). Furthermore, the observation of non-conjugate large-scale transpolar arcs by Østgaard et al. (2003, 2007) (discussed in Sect. 3.5) seems in contradiction with a closed or an open field line topology and remains to be explained.

5.3 Source of Polar Cap Arc Plasmas

In situ observations of plasma structure associated with polar cap arcs made by spacecraft at higher altitudes (>8 R_E) have been relatively rare until recently due to the lack of regular coverage of the high-altitude lobe. The only pre-2000 high-altitude in situ observation we are aware of was reported by Huang et al. (1987, 1989), who showed ISEE observations (at 22 R_E down-tail and $Z_{GSM} = -8$ R_E) of 'filamentary' structures of energetic plasma in the lobes, whilst the lower-altitude Dynamics Explorer 1 satellite (apogee 3.65 R_E) observed a transpolar arc. The authors interpreted their observations as spatial filaments somehow protruding from the plasma sheet. The launch of Cluster has transformed this position, providing regular observations of the lobe in an era in which global-scale imaging of the polar cap region has been made possible by Polar, IMAGE and DMSP. Given the lifetimes of the Polar and IMAGE missions in particular, data from the period 2000–2005 is particularly rich. Nonetheless, it took a while for such data to be exploited. Shi et al. (2013) reported observations of what they termed 'solar wind entry events' which were similar in nature to those observed by Huang et al. (1987, 1989). A statistical analysis revealed that these 'entry events' occurred preferentially under northward IMF conditions; coupled with the fact that the pitch angle distribution was nearly isotropic and the proton energy was similar to that typically observed in the cusp or solar wind, Shi et al. (2013) interpreted their observations in the framework of solar wind plasma entering through the mechanism of lobe reconnection. Fear et al. (2014) presented observations of uncharacteristically hot plasma observed in the southern hemisphere magnetotail lobe; the location of the spacecraft mapped to a transpolar arc observed by IMAGE. They interpreted their observations differently from preceding studies in two respects: firstly, they interpreted the observation of multiple instances of the hot plasma as due to the back-and-forth motion of the transpolar arc in response to changing IMF B_y conditions (consistent with the IMAGE observations, and distinct from the interpretation of spatially separated filaments by Huang et al. 1987, 1989); secondly, they interpreted the hot plasma distribution as being essentially a plasma sheet population formed as a result of the magnetotail reconnection process advocated by Milan et al. (2005) (supported by the observation of a double loss cone in the plasma distribution, and contrary to the Shi et al. (2013) interpretation of direct solar wind entry).

Several subsequent studies have investigated this phenomenon further. Some have adopted the interpretation of direct solar wind entry proposed by Shi et al. (2013): Mailyan et al. (2015) reported a conjunction between plasma populations of the type observed by Shi et al. (2013) and a transpolar arc, interpreted in the framework of direct solar wind entry, implicitly placing the transpolar arc on open magnetic field lines, and Gou et al. (2016) performed a statistical analysis, arguing that observed dependencies could be explained by IMF control of the high latitude reconnection process. Others have sided with a closed field line topology formed by magnetotail reconnection: Fear et al. (2015) argued that if transpolar arcs are formed by magnetotail reconnection, then an observed interaction between a transpolar arc and the cusp spot (Frey et al. 2002) indicated that the high latitude reconnection process (indicated by the presence of the cusp spot) was actually opening the closed magnetic flux associated with the transpolar arc. More recently, Nowada et al. (2018) has provided in situ observations of the cusp region close to a transpolar arc. However, these subsequent studies have assumed one interpretation or the other—there is certainly scope for testing these two competing hypotheses further. If, as appears to be the case from multiple case studies, the appearance of hot plasma-sheet-like or solar-wind-like plasma in the lobes is associated with transpolar arcs, then the question of whether these populations are caused by direct solar wind entry or magnetotail reconnection is intrinsically linked to whether the corresponding transpolar arcs are on open or closed magnetic field lines (Sect. 5.2).

5.4 Connection with Substorm Triggering

The increased probability of observing polar cap arcs during periods of northward IMF leads to the presumption that polar cap arcs are associated with an inactive state of the magnetosphere. It is therefore surprising to see that polar cap arcs can in fact coexist with substorm disturbances. The two features can even be optically connected to each other. Whether this suggests a physical relation, or is merely a coincidence of independent processes, has been a subject of a long debate and still remains a puzzle. The complexity is partially due to the variable relation between the evolution sequences of polar cap arcs and substorms as reviewed below, and partially due to the lack of understanding of the magnetotail configuration with the presence of polar cap arcs.

Murphree et al. (1987, 1993, 1994) identified events where polar cap arcs are connected to the nightside auroral oval and found that the connection points are often associated with substorm-like intensifications. The connection points tend to be located at one end of the intensifications, rather than at any point along the intensifications. As the intensifications move eastward or westward, polar cap arcs move with them. Polar cap arcs can maintain their forms well into the substorm expansion phase, and then fade or rotate to join the oval. Henderson et al. (1996) found the substorm bulge does not necessarily occur on one side of polar cap arcs and can expand across the arcs. Again polar cap arcs persist well into the expansion phase and start to fade during the expansion phase.

On the other hand, Shepherd et al. (1990) and Nielsen et al. (1993) found that polar cap arcs fade or even disappear before substorm onset, and can re-brighten or reappear after the onset. The timing of the onset does not seem to have a very clear relation with the timing of arc fading. Instead, arcs fade around the start of substorm growth phase (Shepherd et al. 1990), or minutes before the onset when there is an overall dimming of the nightside auroral oval (Nielsen et al. 1993). Wu et al. (1991) found that polar cap arcs briefly intensify before fading. The intensification and/or the onset of fading occurs during the growth phase or near the start of the expansive phase of substorms. The fading stops at the beginning of the recovery phase. Spann et al. (1998) observed several polar arcs developing following a pseudo-breakup or a small substorm while the IMF B_z was strongly negative. While most of them dissipate by the time of the onset of a following substorm, one persists into the onset. Craven et al. (1986) noted a polar cap arc newly forming during a substorm, appearing first at local midnight and lengthening sunward to noon. Gusev and Troshichev (1990) also detected a polar cap arc formed at the beginning of a substorm. According to Kullen et al. (2002), TPAs typically fade some 10 s of minutes up to 1–2 hours after the onset of substorms.

A statistical examination of the time variation of polar cap arc intensity relative to substorm onset was conducted by Weygand et al. (2001). They categorized their database based on the location of polar cap arcs, dawnside or duskside. The intensity was found to vary significantly. The intensity of the duskside arcs decreased until onset, while that of the dawnside arcs showed a minimum intensity about 850 s before onset and then increased. The authors hence infer a weak interaction between polar cap arcs and substorms and suggest that the disruption of polar cap arcs is most likely due to an indirect disruption of the magnetotail convection caused during the substorm growth phase.

Recently, Nishimura et al. (2013) and Zou et al. (2015) studied the polar cap arc-substorm relation from a different perspective, where they examined the timing when polar cap arcs contact the nightside auroral oval and the onset of substorm-like intensifications (see Fig. 23). They took advantage of a network of ground-based imagers which could resolve the structures of polar cap arcs and the auroral oval in more detail than the previous spaceborne imaging, without sacrificing the spatial coverage. They found that as polar cap arcs

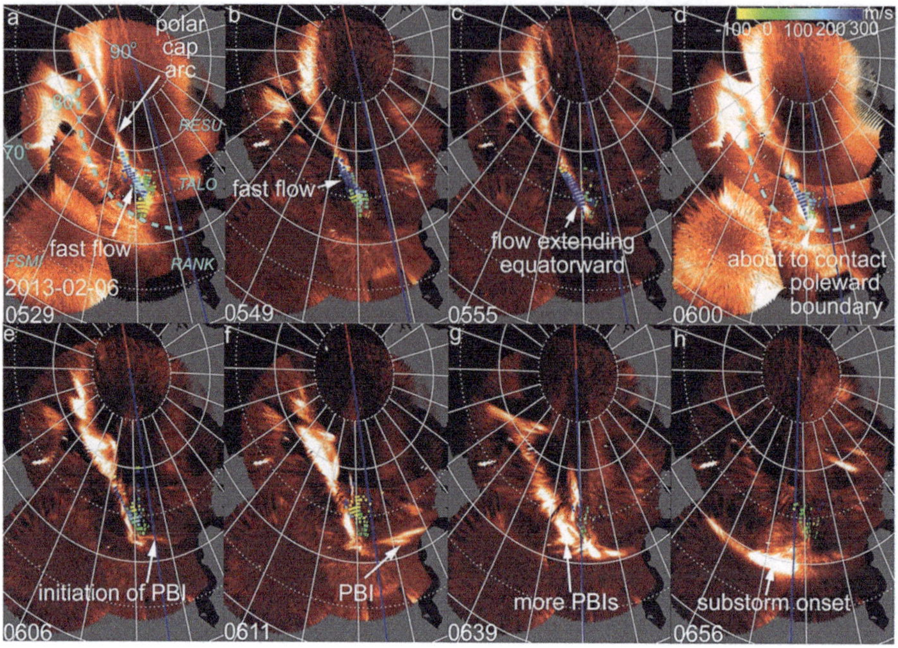

Fig. 23 (**a**–**h**) Snapshots (630.0 nm) showing the association of a polar cap arc to PBI activity within a thick auroral oval on 6 February 2013. White lines are isocontours every 5° in magnetic latitude and every 15° longitude in Altitude Adjusted Corrected Geomagnetic Coordinates. The blue and red lines mark the midnight and noon meridian. A running median intensity over 2 h has been subtracted except for (**a**) and (**d**). (**a**) and (**d**) show absolute intensity and retain the faint and continuous emission from the nightside auroral oval, and the poleward boundary of the oval is identified and marked as the dashed light blue curve. Radar measurements (0–20 range gates) from SuperDARN at RKN station are also shown (Zou et al. 2015)

extend equatorward from high latitude, their contact with the nightside auroral poleward boundary is associated with new and substantial intensifications within the oval for 85% of the time. These intensifications statistically occur within 10 min and ±1 h MLT from the contact. They appear as poleward boundary intensifications (PBIs) in a thick auroral oval, which can be followed by subsequent substorm activity. Or they appear as an intensification of the only resolvable arc within a thin oval, which can also exhibit substantial substorm-like poleward expansion. Since the studied polar cap arcs are associated with enhanced equatorward-directed plasma flows in the polar cap, the authors interpret the sequence as fast equatorward lobe flows compressing the plasma sheet and triggering disturbances within the plasma sheet.

6 Summary

Since the previous review paper by Zhu et al. (1997) we have obtained new understanding about polar cap arcs by investigating observations of various space-based and ground-based instruments. Still, however, there are several unclarified issues regarding the generation, classification and dynamics of polar cap arcs, which are summarized as follows:

✓ **Classification of polar cap arcs:** In this review paper, the polar cap arcs are classified into three categories, large-scale TPA, small-scale Sun-aligned arcs, and polar rain aurora. In addition to these three, we introduced horse-collar aurora in possible relation to Sun-aligned arcs. However, it is still unclear if this classification is appropriate when we consider the source mechanisms of the arcs. For example, the Sun-aligned arc detected by ground-based all-sky cameras may simply be a small version of TPA seen in global imaging from satellites. Not only that, it has not yet been clarified what differentiate small-scale Sun-aligned arcs, polar rain aurora and horse-collar aurora, which could be revealed by combining ground-based observations of Sun-aligned arcs with those of polar rain aurora and Horse-collar aurora from space-based observations, for example by the DMSP SSUSI instrument. The classification of the polar cap arcs should be discussed in close association with their source mechanisms.

- ✓ **Generation mechanism:** The review paper introduces 5 different generation mechanisms of polar cap arcs, 1) IMF B_y switching and resultant twist of the magnetotail, 2) intrusion of closed flux into the polar cap by transient nightside reconnection (TRINNI), 3) interchange instability, 4) plasma sheet thickening, and 5) dayside reconnection driving processes. Most of these mechanisms have been proposed since the previous review by Zou et al. (1997), indicating that our understanding of the generation of polar cap arcs has progressed in the last two decades. From this stage, we should discuss if one of these processes can explain all the types of polar cap arcs, or polar cap arcs can be produced by multiple different processes, based on the previous observational evidence.
- ✓ **Dawn-dusk motion:** The well-known IMF B_y control of the dawn-dusk motion of polar cap arcs can be explained by the model of Milan et al. (2005) based on the flux transportation by the lobe reconnection. However, some fractions of the arcs on the dawnside are known to move poleward irrespective of the sign of IMF B_y. The mechanism of such B_y-independent arcs should be discussed to fully understand the dynamics of polar cap arcs. In addition, it should be examined how the small-scale plasma flow (i.e., localized shear flow) in the vicinity of the arcs is embedded within the large-scale convection system, for example visualized by SuperDARN, during northward IMF intervals.
- ✓ **Is the polar cap field-line open or closed?:** There is clear evidence for polar cap arcs occurring on open field lines and for polar cap arcs occurring on closed field lines which could easily be explained by different formation mechanisms. However, there exists a grey zone, with polar cap arcs for which the magnetic field line configuration cannot be determined experimentally. This remains one of the key issues associated with polar cap arcs. The knowledge of the magnetic field line topology is of prime importance to understand the formation mechanism of polar cap arcs. The main questions surrounding this issue are: If the arcs are occurring on open field lines, how are particles being accelerated to such high energies? And, if they are occurring on closed field lines, what is the configuration of the Earth's magnetotail? These questions should be addressed when we discuss the generation process of polar cap arcs.
- ✓ **Relation to the substorm triggering process:** Recent observations indicated that intensifications of the poleward boundary of the main auroral oval (PBI) occur when equatorward progressing polar cap arcs arrive at the poleward boundary and that those PBIs are followed by the substorm-like activity. However, an actual mechanism occurring in the magnetotail probably near the interface between the lobe and plasma sheet is still unclear. We should tackle this problem by combining a network of ground-based observations with in-situ satellite observations and MHD simulations having sufficient spatial resolution.

To answer all the questions summarized here, it is needed to combine all the available observations and numerical simulations. Now we are able to use several ground-based all-sky cameras in North America, Scandinavia and Antarctica, global convection measurements by

SuperDARN, localized high-resolution observations of ionospheric plasma by several incoherent scatter radars. We can also utilize space-based measurements by the DMSP, Cluster, THEMIS (Time History of Events and Macroscale Interactions) and MMS (Magnetospheric Multiscale) satellites for observing optical and particle characteristics of polar cap arcs. Combination of those observations with recently developing high spatial resolution MHD simulations will lead us to the further understanding of the nature and origin of polar cap arcs. One of the big difficulties in reaching complete understanding of the nature of polar cap arcs is the lack of continuous interhemispheric observations in the polar cap region. We suggest that the development of a twin-spacecraft mission to provide interhemispheric observations of the auroras is a high priority to advance auroral research in general, and the understanding of polar cap auroras in particular. What we really need is instruments constantly viewing the entire auroral region (i.e. global auroral imagers) with as high resolution as possible in time and space, that stay for several years above both the northern and southern polar cap.

Acknowledgements K.H. is supported by JSPS Kakenhi (26302006). K.H., A.K., S.M., R.M., R.F. are supported by the International Space Science Institutes (ISSI) International Team program. J.R. was supported by Natural Environment Research Council studentship NE/L002531/1, R.F. by Science and Technology Facilities Council Consolidated Grant ST/R000719/1, and S.M. by STFC Consolidated Grant ST/N000749/1. Y.Z. is supported by UCAR's Cooperative Programs for the Advancement of Earth System Science (Jack Eddy Postdoctoral Fellowship), and by National Science Foundation (AGS-1664885). H.F. was supported by NSF award AGS-1004736 and NASA's Explorers Program through contracts NNG12FA45C and NNG12FA42I.

References

M. Ashrafi, ASK: Auroral structure and kinetics in action. Astron. Geophys. **48**(4), 4.35–4.37 (2007)

D.N. Baker, S.J. Bame, W.C. Feldman, J.T. Gosling, R.D. Zwickl, J.A. Slavin, E.J. Smith, Strong electron bidirectional anisotropies in the distant tail-ISEE 3 observations of polar rain. J. Geophys. Res. **91**, 5637–5662 (1986). https://doi.org/10.1029/JA091iA05p05637

F.T. Berkey, L.L. Cogger, S. Ismail, Y. Kamide, Evidence for a correlation between Sun-aligned arcs and the interplanetary magnetic field direction. Geophys. Res. Lett. **3**, 145–147 (1976). https://doi.org/10.1029/GL003i003p00145

J. Bonnell, R.C. Elphic, S. Palfery, R.J. Strangeway, W.K. Peterson, D. Klumpar, C.W. Carlson, R.E. Ergun, J.P. McFadden, Observations of polar cap arcs on FAST. J. Geophys. Res. **104**, 12,669 (1999)

W.J. Burke, M.S. Gussenhoven, M.C. Kelley, D.A. Hardy, F.J. Rich, Electric and magnetic field characteristics of discrete arcs in the polar cap. J. Geophys. Res. **87**(A4), 2431–2443 (1982). https://doi.org/10.1029/JA087iA04p02431

H.C. Carlson, S.W.H. Cowley, Accelerated polar rain electrons as the source of Sun-aligned arcs in the polar cap during northward interplanetary magnetic field conditions. J. Geophys. Res. **110**, A05302 (2005). https://doi.org/10.1029/2004JA010669

H.C. Carlson, R.A. Heelis, E.J. Weber, J.R. Sharber, Coherent mesoscale convection patterns during northward interplanetary magnetic field. J. Geophys. Res. **93**(A12), 14501–14514 (1988). https://doi.org/10.1029/JA093iA12p14501

J.A. Carter, S.E. Milan, R.C. Fear, A. Kullen, M.R. Hairston, Dayside reconnection under interplanetary magnetic field B_y-dominated conditions: The formation and movement of bending arcs. J. Geophys. Res. Space Phys. **120**, 2967–2978 (2015). https://doi.org/10.1002/2014JA020809

J.A. Carter, S.E. Milan, R.C. Fear, M.-T. Walach, Z.A. Harrison, L.J. Paxton, B. Hubert, Transpolar arcs observed simultaneously in both hemispheres. J. Geophys. Res. Space Phys. **122**(6), 6107–6120 (2017). https://doi.org/10.1002/2016JA023830

S.-W. Chang et al., A comparison of a model for the theta aurora with observations from Polar Wind and SuperDARN. J. Geophys. Res. **103**, 17,367 (1998)

G. Chisham, M. Lester, S.E. Milan, M.P. Freeman, W.A. Bristow, A. Grocott, K.A. McWilliams, J.M. Ruohoniemi, T.K. Yeoman, P.L. Dyson, R.A. Greenwald, T. Kikuchi, M. Pinnock, J.P.S. Rash, N. Sato, G.J. Sofko, J.-P. Villain, A.D.M. Walker, A decade of the Super Dual Auroral Radar Network (SuperDARN): Scientific achievements, new techniques and future directions. Surv. Geophys. **28**, 33–109 (2007). https://doi.org/10.1007/s10712-007-9017-8

S.W.H. Cowley, Magnetospheric asymmetries associated with the y-component of the IMF. Planet. Space Sci. **29**, 79 (1981)

J.D. Craven, L.A. Frank, C.T. Russell, E.J. Smith, R.P. Lepping, Global auroral responses to magnetospheric compressions by shocks in the solar wind: two case studies, in *Solar Wind–Magnetosphere Coupling*, Y. Kamide, J.A. Slavin (Terra Scientific, Tokyo, 1986), p. 367. 1986

J.D. Craven, J.S. Murphree, L.L. Cogger, L.A. Frank, Simultaneous optical observations of transpolar arcs in the two polar caps. Geophys. Res. Lett. **18**, 2297–2300 (1991). https://doi.org/10.1029/91GL02308

N.U. Crooker, F.J. Rich, Lobe cell convection as a summer phenomenon. J. Geophys. Res. **98**, 13,403–13,407 (1993)

J.A. Cumnock, High-latitude aurora during steady northward interplanetary magnetic field and changing IMF B_y. J. Geophys. Res. **110**, A02304 (2005). https://doi.org/10.1029/2004JA010867

J.A. Cumnock et al., Evolution of the global aurora during positive IMF B_z and varying IMF B_y conditions. J. Geophys. Res. **102**, 17,489–17,497 (1997). https://doi.org/10.1029/97JA01182

J.A. Cumnock, J.R. Sharber, R.A. Heelis, L.G. Blomberg, G.A. Germany, J.F. Spann, W.R. Coley, Interplanetary magnetic field control of theta aurora development. J. Geophys. Res. (2002). https://doi.org/10.1029/2001JA009126

J.A. Cumnock, L.G. Blomberg, I.I. Alexeev, E.S. Belenkaya, S.Y. Bobrovnikov, V.V. Kalegaev, Simultaneous polar aurorae and modelled convection patterns in both hemispheres. Adv. Space Res. **38**, 1685–1693 (2006)

J.A. Cumnock, L.G. Blomberg, A. Kullen, T. Karlsson, K.T. Sundberg, Small-scale characteristics of extremely high latitude aurora. Ann. Geophys. **27**, 3335–3347 (2009). https://doi.org/10.5194/angeo-27-3335-2009

H. Dahlgren, G. Perry, J.-P. St-Maurice, T. Sundberg, K. Hosokawa, J.L. Semeter, M.J. Nicolls, K. Shiokawa, 3D imaging reveals electrodynamics of polar cap aurora. News Rev. Astron. Geophys. **55** (2014). https://doi.org/10.1093/astrogeo/atu215

H. Dahlgren, B.S. Lanchester, N. Ivchenko, D.K. Whiter, Electrodynamics and energy characteristics of aurora at high resolution by optical methods. J. Geophys. Res. Space Phys. **121**, 5966–5974 (2016). https://doi.org/10.1002/2016JA022446

R.D. Elphinstone, K. Jankowska, J.S. Murphree, L.L. Cogger, The configuration of the auroral distribution for interplanetary magnetic field B_z northward, 1, IMF B_x and B_y dependencies as observed by the Viking satellite. J. Geophys. Res. **95**, 5791 (1990)

S. Eriksson, J.B.H. Baker, S.M. Petrinec, H. Wang, F.J. Rich, M. Kuznetsova et al., On the generation of enhanced sunward convection and transpolar aurora in the high-latitude ionosphere by magnetic merging. J. Geophys. Res. **110**, A11218 (2005). https://doi.org/10.1029/2005JA011149

G.J. Fasel, Dayside poleward moving auroral forms: A statistical study. J. Geophys. Res. **100**, 11,891–11,905 (1995). https://doi.org/10.1029/95JA00854

R.C. Fear, The northward IMF magnetosphere, in *Magnetospheres in the Solar System*, ed. by R. Maggiolo, N. André, H. Hasegawa, D. Welling. AGU Centennial Monograph (Wiley, New York, 2019), in press

R.C. Fear, S.E. Milan, The IMF dependence of the local time of transpolar arcs: Implications for formation mechanism. J. Geophys. Res. **117**, A03213 (2012a). https://doi.org/10.1029/2011JA017209

R.C. Fear, S.E. Milan, Ionospheric flows relating to transpolar arc formation. J. Geophys. Res. **117**, A09230 (2012b). https://doi.org/10.1029/2012JA017830

R.C. Fear, S.E. Milan, R. Maggiolo, A.N. Fazakerley, I. Dandouras, S.B. Mende, Direct observation of closed magnetic flux trapped in the high-latitude magnetosphere. Science **346**, 1506–1510 (2014). https://doi.org/10.1126/science.1257377

R.C. Fear, S.E. Milan, J.A. Carter, R. Maggiolo, The interaction between transpolar arcs and cusp spots. Geophys. Res. Lett. **42**, 9685–9693 (2015). https://doi.org/10.1002/2015GL066194

Y.I. Feldstein, P.T. Newell, I. Sandahl, J. Woch, S.V. Leontjev, V.G. Vorobjev, Structure of auroral precipitation during a theta aurora from multisatellite observations. J. Geophys. Res. **100**(A9), 17429–17442 (1995). https://doi.org/10.1029/95JA00379

L.A. Frank, J.D. Craven, Imaging results from Dynamics Explorer 1. Rev. Geophys. **26**, 249 (1988)

L.A. Frank, J.D. Craven, K.L. Ackerson, M.R. English, R.H. Eather, R.L. Carovillano, Global auroral imaging instrumentation for the Dynamics Explorer Mission. Space Sci. Instrum. **5**, 369–393 (1981)

L.A. Frank, J.D. Craven, J.L. Burch, J.D. Winningham, Polar views of the Earth's aurora with dynamics explorer. Geophys. Res. Lett. **9**, 1001–1004 (1982). https://doi.org/10.1029/GL009i009p01001

L.A. Frank, J.D. Craven, D.A. Gurnett, S.D. Shawhan, J.L. Burch, J.D. Winningham et al., The theta aurora. J. Geophys. Res. **91**, 3177–3224 (1986). https://doi.org/10.1029/JA091iA03p03177

H.U. Frey, Localized aurora beyond the auroral oval. Rev. Geophys. **45**, RG1003 (2007). https://doi.org/10.1029/2005RG000174

H.U. Frey, S.B. Mende, T.J. Immel, S.A. Fuselier, E.S. Claflin, J.-C. Gerard, B. Hubert, Proton aurora in the cusp. J. Geophys. Res. **107**, 1091 (2002). https://doi.org/10.1029/2001JA900161

H.U. Frey et al., Dayside aurora. Space Sci. Rev. (2019, this issue). https://doi.org/10.1007/s11214-019-0617-7

I.V. Golovchanskaya, A. Kullen, Y.P. Maltsev, H. Biernat, Ballooning instability at the plasma sheet-lobe interface and its implications for polar arc formation. J. Geophys. Res. **111**, A11216 (2006). https://doi.org/10.1029/2005JA011092

X.C. Gou, Q.Q. Shi, A.M. Tian, W.J. Sun, M.W. Dunlop, S.Y. Fu, Q.G. Zong, G. Facsko, M. Nowada, Z.Y. Pu, B. Mailyan, T. Xiao, X.C. Shen, Solar wind plasma entry observed by Cluster in the high-latitude magnetospheric lobes. J. Geophys. Res. **121**, 4135–4144 (2016). https://doi.org/10.1002/2015JA021578

A. Goudarzi, M. Lester, S.E. Milan, H.U. Frey, Multi-instrumentation observations of a transpolar arc in the northern hemisphere. Ann. Geophys. **26**, 201–210 (2008). https://doi.org/10.5194/angeo-26-201-2008

A. Grocott, S.W.H. Cowley, J.B. Sigwarth, Ionospheric flow during extended intervals of northward but B_y-dominated IMF. Ann. Geophys. **21**, 509–538 (2003)

A. Grocott, S.V. Badman, S.W.H. Cowley, T.K. Yeoman, P.J. Cripps, The influence of IMF B_y on the nature of the nightside high-latitude ionospheric flow during intervals of positive IMF B_z. Ann. Geophys. **22**, 1755–1764 (2004)

M.G. Gusev, O.A. Troshichev, Hook-shaped arcs in dayside polar cap and their relationship to the IMF. Planet. Space Sci. **34**, 489 (1986)

M.G. Gusev, O.A. Troshichev, Relation of sun-aligned arcs to polar cap convection and magnetic disturbances. J. Atmos. Sol.-Terr. Phys. **38**, 1–11 (1990)

M.S. Gussenhoven, Extremely high latitude auroras. J. Geophys. Res. **87**, 2401–2412 (1982). https://doi.org/10.1029/JA087iA04p02401

D.A. Hardy, Intense fluxes of low-energy electrons at geomagnetic latitudes above 85°. J. Geophys. Res. **89**, 3883–3892 (1984). https://doi.org/10.1029/JA089iA06p03883

D.A. Hardy, W.J. Burke, M.S. Gussenhoven, DMSP optical and electron measurements in the vicinity of polar cap arcs. J. Geophys. Res. **87**, 2413–2430 (1982). https://doi.org/10.1029/JA087iA04p02413

D.A. Hardy, M.S. Gussenhoven, K. Riehl, R. Burkhardt, N. Heinemann, T. Schumaker, The Characteristics of polar cap precipitation and their dependence on the interplanetary magnetic field and the solar wind, in *Solar Wind Magnetosphere Coupling, Astrophysics and Space Science Library*, vol. 126, ed. by Y. Kamide, J.A. Slavin (Reidel, Dordrecht, 1986), pp. 575–604

M.G. Henderson, J.S. Murphree, J.M. Weygand, Observations of auroral substorms occurring together with preexisting "quiet time" auroral patterns. J. Geophys. Res. **101**, 24,621 (1996)

E.W. Hones Jr., J.D. Craven, L.A. Frank, D.S. Evans, P.T. Newell, The horse-collar aurora: A frequent pattern of the aurora in quiet times. Geophys. Res. Lett. **16**, 37 (1989)

K. Hosokawa, J.I. Moen, K. Shiokawa, Y. Otsuka, Motion of polar cap arcs. J. Geophys. Res. **116**, A01305 (2011). https://doi.org/10.1029/2010JA015906

C.Y. Huang, L.A. Frank, W.K. Petersen, D.J. Williams, W. Lennartsson, D.G. Mitchell, R.C. Elphic, C.T. Russell, Filamentary structures in the magnetotail lobes. J. Geophys. Res. **92**, 2349–2363 (1987)

C.Y. Huang, J.D. Craven, L.A. Frank, Simultaneous observations of a theta aurora and associated magnetotail plasmas. J. Geophys. Res. **94**, 10,137 (1989)

S. Ismail, C.-I. Meng, A classification of polar cap auroral arcs. Planet. Space Sci. **30**, 319 (1982)

S. Ismail, D.D. Wallis, L.L. Cogger, Characteristics of polar cap sun-aligned arcs. J. Geophys. Res. **82**, 4741 (1977)

P.T. Jayachandran, K. Hosokawa, J.W. MacDougall, S. Mushini, R.B. Langley, K. Shiokawa, GPS total electron content variations associated with a polar cap arc. J. Geophys. Res. **114**, A12304 (2009). https://doi.org/10.1029/2009JA014916

P.T. Jayachandran, K. Hosokawa, K. Shiokawa, Y. Otsuka, C. Watson, S.C. Mushini, J.W. MacDougall, P. Prikryl, R. Chadwick, T.D. Kelly, GPS total electron content variations associated with poleward moving Sun-aligned arcs. J. Geophys. Res. **117**, A05310 (2012). https://doi.org/10.1029/2011JA017423

P.T. Jayachandran, A. Hamza, K. Hosokawa, H. Mezaoui, K. Shiokawa, GPS amplitude and phase scintillation associated with polar cap auroral forms. J. Atmos. Sol.-Terr. Phys. **164**, 185–191 (2017)

Z. Kaymaz, G.L. Siscoe, J.G. Luhmann, R.P. Lepping, C.T. Russell, Interplanetary magnetic field control of magnetotail magnetic field geometry: IMP 8 observations. J. Geophys. Res. **99**(A6), 11113–11126 (1994). https://doi.org/10.1029/94JA00300

A. Koustov, K. Hosokawa, N. Nishitani, T. Ogawa, K. Shiokawa, Rankin Inlet PolarDARN radar observations of duskward moving Sun-aligned optical forms. Ann. Geophys. **26**, 2711–2723 (2008). https://doi.org/10.5194/angeo-26-2711-2008

A.V. Koustov, K. Hosokawa, N. Nishitani, K. Shiokawa, H. Liu, Signatures of moving polar cap arcs in the F-region PolarDARN echoes. Ann. Geophys. **30**, 441–455 (2012). https://doi.org/10.5194/angeo-30-441-2012

A. Kozlovsky, A. Aikio, T. Turunen, H. Nilsson, T. Sergienko, V. Safargaleev, K. Kauristie, Dynamics and electric currents of morningside Sun-aligned auroral arcs. J. Geophys. Res. **112**, A06306 (2007). https://doi.org/10.1029/2006JA012244

A. Kullen, The connection between transpolar arcs and magnetotail rotation. Geophys. Res. Lett. **27**, 73–76 (2000). https://doi.org/10.1029/1999GL010675

A. Kullen, Transpolar Arcs: Summary and Recent Results, in *Auroral Phenomenology and Magnetospheric Processes: Earth And Other Planets*, ed. by A. Keiling, E. Donovan, F. Bagenal, T. Karlsson (2012). https://doi.org/10.1029/2011GM001183

A. Kullen, P. Janhunen, Relation of polar auroral arcs to magnetotail twisting and IMF rotation: a systematic MHD simulation study. Ann. Geophys. **22**, 951–970 (2004)

A. Kullen, M. Brittnacher, J.A. Cumnock, L.G. Blomberg, Solar wind dependence of the occurrence and motion of polar auroral arcs: A statistical study. J. Geophys. Res. **107**, 1362 (2002). https://doi.org/10.1029/2002JA009245

A. Kullen, J.A. Cumnock, T. Karlsson, Seasonal dependence and solar wind control of transpolar arc luminosity. J. Geophys. Res. **113**, A08316 (2008). https://doi.org/10.1029/2008JA013086

A. Kullen, R.C. Fear, S.E. Milan, J.A. Carter, T. Karlsson, The statistical difference between bending arcs and regular polar arcs. J. Geophys. Res. Space Phys. **120**, 10 (2015). https://doi.org/10.1002/2015JA021298

K. Lassen, On the classification of high-latitude auroras. Geophys. Publ. **29**, 87 (1972)

K. Lassen, C. Danielsen, Quiet-time pattern of auroral arcs for different directions of the interplanetary magnetic field in the Y-Z plane. J. Geophys. Res. **83**, 5277 (1978)

K.M. Laundal, N. Østgaard, K. Snekvik, H.U. Frey, Interhemispheric observations of emerging polar cap asymmetries. J. Geophys. Res. **115**, A07230 (2010). https://doi.org/10.1029/2009JA015160

K.M. Laundal, J.P. Reistad, C.C. Finlay, N. Østgaard, P. Tenfjord, K. Snekvik, A. Ohma, Interplanetary magnetic field B_x component influence on horizontal and field-aligned currents in the ionosphere. J. Geophys. Res. **123**, 3360–3379 (2018). https://doi.org/10.1002/2017JA024864

K. Liou, J.M. Ruohoniemi, P.T. Newell, R. Greenwald, C.-I. Meng, M.R. Hairston, Observations of ionospheric plasma flows within theta auroras. J. Geophys. Res. **110**, A03303 (2005). https://doi.org/10.1029/2004JA010735

R. Maggiolo, J.A. Sauvaud, D. Fontaine, A. Teste, E. Grigorenko, A. Balogh, A. Fazakerley, G. Paschmann, D. Delcourt, H. Rème, A multi-satellite study of accelerated ionospheric ion beams above the polar cap. Ann. Geophys. **24**(6), 1665–1684 (2006). https://doi.org/10.5194/angeo-24-1665-2006

R. Maggiolo, M. Echim, J. de Keyser, D. Fontaine, C. Jacquey, I. Dandouras, Polar cap ion beams during periods of northward IMF: Cluster statistical results. Ann. Geophys. **29**, 771–787. https://doi.org/10.5194/angeo-29-771-2011 (2011)

R. Maggiolo, M. Echim, C. Simon Wedlund, Y. Zhang, D. Fontaine, G. Lointier, J.-G. Trotignon, Polar cap arcs from the magnetosphere to the ionosphere: kinetic modelling and observations by Cluster and TIMED. Ann. Geophys. **30**, 283–302 (2012). https://doi.org/10.5194/angeo-30-283-2012

B. Mailyan, Q.Q. Shi, A. Kullen, R. Maggiolo, Y. Zhang, R.C. Fear, Q.-G. Zong, S.Y. Fu, X.C. Gou, X. Cao, Z.H. Yao, W.J. Sun, Y. Wei, Z.Y. Pu, Transpolar arc observation after solar wind entry into the high-latitude magnetosphere. J. Geophys. Res. **120**, 3525–3534 (2015). https://doi.org/10.1002/2014JA020912

K. Makita, C.I. Meng, S.I. Akasofu, Transpolar auroras, their particle precipitation, and IMF B_y component. J. Geophys. Res. **96**, 14,085 (1991)

D. Mawson, Auroral observations at the Cape Royds station, Antarctica. Trans. R. Soc. S. Aust. **XL**, 151–212 (1916)

D. Mawson, Records of the Aurora Polaris. Australasian expedition 1911–1914. Sci. Rep. B **11**, 5758367 (1925)

N.C. Maynard et al., Response of the open–closed field line boundary in the evening sector to IMF changes: A source mechanism for Sun-aligned arcs. J. Geophys. Res. **108**(A1), 1006 (2003). https://doi.org/10.1029/2001JA000174

N.C. Maynard, C.J. Farrugia, W.J. Burke, D.M. Ober, F.S. Mozer, H. Rème, M. Dunlop, K.D. Siebert, Cluster observations of the dusk flank magnetopause near the sash: Ion dynamics and flow-through reconnection. J. Geophys. Res. **117**, A10201 (2012). https://doi.org/10.1029/2012JA017703

D.J. McEwen, Y. Zhang, A continuous view of the dawn-dusk polar cap. Geophys. Res. Lett. **27**, 477 (2000)

S.B. Mende, H. Heetderks, H.U. Frey, M. Lampton, S.P. Geller, S. Habraken, E. Renotte, C. Jamar, P. Rochus, J. Spann, S.A. Fuselier, J.-C. Gerard, R. Gladstone, S. Murphree, L. Cogger, Far ultraviolet imaging from the IMAGE spacecraft: 1. System design. Space Sci. Rev. **91**, 243–270 (2000). https://doi.org/10.1023/A:1005271728567

C.-I. Meng, Polar cap arcs and the plasma sheet. Geophys. Res. Lett. **8**, 273 (1981)

C.-I. Meng, H.W. Kroehl, Intense uniform precipitation of low-energy electrons over the polar cap. J. Geophys. Res. **82**, 2305–2313 (1977). https://doi.org/10.1029/JA082i016p02305

S.E. Milan, M. Lester, S.W.H. Cowley, M. Brittnacher, The convection and auroral response to a southward turning of the IMF: Polar UVI, CUTLASS, and IMAGE signatures of transient magnetic flux transfer at the magnetopause. J. Geophys. Res. **105**, 15741–15756 (2000)

S.E. Milan, B. Hubert, A. Grocott, Formation and motion of a transpolar arc in response to dayside and nightside reconnection. J. Geophys. Res. **110**, A01212 (2005). https://doi.org/10.1029/2004JA010835

P.F. Mizera, D.J. Gorney, D.S. Evans, On the conjugacy of the aurora: High and low latitudes. Geophys. Res. Lett. **14**(3), 190–193 (1987). https://doi.org/10.1029/GL014i003p00190

J.S. Murphree, L.L. Cogger, Observed connections between apparent polar cap features and the instantaneous diffuse auroral oval. Planet. Space Sci. **29**, 1143 (1981)

J.S. Murphree, C.D. Anger, L.L. Cogger, The instantaneous relationship between polar cap and oval auroras at times of northward interplanetary magnetic field. Can. J. Phys. **60**, 349 (1982)

J.S. Murphree, L.L. Cogger, C.D. Anger, D.D. Wallis, G.G. Shepherd, Oval intensifications associated with polar arcs. Geophys. Res. Lett. **14**, 403–406 (1987)

J.S. Murphree, R.D. Elphinstone, M.G. Henderson, L.L. Cogger, D.J. Hearn, Interpretation of optical substorm onset observations. J. Atmos. Sol.-Terr. Phys. **55**(8), 1159–1170 (1993). https://doi.org/10.1016/0021-9169(93)90044-Y

J.S. Murphree, J.B. Austin, D.J. Hearn, L.L. Cogger, R.D. Elphinstone, J. Woch, Satellite observations of polar arcs. J. Atmos. Sol.-Terr. Phys. **56**(2), 265 (1994)

S.M. Naehr, F.R. Toffoletto, Quantitative modeling of the magnetic field configuration associated with the theta aurora. J. Geophys. Res. **109**, A07202 (2004). https://doi.org/10.1029/2003JA010191

P.T. Newell, C.-I. Meng, Creation of theta-auroras: the isolation of plasma sheet fragments in the polar cap. Science **270**, 1338 (1995)

P.T. Newell, D. Xu, C.-I. Meng, M.G. Kivelson, Dynamical polar cap: a unifying approach. J. Geophys. Res. **102**, 127 (1997). 1997

P.T. Newell, K. Liou, C.-I. Meng, M.J. Brittnacher, G. Parks, Dynamics of double-theta aurora: Polar UVI study of January 10-11. J. Geophys. Res. Space Phys. **104**(A1), 95–104 (1999). 1997

P.T. Newell, K. Liou, G.R. Wilson, Polar cap particle precipitation and aurora: Review and commentary. J. Atmos. Sol.-Terr. Phys. **71**, 199–215 (2009). https://doi.org/10.1016/j.jastp.2008.11.004

E. Nielsen, J.D. Craven, L.A. Frank, R.A. Heelis, Ionospheric flows associated with a transpolar arc. J. Geophys. Res. **95**(A12), 21,169–21,178 (1990). https://doi.org/10.1029/JA095iA12p21169

E. Nielsen, R. Elphinstone, D. Hearn, J.S. Murphree, T. Potemra, Oval intensification event observed by STARE and Viking. J. Geophys. Res. **98**, 6163 (1993)

Y. Nishimura, L.R. Lyons, K. Shiokawa, V. Angelopoulos, E.F. Donovan, S.B. Mende, Substorm onset and expansion phase intensification precursors seen in polar cap patches and arcs. J. Geophys. Res. **118**, 2034–2042 (2013). https://doi.org/10.1002/jgra.50279

M. Nowada, R.C. Fear, A. Grocott, Q.-Q. Shi, Q.-G. Zong, Y. Wei, S.-Y. Fu, Z.-Y. Pu, B. Mailyan, H. Zhang, Subsidence of Ionospheric Flows Triggered by Magnetotail Magnetic Reconnection During Transpolar Arc Brightening. J. Geophys. Res. **123**, 3398–3420 (2018). https://doi.org/10.1029/2017JA024701

T. Obara, M. Kitayama, T. Mukai, N. Kaya, L. Cogger, S. Murphree, Simultaneous observations of Sun-aligned polar cap arcs in both hemispheres by EXOS-C and Viking. Geophys. Res. Lett. **15**(7), 713–716 (1988). https://doi.org/10.1029/GL015i007p00713

N. Østgaard, S.B. Mende, H.U. Frey, L.A. Frank, J.B. Sigwarth, Observations of non-conjugate theta aurora. Geophys. Res. Lett. **30**, 2125 (2003). https://doi.org/10.1029/2003GL017914

N. Østgaard, S.B. Mende, H.U. Frey, J.B. Sigwarth, Simultaneous imaging of the reconnection spot in the opposite hemispheres during northward IMF. Geophys. Res. Lett. **32**, L21104 (2005). https://doi.org/10.1029/2005GL024491

N. Østgaard, S.B. Mende, H.U. Frey, J.B. Sigwarth, A. Åsnes, J.M. Weygand, Auroral conjugacy studies based on global imaging. J. Atmos. Sol.-Terr. Phys. **69**(3), 249–255 (2007). https://doi.org/10.1016/j.jastp.2006.05.026

C.J. Owen, J.A. Slavin, I.G. Richardson, N. Murphy, R.J. Hynds, Average motion, structure and orientation of the distant magnetotail determined from remote sensing of the edge of the plasma sheet boundary layer with E > 35 keV ions. J. Geophys. Res. **100**, 185 (1995)
L.J. Paxton, D. Morrison, Y. Zhang, H. Kil, B. Wolven, B.S. Ogorzalek et al., Validation of Remote Sensing Products Produced by the Special Sensor Ultraviolet Scanning Imager (SSUSI): A far UV Imaging Spectrograph on DMSP F16, in *Optical spectroscopic techniques, remote sensing, and instrumentation for atmospheric and space research IV*, ed. by A.M. Larar, M.G. Mlynczak. Proc. SPIE, vol. 4485 (SPIE, San Diego, 2002), pp. 338–348. https://doi.org/10.1117/12.454268
G.W. Perry, H. Dahlgren, M.J. Nicolls, M. Zettergren, J.-P. St.-Maurice, J.L. Semeter, T. Sundberg, K. Hosokawa, K. Shiokawa, S. Chen, Spatiotemporally resolved electrodynamic properties of a Sun-aligned arc over Resolute Bay. J. Geophys. Res. **120**, 9977–9987 (2015). https://doi.org/10.1002/2015JA021790
W.K. Peterson, E.G. Shelley, Origin of the plasma in a cross-polar cap auroral feature (theta aurora). J. Geophys. Res. **89**, 6729 (1984)
R.L. Rairden, S.B. Mende, Properties of 6300-Å auroral emissions at South Pol. J. Geophys. Res. **94**(A2), 1402–1416 (1989). https://doi.org/10.1029/JA094iA02p01402
J.A. Reidy, R.C. Fear, D.K. Whiter, B.S. Lanchester, A.J. Kavanagh, L.J. Paxton et al., Multi-instrument observation of simultaneous polar cap auroras on open and closed magnetic field lines. J. Geophys. Res. Space Phys. **122**, 4367–4386 (2017). https://doi.org/10.1002/2016JA023718
J. Reidy, R.C. Fear, D. Whiter, B.S. Lanchester, A.J. Kavanagh, S.E. Milan et al., Interhemispheric survey of polar cap aurora. J. Geophys. Res. Space Phys. **123**, 7283–7306 (2018). https://doi.org/10.1029/2017JA025153
B.V. Rezhenov, A possible mechanism for theta aurora formation. Ann. Geophys. **13**, 698 (1995)
R.M. Robinson, R.R. Vondrak, K. Miller, T. Dabbs, D. Hardy, On calculating ionospheric conductances from the flux and energy of precipitating electrons. J. Geophys. Res. **92**(A3), 2565–2569 (1987). https://doi.org/10.1029/JA092iA03p02565
J.M. Ruohoniemi, R.A. Greenwald, Statistical patterns of high-latitude convection obtained from Goose Bay HF radar observations. J. Geophys. Res. **101**, 9494–9510 (1996). https://doi.org/10.1029/96JA0158
M. Samara, R.G. Michell, Optical and radar observations of small-scale polar cap auroral structures. J. Atmos. Sol.-Terr. Phys. **105–106**, 308–312 (2013)
G.G. Shepherd, A. Steen, J.S. Murphree, Auroral boundary dynamics observed simultaneously from the Viking spacecraft and from the ground. J. Geophys. Res. **95**, 5845 (1990)
Q.Q. Shi, Q.-G. Zong, S.Y. Fu, M.W. Dunlop, Z.Y. Pu, G.K. Parks, Y. Wei, W.H. Li, H. Zhang, M. Nowada, Y.B. Wang, W.J. Sun, T. Xiao, H. Reme, C. Carr, A.N. Fazakerley, E. Lucek, Solar wind entry into the high-latitude terrestrial magnetosphere during geomagnetically quiet times. Nat. Commun. **4**, 1466 (2013). https://doi.org/10.1038/ncomms2476
I. Shinohara, S. Kokubun, Statistical properties of particle precipitation in the polar cap during intervals of northward interplanetary magnetic field. J. Geophys. Res. **101**, 69–82 (1996). https://doi.org/10.1029/95JA01848
K. Shiokawa, K. Yumoto, K. Hayashi, T. Oguti, D.J. McEwen, A statistical study of the motions of auroral arcs in the high-latitude morning sector. J. Geophys. Res. **100**, 21,979–21,985 (1995). https://doi.org/10.1029/95JA01564
K. Shiokawa, K. Yumoto, N. Nishitani, K. Hayashi, T. Oguti, D.J. McEwen, Y. Kiyama, F.J. Rich, T. Mukai, Quasi-periodic poleward motions of Sun-aligned auroral arcs in the high-latitude morning sector: A case study. J. Geophys. Res. **101**, 19,789–19,800 (1996). https://doi.org/10.1029/96JA01202
K. Shiokawa, T. Ogino, K. Hayashi, D.J. McEwen, Quasi-periodic poleward motions of morningside Sun-aligned arcs: A multi-event study. J. Geophys. Res. **102**, 24,325–24,332 (1997). https://doi.org/10.1029/97JA02383
J.-H. Shue, P.T. Newell, K. Liou, C.-I. Meng, The quantitative relationship between auroral brightness and solar EUV Pedersen conductance. J. Geophys. Res. **106**, 5883 (2001)
S.P. Slinker, J.A. Fedder, D.J. McEwen, Y. Zhang, J.G. Lyon, Polar cap study during northward interplanetary magnetic field on 19 January 1998. Phys. Plasmas **8**, 1119 (2001)
J.F. Spann, M. Brittnacher, R. Elsen, G.A. Germany, G.K. Parks, Initial response and complex polar cap structures of the aurora in response to the January 10, 1997 magnetic cloud. Geophys. Res. Lett. **25**, 2577 (1998)
T.J. Stubbs, R.R. Vondrak, N. Østgaard, J.B. Sigwarth, L.A. Frank, Simultaneous observations of the auroral ovals in both hemispheres under varying conditions. Geophys. Res. Lett. **32**, L03103 (2005). https://doi.org/10.1029/2004GL021199
T. Tanaka, T. Obara, M. Kunitake, Formation of the theta aurora by a transient convection during northward interplanetary magnetic field. J. Geophys. Res. **109**, A09201 (2004). https://doi.org/10.1029/2003JA010271

T. Tanaka, T. Obara, M. Watanabe, S. Fujita, Y. Ebihara, R. Kataoka, Formation of the Sun-aligned arc region and the void (polar slot) under the null-separator structure. J. Geophys. Res. **122**, 4102–4116 (2017). https://doi.org/10.1002/2016JA023584

T. Tanaka, T. Obara, M. Watanabe, S. Fujita, Y. Ebihara, R. Kataoka, M. Den, Cooperatives Roles of Dynamics and Topology in Generating the Magnetosphere-Ionosphere Disturbances: Case of the Theta Aurora. J. Geophys. Res. Space Phys. **123**(12), 9991–10008 (2018)

A. Teste, D. Fontaine, J.-A. Sauvaud, R. Maggiolo, P. Canu, A. Fazakerley, CLUSTER observations of electron outflowing beams carrying downward currents above the polar cap by northward IMF. Ann. Geophys. **25**, 953–969 (2007). https://doi.org/10.5194/angeo-25-953-2007

M.R. Torr, D.G. Torr, M. Zukic, R.B. Johnson, J. Ajello, P. Banks, K. Clark, K. Cole, C. Keffer, G. Parks, B. Tsurutani, J. Spann, A far ultraviolet imager for the International Solar-Terrestrial Physics mission. Space Sci. Rev. **71**(1–4), 329–383 (1995)

N.A. Tsyganenko, A magnetospheric magnetic field model with a warped tail current sheet. Planet. Space Sci. **37**, 5 (1989)

N.A. Tsyganenko, D.H. Fairfield, Global shape of the magnetotail current sheet as derived from Geotail and Polar data. J. Geophys. Res. **109**, A03218 (2004). https://doi.org/10.1029/2003JA010062

C.E. Valladares, H.C. Carlson Jr., The electrodynamic, thermal, and energetic character of intense Sun-aligned arcs in the polar cap. J. Geophys. Res. **96**(A2), 1379–1400 (1991). https://doi.org/10.1029/90JA01765

C.E. Valladares, H.C. Carlson Jr., K. Fukui, Interplanetary magnetic field dependency of stable Sun-aligned polar cap arcs. J. Geophys. Res. **99**, 6247–6272 (1994). https://doi.org/10.1029/93JA03255

C. van der Meeren, K. Oksavik, D.A. Lorentzen, L.J. Paxton, L.B.N. Clausen, Scintillation and irregularities from the nightside part of a Sun-aligned polar cap arc. J. Geophys. Res. **121**, 5723–5736 (2016). https://doi.org/10.1002/2016JA022708

R.J. Walker, R.L. Richard, T. Ogino, M. Ashour-Abdalla, The response of the magnetotail to changes in the IMF orientation: The magnetotail's long memory. Phys. Chem. Earth Part C **24**, 221 (1999)

M. Watanabe, M.R. Hairston, Observation of a unipolar field-aligned current system associated with IMF B_y-triggered theta auroras. J. Geophys. Res. **121**, 4483–4497 (2016). https://doi.org/10.1002/2015JA022016

E.J. Weber, J. Buchau, Polar cap F-layer auroras. Geophys. Res. Lett. **8**(1), 125–128 (1981). https://doi.org/10.1029/GL008i001p00125

J.M. Weygand, J.S. Murphree, M.G. Henderson, G.A. Enno, Simultaneous closed magnetic field line polar arcs and substorms. J. Atmos. Terr. Phys. **63**, 643–655 (2001). https://doi.org/10.1016/S1364-6826(00)00184-X

M. Wiltberger, R.S. Weigel, W. Lotko, J.A. Fedder, Modeling seasonal variations of auroral particle precipitation in a global-scale magnetosphere-ionosphere simulation. J. Geophys. Res. **114**, A01204 (2009). https://doi.org/10.1029/2008JA013108

J.D. Winningham, W.J. Heikkila, Polar cap auroral electron fluxes observed with Isis 1. J. Geophys. Res. **79**(7), 949–957 (1974). https://doi.org/10.1029/JA079i007p00949

Q. Wu, T.J. Rosenberg, F.T. Berkey, R.H. Eather, Intensification and fading of auroral arcs in the dusk-midnight sector of the polar cap. J. Geophys. Res. **96**(A5), 7709–7719 (1991). https://doi.org/10.1029/91JA00256

J. Xing, Q. Zhang, D. Han, Y. Zhang, N. Sato, S. Zhang et al., Conjugate observations of the evolution of polar cap arcs in both hemispheres. J. Geophys. Res. Space Phys. **123**, 1794–1805 (2018). https://doi.org/10.1002/2017JA024272

T. Yamamoto, M. Ozaki, A numerical model of the dayside aurora. J. Geophys. Res. **110**, A05215 (2005). https://doi.org/10.1029/2004JA010786

D.Y. Yeager, L.A. Frank, Low-energy electron intensities at large distances over the Earth's polar cap. J. Geophys. Res. **81**, 3966–3976 (1976). https://doi.org/10.1029/JA081i022p03966

Y. Zhang, D.J. McEwen, N. Kong, J.J. Sojka, L. Zhu, Horizontal electrojet associated with the sun-aligned arcs. J. Atmos. Sol.-Terr. Phys. **61**, 1249–1257 (1999)

Y. Zhang, L.J. Paxton, P.T. Newell, C.I. Meng, Does the polar cap disappear under an extended strong northward IMF? J. Atmos. Sol.-Terr. Phys. **71**(17–18), 2006–2012 (2009)

Y. Zhang, L.J. Paxton, Q.-H. Zhang, Z. Xing, Polar cap arcs: sun-aligned or cusp-aligned? J. Atmos. Sol.-Terr. Phys. **146**, 123–128 (2016)

L. Zhu, C.E. Valladares, J.J. Sojka, R.W. Schunk, D.J. Crain, Model-observation comparison study of multiple polar cap arcs. J. Geophys. Res. **101**(A1), 323–333 (1996). https://doi.org/10.1029/95JA02758

L. Zhu, R.W. Schunk, J.J. Sojka, Polar cap arcs: A review. J. Atmos. Sol.-Terr. Phys. **59**, 1087–1126 (1997). https://doi.org/10.1016/S1364-6826(96)00113-7

Y. Zou, Y. Nishimura, L.R. Lyons, E.F. Donovan, K. Shiokawa, J.M. Ruohoniemi, K.A. McWilliams, N. Nishitani, Polar cap precursor of nightside auroral oval intensifications using polar cap arcs. J. Geophys. Res. **120**, 10,698–10,711 (2015). https://doi.org/10.1002/2015JA021816

Space Sci Rev (2019) 215:53
https://doi.org/10.1007/s11214-019-0616-8

Airglow Patches in the Polar Cap Region: A Review

Keisuke Hosokawa[1] · Ying Zou[2,3] · Yukitoshi Nishimura[2]

Received: 30 April 2019 / Accepted: 18 October 2019 / Published online: 18 November 2019

Abstract Polar cap airglow patches have been known as regions of enhanced 630.0 nm airglow detected by ground-based all-sky imagers at the polar cap latitudes well inside the main auroral oval. Although they were already recognized almost four decades ago as counterparts of polar cap (plasma density) patches, such airglow observations had not been utilized extensively for the studies of ionospheric structures and/or magnetosphere-ionosphere coupling processes in the polar cap. In the last two decades, following the development of highly-sensitive airglow imagers equipped with cooled CCD (Charge Coupled Device) cameras, it has become possible to visualize the dynamical temporal evolution and complicated spatial structure of airglow patches with improved signal-to-noise ratio. Such a progress has enabled us not only to use airglow patches as tracers for plasma convection in the polar cap but also to understand the generation of small-scale plasma irregularities in the ionospheric F region. In addition, recent observations demonstrated a case in which an airglow patch was accompanied by an intense flow channel and corresponding field-aligned current structure along its edges. This implies that airglow patches can signify magnetosphere-ionosphere coupling process in the region of open field lines at the polar cap latitudes, serving as a remote sensing tool just like auroras do. Further studies showed an association of airglow patches with the intensification of aurora on the nightside (Poleward Boundary Intensification: PBI and/or streamer) leading to the expansion phase onset of substorms. This paper

Auroral Physics
Edited by David Knudsen, Joe Borovsky, Tomas Karlsson, Ryuho Kataoka and Noora Partmies

✉ K. Hosokawa
keisuke.hosokawa@uec.ac.jp

Y. Zou
yz0025@uah.edu

Y. Nishimura
toshi16@bu.edu

1 Department of Communication Engineering and Informatics, University of Electro-Communications, Tokyo, Japan

2 Center for Space Physics, Boston University, Boston, USA

3 Department of Space Science, University of Alabama in Huntsville, Huntsville, USA

reviews such recent progresses in the researches of airglow patches obtained by combining data from all-sky airglow imagers, radars and low-altitude satellite observations in the polar cap.

Keywords Airglow · Polar cap · Ionospheric convection · Substorm · Field-aligned currents

1 Introduction

1.1 History and Definition

Airglow patches in the polar cap region were first detected by Weber et al. (1984) using all-sky intensified photometers (ASIPs). They were identified in the all-sky images as regions of enhanced airglow at 630.0 nm wavelength, which is known to be proportional to the electron density at the F region altitudes (see Sect. 1.2 for detail). Figure 1 shows a sequence of 630.0 nm airglow patches travelling across the field-of-view of the ASIP. The drift velocity of the airglow patches was anti-sunward (left to right in Fig. 1) and the speed was $\sim 700\ \mathrm{m\,s^{-1}}$, which is consistent with anti-sunward convection in the polar cap during southward IMF conditions. In this example, the 630.0 nm emission intensity was elevated by $\sim$ 200 Rayleigh, which is sub-visible and much darker than typical reddish aurora in the polar cap (Weber and Buchau 1981). Weber et al. (1984) also demonstrated that such enhancements of the 630.0 nm airglow cannot be explained by impact excitation due to soft electron precipitation in the polar cap region. Then, they concluded that the observed airglow patches are not produced by direct particle precipitation, but are primarily airglow structures produced by patches of enhanced F region ionization. In fact, simultaneous ionosonde observations showed that the F region electron density was enhanced by a factor of $\sim$ 10 inside the airglow patches.

Before the first observations of airglow patches by Weber et al. (1984), there had been several radio observations (e.g., ionosonde, total electron content, scintillation etc.) showing periodic enhancements of FoF2, one of the proxies for the F region peak density, at polar cap latitudes (e.g., Weber et al. 1986). Weber et al. (1986) also identified corresponding increases in GPS TEC and amplitude scintillation in satellite signals at VHF and UHF frequencies. These electron density enhancements were named "polar cap patches" which are regions of enhanced electron density travelling from the dayside sunlit region across the central polar cap. The origin of the newly observed airglow enhancements by Weber et al. (1984) was discussed in close connection with the passage of patches of high-density plasma. Specifically, the airglow patches are optical manifestations of polar cap electron density patches. Since the early observations by Weber et al. (1984, 1986), there have been several optical observations in the polar cap showing the two-dimensional structure of polar cap patches (e.g., Fukui et al. 1994; McEwen and Harris 1996; Pedersen et al. 2000; Zhang et al. 2003). Later development of cooled-CCD cameras in the late 90's enabled us to visualize the detailed spatial structure of airglow patches with much improved signal-to-noise ratio (Dandekar and Bullett 1999; Hosokawa et al. 2006; Carlson et al. 2004). Such recent observations will be introduced in Sect. 1.3.

The definition of polar cap electron density patches is "a region of dense plasma at F region altitudes whose electron density is at least 2 times larger than the surrounding region" (Crowley 1996). This definition has widely been used for detecting electron density patches by using the electron density data from the low-altitude polar orbiting satellites such as DE-2, DMSP, CHAMP and Swarm (Coley and Heelis 1998; Noja et al. 2013; Spicher

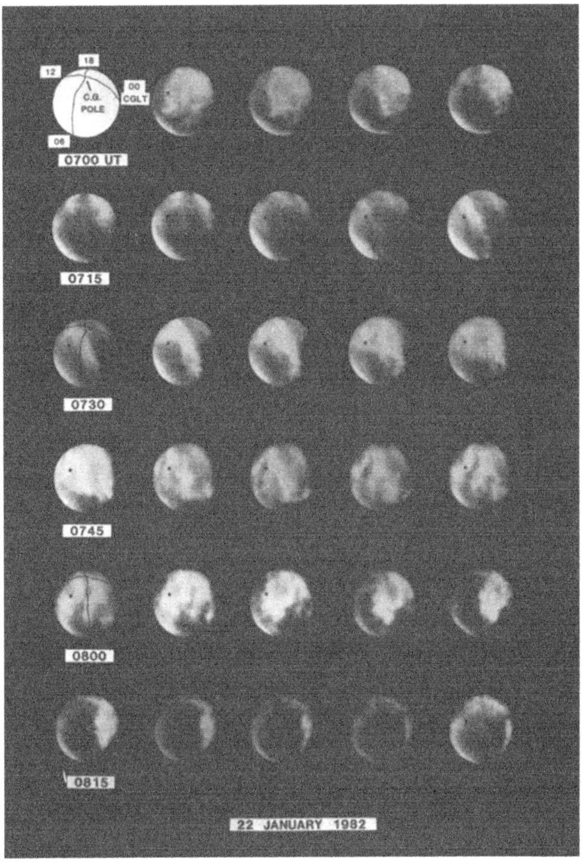

Fig. 1 Successive 630.0 nm airglow images obtained from Thule, Greenland by an ASIP instrument (Weber et al. 1984). The white regions indicate areas of enhanced electron density, which are signatures of polar cap airglow patches. Airglow patches drifted from left (12 MLT) to right (00 MLT) which is anti-sunward direction

et al. 2017). In contrast to electron density patches, there has been no quantitative definition of airglow patches that is used universally. But identification is often easy due to patches' marked difference from auroras in several respects, such as spatial form, characteristic motion, recurrence pattern, and emission wavelengths. Hosokawa et al. (2009b) designed an automatic identification procedure where airglow patches are identified as structures with 630.0 nm emission intensity $>$ 30 R higher than the background, and with 557.7 nm emission intensity $<$ 20 R higher (to avoid auroras).

Airglow patches are not auroras, but they are natural optical emissions coming from the atmosphere just like auroras, and their dynamics can be closely related and/or analogized to auroras as seen below. This chapter is mainly focused on progresses made based on optical observations of airglow patches.

1.2 Airglow Mechanism

Airglow patches have been observed by using 630.0 nm airglow whose central emission altitude is around 250 km. The process producing this emission can be described by the following two steps reactions (Hays et al. 1978, and references therein):

$$O^+ + O_2 \rightarrow O_2^+ + O$$

$$O_2^+ + e \rightarrow O^*(^1S, ^1D) + O^*(^1S, ^1D)$$

Fig. 2 Schematic illustration showing the altitude profile of the 630.0 nm airglow emission (red curve). The altitude profile of the densities of atomic oxygen ion (∼ electron density in the F region) and molecular oxygen (black curves)

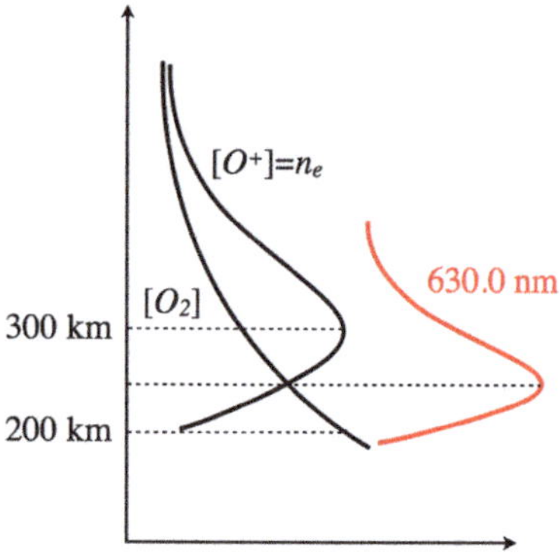

The excited state oxygen atom at 1D state was produced by the dissociative recombination through the molecular oxygen atom, which in turn produces emission when it transits to the ground state at ^{3}P through the following equation:

$$O^*(^1D) \rightarrow O(^3P) + hv.$$

The emission produced by this process is determined to be 630.0 nm, which is often called red line airglow emission (OI6300).

By looking at the first reaction describe above, the intensity of the emission depends not only on the O^+ density but also on the density of molecular oxygen. The O^+ density is maximized near the peak of F region at around 300 km altitude because they are the dominant ion species at that altitude. The density of neutral oxygen molecule decreases with increasing altitude. This makes the peak of the red line emission slightly below the F region peak altitude (Fig. 2). It depends on the background neutral density conditions (season, local time, and solar wind conditions) as described in Sakai et al. (2014). However, it is mostly located at around 250 km. The emission intensity of 630.0 nm airglow has simply been used as a proxy for the F region electron density (Barbier et al. 1962). As mentioned above, however, the observed emission intensity is also associated with the height distribution of airglow patches. To reduce this effect, much weaker oxygen emission at 777.4 nm (Makela et al. 2001) is sometimes employed for identifying the structure of airglow patches (Carlson et al. 2006). The use of 777.4 nm emission allows us to ignore the effect of molecular oxygen; thus, it provides more direct observations of the electron density at the peak altitude of F region.

In addition to the emission produced through the recombination of cold oxygen ions describe above, thermal excitation is also known to produce the 630.0 nm airglow emission by heat applied to atomic oxygen (Carlson et al. 2013; Kwagala et al. 2017, 2018a, 2018b). Thus, when we examine the 630.0 nm images from ground-based airglow imagers, it is important to distinguish the origins of the airglow emission (recombination or thermal excitation). At least, care must be taken to confirm that the emission is not due to direct impact excitation of oxygen by electrons precipitating from the magnetosphere.

1.3 Optical Signature

Optical observations enable a direct visualization of the 2D structure of patches. Patches are often 100–2000 km in size and appear as a circular or cigar-shaped structures. The cigar shape is sometimes elongated in the dawn-dusk direction (Fig. 3a) (Carlson 2003; Hosokawa et al. 2010a, 2013, 2014), and sometimes in the noon-midnight direction (Fig. 3b)

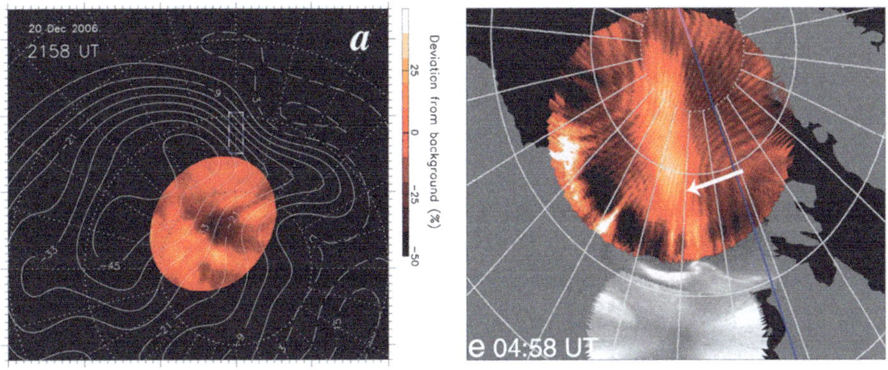

Fig. 3 Left: an example of dawn-dusk elongated airglow patches in MLT/magnetic latitude coordinates (Hosokawa et al. 2009a). The patch is recorded at 630 nm emission by the OMTI imager at Resolute Bay. Airglow intensity is color-coded according to the percentage deviation from the 1 h running average. Superimposed contours give a pattern of the high-latitude electrostatic potential derived from SuperDARN. Right: an example of noon-midnight elongated airglow patches as marked with the white arrow (Nishimura et al. 2013). The blue and pink lines mark magnetic midnight and noon. The grey-scale images equatorward of the 630.0-nm image are white-light aurora measurements from THEMIS ASIs

(Nishimura et al. 2013; Zou et al. 2015). The specific elongation direction is found to vary with the IMF conditions (Zou et al. 2015), the dawn-dusk elongation occurring under strong IMF B_z and the noon-midnight elongation under strong IMF B_y. Factors affecting the elongation direction include how wide the dayside source region of patches is and how much patches rotate within the polar cap (Oksavik et al. 2010). Patches generally have a luminosity of tens to a few hundreds of Rayleighs in the 630.0 nm emission. The luminosity on average decreases as the solar activity declines (McEwen and Harris 1996). It also decreases as patches move away from the dayside cusp (McEwen and Harris 1996). The luminosity has an asymmetric spatial distribution, where the gradient at the patch leading edge is 2–3 times steeper than that at the trailing edge (Hosokawa et al. 2016a). The trailing edge is found to exhibit fingerlike undulations which are likely manifestations of plasma structuring through the gradient-drift instability (GDI) (Hosokawa et al. 2016a) (see details in Sect. 3.1). Such an internal structuring process is also seen in the simultaneous observations of the 630.0 nm emission and incoherent scatter radar observations (Dahlgren et al. 2012a, 2012b). Patches also have faint emissions at 577.7 nm wavelength, the intensity of which is $< 40\%$ of that of 630.0 nm (McEwen and Harris 1996).

1.4 Motion of Airglow Patches

The motion of airglow patches is controlled by local plasma convection, as found from simultaneous measurements of patches and convection (Fukui et al. 1994; Thomas et al. 2015; Valladares et al. 2015). When convection is sheared, patches are observed to deform accordingly (Hosokawa et al. 2010a). Since convection in the polar cap is governed by the IMF conditions, patch motion is a function of IMF. Statistical analysis finds that the speed of patches increases with the magnitude of the southward IMF B_z component, and the dawn-dusk component of the patch speed correlates with the IMF B_y component (Fig. 4) (Zhang et al. 2003; Hosokawa et al. 2009b). When the IMF has a temporal variation, for example turning north from south, patches respond accordingly, stagnating and decaying within the polar cap (Hosokawa et al. 2011).

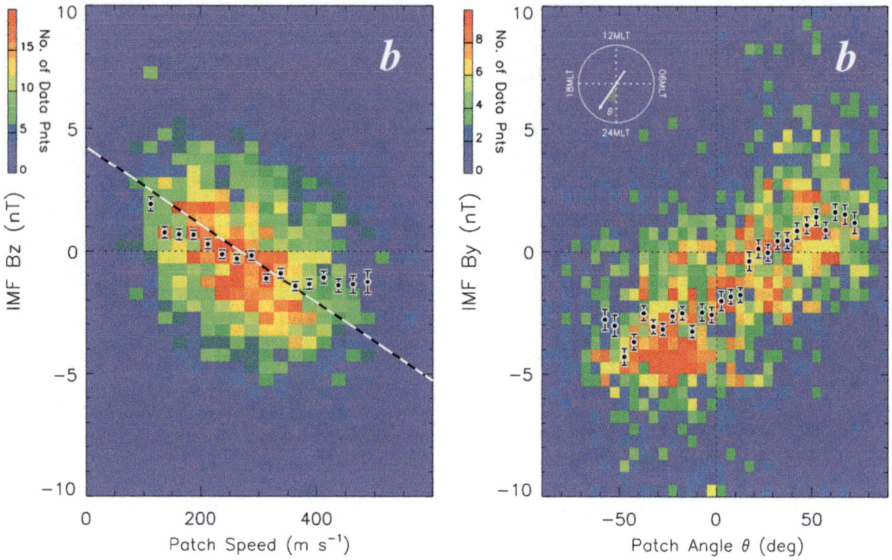

Fig. 4 Left: Distribution of the speed of patches as a function of the IMF B_Z. The density of data points is self-normalized and shown on a linear scale. The dots and bars represent the mean and standard error of the distributions, respectively, in 25 $m\,s^{-1}$ wide bins of patch speed. Right: Distribution of the moving direction of patches (θ: angle of azimuth relative to the noon-midnight line) as a function of the IMF B_Y. The dots and bars represent the mean and standard error of the distributions, respectively, in 5° wide bins of patch speed (Hosokawa et al. 2009b)

One recent finding is that convection in the central polar cap is spatially structured rather than smooth as traditionally assumed. In fact, quite often convection consists of fast-propagating mesoscale structures embedded in a slower convection background and the patch motion is closely related to these mesoscale structures (Nishimura et al. 2014; Zou et al. 2015; Wang et al. 2016; Lyons et al. 2016a). Details of the mesoscale convection will be addressed in Sect. 3.2.

Patches also have vertical motion because the ExB convection has a finite vertical projection at regions where magnetic field lines are inclined. Patches thus move upward when convecting towards the pole, and downward when away from the pole. The vertical motion has a direct impact on the luminosity of the patches, causing the luminosity to increase and then decrease during the course of transpolar motion (Perry et al. 2013). Valladares et al. (2015) also showed an example of airglow patches in which their luminosity changed during their transit across the field-of-view of the imager. By reproducing such a situation using Global Theoretical Ionospheric Model (GTIM), they showed that downward neutral wind associated with gravity waves could produce the observed variation of the 630.0 nm emission intensity.

2 Dayside Observations

2.1 Generation Process

One of the possible sources of airglow patches is high-density plasma in the dayside sunlit region. The high-density plasmas in the source region are captured by the high-latitude convection, in particular the anti-sunward flow near the cusp. During very disturbed conditions, for example during large magnetic storms, the high-latitude convection expands towards lower latitudes and overlaps with the high-density source plasma in the sunlit region. In such cases, high-density plasma, that is as dense as the dayside sunlit region, can be captured and transported towards the central polar cap region as so-called tongue of ionization (TOI: e.g., Sato 1959; Foster et al. 2005). Such a large-scale structure is also observed by an all-sky imager as a huge plume of enhanced 630.0 nm airglow extending in the noon-midnight direction (Hosokawa et al. 2010b). Van der Meeren et al. (2014) also introduced an example of TOI in the 630.0 nm airglow data which stretched all the way across the polar cap from the Canadian dayside sector to Svalbard on the nightside.

The TOI structure has a more continuous spatial shape than patches and are mainly elongated in the noon-midnight direction. However, most of the airglow patches seen in the central polar cap during disturbed conditions are known to be cigar-shaped with elongation more in the dusk-to-dawn direction (Hosokawa et al. 2014). Such a patchy and individual spatial distribution is associated with the temporal variation of the convection on the dayside. In the last 2–3 decades, numerous mechanisms have been proposed for chopping the dayside sunlit plasma into patches: (1) the large-scale direction change in the plasma convection near the cusp due to changes in the IMF B_y and B_z (Anderson et al. 1988), (2) plasma density reduction due to convection jets through frictional heating (Valladares et al. 1996); (3) repetitive capturing of the dayside sunlit plasma due to the expansion/contraction of the open/closed field line boundary by transient magnetopause reconnection (Lockwood and Carlson 1992; Carlson et al. 2004, 2006); (4) cutting of high-density plasma by the downward Birkeland current sheet at the equatorward boundary of the flow disturbance (Moen et al. 2006). These proposed mechanisms actually differ from each other. However, all the mechanisms need temporal variations of convection in the vicinity of the dayside inflow region near the cusp. In addition to these mechanisms, impact ionization by soft particle precipitation could also be one of the possible generation mechanisms of polar cap patches (Millward et al. 1999). In particular, it has been pointed out that transient aurora, known as Poleward Moving Auroral Forms (PMAFs: Fasel 1995), plays an important role in producing airglow patches (documented in detail in the next section).

2.2 Connection with PMAF

Evidence shows that at least some airglow patches are formed from poleward moving auroral forms (PMAFs). PMAFs are auroral structures that brighten at the equatorward boundary of the dayside auroral oval and propagate poleward into the polar cap (e.g. Sandholt et al. 1986). They are accompanied by fast poleward-directed plasma flows (Milan et al. 1999; Thorolfsson et al. 2000; Oksavik et al. 2004) and are thought to be the optical signature of pulsed reconnection on the dayside magnetopause (Southwood 1987). PMAFs often occur as a series of events separated by a few to ten minutes (Fasel 1995). The occurrence of airglow patches also exhibits periodicity. The periodicity occurs at two distinct periods, 40 min and at 5–12 min (Hosokawa et al. 2013). The larger-period modulations are postulated to result from large-scale reconfiguration of the dayside convection pattern while the shorter-period ones relate to dayside magnetopause reconnection occurring at the same rate as PMAFs.

Observations also directly show that airglow patches can emerge from the most poleward location of PMAFs and continue to propagate poleward (Fig. 5) (Lorentzen et al. 2010; Hosokawa et al. 2016b), even reaching the nightside polar cap and auroral zone (Nishimura et al. 2014). Those patches appear faint, having an intensity less than 150 R in comparison to

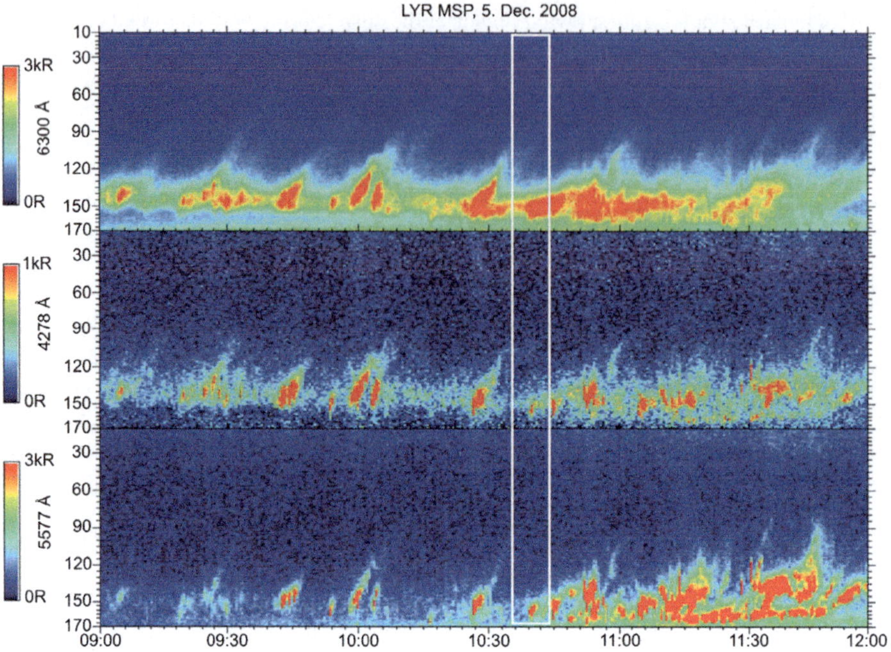

Fig. 5 3 wavelength (630.0 nm, 427.8 nm and 557.7 nm) photometer observation of PMAFs and newly created patches (Lorentzen et al. 2010). Only in the 630.0 nm observation (top panel), faint traces of airglow enhancement are seen to emerge from the poleward edge of PMAFs, indicating a nearly one-to-one correspondence between the two features

the typical hundreds-of-Rayleigh patches seen in Hosokawa et al. (2006, 2009b). This fact indicates that the density enhancement within the PMAF-created patches is low. The density enhancement is likely to arise from particle precipitation and/or upwelling of plasma due to Joule heating at lower altitude at the cusp (Lorentzen et al. 2010; Hosokawa et al. 2016b).

PMAF-created patches are known to cause severe scintillation and resultant loss-of-lock of trans-ionospheric GNSS navigation signals when they are produced on the dayside (Oksavik et al. 2015). Similar effects on the GNSS signals are detected even when such airglow patches encounter the nightside auroral oval after their travel across the pole (Jin et al. 2014; Van der Meeren et al. 2015). Thus, PMAF-created patches are of particular importance in the framework of space weather.

3 Central Polar Cap Observations

3.1 Plasma Structuring

Polar cap patches have been known to be sources of plasma irregularities in the polar cap region. Such small-scale irregularities in the electron density are mainly observed by using coherent scatter radar such as Super Dual Auroral Radar Network (SuperDARN, Chisham et al. 2007). In this case, irregularities are detected as scattering targets of such coherent radars. For example, Milan et al. (2002) showed characteristic poleward moving blobs of backscatter echoes as signatures of polar cap patches. Later, Hosokawa et al. (2009a) demonstrated

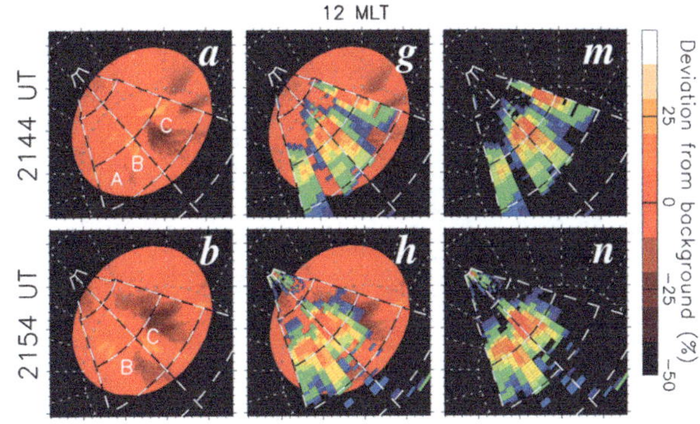

Fig. 6 A sequence of 630.0 nm airglow patches (left panels) and SuperDARN radar backscatter power maps (right panels). In the central panels, superposition of airglow and radar backscatter are shown for comparison (Hosokawa et al. 2009a). The radar echoes are distributed in the entire part of the airglow patches

that radar backscatter echoes were only detected within simultaneously observed airglow patches (Fig. 6). This proves that polar cap patches are active sources of deca-meter scale plasma irregularities.

Past research has suggested plasma irregularities within patches are formed through the so-called gradient-drift instability (GDI: Keskinen and Ossakow 1982). GDI is operative at F region altitudes when the background convection is parallel to the gradient in the electron density. Thus, GDI works only in the trailing edge of patches. In fact, Milan et al. (2002) indicated that the radar backscatter power was relatively larger in the trailing edge of the echo region, which is consistent with the preference of the growth of GDI on the trailing edge. From these observations, GDI has been considered as one of the mechanisms for the internal structuring of plasma within airglow patches. Later, Moen et al. (2012) and Oksavik et al. (2012) demonstrated, by using in-situ electron density measurement by a sounding rocket, that GDI is the primary structuring mechanism of patches in the dayside cusp region. At the same time, Oksavik et al. (2012) pointed out the importance of structured particle precipitation as a source of larger scale "seed" irregularities, which are then efficiently broken down into decameter-scale irregularities by the GDI mechanism. More recent in-situ electron density observations by Goodwin et al. (2015) using the Swarm constellation have suggested that the small-scale internal structure of airglow patches are the product of particle impact ionization in the cusp. On larger spatial scales, GDI can also affect the entire shape of airglow patches. Recently, Hosokawa et al. (2013, 2016a) identified wavy/undulating structure along the trailing edge of airglow patches imaged by high time resolution EMCCD airglow camera (Fig. 7). Such wavy structures are considered to be an indication of GDI as a structuring mechanism of airglow patches.

3.2 Flow Channel, Precipitation and FAC

As noted in Sect. 1.4, airglow patches are recently found to be collocated with mesoscale convection structures that are embedded in a slower convection background (Nishimura et al. 2014; Zou et al. 2015; Wang et al. 2016; Lyons et al. 2016a). These patches occur during non-storm time especially under B_y-dominated IMF conditions. They tend to have a cigar shape that is elongated along or at a small angle from the noon-midnight meridian direction. Since the collocated convection structures have a similarly elongated shape, they are often referred to as flow channels. The flow channels have sharp velocity gradients, particularly at the leading edge of patches. The velocities within the flow channels are usually > 200 m/s larger than the surrounding plasma. Different from the background convection, the flow channels exhibit dynamic evolution even during steady IMF conditions. They may transport

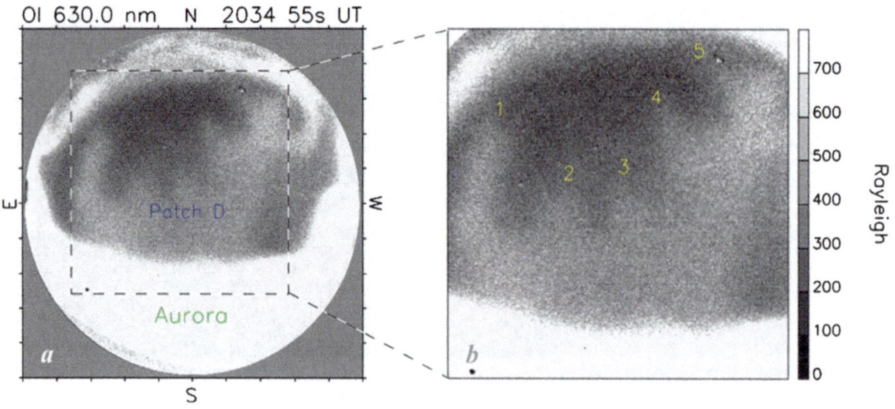

Fig. 7 630.0 nm all-sky image of an airglow patch obtained by an all-sky EMCCD camera in Longyearbyen, Norway (Hosokawa et al. 2016b). This airglow patch drifted anti-sunward (from top to bottom in the all-sky field-of-view). In the trailing edge of the airglow patch (top side), intermediate scale (a few tens of kilo-meter) finger-like structures (numbered from 1 to 5) are seen

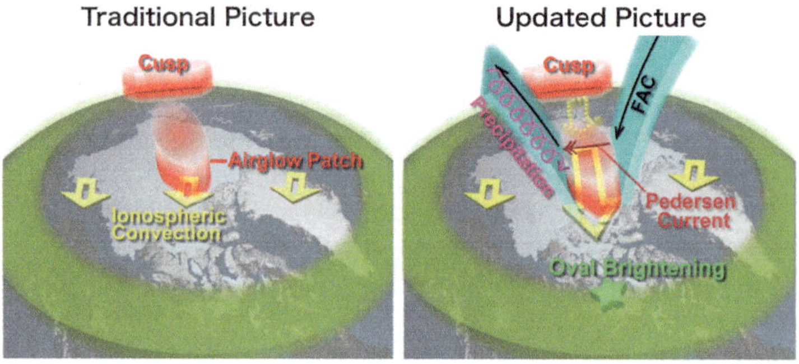

Fig. 8 Traditional and updated pictures of convection associated with airglow patches (Nishimura and Zou 2017). The green oval denotes the auroral oval and the encircled region is the polar cap. Traditional convection (yellow vectors) is assumed to be smooth and airglow patches (red column) follow large-scale convection. Now patches are realized to be often transported by fast flow channels (larger yellow vectors than surrounding). The channels are associated with FACs (blue sheets) and precipitation (purple helix). FACs are closed through Pedersen currents (red double-arrow). As the channel arrives at nightside auroral oval, oval brightening occurs at the same longitude (Sect. 4). Note that the actual distribution of FACs and precipitation is more complex than shown. There are multiple sheets and the upward FACs and precipitation are often (not always) spread over the patch leading edge

patches all the way to the nightside auroral zone, or may disappear during the course of patch propagation.

The driver of the flow channels is inferred to lie in the magnetospheric lobe based on the fact that they are associated with field-aligned currents (FAC) and soft electron precipitation as shown by the illustration in the right panel of Fig. 8. The left panel presents the scenario if patches follow the smooth background convection. The currents associated with the flow channels are generally weak, being 0.1–0.2 $\mu A/m^2$, and are likely carried by electrons of a few hundred eV (Zou et al. 2016, 2017). They usually exhibit a Region 1 sense, i.e., a downward FAC lying dawnward of an upward FAC, and can close through Pedersen currents in the ionosphere. The Region 1 sense of FACs is consistent with the locally

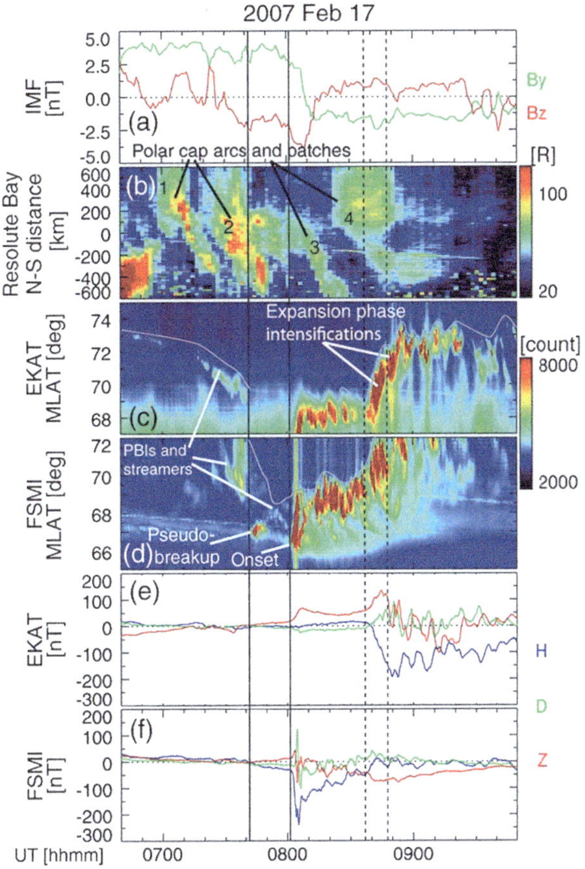

Fig. 9 (**a**) IMF B_Y and B_Z, keograms of the (**b**) Resolute Bay, (**c**) RANK, and (**d**) SNKQ imagers, and (**e**) SNKQ magnetometer data (Nishimura et al. 2013). As airglow patches reach the poleward boundary of nightside auroral oval, auroral intensifications subsequently occur

enhanced dawn-dusk electric field across patches. The presence of FACs and particle precipitation implies that the flow channels seen in the ionosphere are driven by processes in the magnetosphere. Specifically, the flow channels are postulated to form during dayside magnetic reconnection. As the reconnected magnetic flux tube propagates anti-sunward away from the dayside reconnection site to the magnetotail, its ionospheric footprint, which is the flow channel, entrains patches across the polar cap. In other words, the evolution of airglow patches corresponds to the evolution of flux tubes formed by the dayside reconnection.

It is noteworthy that not all airglow patches are associated with flow channels and those that are not tend to follow the background convection. It is also likely that not all flow channels are associated with patches. The generation and evolution of the flow channels is not yet understood and requires further investigation.

4 Nightside Observations

Airglow patches exit the nightside polar cap mainly and nearly symmetrically around magnetic midnight (Moen et al. 2007, 2015). One interesting feature recently noted is that as patches exit, auroras within the nightside auroral oval intensify (Lorentzen et al. 2004; Nishimura et al. 2013, 2014). One example is given in Fig. 9. In Fig. 9, four polar cap structures (which are a mixture of airglow patches and small-scale polar cap arcs) propagated anti-sunward at a speed of $\sim 500\ \mathrm{m\,s^{-1}}$. The first two arrived at the auroral poleward boundary in association with PBI and streamer intensification, which was followed by a pseudobreakup and a substorm marked by the solid vertical lines, respectively. The third and fourth polar cap structures propagated toward the expansion-phase active aurora and they concurred with two major poleward expansions of the auroral oval marked by the dashed vertical lines. This phenomenon echoes observations made with radars where as fast flow channels arrive at the poleward boundary of the nightside auroral oval from the polar cap, poleward boundary intensifications (PBIs) occur (Nishimura et al. 2010; Lyons et al. 2011, 2016b, 2018; Zou et al. 2014). Since airglow patches are commonly associated with flow channels as reviewed in Sect. 3.2, the two observational findings have likely reflected the same physical process.

One key to understanding the relation between patches and auroral intensifications is the timing of intensifications relative to the arrival of patches. This, however, cannot be precisely determined using optical imagers as patches are faint and diffuse. Studies using radars suggest that flow channels statistically arrive right before or simultaneously with the auroral intensifications (Zou et al. 2014). The order of sequence implies that flow channels might trigger auroral intensifications. Currently two different models have been proposed to interpret the causality. One is based on dynamics in the magnetosphere. It is believed that the flow channels correspond to fast flows in the lobe that are 2–3 R_e wide and that as the lobe flows arrive at the nightside plasma sheet they locally compress the plasma sheet. Magnetotail reconnection then develops and the associated precipitation illuminates auroras (Nishimura and Lyons 2016). The other only concerns ionospheric electrodynamics. As the flow channel approaches the auroral oval, upward and downward FACs are induced at the poleward boundary owing to the sharp gradient of ionospheric conductance. The upward FACs correspond to the observed auroral intensifications (Ohtani and Yoshikawa 2016; Ohtani et al. 2018).

The relation between patches and auroral intensifications has important applications in space weather research, as it is the coexistence of the two that induces strong scintillations in GPS signals (Van der Meeren et al. 2015). One example is given in Fig. 10, which presents the level of phase scintillation at various time instances of a substorm disturbance. Before the substorm onset (Figs. 10a and 10e) phase scintillation was low ($\sigma_\phi < 0.2$) across the entire area under monitoring, including the auroral oval and the southward-propagating airglow patch. As the southwestern edge of the patch approaching the auroral oval, the substorm initiated and the oval expanded poleward. At regions where the patch collocated with the intense auroral emissions, phase scintillation increased to values of $\sigma_\phi > 1$ (Figs. 10b–10c and 10f–10g). Scintillation at the other portions of the oval remained low. After the expansion subsided and the aurora faded, phase scintillation returned to $\sigma_\phi < 0.2$.

5 Summary

Polar cap airglow patches were already recognized almost four decades ago as counterparts of polar cap (plasma density) patches. They are the characteristic optical feature of the polar cap under southward IMF condition. However, such airglow observations had not been utilized extensively for the studies of ionosphere in the polar cap, probably due to the limitation in both the spatial and temporal resolutions of optical measurements. In the last two decades, following the development of highly-sensitive airglow imagers equipped with cooled CCD cameras, it has become possible to visualize the dynamical temporal evolution and complicated spatial structure of airglow patches with improved signal-to-noise ratio. Such a

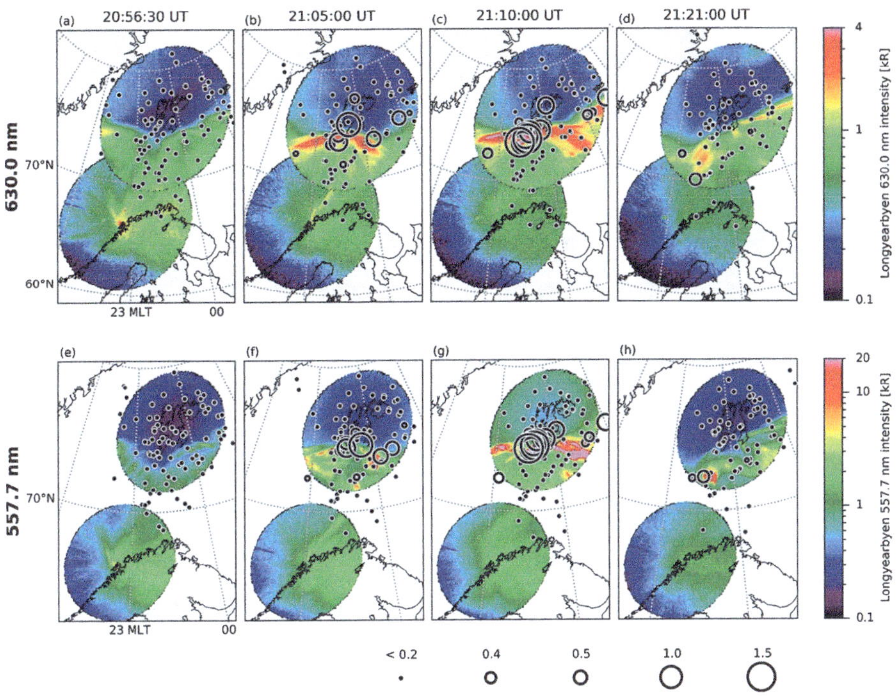

Fig. 10 Optical emissions and phase scintillation (**a**, **e**) before, (**b**–**c** and **f**–**g**) during, and (**d**, **h**) after the substorm expansion phase. Panels (**a**–**d**) and (**e**–**h**) show red line (630.0 nm) and green line (557.7 nm) emissions, respectively. The GNSS ionospheric piercing points are shown as circles, with the size representing the intensity of the phase scintillation. The arriving patch can be seen in Fig. 3a as a light blue area stretching from west to northeast of Svalbard (Van der Meeren et al. 2015)

progress has enabled us not only to use airglow patches as tracers for plasma convection in the polar cap but also to understand the generation of small-scale plasma irregularities, which are known as backscatter targets for coherent radars and sources of scintillation in the trans-ionospheric satellite signals, through plasma instabilities. In addition, recent observations demonstrated a case in which an airglow patch was accompanied by an intense flow channel and corresponding field-aligned current structure along its edges. This implies that airglow patches can signify magnetosphere-ionosphere coupling process in the region of open field lines at the polar cap latitudes, serving as a remote sensing tool just like auroras do. Further studies showed an association with airglow patches with the intensification of aurora on the nightside (PBI and/or streamer) leading to the expansion phase onset of substorms. In this sense, airglow patches may provide critical clues for discussing the coupling processes between polar cap and the nightside auroral oval, and the onset mechanism of substorms and auroral breakups. In order to facilitate such studies connecting ionospheric and magnetospheric physics, it is highly demanded to combine data from all-sky airglow imagers and radars in the polar cap with those in the auroral zone, and then coordinate multi-instrumental experiments including space-based measurements.

Acknowledgements K.H. is supported by JSPS Kakenhi (26302006). Y.Z. is supported by UCAR's Cooperative Programs for the Advancement of Earth System Science (Jack Eddy Postdoctoral Fellowship), and by National Science Foundation (AGS-1664885). Y.N. is supported by NASA grant NNX17AL22G and 80NSSC18K0657, NSF grants PLR-1341359, AGS-1737823, and AFOSR FA9559-16-1-0364.

Publisher's Note Springer Nature remains neutral with regard to jurisdictional claims in published maps and institutional affiliations.

References

D.N. Anderson, J. Buchau, R.A. Heelis, Origin of density enhancements in the winter polar cap ionosphere. Radio Sci. **23**, 513–519 (1988). https://doi.org/10.1029/RS023i004p00513

D. Barbier, F.E. Roach, W.R. Steiger, The summer intensity variations of [OI] 6300 Å in the tropics. J. Res. Natl. Bur. Stand. D, Radio Propag. **66D**(1), 145–152 (1962)

H.C. Carlson, A voyage of discovery into the polar cap and Svalbard, in *Proceedings of the Egeland Symposium on Auroral and Atmospheric Research*, Dep. of Phys., Univ. of Oslo, Norway, ed. by J. Moen, J.A. Holtet (2003), pp. 33–54. ISBN 82-91853-09-6

H.C. Carlson, K. Oksavik, J. Moen, T. Pedersen, Ionospheric patch formation: direct measurements of the origin of a polar cap patch. Geophys. Res. Lett. **31**, L08806 (2004). https://doi.org/10.1029/2003GL018166

H.C. Carlson, J. Moen, K. Oksavik, C.P. Nielsen, I.W. McCrea, T.R. Pedersen, P. Gallop, Direct observations of injection events of subauroral plasma into the polar cap. Geophys. Res. Lett. **33**, L05103 (2006). https://doi.org/10.1029/2005GL025230

H.C. Carlson, K. Oksavik, J.I. Moen, Thermally excited 630.0 nm $O(^1D)$ emission in the cusp: a frequent high-altitude transient signature. J. Geophys. Res. Space Phys. **118**, 5842–5852 (2013). https://doi.org/10.1002/jgra.50516

G. Chisham et al., A decade of the Super Dual Auroral Radar Network (SuperDARN): scientific achievements, new techniques and future directions. Surv. Geophys. **28**, 33–109 (2007). https://doi.org/10.1007/s10712-007-9017-8

W.R. Coley, R.A. Heelis, Seasonal and universal time distribution of patches in the northern and southern polar caps. J. Geophys. Res. **103**(A12), 29,229–29,237 (1998). https://doi.org/10.1029/1998JA900005

G. Crowley, Critical review of ionospheric patches and blobs, in *The Review of Radio Science 1992–1996*, ed. by W. Ross Stone (Oxford University Press, Oxford, 1996)

H. Dahlgren, J.L. Semeter, K. Hosokawa, M.J. Nicolls, T.W. Butler, M.G. Johnsen, K. Shiokawa, C. Heinselman, Direct three-dimensional imaging of polar ionospheric structures with the Resolute Bay Incoherent Scatter Radar. Geophys. Res. Lett. **39**, L05104 (2012a). https://doi.org/10.1029/2012GL050895

H. Dahlgren, G.W. Perry, J.L. Semeter, J.-P. St-Maurice, K. Hosokawa, M.J. Nicolls, M. Greffen, K. Shiokawa, C. Heinselman, Space-time variability of polar cap patches: direct evidence for internal plasma structuring. J. Geophys. Res. **117**, A09312 (2012b). https://doi.org/10.1029/2012JA017961

B.S. Dandekar, T.W. Bullett, Morphology of polar cap patch activity. Radio Sci. **34**(5), 1187–1205 (1999). https://doi.org/10.1029/1999RS900056

G.J. Fasel, Dayside poleward moving auroral forms: a statistical study. J. Geophys. Res. **100**(A7), 11891–11905 (1995). https://doi.org/10.1029/95JA00854

J.C. Foster et al., Multiradar observations of the polar tongue of ionization. J. Geophys. Res. **110**, A09S31 (2005). https://doi.org/10.1029/2004JA010928

K. Fukui, J. Buchau, C.E. Valladares, Convection of polar cap patches observed at Qaanaaq, Greenland during the winter of 1989–1990. Radio Sci. **29**(1), 231–248 (1994). https://doi.org/10.1029/93RS01510

L.V. Goodwin, B. Iserhienrhien, D.M. Miles, S. Patra, C. van der Meeren, S.C. Buchert, J.K. Burchill, L.B.N. Clausen, D.J. Knudsen, K.A. McWilliams, J. Moen, Swarm in situ observations of F region polar cap patches created by cusp precipitation. Geophys. Res. Lett. **42**, 996–1003 (2015). https://doi.org/10.1002/2014GL062610

P.B. Hays, D.W. Rusch, R.G. Roble, J.C.G. Walker, The OI (6300 Å) airglow. Rev. Geophys. **16**(2), 225–232 (1978). https://doi.org/10.1029/RG016i002p00225

K. Hosokawa, K. Shiokawa, Y. Otsuka, A. Nakajima, T. Ogawa, J.D. Kelly, Estimating drift velocity of polar cap patches with all-sky airglow imager at Resolute Bay, Canada. Geophys. Res. Lett. **33**, L15111 (2006). https://doi.org/10.1029/2006GL026916

K. Hosokawa, K. Shiokawa, Y. Otsuka, T. Ogawa, J.-P. St-Maurice, G.J. Sofko, D.A. Andre, Relationship between polar cap patches and field-aligned irregularities as observed with an all-sky airglow imager at Resolute Bay and the PolarDARN radar at Rankin Inlet. J. Geophys. Res. **114**, A03306 (2009a). https://doi.org/10.1029/2008JA013707

K. Hosokawa, T. Kashimoto, S. Suzuki, K. Shiokawa, Y. Otsuka, T. Ogawa, Motion of polar cap patches: a statistical study with all-sky airglow imager at Resolute Bay, Canada. J. Geophys. Res. **114**, A04318 (2009b). https://doi.org/10.1029/2008JA014020

K. Hosokawa, J.-P. St-Maurice, G.J. Sofko, K. Shiokawa, Y. Otsuka, T. Ogawa, Reorganization of polar cap patches through shears in the background plasma convection. J. Geophys. Res. **115**, A01303 (2010a). https://doi.org/10.1029/2009JA014599

K. Hosokawa, T. Tsugawa, K. Shiokawa, Y. Otsuka, N. Nishitani, T. Ogawa, M.R. Hairston, Dynamic temporal evolution of polar cap tongue of ionization during magnetic storm. J. Geophys. Res. **115**, A12333 (2010b). https://doi.org/10.1029/2010JA015848

K. Hosokawa, J.I. Moen, K. Shiokawa, Y. Otsuka, Decay of polar cap patch. J. Geophys. Res. **116**, A05306 (2011). https://doi.org/10.1029/2010JA016297

K. Hosokawa, S. Taguchi, Y. Ogawa, T. Aoki, Periodicities of polar cap patches. J. Geophys. Res. Space Phys. **118**, 447–453 (2013). https://doi.org/10.1029/2012JA018165

K. Hosokawa, S. Taguchi, K. Shiokawa, Y. Otsuka, Y. Ogawa, M. Nicolls, Global imaging of polar cap patches with dual airglow imagers. Geophys. Res. Lett. **41**, 1–6 (2014). https://doi.org/10.1002/2013GL058748

K. Hosokawa, S. Taguchi, Y. Ogawa, Edge of polar cap patches. J. Geophys. Res. Space Phys. **121**, 3410–3420 (2016a). https://doi.org/10.1002/2015JA021960

K. Hosokawa, S. Taguchi, Y. Ogawa, Periodic creation of polar cap patches from auroral transients in the cusp. J. Geophys. Res. Space Phys. **121**, 5639–5652 (2016b). https://doi.org/10.1002/2015JA022221

Y. Jin, J.I. Moen, W.J. Miloch, GPS scintillation effects associated with polar cap patches and substorm auroral activity: direct comparison. J. Space Weather Space Clim. **4**, A23 (2014). https://doi.org/10.1051/swsc/2014019

M.J. Keskinen, S.L. Ossakow, Nonlinear evolution of plasma enhancements in the auroral ionosphere. 1. Long wavelength irregularities. J. Geophys. Res. **87**, 144 (1982)

N.K. Kwagala, K. Oksavik, D.A. Lorentzen, M.G. Johnsen, On the contribution of thermal excitation to the total 630.0 nm emissions in the northern cusp ionosphere. J. Geophys. Res. Space Phys. **122**, 1234–1245 (2017). https://doi.org/10.1002/2016JA023366

N.K. Kwagala, K. Oksavik, D.A. Lorentzen, M.G. Johnsen, How often do thermally excited 630.0 nm emissions occur in the polar ionosphere? J. Geophys. Res. Space Phys. **123**, 698–710 (2018a). https://doi.org/10.1002/2017JA024744

N.K. Kwagala, K. Oksavik, D.A. Lorentzen, M.G. Johnsen, K.M. Laundal, Seasonal and solar cycle variations of thermally excited 630.0 nm emissions in the polar ionosphere. J. Geophys. Res. Space Phys. **123**, 7029–7039 (2018b). https://doi.org/10.1029/2018JA025477

M. Lockwood, H.C. Carlson, Production of polar cap electron density patches by transient magnetopause reconnection. Geophys. Res. Lett. **19**, 1731–1734 (1992). https://doi.org/10.1029/92GL01993

D.A. Lorentzen, N. Shumilov, J. Moen, Drifting airglow patches in relation to tail reconnection. Geophys. Res. Lett. **31**, L02806 (2004). https://doi.org/10.1029/2003GL017785

D.A. Lorentzen, J. Moen, K. Oksavik, F. Sigernes, Y. Saito, M.G. Johnsen, In situ measurement of a newly created polar cap patch. J. Geophys. Res. **115**, A12323 (2010). https://doi.org/10.1029/2010JA015710

L.R. Lyons, Y. Nishimura, H.-J. Kim, E. Donovan, V. Angelopoulos, G. Sofko, M. Nicolls, C. Heinselman, J.M. Ruohoniemi, N. Nishitani, Possible connection of polar cap flows to pre- and post-substorm onset PBIs and streamers. J. Geophys. Res. **116**, A12225 (2011). https://doi.org/10.1029/2011JA016850

L.R. Lyons, Y. Nishimura, Y. Zou, Unsolved problems: mesoscale polar cap flow channels' structure, propagation, and effects on space weather disturbances. J. Geophys. Res. Space Phys. **121**, 3347–3352 (2016a). https://doi.org/10.1002/2016JA022437

L.R. Lyons et al., The 17 March 2013 storm: synergy of observations related to electric field modes and their ionospheric and magnetospheric effects. J. Geophys. Res. Space Phys. **121**, 10,880–10,897 (2016b). https://doi.org/10.1002/2016JA023237

L.R. Lyons, Y. Zou, Y. Nishimura, B. Gallardo-Lacourt, V. Angelopoulos, E.F. Donovan, Stormtime substorms: occurrence and flow channel triggering. Earth Planets Space **70**, 81 (2018)

J.J. Makela, M.C. Kelley, S.A. González, N. Aponte, R.P. McCoy, Ionospheric topography maps using multiple-wavelength all-sky images. J. Geophys. Res. **106**(A12), 29161–29174 (2001). https://doi.org/10.1029/2000JA000449

D.J. McEwen, D.P. Harris, Occurrence patterns of *F* layer patches over the north magnetic pole. Radio Sci. **31**(3), 619–628 (1996). https://doi.org/10.1029/96RS00312

S.E. Milan, M. Lester, S.W.H. Cowley, J. Moen, P.E. Sandholt, C.J. Owen, Meridian-scanning photometer, coherent HF radar, and magnetometer observations of the cusp: a case study. Ann. Geophys. **17**, 159–172 (1999)

S.E. Milan, M. Lester, T.K. Yeoman, HF radar polar patch formation revisited: summer and winter variations in dayside plasma structuring. Ann. Geophys. **20**, 487 (2002)

G.H. Millward, R.J. Moffett, H.F. Balmforth, A.S. Rodger, Modeling the ionospheric effects of ion and electron precipitation in the cusp. J. Geophys. Res. **104**(A11), 24603–24612 (1999). https://doi.org/10.1029/1999JA900249

J. Moen, H.C. Carlson, K. Oksavik, C.P. Nielsen, S.E. Pryse, H.R. Middleton, I.W. McCrea, P. Gallop, EISCAT observations of plasma patches at sub-auroral cusp latitudes. Ann. Geophys. **24**(9), 2363–2374 (2006)

J. Moen, N. Gulbrandsen, D.A. Lorentzen, H.C. Carlson, On the MLT distribution of *F* region polar cap patches at night. Geophys. Res. Lett. **34**, L14113 (2007). https://doi.org/10.1029/2007GL029632

J. Moen, K. Oksavik, T. Abe, M. Lester, Y. Saito, T.A. Bekkeng, K.S. Jacobsen, First in-situ measurements of HF radar echoing targets. Geophys. Res. Lett. **39**, L07104 (2012). https://doi.org/10.1029/2012GL051407

J. Moen, K. Hosokawa, N. Gulbrandsen, L.B.N. Clausen, On the symmetry of ionospheric polar cap patch exits around magnetic midnight. J. Geophys. Res. Space Phys. **120**, 7785–7797 (2015). https://doi.org/10.1002/2014JA020914

Y. Nishimura, L.R. Lyons, Localized reconnection in the magnetotail driven by lobe flow channels: global MHD simulation. J. Geophys. Res. Space Phys. **121**, 1327–1338 (2016). https://doi.org/10.1002/2015JA022128

Y. Nishimura, Y. Zou, *Studying the Auroras and What Makes Them Shine* (Scientia, Bristol, 2017)

Y. Nishimura et al., Preonset time sequence of auroral substorms: coordinated observations by all-sky imagers, satellites, and radars. J. Geophys. Res. **115**, A00I08 (2010). https://doi.org/10.1029/2010JA015832

Y. Nishimura, L.R. Lyons, K. Shiokawa, V. Angelopoulos, E.F. Donovan, S.B. Mende, Substorm onset and expansion phase intensification precursors seen in polar cap patches and arcs. J. Geophys. Res. Space Phys. **118**, 2034–2042 (2013). https://doi.org/10.1002/jgra.50279

Y. Nishimura et al., Day-night coupling by a localized flow channel visualized by polar cap patch propagation. Geophys. Res. Lett. **41**, 3701–3709 (2014). https://doi.org/10.1002/2014GL060301

M. Noja, C. Stolle, J. Park, H. Lühr, Long-term analysis of ionospheric polar patches based on CHAMP TEC data. Radio Sci. **48**, 289–301 (2013). https://doi.org/10.1002/rds.20033

S. Ohtani, A. Yoshikawa, The initiation of the poleward boundary intensification of auroral emission by fast polar cap flows: a new interpretation based on ionospheric polarization. J. Geophys. Res. Space Phys. **121**, 10,910–10,928 (2016). https://doi.org/10.1002/2016JA023143

S. Ohtani, T. Motoba, J.W. Gjerloev, J.M. Ruohoniemi, E.F. Donovan, A. Yoshikawa, Longitudinal development of poleward boundary intensifications (PBIs) of auroral emission. J. Geophys. Res. Space Phys. **123**, 9005–9021 (2018). https://doi.org/10.1029/2017JA024375

K. Oksavik, J. Moen, H.C. Carlson, High-resolution observations of the small-scale flow pattern associated with a poleward moving auroral form in the cusp. Geophys. Res. Lett. **31**, L11807 (2004). https://doi.org/10.1029/2004GL019838

K. Oksavik, V.L. Barth, J. Moen, M. Lester, On the entry and transit of high-density plasma across the polar cap. J. Geophys. Res. **115**, A12308 (2010). https://doi.org/10.1029/2010JA015817

K. Oksavik, J. Moen, M. Lester, T.A. Bekkeng, J.K. Bekkeng, In situ measurements of plasma irregularity growth in the cusp ionosphere. J. Geophys. Res. **117**, A11301 (2012). https://doi.org/10.1029/2012JA017835

K. Oksavik, C. Meeren, D.A. Lorentzen, L.J. Baddeley, J. Moen, Scintillation and loss of signal lock from poleward moving auroral forms in the cusp ionosphere. J. Geophys. Res. Space Phys. **120**, 9161–9175 (2015). https://doi.org/10.1002/2015JA021528

T.R. Pedersen, B.G. Fejer, R.A. Doe, E.J. Weber, An incoherent scatter radar technique for determining two-dimensional horizontal ionization structure in polar cap F region patches. J. Geophys. Res. **105**(A5), 10637–10655 (2000). https://doi.org/10.1029/1999JA000073

G.W. Perry, J.-P. St-Maurice, K. Hosokawa, The interconnection between cross-polar cap convection and the luminosity of polar cap patches. J. Geophys. Res. Space Phys. **118**, 7306–7315 (2013). https://doi.org/10.1002/2013JA019196

J. Sakai, K. Hosokawa, S. Taguchi, Y. Ogawa, Storm time enhancements of 630.0 nm airglow associated with polar cap patches. J. Geophys. Res. Space Phys. **119**, 2214–2228 (2014). https://doi.org/10.1002/2013JA019197

P.E. Sandholt, C.S. Deehr, A. Egeland, B. Lybekk, R. Viereck, G.J. Romick, Signatures in the dayside aurora of plasma transfer from the magnetosheath. J. Geophys. Res. **91**(A9), 10063–10079 (1986). https://doi.org/10.1029/JA091iA09p10063

T. Sato, Morphology of ionospheric F2 disturbances in the polar regions. Rep. Ionos. Space Res. Jpn. **131**, 91 (1959)

D.J. Southwood, The ionospheric signature of flux transfer events. J. Geophys. Res. **92**(A4), 3207–3213 (1987). https://doi.org/10.1029/JA092iA04p03207

A. Spicher, L.B.N. Clausen, W.J. Miloch, V. Lofstad, Y. Jin, J.I. Moen, Interhemispheric study of polar cap patch occurrence based on Swarm in situ data. J. Geophys. Res. Space Phys. **122**, 3837–3851 (2017). https://doi.org/10.1002/2016JA023750

E.G. Thomas et al., Multi-instrument, high-resolution imaging of polar cap patch transportation. Radio Sci. **50**, 904–915 (2015). https://doi.org/10.1002/2015RS005672

A. Thorolfsson, J.-C. Cerisier, M. Lockwood, P.E. Sandholt, C. Senior, M. Lester, Simultaneous optical and radar signatures of poleward-moving auroral forms. Ann. Geophys. **18**, 1054 (2000)

C.E. Valladares, D.T. Decker, R. Sheehan, D.N. Anderson, Modeling the formation of polar cap patches using large plasma flows. Radio Sci. **31**(3), 573–593 (1996). https://doi.org/10.1029/96RS00481

C.E. Valladares, T. Pedersen, R. Sheehan, Polar cap patches observed during the magnetic storm of November 2003: observations and modeling. Ann. Geophys. **33**, 1117–1133 (2015). https://doi.org/10.5194/angeo-33-1117-2015

C. Van der Meeren, K. Oksavik, D. Lorentzen, J.I. Moen, V. Romano, GPS scintillation and irregularities at the front of an ionization tongue in the nightside polar ionosphere. J. Geophys. Res. Space Phys. **119**, 8624–8636 (2014). https://doi.org/10.1002/2014JA020114

C. Van der Meeren, K. Oksavik, D.A. Lorentzen, M.T. Rietveld, L.B.N. Clausen, Severe and localized GNSS scintillation at the poleward edge of the nightside auroral oval during intense substorm aurora. J. Geophys. Res. Space Phys. **120**, 10,607–10,621 (2015). https://doi.org/10.1002/2015JA021819

B. Wang, Y. Nishimura, L.R. Lyons, Y. Zou, H.C. Carlson, H.U. Frey, S.B. Mende, Analysis of close conjunctions between dayside polar cap airglow patches and flow channels by all-sky imager and DMSP. Earth Planets Space **68**(1), 150 (2016). https://doi.org/10.1186/s40623-016-0524-z

E.J. Weber, J. Buchau, Polar cap F-layer auroras. Geophys. Res. Lett. **8**(1), 125–128 (1981). https://doi.org/10.1029/GL008i001p00125

E.J. Weber, J. Buchau, J.G. Moore, J.R. Sharber, R.C. Livingston, J.D. Winningham, B.W. Reinisch, F layer ionization patches in the polar cap. J. Geophys. Res. **89**, 1683 (1984)

E.J. Weber, J.A. Klobuchar, J. Buchau, H.C. Carlson Jr., R.C. Livingston, O. de la Beaujardiere, M. McCready, J.G. Moore, G.J. Bishop, Polar cap F layer patches: structure and dynamics. J. Geophys. Res. **91**(A11), 12121–12129 (1986). https://doi.org/10.1029/JA091iA11p12121

Y. Zhang, D.J. McEwen, L.L. Cogger, Interplanetary magnetic field control of polar patch velocity. J. Geophys. Res. **108**, 1214 (2003). https://doi.org/10.1029/2002JA009742

Y. Zou, Y. Nishimura, L.R. Lyons, E.F. Donovan, J.M. Ruohoniemi, N. Nishitani, K.A. McWilliams, Statistical relationships between enhanced polar cap flows and PBIs. J. Geophys. Res. Space Phys. **119**, 151–162 (2014). https://doi.org/10.1002/2013JA019269

Y. Zou, Y. Nishimura, L.R. Lyons, K. Shiokawa, E.F. Donovan, J.M. Ruohoniemi, K.A. McWilliams, N. Nishitani, Localized polar cap flow enhancement tracing using airglow patches: statistical properties, IMF dependence, and contribution to polar cap convection. J. Geophys. Res. Space Phys. **120**, 4064–4078 (2015). https://doi.org/10.1002/2014JA020946

Y. Zou et al., Localized field-aligned currents in the polar cap associated with airglow patches. J. Geophys. Res. Space Phys. **121**, 10,172–10,189 (2016). https://doi.org/10.1002/2016JA022665

Y. Zou, Y. Nishimura, L.R. Lyons, K. Shiokawa, Localized polar cap precipitation in association with nonstorm time airglow patches. Geophys. Res. Lett. **44**, 609–617 (2017). https://doi.org/10.1002/2016GL071168

GPSR Compliance

The European Union's (EU) General Product Safety Regulation (GPSR) is a set of rules that requires consumer products to be safe and our obligations to ensure this.

If you have any concerns about our products, you can contact us on ProductSafety@springernature.com

In case Publisher is established outside the EU, the EU authorized representative is:

Springer Nature Customer Service Center GmbH
Europaplatz 3
69115 Heidelberg, Germany

Batch number: 10731189

Printed by Printforce, the Netherlands